Poisonous Mushrooms of the Northern United States and Canada

Cover: Watercolor by Henry A. C. Jackson, reproduced courtesy of the National Gallery of Canada. The painting is part of a collection donated to the Gallery by Dr. Naomi Jackson Groves. *Boletus subvelutipes* is in the *Luridi* group, all of which should be avoided. Chapter 13 gives a discussion of the genus; the species is treated under *B. erythropus*.

POISONOUS MUSHROOMS of the Northern United States and Canada

JOSEPH F. AMMIRATI
University of Washington, Seattle, Washington

JAMES A. TRAQUAIR
Agriculture Canada, Harrow, Ontario

PAUL A. HORGEN
University of Toronto, Toronto, Ontario

University of Minnesota Press
Minneapolis

Poisonous Mushrooms of the Northern United States and Canada

Published in the United States of America by the University of Minnesota Press, 2037 University Avenue Southeast, Minneapolis MN 55414

The University of Minnesota is an equal-opportunity educator and employer.

Published in Canada by Fitzhenry & Whiteside Limited, 195 Allstate Parkway, Markham, Ontario L3R 2T8, in cooperation with Agriculture Canada and the Canadian Government Publishing Centre, Supply & Services Canada.

Library of Congress Cataloging in Publication Data

Ammirati, Joseph F.

Poisonous mushrooms of the northern United States and Canada

Includes bibliographies and index.
1. Mushroms, Poisonous — United States. 2. Mushrooms, Poisonous — Canada. I. Traquair, James Alvin. II. Horgen, Paul A. III. Title.

QK605.A4 589.2'0469 84-20979
ISBN 0-8166-1407-5

Color separations and litho prep by Graphitech Inc.
Printed by Ashton-Potter
Bound by T.H. Best Printing Company
Printed and bound in Canada

Contents

Preface

People in the United States and Canada have recently shown increased interest in collecting wild mushrooms and other large fungi. The activities and inquiries of natural history groups and private citizens reflect this interest, and many new mycological societies and clubs are now being organized. Most mushroom collectors are interested in edibility of mushrooms. However, artists and photographers are interested in the variety of shapes, colors, and forms of mushrooms; biologists in their role as decomposers, parasites, and symbionts in nature as well as their use in general research; and medical researchers in seeking sources of new drugs. Some people collect mushrooms that have hallucinogenic properties; however, this practice is illegal in Canada and in many parts of the United States, and is ill-advised for health reasons.

Europeans are avid mushroom hunters. They have a tradition of foraging for mushrooms for food. They learn to recognize edible species through personal experience or from information passed from generation to generation. Many mushrooms that are collected in Europe are also found in North America. Others are merely close relatives, and their edibility is often not known with any certainty.

North Americans, particularly those of British descent, are cautious about eating mushrooms. This concern stems partly from the many superstitions that have surfaced to explain the unusual habits of mushrooms and their life history. The rapid development, ephemeral nature, and common association of mushrooms with decaying material have prompted people to attribute magical and mysterious properties to wild mushrooms. In fact, certain mushrooms were key elements of many religious cults in ancient times and still play an important role in primitive societies today.

Many edible mushrooms are easily confused with species that are deadly poisonous. Poisonings frequently result when mushroom hunters are careless and incorrectly identify the particular species that they are seeking. Indiscriminant tasting based on hearsay advice or accidental nibbling by children and pets also result in poisoning. When poisonings occur, biologists and medical practitioners are faced with the task of identifying the mushrooms implicated. When a positive determination is not possible, a professional mycologist must be consulted.

This publication is designed mainly to help medical practitioners, mycologists, biologists, and mushroom foragers identify poisonous mushrooms quickly and accurately. However, it is also designed to provide enough basic information on poisonous fungi that research in this area can be furthered. To accommodate both the general and technical audiences for which this book is intended, some general information given in certain sections of the book is repeated in greater detail in later, more technical, sections.

The first four chapters of Part One of this book provide general information on mushrooms and other large fungi and present methods for studying them. These chapters should be used in conjunction with the Glossary and the illustrations. Chapter 5 contains taxonomic keys and family descriptions of poisonous mushrooms and related fungi. There are two types of taxonomic keys: the general keys are based on macroscopic features, and the technical keys are based on a combination of macroscopic and microscopic features. The technical keys are more accurate than the general keys and should be used whenever possible. The general keys are designed to be used by the educated amateur, whereas the technical keys are intended primarily for use by the mushroom specialist.

Part Two of this book deals specifically with fungal poisonings. Chapter 6 provides an overview of the subject. It sets out guidelines for safe collecting and presents a list of poisonous and edible mushrooms that look alike. However, its most important feature is a key to the seven types of fungal poisonings based on symptomatology.

The seven remaining chapters each deal with one of these types of poisoning, labeled by toxin group. Each chapter contains a discussion of the toxicology and symptomatology of the toxin group and provides a list of the genera and species that are known to contain the toxin or toxins responsible. General descriptions of the symptom-producing genera likely to occur in the northern United States and Canada are provided, along with general and technical treatments of Canadian species, including keys to species when appropriate. The general information contained in each species description is supplemented by detailed technical data included for mycologists interested in further study. Colored photographs of most of the species treated are included, along with line drawings illustrating microscopic characteristics. Useful references, when available, are listed for each species. Medical treatments for poisoning are not presented here. Other references, such as *Toxic and Hallucinogenic Mushroom Poisoning* by G. Lincoff and D. Mitchel (New York: Van Nostrand Reinhold Co.; 1972), should be consulted for a thorough coverage of toxicology, symptomatology, and treatment.

This book offers two methods for identifying the genus and species of a specimen of toxic fungus involved in a poisoning case. One approach is to identify the genus by following the general and technical keys in Chapter 5. Then refer to the appropriate chapter in Part Two, as indicated in parentheses after the genus name, and locate the detailed description of that genus there. The genera appear alphabetically in each chapter of Part Two to facilitate the search. By studying the comprehensive descriptions of the poisonous species of the genus provided there, and by examining the corresponding illustrations, you should be able to make a positive identification. Alternatively, with the key in Chapter 6 you can identify the specific toxin group involved in a poisoning from the symptoms exhibited by the patient. Then turn to the chapter dealing with that particular toxin group and identify the genus and species involved by looking at the illustrations and reading all the general and technical descriptions that seem appropriate. The toxin groups and the photographs are color coded for quick reference. It is best to cross-check your identification using both methods.

The collections used for most of the descriptions and illustrations in this book are listed in Appendix 1. All microscopic studies were made from dried specimens using potassium hydroxide or Melzer's reagent as the mounting medium.

Only species known to be poisonous are formally described in Part Two of this book. These species have been selected on the basis of chemical analysis of the toxic substances or because of documented or consistent reports of poisonings. The lists of poisonous fungi contained in this book may not be complete.

Do not assume that any mushroom not described or illustrated in this book is edible. Never eat a mushroom unless you are certain of its identity.

Acknowledgments

During the writing of this book many people contributed information and many hours of hard work. We want to thank everyone involved in the project but especially those who reviewed part of or the entire manuscript. We are extremely grateful to Dr. H. E. Bigelow, University of Massachusetts, Amherst, MA; Dr. K. A. Harrison, Kentville, N.S.; Dr. D. Jenkins, University of Alabama, Birmingham, AL; Dr. C. Marr, State University of New York, College of Education, Oneonta, NY; Dr. S. J. Mazzer, Kent State University, Kent, OH; Dr. D. M. Simons, Wilmington, DE; Dr. A. H. Smith, University of Michigan, Ann Arbor, MI; Dr. D. E. Stuntz, University of Washington, Seattle, WA; Dr. W. J. Sundberg, Southern Illinois University, Carbondale, IL; and Dr. H. D. Thiers, San Francisco State University, San Francisco, CA. Drs. R. Halling and N. Weber also reviewed short sections of the manuscript and C. L. Ovrebo and Donna-Leigh Rendall helped prepare the sections on *Tricholoma* and *Lactarius*, respectively.

Most of the line drawings were prepared by Audra Geras, Toronto, Ont.; the remainder (Figs. 1–3, 11–16, 99, 131, and 157–161) were drawn by Dr. J. A. Traquair. The original manuscript was typed by Mrs. Patricia Orosz, Waterdown, Ont. This book was edited at Agriculture Canada by Sharon Rudnitski, Research Program Service.

We are also very grateful to Dr. Donald J. S. Barr, Biosystematics Research Institute, Ottawa, Ont., for his help and extreme patience throughout this project.

An important part of this book is the color photographs of individual species. Thanks are extended to all those who contributed photographs, credited below.

Dr. Joe Ammirati
Figs. 46, 47, and 220

Catherine Ardrey
Figs. 73, 75, 85, 181, 190, 193, 201, 206, and 230

Harley Barnhart
Figs. 191 and 196

Dr. Michael Beug
Figs. 38, 53, 69, 84, 88, 89, 105, 107, 108, 110–113, 115, 197, 215, and 232

Mary Ferguson
Fig. 208

Dr. Ken Harrison
Figs. 45, 52, 68, 76, 180, 188, 194, 195, 199, 202, and 205

Ken Harrison, Jr.
Figs. 17, 34, 192, and 233

Dr. Dick Homola
Figs. 183, 218, 228, and 234

Dr. David Largent
Figs. 39, 71, 74, 179, and 209

Dr. David Malloch
Fig. 225

Dr. Currie Marr
Figs. 221 and 222

Margaret McKenny
Figs. 72, 186, 203, and 212

William McLennon
Figs. 184 and 227

Kit Scates
Figs. 35, 70, 86, 106, 109, 114, 116, 177, 216, 219, 223, 229, and 231

Dr. Don Simons
Figs. 36, 37, and 83

Dr. Alexander Smith
Figs. 40, 189, 200, and 204

Dr. Walt Sundberg
Figs. 48, 49, 198, 207, 213, and 224

Dr. Leo Tanghe
Figs. 54, 178, and 210

Dr. Harry Thiers
Figs. 185, 187, 211, 214, and 217

Dr. Jim Traquair
Fig. 226

Ellen Trueblood
Figs. 41 and 87

Ben Woo
Fig. 182

Part One The fungus fruit body

1

INTRODUCTION

Fruit Body

The fruit body of a fungus is the key to its identification and the part collected for food. Thousands of fungi produce fruit bodies. Some fruit bodies are minute, barely visible to the naked eye; others are very large, weighing as much as 70 kg. Fruit bodies not only differ in size and weight but also in shape, texture, color, and other features.

Most of the fungi that produce fruit bodies belong to either the division Basidiomycota or the division Ascomycota. A third division, Zygomycota, also has some members that produce small to minute fruit bodies. The Zygomycota usually develop underground and are difficult to find; therefore they are not frequently encountered by mushroom collectors and are not treated further here. These three divisions correspond to the classes Ascomycetes, Basidiomycetes, and Zygomycetes of earlier literature.

The Basidiomycota produce their sexual spores on a special club-shaped cell called a **basidium** (pl. **basidia**) (Fig. 1B). The Ascomycota produce sexual spores in a special sac-like cell called an **ascus** (pl. **asci**) (Fig. 2C). The suffix -mycota means fungi; therefore the Basidiomycota are fungi that produce spores on a basidium, whereas the Ascomycota are fungi that produce spores in an ascus.

Various common names have been applied to the fruit bodies produced by fungi. The mushroom, the most common form of fruit body produced by the Basidiomycota, is the name familiar to most people. The term toadstool is often used for poisonous fungi.

Kinds of fruit bodies produced by the Basidiomycota include gill mushrooms, boletes, hedgehog fungi, chanterelles, polypores (sometimes called conks or bracket fungi), coral fungi, stink horns, puffballs, and several others. Rusts and smuts, also members of the Basidiomycota, are plant parasites that do not form fruit bodies.

Some Ascomycota commonly collected for food include cup fungi, morels, false morels, and earthtongues. These are in the class Discomycetes. Other Ascomycota produce minute fruit bodies too small to be seen with the naked eye or so small that they go unnoticed by the average collector. Because these are not normally collected for food, they are not considered further here.

The fruit body of the fungus is technically called a sporocarp, a term that means a fruit that produces spores. Basidiocarp, a term combining basidium and carp, is the technical name for the fruit bodies produced by the Basidiomycota, and ascocarp, a term combining ascus and carp, is the technical name for the fruit bodies of the Ascomycota.

Fungi are sometimes divided into two groups according to the texture of the fruit bodies. Fungi that produce soft, fleshy fruit bodies that decay readily are called fleshy fungi. Members of the Basidiomycota and the Ascomycota are found in this group. The second group of fungi, mainly members of the Basidiomycota, produce fruit bodies with a tough to leathery or woody consistency. They usually do not easily break and do not readily decay; in fact some may persist for several growing seasons. Some fungi are not truly fleshy nor are they woody to tough in texture. Instead they have a consistency intermediate between fleshy and woody. Smith and Smith (1973) call these intermediates the almost fleshy fungi.

In this treatment we are concerned primarily with the fleshy and almost fleshy fungi, because these are the ones collected for food. The tough to woody species, which are primarily polypores such as bracket fungi and conks, are not usually collected for food because their texture makes them inedible. Known exceptions are discussed in Chapter 13.

Life Cycle

The fungus fruit body is the spore-producing stage of the life cycle. Most fungi reproduce by spores and the fruit bodies have developed specifically for the production and release or dispersal of spores. The spores produced by fruit bodies are usually the result of sexual reproduction.

The fruit body is the visible part of the growing fungus. It is supported by and develops from an extensive network of thread-like filaments called **hyphae** (sing. **hypha**). Hyphae are often collectively termed the **mycelium**; the food-absorbing part of the fungus as opposed to the spore-producing fruit body of the fungus is called the **vegetative mycelium.** The individual hyphae that compose the mycelium absorb nutrients and water from the substratum in which they are growing. When the nutrient supply is adequate and environmental conditions are favorable, some fungi may grow in the same location for several years. Fungi cannot make their own food, namely carbohydrates, as can green plants. Some are saprophytic, obtaining food

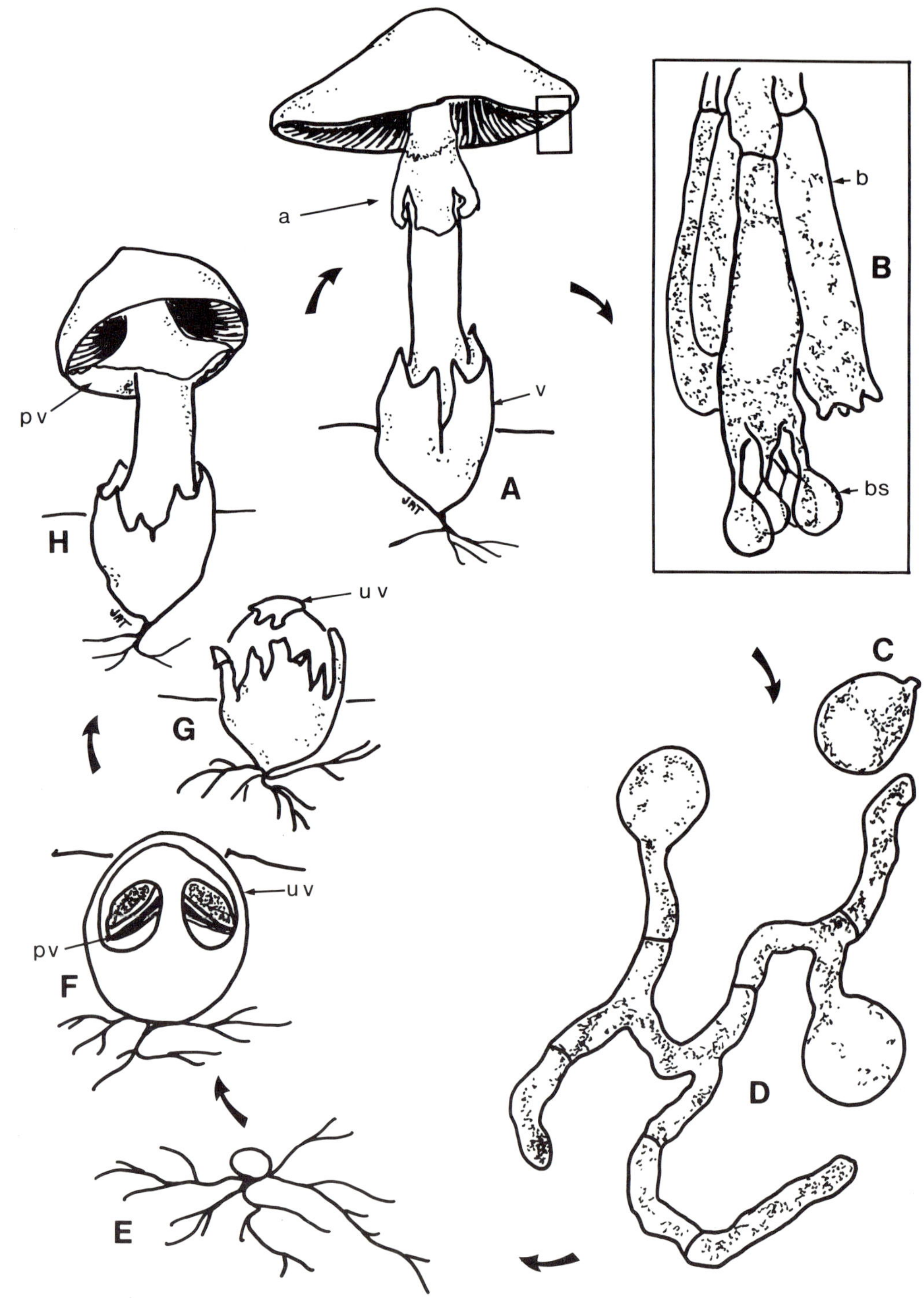

Fig. 1. The life cycle of a member of the Basidiomycota, as represented by *Amanita phalloides*: A, mature basidiocarp with annulus (*a*) and volva (*v*); B, section of the surface of the hymenium of the gill showing the basidia (*b*) and developing basidiospores (*bs*); C, basidiospore; D, hyphal fusion of a primary mycelium derived from germinated basidiospores; E, secondary mycelium with the primordial basidiocarp, the button stage; F, section of a primordial basidiocarp, the egg-like stage, showing the gills covered by a membranous partial veil (*pv*) and the whole basidiocarp surrounded by a universal veil (*uv*); G, emergence of the cap and fragmentation of the universal veil; H, elongation of the stalk, expansion of the cap, and rupture of the partial veil to form the annulus.

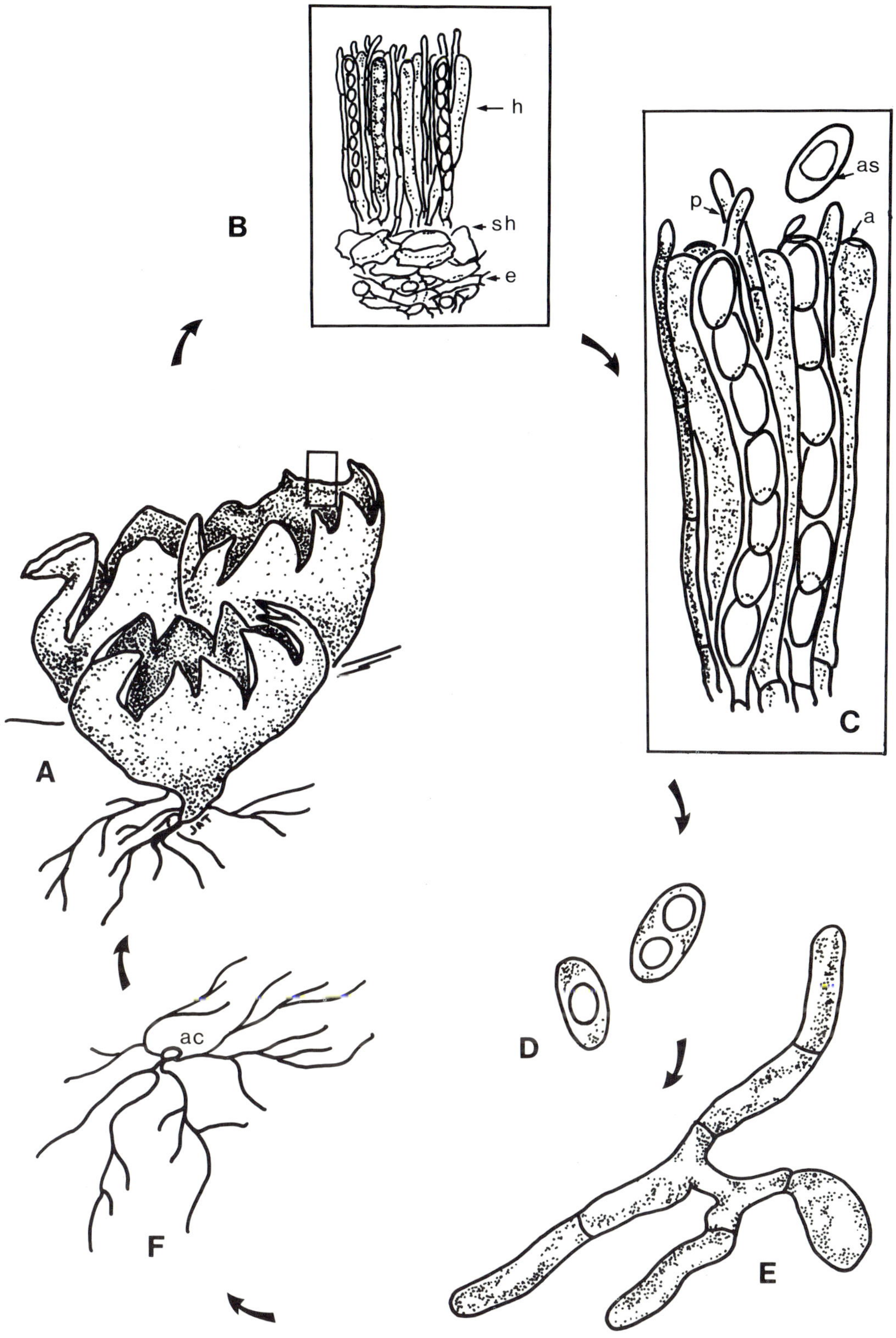

Fig. 2. The life cycle of a member of the Ascomycota, as represented by *Sarcosphaera crassa*: A, mature ascocarps; B, section showing the hymenium (*h*), subhymenium (*sh*), and excipulum (*e*); C, section of the hymenium showing the operculate ascus (*a*), paraphyses (*p*), and an ascospore (*a*); D, ascospores; E, primary mycelium derived from a germinated ascospore; F, mycelial network attached to a primordial ascocarp (*ac*).

from dead organic material, whereas others are parasitic on living plants or animals or even on other fungi. Many fungi, especially gill mushrooms and boletes, have an extensive mycelium that lives in association with the roots of green plants. This association, which is beneficial to both the fungus and host plant, is termed a **mycorrhiza** (pl. **mycorrhizae**).

When conditions are favorable and the mycelium is at the proper stage of development, one or more fruit bodies are produced by the fungus. The actual conditions necessary for fruit body formation and spore production are not clearly understood. Humidity, light, temperature, aeration, and nutrition are commonly involved. The genetic makeup and the general physiology of the fungus hyphae are also important in the initiation and formation of young fruit bodies and their development to a mature stage. The spores produced by a fruit body are released when it is mature. When they land in a suitable environment, the spores germinate and the hyphae grow to initiate the life cycle anew. The life cycle of a typical member of the Basidiomycota is presented in Fig. 1; Fig. 2 illustrates the life cycle of a member of the Ascomycota. Macroscopic and microscopic features of the fungus fruit body are used extensively in the description, classification, and identification of fungus species. The vegetative mycelia are often similar among species and to date have not been particularly helpful in identification. Important characteristics of fungus fruit bodies are described and discussed in Chapters 2 and 3 and should be learned and understood before attempting to identify fungi to genus (pl. genera) or species (pl. species).

In the following chapters the fruit bodies of the Basidiomycota and the Ascomycota are discussed separately. Macroscopic features, those that can be seen with the naked eye or hand lens, are dealt with in Chapter 2. Microscopic features, those which can be determined only through a microscope, are discussed in Chapter 3. Terminology and definitions pertinent to understanding the structure and descriptions of fruit bodies are presented in the Glossary, and line drawings illustrating these features are interspersed and cited appropriately throughout. Certain illustrations and definitions of terms in the following chapters are taken or adapted from works by Largent (1977) and Largent et al. (1977).

Literature

Largent, D. L. How to identify mushrooms to genus I: Macroscopic features. 2nd ed. Eureka, CA: Mad River Press; 1977.

Largent, D. L.; Johnson, D.; Watling, R. How to identify mushrooms to genus III: Microscopic features. Eureka, CA: Mad River Press; 1977.

Smith, H. V.; Smith, A. H. How to know the non-gilled fleshy fungi. Dubuque, IA: William C. Brown Co.; 1973.

2

MACROSCOPIC FEATURES

Basidiomycota

The Basidiomycota produce several types of fruit bodies. Mushroom-like forms include gill mushrooms, chanterelles, boletes, certain hedgehog fungi, and some polypores. Others include coral fungi, conks, bracket fungi, stinkhorns, and puffballs. Gill mushrooms are the most common of these forms, and the terminology used to describe the macroscopic features of gill mushrooms is presented first. These terms, however, are not exclusive to gill mushrooms; they can be applied equally to other mushroom-like fruit bodies. The discussion of these most commonly used terms is followed by explanations of the terminology used exclusively to describe the special features of fruit bodies of forms other than gill mushrooms and finally of puffballs and related fungi.

GILL MUSHROOMS

The typical gill mushroom (Figs. 1A, 3A) consists of an umbrella-shaped **cap**, technically called the pileus; **gills**, technically called lamellae; and a **stalk**, technically called the stipe (Fig. 3A). The blade-like gills comprising the **hymenophore**, which are on the underside of the cap, are oriented so that the edge of each gill is perpendicular to the substratum directly beneath the cap. The surface layer of each gill, which produces the spore-producing basidia, is called the **hymenium** (Fig. 1B). The surface of the cap and stalk, and the interior **flesh** of the fruit body, are usually sterile in mushrooms, that is, they do not produce spores. The stalk and cap of certain fruit bodies are modified; sometimes the stalk is even lacking. For example, species in the mushroom genera *Pleurotus* and *Crepidotus* sometimes lack a distinct stalk (Fig. 9B).

Size and stature

The size of fruit bodies is important, especially the range in size for young to mature fruit bodies. The cap diameter is perhaps the most useful measurement in identification, but the height of the cap may also be helpful, particularly when the shape is elongated. The length of the stalk including the belowground portion, its width at the apex, and its width at the broadest point in the case of clavate, ventricose, or bulbous stalks are also helpful (see this chapter, "Stalk").

General stature and the size of the cap in relation to that of the stalk usually help with identification. Photographs or a simple line drawing showing the shape and relative size of the cap and stalk are useful.

Cap

Shape

The cap is divided into two main areas: the **disc**, which comprises the center of the cap; and the **margin**, the area from the disc to the edge (Fig. 4H). The general shape of the cap, as well as that of the disc and margin, changes as the fruit body matures; therefore, a range of shapes for young to mature fruit bodies is usually recorded. General cap shape may be somewhat characteristic of a species, but this feature alone is generally of limited help in identification. Terms used to describe cap shape include **campanulate, conic, convex, parabolic,** and **plane.** These shapes are illustrated in Fig. 4 A–I and are described in the Glossary.

The disc of the cap may be shallowly to deeply depressed (Fig. 5 D–F) or funnel shaped (Fig. 4I). The depression may be small and deep or broad and more or less shallow. A raised knob in the center of the cap is called an **umbo** (Fig. 5B). An umbo may be broad and rounded or sharply pointed (Fig. 5C). Caps with an umbo are described as **umbonate.** Caps with a sharply pointed umbo are termed **acutely umbonate**; **broadly umbonate** refers to caps with a rounded umbo (compare Figs. 5 A–C, 7 A–C). When a depressed cap has an umbo in the depression, the condition is called **umbilicate** (Fig. 5F).

The shape of the cap margin, as seen in a cross section of the cap, may sometimes help to separate genera, for example to differentiate *Mycena* from *Collybia.* Common margin shapes include **decurved, enrolled, incurved, recurved,** or **upturned.** See Fig. 5 G–K and the Glossary for descriptions of these terms.

In surface view the extreme edge of the margin varies in outline. When the edge is even it is termed **entire**, but sometimes it is scalloped, eroded, or split. The cap margin sometimes has pieces of partial veil attached to its edge, a condition described as **appendiculate** (Figs. 107, 109).

The surface of the margin may also vary. In some cases it has radially oriented lines of varying lengths, radiating from the edge of the margin inward toward the center. Various terms such as **plicate, striate, sulcate,** and **translucent–striate** are used to describe this feature. Refer to the Glossary for descriptions of these terms.

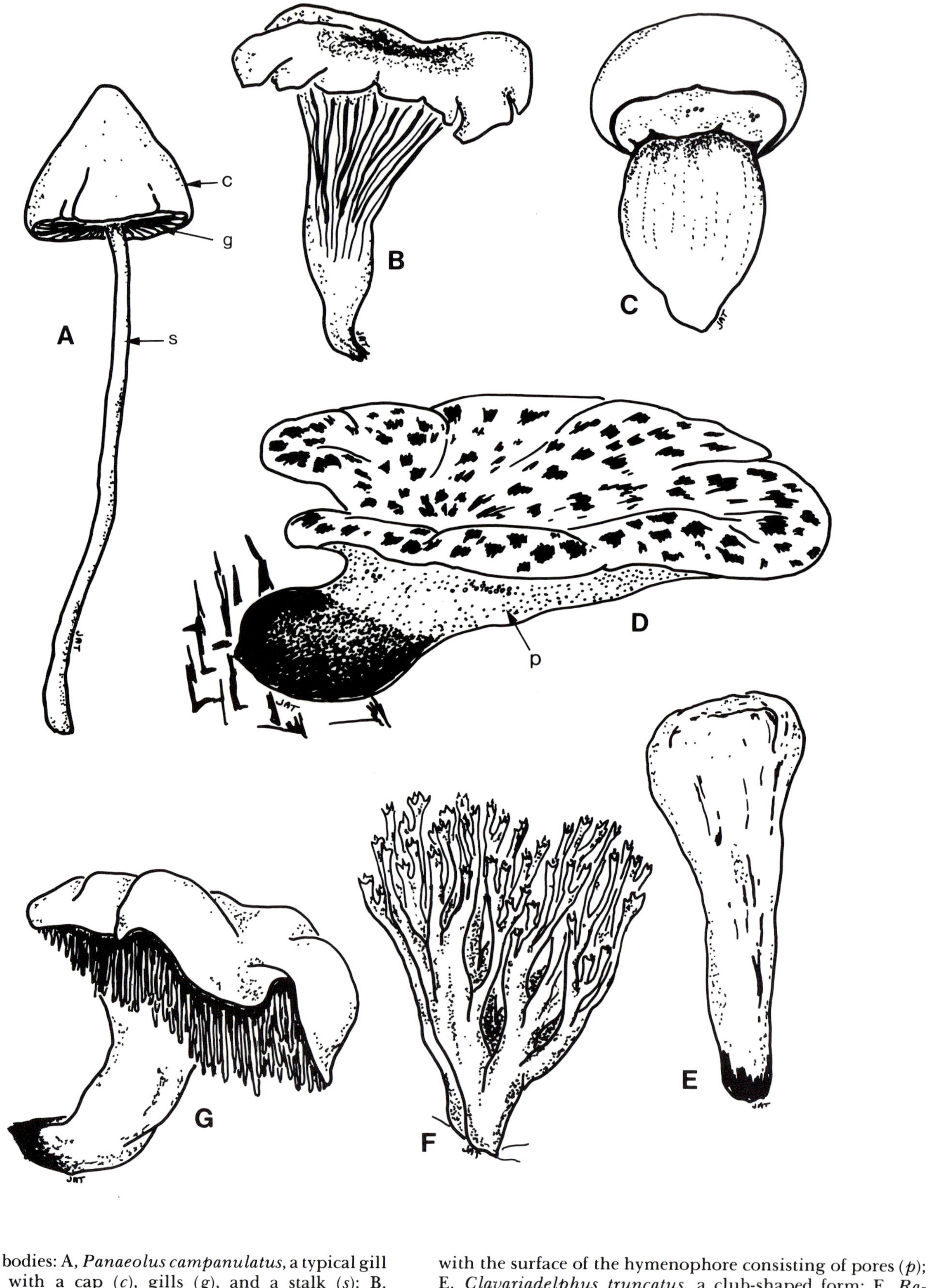

Fig. 3. Fruit bodies: A, *Panaeolus campanulatus*, a typical gill mushroom, with a cap (*c*), gills (*g*), and a stalk (*s*); B, *Cantharellus cibarius*, with thick, gill-like ridges; C, *Boletus edulis*, with a fleshy tube layer (*t*); D, *Polyporus squamosus*, with the surface of the hymenophore consisting of pores (*p*); E, *Clavariadelphus truncatus*, a club-shaped form; F, *Ramaria* sp., a coral-like form; G. *Dentinum repandum*, with spines on the undersurface of the cap.

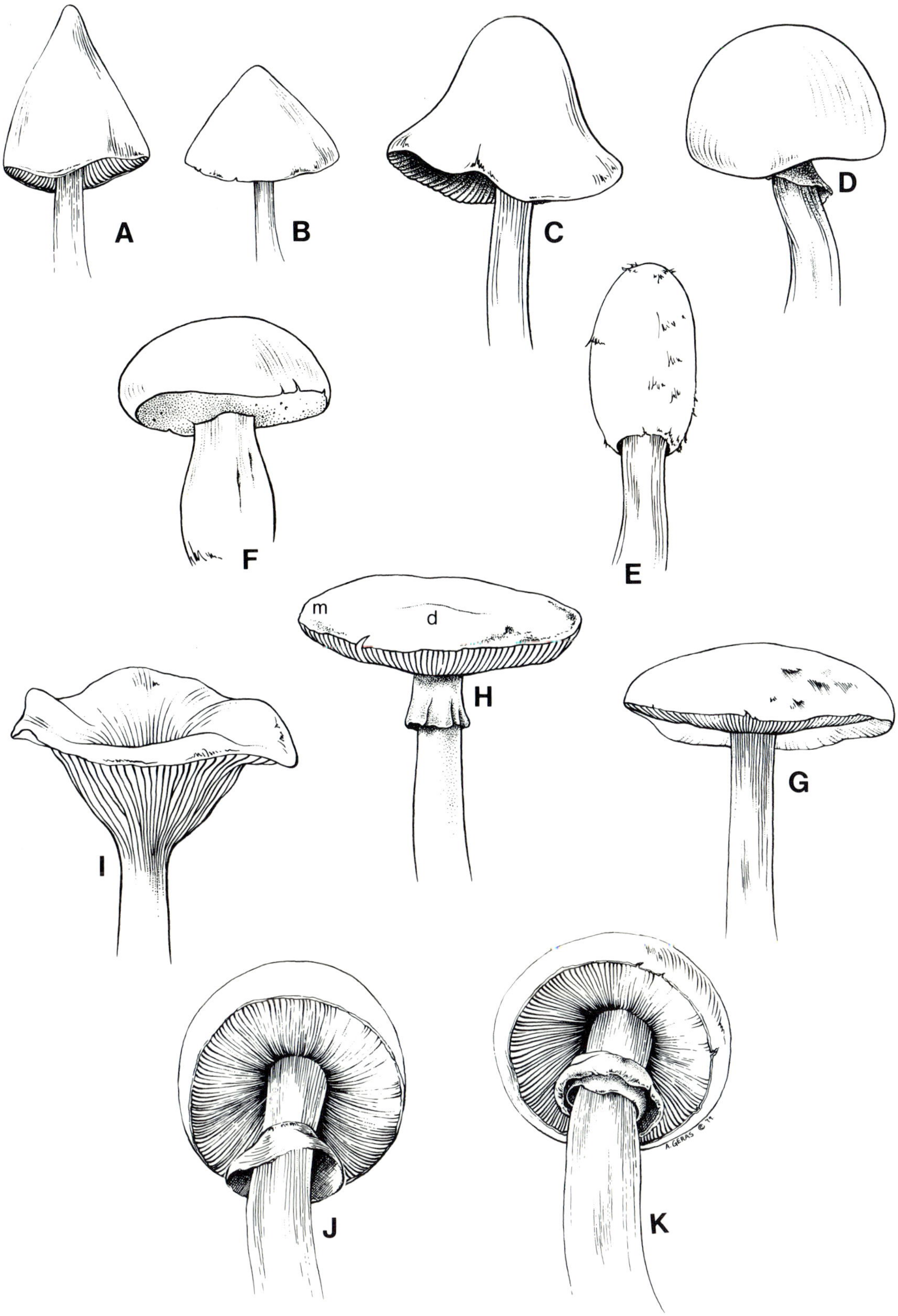

Fig. 4. Cap shapes (A–I) and annulus structures (J, K). Caps: A, conic; B, obtusely conic; C, campanulate; D, parabolic; E, narrowly parabolic; F, cushion-shaped; G, convex; H, plane, noting the central disc (*d*) and the margin (*m*); I, funnel-shaped. Annuli: J, single; K, double.

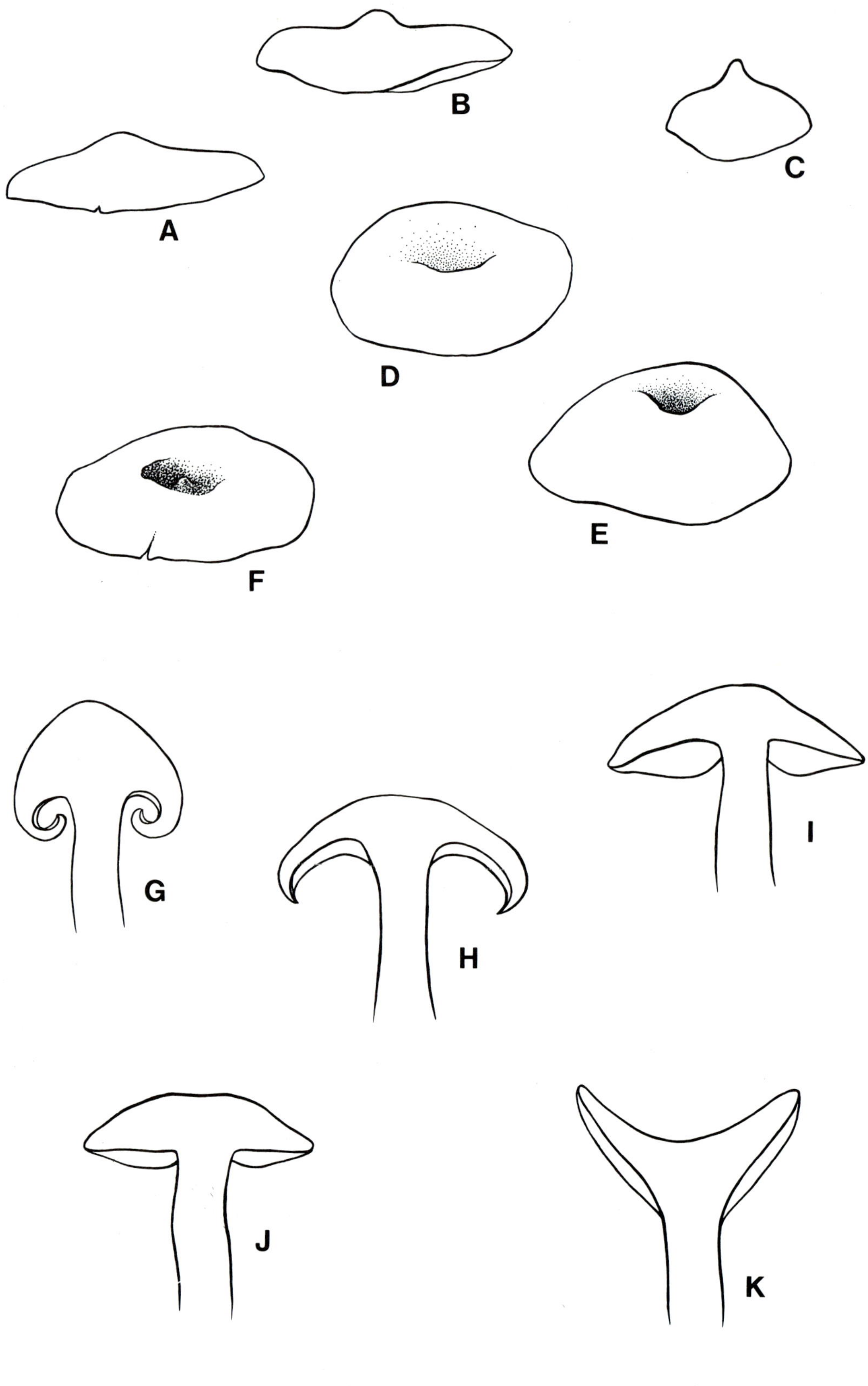

Fig. 5. Cap shapes (A–F) and margin types (G–K). Caps: A and B, umbonate; C, acutely umbonate; D, shallowly depressed; E, deeply depressed; F, umbilicate. Margins: G, enrolled; H, incurved; I, slightly decurved; J, straight; K, upturned.

Surface

The surface of the cap of mushrooms and other fruit bodies has a variety of features that are used in species identification and description. These features often can be seen with the naked eye or a hand lens. Other characteristics such as the presence of a sticky or moist surface are detected by lightly touching the surface.

The cap surface may appear dull or shiny to silky. It may be dry, moist, or **lubricous**, as if covered with a thin layer of oil. Some caps have a tacky to sticky feeling when moist due to a thin layer of jelly-like material called **gluten** on the surface, a condition described as **viscid.** When the cap surface has a distinct, visible coating of gluten, it is described as **glutinous.** Viscid and glutinous caps often have debris stuck to their surfaces.

In many fruit bodies the cap surface is moist when fresh but becomes dry as the fruit bodies mature or after they have been collected. When the color changes markedly as the surface dries from what it was in fresh caps, the condition is called **hygrophanous**. The color change may occur rapidly, especially in small fruit bodies, or it may take several hours. Caps that gradually change color over 24 hours or longer are not usually considered hygrophanous.

When the surface of the cap is smooth, without cracks, wrinkles, or depressions, it is termed **even**. A split surface may be described as **laciniated** or **rimose**, a cracked surface as **areolate**, and a wrinkled surface as **corrugate** or **rivulose**. See the Glossary for more precise descriptions of these terms.

The hyphae on the cap surface are arranged and associated in various ways. Often the cap is bald or naked, a condition described as **glabrous**. When the hyphae on the cap are associated in clusters, the surface may appear to be covered with fibrils, hairs, small to coarse scales, granules, or powder. These processes may easily be removed or may be present only on young specimens, disappearing with age. However, such processes sometimes last throughout the development of the cap and appear to be an actual part of the cap surface.

Numerous terms are used to describe the nature of the cap surface. Some of the important and more commonly used terms include **appressed-fibrillose, floccose**, **hirsute**, **pruinose**, **pubescent**, **scurfy**, **squamous, squamulose, squarrose, tomentose,** and **villous.** These and other terms are illustrated in Fig. 6 and described in the Glossary.

When describing the surface of the cap it is important to note the condition of the fruit body. The age of the fruit body and the effects of weathering often change or modify the cap surface. Describe as many stages of development as possible and note the weather conditions under which the fruit bodies are growing.

Color

Although variable and often difficult to describe, color is a very important characteristic. It is often affected by the age of the fruit body or by weather conditions, habitat, or even geographical location. Take care to note the range of color among young and old specimens and observe any color changes or staining that might take place with aging or from handling or bruising. Colors should be recorded as soon after collecting as possible because they may change as the fruit bodies begin to dry or decay. Note the conditions under which the colors are described, because colors seen in the sunlight appear slightly different when seen under artificial light.

Mushrooms and other fruit bodies exhibit a variety of colors, as varied as those in flowering plants. People tend to use common color terms differently, so it is best to refer to a color guide or color chart when describing the color of fruit bodies. The *Methuen Handbook of Colour* (Kornerup and Wanscher 1978) is a fairly good color guide, available at a reasonable price. When accurate color notes cannot be taken, a color photograph or slide is the best alternative.

Many of the color terms used in the species descriptions in Part Two of this book are general descriptions of shades selected to convey a general impression of the color range for a wide audience. Some of these general terms, however, are quite selective in meaning. A few of the more obscure terms used fairly often in the book, such as **agate**, **alutaceous**, **fawn**, **fulvous**, and **rufous**, are defined in the Glossary. Other color terms, however, are also used, mainly in the technical descriptions of Part Two, as defined in *Color Standards and Color Nomenclature* (Ridgway 1912). A list of the technical color terms from Ridgway's color guide used in this book appears in Appendix 3.

Besides noting cap color at various stages, see whether the cap is uniformly colored or if two or more regions are differently colored. For example, the central region of the cap is often colored differently than the margin. Sometimes caps have zones of color or are complexly colored or mottled. The surface may also stain or discolor when the fruit body is bruised or handled; check for this feature by simply rubbing or deliberately bruising the cap surface.

When fibrils, scales, or other processes occur on or as part of the cap surface, accurately describe the color and note whether the background color differs.

Flesh

The cap flesh should be carefully described. Note the following features:

- color and any color changes
- thickness—at the center and margin
- consistency—soft, hard, firm, turgid, or fragile; tough, woody, and leathery are terms that are useful for describing polypores

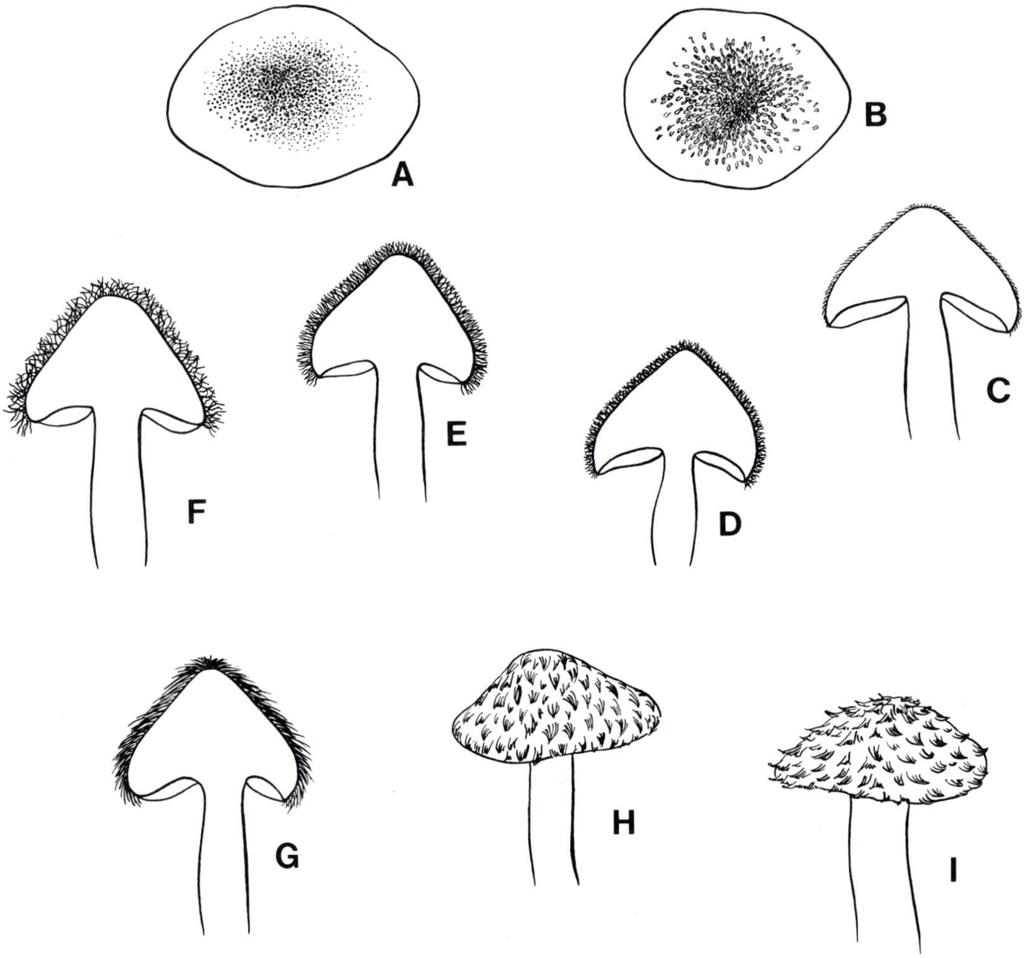

Fig. 6. Cap surfaces: A, pruinoise; B, scurfy; C, pubescent; D, floccose; E, tomentose; F, somewhat matted-fibrillose; G, villous; H, squamulose; I, recurved-squamulose.

- taste and odor—these characteristics are important in distinguishing species and sometimes in identifying genera; however, tastes and odors are difficult to describe and not all people can differentiate well among them.

To determine odor, crush a small piece of flesh; sometimes simply smelling the flesh is adequate. Often the odor is not distinctive or is simply fungus-like. Some distinctive adjectives, however, used to describe odors are unpleasant, disagreeable, fragrant, sweet, anise-like, radish-like, bean-like, like green corn, or like fresh meal.

Taste also varies. Sometimes it is simply mild, indistinctive, or fungus-like; in other cases it is peppery or acrid, leaving a burning sensation in the mouth, like fresh meal, bitter, anise-like, creosote-like, or like phenol. The flesh of some fungi is mild when first tasted, becoming distinctive, for example peppery, after a few seconds. Taste is not always an indication of edibility. Some of the most poisonous fungi have a mild or even a pleasant taste. When tasting the flesh or some other part of a fruit body for identification purposes, be sure not to swallow it.

The flesh of some fungi produces a milk-like or watery substance referred to as **latex**. When the flesh of the cap, gills, or stalk is cut or broken, the latex usually appears on the exposed surface. Latex is characteristic of the genus *Lactarius* but also occurs in other fungi. Note the presence or absence of latex in the flesh. When it is present, note its color, whether the color changes, whether it stains the broken flesh or surface of the fruit body, and, if possible, its taste.

Gills

The gills of a gill mushroom produce the spores. In fruit bodies other than gill mushrooms, the gills are replaced by tubes, spines, or teeth, slightly wrinkled ridge-like folds, or some other type of spore-producing surface.

One of the most important features of the gills is the way they are attached to the stalk or their position in relation to the stalk. The gills usually run in a radial direction from the stalk to the edge of the cap. Look at the gills and follow them from the edge of the cap inward to the stalk to determine the type of gill

Fig. 7. Types of gill attachment (A-I) and spacing (J-M). Attachments: A and B, free; C, seceding; D, adnate; E, deeply adnexed; F, deeply sinuate; G, uncinate; H, adnexed; I, decurrent. Spacing: J, distant; K, subdistant; L, close; M, crowded.

attachment. The gills may be attached in various ways or may not be attached at all. The various kinds of gill attachment, such as **adnate, adnexed, arcuate, arcuate–decurrent, decurrent, free,** and **subdecurrent,** are listed in the Glossary and the most common ones are illustrated in Fig. 7 A–I. In some mushrooms the gills approach the stalk apex, often very closely, but do not actually touch or contact the stalk; such gills are described as free (Fig. 7 A and B). Care should be taken in determining the type of gill attachment because the gills of some mushrooms pull or break away from the stalk with maturity, a condition described as **seceding** (Fig. 7C). When the gills have seceded, small lines representing the point of gill attachment can usually be seen on the stalk apex.

Spacing of the gills is also of considerable importance. Spacing refers to the distance between gills. The terms **close, crowded, distant,** and **subdistant** are used to describe spacing. See Fig. 7 J–M and the Glossary for clarification of these terms.

The width of the gills, namely the distance from the gill edge to where it attaches to the cap, should also be noted. The terms narrow, moderately broad, and broad are commonly used. Gills that are broadest in the middle are called **ventricose**.

Gill color is very important, especially the color of young gills before the spores mature. The color of mature gills sometimes helps to indicate the color of the spores, but as the colors do not always correspond, caution should be used in relating mature gill color to spore deposit color. A spore deposit should always be taken to ascertain the spore color. Also note whether the gill **edge** is the same color as the gill **faces** and record the color of the gill edge when it differs.

The gills of some mushrooms contain latex. When these gills are cut or broken, the latex appears as small droplets on the cut surface. Describe the characteristics of the latex of the gills the same as you would for latex associated with the cap.

The general outline of the gill edge should also be noted. The gill edge is often entire (Fig. 8A), but it may be minutely fringed, **eroded**, or finely curled to wavy (Fig. 8 B–E). At times the gill edge is toothed, like the edge of a saw. When the teeth are fairly coarse the edge is called **serrate** (Fig. 8G); when the teeth are small the edge is described as **serrulate** (Fig. 8F).

Some gills do not extend the full distance from the cap edge to the stalk. Short gills are called **lamellulae** (Fig. 9B). Lamellulae may be of various lengths on a single cap; each length category is called a **series**. For example, the gills of a mushroom may be described as adnate with three series of lamellulae.

Some mushrooms produce gills that are branched or forked, a feature that may help in identifying species or even genera. Also, short veins may connect the face of one gill to another or extend into the interspace between two gills. Gills of this nature are described as **interveined**.

The gills of some mushrooms, mainly those in the genus *Coprinus*, decompose into an inky fluid when the fruit body is mature. The gills of such mushrooms are said to be **deliquescent**. The flesh of the cap in these mushrooms is often deliquescent also.

Stalk

Many features of the stalk are important in the description and identification of species. The length of

Fig. 8. Gill edges: A, entire; B, crenate; C, wavy; D, eroded; E, crisped; F, serrulate; G, serrate.

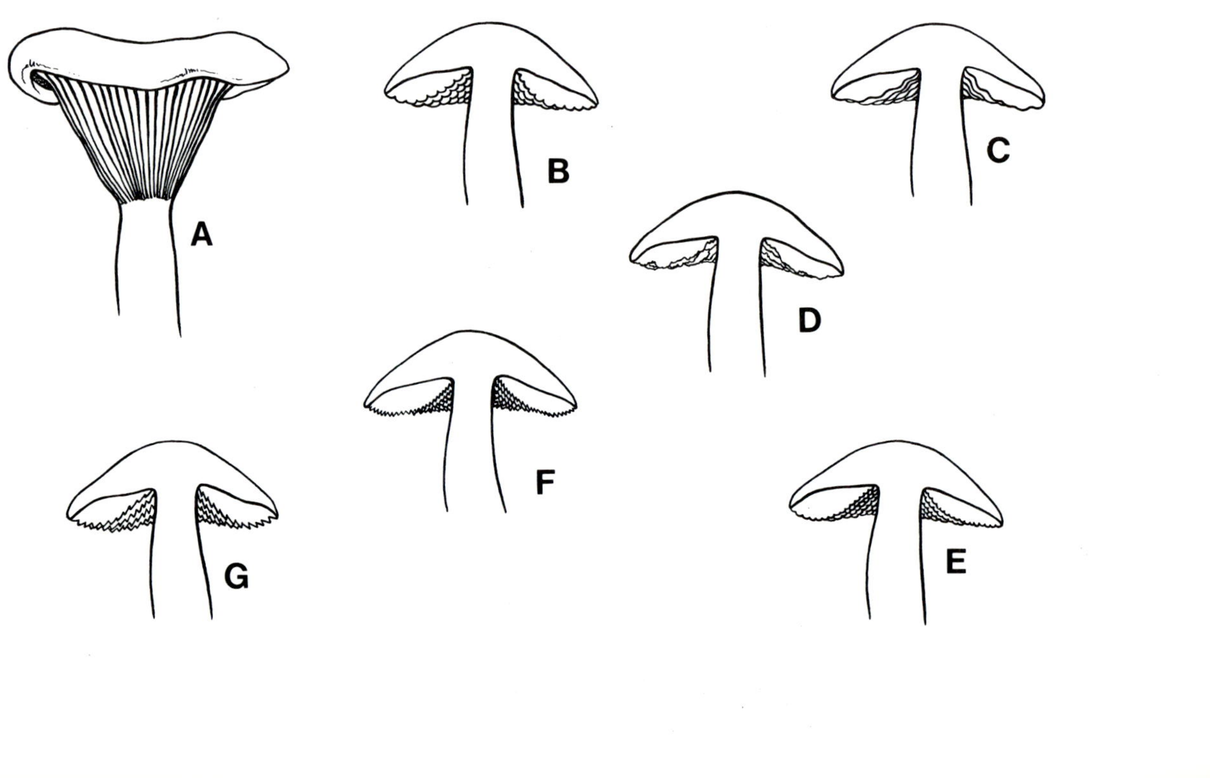

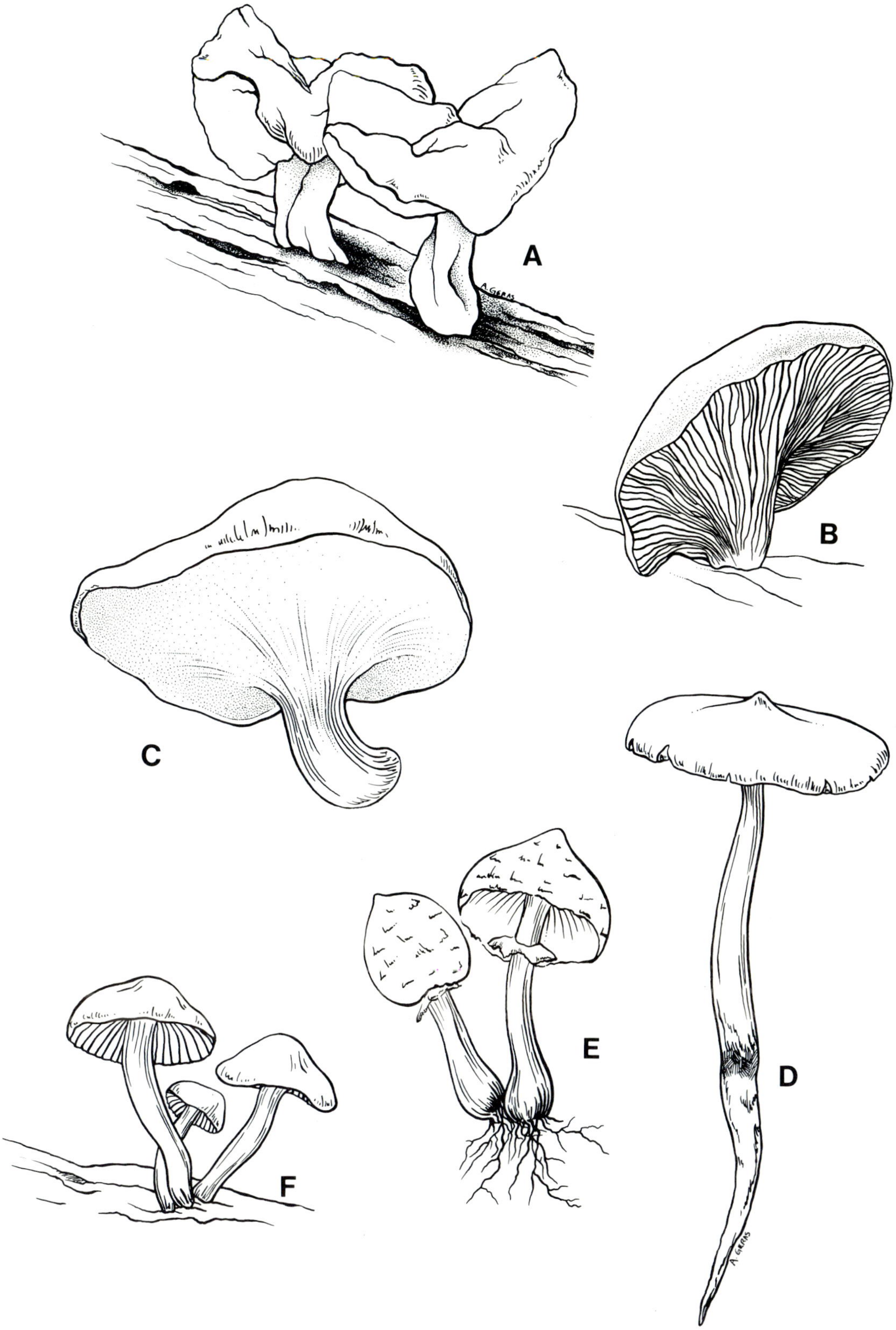

Fig. 9. Stalk attachments (A–C) and bases (D–F): A, central; B, lateral; C, eccentric; D, rooting; E, exhibiting rhizomorphs; F, inserted.

the stalk and the width at the apex and base or at the thickest portion should be recorded. When the stalk is attached in the center of the cap the attachment is termed **central** (Figs. 3A, 9A); but at times the attachment is **lateral**, with the stalk attached to the margin of the cap (Fig. 9B), or **eccentric**, with the attachment intermediate in position between lateral and central (Fig. 9C).

The stalk base is attached to the vegetative mycelium. Where contact is made with the substratum there is often a superficial mass of hyphae called the **basal mycelium**, or when the hyphae are densely matted or wooly, the **basal tomentum**. Hyphae on the stalk base that form distinct strands are called **rhizoids**, whereas a thicker, cord-like strand of hyphae is termed a **rhizomorph** (Fig. 9E). When the hyphae are large, long, stiff, and hair-like, the stalk base is described as **strigose**. Infrequently the stalk base is naked and directly attached to the substratum; if so, it is said to be **inserted** (Fig. 9F).

In some fruit bodies the stalk base continues into the substratum as a root-like process. The stalk is said to be **rooting** and the root-like base is called a **pseudorhiza** (Fig. 9D).

When the stalk is round in cross section it is termed **terete**, and when it is somewhat flattened it is called **compressed**. When the stalk is the same diameter from apex to base in longitudinal view, it is described as **equal** (Fig. 10A); or it may be described as **tapered** in one direction or another, often toward the base (Fig. 10B). When the stalk is enlarged below and club-shaped, it is termed **clavate** (Fig. 10 C and D); when it appears swollen in the middle, **ventricose**; or when thickened only at the very base, like the base of a green onion, **bulbous** (Fig. 10 E–G). The bulb at the stalk base may be evenly rounded or it may have a rim or margin above (Fig. 10G).

The stalk surface may be modified in various ways. When describing it, compare the apex, middle, and base of the stalk. Also, check for remains of the partial or universal veil. General terms for describing the stalk surface are similar to those used to describe the cap surface. However, there are some features specifically related to the stalk that require some discussion. Ridges or coarse ribs may be present on the stalk surface. At times very coarse, sharp ridges develop on the stalk giving it a channeled appearance, a condition termed **lacunary** (Fig. 10H). Boletes often have colored spots on the stalk surface called **glandular dots** (Fig. 10J). When the surface of the stalk is roughened with small pointed scales, it is described as **scabrous**, or when marked with small dots, **punctate** (Fig. 10I); when there is a distinct net-like pattern on the surface, the condition is described as **reticulate** (Fig. 10K).

The color of the stalk should be recorded as for the cap, clearly noting any color changes occurring with age or caused by bruising.

The interior of the stalk may be solid or it may be hollow with a more or less distinct longitudinal channel through the center; when, however, the central part of the stalk is filled with soft, cottony material, the stalk is referred to as **stuffed**. The actual texture of the stalk is helpful in identifying some genera and even some species. A common stalk texture is **cartilaginous**, a condition in which the stalk is usually fairly thin, often less than 5 mm, firm and tough but readily bent, and often cleanly splitting when bent between the fingers. Another common stalk texture is a condition termed **fleshy–fibrous**; the stalk in this case is usually thicker than 5 mm, leaving a more or less ragged edge when broken in half. In some mushrooms, for example the genera *Russula* and *Lactarius*, the stalk breaks like chalk into small chunks. Other stalks have a woody, corky, or leathery texture. These textures are common among some of the hedgehog fungi and the polypores. Check the flesh of the stalk for the presence of latex.

Universal and partial veils

The **universal veil** and the **partial veil** occur in certain gill mushrooms and in some boletes. Because the characteristics of the veils are similar in boletes and gill mushrooms, this discussion applies equally to both forms. The tubes referred to below perform the same function in boletes that the gills do in gill mushrooms. The tubes and other distinguishing characteristics of boletes, as well as interesting features of other mushroom-like forms of Basidiomycota, are discussed fully later in this chapter.

The universal veil covers the entire young fruit body, including the cap, stalk, and gills or tubes (Fig. 1 F and G). The partial veil covers only the developing gills or tubes (Fig. 1 F and H).

The universal veil and its remains

During early development, the **button** stage, certain mushrooms are covered by the universal veil. In the early stages the fruit body grows fairly slowly and uniformly, and the growth of the veil usually keeps pace with it. At this stage the fruit body resembles an egg or a young puffball; but if it is cut longitudinally, a cap, stalk, and gills or tubes are evident (Fig. 1F). Eventually, the mushroom begins to enlarge and elongate more rapidly than does the universal veil, and the emerging cap and stalk break through the veil (Fig. 1 G and H).

The texture of the veil greatly determines where, and in what form, the remains of the veil are found on the fruit body. To determine if a mature mushroom was originally enclosed in a universal veil, look for one or more of the following:

- a membranous cup around the base of the stalk—if delicate, this tissue may remain in the ground when the mushroom is collected; this structure is called a **volva** (Figs. 1A, 36)
- patches of tissue at the base of the stalk (Fig. 85)
- a dry, powdery coating of separable granules, often easily removed, on the cap or the stalk or both

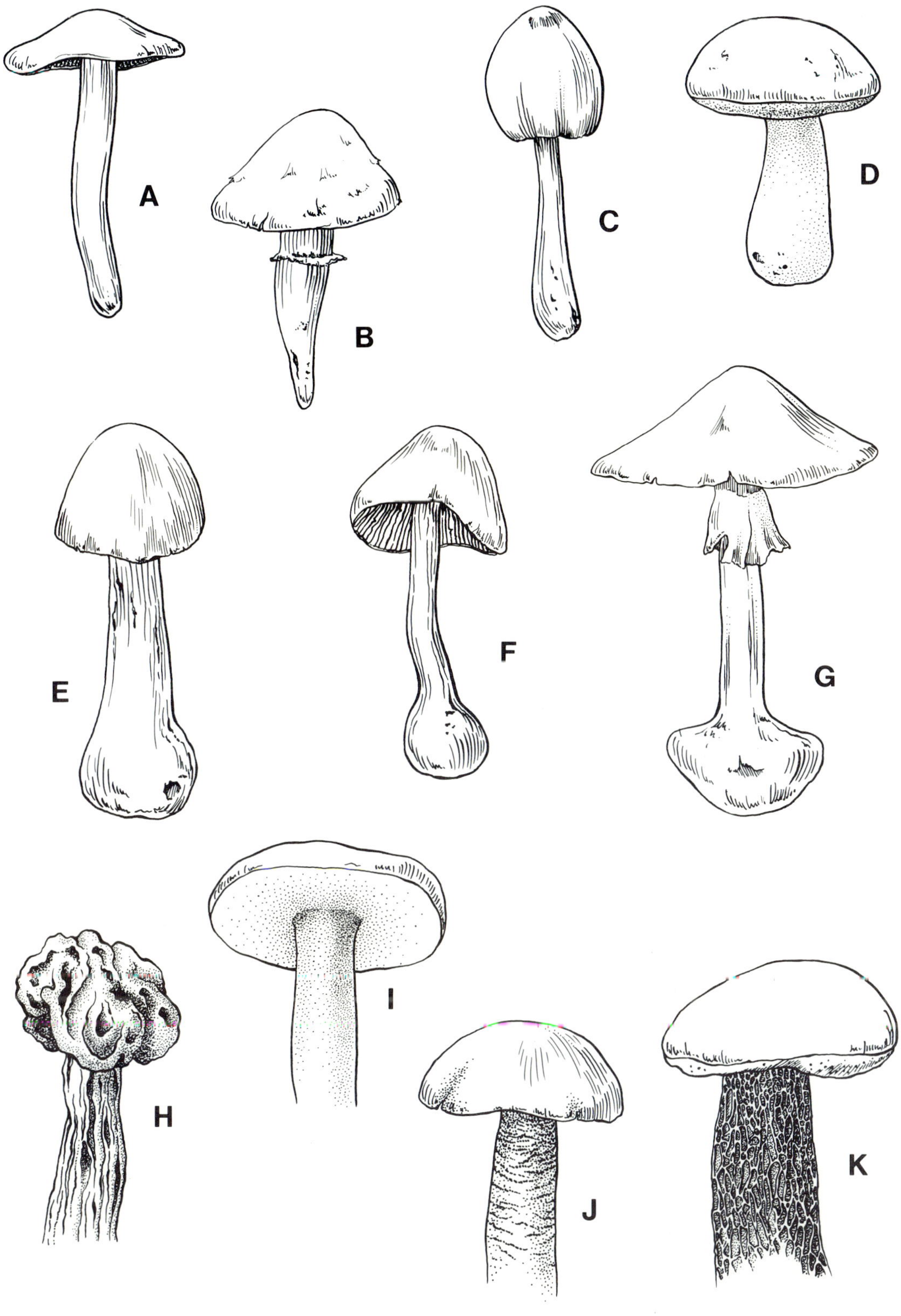

Fig. 10. Stalk shapes (A–G) and surfaces (H–K): A, equal; B, tapered downward; C, narrowly clavate; D, clavate; E and F, bulbous; G, abruptly bulbous; H, lacunary; I, punctate; J, exhibiting glandular dots; K, reticulate.

- membranous or wart-like patches of tissue on the cap surface, sometimes colored differently from the cap and usually cleanly separable from it (Fig. 85).

The partial veil and its remains

Certain fruit bodies have both a universal and partial veil; others may have only one or the other (Figs. 1A, 4H). Many lack both types of veils. A partial veil in buttons usually covers only the developing gills or tubes. It extends from the cap edge to the stalk, usually near the apex. As the cap expands, especially in a radial direction, the partial veil is broken (Fig. 1H). A partial veil may be membranous or composed of fine fibrils. When it breaks or is torn, veil fragments may remain attached to the edge of the cap, to the stalk, or to both. To determine if a partial veil is present, look for at least one of the following features:

- a more or less membranous ring on the stalk, called the **annulus** (Fig. 4 H and J)—In some fruit bodies the annulus may become free and fall to the base of the stalk. This situation is to be distinguished, however, from a **basal annulus**, in which the ring is actually formed toward the base of the stalk. The annulus is often composed of a single layer but when it has a tough or cottony outer coating and a membranous inner one, a double annulus may result (Fig. 4K). In contrast, the annulus may be **evanescent**, so thin and delicate that it quickly wears away, leaving at most a faint trace on the stalk, the annular zone
- a zone of fibrils on the stalk apex representing the remains of a fibrillose veil, called a **cortina**
- ragged or tooth-like patches of tissue or fibrils on or hanging from the cap edge (Figs. 107, 109). The cap margin when hung with pieces of the partial veil is described as **appendiculate**.

OTHER MUSHROOM-LIKE FRUIT BODIES

The typical form and shape exhibited by the gill mushroom, namely the umbrella-shaped cap on a stalk, is also characteristic of fruit bodies of other Basidiomycota such as boletes, some polypores, certain hedgehog fungi, and chanterelles. The main difference between gill mushrooms and these other fruit bodies is in the nature or configuration of the spore-producing surface, the hymenophore. Secondly, the texture of the fruit body may be somewhat different in these forms. Finally, as in some gill mushrooms, the stalk or cap or both may be lost or modified in some of these other forms.

Boletes and polypores produce their spores inside **tubes** that cover the underside of the cap (Fig. 3 C and D). The flesh and the tubes in boletes are soft to firm and fleshy, whereas in polypores such as bracket fungi and conks, the tubes and flesh are usually leathery, tough to fibrous, or woody.

The fleshy nature of boletes relates them to gill mushrooms. Macroscopic features of the cap and to a degree the stalk are similar to those for gill mushrooms. The tubes on the underside of the cap form a soft to spongy layer that can often be easily separated from the flesh of the cap (Fig. 11 A and B). The spore-producing tissue, the hymenium, lines the inside of each tube. The tubes are arranged vertically, that is, perpendicular to the substratum (Fig. 11 E); the opening of each tube is called a **pore**.

The following characteristics of the tubes are important for identification:

- length—measure their maximum length and their length near the stalk; the tubes are sometimes shorter in the area around the stalk
- color—note also any staining or discoloration
- pores—note the number of pores per square millimetre, their shape, for example round or angular, and their color and any staining or discoloration; remember that the pores are not always the same color as the tubes.

Polypores such as bracket fungi or conks also produce spores on the inside of tubes (Fig. 11C). The tubes usually are not fleshy but have a tougher consistency than do those of boletes. The tubes should be described as for boletes. Sometimes a somewhat special terminology is used in descriptions of polypores. If interested in this group, one should obtain references specifically related to polypores (e.g. Overholts 1953).

The hedgehog fungi produce **spines** in place of gills (Fig. 3G). The spines cover the undersurface of the cap, and the surface of each spine produces spores. As in describing the gills of a mushroom, the color and texture of the spines are important features. Also note their length; whether or not they are decurrent on the stalk; the number of spines per square millimetre; and the nature of the tip of the spines, for example whether they are blunt or sharply pointed. The consistency of many hedgehog fungi is tough to fibrous or even woody, for example as in the genus *Hydnellum*; in others, such as in the genus *Dentinum*, the consistency is more or less fleshy. There is no good evidence of poisoning by hedgehog fungi.

The stalk and cap of certain fruit bodies in the Basidiomycota are highly modified or even lacking. For example, many polypores lack a stalk and grow laterally attached to the side of a log or tree, the feature from which they derive the name bracket fungi (Fig. 11C). In some instances the fruit body consists of a simple, vertical, cylindrical to club-shaped stalk with the spore-producing surface occurring on the upper portion of the stalk; this form is typical of some coral fungi (Fig. 3E). Other coral fungi have branched fruit bodies consisting of more or less vertical branches that originate from a single, often short stalk (Fig. 3F). In these forms the spores are produced on the branches. Finally, some fruit bodies lack both a distinct stalk and cap; for example, certain hedgehog fungi produce branches

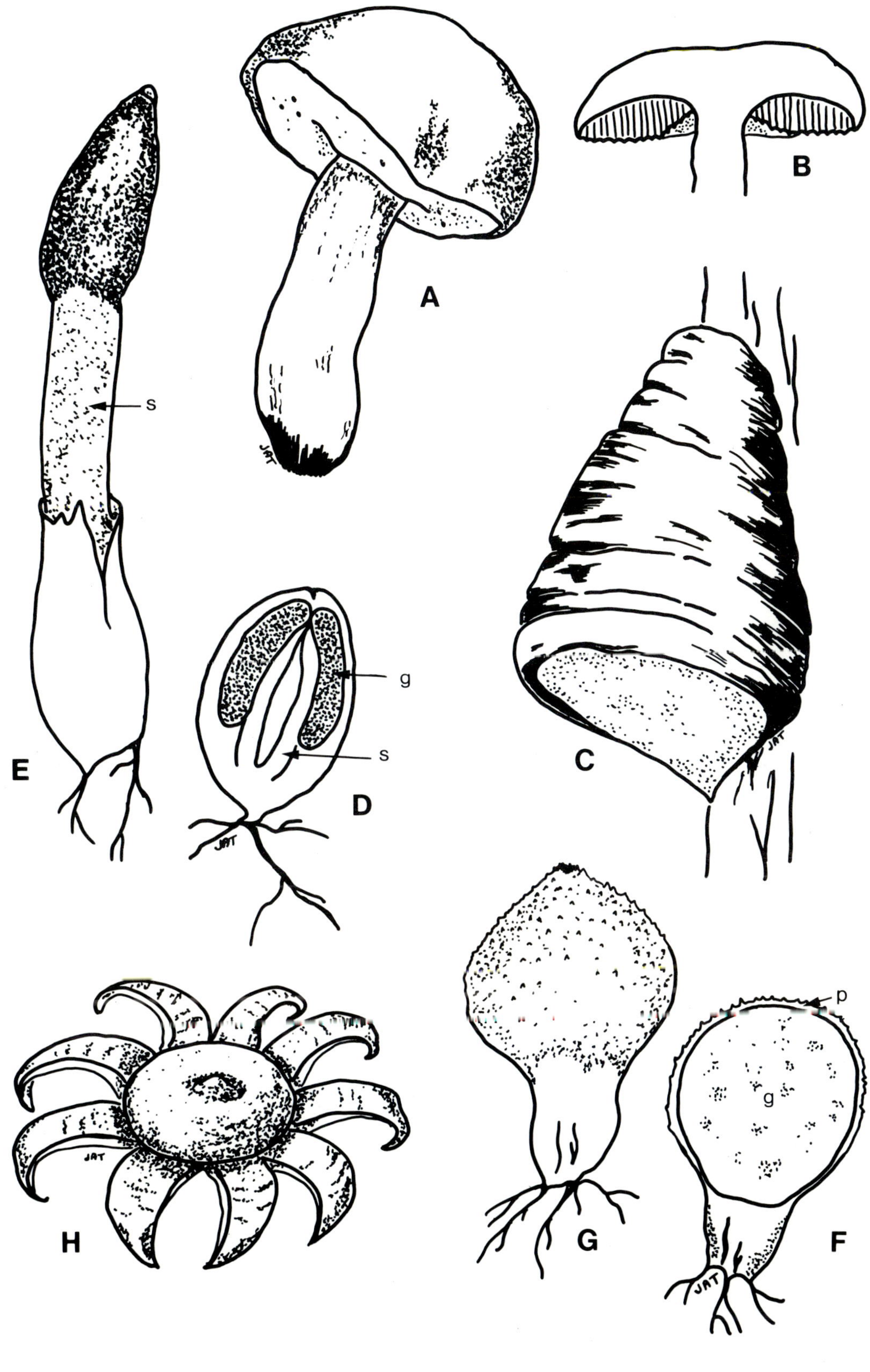

Fig. 11. Fruit bodies of the Basidiomycota: A, *Boletus* sp., with a fleshy tube layer, exhibiting a porous cap undersurface; B, *Boletus* sp., showing the tube structure in cross section; C, *Fomes fomentarius*, a bracket or conk form, with the surface of the hymenium porous; D, *Mutinus* sp., a stinkhorn at the egg-like stage, showing the internal gleba (*g*) and the stalk (*s*); E, *Mutinus* sp., a mature stinkhorn; F, *Lycoperdon* sp., a puffball, showing the central gleba (*g*) and surrounding peridium (*p*) in cross section; G, *Lycoperdon* sp., a small puffball; H, *Geastrum* sp., an earthstar.

that are more or less covered with downwardly directed spines. The spores are produced on the surface of each spine.

PUFFBALLS AND RELATED FUNGI

Puffballs and their allies are also members of the Basidiomycota. Besides puffballs, this group includes earthstars and stinkhorns. This entire group of fungi produce their spores inside the fruit body, which is called the basidiocarp as in other Basidiomycota. The spores usually remain enclosed until maturity, or they are exposed just at the time of maturation, as in stinkhorns.

In the following discussion macroscopic features of puffballs and stinkhorns are emphasized because the fruit bodies of these fungi are the ones most likely to be seen when one is collecting fungi for food.

Puffballs are commonly collected for food, so the characteristics of this group of fungi should be kept clearly in mind. The fruit body is often a more or less spherical to globose or pear-shaped structure. There is typically no true stalk but in some genera there is a **sterile base** that may be stalk-like in appearance (Fig. 11G). The spore tissue, called the **gleba**, is produced inside the fruit body and is exposed or released at maturity or soon after maturity. The gleba is enclosed by a distinct wall, the **peridium** (Fig. 11G). The peridium is often two-layered, although the outer layer may be cracked or even entirely lost at maturity. The peridium varies in thickness; in *Calvatia* and *Lycoperdon* it is fairly thin, whereas in *Scleroderma* it is relatively thick. The surface of the fruit body may be smooth to wrinkled or covered with warts, spines, or scales. In some puffballs scales or warts are formed by a cracking of the outer layer of the peridium as the fruit body expands.

The gleba of many puffballs is soft or somewhat fleshy to firm. It is pure white when young, but it gradually changes from white to some shade of yellow, brown, or lilac as it matures. With maturity the spores often gradually turn into a dry, powdery mass. Puffballs that have a more or less soft, white gleba when immature typically have a thin peridium. Examples of this group are *Lycoperdon, Bovista,* and *Calvatia.* In the genus *Scleroderma* the gleba is firm and very hard and is usually somewhat violet or purple; however, some pale or white mottling may be present, and very young fruit bodies of *Scleroderma* sometimes have a pale to white gleba. The gleba of *Scleroderma* at maturity forms a powdery mass of spores that is blackish to blackish brown or purple brown. Besides these features, this genus has a characteristic thick peridium. *Scleroderma* are commonly called thick-skinned puffballs. They should not be eaten.

The powdery mass of spores that is characteristic of puffballs is released from the interior of the fruit body in various ways. For example in *Lycoperdon* a small pore develops in the top of the fruit body through which the spores gradually escape, in puffs. In others such as certain species of *Scleroderma* the peridium splits open in an irregular to star-like pattern allowing the spores to escape. In still others the peridium breaks irregularly as a result of mechanical damage or disintegration, thereby releasing the spores.

When describing puffballs, describe both young and mature stages; the characteristics of mature specimens, however, must be ascertained for accurate identification. The following characteristics should be noted:

- size and shape
- presence or absence of a sterile base
- color of the surface and any color changes occurring with age or when bruised or handled
- the nature of the surface, for example whether it is smooth, wrinkled, or warty
- the characteristics of a median longitudinal section through the fruit body, such as the thickness and color of the peridium and the consistency and color of the gleba, comparing young and mature specimens and noting the color of the mature spore mass
- the characteristics of the surface of the peridium at maturity, noting how it breaks up or changes with age and how the powdery spore mass escapes from the fruit body.

Earthstars are closely related to puffballs and should be mentioned here. Like puffballs they have a two-layered peridium enclosing the gleba. The outer layer, which is often thicker than the inner layer, splits radially in a star-like pattern; the resulting rays curve back exposing the inner peridium which surrounds the spore mass (Fig. 11H). The inner peridium is sometimes raised on a short stalk and typically has a pore at the top. The spore mass, which is dry and powdery at maturity, eventually escapes through the pore. Descriptions of these fungi are similar to those for puffballs.

In stinkhorns the gleba (Fig. 11D) is enclosed in the fruit body until maturity. The young fruit body, called the egg stage, is attached to the substratum by a rhizomorph (Fig. 11E). The egg has a thin, papery outer layer and a second layer of jelly-like material that gives the egg a rubbery to jelly-like texture. These layers make up the peridium that encloses the gleba; the gleba almost completely surrounds a hollow stalk, sometimes called the **receptacle** (Fig. 11E). At maturity, the stalk elongates rapidly and breaks out of the egg, exposing the spore mass. Usually, remnants of the egg remain as a cup-like structure around the base of the stalk.

The spore-producing area may occur at the apex of the stalk as in the genera *Mutinus* and *Phallus* (Fig. 11E) or on stout columns, on arms, or on a chambered to lattice-like head as in the genera *Clathrus* and *Linderia.* The spore mass is always slimy and has a fetid odor at maturity.

In describing stinkhorns note at least the following:

- size, shape, and external color of the egg
- size, shape or configuration, texture, and color of the stalk
- position and color of the spore mass.

Few people collect stinkhorns because of the extremely strong, fetid odor they produce when mature. However, Farlow (1890) reported that a child was poisoned by *Clathrus columnatus* Bosc and that several hogs died from eating this species.

Ascomycota

The Ascomycota treated here are in the class commonly called the Discomycetes. Most of the Ascomycota that produce fruit bodies large enough to be of interest to those collecting fungi for food belong to the Discomycetes. Some of the larger and more frequently encountered Discomycetes include the cup fungi, the highly prized morels, and the false morels. The name cup fungi refers to species that produce a more or less cup-shaped fruit body. **Apothecium** (pl. **apothecia**) is the name applied to the fruit body of the Discomycetes.

APOTHECIA

The terminology used to describe apothecia is similar to that used to describe mushrooms. Refer to the appropriate sections of this treatment and the Glossary for general descriptive terms and definitions.

Many apothecia are cup-shaped, lacking a distinct stalk or cap (Figs. 2A, 12A). The inner surface of the cup produces the spores. The spore-producing surface is called the hymenium, as it is in the Basidiomycota. The exterior of the cup is usually simply called the outer surface. A description of an apothecium of this kind should include shape, size of all stages, color of the hymenium and the outer surface, surface features of the hymenium and outer surface, such as whether they are smooth or glabrous, and the consistency and color of the flesh of the cup. The flesh of the apothecium has a specialized name, the **excipulum**.

Many of the larger apothecia, such as false morels, have a distinct stalk and a distinct cap or cap-like region (Fig. 46). The cap may simply consist of a cup raised on a stalk (Fig. 12B). Often, however, the cap loses its typical cup shape and is instead saddle shaped (Fig. 12C), globose to ovoid (Fig. 10H), or more irregularly shaped to lobed. As the shape of the cap changes, the configuration of the surface may also change and become wrinkled to convoluted (Fig. 10H) or pitted to honeycombed (Fig. 12D).

As the cap becomes more irregular to folded, the outer surface, namely the underside of the cap, often becomes less distinct and the margin of the cap, instead of being free (Fig. 234), becomes enrolled (Figs. 46, 49) or fused with the tissue of the stalk as in *Morchella* (Fig. 210). The fusion may be partial or the cap surface may be fused with the stalk along its entire length.

Note these important features when describing the apothecia:

- shape and configuration—for example whether the cap is convoluted or pitted
- size, width, and height
- consistency of the cap—whether or not it is fragile
- color of the hymenium of young and mature specimens and tendency to any bruising
- color and shape of the outer or under surface of the cap.

Clearly indicate whether the cap is fused with the stalk; the degree of fusion can be more easily seen if a median longitudinal section is made through the cap and stalk.

The stalk of the apothecium may be narrow and cylindrical with an even surface, but often it is modified both in shape and size. As in other mushroom forms previously discussed, the stalk may be equal, tapered downward or upward, clavate, or more or less bulbous. Although some stalks are fairly narrow, others are thick and sometimes almost as wide as the diameter of the cap. The surface may have longitudinal ridges or folds with corresponding grooves, or distinct longitudinal or network-like ribs, the condition described as lacunary (Fig. 10H).

Often the stalk is hollow and has a single longitudinal channel, or it is strongly folded with corresponding deep furrows or cavities. To see this feature make one or more cross sections of the stalk. Thick stalks that are strongly folded tend to have more than one channel that may or may not extend the full length of the stalk.

When describing the stalk note the following:

- length and width at the apex and base
- shape
- configuration of the surface
- whether or not the surface is glabrous, pubescent, or otherwise modified
- color of the surface
- nature of the stalk interior—whether it is hollow and the number and length of any cavities that may be present.

The flesh of the stalk, like the cap, is technically called the excipulum. Make a median longitudinal section through the cap and stalk to determine the color of the excipulum; note any color changes that occur after cutting the flesh. Also note the consistency of the excipulum, and its odor and taste.

Habitat and Habit of Growth

The habitat of a particular specimen, namely the environment in which it grows, is often helpful in identification. Some fungi have very specific environmental and substrate requirements, whereas others have more general requirements.

Make note of the general environment in which a fungus is found, for example whether it is growing in a grassy field, a meadow, or a lawn; in well-manured soil; or in pine woods, oak woods, or beech-maple woods. When a fungus grows in a wooded area or where there are scattered trees or shrubs, the fungus may form

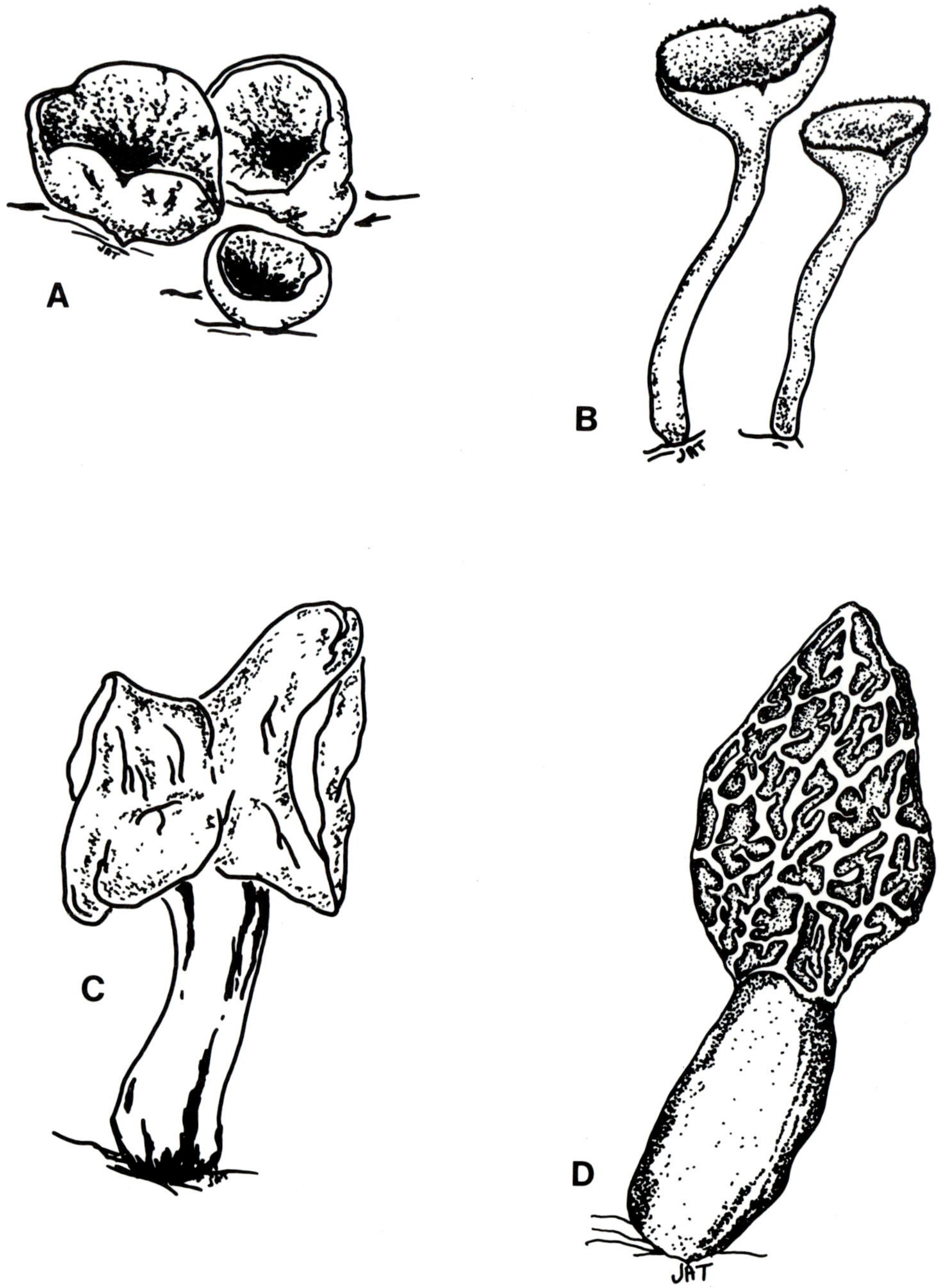

Fig. 12. Fruit bodies of the Ascomycota: A, *Peziza* sp., a cup-like, stalkless form; B, *Helvella macropus*, a stalked, cup-like structure; C, *Helvella* sp., exhibiting a saddle-like, lobed cap; D, *Morchella* sp., with a pitted, sponge-like cap.

mycorrhizae, that is, the hyphae may grow in association with the roots of green plants. In this situation note also the shrubs and trees growing in the vicinity of the fungus.

The substrate from which the fungus is fruiting is one of the most helpful habitat characteristics. Many fungi grow on the ground, often in a mixture of soil and dead organic material, and are simply called **terrestrial**. Other fungi fruit directly from deep accumulations of leaf litter and other dead organic material and are described as **humicolous**. Fungi that grow on wood, usually on rotting wood, are called **lignicolous**, and those that grow on dung or manure, **coprophilous**.

The habit of growth refers to the spacing between individual fruit bodies as they occur in their natural habitats. Habit of growth is an important feature because the number of fruit bodies produced in a fruiting is an indication of the extent of mycelium involved. **Solitary** means the occurrence of one fruit body at a time. **Scattered** indicates that the fruit bodies are widely separated, 30–60 cm apart. **Gregarious** means that the fruit bodies occur close together or in groups. **Caespitose** is used when the fruit bodies are growing very close together and are joined at their bases (Fig. 9E). Fruit bodies that are almost caespitose are called **subcaespitose. Connate** indicates a caespitose habit where the stalks are grown together for a considerable distance from the base upward.

Fungi do not always have just one habit of fruiting; one species may, for example, grow scattered to gregarious; another may be solitary to scattered.

Distribution and Seasonal Occurrence

The geographic distribution of some fungi is limited; others are widely distributed. *Phaeolepiota aurea*, for example, occurs primarily in western North America, whereas *Armillariella mellea* is widely distributed. However, the actual distribution of many species is not known. More collecting and study are needed before distribution maps for most fungi can be made.

The seasonal occurrence of fungi varies from species to species. Certain species fruit at a specific time of the year, as for example *Gyromitra esculenta*, which fruits in the spring; others fruit during the entire growing season when growth conditions are favorable, as does *Collybia dryophila*. *Cortinarius* and *Tricholoma* fruit primarily in the fall. *Russula*, certain *Lactarius*, and several species of *Amanita* and *Boletus* commonly fruit in the summer. *Amanita muscaria* and *A. pantherina* sometimes fruit in June and July and again in the fall. Environmental conditions, especially temperature, rainfall, and geographic location, influence the time of fruiting in fungi. The abundance of a species varies from year to year. The factors controlling crop size are poorly known, but abundance seems to be somewhat dependent on weather conditions and location.

Literature

Farlow, W. G. Poisonous action of *Clathrus columnatus*. Bot. Gaz. 15:45–46; 1890.

Kornerup, A.; Wanscher, J. H. Methuen handbook of colour. 3rd ed. London: Methuen & Co., Ltd.; 1978.

Overholts, L. O. The Polyporaceae of the United States, Alaska, and Canada. Ann Arbor, MI: University of Michigan Press; 1953.

Ridgway, R. Color standards and color nomenclature. Washington, DC: R. Ridgway; 1912.

3

MICROSCOPIC FEATURES

Most of the following discussion in this chapter concerns the hyphal structure and spores of the Basidiomycota. The two classes of Basidiomycota treated in this book are discussed separately here. The class Hymenomycetes, in which the spore-producing structures are located exteriorly, is the more important of the two and is discussed first. Emphasis in the treatment of the Hymenomycetes is placed on gill mushrooms, with references to other forms such as boletes, coral fungi, and polypores. Largent et al. (1977) provide further details of this group. Following the treatment of the Hymenomycetes is a brief description of the hyphal structure of puffballs and stinkhorns of the class Gasteromycetes, in which the spores are enclosed within the fruit body. The chapter is completed by a section on the hyphae and spores of the class Discomycetes of the Ascomycota. The discussion of the Discomycetes is simplified and restricted in scope.

Basidiomycota

Although macroscopic characteristics are important in describing and identifying fungi, an accurate account of a species or a correct identification often depends on close study of the microscopic features. Both the vegetative mycelium and the fungus fruit body are composed of hyphae. Although microscopic studies of the vegetative mycelium are not very helpful, studying the hyphal structure of the fruit body under a microscope provides numerous characteristics and differences that aid in identification. The fruit body of a typical mushroom is composed of three main parts: the cap; the hymenophore, consisting of gills, tubes, or spines, on the underside of the cap; and the stalk. These structures have a surface layer as well as internal tissue called the **trama**. Although the trama provides some important microscopic characteristics, many more important features are found on the surface of the fruit body. As distinguished microscopically, the surface of the cap and, in particular, the surface of the hymenophore, namely the hymenium, are especially important in identification.

The spores, which are produced on basidia in the Basidiomycota, are called **basidiospores**. They are microscopic in size and provide many important characteristics for identifying fungi.

HYPHAL STRUCTURE OF MUSHROOMS AND OTHER HYMENOMYCETES

Surface of the cap

The surface layer of the cap is usually differentiated from the flesh of the cap. The interior flesh of the cap is called the cap trama. The surface layer in a microscopic description is commonly called the **cuticle**. Gill mushrooms, boletes, polypores, chanterelles, and hedgehog fungi with a cap and stalk have a similar structure. Coral fungi occur as single stalks or highly branched fruit bodies that likewise have a trama and extensive surface area, but no distinct cap.

The macroscopic features of the cap surface, that is, whether it is smooth, appressed, fibrillose, scaly, or otherwise, are related to the arrangement and association of the hyphae. If the hyphae are studied carefully with a microscope, much of what is seen macroscopically can be described and explained at the microscopic level (see Ch. 4, "Microscopic Techniques").

The hyphae of the cap cuticle vary greatly in size and in shape. The width of the hyphae, measured in micrometres (μm), range in size from a few micrometres to in excess of 25 μm. The shape of the hyphae varies from thread-like or cylindrical, a condition termed **filamentous** (Figs. 28D, 162C), to greatly inflated, to more or less spherical (Fig. 23E). The hyphae comprising the cap cuticle may be **interwoven** or they may be arranged more or less **parallel** to one another.

The hyphae of the cap cuticle may occur in one or more layers and are arranged in various ways ranging from parallel to the cap surface to perpendicular to it, with various intermediate orientations between the two extremes. The arrangement of the surface hyphae is best seen in a tangential or radial section of the cap. See Chapter 4, "Microscopic Techniques", for the method.

When the hyphae are arranged more or less parallel to the cuticle, the cap surface is called a **cutis** (Figs. 13A, 28D). The hyphae are often appressed to the surface, but sometimes they are loosened or upturned and appear macroscopically as scales or other structures. The terminal portion of each hypha, sometimes called the end cell, is often free, but sometimes several end cells are clustered together, agglutinated, or upturned as viewed under a microscope. When individual end cells become modified or specialized, they are called **pileocystidia**.

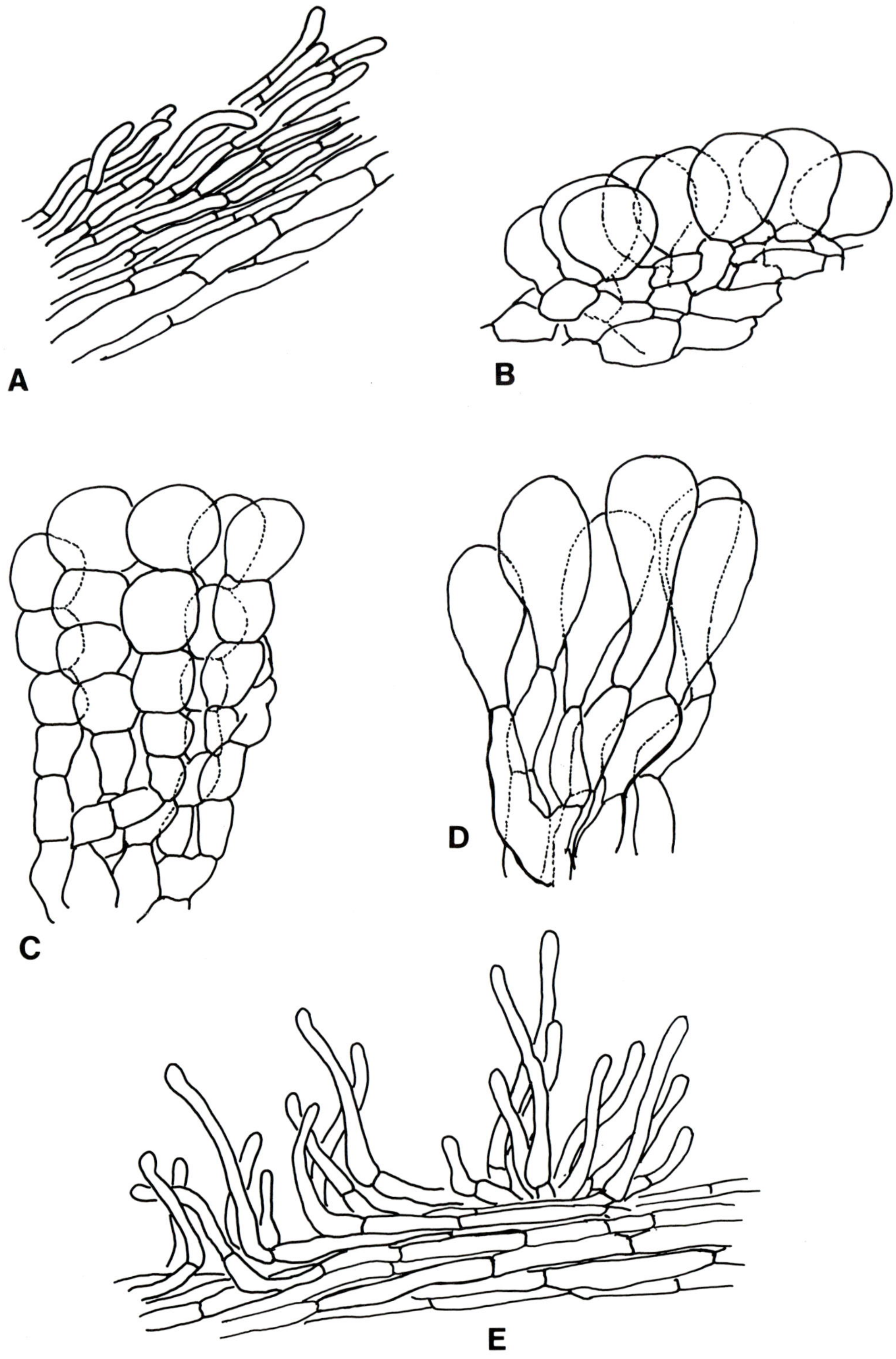

Fig. 13. Basic cuticles: A, cutis; B, cellular type; C, epithelium; D, hymeniform type; E, trichoderm.

The most highly modified and complex cap cuticles are composed of hyphae that are oriented perpendicularly to the surface. A complete discussion of these types of cuticles is beyond the scope of this publication. Technical papers on this subject should be consulted (Largent et al. 1977). Only a few basic types are discussed here.

When the hyphae of the cuticle are oriented perpendicularly to the cap surface and are more or less parallel to one another as seen in radial section, the cuticle is called a **derm**. The hyphae, as viewed in scalp section, from above, appear as a series of oval cells, like a series of bubbles. See Chapter 4, "Microscopic Techniques", for the methods of obtaining these sections. When the hyphae are more or less filiform to cylindrical, the surface is called a **trichoderm** (Fig. 13E).

A derm may also be composed of more or less isodiametric cells, having globose, subglobose, or vesiculose shapes. When the cells occur in a single layer, the cap surface is described as **cellular** (Fig. 13B). When the cells are highly inflated, and more importantly when such cells occur in many layers, the cuticle is called an **epithelium** (Fig. 13C). When the cuticle of the cap is covered with cells that are basidium-like in shape and size, it is described as **hymeniform** (Figs. 13D, 23E).

The hyphae of the cap cuticle may be imbedded in or may produce gelatinous material. In this case the prefix ixo-, meaning viscid, is used with the term describing the structure of the cap surface, for example **ixocutis** or **ixotrichoderm**.

When the cap cuticle is composed of two distinct layers, the outermost one is called the **epicutis** and the inner one is called the **subcutis**. The hyphae in the subcutis may be differently pigmented or may differ in size, shape, or arrangement in any combination, as compared with the hyphae of the epicutis and the cap trama.

The hyphae making up the cuticle of the cap may be thin walled, 0.5 μm or less, or thick walled. The hyphal walls may be smooth or they may be encrusted due to the deposition of pigments or other materials on the surface of the hyphae. These deposits may occur in many forms such as spirals, pegs, or spines.

The color of the hyphae as viewed under the light microscope is also important. The color usually represents pigment on the hyphal surface, in the hyphal wall, or within the hyphae. The mounting medium used for observation of the hyphae may change the color or remove the pigment entirely, thus making the hyphae appear colorless, a condition described as **hyaline**. Therefore, be sure to report the medium in which the hyphae are studied (see Ch. 4, "Chemical Reagents").

Trama of the cap and stalk

The trama of the cap is continuous with the trama of the stalk and is closely associated with the trama of the gills. The microscopic structure of the cap trama may be studied by making either radial or tangential sections of the cap, preferably both (see Ch. 4, "Microscopic Techniques"). The arrangement, size, shape, and color of the hyphae and the nature of the hyphal walls of the trama are studied as described for the cap cuticle.

The hyphae of the cap trama are frequently interwoven and variously branched. They sometimes grade gradually into the cap surface, but often a change in the orientation, size, or coloration of the hyphae indicates the beginning of the cap cuticle. The hyphae of the cap trama turn downward to form the trama of the gills. If you follow the hyphae of the cap trama downward into the tissue of the stalk, you can usually see where the stalk trama begins because the hyphae usually become more or less longitudinally arranged, parallel to the long axis of the stalk. The hyphae themselves may be parallel to one another or somewhat interwoven.

The hyphae of the cap and stalk trama vary from cylindrical to more or less inflated. In some genera, for example *Lactarius* and *Russula,* there are groups of inflated cells among the cylindrical hyphae. These cells are called **sphaerocysts** (Fig. 166E), and they account for the brittle to chalk-like consistency of the flesh in these fungi.

The trama may have other types of hyphae, such as **oleiferous** hyphae, **gloeoplerous** hyphae, or **lactiferous** hyphae. Oleiferous hyphae are often refractive and have oil-like contents; gloeoplerous hyphae have resinous or granular contents; and lactiferous hyphae contain a milky or latex-like substance. At times the presence or absence of such hyphae is helpful in distinguishing genera or groups of species, as for example in *Ramaria.* Usually, however, the feature is not helpful in separating species. Various mounting media such as Melzer's reagent or Cotton Blue are used to study these specialized hyphae. Depending on the medium, these hyphae may give characteristic color reactions (see Ch. 4, "Chemical Reagents").

Surface of the stalk

The microscopic structure of the surface layer of the stalk, which is referred to as the stalk cuticle, is helpful in identification but is usually not as important as that of the cap. Studies here should include noting the arrangement of the hyphae as well as other features as described for the cap surface. Often the hyphae of the stalk cuticle, like those of the cap cuticle, have cystidia, particularly at the apex, that are modified in size, shape, color, or otherwise. These cystidia are called **caulocystidia**, the suffix caulo- referring to the stalk.

Trama of the hymenophore

The hymenophore, which takes the form of gills, tubes, or spines in most mushroom-like Basidiomycota, consists interiorly of sterile tissue, the trama, which supports and produces the hymenium, namely the outer layer of the gills or spines or the inner surface of the tubes. The microscopic features of gills, spines, and tubes are similar. A description of the gill trama only is given in this section, but the features discussed apply generally to the tube and spine trama also.

The gill trama gives the gills their basic shape and greatly determines their size. The gill trama is composed of hyphae that grow downward from the cap trama toward the gill edge. Some of the tramal hyphae in the gill branch off to produce the hymenium on the faces of the gills; the remainder extend to the gill edge, where they may also terminate as a hymenium.

The arrangement of the hyphae in the gill trama is extremely important in distinguishing mushroom families and even genera. The arrangement and organization of the hyphae in the gill trama is determined by studying cross sections of the gills (see Ch. 4, "Microscopic Techniques").

The arrangement of the hyphae in the gill trama is classified in four ways: **parallel** to **subparallel**, **interwoven**, **divergent**, and **convergent**. Mature specimens are usually selected for sectioning; however, for some mushrooms such as species in the family Hygrophoraceae, you have to section young specimens to determine the arrangement of the hyphae. The four basic patterns of arrangement are described in the Glossary and illustrated in Fig. 14 A–D.

Record also the width of the tramal hyphae, noting both the narrowest and the broadest measurements; their shape; the thickness of the hyphal walls; the color of the hyphae; and the nature of the hyphal surface, such as whether it is smooth or encrusted.

The **subhymenium** (pl. **subhymenia**) is a narrow zone of hyphae between the gill trama and the hymenium. It produces the basidia, basidioles, and cystidia discussed below. There are various types of subhymenia, but a discussion of these is beyond the scope of this treatment. Largent et al. (1977) provide descriptions and illustrations of various subhymenia.

Hymenium

The spore-producing layer of the hymenophore, the hymenium, is actually composed of end cells of the hyphal system of the fruit body. This structure can generally be seen by tracing the hyphae of a fruit body up the stalk into the cap trama and downward into the gill trama, where they eventually terminate as end cells in the hymenium. Of course not all the hyphae terminate in the hymenium, but this pattern holds for most fruit bodies, especially those that are mushroom-like.

Three basic types of cells can be distinguished in the hymenium: **basidia** (sing. **basidium**), **basidioles**, and **cystidia** (sing. **cystidium**). Basidia are the cells that produce the sexual spores, called basidiospores in the Basidiomycota. Basidioles are usually considered to be immature or aborted basidia. These two types of cells are always present and are discussed here first. The cystidia, which are sterile, differentiated end cells (Fig. 29C), are only present in some species. Their description follows that of the basidia and basidioles.

Basidia and basidioles

Basidia are produced by the subhymenium. They vary in size, shape, and structure depending on the group of Basidiomycota in which they occur. The Basidiomycota in this treatment all have a single-celled basidium that is typically thin walled and club shaped when mature. The basidium when immature is narrowly clavate. As the cell enlarges and attains its mature, more or less clavate shape, a series of complex nuclear changes called meiosis occurs, resulting in the production of four nuclei. The nuclei gradually migrate toward the tip of the basidium, where four narrow, awl-shaped protrusions called **sterigmata** (sing. **sterigma**) develop. One spore develops at the tip of each sterigma and usually one nucleus migrates into each spore (Figs. 1B, 56B). Some basidia produce less than four or even more than four such spores, but most commonly the number produced is four. The spores are formed externally on the surface of the basidium. When they are mature, they are forcibly ejected from the basidium.

When studying basidia, note the following:

- number of sterigmata—basidia with two and four sterigmata may occur on the same gill
- shape—narrowly clavate, clavate, or broadly clavate
- size—length and width at the broadest point, recording a range of size
- color—note the mounting medium and the contents of the basidia.

Like the hyphae, the color of a basidium may be in the wall or inside the cell and sometimes varies with the mounting medium. Often the basidia are simply colorless, the condition described as hyaline, and without contents, except perhaps for small granules and droplets.

The characteristics of the basidia in gill mushrooms and related fungi are sometimes helpful in identifying genera, as for example the presence of siderophilous granulation in basidia of *Lyophyllum* (see Ch. 4, "Chemical Reagents"). Basidia function mainly in the production of basidiospores, which provide some of the best characters for identification.

Basidioles typically originate from the subhymenium. They are similar in shape to basidia and may be considered immature, aborted, or developmental stages of basidia. They are of little taxonomic value.

Cystidia

Cystidia may occur on the surface of the cap, the stalk, and the gills including the gill edge, but do not occur in all species of mushrooms. The cystidia that

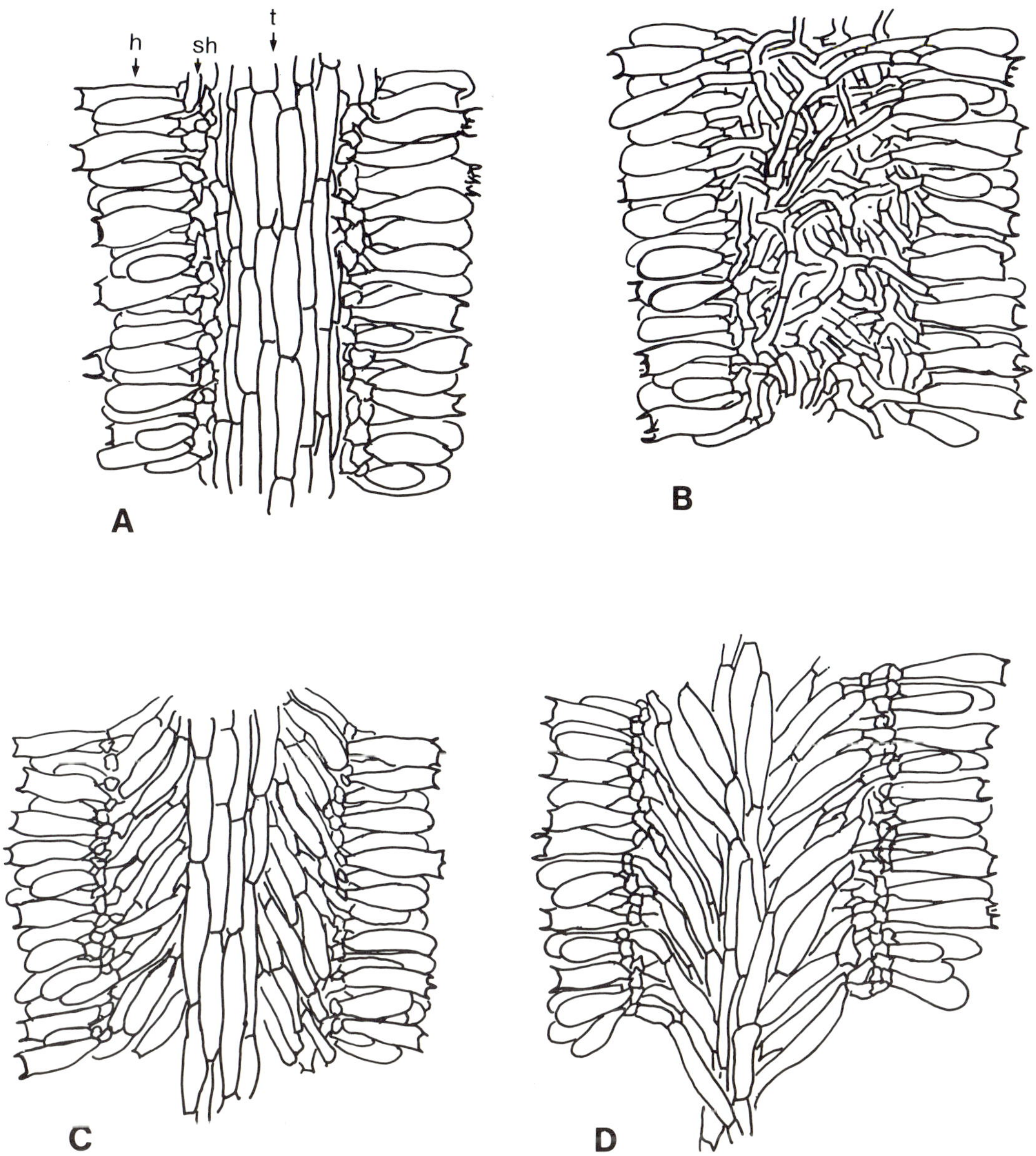

Fig. 14. Basic gill tramae: A, regular; B, interwoven; C, divergent; D, convergent.

occur on the cap, pileocystidia, and on the stalk, caulocystidia, have already been mentioned. Two types of cystidia occur on the gills. **Cheilocystidia** are those produced on the edges of the gill, whereas those produced on the gill faces are called **pleurocystidia**. Some mushrooms have both cheilocystidia and pleurocystidia; others have only cheilocystidia. Cheilocystidia are also found on the pores of the tubes in boletes and polypores and on the apex of the spines of hedgehog fungi. Pleurocystidia are found on the inner surface of the tubes of boletes and polypores and on the sides of spines in hedgehog fungi. This classification is based on the location at which the cystidia occur.

However, cystidia may also be described on the basis of other features such as morphology, function, content, or origin. Basic types include the **leptocystidium**, the **metuloid**, the **pseudocystidium** and **macrocystidium**, the **gloeocystidium**, the **chrysocystidium** (a type of gloeocystidium), and the **broom cell**. Some cystidia contain pigment; others are encrusted or ornamented with crystals of amorphous deposits or are covered with appendages or protuberances.

Cystidia are variously shaped. Their shape is especially important because it may be characteristic of a species or perhaps a group of species. Common shapes include **capitate, filiform, fusiform, mucronate, ventricose,** and **vesiculate,** to name a few. Terminology and illustrations for these cystidial features are given in the Glossary.

Clamp connections

A **clamp connection** is a special type of hyphal branch, present in some Basidiomycota. This small, semicircular hyphal branch is laterally attached to the walls of two adjoining cells and arches over the septum between them (Figs. 18C, 50D, 154C). Not all hyphae in a given fruit body have clamp connections, so check the septa of hyphae from various parts of the fruit body to be sure of their presence or absence. In some fungi clamp connections are abundant; in others they are rare. Some clamp connections are small and difficult to observe; others are large and easy to observe.

Veils

The partial and universal veils, including their remnants, are composed of hyphae. Usually it is possible to differentiate the tissue of a veil from that of the cap or stalk, but not always.

When describing the hyphae of the veils or of their remains, such as the annulus formed by the partial veil or the volva formed by the universal veil, note the following:

- arrangement of the hyphae
- size and shape of the hyphae—look especially for sphaerocysts or other special cells
- thickness of the hyphal walls
- any encrustations or other structures on the hyphal walls
- color and contents of the hyphae.

HYPHAL STRUCTURE OF PUFFBALLS AND STINKHORNS

Puffballs and stinkhorns, of the class Gasteromycetes, produce their spores on basidia that are formed inside the fruit body. The spore-producing tissue, the gleba, is extensive in most puffballs, comprising most of the fruit body. In stinkhorns it is less extensive. The peridium, the wall that encloses the gleba, protects the spore-producing tissue, at least until maturity. Below is a brief discussion of the microscopic structure of the gleba and the peridium. The microscopic features of the peridium are not extensively used in species descriptions.

Gleba

The mature gleba is primarily composed of spores. However, the gleba of young stages consists of a distinct hymenium with supporting tissue, visible in most puffballs. Often the basidia are so closely packed that the gleba appears to be almost continuous, and the hymenium is difficult to differentiate as a distinct layer. A thorough microscopic description of the trama and basidia is desirable, but often not essential, in identifying species.

In certain puffballs such as the genus *Lycoperdon*, the mature spore mass is associated with sterile, thread-like hyphae called **capillitia** (sing. **capillitium**). The shape, size, branching pattern, and color of capillitia are important features in identifying certain species or even genera. The presence of capillitia alone may help to identify the genus.

Peridium

When the peridium is composed of more than one layer, describe each layer individually. Determine the way the hyphae are arranged by making thin sections of the peridium. Note the shape and width of the hyphae, their color, the thickness of the hyphal walls, and whether they are smooth or encrusted.

BASIDIOSPORES

The characteristics of basidiospores are extremely important in classifying and identifying the Basidiomycota. The color of spores in mass, ascertained by taking a spore deposit, is the only feature of spores that can be determined without a microscope. This feature is nevertheless best considered in this section, along with the discussion of the other characteristics of spores that are determined microscopically.

The microscopic study of spores can be made in water, an aqueous solution of potassium hydroxide, Melzer's reagent, or some other chemical reagent (see Ch. 4, "Chemical Reagents"). Always indicate the medium in which the spores are studied because some of the features differ in various media.

Much of the terminology used when discussing the spores of the Basidiomycota can also be applied to the spores of the Ascomycota. Therefore, reference is occasionally made to this section elsewhere in this book during discussions of the spores of the Ascomycota.

Spore deposits

A spore deposit can usually be obtained from members of the Basidiomycota that forcibly eject their spores. Gill mushrooms, boletes, hedgehog fungi, chanterelles, and coral fungi all give a spore deposit when the fruit body is fresh and mature. Young fruit bodies sometimes produce a light deposit, but often they are too immature to produce spores. Polypores produce a spore deposit under the proper conditions, but sometimes a spore deposit cannot be obtained for these fungi.

Spore deposits should be taken soon after the fruit body is collected, either in the field or immediately after returning home. The true color is analyzed most accurately when white paper is used to obtain the deposit. However, deposits taken on glass slides can also be used to ascertain true color. Use the following procedure. Cut the stalk from the fruit body just below the attachment of the gills, tubes, or spines. Place the cap with the spore-producing surface down on a piece of white paper or on a glass slide. Wrap the cap and the white paper or slide in waxed paper; do not use plastic bags. Spore deposits from coral fungi are obtained by laying the branches of the fruit body on a piece of white paper or on a glass slide and wrapping the specimen as described.

When spore deposits are set up in the field, place the wrapped specimens in your collecting basket or in a small container to prevent them from being disturbed. At home or in the laboratory, the wrapped specimens can be placed on a tabletop or bench. Allow several hours for the spore deposit to form. A heavy deposit is required to obtain the true color of the spores in mass. After a sufficient wait, unwrap the specimen, remove the cap from the paper, and analyze the color of the spore deposit, preferably with the aid of a color chart or guide.

Retain the spore deposit with the collection. Use the spores from the deposit to obtain measurements of spore size. Mature spores are needed for accurate measurements.

Because puffballs and their relatives do not forcibly eject their spores from the basidia at maturity, an actual spore deposit as just described cannot be obtained for these forms. The color of the mature gleba, which is often a powdery mass of spores, is usually given as the color of the spores in mass for puffballs and their relatives. The color of mature spores, and occasionally the changes in color of the gleba as the spores mature, are important features in identifying puffballs.

Color

The color of individual spores under the microscope is usually recorded immediately after they are mounted for study. Often the spores are hyaline as viewed when mounted in water or potassium hydroxide, but sometimes they are somewhat pigmented. Some spores change color when placed in Melzer's reagent or potassium hydroxide, as described in the section "Chemical Reactions of Spores", later in this chapter. It therefore may be desirable to check the spore color again, after some time has lapsed.

Size

Spores should be measured with an oil-immersion lens at a magnification of 1000 times. The most accurate spore measurements are obtained from mature spores taken from a spore deposit, but spores from the hymenophore can also be measured. In some fruit bodies, spores may be obtained from natural deposits on the stalk surface or on the surface of an adjacent cap.

To obtain accurate spore measurements, measure the length and width of at least 10 spores selected at random. Spore measurements of length and width are usually expressed as a range, for example 7–9 × 4–5 μm, with the length given first. When some spores occasionally fall outside this range, for example some shorter and some wider than the measurements given above, the range is expressed thus: (6–)7–9 × 4–5(–5.5) μm. Average spore length and width may also be helpful.

To obtain consistent, accurate measurements, the spores should be measured in **side view**. To determine when a spore is in side view or in **face view** requires some understanding of how spores are shaped and their position on the basidium. See Fig. 15 for a comparison of these features.

Attachment and shape

The end of the spore that is attached to the sterigma is called the **proximal end** (Fig. 15); the actual point of attachment is called the **apiculus** (pl. **apiculi**) (Fig. 15). When the spore is detached from the sterigma, the apiculus usually appears as a more or less small to minute sharp point at the proximal end of the spore (Fig. 15). At times a flattened area or a slight depression commonly called the **plage** (Fig. 27A) occurs just above the apiculus on the dorsal side of the spore. This feature only occurs in certain genera of mushrooms, for example, the genus *Galerina*. The top of the spore as it sits on the basidium (Fig. 15) is the **apex** of the spore.

In mushrooms and other fungi that forcibly eject their spores, each spore is attached obliquely to the sterigma (Fig. 15). In puffballs and its relatives each spore is centrally positioned on the sterigma, so that the longitudinal axis of the spore corresponds to that of the sterigma (Fig. 16). Some puffballs and stinkhorns produce very short sterigmata, or the sterigmata are entirely lacking, in which case the spores are described as **sessile** (Fig. 16). The spores of other puffballs, however, for example those produced by certain species of *Lycoperdon*, often have a well-developed **pedicel,** namely a modified sterigma, that remains attached to the spore when it is dislodged or falls from the basidium.

Spore shape is usually determined in side view and face view. Some of the more commonly used terms to describe shapes include **angular, ellipsoidal, fusiform, globose, nodulose,** and **subglobose.** These terms are defined in the Glossary.

Ornamentation

The surface of many basidiospores is smooth and the outline usually even (Fig. 19A). Some basidiospores, however, are covered with ornamentation, usually a modification of the surface layer of the spore wall. There are various types of ornamentation and at times it is difficult to determine the exact details of the ornamentation with a light microscope. Spore ornamentation should be studied using an oil-immersion lens at a magnification of 1000 times. Always focus on the spore surface to determine the pattern and nature of spore ornamentation. Some

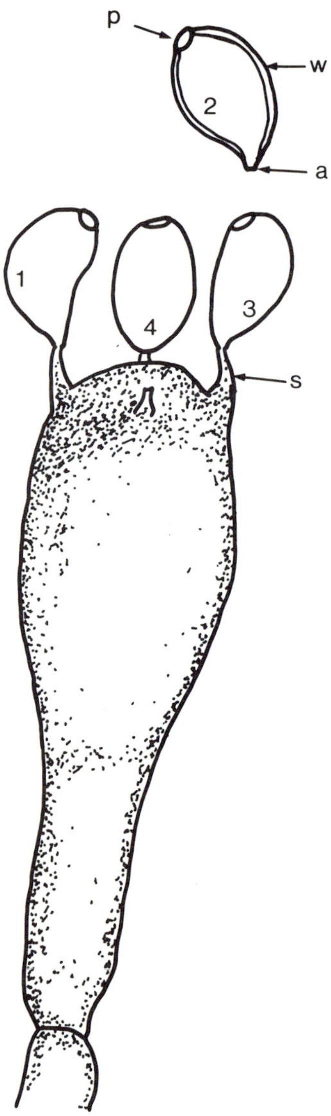

15

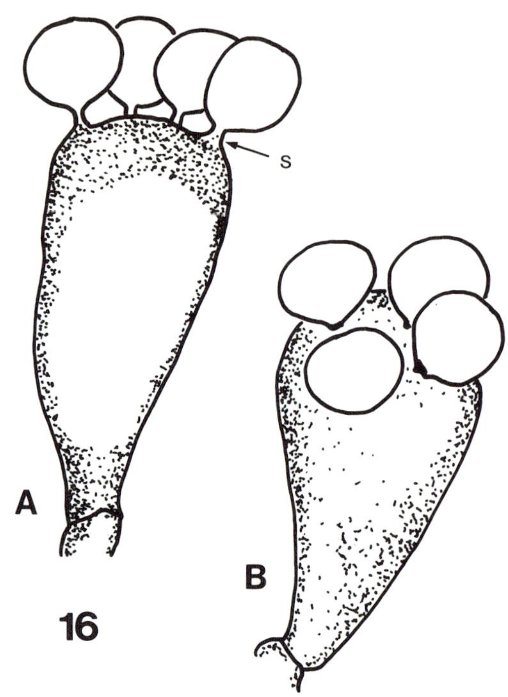

16

Fig. 15. Basidium and basidiospores of the typical gill mushroom. Note the obliquely placed spores borne on long, horn-like sterigmata (*s*). Spores 1–3 are shown in side view and spore 4 in face view. An apical germ pore (*p*) in the thick wall (*w*) is found at the distal end of the spore and an apiculus (*a*) is found at the proximal end.

Fig. 16. Basidia of puffballs: A, basidium with a short sterigmata and symmetrical spores; B, basidium with no sterigmata and sessile, symmetrical spores.

commonly encountered types of spore ornamentation include **calyptrate, echinulate, punctate, reticulate, striate, tuberculate,** and **verrucose.** These terms are defined in the Glossary.

Other features

Apical germ pore

The apex of some spores has an **apical germ pore.** This pore is actually a thin spot through which the spore may germinate (Fig. 104A). At times the apical germ pore is so broad it causes the apex of the spore to appear flattened.

Guttulate spores

When one or more oil globules are present in the spore, the spores are described as **guttulate** (Fig. 42 A and B). When they are absent, the spores are termed **eguttulate.**

Spore wall

The spore wall may be composed of up to five layers. Depending on the number and thickness of these layers, the wall is described as either thin or thick. Thin-walled spores usually have a wall thickness of 0.5 μm or less. Actual measurements of spore walls are not usually recorded in descriptions; the spores are simply described as thin-walled or thick-walled.

Chemical reactions of spores

Spores mounted in water usually do not change color. In an aqueous solution of potassium hydroxide, however, pigmented spores may fade somewhat or the color may darken; hyaline spores in this solution usually remain colorless.

There are several chemicals used to study spores because of the specific reactions they produce with certain spores. Perhaps the most widely used mounting medium is Melzer's reagent, an iodine solution. Many spores do not react with Melzer's reagent and are called nonamyloid. If, however, the spore wall or the ornamentation turns blue, black, or gray in Melzer's reagent the spore is termed amyloid; if the spores turn reddish or purplish brown, the spore is termed dextrinoid.

Cotton Blue as well as other chemicals are used to stain spores. A cyanophilous spore is one in which the wall turns dark blue with Cotton Blue. When the reaction does not occur, the spores are called acyanophilous.

See Chapter 4, "Chemical Reagents", for a more complete discussion of these terms.

Ascomycota

A knowledge of the anatomy and hyphal structure of the apothecium, the fruit body of the Ascomycota, as well as a microscopic study of the spores, called **ascospores** in the Ascomycota, are essential for an accurate identification of Discomycetes, the only class of the Ascomycota treated in this book.

A section through a typical apothecium reveals several layers of tissue (Fig. 2 B and C). The **excipulum,** composed of hyphae, is the sterile tissue making up the fleshy part of the apothecium. It is divided into two layers: the **ectal excipulum** and the **medullary excipulum.** The ectal excipulum is the outer surface of a cup-like apothecium or the undersurface of the cap of a stalked apothecium. The medullary excipulum comprises the inner, fleshy portion of the apothecium. The subhymenium, the layer that produces and supports the hymenium, lies next to the medullary excipulum. The details of the subhymenium are not critical for the purposes of this treatment and are not dealt with further here. The hymenium, the spore-producing layer, covers the upper surface of the cap or lines the inner surface of the cup.

HYPHAL STRUCTURE OF APOTHECIA

Excipulum

A knowledge of the excipulum is essential for the accurate identification of many Discomycetes. In recent years there has been an attempt to name and define the various kinds of tissues composing the excipulum of apothecia (Korf 1973 and references cited therein). However, a detailed account of the structure of the excipulum is beyond the scope of this treatment.

The orientation, shape, and size of the hyphae composing the excipulum are probably the most important features. Note the color of the hyphae, the thickness of the hyphal wall, and whether the hyphal wall surface is smooth or encrusted or otherwise ornamented.

The flesh of the stalk, when a stalk is present, is also called the excipulum. There is usually an outer surface layer and an inner fleshy layer of excipulum making up the stalk. Make a longitudinal section of the stalk to study its structure.

Hymenium

The spore-producing layer of the apothecium, the hymenium, is actually composed of end cells of the hyphal system of the fruit body, as is the case in the Basidiomycota. Ascospores are formed in special cells called asci. Usually there are eight spores in each ascus (Fig. 2C), but there are sometimes fewer or more than eight. Besides asci, there are sterile cells or filaments present in the hymenium called **paraphyses** (sing. **paraphysis**). Paraphyses are usually narrow, cylindrical structures, simple or branched (Figs. 2C, 44B), and sometimes divided by cross walls, a condition termed **septate.** They are interspersed among the asci and are more or less perpendicular to the surface of the hymenium, as are the asci.

In describing the hymenium, the length and width, shape, and color of the paraphyses and asci are recorded. The thickness of the walls of the paraphyses and the asci may also be measured. Note especially the number of spores in each ascus. Often the asci are studied in Melzer's reagent, an iodine solution, to determine if the tip or wall of the ascus turns blue. This feature is helpful in identifying genera; for example, *Peziza* and its relatives usually have asci that turn blue at the tip or throughout the ascus wall, whereas *Morchella* and *Gyromitra* and relatives give a negative reaction in Melzer's reagent.

ASCOSPORES

Ascospores are one of the most important microscopic features used in identifying Discomycetes. A spore deposit can be obtained for certain Discomycetes, as it can for mushroom-like forms of Basidiomycota, to determine the color of the spores in mass. This character is not an important feature in the identification of Discomycetes, however, so usually a spore deposit is not taken. Nevertheless, spore deposits of Discomycetes are discussed briefly here, simply as a matter of interest, along with a description of the more important microscopic features of ascospores.

Spore deposits

The asci of most Discomycetes forcibly eject their spores. For those species treated here, there is a small lid, the **operculum,** near the apex of the ascus, which pops open when the ascus is mature (Fig. 2C). The spores in each ascus are discharged when the pressure builds up sufficiently within the ascus to force them out. Often, many asci release their spores simultaneously, producing a cloud of spores over the surface of the hymenium.

Place a sheet of white paper or a glass slide under the hymenium of a mature fruit body. Wrap the specimen in wax paper and leave it undisturbed for several hours. A spore deposit usually develops as a result of an accumulation of spores shot out of the asci over a period of time. The spore deposit is usually white to slightly buff, pale yellowish, or yellowish brown.

Color

Spore color under the microscope is important and varies with the mounting medium. Note whether the spore wall is smooth or ornamented, and describe the type of ornamentation.

Size

Mature spores must be obtained for accurate measurement. Select spores from mature or old apothecia, or take them from a spore deposit. Apothecia collected for food are often too young to be useful for spore measurement. Record a range for spore length and width as described for basidiospores and note the shape of the spores.

Other features

Oil globules

The presence or absence of internal oil globules in each spore and the number of globules, when present, are important features in identifying certain genera such as *Helvella* and *Gyromitra.* As in basidiospores, when such globules are present in ascospores the spores are described as guttulate, and in their absence as eguttulate.

Droplets

A special feature of fresh spores of *Morchella* should be noted here. In fresh specimens of *Morchella* a crown of tiny droplets at the ends of each spore is visible. These droplets are lacking in the genera *Gyromitra* and *Helvella.*

Apiculi

The spores of some species of *Gyromitra* have an apiculus. This short, blunt to rounded or sharp point occurs at each end of the ascospore. Ascospores having apiculi are described as being apiculate.

Literature

Korf, R. P. Discomycetes and Tuberales. Ainsworth, G. C.; Sparrow, F. K.; Sussman, A. S., eds. The fungi. Vol. 4a. New York, NY: Academic Press; 1973.

Largent, D. L.; Johnson, D.; Watling, R. How to identify mushrooms to genus III: Microscopic features. Eureka, CA: Mad River Press; 1977.

4

METHODS OF STUDY

Collecting

The following equipment is essential for collecting mushrooms. A sturdy container is required to carry the specimens. A fairly shallow, wide market basket or picnic basket is preferable. It protects the specimens from external mechanical damage and allows good air circulation through the slits in the walls and bottom of the basket. A hunting knife with a thick blade or a similar instrument is needed for digging specimens from the ground or prying them from wood. A hand lens is useful for examining specimens in detail. Always carry a magnetic compass for determining the direction back to your starting point. Finally, a notebook, a pencil, and paper for taking spore deposits are handy.

When you make a collection, give it a number for reference and record the location, habitat, and other pertinent information. See this chapter, "Macroscopic Techniques", for a sample form for recording such information. When a spore deposit is taken in the field, wrap it separately and give it the same number as the collection so you can match the two when you arrive home or at the laboratory. The procedure for taking a spore deposit is given in the previous chapter.

When collecting, carefully remove each specimen from the substratum. Study each specimen as you collect it to be certain you are including only one species in each collection. Look for notable features such as the presence or absence of a volva, and record any other special characteristics you notice. Also, make a sketch or a mental note of the general appearance of the fungus for later reference. Good notes will help you later with identification.

Waxed paper is best for wrapping individual specimens or collections. Never use plastic bags, especially for specimens of fleshy fungi. They cause moisture to accumulate in the collection, which increases the rate of decay or deterioration of the specimens. To wrap a specimen or a collection, select a piece of waxed paper of appropriate size. Lay the specimens with field notes on the waxed paper, roll the waxed paper around the specimens in the shape of a cylinder, and twist the paper tight at both ends. Always wrap collections of different fungi separately. Place the specimens carefully in the basket; do not put large or heavy specimens on smaller, fragile ones. Some rearrangement of the specimens may therefore be required as you collect.

Some collectors take color photographs or even black-and-white prints of each collection. They are helpful for identification and are an excellent source of reference as they accumulate over the years. Color slides are especially useful because they can be used for lectures and are easily and cheaply duplicated.

Collecting mushrooms for food is a somewhat different proposition than collecting them for study. Usually when gathering mushrooms for the table, you collect only one or two specific kinds with which you are fully familiar. On the other hand, when collecting mushrooms for study, you usually include several unknown fungi. Never collect fungi for food unless you are absolutely sure of the identification.

When transporting collections for food or study, do not pile the specimens on top of one other. Those on the bottom will become smashed or damaged, especially when they are transported long distances. Do not leave specimens in a closed car for even a short time, especially when it is a hot, sunny day. They will overheat rapidly and decay.

On arriving home or at the laboratory, put the specimens in a cool place. Set up spore deposits if you did not do so in the field. Then select specimens for study. Study the small or delicate specimens first because they tend to deteriorate more quickly than larger ones. If you have collected *Coprinus*, however, work on these specimens first instead, because they quickly deteriorate into an inky mass shortly after they are collected.

Fungi collected for food should be taken care of immediately after you return from the field. Double-check all specimens to be sure they are all the same species.

Macroscopic Techniques

Taking accurate and useful notes on mushrooms and other fungi requires practice and a knowledge of various descriptive terms. The terminology required for describing the macroscopic features of fruit bodies is given in Chapter 2 of this book. Refer to the Glossary and the accompanying illustrations while studying Chapter 2, to fully understand the specialized meanings of terms used in describing fungi.

The materials needed for taking notes on fungi are few. Take your notes in pencil or pen, preferably with permanent ink, on 20 × 28 cm sheets of paper or 10 × 15 cm cards, or on special forms. A sample form is presented on the next page. A ruler marked off in

MACROSCOPIC NOTES ON FUNGI

NAME | Collected by:

COLLECTION NO. | Date:

LOCALITY

HABITAT

Environment:

Substrate:

HABIT OF GROWTH

CAP

Size (mm) | Shape

Color

young: | old:

Margin

shape: | form:

Nature of surface

Flesh

taste: | odor:

consistency: | color:

color changes:

Latex

presence or absence: | taste:

color: | color change:

SPORE-PRODUCING STRUCTURES

Type

Type of attachment to stalk

Width (mm)

Length (mm)

Color

young: | old:

Color changes

Nature of gill edges or pores

Lamellulae

number:

STALK

Shape

Width (mm)

at apex: at base:

Length (mm)

Color

young: old:

Color changes

Surface features

Flesh

texture: color-young:

color-old: color changes:

PARTIAL VEIL

Presence or absence

Structure

Annulus

presence or absence: color:

color changes: position:

attachment: no. of layers:

UNIVERSAL VEIL

Presence or absence

Type of remnants on stalk

shape: color:

Type of remnants on cap

color: color changes:

SPORE DEPOSIT COLOR

COMMENTS

millimetres and a razor or knife for cutting or splitting fruit bodies are essential. A hand lens is also useful.

One of the most important macroscopic features of fruit bodies is the color. Although a general indication of color is helpful, a color guide or set of color charts is best. A color guide allows you to take more accurate color readings and is a permanent record that can be referred to by other workers, thus making it easy to interpret and communicate fruit body colors.

For some fungi, determining the reaction of the surface and flesh of the fruit body to one or more chemical reagents is important in their identification. Some of the more common of the many reagents used include strong alkalis, strong acids, and iron salts. See this chapter, "Chemical Reagents", for the preparation of these chemicals and the reactions of fungal materials to them.

Preserving and Storing Specimens

Preserved specimens have several uses. They can be retained for reference, for keeping records from season to season, and for exchange with other collectors. Drying is the best method for preserving specimens. Specimens preserved in liquid are not very useful for identification purposes.

Fruit bodies can be dried on screens over low heat. A temperature of about 45°C is satisfactory. The warm air must be circulated around the specimens and the moisture removed as the specimens dry. One of the best-designed driers is that used by Dr. A. H. Smith and many of his students (Smith and Smith 1973). This drier is constructed of several removable screen shelves enclosed by a fireproof canvas. The heat source, such as hot plates or light bulbs, is placed beneath the shelves, and the enclosing canvas forms a chimney through which the warm air travels upward and around the specimens. The specimens slowly dry as the moisture is gradually removed. A kitchen oven is normally not suitable because the moisture does not escape readily and the specimens cook before they dry. When specimens are large, split them lengthwise before placing them on the drier.

Once the specimens are dried, store them in small cardboard or plastic boxes. Paper packets are used by some people but these do not protect the specimens from damage as well as boxes do. Most fruit bodies, if not fumigated, are eaten by insects, especially small beetles or weevils. Moth balls, with naphthalene as the active ingredient, are excellent protection against insects. Add a few crystals to each box. *p*-Dichlorobenzene can also be used in place of naphthalene.

A collection of mushrooms or other fungi is called a herbarium. If you maintain a personal herbarium, place the specimens in a dry place. The insect repellent produces considerable odor, so store the specimens away from the main household. It is also not a good idea to breathe the repellents for extended times. Remember that each collection you put in the herbarium should be properly labeled and given a number for reference. Notes on collections can be kept in a file and given numbers that correspond to the collection numbers.

Some workers use silica gel to dry small specimens. Activated silica gel, mesh size 6-16, is best for drying mushrooms. Silica gel can often be obtained from florists and nursery supply houses.

Microscopic Techniques

The microscopic study of fungus fruit bodies is essential for accurate identification. It involves a detailed study of the spores and the hyphal structure of the fruit body. A study of microscopic characters may seem difficult to the novice but is no more difficult than studying macroscopic features. Once the study is initiated, most people find it interesting and rewarding.

Every macroscopic feature of a fungus fruit body can be correlated with a microscopic feature. If a particular species has a scaly cap surface, a study of the individual scales shows how the hyphae are arranged to form the scales. However, studies with the light microscope also reveal characteristics that cannot be seen with the naked eye, for example whether the spores are thick walled or thin walled.

Microscopic studies can be made from fresh or preserved specimens. Most workers do microscopic studies from preserved specimens. When the specimens are properly preserved and the correct procedures for reviving and studying them are followed, good results can be obtained.

Equipment

The biggest drawback to studying fruit bodies microscopically is usually the availability of a good-quality compound microscope, with an oil-immersion lens capable of magnification at 100 times. The best place to gain access to one is the biology department of a college or university or even a highschool.

The following equipment and materials are needed for the microscopic study of fruit bodies:

- microscopic slides—standard glass slides
- cover glasses—thin cover glasses numbered 0-1 are best for work with an oil-immersion lens
- razor blades—single- or double-edged. High-quality double-edged blades are excellent for making sections, but be careful not to cut your fingers. To prevent cutting your fingers tape one side of the double edge with masking tape
- forceps—for handling small pieces of fungus tissue
- dissecting needles—these instruments have a long, needle-like end attached to a wooden handle. They are handy for manipulating sections and small pieces of tissue

- lens paper—soft, nonabrasive tissue for cleaning microscopic lenses. A box of facial tissue or similar absorbent material is also handy, for example for absorbing excess reagents
- dropper bottles—for storage of reagents such as water
- immersion oil—good-quality oil for an oil-immersion lens. Note that some are carcinogenic, so select a safe one
- lens-cleaning compound—xylene is good for removing oil and other debris from lenses
- eyepiece and stage micrometer—for hyphal and spore measurements. Take measurements with an oil-immersion lens, at a magnification of 100 times
- chemical reagents—various reagents are used. A list of the more commonly used chemicals with instructions on their preparation and descriptions of their reactions with fungal material is presented in this chapter, "Chemical Reagents".

Techniques

The study of ascospores and basidiospores with the compound microscope presents no real problem. Spores for microscopic study are best taken from spore deposits. Scrape the spores onto a drop of mounting medium on a glass slide or touch the wetted tip of a dissecting needle to the spore mass, and place the adhering spores in a drop of mounting medium. Then place a cover glass over the preparation, and the slide is ready for viewing under the microscope. You can also study spores taken from small pieces or thin sections of the spore-producing structures. Simply take a piece or a section of the material and mount it as you would the spores.

Studying the surface or the trama of a fruit body is more difficult than studying the spores, however, and requires some special techniques. You need thin pieces of tissue, one to a few cells thick, to place under the microscope. One method for obtaining thin samples for study is to prepare a squash mount. Place a small piece of fruit body tissue on a slide in a drop of mounting medium. Then cover the tissue with a cover glass, and press gently on the cover glass to spread the tissue out for viewing with the microscope. Squash mounts are of some help in microscopic studies, but hand sections of fruit body tissue are more widely used, especially for studying the arrangement of the hyphae.

Making hand sections of fungal fruit bodies requires the use of a sharp razor blade. Sections are made from a small piece of fruit body tissue. The idea is to slice a section of the tissue thin enough so that the arrangement, shape, and size of individual hyphae can be studied.

Sections of fungus tissue can be made in the following manner. Hold a piece of tissue between the thumb and forefinger. Place the razor blade face down on the forefinger with the cutting edge toward the fungus tissue. Pass the blade through the tissue toward the thumb, using the thumb as a backstop. You may want to place a piece of tape around the end of your thumb to prevent you from cutting into it with the razor blade. Another method involves the use of elderberry pith. Take a piece of pith about 50 mm long and cut it longitudinally into two halves. Place the tissue to be sectioned between the two halves near one end of the pith. Be sure the material is placed so that the plane of sectioning is parallel to the end of the pith. Place the razor blade near the end of the pith stick and pass the blade through the pith and the tissue of the fungus. Make a rather thick section the first time to obtain a good cutting surface. Then pass the razor blade several times quickly through the pith and fungus tissue, making thin sections. Wet the razor blade slightly before sectioning; sectioning becomes easier with a wet blade and the sections tend to accumulate on the wet surface. The sections can then easily be scraped from the razor blade with a wetted dissecting needle and placed in a drop of mounting medium. When elderberry pith is not available, a piece of carrot 10–20 mm thick and about 50 mm long can be split longitudinally and used in its place.

The fruit body of a mushroom usually has three parts: a cap, gills, and a stalk. Each may be sectioned in three basic ways: radially, tangentially, and transversely.

Any section that follows a plane parallel to the radius of a circle is a **radial section**. A radial section of a cap runs from the edge of the cap to or through the center. A longitudinal section of the stalk passing through the center of the stalk is also a radial section, specifically termed a **median longitudinal section**.

When the cap is cut from margin to margin without passing through the center, a **tangential section** is obtained. Thin sections for study are often made parallel to the plane of this section. A longitudinal section of the stalk that does not pass through the center is also a tangential section.

A transverse section of the stalk is made perpendicular to the long axis of the stalk. Such a section is referred to as a **cross section**. For microscopic study thin sections are made parallel to the plane of the cross section.

A fourth kind of section is the **scalp section**, sometimes useful for viewing the arrangement of the hyphae on the surface of the stalk or cap. A scalp section is a thin section made along the surface of the stalk in a longitudinal direction or along the slope of the cap.

To obtain sections of a gill or gills requires a slightly different technique. Using a sharp razor blade or scalpel remove a small wedge of tissue from the cap with several gills attached, just as if you were cutting a piece of pie from a round pie plate. Cut away a portion of the margin. Make the cut so the razor passes through the flesh of the cap and across each gill. When looking at the cut surface, each gill should appear more or less like a tall, narrow isosceles triangle, with the base of the triangle attached to the flesh of the cap. This view presents a cross section of the gills. Hold the wedge of tissue between the thumb and forefinger, or place it between two pieces of pith, and slice thin sections parallel to the plane of the prepared surface. The thin sections present tangential sections of the cap surface and trama and cross sections of the gills. This technique is the most commonly used, and the cross sections of the gills so obtained are used particularly to determine the arrangement of the hyphae of the gills, namely whether they are parallel, interwoven, divergent, or convergent. This technique is also used to make sections of the tubes in boletes and polypores. Spines are difficult to section in this manner. To study the hyphal arrangement of spines, lay the spines on a slide and make a squash mount, by gently applying pressure on the cover glass.

Another technique is sometimes used for studying gill edges to determine the presence or absence of cheilocystidia. Simply remove a gill or piece of gill with the edge intact. Lay it face down on a slide in a drop of mounting medium and cover it with a cover glass. Place the preparation under the microscope and study the gill edge.

Remember that too much pressure on the cover glass can distort the arrangement of the hyphae when looking at sections or gill edges, so handle the material carefully and apply pressure gently.

To study dried material, place a piece of dried tissue in a small amount of 95% ethanol for about 30 seconds. Be sure to wet the tissue entirely. Transfer the piece of tissue to tap water and allow the tissue to soak until it becomes completely pliable. Remove the tissue from the water and blot it dry. Then proceed with sectioning.

Another method can be used for reviving dried tissue. Obtain a Petri dish or a small closed container. Place in it a few layers of paper towels that have been moistened with tap water. Put a piece or pieces of fungus tissue on or between the towel layers, close the container, and let the tissue slowly absorb the water until it is pliable. Allow 20 to 30 minutes. This method is slower than the one above but gives good results and does not require the use of ethanol.

Chemical Reagents

Many chemical reagents and solutions have been used to study the Ascomycota and the Basidiomycota. Some have very specific uses; others are commonly used in general studies. The most common as well as some of the most useful reagents are listed below, with methods for their preparation. Handle all reagents with care because some are extremely toxic. Always store reagents in proper containers. Use dropper bottles for everyday use and never contaminate or mix reagents. Keep all solutions out of the reach of children, and label each container, clearly indicating the contents by name and concentration. Largent et al. (1977) provides further information.

Water

Water is commonly used for fresh material and can be used for revived tissue. Use tap or distilled water. Water is used as the accepted standard for comparison of the effects of other media on spores and hyphae.

Potassium hydroxide

Make a 3–5% aqueous solution. Dissolve 3–5 g of potassium hydroxide in 97–95 mL of water. Stains in aqueous solution such as Congo Red and Phloxine can be combined with potassium hydroxide. Potassium hydroxide can be used as a mounting medium for microscopic studies and also for testing macroscopic color reactions. It is also good for reviving sections made directly from dried tissue. Sodium hydroxide or ammonium hydroxide can be substituted for potassium hydroxide. Remember that spore color may change with these reagents.

Iron salts

A 10% aqueous solution of ferrous sulfate ($FeSO_4$) is normally used for testing macroscopic color reactions of the surface and flesh of fruit bodies. Fresh material should be used for this test, and the reagent should be freshly prepared for accurate results. Three reactions are possible: no change, a negative reaction; a change to olive, grayish olive, green, or blackish green; or a change to pinkish or red. The color change is sometimes slow to develop.

Cotton Blue

This dye is used in solution with lactic acid or a mixture of lactic acid and phenol (Largent et al. 1977). One procedure is simply to use a 1% lactic acid solution of Cotton Blue, made by dissolving 1 g of Cotton Blue in 99 mL of lactic acid. Place the material to be studied in one drop of Cotton Blue solution on a glass slide; cover with a cover glass and study. Heat the slide gently for better results. A positive reaction may take up to an hour to develop.

Cotton Blue is often used to study spores; the spore wall or the ornamentation becomes strongly colored blue when the reaction is positive. Such a spore is described as **cyanophilous**, and spores in which a negative reaction is observed are termed **acyanophilous**. Some chrysocystidia stain blue in Cotton Blue.

Cresyl Blue

A 0.5–1.0% aqueous solution is normally used. Dissolve 0.5–1.0 g of Cresyl Blue in 99.5–99.0 mL of water. Allow the preparation to stand for 5-10 minutes and filter out the excess dye. Place material to be studied in a drop of Cresyl Blue solution on a glass slide; cover with a cover glass and study.

Certain spore walls and hyphae turn red to violet in Cresyl Blue. A structure that gives a positive reaction is described as **metachromatic**. The reagent can also be used to detect gelatinous areas in fruit bodies.

Melzer's reagent

Melzer's reagent is essential for the study of both the Ascomycota and the Basidiomycota. It gives distinctive color reactions with certain spores and hyphae and is useful for studying the fine details of spores.

Melzer's reagent is prepared by adding 1.5 g of iodine crystals, 5.0 g of potassium iodine, and 100 g (mL) of chloral hydrate to 100 mL of water. Warm the solution but do not allow it to boil. The solution keeps for a long time. Do not combine Melzer's reagent with any type of alkali because a cloudy precipitate develops.

Place the material to be studied in a drop of Melzer's reagent on a glass slide, cover with a cover glass, and study. Usually a positive reaction occurs almost immediately. When the reaction is doubtful leave the material in solution for about 30 minutes. Fresh material sometimes gives a slower reaction than dried material. Depending on the reaction of the spores or hyphae in Melzer's reagent, the material may be described as **nonamyloid**—no change, remaining the yellowish color of the reagent or hyaline; **amyloid**—turning blue, black, violet, violet black, or blue gray; or **dextrinoid**—turning brown to reddish or purplish brown.

Melzer's reagent is also good for reviving dried tissues. The hyphae typically regain their normal shape and size.

Phloxine

Use a 1% aqueous solution. Dissolve 1 g of Phloxine in 99 mL of water. Filter out excess dye. Phloxine is commonly used with aqueous potassium hydroxide. It is taken up by the diffuse cytoplasm within the hyphae and is especially good for studying fresh material. See the next section, "Congo Red", for the recommended procedure and use.

Congo Red

This dye can be used as a 1% aqueous solution, made by dissolving 1 g of Congo Red in 99 mL of water and filtering out the excess dye, or with ammonium hydroxide, made by saturating concentrated ammonium hydroxide with Congo Red.

Congo Red is often used in combination with Phloxine and can be added to potassium hydroxide. Congo Red stains the walls of hyphae or spores.

A good procedure is to place the material for study in a mixture of one drop of 1% Congo Red, one drop of 1% Phloxine, and one drop of 3% potassium hydroxide or ammonium hydroxide. Excess reagent can be removed with an absorbent tissue by placing it on one side of the cover glass. With this procedure the walls stain red due to their reaction with Congo Red and the contents stain pinkish to reddish by reacting with Phloxine.

Acetocarmine

Basidia of certain mushrooms, for example the genus *Lyophyllum*, contain particles that are blackish purple or violet black in acetocarmine. These basidia, and often the granules, are described as **siderophilous**.

The formula given here is taken from the work of Largent et al. (1977). Grund and Marr (1965) describe other procedures.

To prepare an acetocarmine solution, first prepare a 45% aqueous acetic acid solution by adding 45 mL of glacial acetic acid to 55 mL of water. Then prepare a 45% aqueous ethanol solution by adding 45 mL of ethanol to 55 mL of water. Boil the aqueous acetic acid with an excess of carmine for 30 minutes. Filter out the excess carmine and dilute to half strength with the aqueous ethanol. Add 2-4 drops of 5% aqueous ferric chloride ($FeCl_3.6H_2O$).

To observe the reaction of basidia to acetocarmine, place a small piece of gill in a few drops of acetocarmine solution on a glass slide and heat over a direct flame. When the liquid begins to evaporate and a film forms, remove the piece of gill and place it in a few more drops of acetocarmine solution and heat again. Repeat this procedure two or three times. Then cover the gill piece with a cover glass and observe immediately.

Literature

Grund, D. W.; Marr, C. D. New methods for demonstrating carminophilous granulation. Mycologia 57(4):583–587; 1965.

Largent, D. L.; Johnson, D.; Watling, R. How to identify mushrooms to genus III: Microscopic features. Eureka, CA: Mad River Press; 1977.

Smith, H. V.; Smith, A. H. How to know the non-gilled fleshy fungi. Dubuque, IA: William C. Brown Co.; 1973.

5

TAXONOMY

Fungi can be defined as organisms that reproduce by spores and lack chlorophyll, being either saprobic, parasitic, or symbiotic. Some people recognize fungi as part of the kingdom Protista, whereas others believe that fungi should be a division of the kingdom Plantae, the green plants. Here, these organisms are placed in a separate kingdom, Fungi.

Nomenclature

Unfortunately for the nonscientist and amateur mushroom collector, there are few common names for mushrooms as compared with the many names for birds and flowering plants. Moreover, many of the common names used are imprecise and somewhat confusing to anyone other than their authors. To facilitate communication among students of fungi and other plants, a system of naming has been developed. The rules are set down in the *International Code of Botanical Nomenclature,* prepared by experienced taxonomists (Stafleu 1978).

The scientific name of a species, its taxon (pl. taxa), is a two-word combination called a binomial, consisting of the name of the genus (e.g. *Clitocybe*) followed by a specific epithet (e.g. *C. dealbata*), which indicates a particular kind of fungus in the genus. The names of genera and higher ranks may stand alone, but the species name is not complete without the appropriate generic name. Genus and species names appear in italics in printed text. The binomial system of nomenclature was initiated by Linnaeus, who gave each plant a name in Latin, the scientific language of his day and now the traditional language of the taxonomist. In scientific publications the species name is followed by a citation of its author, for example *Amanita ocreata* Peck. When a fungus has been reclassified in a different genus from the one in which it was originally described, the name of the author of the species epithet is placed in parentheses and is followed by the author of the new combination, for example *Clitocybe dealbata* (Fr.) Kummer. Standard abbreviations are used for many author's names. A list of those used in this book is given in Appendix 2.

In order to standardize the naming of fungi, a starting point has been assigned and rules of priority have been established. The starting point for most of the fungi is the *Systema Mycologicum,* published in 1821 by E. M. Fries in Sweden. Names assigned before the *Systema Mycologicum* was published are not considered to be valid. To be validly published in subsequent years, the name must be published with a description and, since 1935, must also be accompanied by a Latin diagnosis. When more than one name is validly published for a species, priority is given to the earliest name.

A type concept is used to stabilize the specific epithet. The author of a new species of fungus must designate a type specimen upon which the description is based. Future scientists can then compare their taxon to the type.

Classification

Classification is the grouping of related elements. As is the case with vascular plants, relationships are recognized by comparing morphological features, usually of reproductive structures. Traditionally, in fungi the morphological features studied were macroscopic. Now microscopic features and developmental studies are used to aid in the construction of a more natural classification of the fungi.

Closely related species are grouped in genera, genera are grouped in families, families in orders, orders in classes, classes in subdivisions (if used), subdivisions in divisions, and divisions in the kingdom Fungi. In this way a hierarchy is constructed. For example, the species *Clitocybe dealbata* is classified as follows:

kingdom Fungi
division Basidio**mycota**
class Hymeno**mycetes**
order Agaric**ales**
family Tricholomat**aceae**
genus *Clitocybe*
species *C. dealbata*

The name of each group, except for the genus and species, has a characteristic ending, which is distinguished by boldface in the example. The genus and species names are always italicized in printed text. Only the specific epithet has the initial letter lower case. The names of taxa above the rank of genus are always plural in form and require a plural verb.

Traditionally, the Fungi were divided into four main classes, the Phycomycetes, the Ascomycetes, the Basidiomycetes, and the Fungi Imperfecti. More recently, mycologists are recognizing several classes of fungi that are included in the following five divisions:

- Mastigomycota—the water molds or aquatic fungi, characterized by motile spores
- Zygomycota—terrestrial, characterized by nonmotile spores and zygospores
- Ascomycota—primarily terrestrial, characterized by sexual reproduction involving an ascus
- Basidiomycota—primarily terrestrial, characterized by sexual reproduction involving a basidium
- Deuteromycota—primarily terrestrial, characterized by asexual, nonmotile spores.

The fungi we are considering here belong primarily to the division Basidiomycota. A few belong to the division Ascomycota, in the class Discomycetes.

The division Basidiomycota is recognized by clavate basidia on which spores are produced. The two classes of the Basidiomycota dealt with in Part Two are the Hymenomycetes and the Gasteromycetes.

In the class Hymenomycetes the basidia line the interior of tubes or cover the surface of gills, spines, or some other reproductive structure on the exterior of the fruit body. The two orders of Hymenomycetes treated here are the Agaricales and the Aphyllophorales. All members of the order Agaricales are fleshy mushrooms, the agarics and the boletes, whereas the order Aphyllophorales includes tough and leathery fungi such as the polypores and clavaroid fungi.

The order Agaricales contains numerous families such as the Agaricaceae, Tricholomataceae, Coprinaceae, Strophariaceae, and Cortinariaceae. These families are in part recognized on the basis of spore color, the presence or absence of veils, and type of gill attachment.

Several families are included in the modern concept of the order Aphyllophorales, but only a few are eaten because most are too tough, fibrous, or woody. The more fleshy representatives of this order are found in the families Polyporaceae, for example the genera *Albatrellus*, *Grifola*, and *Laetiporus*; the Clavariaceae, particularly the genus *Ramaria*; and the Cantharellaceae, especially the genera *Cantharellus*, *Gomphus*, and *Craterellus*.

In the class Gasteromycetes, the puffballs and relatives, the basidia are produced in the gleba, which is enclosed by the peridium. Puffballs in the order Sclerodermatales, the only order in this class that contains species formally described in Part Two, have a tough peridium that protects a purplish black gleba. The spore are powdery at maturity and are released when the peridium is ruptured.

The division Ascomycota is recognized by sac-like or cylindrical asci, within which spores are produced. Only a few members of the Pezizales, in the class Discomycetes, are formally described in Part Two. The fruit bodies of fungi in this class, the apothecia, are typically cup-like structures (e.g. *Sarcosphaera*), with the spore-producing surface, which lines the interior of the cup, exposed. In morels and false morels the apothecium has a distinct stalk; in morels the spore-producing surface is pitted, whereas in false morels it is often convoluted.

A list of all the families and selected genera of the Basidiomycota and the Ascomycota treated in this book is given in Table 1. Genera that are listed in Part Two of this book as containing species suspected or known to be poisonous appear in boldface type. The genera that appear in lightface type are not described formally in Part Two; however, do not assume that all the species belonging to these genera are edible. Enough information may simply not have been available in some cases to determine edibility.

Keys and Family Descriptions

The first key provided here separates the two major divisions of fungi, the Basidiomycota and the Ascomycota. A set of general and technical keys are then presented for the Basidiomycota, followed by a corresponding set of keys for the Ascomycota.

The general keys are based on macroscopic characteristics of fruit bodies and on spore deposit color, where applicable. In some instances a hand lens is helpful, but most characteristics can be seen with the unaided eye. Keys based primarily on macroscopic characteristics do not always lead to the correct family or genus, or several attempts may be necessary before a correct decision is reached. The technical keys combine macroscopic and microscopic characters to separate the taxa. Use the technical keys, whenever possible, to confirm your preliminary identification.

The general treatment of the Basidiomycota consists of a general key that ultimately separates the families of the division. Also provided are general descriptions for each of the families that contain the species described formally in Part Two, as well as descriptions for two families of the order Agaricales, Gomphidiaceae and Hygrophoraceae, that at present are not known to contain any poisonous species. Descriptions for these two families are included to facilitate the study of this important order as new information on toxicity becomes available. The family descriptions include, whenever helpful, general keys to important genera of these families. The appropriate chapter references to Part Two are provided in the family descriptions to help you locate the various genus descriptions, which are arranged alphabetically within chapters.

The technical treatment of the Basidiomycota includes a technical key separating the families of the order Agaricales and technical keys to selected genera of families (Shaffer 1968, Smith 1973, Stuntz 1977). Technical keys are provided whenever separation on the basis of macroscopic features alone may not be conclusive.

Table 1 Important Taxa of Fleshy and almost Fleshy Fungi

Order	Family	Genus (genera listed in Part Two are in boldface)
DIVISION BASIDIOMYCOTA, CLASS HYMENOMYCETES		
Agaricales	Agaricaceae	***Agaricus*** (Ch. 13) *Cystoagaricus* *Melanophyllum*
	Amanitaceae	***Amanita*** (Chs. 7, 10, 11, 13) *Limacella*
	Bolbitiaceae	*Agrocybe* *Bolbitius* ***Conocybe*** (Chs. 7, 12)
	Boletaceae	*Boletellus* ***Boletus*** (Chs. 10, 13) *Fuscoboletinus* *Gastroboletus* *Gyrodon* *Gyroporus* *Leccinum* *Pulveroboletus* *Strobilomyces* *Suillus* ***Tylopilus*** (Ch. 13)
	Coprinaceae	***Coprinus*** (Ch. 9) ***Panaeolus*** (Ch. 12) *Psathyrella*
	Cortinariaceae	*Alnicola* ***Cortinarius*** (Ch. 7) *Crepidotus* ***Galerina*** (Ch. 7) ***Gymnopilus*** (Ch. 12) ***Hebeloma*** (Chs. 10, 13) ***Inocybe*** (Ch. 10) *Phaeocollybia* ***Phaeolepiota*** (Ch. 13) *Phaeomarasmius* ***Pholiota*** (Ch. 13) *Rozites* *Simocybe* *Tubaria*
	Gomphidiaceae	*Chroogomphus* *Gomphidius*
	Hygrophoraceae	*Hygrophorus*
	Lepiotaceae	***Lepiota*** (Chs. 7, 13)
	Paxillaceae	***Paxillus*** (Ch. 13) *Phylloporus*
	Pluteaceae (=Volvariaceae)	*Chamaeota* ***Pluteus*** (Ch. 12) *Volvariella*
	Rhodophyllaceae (=Entolomataceae)	*Alboleptonia* *Claudopus* *Clitopilus* ***Entoloma*** (Ch. 13) (see rhodophylloid mushrooms) *Leptonia*

(Continued)

Table 1 Important Taxa of Fleshy and almost Fleshy Fungi (*Continued*)

Order	Family	Genus (genera listed in Part Two are in boldface)
DIVISION BASIDIOMYCOTA, CLASS HYMENOMYCETES		
Agaricales (*continued*)	Rhodophyllaceae (=Entolomataceae) (*continued*)	***Nolanea*** (Ch. 13) (see rhodophylloid mushrooms)
		Pouzarella
		Rhodocybe
		Rhodophyllus
	Russulaceae	***Lactarius*** (Ch. 13)
		Russula (Chs. 10, 13)
	Strophariaceae	*Melanotus*
		Naematoloma (Ch. 7)
		Pholiota (see Cortinariaceae)
		Psilocybe (Ch. 12)
		Stropharia (Chs. 12, 13)
	Tricholomataceae	*Armillaria*
		Armillariella[1] (Ch. 13)
		Baeospora
		Cantharellula
		Catathelasma
		Clitocybe (Chs. 9, 10, 13)
		Clitocybula
		Collybia (Ch. 13)
		Crinipellis
		Cystoderma
		Dermoloma
		Flammulina
		Hohenbuehelia
		Hygrophoropsis[2] (Ch. 10)
		Hypsizygus
		Laccaria
		Lentinellus
		Lentinus
		Leucopaxillus
		Lyophyllum
		Marasmius
		Melanoleuca
		Mycena (Chs. 10, 13)
		Omphalina
		Omphalotus (Chs. 10, 13)
		Oudemansiella
		Panellus
		Panus
		Phyllotopsis
		Pleurocybella
		Pleurotus
		Resupinatus
		Schizophyllum
		Strobilurus
		Tricholoma (Ch. 13)
		Tricholomopsis (Ch. 13)
		Xeromphalina
		Xerulina

(*Continued*)

Table 1 Important Taxa of Fleshy and almost Fleshy Fungi (*Concluded*)

Order	Family	Genus (genera listed in Part Two are in boldface)
DIVISION BASIDIOMYCOTA, CLASS HYMENOMYCETES		
Aphyllophorales	Cantharellaceae	*Cantharellus* *Craterellus* ***Gomphus*** (Ch. 13)
	Clavariaceae	*Clavaria* *Clavariadelphus* *Clavicorona* *Clavulina* *Clavulinopsis* ***Ramaria*** (Ch. 13) *Ramariopsis*
	Hydnaceae	*Dentinum* *Hydnellum*
	Polyporaceae	***Albatrellus*** (Ch. 13) (see polypores) *Grifola* ***Laetiporus*** (Ch. 13) (see polypores)
DIVISION BASIDIOMYCOTA, CLASS GASTEROMYCETES		
Lycoperdales	Lycoperdaceae	*Bovista* *Calvatia* *Lycoperdon*
Phallales	Phallaceae	*Clathus* *Linderia* *Mutinus* *Phallus*
Sclerodermatales	Sclerodermataceae	***Scleroderma*** (Ch. 13)
DIVISION ASCOMYCOTA, CLASS DISCOMYCETES		
Pezizales	Helvellaceae	*Discina* ***Gyromitra*** (Ch. 8) ***Helvella*** (Ch. 13) *Neogyromitra* *Rhizina* *Underwoodia*
	Morchellaceae	***Disciotis***[2] (Ch. 8) ***Morchella***[1] (Ch. 13) *Ptychoverpa* ***Verpa***[1] (Ch. 13)
	Pezizaceae	*Aleuria* *Caloscypha* *Peziza* ***Sarcosphaera*** (Chs. 8, 13)
	Sarcoscyphaceae	*Sarcoscypha*

[1]Commonly eaten, but cannot be tolerated by some individuals.
[2]Possibly contains poisonous species, but not proven conclusively.

The treatment of the Ascomycota is not as extensive as that of the Basidiomycota because there are only a relatively few genera of importance to mushroom collectors in this division. There is a general key separating the important families of the Discomycetes, the class which contains all the poisonous species formally described in Part Two, followed by descriptions and general keys to selected genera of these families. Also included is a general description of the family Sarcoscyphaceae, which contains only one genus sometimes encountered by mushroom collectors. As this genus is not known to contain poisonous species, it is not formally described in Part Two. A parallel set of technical keys is also presented.

Taxa that contain the poisonous species treated in Part Two of this book are printed in boldface type in the keys. The appropriate page or chapter to which you must refer for further information follows in parentheses. Taxa that appear in lightface in the keys are not formally described in Part Two, either because they are not known to contain poisonous species or because the species are usually considered edible. Always refer to generic descriptions and illustrations in Part Two to verify your determination. Study Chapters 2 and 3 thoroughly before attempting to use these keys, to familiarize yourself with the terminology.

Remember that these keys do not include all families and genera currently known. If you cannot identify your specimen with the help of this book, it is not necessarily safe to eat. Never eat any mushroom you are not able to identify.

KEY TO THE MAJOR DIVISIONS

Use this key to determine the major division to which your specimen belongs. Then turn to the appropriate page, indicated in parentheses, to determine its family.

1. Fruit body with or without a stalk. Shape: with a distinct cap that has pores, spines, or gills on the undersurface; club-shaped, vase-shaped, or much-branched; ball-shaped and opening by irregular cracks or openings, by a pore, or like a star; or bearing a slimy, ill-smelling mass of spores on a stalk or on branches. Sexual spores produced on a club-shaped cell called a basidium .... **Basidiomycota** (p. 48)

   Fruit body with or without a stalk. Shape: cup-shaped; saddle-shaped; or with the cap or head convoluted, pitted, or ridged to relatively smooth. Sexual spores produced in a sac-shaped to club-shaped cell called an ascus................................ .... **Ascomycota** (p. 67)

BASIDIOMYCOTA: GENERAL

Key to families

Use the key in this section ultimately to identify the particular family of the Basidiomycota to which your specimen belongs, based on its macroscopic features. Use the technical key on page 60, which incorporates the microscopic characteristics, to confirm your separation of the families of the order Agaricales. Then turn to the appropriate page, indicated in parentheses after the family name, to determine the genus of your specimen. The species of these genera that are known to be poisonous are treated in Part Two, where genera and species descriptions are arranged alphabetically. The chapters where the species descriptions can be located are given in parentheses after the genus names.

1. Spores produced within the fruit body, and released, not forcibly ejected, when or after the fruit body matures; spore tissue, the gleba, typically enclosed by a distinct wall, the peridium, as it develops .... **Gasteromycetes** 2

   Spores produced on the surface of gills, spines, wrinkles, ridges, or erect columns or branches, or on the inner surface of tubes, and forcibly ejected upon maturity; spore tissue not enclosed by a distinct wall as it develops .... **Hymenomycetes** 6

2. Fruit body small, less than 10 mm across when mature, resembling a vase or miniature nest with small pill-like structures inside....... Nidulariales
   bird's-nest fungi; too small to be useful as food

   Fruit body usually larger than 10 mm across and not as above 3

3. Mature fruit body consisting of a single stalk, several more or less erect arms, or a lattice-like column. Spores produced in a foul to disagreeable slimy mass, dispersed by insects. Immature fruit body developing from a fleshy to a gelatinous or rubbery egg .... Phallales
   stinkhorns; not normally collected for food; too distinctive to be inadvertently eaten (see Ch. 2)

   Fruit body not as above. Spore mass often dry and powdery at maturity 4

4. Fruit body more or less resembling an unexpanded gill mushroom or subterranean and truffle-like; stalk often running into or through the spore mass, but sometimes rudimentary .... Hymenogastrales
   includes false truffles; little is known of their edibility

   Fruit body not mushroom-like, fruiting aboveground, and appearing puffball-like................. 5

5. Immature puffball when cut in half usually having a firm purple to purple brown gleba, but at times having the gleba mottled pallid or whitish; peridium thick; stalk, when present, fibrous. Capillitium usually absent (use microscope) .... **Sclerodermataceae** (Sclerodermatales) (p. 57)

Immature puffball when cut in half having a soft to firm white gleba; peridium usually thin; gleba composed of irregular spore-producing chambers; stalk when present represented by a sterile base that is firm to soft and usually chambered. Capillitium usually present (use microscope).... .... Lycoperdaceae (Lycoperdales)
edible puffballs, e.g. *Lycoperdon*, *Calvatia*, and *Bovista*

6. Fruit body tough, leathery, rubbery, woody, or semifleshy; stalk present or absent. Spore-producing surface in the form of tubes that are visible as pores on the undersurface of the cap. Fungi often growing on wood.......................... .... **Polyporaceae** (Aphyllophorales) (p. 56)

Fruit body of various textures, from fleshy to woody. Spore-producing surface relatively smooth or in the form of gills, spines, folds, wrinkles, or tubes; tubes when present easily separable from the flesh of the cap and having a soft spongy to gelatinous texture.................... 7

7. Spore-producing surface in the form of distinct spines that hang downward from the undersurface of the cap, from branches, or from a central mass of tissue that is more or less laterally attached to the substratum................................ .... Hydnaceae (Aphyllophorales)
not known to contain any poisonous species

Spore-producing surface not composed of spines but of gills, tubes, wrinkles, or ridges, or nearly smooth 8

8. Fruit body in the form of a single upright club or cylinder, or sparsely to profusely branched with a fairly well-developed base or stalk and with rounded to flattened branches directed upward or outward.. .... **Clavariaceae** (Aphyllophorales) (p. 53)

Fruit body with a distinct cap, and often with a distinct stalk 9

9. Fruit body typically fleshy to semifleshy, occasionally somewhat tough to fibrous or leathery. Spore-producing surface in the form of gills..... .... **several families of Agaricales** 11
when fruit body extremely tough to woody, check polypores, Ch. 13

Fruit body lacking gills.................................... 10

10. Fruit body typically with a distinct stalk and cap. Spore-producing surface nearly smooth or wrinkled, with veins, ridges, or thickened gill-like structures with blunt edges on the undersurface of the cap .... **Cantharellaceae** (Aphyllophorales) (p. 51)

Fruit body typically fleshy to semifleshy. Spore-producing surface in the form of tubes visible as pores on the undersurface of the cap; tubes soft and spongy to gelatinous in texture.................. .... **Boletaceae** (Agaricales) (p. 51)

11. Universal veil present, leaving a volva around the stalk base or warts or patches of tissue on the stalk base; cap surface glabrous or with patches or warts of universal veil tissue. Gills free or nearly so from the stalk apex.................................. 12

Universal veil usually absent, but when present leaving no volva, warts, or patches of tissue on the stalk or cap. Gills attached to or free from the stalk apex 13

12. Annulus or annular zone lacking. Spore deposit pink, and gills with a pink to salmon color when mature .... **Pluteaceae (=Volvariaceae)** (p. 56)

Annulus or annular zone often but not always present. Spore deposit usually white to cream color .... **Amanitaceae** (p. 51)

13. Gills free or nearly so from the stalk apex (check apex of stalk carefully to see that lines formed by the gills are lacking)...................................... 14

Gills clearly attached to the stalk apex (e.g. adnate, adnexed, sinuate, or decurrent) 17

14. Partial veil lacking, hence no annulus or annular zone. Spore deposit pink, and gills shaded pink to salmon color when mature .... **Pluteaceae (=Volvariaceae)** (p. 56)

Partial veil often present, but when lacking, then fruit body not with the above combination of characteristics.. 15

15. Spore deposit dark brown, chocolate brown, or purplish brown; gills similarly colored when mature ... **Agaricaceae (*Agaricus*)** (p. 51; Ch. 13)
in *Melanophyllum* Velen. the spore deposit is green to olive at first, becoming chocolate to blackish brown

Spore deposit white buff, cream color, pinkish, or greenish (and remaining so); gills not becoming dark brown from spores with age (three choices below) 16

16. Spore deposit white. Stalk often coated with a layer of viscid or glutinous material up to a more-or-less distinct annular zone ... **Amanitaceae** (p. 51)

Spore deposit white, buff, cream color, pinkish, or greenish. Stalk never with a distinctly viscid or gelatinous layer.................. **Lepiotaceae** (p. 55)

Spore deposit white. Fruit body small, with deeply adnexed to free gills. Fungi found in grass........ .... Tricholomataceae (*Marasmius oreades* (Bolt. ex Fr.) Fr.)
not known to be poisonous, but other poisonous species are sometimes confused with it (Ch. 10, *Clitocybe dealbata*)

17. Spore deposit vinaceous brown, chocolate, fuscous, or blackish brown to black. Gills of mature fruit body deliquescent, namely disintegrating by self-digestion into a black to blackish inky mass..... .... **Coprinaceae (*Coprinus*)** (p. 53; Ch. 9)

Gills usually not as above, but when somewhat deliquescent, then the spore deposit not dark brown to black 18

18. Gills thick and waxy. Spore deposit white .... Hygrophoraceae (p. 55)
not known to contain any poisonous species, but see Chapter 6; when gills have a distinct pinkish to purplish color, it may be *Laccaria* Berk. & Br. (Tricholomataceae)
Gills not typically thick and waxy, but when thick, then the spore deposit some shade of gray, brown, or olive 19

19. Spore deposit white, cream color, or yellow to orange 20
Spore deposit pinkish buff, pink, grayish lilac, vinaceous brown, fawn color, reddish brown to brown, black, olive, or some other dark brownish color 21

20. Fruit body often with a short stalk that is thick in relation to the width of the cap; stalk and cap fragile to brittle, breaking like soft chalk into small pieces when smashed or squeezed .... **Russulaceae** (p. 57)
Fruit body not completely fitting the above description; texture of cap and stalk fleshy, fibrous, pliant, or cartilaginous, but not snapping or breaking like soft chalk .... **Tricholomataceae** (p. 58)

21. Spore deposit pinkish buff to distinctly pink, fleshy brown to brownish pink, vinaceous cinnamon, fawn color, vinaceous buff, or with a grayish lilac tone 22
Spore deposit black, brownish black, chocolate, smoky brown, reddish brown, orange brown, yellow brown, olive brown, or rusty brown, but not pinkish brown or fleshy brown 23

22. Spore deposit distinctly pink or grayish lilac; fungi growing on wood and usually more or less laterally attached. Or spore deposit pinkish buff; fungi terrestrial, with a central stalk .... **Tricholomataceae** (p. 58)
Spore deposit fleshy brown to pinkish brown, or fawn color to vinaceous buff or vinaceous cinnamon; fungi often terrestrial with a distinct central stalk, but at times occurring on wood or woody debris and laterally attached .... **Rhodophyllaceae (=Entolomataceae)** (p. 56)

23. Spore deposit fuscous to almost black, or smoky brown to olive smoky brown. Gills usually decurrent, thick, and distant .... Gomphidiaceae (p. 55)
not known to contain any poisonous species, but see Chapter 6
Spore deposit and gills not fitting the above description 24

24. Gills readily separable as an intact layer from the tissue of the cap, often narrow and frequently with interconnecting ridges and veins, especially near the stalk. Spore deposit clay color to chocolate .... **Paxillaceae** (p. 55)
Gills not with all the above features. Spore deposit variable in color 25

25. Spore deposit yellow brown, orange brown, bright rusty brown, dull rusty brown, cinnamon brown, dull earth brown, or umber 26
Spore deposit purple brown, dusky brown, brownish black, or black (rarely reddish brown)...... 29

26. Fungi lignicolous, occurring on logs, on dead trees, or occasionally on living trees .... **Cortinariaceae** (p. 54)
when on rotten wood, wood chips, or woody debris and when spore deposit is earth brown, it may be *Agrocybe* Fayod (Bolbitiaceae)
Fungi terrestrial, occurring on soil, on humus, or in moss, but not on wood 27

27. Fruit body small; stalk thin; cap hygrophanous. Fungi growing in moss .... **Cortinariaceae (*Galerina*)** (p. 54; Ch. 7)
Fruit body not as above, but when small, fungi not growing in moss 28

28. Fruit body often small. Spore deposit bright rusty brown to orange brown. Stalk thin and often cartilaginous. Young caps conic to bell-shaped with the margin straight; margin plicate–striate in some caps and somewhat deliquescent.......... .... **Bolbitiaceae** (p. 51)
Fruit body variable in size, but when small, then not with the above features .... **Cortinariaceae** (p. 54)
when fruit body has a dull clay color to earth brown spore deposit, is annulate, and is growing in a grassy area, check Bolbitiaceae, p. 51

29. Cap bell-shaped to parabolic, not expanding. Gills mottled with blackish patches of spores when mature. Spore deposit black. Fungi often growing on dung or manured soil ... **Coprinaceae** (p. 53)
Cap bell-shaped to parabolic or conic to convex. Gills not mottled with patches of mature spores 30

30. Basal portion of the stalk and at times the cap staining or turning greenish to bluish green in age or when bruised .... **Strophariaceae** (p. 57)
also check ***Panaeolus*** (Coprinaceae), Ch. 12
Stalk or fruit body not staining or turning bluish green to greenish.. 31

31. Stalk extremely fragile. Cap usually hygrophanous; cap margin usually straight in young specimens .... **Coprinaceae** (p. 53)
Fruit body not with the above features .... **Strophariaceae** (p. 57)

Descriptions of families, with keys to genera when appropriate

Agaricaceae — Agaricales

Fruit body fleshy, with a stalk and cap, and with spores produced on gills on the undersurface of the cap; gills free or nearly so; stalk central; annulus often present from partial veil, but not always; universal veil typically absent, hence volva usually lacking. Spores chocolate to dark brown or blackish in deposit, although sometimes green to olive in deposit at first but soon becoming dark brown, variously shaped (subglobose, ovoid, oblong, ellipsoidal); apical germ pore present or absent; surface smooth. Cheilocystidia often present. Gill trama regular. Cap cuticle usually consisting of filamentous hyphae or sometimes covered by isodiametric cells in the form of an epithelium. Clamp connections present or absent. Fungi growing on soil, dung, or wood.

The main genus is *Agaricus* (Ch. 13). *Melanophyllum* Velen. is related to *Agaricus* and has a green to olive spore deposit that changes to chocolate or blackish brown on standing. Little is known about the edibility of *Melanophyllum*. The above description is based on macroscopic features. Use the technical key incorporating microscopic characteristics on page 61 to separate these genera.

Amanitaceae — Agaricales

Fruit body fleshy, with a stalk and cap, and with spores produced on gills on the undersurface of the cap; gills free or nearly so; stalk central; annulus present or absent; universal veil present, usually leaving a volva or other remains on the stalk and sometimes leaving remains on the cap. Spores usually white to cream color, at times pink or pale green in deposit, variously shaped (globose, subglobose, ovoid, ellipsoidal, or subcylindrical), lacking an apical germ pore, usually smooth but rarely minutely ornamented, usually amyloid or nonamyloid, rarely dextrinoid. Cheilocystidia absent or poorly differentiated; gill edges may be lined with sterile cells from the veil. Gill trama divergent, at least in young fruit bodies. Cap cuticle normally consisting of filamentous hyphae, which at times are poorly developed because of universal veil. Clamp connections present or absent. Fungi saprobic on soil or wood, or mycorrhizal.

The following key is based on macroscopic features. Use the technical key on page 62, which incorporates microscopic characteristics, to confirm your identification.

1. Stalk usually with a volva, but if not, then the remains of the universal veil present as warts or patches on the cap or around or on the stalk base; annulus often present but not always .... ***Amanita* (including *Amanitopsis* Roze)** (Chs. 7, 11, and 13)

   Stalk lacking a volva and lacking patches or warts of veil tissue associated with the cap or stalk; stalk either dry or coated with a viscid or glutinous veil, sometimes with an annular zone above .... *Limacella* Earle

Bolbitiaceae — Agaricales

Fruit body fleshy or fleshy-soft, at times somewhat fragile, typically with a stalk and cap, and with spores produced on gills on the undersurface of the cap; gills free or attached; stalk central, annulate or not, without a universal veil in our material, hence lacking a volva. Spores buff, rust brown, rich brown, hazel brown, or dark brown in deposit, often ellipsoidal, usually with a distinct apical germ pore; surface smooth to roughened; spore wall thin or thick. Cheilocystidia and sometimes pleurocystidia present. Gill trama regular. Cap cuticle consisting of more or less isodiametric to pear-shaped cells that form a hymeniform surface or an epithelium, sometimes also with filamentous hyphae. Clamp connections present or absent. Fungi saprobic on soil, dung, wood, or other plant remains.

The following key is based on macroscopic features. Use the technical key on page 62, which incorporates microscopic characteristics, to confirm your identification.

1. Cap with a viscid layer when fresh. Spore deposit rusty brown 2

   Cap usually not viscid, but if so, then spore deposit earth brown to dark dingy brown 3

2. Cap very thin-fleshed and fragile, often collapsing and almost deliquescent; margin plicate–striate .... *Bolbitius* Fr.

   Cap moderately thick-fleshed; margin not plicate–striate......................... ***Conocybe*** (Chs. 7 and 12)

3. Spore deposit dark to dull brown, either fuscous, earth brown, dark dingy brown, or dark milky chocolate. Stalk 3 mm or more thick, fleshy-fibrous .... *Agrocybe* Fayod

   Spore deposit bright rusty brown; stalk seldom more than 2–3 mm thick, fragile to brittle-cartilaginous .... ***Conocybe*** (Chs. 7 and 12)

Boletaceae — Agaricales

Fruit body fleshy, with a stalk and cap, and with spores typically produced on the inner surface of tubes, though rarely on gills; stalk central or at times eccentric, rarely lateral, annulate or not, not volvate, though universal veil sometimes present. Spores yellow, golden yellow, ocher, pinkish cinnamon, vinaceous brown, ferruginous, dark brown, cinnamon, clay color, olive brown, fuscous, or black in deposit, variously shaped (globose, subglobose, ovoid, ellipsoidal, almond-shaped, subfusiform, subcylindrical), usually lacking an apical germ pore, and smooth or variously ornamented. Cystidia usually present. Tube trama divergent at least when young, often gelatinous. Cap cuticle various, often consisting of filamentous hyphae variously arranged, and sometimes consisting of more or less isodiametric cells. Clamp connections present or absent. Fungi saprobic on soil or wood, mycorrhizal, or parasitic on the fruit bodies of other fungi.

The following key is based on macroscopic features. Use the technical key on page 62 (Smith and Thiers 1971), which incorporates microscopic characteristics, to confirm your identification.

1. Tubes contorted or arranged in such a way that spores cannot fall free from them; spores not discharged from basidia as in other boletes; hence, no spore deposit can be obtained (mostly western species) .... *Gastroboletus* Lohwag

   Tubes vertically arranged, with pores open at maturity; spores discharged from basidia; spore deposit can be obtained from mature specimens, although overmature or young fruit bodies may not give a spore deposit 2

2. Cap and stalk surface woolly to floccose–scaly or hairy. Tubes grayish white when young; spore deposit rusty brown to blackish. Flesh of fresh fruit bodies staining reddish then blackish .... *Strobilomyces* Berk.

   Fruit body not with the above features 3

3. Stalk roughened with small scales or points that are pale to dark gray, dark brown, or blackish, although if scales or points pallid at first, then usually darkening with age or by maturity .... *Leccinum* S. F. Gray

   Stalk not as above 4

4. Fruit body with a delicate floccose–fibrillose sulfur yellow to bright yellow veil that typically leaves a zone on the stalk .... *Pulveroboletus* Murrill

   Fruit body typically lacking a veil, but when present, then not as above 5

5. Pores of tubes uneven, 2–5 mm wide; tubes very shallow, light yellow with tints of green in early stages. Stalk usually eccentric to lateral. Fungi usually associated with ash trees (*Fraxinus*) .... *Gyrodon (Boletinellus) merulioides* (Schw.) Singer

   Tubes not as above. Stalk usually centrally attached 6

6. Spore deposit grayish brown, vinaceous brown, purple brown to purple drab or vinaceous or paler 7

   Spore deposit yellow, greenish yellow, olive, olive brown, cinnamon, or tawny 8

7. Pores of tubes often elongated radially, at least with age. Cap surface usually viscid, but when dry, then distinctly fibrillose. Veil often present, leaving a ring or zone on the stalk. Fungi often associated with larch trees (*Larix*) .... *Fuscoboletinus* Pomerleau & Smith

   compare *Suillus*, couplet 9

   Pores of tubes circular; young tubes pallid to almost whitish to olive buff or grayish at first, often becoming pinkish to vinaceous with age. Cap surface glabrous, unpolished or velvety, not normally viscid. Veil typically lacking .... ***Tylopilus*** (Ch. 13)

8. Stalk usually hollow at maturity. Spore deposit pale yellow .... *Gyroporus* Quél.

   Fruit body not with the above features 9

9. Fruit body with at least two of the following features: (*1*) cap viscid to slimy; (*2*) stalk glandular-dotted; (*3*) veil leaving an annulus on the stalk or adhering to the margin; (*4*) pores of tubes tending to be arranged in radial rows (in this case the cap is usually dry and fibrillose); (*5*) spore deposit pale dingy cinnamon to near clay color to olive .... *Suillus* Mich. ex S. F. Gray (including *Boletinus* Kalchbr.)

   Fruit body not with any two or more of the above features in combination .... ***Boletus*** (Ch. 13)

   Boletellus Murrill also keys out here; the spores must be studied: smooth in *Boletus*; with wings, folds, or striations in *Boletellus*

Cantharellaceae — Aphyllophorales

Fruit body funnel-shaped, tubular, made up of many small caps or with a stalk and cap although often not well-demarcated, fleshy or membranous. Surface of hymenium usually, but not always, on the undersurface of a cap or cap-like structure, often continuing down the stalk for a considerable distance, smooth, radially or reticulately ridged, wrinkled, or sometimes gill-like. Spores white, cream color, ocher, ocher brown, rusty, or light pinkish buff to salmon in deposit, variable in shape, smooth to finely wrinkled to reticulate or tuberculate, thin- or thick-walled, nonamyloid, sometimes cyanophilous. Hymenium usually lacking cystidia. Trama composed of thin- to thick-walled hyphae; clamp connections present or absent; vascular hyphae sometimes present, sometimes terminating in the hymenium. Fungi mainly saprobic and usually terrestrial.

The following key is based on macroscopic features. Only the three most common genera are keyed here. Use the technical key on p. 62 to confirm your identification. For a more complete key, see the work of Petersen (1973).

1. Fruit body fleshy, purplish to purplish brown or violet brown; stalk solid. Spores produced on a wrinkled surface. Several caps sometimes produced from a main stalk .... ***Gomphus*** (Ch. 13)

   Fruit body not with the above combination of characters 2

2. Cap surface and flesh often breaking up into coarse scales; stalk becoming hollow. Spore deposit usually ocher to ocher brown or rusty, less commonly cream color .... ***Gomphus*** (Ch. 13)

Cap not as above. Spore deposit usually white to cream color or light pinkish buff to salmon, but when it is somewhat ocher to ocher brown, then cap not breaking into coarse scales 3

3. Fruit body funnel-shaped, hollow, thin-fleshed, grayish brown to blackish brown or bluish black .... *Craterellus* Pers.

Fruit body not with the above combination of characters, usually with strong ocherous colors or sometimes reddish to orangish or even purplish .... *Cantharellus* Adans. ex Fr.

the thin to membranous species of *Craterellus* may be difficult to separate from *Cantharellus*, and the larger, fleshy forms of *Cantharellus* may resemble *Gomphus*; therefore, use the technical key, when possible, to separate these genera

Clavariaceae — Aphyllophorales

Fruit body erect, cylindrical, fusiform, club-shaped to top-shaped, or branched to coral-like; consistency fleshy, fibrous, leathery, sometimes fragile; surface of hymenium smooth to longitudinally ridged, produced all around the upper surface of the fruit body or branches. Spores white, creamy, yellow, ocher, or brown in deposit, variously shaped, smooth, longitudinally wrinkled, tuberculate or echinulate, usually nonamyloid, sometimes cyanophilous. Cystidia sometimes present. Trama composed of thin- and sometimes thick-walled hyphae; clamp connections present or absent. Fungi mycorrhizal, or saprobic on soil or wood.

The following key is based on macroscopic characters. However, as clavarioid genera are difficult to separate without microscopic features, use the technical key on p. 62 whenever possible.

1. Fruit body branched in a coral-like manner...... 2

Fruit body unbranched or at most with one or two side branches, never with coral-like branching... 6

2. Branching more or less whorled; branch tips box-like or squared off, each often with a fringe of short tooth-like projections that make the fruit body appear crown-like. Fruit body growing on wood .... *Clavicorona* Doty

Branching irregular or sometimes occurring in pairs; branch tips blunt to pointed, not arranged in a crown-like structure 3

3. Fruit body profusely branched. $FeSO_4$ on fresh branches turning them dull olive to blue or bright green. Spore deposit yellow, ocher, or yellow brown to brown.................... ***Ramaria*** (Ch. 13)

Fruit body not with the above combination of characters 4

4. Spore deposit white. Fruit body sparsely to somewhat profusely branched; $FeSO_4$ producing no reaction on fresh branches .... *Ramariopsis* (Donk) Conner

Fruit body not with all of the above features 5

5. Spore deposit white to cream color. $FeSO_4$ turning branches gray green.. .... *Clavulinopsis* van Overeem

Spore deposit usually white. $FeSO_4$ not turning branches gray green with $FeSO_4$........................ .... *Clavaria* Fr. *Clavulina* Schroet.

6. Fruit body simple, broadly club-like or turban-like; interior in the apex spongy or soft and punky, usually 5 mm or more thick at the widest part .... *Clavariadelphus* Donk

Fruit body not with the above combination of characters 7

7. Fruit body or branch tip or tips of fruit body cup-like; rim of cup usually crown-like, with erect tooth-like projections........... *Clavicorona* Doty

Fruit body or branch tip or tips of fruit body acute to rounded but not as described above 8

8. Spore deposit usually white. $FeSO_4$ not turning branches of fruit body green to gray green .... *Clavaria* Vaill. ex Fr.

Spore deposit white to cream color. $FeSO_4$ often turning branches of fruit body green to gray green .... *Clavulinopsis* van Overeem

Clavulinopsis and *Clavariadelphus* are often difficult to separate. *Clavariadelphus* has fruit bodies that are usually 5 mm or more thick in the widest portion, with the interior in the apical portion typically spongy to soft and punky. *Clavaria* may also be difficult to separate from *Clavulinopsis*, especially using a general key. Usually *Clavaria* species are very fragile and more easily broken than *Clavulinopsis* species; this relative character, however, is not easily interpreted without some experience.

Coprinaceae — Agaricales

Fruit body membranous to fleshy, at times fairly fragile, sometimes partially deliquescent, with a stalk and cap, and with spores produced on gills on the undersurface of the cap; gills free to variously attached; stalk central, annulate or not, usually not volvate. Spores black to fuscous to olivaceous black or at times purplish fuscous to chocolate brown or date brown in deposit, variously shaped, often with a conspicuous apical germ pore; surface smooth or variously ornamented. Gill trama regular. Cystidia sometimes present. Cap cuticle, at least the outer layer, composed of isodiametric to short broad cells that are often arranged in a hymeniform layer or epithelium, less often composed of filamentous hyphae. Clamp connections usually present. Fungi usually saprobic on soil, dung, wood, or other plant remains.

The following key is based on macroscopic features. Use the technical key on page 63, which incorporates microscopic features, to confirm your identification.

1. Gills of mature fruit body deliquescent, disintegrating by self-digestion into a black slimy mass .... ***Coprinus*** (Ch. 9)

   Gills not deliquescent.. 2

2. Cap bell-shaped to parabolic, not usually expanding. Gills mottled black, with patches of mature spores .... ***Panaeolus* (including *Panaeolina* Maire)** (Ch. 12)

   Cap shape various, but when shaped as above, then gills not mottled.. 3

3. Fruit body small. Cap flesh extremely thin, usually 1 mm or less at the center; cap margin usually conspicuously striate to plicate-sulcate. Stalk extremely fragile .... **Coprinus** (Ch. 9)

   Fruit body small to medium-sized or at times larger. Cap flesh usually 1 mm or more thick at the center; cap margin usually straight; cap often hygrophanous, quickly fading in color after collecting .... *Psathyrella* (Fr.) Quél. (including *Lacrymaria* (Pat.) Smith & Singer in Singer)

Cortinariaceae — Agaricales

Fruit body membranous to fleshy, usually with a stalk and cap but sometimes with the stalk much reduced to lacking, and with spores produced on gills on the undersurface of the cap; gills variously attached, rarely free; stalk, when present, central, eccentric, or lateral, annulate or not, rarely volvate. Spores usually some shade of brown ranging from ocher, yellow brown, ferruginous, clay color, cinnamon brown, deep brown, vinaceous brown, brick red, chocolate, fuscous, or blackish brown in deposit, variously shaped, typically lacking an apical germ pore (compare Strophariaceae), smooth to ornamented or roughened, thin- or thick-walled, nonamyloid. Gill trama usually regular to irregular. Cheilocystidia and pleurocystidia sometimes present. Cap cuticle often consisting of filamentous hyphae or less commonly composed of more or less isodiametric to pedicellate cells, forming a hymeniform layer or an epithelium. Clamp connections present or absent. Fungi saprobic on soil, dung, mosses, wood, or other plant remains, or mycorrhizal.

The following key is based on macroscopic features. Use the technical key on page 63, which incorporates microscopic features, to confirm your identification.

1. Cap lacking a distinct stalk, attached to the substratum laterally, shelf- or bracket-like; point of attachment sometimes with a very short stub or knob. Fungi occurring on wood........................ .... *Crepidotus* (Fr.) Quél.

   Cap with a distinct stalk that is usually centrally attached.. 2

2. Fungi occurring on wood, namely on limbs or logs of standing dead or living trees or at times on buried wood, rarely occurring on humus........ 3

   Fungi terrestrial, occurring in soil, humus, or moss beds, rarely on wood 5

3. Spore deposit rusty orange brown to rusty orange .... ***Gymnopilus*** (Ch. 12)

   Spore deposit dull to dark brown, cinnamon brown, or rusty brown to yellow brown or more ocherous... 4

4. Cap usually viscid and scaly, but when cap glabrous, then stalk with patches or scales below the annulus or annular zone.. ***Pholiota*** (Ch. 13)
   Pholiota is sometimes considered a member of the Strophariaceae

   Cap or stalk or both not as above........................ .... ***Galerina*** (Ch. 7)
   Pholiota (Ch. 13)
   when on wood chips and with an earth brown spore deposit, it also may be *Agrocybe* Earle; when on rotten wood, check ***Inocybe***, Ch. 10

5. Base of stalk deeply rooting, with fruit bodies in small to large clusters or sometimes scattered. Cortina absent even in young specimens .... *Phaeocollybia* Heim (=*Naucoria* (Fr.) Quél.)

   Fruit body not as above, that is, when cortina absent, stalk usually not rooting 6

6. Cap surface and stalk below annulus with a granular to powdery coating; stalk with a flaring annulus; large pieces of veil tissue present on cap margin; cap and stalk orange to orange tan .... ***Phaeolepiota*** (Ch. 13)

   Not with all the above features 7

7. Stalk 10–25 mm thick at the apex; a distinct membranous annulus present on the stalk apex........ .... *Rozites* Karst.

   Stalk thin to thick, but not with a distinct annulus 8

8. Spore deposit usually dull brown, namely dull tawny to clay color, rarely reddish brown to vinaceous brown. Gill edges whitish to white, crenulate, and often beaded with droplets. Cap sticky to viscid when fresh. Partial veil often absent, but when present, then composed of fibrils and usually not persistent....................... .... ***Hebeloma*** (Ch. 13)
   compare ***Inocybe***, Ch. 10

   Fruit body not with the above features 9

9. Fruit body usually small, with slender stalk 1–3 mm thick. Cap typically conic to parabolic or bell-shaped; margin often straight in early stages. Fungi typically occurring in moss beds or associated with mosses, at times terrestrial or occurring on wood................................. ***Galerina*** (Ch. 7)
   when fruit bodies are on small sticks, moist leaves, or litter, it may be *Tubaria* (W. G. Smith) Gill.

   Fruit body not with all the above features....... 10

10. Cortina a cobwebby veil of fibrils, usually well-developed with its remains usually evident on the stalk or cap edge. Spore deposit often bright orange brown to rusty brown, but at times duller brown. Cap surface variable (dry, moist, viscid, or hygrophanous)................ ***Cortinarius*** (Ch. 7)
when in clusters on soil, check ***Pholiota***, Ch. 13

Cortina lacking or poorly developed, sometimes present and evident in unexpanded fruit body but usually not evident in expanded fruit body. Spore deposit dull brown to grayish brown or dull grayish umber. Cap surface dry, only slightly tacky in wet weather, never viscid .... ***Inocybe*** (Ch. 10)
compare ***Hebeloma***, Ch. 13, or when growing in clusters, check ***Pholiota***, Ch. 13

Gomphidiaceae — Agaricales

Fruit body fleshy, with a stalk and cap, and with spores produced on gills on the undersurface of the cap; gills attached, decurrent, and thick; stalk relatively central, not annulate, but sometimes with a partial veil that is glutinous, composed of fibrils, or powdery, and without a universal veil. Spores smoky gray brown to fuscous or black in deposit, usually 15 μm long or more; shape ovoid, ellipsoidal, or subfusiform; apical germ pore absent; surface smooth. Gill trama divergent, at least in young fruit bodies. Cystidia large, thin- or thick-walled. Cap cuticle composed of filamentous hyphae. Clamp connections typically absent. Fungi saprobic on soil or mycorrhizal.

The family contains two genera, *Gomphidius* and *Chroogomphus*, neither of which contain any species known to be poisonous. Except for the key given below which separates the genera on the basis of macroscopic features, and the technical key based on microscopic characters given on p. 64 with which you can confirm your identification, this family is not treated further.

1. Cap surface gelatinous, covered with a layer of gluten, to distinctly viscid; flesh of cap white ... . *Gomphidius* Fr.

Cap surface dry to viscid or tacky; flesh of cap colored, light orange or light yellowish, or near the surface light salmon to pinkish in some species .... *Chroogomphus* (Singer) O. K. Miller

Hygrophoraceae — Agaricales

Fruit body with a stalk and cap, fleshy, with spores produced on gills on the undersurface of the cap; gills attached although sometimes seceding, often rather thick and appearing waxy; stalk central or nearly so, not annulate but sometimes with a veil of gluten or fibrils, without a volva. Spores white in deposit, variously shaped, lacking an apical germ pore, usually smooth, usually nonamyloid. Gill trama regular, irregular, or divergent. Basidia typically long and narrow, about six times the length of the spores or longer. Cystidia present or absent. Cap cuticle various, composed of filamentous hyphae to form a cutis or a trichoderm or of more or less isodiametric cells to form a hymeniform layer or an epithelium. Clamp connections present or absent. Fungi saprobic on soil or mycorrhizal.

Only the genus *Hygrophorus* is used here. Some workers separate its species into two genera, the other being *Hygrocybe*. Others divide the genus even further (Smith 1973, Stuntz 1977). However, these genera are combined under the one name *Hygrophorus* in this book, making the inclusion of keys for this family unnecessary. As this genus is not known at present to contain any poisonous species, it is not treated further here, except for a brief citation in Chapter 6, "Scope".

Lepiotaceae — Agaricales

Fruit body with a stalk and cap, fleshy, with spores produced on gills on the undersurface of the cap; gills free; stalk central, usually annulate, without a universal veil, hence not volvate. Spores white, cream color, pale buff (occasionally pinkish), or light greenish in deposit, variously shaped, with or without an apical germ pore, smooth, amyloid, dextrinoid, or nonamyloid. Gill trama regular to irregular. Cystidia sometimes present. Cap cuticle various, filamentous forming a cutis or a trichoderm or composed of more or less isodiametric cells forming a hymeniform layer or an epithelium. Clamp connections present or absent. Fungi saprobic on soil, wood, or other plant remains.

Only the genus *Lepiota* (Chs. 7 and 13) is used here. Some workers separate its species into several genera. Some of these besides *Lepiota* are: *Chlorophyllum* Massee (greenish spore deposit), *Cystolepiota* Singer, *Leucoagaricus* (Locquin) Singer, *Leucocoprinus* Pat., and *Macrolepiota* Singer. These workers use microscopic features, at least in part, to separate these genera. However, these genera are combined under the one name *Lepiota* in this book, making the inclusion of keys for this family unnecessary. Stuntz (1977) provides further information.

Paxillaceae — Agaricales

Fruit body with a cap, with or without a stalk, fleshy, and with spores produced on the surface of gills on the undersurface of the cap; gills decurrent (when a stalk is present), often easily and cleanly separable as a unit from the cap trama; stalk, when present, central, eccentric, or lateral, neither annulate nor volvate. Spores dark buff to medium brown, olivaceous brown, or dark brown (chocolate brown to coffee brown) in deposit, ovoid to ellipsoidal or oblong, lacking an apical germ pore, and smooth. Gill trama divergent, at least in young fruit bodies. Cystidia sometimes present. Clamp connections usually present. Fungi mycorrhizal, or saprobic on wood or soil.

Paxillus (Ch. 13) is the main genus, although the family contains several other genera, including *Phylloporus* Quél. Others place *Phylloporus* in the Boletaceae. It differs from *Paxillus* microscopically and also has more or less waxy, brightly colored gills, often yellow, which commonly have interconnecting veins and ridges that may form large to irregularly shaped pores. *Phylloporus* does not contain any known poisonous species. Use the technical key on page 64, which incorporates microscopic characteristics, to separate these two genera.

Pluteaceae (=Volvariaceae) — Agaricales

Fruit body with a stalk and cap, fleshy, often soft, with spores produced on gills on the undersurface of the cap; gills free; stalk central, with or without an annulus or volva. Spores sordid pink to brownish pink or vinaceous brown in deposit, variously shaped, lacking an apical germ pore, smooth, and nonamyloid. Gill trama convergent, at least in young fruit bodies. Cystidia often present. Cap surface composed of filamentous hyphae forming a cutis or a trichoderm or of more or less isodiametric cells forming a hymeniform layer. Clamp connections present or absent. Fungi saprobic on soil, wood, or other plant remains, rarely parasitic on the fruit bodies of other fungi.

The two genera of this family most often found in North America are *Volvariella* and *Pluteus*. *Volvariella* is not known to contain any poisonous species, but *Pluteus* (Ch. 12) is thought to contain a hallucinogenic species. A macroscopic key to separate the two genera is given below. Use the technical key based on microscopic characters on p. 64 to confirm your identification. A third genus, *Chamaeota* (W. G. Smith) Earle, can be distinguished from the other two by the presence of an annulus. However, it is rare and is not commonly encountered in North America.

1. Volva present; annulus absent.. *Volvariella* Speg.
   Volva and annulus absent .... ***Pluteus*** (Ch. 12)

Polyporaceae — Aphyllophorales

Fruit body variable in shape, semicircular, fan-shaped, shell-shaped (as half a bivalve), hoof-shaped, with a cap and stalk, flat on the surface of the substratum with the spore-producing surface facing outward, or spread out over the substratum and turned back at the margin to form a narrow cap; consistency usually leathery, corky, or woody, but sometimes membranous, spongy, fibrous, or fleshy. Surface of hymenium on the undersurface of the cap or facing outward when the fruit body is flat on the surface of the substratum, usually consisting of pores or sometimes gill-like to elongate or labyrinthine, at times forming flattened teeth. Spores white, yellow, or brown in deposit, variously shaped, often smooth, at times echinulate to tuberculate, thin- to thick-walled, usually nonamyloid. Hymenium with or without cystidia. Trama composed of thin- and often thick-walled hyphae. Clamp connections present or not. Fungi saprobic on wood, sometimes terrestrial, also parasitic on trees.

So few species of this large family are fleshy enough to be collected by foragers looking for edible mushrooms that a key to the genera is not feasible for the purposes of this book. The main genera *Albatrellus* and *Laetiporus* are treated in Chapter 13 alphabetically under the name polypores.

Rhodophyllaceae (=Entolomataceae) — Agaricales

Fruit body with a cap, usually with a stalk, fleshy to membranous, with spores produced on gills on the undersurface of the cap; gills attached (when stalk is present), sometimes seceding; stalk, when present, central to eccentric, without an annulus or a volva. Spores grayish pink to pinkish cinnamon or vinaceous buff, often angular or nodulose in side view or in end view or both, sometimes minutely angular–roughened, lacking an apical germ pore, and nonamyloid. Gill trama regular to irregular. Cystidia sometimes present. Cap surface composed of filamentous hyphae forming a cutis or a trichoderm; end cells sometimes forming cystidia. Clamp connections present or absent. Fungi mycorrhizal, or saprobic on soil, wood, or other plant remains.

Some workers place all species in one genus, *Rhodophyllus* Quél.; others place them all in the genus *Entoloma* (Fr.) Quél. In this treatment, however, the family is divided into the genera *Entoloma* and *Nolanea* (Ch. 13), as well as *Leptonia*, *Alboleptonia*, *Claudopus*, *Clitopilus*, *Pouzarella*, and *Rhodocybe* (see the key below). The genus *Eccilia* (Fr.) Kummer may also be used but is not included here. Microscopic features are required for correctly separating at least some of these genera. The following key, based on macroscopic features, is adapted from that of Largent (1973). Use the technical key on page 64, which incorporates microscopic characteristics, to separate the genera further and to confirm your identification. The technical key is adapted from that of Stuntz (1977).

1. Cap sessile, laterally attached and shelving, or attached by a short stub or knob, and without a distinct stalk .... *Claudopus* (W. G. Smith) Gill.
   Cap with a distinct stalk that is usually, but not always, centrally attached to the cap 2
2. Stalk eccentric to lateral, well-developed 3
   Stalk centrally attached...................................... 4
3. Gills distinctly decurrent... *Clitopilus* (Fr.) Quél.
   Gills adnexed to adnate ....
   .. ***Entoloma* (rhodophylloid mushrooms)** (Ch. 13)
4. Stalk slender, usually 1–3 mm thick but never more than 5 mm thick, often fibrous–pliant, or fragile, or brittle; cap trama typically thin to membranous (three choices)................................... 5

Stalk stouter, usually 5 mm or more thick, fleshy-fibrous and soft, or pulpy, or with a firm cartilaginous rind and a soft interior; cap trama usually 3–5 mm thick, rarely thin enough to be called membranous .... ***Entoloma* (rhodophylloid mushrooms)** (Ch. 13)

5. Mature caps parabolic to bell-shaped, or at least strongly convex; margin straight in expanded caps .... ***Nolanea* (rhodophylloid mushrooms)** (Ch. 13)

   Mature caps plane to shallowly convex; margin enrolled in unexpanded caps .... *Leptonia* (Fr.) Quél. (including *Alboleptonia* Largent & Benedict)

   When stalk base strigose, coated with fairly long hairs .. *Pouzarella* Mazzer (=*Pouzaromyces* Pilát)

Russulaceae — Agaricales

Fruit body with a stalk and cap, fleshy and typically brittle, and with spores produced on gills on the undersurface of the cap; gills attached or free; stalk usually central, usually not annulate, not volvate. Spores white, cream color, yellow, pinkish buff, or orange in deposit, globose to subglobose, ovoid or obovoid to ellipsoidal, lacking an apical germ pore, and ornamented with ridges or wings that are usually amyloid. Gill trama composed of filamentous hyphae and groups of more or less globose cells called sphaerocysts or composed of filamentous interwoven hyphae. Cystidia often present. Cap cuticle various, often composed only of filamentous hyphae forming a cutis or a trichoderm. Cap trama often composed of filamentous hyphae and groups of sphaerocysts. Clamp connections absent. Fungi mycorrhizal or saprobic on wood or soil.

The following key is based on macroscopic features. Use the technical key on page 65, which incorporates microscopic characteristics, to confirm your identification.

1. Fruit body exuding a watery juice or latex-like substance, especially in young fresh specimens when cut (cut stalk apex with a razor to check this character) .... ***Lactarius*** (Ch. 13)

   Fruit body not exuding a watery juice or latex-like substance when cut .... ***Russula*** (Ch. 13)

Sclerodermataceae — Sclerodermatales

Fruit body usually aboveground, nearly spherical to ovoid or pear-shaped, often yellow to yellow brown when mature, sessile or on a false stalk consisting of fibrous mycelia binding together soil, sand, and debris. Peridium thick, usually not differentiated into layers, opening by breaking irregularly or into lobes or by disintegration. Gleba often violaceous or dark-colored and mottled with whitish to pallid veins in early stages although sometimes whitish to pallid throughout when very young, powdery when mature. Spores ornamented, yellow brown to brown or blackish when mature. Capillitium absent to rudimentary. Fungi mycorrhizal, or saprobic on soil or rotten wood.

The main genus *Scleroderma*, which is characterized as above, is treated in Chapter 13. As there is only one genus of importance in this family, there is no need to include keys.

Strophariaceae — Agaricales

Fruit body with a cap and usually with a stalk, fleshy to thin, with spores produced on gills on the undersurface of the cap; gills attached (sinuate, adnate, adnexed, or decurrent); stalk central to eccentric or lateral; partial veil present or absent, sometimes annulate, not volvate. Spores dark brown to fuscous or purplish fuscous to violaceous black, variously shaped, usually smooth to rarely slightly ornamented, usually with a distinct apical germ pore but sometimes only thin to modified at the apical end. Gill trama regular to irregular. Cystidia present or absent. Cap cuticle usually composed of filamentous hyphae forming a cutis or less commonly a trichoderm but rarely composed of more or less isodiametric cells forming an epithelium. Clamp connections usually present. Fungi saprobic, terrestrial or occurring on wood, although sometimes occurring on moss, dung, or other plant remains.

The following key is based on macroscopic features. Use the technical key on page 65, which incorporates microscopic characteristics, to confirm your identification. There are two technical keys for separating the genera of this family. In Key I, *Psilocybe* includes the genera *Stropharia* and *Naematoloma* (=*Geophila* Quél. and *Hypholoma* (Fr.) Quél. of some workers). In Key II, *Stropharia* and *Naematoloma* are separated from *Psilocybe*.

1. Spore deposit dull brown to rusty brown. Stalk central. Fungi often occurring on wood, but not always .... ***Pholiota*** (Cortinariaceae) (Ch. 13)

   Spore deposit darker, purple brown, violaceous brown, or dark grayish brown to blackish brown, rarely brownish red to chocolate 2

2. Stalk typically eccentric, shorter than the diameter of the cap .... *Melanotus* Pat.

   Stalk central, and longer than above 3

3. Basal portion of the stalk and at times the cap staining or turning greenish to bluish green or bluish when bruised or with age .... ***Psilocybe*** (Ch. 12)

   Stalk not as above 4

4. Annulus present 5

   Annulus absent 7

5. Annulus dry, and membranous, floccose, or fibrillose; cap viscid or not; stalk not viscid below the annulus. Gills decurrent or otherwise attached.. 6

Annulus viscid and forming a viscid zone on the stalk; cap and stalk below the annulus viscid when fresh. Gills adnexed, adnate, or uncinate, not decurrent .... ***Stropharia*** (Chs. 12 and 13)

6. Annulus poorly developed to thin; remnants of partial veil often present on the cap margin. Fungi usually occurring on wood .... ***Naematoloma*** **(=*Hypholoma* (Fr.) Quél.)** (Ch.7)

   Annulus well-developed and membranous; when the veil remnants adhere to the cap margin, then fungi terrestrial....... ***Stropharia*** (Chs. 12 and 13)

7. Stalk slender, usually 1-3 mm wide but occasionally up to 5 mm thick, often fibrous-pliant, or fragile, or brittle. Cap flesh very thin to membranous .... ***Psilocybe*** (Ch. 12)

   Stalk thicker and stouter, 5 mm or more thick at the apex, fleshy-fibrous and soft, or pulpy, or with a firm cartilaginous rind and a softer interior. Cap trama thicker than above, at least 3-5 mm thick at the center, rarely membranous .... ***Psilocybe*** (Ch. 12)

   also check ***Stropharia,*** Chs. 12 and 13, and ***Naematoloma,*** Ch. 7. Some workers now include these genera in *Psilocybe*

Tricholomataceae — Agaricales

Fruit body with a cap and usually with a stalk, membranous, fleshy, or fleshy-tough, with spores produced on gills on the undersurface of the cap; gills attached (when stalk is present) or sometimes free; stalk when present central, eccentric, or lateral, annulate or not (partial veil present or absent), usually without a universal veil (not volvate). Spores white, cream color, pinkish, vinaceous fawn, pale lilaceous or pale greenish in deposit, variously shaped (rarely annular in outline, but if so, then spore deposit white to cream color), lacking an apical germ pore, smooth to variously ornamented, thin- to thick-walled, amyloid, dextrinoid, or nonamyloid. Basidia typically less than six times as long as the spores. Gill trama regular, irregular, or divergent. Cystidia present or absent. Cap cuticle composed of filamentous hyphae forming a cutis or a trichoderm or of more or less isodiametric cells forming a hymeniform layer or an epithelium. Clamp connections present or absent. Fungi mycorrhizal; saprobic on soil, wood, or other plant remains; or parasitic on vascular plants or other fungi.

The following key is based on macroscopic features. Use the technical key on page 65, which incorporates microscopic characteristics, to confirm your identification.

1. Cap lacking a distinct stalk, attached to the substratum laterally, like a shelf or bracket, or basally (with a funnel-shaped cap); point of attachment sometimes with a very short stub or knob 2

   Cap with a distinct stalk; stalk usually centrally attached to the cap, occasionally eccentric or lateral 6

2. Spore deposit pink. Fruit body orange, with a strong repulsive odor. Cap densely tomentose. Fungi occurring on rotting wood...................... .... *Phyllotopsis* Gilbert & Donk ex Singer

   Fruit body not with all the above features......... 3

3. Fruit body thin and tough to leathery. Gills split longitudinally along their edges and rolled back toward their faces. Spore deposit white. Fungi usually occurring on wood.. *Schizophyllum* Fr.

   Fruit body not with longitudinally split gills, though at times with one or more of the other features... 4

4. Gill edges very uneven to serrate. Flesh often tough to rubbery, usually tasting hot and peppery .... *Lentinellus*

   Gill edges even. Flesh tough and pliable, or fibrous, or soft and fleshy... 5

5. Flesh of the cap usually soft and fleshy; cap color white to dull grayish or dull brownish .... *Pleurotus* (Fr.) Quél.

   related genera such as *Pleurocybella* Singer or *Hohenbuehelia* Schulz. may key out here

   Flesh of the cap usually tough to pliable or fibrous; cap often with pink vinaceous reddish tones .... *Panus* Fr.

   when cap is greenish and gills orange to yellowish, it may be *Panellus (Pleurotus) serotinus* (Pers. in Hoffm. ex Fr.) Kühner

6. Fruit body small. Cap tough to fleshy-leathery; color orange brown to brown; taste bitter. Fungi occurring on wood .... *Panellus* Karst.

   Fruit body not fitting all the above features...... 7

7. Fungi occurring on wood (limbs, branches, logs, dead trees, or occasionally living trees) or growing from buried wood, around stumps, or around the base of living trees.. 8

   Fungi terrestrial (e.g. on soil, humus, or litter, or in moss beds), not typically growing on wood... 19

8. Partial veil present, leaving an annulus or annular zone on the stalk 9

   Partial veil lacking, hence no annulus or annular zone... 10

9. Gill edges distinctly serrate. Cap trama and stalk tough and pliable. Fruit bodies occurring singly, scattered, in groups, or in clusters. *Lentinus* Fr.

   Gill edges even. Stalk base at times more or less bulbous. Cap surface often with small pointed scales. Fruit bodies often in clusters .... ***Armillariella*** (Ch. 13)

10. Stalk usually centrally attached and well developed, but sometimes as below 11

Stalk usually eccentric to laterally attached .... *Pleurotus* (Fr.) Quél. and relatives

11. Cap, stalk, and gills yellow to orange; gills narrow, decurrent, thin, not forked. Fruit bodies often occurring in large clusters .... ***Omphalotus*** (Ch. 13)
    in some areas ***Armillariella tabescens,*** Ch. 13, may key out here

    Fruit body not with all of the above features .. 12

12. Fungi scattered to gregarious, occurring on logs or other pieces of wood 13

    Fungi occurring usually in small to large clusters, although some may be densely gregarious 16

13. Stalk tough to elastic or of a more or less horny consistency and rigid. Cap trama pliable and rather tough. Fruit body usually reviving to natural size and color when moistened, not rotting when drying 14

    Stalk soft and fleshy, or brittle, or fragile. Cap trama more or less easily torn or broken when fresh and moist 15

14. Fruit body brightly colored, yellow to orange .... *Xerulina* Singer
    when fruit bodies are small, orange colors are dull, and the stalk base is coated with orange to tawny hairs, it may be *Xeromphalina* Kühner & Maire

    Fruit body variously colored, but not usually bright yellow or orange .... *Marasmius* Fr.

15. Cap often convex to plane, at times broadly rounded; edge of cap margin enrolled in young fruit bodies. Gills attached, but not decurrent .... check ***Collybia*** (Ch. 13)
    it may also be ***Tricholomopsis,*** Ch. 13, and relatives

    Cap often parabolic to bell-shaped or strongly convex; edge of cap margin straight, not enrolled, when young. Gill attachment various (e.g. adnate or adnexed) .... ***Mycena*** (Ch. 13)
    when fruit bodies have decurrent gills, check ***Clitocybe,*** Chs. 9, 10, and 13, or it may be *Omphalina* Quél.

16. Cap surface glutinous to sticky. Stalk velvety. Fungi often occurring on elm, but also on other hardwoods .... *Flammulina velutipes* (Curt. ex Fr.) Singer

    Not fitting all of the above description 17

17. Cap plane to somewhat convex; edge of cap margin enrolled in unexpanded caps. Gills attached but not typically decurrent .... ***Collybia*** (Ch. 13)
    also ***Tricholomopsis,*** Ch. 13, and when stalk is long rooted, *Oudemansiella* Speg.

    Fruit body not as above 18

18. Fruit body usually small. Flesh of cap thin; cap margin striate. Gills arcuate-decurrent to decurrent. Stalk thin with a horny consistency, dark colored and with a tuft or mass of orangish to tawny hairs at the base .... *Xeromphalina* Kühner & Maire

    Fruit body usually small. Flesh of cap thin; cap margin often striate, usually straight in unexpanded caps. Gills variously attached (e.g. adnate or adnexed). Stalk not as above .... ***Mycena*** (Ch. 13)

19. Cap usually orange yellow to apricot color, sometimes yellowish brown; cap margin incurved. Gills deep orange to pale yellow, crowded, decurrent, and often forked. Fungi terrestrial or sometimes occurring on rotten wood .... *Hygrophoropsis aurantiaca* (Wulfen ex Fr.) Maire
    suspected of containing muscarine (Ch. 10); also check ***Clitocybe,*** Chs. 9, 10, and 13

    Not with the above features 20

20. Gills thick, appearing more or less soft and waxy or thick and fairly brittle, often subdistant, relatively widely spaced 21

    Gills typically thin, although rarely thickish, but then not waxy in appearance, usually close to crowded but occasionally subdistant 22

21. Gills purplish to lilac or vinaceous to reddish or pinkish flesh color, often thick and rather hard and brittle .... *Laccaria* Berk. & Br.

    Gills usually otherwise colored, typically soft and waxy in appearance .... *Hygrophorus* Fr. (Hygrophoraceae)

22. Cap dry and surface more or less granulose to powdery. Stalk often with a dry, granulose to powdery coating below the annular zone or on the lower two-thirds .... *Cystoderma* Fayod

    Cap and stalk not as above 23

23. Annulus or annular zone present 24

    Annulus or annular zone lacking 25

24. Stalk with a sheath-like or fibrillose coating over the lower half or two-thirds; sheath often ending in a ring or zone at approximately midstalk or above .... *Armillaria* (Fr.) Quél.
    it may also be *Catathelasma* Lovej., or check ***Tricholoma,*** Ch. 13

    Stalk usually without a distinct sheath-like coating over the lower surface; annulus in the form of a distinct ring or at times only a faint zone .... ***Tricholoma*** (Ch. 13)
    Armillaria and *Tricholoma* are difficult to separate macroscopically; either genus may key out here or in the above choice

25. Flesh of cap dry and firm, often with a disagreeable to bitter taste. Base of stalk attached to an extensive mat of white mycelia around the stalk base .... *Leucopaxillus* Boursier
    compare ***Tricholoma,*** Ch. 13, and ***Clitocybe,*** Chs. 9, 10, and 13

    Fruit body not with all the above features 26

26. Gills decurrent to subdecurrent........................ 27
    Gills adnate, adnexed, or sinuate, not distinctly decurrent.. 31
27. Stalk slender, 1-3 mm or up to 5 mm thick, often fibrous-pliant, or brittle, or fragile. Flesh of cap often thin to membranous............................. 28
    Stalk stouter than above, 5 mm or more thick, fleshy-fibrous and soft, or pulpy, or with a firm cartilaginous rind and a soft interior. Flesh of cap usually at least 3-5 mm thick in the center, rarely membranous... 29
28. Stalk thin, dark colored, rigid and horny in consistency, with a tuft of orangish to yellow brown hairs at the base...................................... .... *Xeromphalina* Kühner & Maire
    Not as above in all features............................. 30
29. Stalk tough to elastic or somewhat rigid. Cap trama pliable and rather tough. Fruit body usually reviving to more or less its natural size and color when moistened, not rotting when dryed.. .... *Marasmius* Fr.
    Stalk soft and fleshy, or brittle, or fragile. Cap trama easily torn or broken when fresh and moist .... ***Clitocybe*** (Chs. 9, 10, and 13) ***Mycena*** (Ch. 13)
    when associated with moss and algae, it may be *Omphalina* Quél.
30. Gills forked, white to grayish, at times spotted reddish brown. Cap gray to grayish olive, with a distinct umbo in the center. Fungi often growing in moss beds.. *Cantharellula umbonata* (Gemlin ex Fr.) Singer
    Fruit body not as above and fungi usually not associated with moss... .... ***Clitocybe*** (Chs. 9, 10, and 13)
    when gills are thick with blunt edges or ridge-like, check **Cantharellaceae,** p. 52, or it may be *Hygrophorus* Fr. (Hygrophoraceae)
31. Stalk slender, 1-3 mm or up to 5 mm thick, often fibrous-pliant, or brittle, or fragile. Flesh of cap often membranous; cap margin often striate.. 32
    Stalk stouter than above, 5 mm or more thick, fleshy-fibrous and soft, or pulpy, or with a firm cartilaginous rind and a soft interior. Flesh of cap usually at least 3-5 mm in the center, rarely membranous... 34
32. Stalk tough to elastic or somewhat rigid. Flesh of cap pliable and rather tough. Fruit body usually reviving to a more or less natural size and color when moistened, not decaying readily when dried .... *Marasmius* Fr.
    Stalk soft and fleshy, or brittle, or fragile. Flesh of cap easily torn or broken when fresh or moist... 33
33. Cap often convex to plane, at times broadly rounded; edge of cap margin enrolled in young stages. Gills not at all decurrent........................ .... ***Collybia*** (Ch. 13)
    Cap often parabolic to bell-shaped or cone-shaped to strongly convex; edge of cap margin straight, not enrolled when young. Gills sometimes slightly decurrent .... ***Mycena*** (Ch. 13)
34. Fungi occurring in clusters of several to many fairly large fruit bodies ... possibly *Lyophyllum* Karst.
    Fungi not as above, but when occurring in clusters, then the clusters usually consisting of only a few fruit bodies 35
35. Cap glutinous to viscid. Stalk long in relation to the width of the cap and deeply rooting into the substratum. Fungi solitary, scattered, or gregarious, sometimes occurring around rotten stumps or growing from buried wood.......................... .... *Oudemansiella* Speg.
    Not fitting all the above features..................... 36
36. Cap flattened to broadly convex, often with a distinct umbo in the center; cap surface soft and smooth, like leather. Gills crowded and usually pure white. Stalk long in relation to the width of the cap and snapping easily, leaving a clean sharp break..................................... *Melanoleuca* Pat.
    Fruit body not as above; stalk usually thicker in relation to the width of the cap and fibrous-fleshy to fleshy, not snapping cleanly.......................... .... ***Tricholoma*** (Ch. 13)
    when the fruit body is in general grayish to blackish or grayish brown to pallid and dull-colored, it may be *Lyophyllum* Karst.; when the cap margin has an enrolled edge when young, check ***Collybia*** (Ch. 13); when the fruit body has a buff to pinkish buff spore deposit and slightly decurrent gills, or when the fruit body is pinkish to lilac or bluish colored, or both, check ***Clitocybe*** **(*Lepista* (Fr.) Gill. group)** (Ch. 13)

BASIDIOMYCOTA: TECHNICAL

Key to the families of the order Agaricales: gill mushrooms and boletes

1. Trama of the cap and stalk with nests of sphaerocysts among filamentous hyphae. Spores with amyloid ornamentation .... **Russulaceae** (p. 65)
   Trama of the cap and stalk lacking sphaerocysts. 2
2. Spores produced on distinct blade-like gills with acute edges.. 3
   Spores produced on wrinkles, veins, ridges, or obtuse-edged gills .... **Cantharellaceae** (Aphyllophorales) (p. 62)
3. Spore deposit flesh brown to pinkish brown, pinkish, vinaceous or reddish cinnamon; spores angular (Fig. 89) or longitudinally striate .... **Rhodophyllaceae (=Entolomataceae)** (p. 64)

Spores not angular or, if so, then spore deposit white or rusty brown to earth brown in deposit 4

4. Gills thick and waxy with fairly acute edges, relatively soft; basidia usually long in relation to width at the apex, at least six times longer than the spores; spore deposit white to whitish .... Hygrophoraceae
   Gills not as above, usually thin and not waxy.. 5
   when thick and waxy to brittle, and spores echinulate, it may be *Laccaria* Berk. & Br. (Tricholomataceae)

5. Gills free, or nearly so, from stalk; gill trama divergent. Universal veil typically present, sometimes in the form of a layer of slime; partial veil often present and leaving an annulus on the stalk. Spores hyaline under the microscope (in water); spore deposit usually white to cream color........ .... **Amanitaceae** (p. 62)
   Not with the above features 6

6. Gills free or practically so; gill trama convergent. Universal veil present or absent. Spore deposit pink to vinaceous or reddish cinnamon............ .... **Pluteaceae (=Volvariaceae)** (p. 64)
   Not with all the above features 7

7. Spore deposit white, pale yellow, pinkish buff or greenish, but when greenish, then not changing to blackish brown or chocolate brown 14
   Spore deposit ocherous brown, yellow brown, clay color, earth brown, umber, rusty brown, orange brown, or some shade of blackish brown, chocolate brown, to black, rarely reddish to olive or greenish, but when greenish, then changing to blackish brown or chocolate brown on drying 8

8. Spore deposit fuscous to olive fuscous; spores large and bolete-like in shape (Fig. 124) and dark brown under the microscope (in water); gills thick, usually distant and decurrent .... Gomphidiaceae (p. 64)
   Characters not as above, but when spore deposit blackish to dark brown, then gills thin........... 9

9. Spore deposit black, cocoa brown to chocolate brown, to dull cinnamon brown, rarely vinaceous to brick red; spores with an apical germ pore; cap surface in section usually cellular to hymeniform .... **Coprinaceae** (p. 63)
   but when composed of filamentous hyphae and the gills deliquescent, follow this choice
   Not as above, but when spores and cap surface as above, then spore deposit ocherous, bright rusty brown, rusty brown, or clay color 10

10. Gills free or slightly attached to the stalk apex. Annulus usually present. Spore deposit blackish, fuscous, olive fuscous, or some shade of chocolate brown; spores typically lacking a distinct apical germ pore. Stalk easily and cleanly separable from the cap................................. **Agaricaceae** (p. 61)
   when spore deposit greenish at first then changing to blackish brown or chocolate brown on drying, follow this choice
   Fruit bodies not with all the above features 11

11. Gills readily separable as an intact layer from the trama of the cap, often narrow and frequently with interconnecting ridges and veins especially near the stalk. Veil absent. Spore deposit clay color to chocolate .... **Paxillaceae** (p. 64)
   Gills not with all the above features 12

12. Spores with an apical germ pore; spore deposit bright rusty brown to clay color. Cap cuticle cellular to hymeniform........ **Bolbitiaceae** (p. 62)
   Not with the above features in combination ... 13

13. Spore deposit rusty brown to clay color; spore surface often roughened, finely wrinkled to warted; spores lacking an apical germ pore....... .... **Cortinariaceae** (p. 63)
   Spore deposit dull yellow brown to dark yellow brown, blackish brown, violaceous brown or more violaceous; spores typically with an apical germ pore (check spores using 100× oil-immersion lens); spore surface usually smooth .... **Strophariaceae** (p. 65)

14. Cap and stalk readily separable from each other. Gills free; gill trama regular. Partial veil usually leaving an annulus. Spore deposit white to off-white, rarely pinkish or greenish to olive green, but when olive or greenish, then not darkening on drying................................. **Lepiotaceae** (p. 55)
   Cap and stalk not easily and cleanly separable. Gills usually attached, rarely free. Spore deposit white, yellowish, buff, pinkish, or vinaceous brown to grayish lilac. Stalk central, eccentric, lateral, or lacking .... **Tricholomataceae** (p. 65)

Keys to important genera

Agaricaceae — Agaricales

1. Spore deposit green to olive when fresh and moist, soon becoming dark brown. Cap cuticle composed of sphaerocysts, forming an epithelium... .... *Melanophyllum*
   Spore deposit lacking green to olive tones when fresh and moist. Cap cuticle forming a cutis or an epithelium 2

2. Spores ovoid to ellipsoidal and smooth. Cap cuticle composed of filamentous more or less repent hyphae, forming a cutis........ ***Agaricus*** (Ch. 13)
   Spores subangular to nodulose in outline. Cap cuticle composed of sphaerocysts, forming an epithelium .... *Cystoagaricus* Singer

Amanitaceae **Agaricales**

1. Universal veil membranous, fleshy or more or less powdery, present in mature fruit bodies as a volva, or as powder, warts, patches, or rings on the cap or on the lower portion of the stalk, or both .... ***Amanita*** (Chs. 7, 11, and 13)

   Universal veil not as above, often glutinous and leaving a layer of slime on the stalk.. *Limacella*

Bolbitiaceae **Agaricales**

1. Cap viscid. Spore deposit bright rust brown 2

   Cap usually not viscid, but when viscid, then spore deposit dark dingy brown to fuscous 3

2. Cap very thin-fleshed and fragile, often collapsing, plicate-striate to plicate-sulcate toward the margin .... *Bolbitius*

   Cap at least moderately thick-fleshed, not plicate-sulcate toward the margin .... ***Conocybe*** (Chs. 7 and 12)

3. Spore deposit bright rust brown. Stalk fragile or brittle-cartilaginous, seldom more than 2-3 mm thick......................... ***Conocybe*** (Chs. 7 and 12)

   Spore deposit darker, dark dingy brown to fuscous or a dark milky chocolate. Stalk fleshy-fibrous, 3 mm or more thick .... *Agrocybe*

Boletaceae **Agaricales**

1. Spores not discharged from basidia, hence spore deposit not obtainable (the tubes often oriented so that the spores cannot fall free from them) ... . *Gastroboletus*

   Spores discharged from basidia, hence spore deposit obtainable from fresh mature fruit bodies 2

2. Cap and often the stalk covered with a coating of coarse, dry, woolly to floccose scales, or fairly hairy. Spore deposit blackish brown or somewhat rusty brown; spores globose to subglobose; spore surface reticulate or verrucose .... *Strobilomyces*

   Cap and stalk not as above. Spores smooth or when elongated then either smooth or ornamented... 3

3. Spores ornamented with wings, folds, or striations. Fruit body like ***Boletus***, Ch. 13, in appearance .... *Boletellus* Murrill

   Spores smooth or ornamented, but ornamentation not as above 4

4. Fruit bodies with two or more of the following features: (*1*) pores of tubes boletinoid at maturity; (*2*) stalk glandular-dotted; (*3*) veil leaving an annulus on the stalk or remnants on the edge of the cap; (*4*) cap glutinous to viscid; (*5*) pleurocystidia occurring in bundles (as revived in potassium hydroxide) with encrusting brown pigment in or around the bundle.................... 5

   Not with two or more of the above features 6

5. Spore deposit grayish brown to wood brown, vinaceous brown, purplish brown, or chocolate brown to purple drab................ *Fuscoboletinus*

   Spore deposit dingy yellow to yellow brown, pale cinnamon tan, olive, olive brown, or a greenish mustard yellow .... *Suillus*

6. Spore deposit pale yellow; spores more or less ellipsoid. Stalk hollow at maturity... *Gyroporus*

   Not as above.. 7

7. Veil dry and floccose to almost powdery in appearance, usually leaving a zone or ring on the stalk when it breaks .... *Pulveroboletus*

   Veil lacking or present, but when present, then not as above 8

8. Spore deposit gray brown, red brown, vinaceous, vinaceous brown, or purple brown .... ***Tylopilus*** (Ch. 13)

   when stalk has dark-colored points or small scales, it may be *Leccinum*

   Stalk scabrous-roughened or covered with small scales or points; scales or points pale to dark gray, dark brown, or blackish, although sometimes scales are pallid or pale in early stages, usually darkening with age or by maturity........ .... *Leccinum*

   Stalk not as above, glabrous and smooth, or when ornamented, then either scurfy, pruinose, squamulose, reticulate, or somewhat lacerated-reticulate .... ***Boletus*** (Ch. 13)

Cantharellaceae **Aphyllophorales**

1. Spores roughened, either finely wrinkled, reticulate, or tuberculate; spore deposit cream to ocher or ocher brown to rusty .... ***Gomphus*** (Ch. 13)

   Spores smooth; spore deposit white, cream, yellow, salmon, or pinkish buff 2

2. Hyphae of the trama with clamp connections .... *Cantharellus*

   Hyphae of the trama lacking clamp connections .... *Craterellus*

Clavariaceae **Aphyllophorales**

1. Fruit body branching in a coral-like manner 2

   Fruit body unbranched or at most with one or two side branches, never with coral-like branching.. 7

2. Branching whorled; branch ends box-like. Spores amyloid. Gloeocystidia present. Fungi often occurring on wood .... *Clavicorona*

   Not with the above combination of characters.. 3

3. Spores white in deposit, globose to subglobose; basidia with two horn-like incurved sterigmata, usually secondarily septate after spore discharge .... *Clavulina*

   Spore deposit color and shape variable; basidia with four sterigmata, not developing a secondary septum 4

4. Spores colored in deposit, yellow to ocher in mass, usually cyanophilous; surface ornamented. $FeSO_4$ turning the hymenium of the upper branches green .... ***Ramaria*** (Ch. 13)

   Spores white to pale cream in deposit, ornamented or not, usually not cyanophilous. $FeSO_4$ reaction positive or negative 5

5. Hyphae of the fruit body trama lacking clamp connections .... *Clavaria*

   Hyphae of the fruit body trama with clamp connections 6

6. Spores usually echinulate, verrucose, or tuberculate. $FeSO_4$ not reacting with the hymenium of the upper branches .... *Ramariopsis*

   Spores smooth. $FeSO_4$ turning the hymenium of the upper branches gray green..... *Clavulinopsis*

7. Branch tips of fruit body cup-like; rim of cup often crown-like, with erect tooth-like projections. Gloeocystidia present in the hymenium. Spores amyloid .... *Clavicorona*

   Not with the above combination of characters.. 8

8. Hyphae of the fruit body trama lacking clamp connections .... *Clavaria*

   Hyphae of the fruit body trama with clamp connections 9

9. Fruit body broadly clavate to top-shaped, often with the interior in the apex soft and punky. Spores smooth, often elongated. $FeSO_4$ turning the hymenium of the upper branches green to gray green or not .... *Clavariadelphus*

   Fruit body usually narrowly clavate, cylindrical, or fusiform, with the interior in the apical portion usually not soft and punky. Spores smooth or angular to angular-nodulose or roughened; shape variable. $FeSO_4$ turning the hymenium of the upper branches green to gray green or not.. .... *Clavulinopsis*

   Clavariadelphus and *Clavulinopsis* are difficult to separate; the spongy to soft or punky interior of the apical part of the fruit body is a helpful character in identifying *Clavariadelphus*, in which the fruit body is also frequently 5 mm or more thick at the widest part

Coprinaceae — Agaricales

1. Gills and often cap trama deliquescent. Gills very thin, often closely packed and with practically parallel faces .... ***Coprinus*** (Ch. 9)

   Gills not deliquescent 2

2. Cap plicate-striate. Hymenium regularly containing brachybasidioles.... *Pseudocoprinus* Kühner

   included as part of ***Coprinus*** in Chapter 9

   Not with both the above characters 3

3. Gills not mottled by patches of maturing spores. Fruit body typically very fragile. Spore deposit black, purple brown, gray brown, or reddish .... *Psathyrella* (Fr.) Quél.

   the spores of ***Panaeolus***, Ch. 12, often retain their normal color in concentrated sulfuric acid; those of *Psathyrella* typically fade or discolor in concentrated sulfuric acid. Compare with spores mounted in water

   Gills distinctly to obscurely mottled with patches of maturing spores 4

4. Gills rusty brown and somewhat mottled at maturity; spores typically ornamented. Pleurocystidia typically in fascicles of two to four .... *Psathyrella* (subgenus *Lacrymaria*)

   Gills evenly fuscous to black at maturity. Pleurocystidia not in fascicles 5

5. Spore surface roughened, ornamented with irregular patches of outer wall material .... ***Panaeolus* (=*Psathyrella*, subgenus *Panaeolina* (Maire) Smith)** (Ch. 12)

   Spore surface smooth; spores in face view often obscurely angular .... ***Panaeolus*** (Ch. 12)

Cortinariaceae — Agaricales

1. Stalk eccentric, lateral, or lacking 2

   Stalk central 3

2. Fruit body sessile or with a short stub-like stalk usually only visible on the undersurface of the cap. Spores lacking an apical germ pore; surface smooth or ornamented .... *Crepidotus*

   Fruit body with a distinct stalk. Spores smooth... .... *Phaeomarasmius* Scherffel

3. Fruit body with a broad flaring membranous annulus whose lower surface, as well as the cuticle of the cap and the stalk, is covered by an epithelial layer composed of sphaerocysts that form a powdery or granular coating .... ***Phaeolepiota*** (Ch. 13)

   Fruit body not as above 4

4. Stalk with a long tapering pseudorhiza; partial veil lacking. Spores ornamented. Cap surface often slimy to more or less viscid .... *Phaeocollybia*

   Not with the above combination of features 5

5. Spores with an apical germ pore, at times inconspicuous (use 100× oil-immersion lens).......... 6

   Spores lacking an apical germ pore 7

6. Pleurocystidia absent. Fruit body with at least one of the following characters: (*1*) cheilocystidia capitate; (*2*) plage of the spores delimited by a

ragged line; or (*3*) fungi associated with mosses .... ***Galerina*** (Ch. 7)

Pleurocystidia present or absent. Fruit bodies lacking the features enumerated above. Fungi often occurring on wood........ ***Pholiota*** (Ch. 13)
Pholiota is sometimes considered a member of the Strophariaceae

7. Spores thin-walled, often collapsed in microscope mounts, nearly hyaline under the microscope ... . *Tubaria* (W. G. Smith) Gill.
when clamp connections are absent, check ***Galerina*** (Ch. 7)

Spores with thicker walls that rarely collapse in microscopic mounts, distinctly colored under the microscope, usually brownish 8

8. Cap typically dry. Stalk 10-20 mm thick, with a distinct membranous annulus .... *Rozites*

Fruiting body not with the above combination of features... 9

9. Spore deposit ocherous tawny to rich cinnamon or fulvous; spores verrucose to finely wrinkled, lacking an apical germ pore. Cheilocystidia often lacking. Partial veil cortinate and always present in early stages. Fungi typically terrestrial .... ***Cortinarius*** (Ch. 7)

Not as above.. 10

10. Spore deposit typically bright rusty orange to orange ocherous; spores verruculose. Cheilocystidia present. Fungi typically occurring on wood, occasionally terrestrial....................................... .... ***Gymnopilus*** (Ch. 12)

Fruit body not with the above combination of features.. 11

11. Spores smooth, angular, nodulose, or rarely coarsely echinate; metuloids often present. Stalk usually 2 mm or more thick. Cap not viscid. Fungi growing on humus, in earth, or on very rotten wood .... ***Inocybe*** (Ch. 10)

Fruit body not as above 12

12. Cap usually viscid when fresh. Stalk more or less fleshy, usually over 4 mm thick. Spores ornamented, i.e. punctate, finely wrinkled, or tuberculate to nearly smooth. Spore deposit often clay color to earth brown, but occasionally brick red to vinaceous brown .... ***Hebeloma*** (Ch. 13)

Not as above. Stalk fragile, usually less than 4 mm thick (four choices below)............................ 13

13. Spores distinctly ornamented, with a sharply defined smooth plage area, or at times appearing almost smooth except for plage line .... ***Galerina*** (Ch. 7)

Not with the above combination of characters (three choices below).................................... 14

14. Cap surface dry to hygrophanous, not viscid; hyphae of the cap cuticle with cystidium-like terminal cells or mixed with sphaerocysts. Spores large, 10 μm or more long. Fungi often associated with alder (*Alnus*) or willow (*Salix*) .... *Alnicola* Kühner

Cap surface lacking small scales; cap cuticle with scattered pileocystidia or with these in groups. Spore deposit often dull fuscous brown or olivaceous umber............................ *Simocybe* Karst.

Cap surface punctate-scaly or floccose-scaly; cells of cap cuticle often encrusted with crystals or pigment. Spore deposit ocherous, yellow brown, or rusty brown......... *Phaeomarasmius* Scherffel

Gomphidiaceae — **Agaricales**

1. Gills, and usually the cap and stalk, dull to bright ocherous in early stages. Cap or gill trama or both containing amyloid hyphae .... *Chroogomphus*

Gills, cap, and stalk not as above; flesh of cap white in early stages. Cap and gill trama containing only nonamyloid or dextrinoid hyphae....... .... *Gomphidius*

Paxillaceae — **Agaricales**

1. Gills brightly colored, waxy, often subdividing and reuniting repeatedly or having more or less evident pores, fairly thick, with distinct pleurocystidia and cheilocystidia; spores usually subfusiform. Stalk central.......... *Phylloporus* Quél.

Gills ocherous, brownish yellow, not particularly thick; spores ellipsoidal, usually less than 10 μm long; cystidia often less distinct. Stalk central to lacking .... ***Paxillus*** (Ch. 13)

Pluteaceae (=Volvariaceae) — **Agaricales**

1. Volva present; annulus lacking.......... *Volvariella*

Volva lacking.. 2

2. Annulus present................................. *Chamaeota*
rare genus, with small number of species

Annulus absent .... ***Pluteus*** (Ch. 12)

Rhodophyllaceae (=Entolomataceae) — **Agaricales**

1. Spores angular in end view only....................... 2

Spores angular in all views 3

2. Spores longitudinally striate in side view............ .... *Clitopilus*

Spores rough to warty in side view...................... .... *Rhodocybe* Maire

3. Stalk eccentric, lateral, or lacking...... *Claudopus*

Stalk central 4

4. Cap minutely tomentose (at least on the disc), squamulose, or distinctly appressed-fibrillose. Fruit body usually small................................. 5

Cap glabrous, hoary, viscid, dry, lubricous, or covered with a layer of minute fibrils and then micaceous. Fruit body small to large 7

5. Stalk base strigose. Hyphae of the cap trama with encrusted distinctly thickened (0.2–0.5 µm or more thick) brownish walls .... *Pouzarella*

   Stalk base naked or covered by cottony mycelia. Hyphae of the cap trama, when encrusted, thin-walled 6

6. Fruit body entirely white to gray-tinged in early stages. Cap cuticle composed of an irregular to entangled layer of hyphae .... *Alboleptonia*

   Cap cuticle usually in the form of a trichoderm or nearly hymeniform, at least on the disc, but when the cuticle is composed of an entangled layer of hyphae, then the fruit body is some color other than white .... *Leptonia*

7. Stalk thick, typically more than 5 mm thick at the apex. Cap trama usually thick; cap shape convex to plane. Basidia usually clamped at the base ... ***Entoloma* (rhodophylloid mushrooms)** (Ch. 13)

   Stalk slender, typically less than 5 mm thick at the apex. Cap trama thin; cap shape often umbilicate or conic to bell-shaped or at least umbonate. Basidia clamped at the base or not 8

8. Stalk usually polished and not longitudinally silky-fibrillose-striate. Cap often umbilicate; odor not like fresh meal, not rancid, nor fishy. Basidia usually not clamped at the base .... *Leptonia*

   Stalk often silky-fibrillose-striate. Cap often conic, bell-shaped, or umbonate; odor often like fresh meal, rancid, or fishy. Basidia clamped at the base or not .... ***Nolanea* (rhodophylloid mushrooms)** (Ch. 13)

Russulaceae — Agaricales

1. Latex present, exuding when the fruit body is cut or broken. Gill trama usually lacking sphaerocysts, except where it grades into the cap trama .... ***Lactarius*** (Ch. 13)

   Latex lacking. Gill trama usually containing sphaerocysts .... ***Russula*** (Ch. 13)

Strophariaceae — Agaricales

Key I

1. Spore deposit dull brown to rusty brown. Stalk central, annulate or not. Chrysocystidia present or absent; spores usually with at least a minute apical germ pore .... ***Pholiota*** (Cortinariaceae) (Ch. 13)

   Spore deposit darker, often with a distinct violaceous to purple brown tone 2

2. Stalk often eccentric, shorter than the diameter of the cap .... *Melanotus*

   Stalk central, annulate or not, usually longer than the diameter of the cap. Chrysocystidia present or not .... ***Psilocybe*** (Ch. 12)

Key II

1. Spore deposit dull brown, cinnamon brown to umber, or olivaceous umber, or rusty brown. Fungi often occurring on wood.................................. .... ***Pholiota*** (Cortinariaceae) (Ch. 13) including *Kuehneromyces* Singer & Smith, which is not known to be poisonous

   Spore deposit deep lilac or purple, or blackish purple or dark sepia, or fuscous 2

2. Stalk eccentric to lateral, or occasionally rudimentary, when present often curved and shorter than the diameter of the cap .... *Melanotus* mainly tropical species

   Stalk central to somewhat eccentric, not curved and short as above 3

3. Chrysocystidia present in the hymenium.......... 4

   Chrysocystidia absent in the hymenium 5

4. Cap cuticle with a cellular subcutis .... ***Naematoloma*** (Ch. 7)

   Cap cuticle lacking a cellular subcutis .... ***Stropharia*** (Chs. 12 and 13)

5. Fruit body staining blue or greenish blue where handled (check the base of the stalk)................. .... ***Psilocybe*** (Ch. 12)

   Fruit body not staining as above....................... 6

6. Stalk viscid when fresh, from a viscid veil........... .... ***Stropharia*** (Chs. 12 and 13)

   Stalk not viscid, although the cap may be viscid .... ***Psilocybe*** (Ch. 12)

Tricholomataceae — Agaricales

1. Stalk typically eccentric to lateral or lacking. Fungi often occurring on wood or woody debris 2

   Stalk typically central and well-developed. Fungi occurring in various habitats such as terrestrial, on wood, or on humus 11

2. Spores amyloid.. 3

   Spores nonamyloid 4

3. Gill edges toothed to serrulate at maturity .... *Lentinellus*

   Gill edges even to uneven or crenulate, but not toothed or serrulate .... *Panellus*

4. Spore deposit pink. Cap surface more or less tomentose. Fruit body typically with a strong unpleasant odor. Fungi occurring on rotting wood .... *Phyllotopsis*

   Not as above... 5

5. Spore deposit white. Gill edges split along their entire length, with the faces diverging or recurving. Fungi occurring on wood or woody debris. Fruit

body fairly small and leathery in texture. .... *Schizophyllum*

Fruit body not with the above combination of characters 6

6. A distinct gelatinous layer in the cap trama or cap with a gelatinous cuticle 7

No gelatinous layer or layers present in or on the cap 8

7. Lamprocystidia present in the hymenium .... *Hohenbuehelia* Schulz.

Lamprocystidia absent in the hymenium .... *Resupinatus* (Nees) S. F. Gray

8. Fruit body tough and readily reviving. Thick-walled hyphae usually present and numerous in the gill trama .... *Panus* and relatives

Fruit body not as above 9

9. Spores cylindrical. Subhymenium well-developed, conspicuous .... *Pleurotus*

Spores globose to short-ellipsoidal. Subhymenium inconspicuous or almost lacking 10

10. Fruit body stipitate. Gill trama parallel or subparallel. Flesh of cap firm or even rather tough .... *Hypsizygus* Singer
when fruit body is orange to yellow or olivaceous, check **Omphalotus**, Ch. 13

Fruit body sessile. Gill trama interwoven. Flesh of cap soft and fragile.......... *Pleurocybella* Singer

11. Spores amyloid.. 12

Spores nonamyloid or dextrinoid.................... 22

12. Fruit body small. Stalk base with orange to fulvous strigose hairs; consistency of stalk tough and horny .... *Xeromphalina*

Fruit body not with the above features 13

13. Stalk (3-)4-10 or more mm thick.................... 14

Stalk 0.3-3(-5) mm thick 20

14. Gills decurrent, repeatedly forking in pairs. Cap margin enrolled in early stages. Fungi often associated with mosses .... *Cantharellula* Singer

Not as above... 15

15. Spores with amyloid ornamentation 16

Spores smooth 17

16. Clamp connections present on the hyphae of the fruit body. Pleurocystidia lacking. Many white mycelia associated with the stalk base and in the surrounding litter and duff......... *Leucopaxillus*

Clamp connections absent. Pleurocystidia or cheilocystidia, or both, often but not always present. Stalk very straight, lacking conspicuous mycelia on and around the base................ *Melanoleuca*

17. Fruit body lacking a veil. Cap cuticle cellular or in the form of a hymeniform layer .... *Dermoloma* (Lange) Singer ex Herink

Fruit body not with the above characters 18

18. Veil of fruit body composed of sphaerocysts, more or less granular, coating the cap surface and the stalk below the annular zone or annulus .... *Cystoderma*

Veil not as above.. 19

19. Stalk massive. Annulus double, with two annuli one below the other. Gills strongly decurrent.... .... *Catathelasma* Lovej.

Stalk not thick and massive. Annulus single. Gills adnexed to emarginate, not decurrent .... *Armillaria* Kummer

20. Cap cuticle lacking layers that are well-differentiated from the underlying cap trama. Spores 5 μm or more long. Fungi occurring on wood, often caespitose..... *Clitocybula* (Singer) Métrod

Not with the above combination of characters..21

21. Cap margin enrolled in early stages. Spores small (<6 μm), subglobose. Fruit body often occurring on decaying conifer cones, sometimes on rotting wood .... *Baeospora* Singer

Cap margin straight in early stages. Spores typically larger. Fruit body occurring in various habitats .... ***Mycena*** (Ch. 13)

22. Spores dextrinoid; gills decurrent, narrow, often forked, colored orange to orange yellow. Fungi terrestrial or occurring on decaying wood .... *Hygrophoropsis* (Schroet.) Maire
suspected of containing muscarine, Ch. 10

Fruit body not with the above combination of characters 23

23. Fruit body tough. Gills with serrate edges .... *Lentinus*

Fruit body not with the above features 24

24. Some evidence of a veil, at least in young fruit bodies. Annulus sometimes present............... 25

Veil absent, or at most rudimentary................ 27

25. Veil more or less granulose. Annulus or some veil remains typically showing on the stalk, composed of sphaerocysts................................ *Cystoderma*

Veil not as above.. 26

26. Fungi growing on wood or woody debris, less commonly terrestrial. Black rhizomorphs associated with fruit body. Veil forming a distinct annulus, visible at least in young specimens. Gills usually decurrent. Gill trama divergent in young specimens....................... ***Armillariella*** (Ch. 13)

Fungi growing on soil or humus. Black rhizomorphs absent. Veil leaving an annulus or annular zone. Gills sinuate to adnexed. Gill trama regular .... ***Tricholoma*** (Ch. 13)

27. Fruit body with whitish to grayish or tan colors, often staining gray to black when injured. Basidia containing siderophilous granules .... *Lyophyllum*

Fruit body not with the above features in combination 28

28. Fruit bodies often large and occurring in large clusters. Gills, cap, and stalk yellow to orange or olivaceous tinted when fresh. Fruit body, especially the gills, luminescent when fresh .... ***Omphalotus*** (Ch. 13)

Fruit bodies not as above 29

29. Gills thickish, somewhat waxy and brittle, pinkish to flesh-colored to vinaceous or violaceous. Spores usually echinulate, but when smooth, then oblong and large (11-22 μm long) .... *Laccaria*

Not with the above combination of features ... 30

30. Cap cuticle in the form of a hymeniform or cellular layer, or composed of broom cells arranged more or less in rows 31

Cap cuticle composed of appressed hyphae, gelatinous or otherwise, or in the form of a trichoderm composed of tangled to irregularly arranged elements 37

31. Stalk typically rooting. Pleurocystidia large and conspicuous. Cap usually viscid when fresh. Gills broad and subdistant .... *Oudemansiella* Speg.

Not with the above combination of characters..32

32. Pleurocystidia absent. Cap cuticle in the form of a hymeniform to cellular layer; hyphae with clamp connections. Fruit body with the aspect of a small *Tricholoma*. Fungi occurring on humus or soil .... *Dermoloma*

Not as above 33

33. Broom cells typically present in the cap cuticle. Fruit body reviving when moistened .... *Marasmius*

Broom cells absent from the cap cuticle 34

34. Fruit body fairly small, tough, yellow to orange, reviving well when moistened and not decaying readily .... *Xerulina*

Fruit body not as above 35

35. Hyphae lacking clamp connections. Fruit body small, often growing on cones of conifers or *Magnolia* .... *Strobilurus* Singer

Not with the above combination of characters..36

36. Margin of cap enrolled in early stages .... ***Collybia*** (Ch. 13)

Margin of cap straight in early stages .... ***Mycena*** (Ch. 13)

37. Cap reddish to orange to yellowish, viscid. Cap cuticle with pileocystidia. Stalk velvety. Fungi caespitose, occurring on wood of elm, aspen, willow, alder, or other hardwoods .... *Flammulina* Karst.

Fruit body not as above 38

38. Stalk fleshy and typically over 5 mm thick, often as thick as 10-30 mm 39

Stalk thin, 0.5-4(-6) mm thick 43

39. Gills sinuate to adnexed. Clamp connections typically absent, but when present, spores are broadly ellipsoidal. Spore deposit white. Fungi terrestrial .... ***Tricholoma*** (Ch. 13)

Not as above 40

40. Gills adnate to usually distinctly decurrent. Clamp connections typically present. Spores smooth or finely ornamented, white, yellowish, pinkish, or pale vinaceous brown in deposit, but spores hyaline under the microscope .... ***Clitocybe*** (Chs. 9, 10, and 13) when black rhizomorphs are associated with the fruit body, check ***Armillariella***, Ch. 13

Not with the above combination of features ... 41

41. Spores globose to subglobose. Basidia containing siderophilous granules. Fungi typically caespitose, sometimes occurring on soil that is rich in lignin such as is found around sawdust piles, but not growing truly on wood .... *Lyophyllum*

Fungi truly lignicolous, not usually growing in large clusters 42

42. Fruit body pale, whitish. Gills broadly adnate to decurrent. Stalk 3-10 mm thick .... ***Clitocybe*** (Chs. 9, 10, and 13)

Fruit body distinctly colored, such as yellow, red, or purple, sometimes grayish to brownish. Cheilocystidia usually distinctive, for example filamentous or enlarged .... ***Tricholomopsis*** (Ch. 13)

43. Cap thin, pliant, and readily reviving when moistened. Cap cuticle with dextrinoid hairs .... *Crinipellis* Pat.

Cap thin to thickish but fragile, not or poorly reviving when moistened 44

44. Fruit body greenish or some shade of yellow to orange, often fairly small. Gills decurrent. Clamp connections absent .... *Omphalina* Quél.

Not as above 45

45. Cap surface scaly to scurfy. Stalk not watery-translucent when fresh. Cheilocystidia enlarged. Fungi occurring on wood or near rotting wood .... ***Tricholomopsis*** (Ch. 13)

Not as above 46

46. Cap more or less convex; margin enrolled in early stages .... ***Collybia*** (Ch. 13)

Cap conic, varying to convex; margin straight to incurved in early stages, not enrolled. Stalk brittle and often watery-translucent when fresh...... .... ***Mycena*** (Ch. 13)

ASCOMYCOTA: GENERAL

Key to important families

Use the following key to identify the particular family of the Ascomycota to which your specimen belongs, based on its macroscopic features. This key separates only the three families of the Ascomycota that contain species which may be collected by people foraging for food. All three families belong to the order Pezizales of the class Discomycetes and are characterized by operculate asci. Use the technical key on page 69, which incorporates the microscopic characteristics, to confirm your identification of family. Then turn to the appropriate page, indicated in parentheses, to determine the genus. The species of these genera that are known to be poisonous are treated in Part Two, where genera and species are arranged alphabetically.

1. Apothecium cup-shaped, urn-shaped, or saucer-shaped, lacking a stalk or occasionally with a short stalk .... **Pezizaceae** (p. 68)

   Apothecium usually not as above but instead saddle-shaped to irregular, or cone-shaped, less commonly cup-like, typically with a distinct cap and stalk 2

2. Apothecium more or less cone-shaped; surface usually pitted or honey-combed, or with distinct longitudinal grooves......... **Morchellaceae** (p. 68)

   Apothecium saddle-shaped, brain-like, wrinkled to irregularly lobed, or cup-like .... **Helvellaceae** (p. 68)

Descriptions of families, with keys to important genera when appropriate

Helvellaceae — Pezizales

Apothecium large, usually with a cap and stalk, disc-like to cup-like or saddle-shaped to brain-like and convoluted or folded and wavy, rarely with the upper portion in the form of a chambered stalk; color gray to black or brown, rarely white to cream color, somewhat fleshy to brittle; stalk simple to chambered. Asci thin-walled, clavate to cylindrical, with an apical operculum, not turning blue in Melzer's reagent. Ascospores usually numbering eight per ascus, smooth or ornamented, ellipsoidal to ovoid or fusiform, hyaline or brown, and guttulate containing one, two, or more rarely three oil globules. Paraphyses usually not forming a network. Fungi saprobic on soil, humus, or rotting wood.

The following key is based on macroscopic features. Use the technical key on page 69, which incorporates microscopic characteristics, to confirm your identification.

1. Stalk 10–50 mm thick or more, often with interior folds or longitudinal grooves. Cap typically some shade of brown (reddish brown to bay brown or yellowish brown) to ocherous; shape saddle-like, brain-like, or wrinkled .... ***Gyromitra*** (Ch. 8)

   Stalk usually 10 mm thick or less, cylindrical or with distinct longitudinal ribs or ridges; cap grayish to grayish brown or blackish, occasionally white to cream color; shape saddle-like, cup-like, or somewhat lobed......... ***Helvella*** (Ch. 13)

Morchellaceae — Pezizales

Apothecium fleshy to somewhat brittle, large, disc-like, stalked and with a sponge-like cap or stalked and with a campanulate pendant cap; color generally buff to brown or grayish; stalk usually simple. Asci thin-walled, clavate to cylindrical, with an apical operculum, not turning blue in Melzer's reagent. Ascospores usually numbering eight per ascus, usually smooth, hyaline, and eguttulate, lacking distinct oil globules, but typically with a crown of tiny droplets at both ends of each spore in fresh specimens mounted in water. Paraphyses usually not forming a network. Fungi saprobic, occurring on humus and soil, fruiting in the spring or into early summer at higher elevations.

The following key is based on macroscopic features. Use the technical key on page 70, which incorporates microscopic characteristics, to confirm your identification.

1. Cap or head of apothecium attached only at or near the stalk apex, the sides hanging free and skirt-like..................................... ***Verpa*** (Ch. 13)

   Cap or head of apothecium attached for most (or at least half) its length to the stalk; surface of cap pitted.................................. ***Morchella*** (Ch. 13)

Pezizaceae — Pezizales

Apothecium usually fairly large, cup-like to disc-like, sessile or with a stalk but not with a distinct cap, fleshy to brittle, usually not tough or gelatinous. Asci thin-walled, clavate to cylindrical, with an apical operculum, turning blue apically or rarely entirely in Melzer's reagent. Ascospores usually numbering eight per ascus, smooth or ornamented, thin-walled, hyaline to brown, spherical to ellipsoidal. Paraphyses usually simple, not forming a network. Fungi saprobic on soil, humus, wood, or dung.

The following key is based on macroscopic features. Use the technical key on page 70, which incorporates microscopic features, to confirm the separation of *Peziza* from *Sarcosphaera*.

1. Apothecium buried in soil and litter in early stages; shape subglobose to spherical; inner surface of fruit body tinted lilac when mature .... ***Sarcosphaera*** (Ch. 13)

   Apothecium not fitting the above description... 2

2. Apothecium lopsided to irregular or split; inner surface, the hymenium, orange to yellow; exterior typically with bluish to greenish tints .... *Caloscypha fulgens* (Pers. ex Fr.) Boud.

   Apothecium disc-shaped to cup-shaped, not typically lopsided or irregular or split; color variable, but not as described above 3

3. Apothecium often with a short stalk; inner surface red to orange red. Fungi often occurring in the spring or early summer .... *Sarcoscypha* (Fr.) Boud. (Sarcoscyphaceae) (p. 69)

   Apothecium usually sessile, lacking a stalk; inner surface not typically red, sometimes yellow orange or orange to scarlet 4

4. Apothecium orange to scarlet, fading to orange yellow. Fungi occurring on bare or newly exposed soil .... *Aleuria aurantia* (Fr.) Fuckel

   Apothecium usually some shade of brown, but at times lilac to violet tinted, cup-shaped, disc-shaped, or shaped like a flattened disc .... *Peziza* Dill. ex Fr.
   compare *Discina* (Fr.) Fr. and ***Disciotis***, Ch. 8, of the Helvellaceae and the Morchellaceae, respectively

Sarcoscyphaceae — Pezizales

Apothecium small to large, usually cup-like, sessile or with a short stalk, tough; hymenium on interior of cup often bright red to orange. Asci fairly thick-walled, cylindrical, operculate, not turning blue in Melzer's reagent; operculum subapical, often oriented obliquely, with a thickened pad surrounding the inner portion of the opening. Ascospores usually numbering eight per ascus, smooth or ornamented, hyaline, thin-walled, globose to ellipsoidal. Paraphyses usually forming a network. Fungi saprobic, occurring on wood or woody debris.

As this family contains only one genus of importance, namely *Sarcoscypha*, which is not known to contain any poisonous species, no keys are provided here.

ASCOMYCOTA: TECHNICAL

Key to important families

1. Asci clavate to cylindrical, thick-walled, with the operculum situated just below the apex and oriented obliquely, not turning blue in Melzer's reagent. Apothecium leathery, tough, or gelatinous, not fleshy, often somewhat stalked. Fungi often occurring on wood 2

   Asci typically clavate to cylindrical, thin-walled, with a terminal operculum, although more rarely broadly clavate to ovoid with walls thickened and sometimes opening by a vertical slit, sometimes turning blue in iodine. Apothecium typically fleshy, or cartilaginous, or brittle 3

2. Apothecium bright-colored, e.g. red, pink, yellow, or orange .... Sarcoscyphaceae (p. 69)
   not known to contain any poisonous species; probably *Sarcoscypha* (Fr.) Boud., but when fruit body brittle and orange, it may be *Aleuria* (Fr.) Fuckel

   Apothecium dark-colored, frequently black to brown but sometimes a violaceous shade .... Sarcosomataceae
   not known to contain any poisonous species

3. Ascus apex intensely blue or whole ascus wall diffusely blue in Melzer's reagent; ascospores hyaline to brown, rarely greenish yellow .... **Pezizaceae** (p. 70)

   Ascus apex not blue in Melzer's reagent 4

4. Apothecium large, always buff to brown or grayish to blackish. Shape variable: either disc-like, or stalked and with a pitted or sponge-like cap, or stalked and with a hanging bell-shaped cap. Ascospores eguttulate, hyaline under a microscope in water, and with a crown of tiny droplets at both ends of the spore in fresh specimens .... **Morchellaceae** (p. 70)

   Apothecium large or small, dull to occasionally brightly colored. Ascospores guttulate or eguttulate, lacking a crown of droplets at both ends of the spore in fresh specimens....................... 5

5. Apothecium large, disc-like to cup-shaped, usually stalked, often saddle-shaped or with a brain-like cap, or with the hymenium on the upper portion of a chambered stalk. Distinct hairs not produced on the outer surface of the apothecium............. .... **Helvellaceae** (p. 69)

   Apothecium usually small, disc-like to cup-shaped, rarely stalked. Hairs present or absent on the outer surface of the apothecium........ Pyronemataceae
   not known to contain any poisonous species

Keys to important genera

Helvellaceae — Pezizales

1. Ascospores ornamented, mostly more than 25 μm long when mature 2

   Ascospores usually smooth, but when ornamented, then mostly less than 25 μm long 4

2. Apothecium column-like to club-shaped............. .... *Underwoodia* Peck

   Apothecium not as above 3

3. Apothecium cup-like, stalked; stalk often short and fluted....................................... *Discina* (Fr.) Fr.
   when cap is disc-like and attached by a rooting process, it may be *Rhizina* Fr.

   Apothecium with a distinct cap and stalk; cap typically convoluted to folded or brain-like....... .... *Neogyromitra* Imai
   sometimes included in *Discina* (Fr.) Fr., see ***Gyromitra***, Ch. 8

4. Apothecium usually with a well-developed stalk; stalk short to long, usually 10 mm thick or less, cylindrical or with distinct longitudinal ribs or ridges. Cap cup-shaped, saddle-shaped, or irre-

gularly saddle-shaped, grayish to grayish brown or blackish, occasionally white to cream. Ascospores smooth or less commonly delicately to coarsely warted..
***Helvella* (including *Paxina* O. Kuntze)** (Ch. 13)

Apothecium stalked; stalk 10–50 mm thick or more, often with interior folds or longitudinal folds or grooves. Cap saddle-shaped to brain-like or convoluted to folded, typically some shade of brown, such as reddish brown to bay brown or yellowish brown, to ocherous. Spores typically smooth, but sometimes faintly reticulate... ***Gyromitra*** (Ch. 8)
sometimes included in *Discina* (Fr.) Fr.

Morchellaceae — Pezizales

1. Apothecium with a pitted to sponge-like cap, stalked, and hollow .... ***Morchella*** (Ch. 13)
   Apothecium not as above, either a large broad cup or with a bell-shaped cap that is free from the stalk except at the very stalk apex 2
2. Apothecium stalked; stalk hollow. Cap bell-shaped, free from the stalk except at the very stalk apex .... ***Verpa*** (Ch. 13)
   species with two-spored asci are sometimes placed in *Ptychoverpa* Boud.
   Apothecium disc-like to cup-shaped. Hymenium often with distinct folds or vein-like ridges. Stalk slight or absent.................................... *Disciotis*
   suspected of containing gyromitrin, Ch. 8

Pezizaceae — Pezizales

1. Apothecium in the form of deep cups, wholly immersed in the soil and litter, and very large. Hymenium, the inner surface of the cup, usually tinted violaceous. Ascospores smooth, ovoid, and biguttulate .... ***Sarcosphaera*** (Ch. 13)
   Apothecium disc-like to cup-shaped, but when cup-shaped, then not a deep cup that is immersed in the soil and litter. Hymenium not usually violaceous, but if so, then usually mixed with brown. Spores of various types ....
   .... *Peziza* Dill. ex Fr.

Literature

Fries, E. M. Systema Mycologicum. Vol. I. Greifswald, Sweden: Moritz; 1821.

Goss, R. D.; Shoop, C. R. A case of mushroom poisoning caused by *Tricholomopsis platyphylla*. Mycologia 72:433–435; 1980.

Korf, R. P. Discomycetes and Tuberales. Ch. 9. Ainsworth, G. C.; Sparrow, F. K.; Sussman, A. S., eds. The fungi. Vol. IVA. New York, NY: Academic Press; 1973.

Largent, D. L. How to identify mushrooms to genus I: Macroscopic features. 2nd ed. Eureka, CA: Mad River Press; 1977 [first printed in 1973].

Petersen, R. Aphyllophorales II: The clavarioid and cantharelloid basidiomycetes. Ch. 20. Ainsworth, G. C.; Sparrow, F. K.; Sussman, A. S., eds. The fungi. Vol. IVB. New York, NY: Academic Press; 1973.

Shaffer, R. L. Keys to the genera of higher fungi. 2nd ed. Ann Arbor, MI: The University of Michigan Biological Station; 1968.

Smith, A. H. Agaricales and related secotioid gastromycetes. Ch. 23. Ainsworth, G. C.; Sparrow, F. K.; Sussman, A. S., eds. The fungi. Vol. IVB. New York, NY: Academic Press; 1973.

Smith, A. H.; Thiers, H. D. The boletes of Michigan. Ann Arbor, MI: The University of Michigan Press; 1971.

Stafleu, F. A., Chairman. International code of botanical nomenclature. Adopted by the 12th International Botanical Congress, Leningrad, July 1975. Utrecht, Holland: Bohn, Scheltema, Holkema; 1978.

Stuntz, D. E. 1977. How to identify mushrooms to genus IV: Keys to families and genera. Eureka, CA: Mad River Press; 1977.

Part Two Fungal Poisoning

6

OVERVIEW

Scope

Part Two of this book contains descriptions of species of fleshy and almost fleshy fungi for which there are either documented cases of poisoning or consistent and repeated reports of poisoning, either in North America or Europe; also included are species from which toxins have been isolated or detected. These fungi produce a variety of symptoms, depending on the species. They can cause digestive upsets ranging from mild to severe, hallucinations, damage to vital organs or protoplasm, or malfunctions of the nervous system. Species such as *Lepiota molybdites,* which is tolerated by some people but is frequently the cause of severe gastrointestinal upset in others, are also formally described in Part Two here.

Some normally edible mushrooms, including the common market mushroom *Agaricus bisporus* (Lange) Moeller & Schaeff. (=*Agaricus brunnescens* Peck), cause digestive upset in some individuals. Such people simply cannot tolerate these mushrooms and exhibit gastrointestinal upset or other symptoms when they eat them. Burrell (1978) used the term mushroom intolerance to describe this type of reaction to *Agaricus bisporus,* observed in one individual by Bergoz (1971). Such fungi, however, cannot be classified as truly poisonous and in general they are not formally described in Part Two.

Gastrointestinal upset such as diarrhea from ingestion of mushrooms has sometimes been termed an allergic reaction; however, this type of reaction probably should not be classed as such. Misuse of the term allergy in such cases is clearly discussed in an excellent short article by Burrell (1978).

A true allergic reaction or hypersensitivity to mushrooms is one caused by simple contact with mushrooms, producing symptoms such as an itch or a rash. For example, one woman reported the development of a rash on her face caused by touching her face with her hands after handling various species of mushrooms. *Chroogomphus vinicolor* (Peck) O. K. Miller, also, has been reported to have caused an uncomfortable burning sensation in the eyes of one individual and skin eruptions on his eyelids from rubbing his eyes after handling this species (H. D. Thiers, Personal commun.). Other types of sensitivity that can be attributed to allergic reactions have also been documented. Chilton (1979) reported that workers at the University of Washington Dental Clinic were able to correlate periods of low saliva flow with ingestion of the store-bought *Agaricus bisporus.* According to Hatfield and Brady (1975), specific hemolysins have been detected in the serum of two previously sensitized patients who developed hemolytic anemia after eating *Paxillus involutus.* Mushrooms causing allergic reactions are generally not, however, formally described in Part Two.

Some mushrooms are reported as being suspected of being poisonous or are not recommended for eating in some literature but are termed nonpoisonous or edible in other references. For example, Groves (1975) lists *Hygrophorus conicus* Fr. as being suspected of being poisonous, whereas Miller (1972) classifies this species as nonpoisonous. In instances where the evidence does not support classifying a particular species as poisonous, it may not be included in this book. More work is needed on such species to determine their edibility.

Some mushroom poisonings are caused by contaminants in the environment that have been absorbed by the mushroom tissue or have been deposited on the mushroom surface. For example, various species of mushrooms can accumulate large quantities of heavy metals, such as mercury. Under certain circumstances eating these normally edible species could lead to human poisoning (Byrne et al. 1976, Seeger and Nutzel 1976). Other examples have been documented of mushrooms accumulating through their mycelium or fruit bodies pesticides, insecticides, or other toxic compounds used as sprays. In one instance a man was poisoned by specimens of *Leccinum,* normally edible, that had been sprayed with a pesticide (H. D. Thiers, Personal commun.). Because these mushrooms are not inherently poisonous, however, they are not formally described in Part Two.

Other mushrooms, such as *Marasmius oreades* (Bolt. ex Fr.) Fr., are known to contain cyanide or related compounds, but the concentration is too low for these mushrooms to be considered poisonous (Göttl 1976; Heineman 1942, 1949).

Mushroom poisoning involving young children requires special mention, because children are likely to consume raw mushrooms found in lawns or play areas. Physicians and pediatricians likely to encounter mushroom poisoning in children should consult Lampe (1977) and the references therein.

Classification of Fungal Poisoning

The actual number of species of fleshy and almost fleshy fungi in North America is not known. Based on taxonomic studies to date, the number exceeds 5000, with some estimates close to 10 000. Based on reports of poisonings and toxin studies, a comparatively small percentage of the total number of species are toxic to humans. However, the actual number of poisonous or edible species is impossible to determine because many fungi are not collected for food and have not been analyzed for toxins.

Many toxins have been found in the poisonous fungi analyzed chemically to date. Information on the chemistry and distribution of the various toxins shows that species in unrelated genera may contain one or more of the same toxins. For example, amatoxins occur in both *Galerina* and *Amanita*. On the other hand, within a single genus, for example *Amanita*, certain species contain one group of toxins while other species contain a completely unrelated group of toxins. A single genus, for example *Tricholoma*, can also contain both edible and poisonous species. Nevertheless a few genera, such as *Inocybe*, are comparatively homogenous, composed of mainly toxic species all containing the same toxin, in this instance muscarine.

Although we have obtained a tremendous amount of information on mushroom toxins in recent years, there are still several poisonous species for which the actual toxin or toxins are unknown. Therefore, it is impossible to set up a classification of poisonous species based entirely on the toxins they contain. Because of the distribution of the various toxins among the families and genera of fungi, as discussed above, it is equally difficult to set up a classification of poisonous species that follows a purely taxonomic scheme.

The most convenient and useful basis for classifying toxic fungi is by the symptoms they produce when eaten by humans. Various schemes have been proposed. Tyler (1971) divided mushroom poisons into four groups: protoplasmic poisons; compounds affecting the nervous system; gastrointestinal irritants; and disulfiram-like constituents, which cause circulatory-respiration difficulties when combined with alcohol. Lampe (1977) presented a more elaborate but useful scheme that he called differential diagnosis of mushroom intoxication by symptoms. Lincoff and Mitchel (1977) divided poisonous mushrooms into seven groups. Their system is based on a combination of known toxic properties of poisonous fungi and symptomatology.

The seven types of fungal poisoning that are used in this book are defined according to the time of onset of symptoms. The similarity and severity of the symptoms are also taken into account. The system is similar to the one advanced by Lincoff and Mitchel (1977). The types of poisoning are labeled by toxin groups. The toxin groups are listed in Table 2. The key to the toxin groups given here was prepared by Dr. D. M. Simons, Chairman of the Toxicology Committee, North American Mycological Association. Each of the seven toxin groups is treated in a separate chapter, in Chapters 7 to 13.

To identify the fungus species involved in a particular poisoning case, follow the key based on symptomatology given below to determine the toxin group involved. Then turn to the appropriate chapter, color-coded and identified by number in parentheses in the key for quick reference. Study the color photographs at the end of the chapter if you have a specimen to compare them with. Then read the descriptions of possible genera and follow the keys to species when applicable to make a final identification. The descriptions of genera appear alphabetically in each chapter, as do the species within each genus. Each chapter also contains a list of symptom-producing species in the toxin group, arranged according to family.

To cross-check your determination, return to Chapter 5, "Taxonomy", and identify the genus a second time by following the appropriate general and technical keys to the genera of families that appear there.

Key to Toxin Groups

1. Illness typically characterized by marked flushing of the face and torso, similar to an alcohol-disulfiram (Antabuse) intoxication, and generally occurring when alcohol is ingested some hours (induction period) after a meal of wild mushrooms .... **coprine** (Ch. 9)

   Illness not associated with an alcohol-mushroom combination or at least not showing an induction period or not characterized by marked flushing 2

2. Onset of illness one-half to three hours after eating mushrooms 3

   Onset of illness six hours or more after eating mushrooms 6

3. Symptoms limited to indications of gastrointestinal irritation: for example nausea, vomiting, diarrhea, or abdominal cramps.......................... .... **gastrointestinal irritants** (Ch. 13)

   Symptoms indicative of peripheral or central nervous system involvement. Gastrointestinal irritation occasionally also present................. 4

4. Symptoms reflecting strong stimulation of the parasympathetic nervous system: for example perspiration, salivation, excessive production of tears, constricted pupils, slow pulse, or abdominal cramps. Symptoms of central nervous system involvement absent.. **muscarine** (Ch. 10)

Symptoms indicative of central nervous system involvement: for example erratic and irrational behavior, confusion, delirium, muscular incoordination, or hallucinations 5

5. Symptoms like those produced by the drug LSD: for example changing moods, altered color perception, or hallucinations .... **hallucinogens** (Ch. 12)

Symptoms not characteristic of a true hallucinogen: for example dizziness, delirium, incoordination, confusion, or withdrawal (like a drunken stupor), with or without hallucinations. Victim sometimes agitated initially and later somnolent.... **ibotenic acid - muscimol** (Ch. 11)

6. Symptoms starting seven to ten or more hours after eating mushrooms, and similar to symptoms of acute hemolytic anemia: for example vomiting, malaise, headache, shaking, chills, fever, aching pains, spasms, rigidity, or jaundice. Symptoms indicative of central nervous system involvement: for example convulsions, delirium, or coma. Liver damage sometimes occurring in some individuals .... **monomethylhydrazine** (Ch. 8)
see also *Gomphus*, Ch. 13, for initial symptoms

Initial symptoms include vomiting, persistent cholera-like diarrhea, and severe abdominal pain, starting most commonly ten to fourteen hours after eating mushrooms. Overt signs and symptoms of toxic hepatitis and toxic nephritis developing two to four days later .... **deadly toxins—amatoxins and others** (Ch. 7)

Guidelines for Safe Collecting and Eating

No single test or rule can be used to determine the edibility of any mushroom or other fungus. Each species must be carefully studied and learned so that edible species can be separated from those that are poisonous. If you are unfamiliar with the structure and identification of fungi, study reliable literature, enroll in a course on mushroom identification, or join a mushroom club or society. As you learn to identify mushrooms and other fungi, have each identification checked by an authority so you know it is correct. In other words learn something about the subject before you start collecting for food. After some of the common species are learned, additional ones can be studied and added to those you already know. In a few seasons you will be able to identify and collect one or more edible species.

Some general tests for determining the edibility of fungi are strictly superstition and are entirely unreliable. One of these is the belief that garlic cloves or silver coins turn blue or black when they are cooked in the same pan with poisonous mushrooms. Nevertheless, there are several useful guidelines that can be followed, which greatly reduce the risk of being poisoned. Some of these guidelines that make collecting fungi for food a less dangerous hobby are listed below.

- Be absolutely sure of the identification of the species you are collecting for food.
- Collect each specimen carefully and remove the entire fruit body from the substratum to check its identification. Several species of mushroom may grow together and it is easy to become careless and make a mistake, especially in a good season (see Fig. 17).
- Collect only one species at a time. If you collect more than one species, keep each in a separate container or package.
- Do not collect overmature or spoiled fruit bodies. Only fresh, young specimens are good for food. As mushrooms age they often decay and are not suitable for eating.
- Keep specimens clean by removing excess soil and debris, but do not cut away parts of the mushroom, or other identifying features. Note the substratum and habitat where the fruit bodies are collected, and be sure of the location.
- Collect and transport specimens in a sturdy, well-aired container, such as a market basket.
- On reaching home put specimens in a cool, dry place, preferably a refrigerator. If specimens are wet or soiled, place or wrap them in a dry, clean package or container.
- Clean and process or eat the specimens as soon as possible after you arrive home. Mushrooms and other fungi deteriorate rapidly and should not be kept longer than 24 hours before processing or eating. As you clean the specimens, cut each lengthwise and check for spoilage or insect damage.
- Do not eat wild mushrooms raw.
- Eat only one kind of mushroom at a time. Do not overeat.
- The first time you sample a species of mushroom, eat only a small amount and observe your reaction to it. Save a few intact fresh specimens for identification in case you become ill.
- Learn as much as you can about the various kinds of poisoning and the related symptoms.
- If you do experience symptoms of poisoning, empty your stomach as soon as possible and seek medical aid.
- Report all poison cases. Your experiences may prevent others from making similar mistakes.

Besides these general guidelines, there are several specific rules for distinguishing edible from poisonous species. Although one usually cannot make generalizations about the edibility of whole genera, these guidelines (Lincoff and Mitchel 1977) are helpful in avoiding certain large groups of species that are dangerous and should help prevent confusion of edible and poisonous species that are similar in color, size, shape, and other features.

- Never eat any white-capped mushroom that you have not positively identified. Some, such as *Amanita virosa,* are deadly and can be easily confused with white *Agaricus* species.
- Beware of any mushroom with an annulus on the stalk. Again confusion of *Amanita* and *Agaricus* is a real possibility, especially when collecting is done carelessly.
- Avoid all little brown mushrooms. They are difficult to identify and are easily confused. Many are poorly known.
- Check carefully those mushrooms that bear remnants of the universal veil, such as a volva in the form of a sac or cup around the stalk base or warts, scales, or patches on the surface of the cap or stalk.
- Species with a swollen or bulbous stalk base should also be carefully checked. Many *Amanita* species have an enlarged stalk base; and in cases where the volva or universal veil material has disintegrated, all that is left is a swollen stalk base.
- Avoid species of *Boletus* that have red to orange tube pores or which stain blue when cut or bruised. Several such are poisonous.
- Many people collect puffballs for food. Cut each puffball in half vertically and examine the interior carefully. The flesh should be pure white and the wall should be thin. A discolored or black to purplish interior means it is either an overripe puffball or a poisonous *Scleroderma* species. When an outline of a mushroom, including cap, gills, and stalk, is visible in vertical section, the specimen is not a puffball; it is instead the button of a mushroom, possibly a deadly *Amanita* species.
- Beware of all fungi with a brain-like to irregularly lobed (*Gyromitra*) or saddle-shaped (*Helvella*) cap. Certain species in this group are dangerous; some can be eaten only when parboiled and then thoroughly cooked. Avoid all of them to be safe.

Poisonous and Edible Species that Look Alike

Except for the intentional ingestion of hallucinogenic mushrooms, most toxic mushrooms are eaten by accident. Small children eat mushrooms simply because they like to put anything they can into their mouths; such occurrences are therefore purely accidental. People collecting mushrooms for food often carelessly collect and eat poisonous species because they confuse edible and poisonous species. This confusion is probably the main cause of mushroom poisoning,

Fig. 17. The commonly eaten mushroom *Agaricus sylvicola* (*left*) and its deadly double *Amanita virosa* (*right*).

Table 2 Poisonous and Edible Species that Look Alike

Toxin group	Poisonous species	Edible double
Deadly toxins—amatoxins and others (Ch. 7)	*Amanita phalloides*	*Amanita calyptroderma* *Amanita fulva* and relatives *Lepiota* spp., large forms such as *L. procera* *Tricholoma flavovirens* (=*T. equestre*) *Tricholoma portentosum* *Tricholoma sejunctum**
	Amanita bisporigera, A. phalloides (pale forms), *A. verna,* and *A. virosa*	*Amanita vaginata,* pale to white forms *Lepiota naucina** *Agaricus* spp., white or pale forms* *Tricholoma* spp., white forms such as *T. resplendens**
	Amanita bisporigera (buttons), *A. verna* (buttons), and *A. virosa* (buttons)	*Agaricus* spp., buttons of white forms Puffballs such as *Lycoperdon perlatum*
	Galerina autumnalis, G. marginata, and *G. venenata*	*Armillariella mellea** *Flammulina velutipes* *Pholiota* spp., small forms such as *P. mutabilis* *Psilocybe stuntzii*†
Monomethylhydrazine (Ch. 8)	*Gyromitra ambigua, G. esculenta, G. infula,* and possibly other *Gyromitra* species	*Gyromitra gigas** *Helvella* spp.* *Morchella* spp.*
Coprine (Ch. 9)	*Coprinus atramentarius, C. insignis,* and *C. variegatus* (=*C. quadrifidus*)	*Coprinus comatus** *Coprinus* spp. such as *C. micaceus* *Lepiota* spp., larger forms such as *L. procera*
Muscarine (Ch. 10)	*Clitocybe dealbata*	*Marasmius oreades*
	Inocybe spp.	*Marasmius oreades* *Tricholoma* spp., small forms such as *T. myomyces*
Ibotenic acid–muscimol (Ch. 11)	*Amanita gemmata*	*Russula* spp., similarly colored forms
	Amanita muscaria, white to pale forms	*Agaricus* spp., white forms*
	Amanita cothurnata	*Lepiota* spp., white forms* *Agaricus* spp., white forms*
	Amanita muscaria, yellow to orange or reddish orange forms	*Amanita caesarea* *Amanita flavorubens* (=*A. flavorubescens*)* *Amanita rubescens* *Armillariella mellea**
	Amanita pantherina	*Amanita* spp., similarly colored nontoxic forms
	Amanita solitaria group	*Agaricus* spp., white forms*
Hallucinogens (Ch. 12)	*Gymnopilus* spp. such as *G. spectabilis*	*Armillariella mellea** *Pholiota* spp., large forms*
	Panaeolus foenisicii	*Psathyrella candolleana* *Agrocybe pediades* *Marasmius oreades*
	Panaeolus spp.	*Coprinus* spp.
Gastrointestinal irritants (Ch. 13)	*Agaricus* spp.	*Agaricus campestris* and others*
	Boletus spp., poisonous forms such as *B. sensibilis*	*Boletus* spp., edible forms
	Entoloma spp.	*Pluteus* spp. such as *P. cervinus*
	Hebeloma spp., large forms	*Rozites caperata*

(*Continued*)

Table 2 Poisonous and Edible Species that Look Alike (*Concluded*)

Toxin group	Poisonous species	Edible double
Gastrointestinal irritants (Ch. 13) (*continued*)	*Lepiota molybdites*	*Lepiota* spp., large forms *Coprinus comatus*
	Naematoloma fasciculare	*Armillariella mellea** *Naematoloma capnoides* *Naematoloma sublateritium*
	Paxillus involutus	*Lactarius* spp.*
	Ramaria spp. such as *R. gelatinosa*	*Ramaria* spp., edible forms
	Scleroderma spp.	Puffballs, edible forms such as *Lycoperdon* spp.

*These species are sometimes not tolerated by certain individuals.
†Hallucinogenic.

especially the severe or fatal cases. Mistakes arise when the collector has a poor knowledge of poisonous species and is channeled into making an error by the many poisonous and edible mushrooms that look alike.

Poisonous and edible species that look alike are a main point of discussion in many modern books on toxic mushrooms (e.g. Lincoff and Mitchel 1977); however, this problem has been recognized for decades and is pointed out in many early European works.

Examples of mushroom poisoning from ingestion of the poisonous doubles of edible species abound. Some years ago, in the Upper Peninsula of Michigan, several golfers mistakenly ate the white form of *Amanita muscaria,* thinking it was an edible species of *Agaricus*. Not only did they confuse the two species but they ate the mushrooms raw while golfing and soon became extremely ill (Ingrid Bartelli, Personal commun.). The confusion of the poisonous *Amanita phalloides* with the similarly colored *Tricholoma flavovirens* (Pers. ex Fr.) Lundell (=*T. equestre* (L. ex Fr.) Kummer) is also well documented. Another example is illustrated in Fig. 17, which compares the edible *Agaricus sylvicola* with its deadly double, *Amanita virosa*. A personal experience (J. F. Ammirati) involved sorting through a bushel of mushrooms collected for eating; besides *Tricholoma myomyces* (Fr.) Lange, the intended edible species, several specimens of *Melanoleuca*, which is not known to be toxic, and a considerable number of *Inocybe* species, which contain muscarine, were found. The collectors could not differentiate between these genera. They knew nothing about mushroom characteristics, such as spore deposits, having learned to recognize the edible species from their parents while growing up in Europe. No poisoning resulted from this incident because the collectors had their specimens checked before eating or preserving them. Later, however, another group of collectors experienced poisoning from *Inocybe* after eating a similar mushroom mix.

Mushroom collectors must learn to recognize both edible and poisonous species, or at least must be aware of the characteristics of poisonous species and know which poisonous doubles may be confused with the edible species they are collecting. Table 2 lists the poisonous doubles of some edible species. This list is adapted from the work of Lincoff and Mitchel (1977), and the mushrooms are arranged according to toxin group. Edible species that have been reported as poisonous to some individuals because of intolerance are marked with an asterisk. As this list is not complete, collectors are advised to compile a personal listing of possible poisonous species that resemble the edible species they collect.

Literature

Bergoz, R. Trehalose malabsorption causing intolerance to mushrooms. Gastroenterology 60:909; 1971.

Burrell, R. Allergy to mushrooms: Fact or fancy? McIlvainea 3(2):11–14; 1978.

Byrne, A. R.; Ravnik, V.; Kosta, L. Trace element concentrations in higher fungi. Sci. Total Environ. 6:65–78; 1976.

Chilton, S. *Agaricus* reaction. Mycena News 29(2):59; 1979.

Göttl, L. Blausäurebildende Basidiomyzeten. Hat Cyanogenese einen taxonomischen Wert? Z. Pilzkd. 48:185–194; 1976.

Groves, J. W. Edible and poisonous mushrooms of Canada. Ottawa, Ont.: Agriculture Canada. Agric. Can. Publ. 1112; 1979 [first printed in 1962].

Hatfield, G. M.; Brady, L. R. Toxins of higher fungi. Lloydia 38(1):36–55; 1975.

Heineman, P. Observations sur les basidiomycetes à acide cyanohydrique. Bull. Trimest. Soc. Mycol. Fr. 58:99–104; 1942.

Heineman, P. Observations sur les basidiomycetes à acide cyanohydrique II. Lejeunia 13:99–100; 1949.

Lampe, K. F. Mushroom poisoning in children updated. Paediatrician 6:289–299; 1977.

Lincoff, G.; Mitchel, D. H. Toxic and hallucinogenic mushroom poisoning—A handbook for physicians and mushroom hunters. New York, NY: Van Nostrand Reinhold Co.; 1977.

Miller, O. K., Jr. Mushrooms of North America. New York, NY: E. P. Dutton & Co., Inc.; 1972.

Seeger, R.; Nutzel, R. Quecksilbergehalt der Pilze. Z. Lebensm. Unters. Forsch. 160:303–312; 1976.

Tyler, V. E. Mushroom poisons. McKenny, M. The savory wild mushroom (revised and enlarged by Stuntz, D. E.). Seattle, WA: University of Washington Press; 1971.

7

DEADLY TOXINS—AMATOXINS AND OTHERS

Toxicology and Symptomatology

A wide variety of fleshy fungi have caused or have been reported to have caused human death. Athough they have been found to contain several toxins, the amatoxins are the ones that are the most widely distributed.

A number of species in four genera of mushrooms, *Amanita, Conocybe, Galerina,* and *Lepiota,* contain levels of amatoxins high enough to cause severe human poisoning. A recent report determined that minute concentrations of amatoxins occur in many other mushrooms including the edible *Boletus edulis* Bull. ex Fr. and *Cantharellus cibarius* Fr. (Faulstich and Cochet-Meilhac 1976). The levels reported for these mushrooms, however, are not active physiologically; for example, *Amanita phalloides* has a concentration of amatoxins about 25 000 times greater than the edible *Cantharellus cibarius* (Faulstich and Cochet-Meilhac 1976).

A lethal dose of amatoxin-containing mushrooms is dependent on the levels of amatoxins present in the mushroom specimens. In the most deadly specimens of *A. phalloides,* as little as 55 g of mushroom consumed by a 68-kg man could be lethal (Lincoff and Mitchel 1977).

Galerina sulciceps (Berk.) Boedijn is perhaps the most toxic mushroom known to man; it has caused the death of several people in just seven hours after ingestion. The toxin or toxins in this fungus until recently were unknown. Bes (1981), however, has identified amatoxins in *G. sulciceps.* Symptoms include stomach spasms and nausea without vomiting or diarrhea, and there is no long period of delay before the onset of the symptoms (Benedict 1972). This species is not found in North America, however, so is not treated further here.

Several other fleshy fungi, mainly mushrooms, contain other toxins that have caused human death. These include *Cortinarius orellanus* (orellanin) and *Naematoloma fasciculare* (toxin unknown), both discussed in this chapter; *Gyromitra esculenta* (monomethylhydrazine), Ch. 8; *Inocybe patouillardii* (muscarine), Ch. 10; *Amanita muscaria* and *Amanita pantherina,* eaten in large quantities (ibotenic acid - muscimol), Ch. 11; and *Entoloma lividum* and *Paxillus involutus* (gastrointestinal irritants), Ch. 13. In addition, deaths in children have been caused by the ingestion of normally nonlethal mushrooms, including species of *Coprinus, Agaricus, Psilocybe, Lepiota* (which includes *Chlorophyllum*), and *Lactarius* (Buck 1969).

The amatoxins comprise one of two major groups of cyclopeptides, the second being the phallotoxins, isolated from fungi. Chemically, both are characterized as bicyclic peptides, composed of a ring of amino acids bridged by a sulfur atom to form a two-ring structure. Of these two cyclopeptides only the amatoxins appear to be involved in human poisonings. There are no data implicating the phallotoxins in human poisoning (Lincoff and Mitchel 1977).

The cellular mechanism for amatoxin poisoning is well understood. The toxins affect the primary step in gene expression, the synthesis of a class of important biological molecules known as messenger ribonucleic acid (mRNA). The toxin specifically inhibits the catalytic action of the enzyme RNA polymerase B, which synthesizes mRNA. Without mRNA, cells eventually stop synthesizing proteins, and without the spectrum of normal cellular proteins, many being enzymes themselves, the cells cease to function.

There is generally a long delay between eating the mushrooms and the appearance of the first symptoms. The delay may be from 5 to 24 hours, with the symptoms appearing in most victims approximately 12 hours after ingestion. The first symptoms are violent and prolonged vomiting, persistent diarrhea, and intense abdominal pain, which often may be associated with cramping in the feet and legs.

When these early symptoms appear, it is extremely important to identify the mushroom ingested. When no fresh material is available, try to identify spores in the vomit. Thin-layer chromatography can be used to detect the presence of amatoxins in the vomit or feces or in the mushroom tissue itself (Becker et al. 1976, Lincoff and Mitchel 1977).

Meixner (1979) has developed a simple test for detecting amatoxins in fresh mushroom tissue. The following procedure was translated and contributed by Dr. D. M. Simons, Chairman, Toxicology Committee, North American Mycological Association.

- Squeeze a drop of juice from fresh tissue onto a piece of pulp paper, such as newsprint. Use a garlic press, if possible. When only a small fragment of tissue is available, mash it onto the paper so that the juice can be absorbed.
- Circle the wet spot with a pencil to mark the location.
- Dry the spot with gentle heat. A hair dryer held at a distance is suitable.

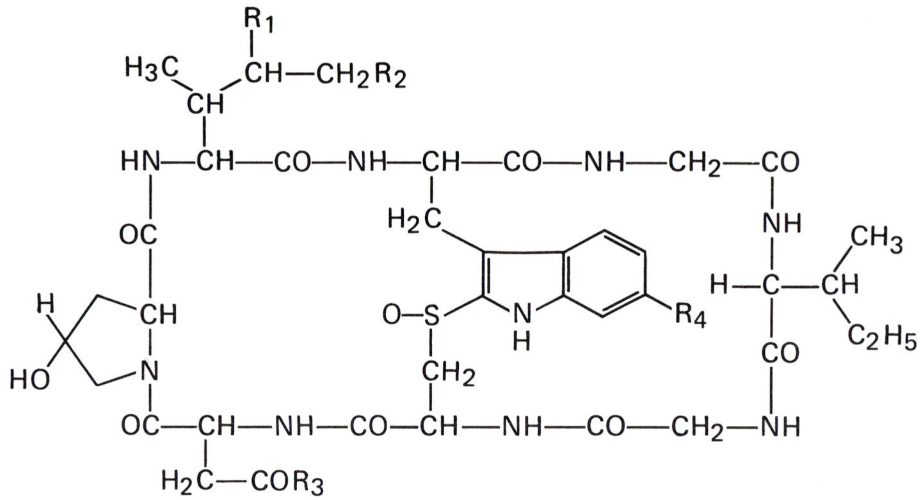

	R_1	R_2	R_3	R_4
α - *amanitin*	OH	OH	NH_2	OH
β - *amanitin*	OH	OH	OH	OH
γ - *amanitin*	H	OH	NH_2	OH

amatoxins

- Add a drop of concentrated hydrochloric acid to the dry spot.

When amatoxins are present, a blue color develops on the paper. The time required for the color to develop is only a minute or two if substantial amounts of amatoxins are present; 10–20 minutes may be required for trace amounts.

The Meixner test is based on an acid-catalyzed reaction of amatoxins with lignin. Cheap pulp paper is therefore suitable because it has a high lignin content. Good-quality cellulose paper does not work. The actual chemistry involved in the reaction is not known. Phallotoxins do not react to form blue products. Vergeer (1979) suggests that the Meixner test be performed indoors, away from direct sunlight and heat. Preliminary tests indicate that pure hydrochloric acid on newsprint turns blue when exposed to sunlight or higher temperatures (63°C and above).

The Meixner test is valuable because it can be carried out rapidly and routinely, even by amateur mycologists with little chemical training. Remember though that concentrated hydrochloric acid should be handled with considerable caution. Be sure to use the Meixner test only for scientific purposes: a negative result can never be taken as proof that the mushroom under test is edible.

The initial violent symptoms are caused by irritation of the gastrointestinal lining by amatoxins. The extremely large losses of body water and salts resulting from the persistent vomiting and diarrhea, left untreated, could result in death from circulatory failure. With proper medical treatment of the electrolyte imbalance and dehydration, shock and circulatory collapse can be avoided (Lincoff and Mitchel 1977).

If the victim survives the first symptoms, there is sometimes a period of apparent improvement. However, this stage is normally short-lived. Severe liver damage usually follows. Standard methods for evaluating the extent of liver damage include monitoring alkaline phosphatase, bilirubin, blood urea nitrogen (BUN), creatinine, lactate dehydrogenase (LDH), serum glutamic oxaloacetic transaminase (SGOT), and serum glutamic pyruvic transaminase (SGPT). The levels of SGOT and SGPT begin to elevate on the third and fourth day after ingesting the mushrooms. It is not uncommon for the white blood cell count to reach 25 000 or greater.

When the victim is fortunate enough to survive the dehydrative shock and acute liver necrosis, he must still contend with another crisis, acute kidney failure. Amatoxins cause direct destruction of the cells in the kidney tubules. All the symptoms that are generally common to uremic patients with acute tubular necrosis are frequently seen: high blood urea, oliguria-anuria, high serum potassium, low serum sodium, acidosis, septicemia, internal bleeding, and pulmonary edema. Proteinuria and slight or moderate hematuria are often observed.

In summary, considering all the symptoms on a time scale: 5-24 hours after consumption, the initial symptoms of vomiting, persistent diarrhea, and abdominal pain appear; 1-3 days after consumption, dehydrative shock sets in; 4-11 days after consumption, liver and kidney dysfunction occurs. After 12 days, complications may result from the earlier symptoms. Secondary dysfunctions such as lesions in the pancreas and lesions in the heart muscle tissue are often observed.

No antidote exists to date for amatoxin poisoning. Thioctic acid has been used to treat cases both in Europe and North America (Lincoff and Mitchel 1977). Supportive measures by the physician and good nursing are of greatest importance to the victim. Physicians should consult poison control centers for further information. Lincoff and Mitchel (1977) provide a detailed discussion of various treatments that have been used and give an evaluation of their success rates.

The above paragraphs are based in part on Lincoff and Mitchel (1977) and unpublished information prepared by Dr. D. M. Simons, Chairman, Toxicology Committee, North American Mycological Association.

Deadly Species

Amanitaceae
- *Amanita*
  - *A. bisporigera*
  - *A. hygroscopica*
  - *A. ocreata*
  - *A. phalloides*
  - *A. suballiacea*
  - *A. tenuifolia*
  - *A. verna*
  - *A. virosa*

Bolbitiaceae
- *Conocybe*
  - *C. filaris*

Cortinariaceae
- *Cortinarius*
  - *C. gentilis*
  - *C. orellanus*
  - *C. speciosissimus*
- *Galerina*
  - *G. autumnalis*
  - *G. marginata*
  - *G. sulciceps*
  - *G. venenata*

Lepiotaceae
- *Lepiota*
  - *L. brunneoincarnata*
  - *L. brunneolilacea*
  - *L. castanea* (suspected)
  - *L. clypeolarioides*
  - *L. felina* (suspected)
  - *L. fuscovinacea*
  - *L. griseovirens* (suspected)
  - *L. heimii*
  - *L. helveola sensu* Huijsman
  - *L. helveola sensu* Josserand
  - *L. locanensis*
  - *L. ochraceofulva* (suspected)
  - *L. pseudohelveola*
  - *L. rufescens*
  - *L. subincarnata*

Strophariaceae
- *Naematoloma*
  - *N. fasciculare*

The species listed above contain or are suspected to contain deadly toxins. Most contain amatoxins. The only exceptions are the species of *Cortinarius,* which contain orellanin; and *Naematoloma fasciculare,* which contains one or more unknown toxins. The list is arranged alphabetically within families. The species in this list, with the exception of the tropical and subtropical species of *Lepiota* and *Galerina sulciceps,* occur or are likely to occur in Canada or in areas of the United States adjacent to Canada.

The remainder of this chapter contains descriptions of these genera arranged fully alphabetically, without regard to family. For each genus, keys to species and species descriptions are included where warranted. Only those species that are commonly occurring in Canada or in the United States adjacent to Canada are treated in detail here. Species treatments are arranged alphabetically within each genus.

Descriptions of Genera and Species

Amanita Pers. ex S. F. Gray — Amanitaceae

Anyone eating mushrooms or dealing with poisonous mushrooms must learn to recognize the genus *Amanita.* Although some species of *Amanita* are edible, many of them are poisonous, and species such as *A. phalloides* are among the most poisonous mushrooms known.

There are four principal morphological characteristics that are shared by most *Amanita* species: the universal veil, the annulus, free or slightly attached gills, and a white spore deposit.

The universal veil completely covers the young mushroom in the button stage and breaks when the stalk elongates and the cap enlarges and expands (Fig. 1 F–H, A). The universal veil breaks in various ways depending on its structure and consistency. When the universal veil is membranous and tough, it often breaks by a slit across the top of the cap, allowing the cap to become free.

In such cases the universal veil is usually left as a cup, the volva, around the stalk base. This type of development is characteristic of *A. phalloides* and its relatives. In some species the universal veil is membranous but more fragile and breaks in an irregular pattern as the mushroom enlarges. In these cases a large piece of universal veil material is often left on the cap, with the remainder forming a cup-like volva around the stalk base. Development of this type is found in *A. calyptroderma* Atk. & Ballen, an edible species, occurring on the west coast of North America. In other species of *Amanita* the universal veil is more fragile and not membranous. For example, part of the universal veil of *A. muscaria* typically breaks into small patches or warts that are variously arranged over the cap surface. The remainder of the universal veil, the volva, is associated with the lower stalk surface and the apex of the bulbous stalk base, where it forms ring-like patches (Fig. 85). In *A. pantherina* the universal veil also breaks into small patches or warts on the cap surface, but the volva typically occurs as a collar-like rim on the apex of the stalk base, indicating where the universal veil broke (Fig. 89). Certain *Amanita* species have a universal veil that crumbles easily into small pieces that range from mealy to more or less powdery. In these species the cap surface has granules or small patches of soft tissue that are easily weathered away, so that the cap of mature specimens often lacks any remains of the universal veil. The volva in these species usually occurs as loosely attached granules or small patches on the lower stalk and at its base.

The volva usually also becomes lost or weathered away with age and at times can be found on the ground around the stalk base. In these species, particularly, both young and mature specimens must be studied. In many *Amanita* species the volva is buried in the soil and litter and may not be evident unless the entire mushroom, including the stalk base, is carefully removed from the soil.

The annulus is formed from the partial veil, a more or less thin membranous layer of tissue that extends from near the apex or middle of the stalk to the edge of the cap (Fig. 1 H and A). In the button stage and other young stages the partial veil covers the young, developing gills. When the mushroom matures, the partial veil typically breaks along the cap edge, leaving a thin, skirt-like ring on the upper part of the stalk. Sometimes the partial veil breaks away from the stalk rather than from the cap edge, leaving a fringe of tissue hanging from the cap edge. Some *Amanita* species do not have a partial veil and therefore have no annulus. These species have traditionally been placed in a separate genus, *Amanitopsis* or *Vaginata.* Now, they are usually placed in the genus *Amanita.*

The gills are often free in *Amanita* but in some species, such as *A. silvicola* Kauffman, they are attached or just reach the stalk (Fig. 7 A and B).

The spore deposit in *Amanita* is usually white. The gills are also usually white or pale colored, but because the spore color is not always the same color as the gills, be sure to take a spore deposit.

Amanita species also have in common a divergent gill trama and a unique stalk trama composed of very large, clavate hyphae and narrow, branching, longitudinally arranged, interwoven hyphae. The divergent gill trama is often used as a key characteristic (Fig. 14C).

Amanita magnivelaris Peck, a species in the *Amanita phalloides* group, is suspected of containing amatoxins but their presence, to our knowledge, has not been verified. *Amanita brunnescens* Atk. has also been suspected but no one has yet verified the presence of lethal amounts of amatoxins in this species. Abdel-Malak (1974) found some yet uncharacterized and unnamed toxins, in undetermined concentrations and toxicity, in the *Amanita* species *A. bisporigera, A. brunnescens, A. flavoconia* Atk., *A. muscaria, A. rubescens* (Pers. ex Fr.) S. F. Gray, and *A. virosa.* Of these only *A. bisporigera* and *A. virosa* are known to contain lethal concentrations of amatoxins. Bartelli (1977) lists *Amanita spreta* (Peck) Sacc. as deadly. We have no reports of poisoning by it. Coker (1917) reports it as harmless, at least in one instance. Ford (1909) found a hemolysin and an additional toxin, in small amounts, in *A. spreta.* Bartelli (1977) provides a photograph and summarizes the main descriptive features.

Several *Amanita* species contain sufficient quantities of amatoxins to cause severe or fatal poisoning. In Canada three of these species, *A. bisporigera*, *A. verna*, and *A. virosa*, have been recorded. *Amanita phalloides*, which to date has not been documented from Canada, occurs in New York State along the shore of Lake Ontario. It has also been reported in the west from Washington and south into California. Therefore, it probably does occur in Canada and you should be on the lookout for it when identifying deadly species. *Amanita ocreata* is a second species not yet reported in Canada but occurring on the Pacific coast in California. Its distribution seems fairly restricted but it has not been thoroughly studied and it may well occur throughout the west, especially where species of oak (*Quercus*) grow. There are at least three other *Amanita* species that contain amatoxins and occur in North America; these are *A. tenuifolia* (Murrill) Murrill, *A. suballiacea* (Murrill) Murrill, and *A. hygroscopica* Coker. However, they appear to grow mainly in the southeastern United States and are not described here.

In summary, deadly *Amanita* species have five characteristics in common. These include the four features previously described, namely a white spore deposit; free or slightly attached gills; a well-developed partial veil typically leaving a membranous annulus; and a bulbous stalk base surrounded by a cup-like to sac-like volva left from the universal veil. The final feature, a microscopic characteristic, is amyloid spores. These five characteristics identify the group to which all of the deadly *Amanita* species belong. When a poisoning results from an *Amanita* in this group, identification of the particular species is academic, because all are extremely dangerous. *Amanita* species in this group occur mainly in forested areas but may be expected to occur wherever there are trees, for example in lawns with scattered trees.

For a discussion of other species of *Amanita* containing toxins other than amatoxins, see Chapters 11 and 13.

Key to selected species

1. Cap dark grayish brown, some shade of olive, greenish, or yellowish green, at times shaded with gray or brown or with blackish streaks, usually darkest in the center....................... ***A. phalloides***

   Cap white to creamy or buff, with the center at times pinkish buff or tannish with age (the colors may be more creamy to yellowish or dingy when old), never distinctly colored as above............. 2

2. Potassium hydroxide not staining the cap surface distinctly yellow. Spores ellipsoidal to broadly ellipsoidal in shape, mostly 10–11 × 6–7.5 μm in size... ***A. verna***

   pale forms of *A. phalloides* may key out here

   Potassium hydroxide staining the cap surface yellow. Spores often broadly ellipsoidal to subglobose, and if ellipsoidal then 9–14 × 7–10 μm in size... 3

3. Basidia predominantly two-spored........................ .... ***A. bisporigera***

   Basidia predominantly four-spored..................... 4

4. Cap white to buff, often with pinkish buff in the center. Spores rarely subglobose, usually 9–14 × 7–10 μm in size. Fungi known only from the Pacific coast................................ ***A. ocreata***

   Cap mainly white but at times with creamy to yellowish tints. Spores more or less globose, mostly 9.2–12 × 8.5–9.5 μm in size ..... ***A. virosa***

Amanita bisporigera Atk.

Figs. 18, 34

Common names. Destroying angel, deadly amanita, white death cap, angel of death.

Distinguishing characteristics. Cap 40–70 mm broad, convex to more expanded; surface dry to somewhat viscid; color white or at times creamy in the center. Gills free, close to crowded, white. Stalk white, 5–10 mm thick; base bulbous. Partial veil leaving a membranous white apical annulus. Universal veil leaving a white membranous conspicuous cup-like volva around the stalk base. Potassium hydroxide staining the cap surface yellow. Basidia two-spored; spores globose to subglobose, 8.5–11 × 8–10.2 μm in size, amyloid.

Observations. The color and general appearance of *A. bisporigera* are similar to those of *A. verna* and *A. virosa* and the three should be carefully compared. The two-spored basidia are a key characteristic to the identification of *A. bisporigera* and the feature forms the basis for its species name. *A. bisporigera* is at times smaller and more slender than either *A. verna* or *A. virosa*, but it varies considerably in size; therefore size should not be used as a diagnostic characteristic.

A. bisporigera appears to be more toxic than other species in this group of *Amanita*, when equal quantities of mushroom tissue are compared. However, a thorough study of the variation in the kinds and amounts of amatoxins in deadly species of *Amanita* has not been made. Because considerable variation occurs in some species, further sampling is required before a conclusion can be reached regarding the toxicity of *A. bisporigera*.

Distribution and seasonal occurrence. *A. bisporigera* is widely distributed across Canada. It is generally less common than *A. virosa* but its occurrence may vary with location and weather. It is most commonly found

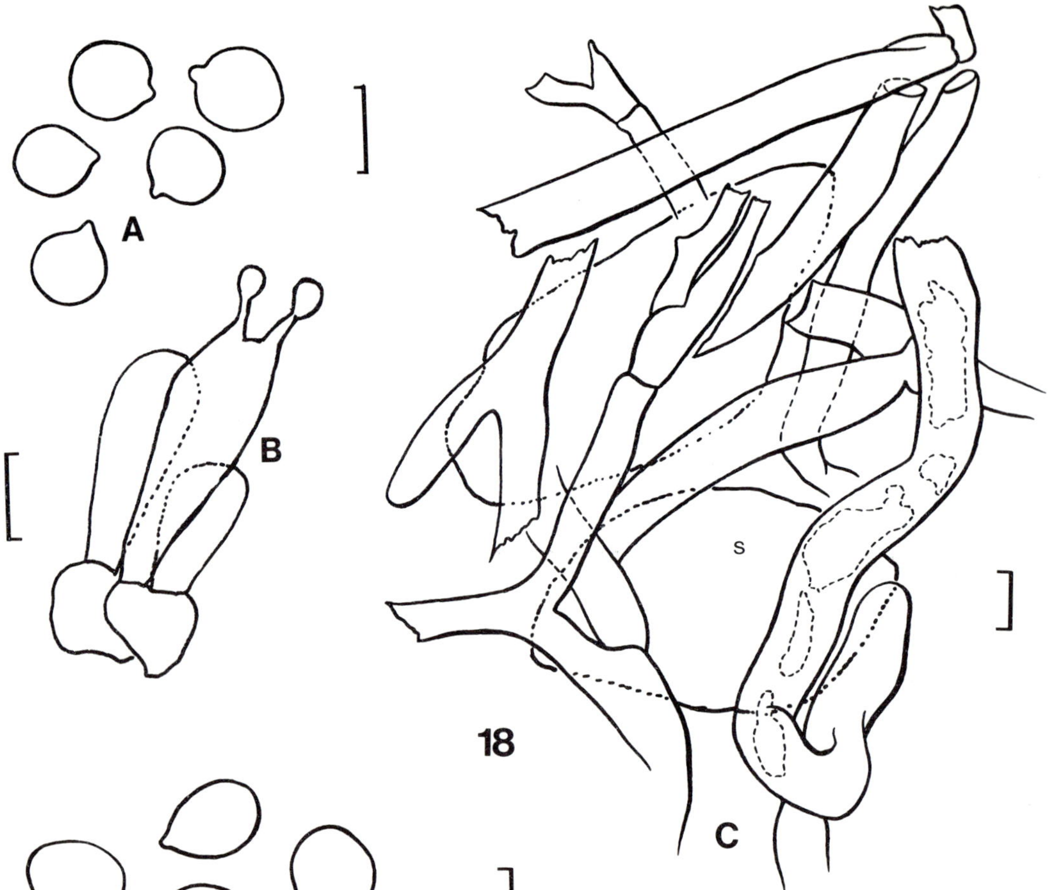

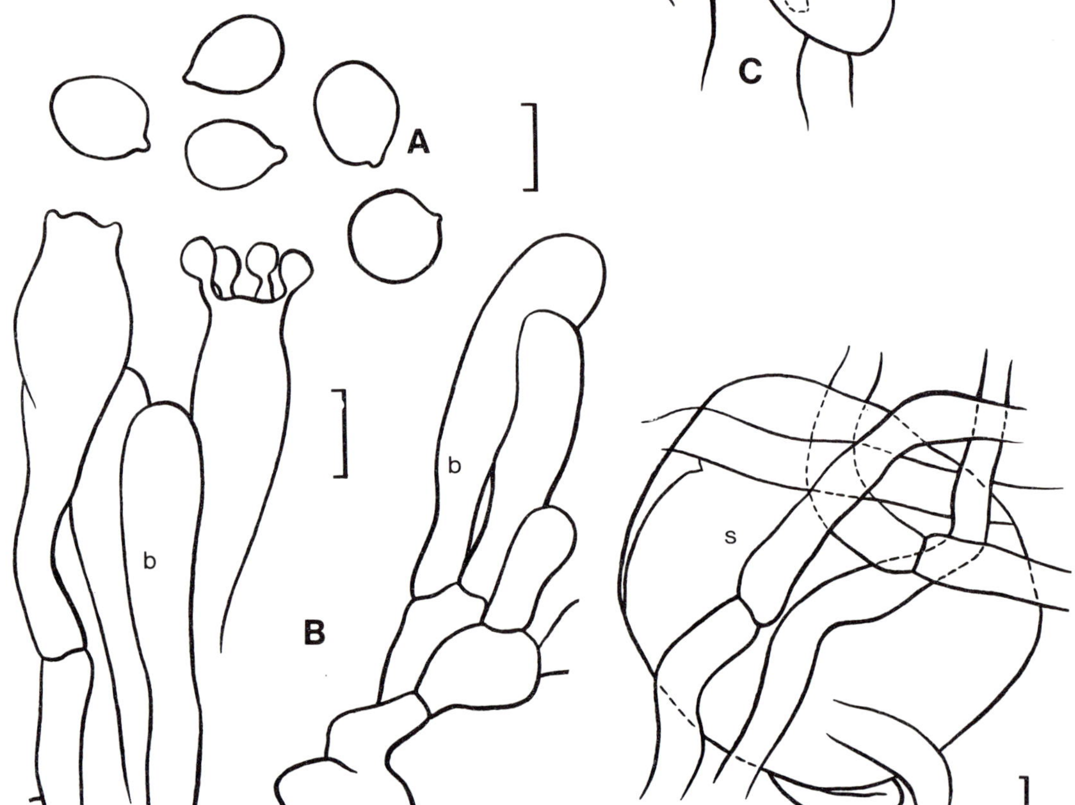

Fig. 18. *Amanita bisporigera*: A, spores; B, basidium; C, hyphae and sphaerocysts (*s*) of the volva (note the absence of clamp connections on the hyphae).

Fig. 19. *Amanita ocreata*: A, spores; B, basidia and basidioles (*b*); C, hyphae and a sphaerocyst (*s*) of the volva.

Scale line 10 μm

in eastern North America and is apparently rare in western North America. There has been one report of it, also, from Mexico (Heim 1957).

Habitat and habit. The fungus is terrestrial under hardwoods or in mixed woods. It occurs fairly commonly under oak (*Quercus*) and has been reported in birch-aspen (*Betula-Populus*) areas in the west. The fruit bodies may grow solitary, scattered, or gregarious.

Technical description. Cap 40-70 mm broad, convex to expanded, somewhat umbonate; margin even; surface dry or viscid, smooth, and glabrous to finely fibrillose; color white to creamy on the disc. Flesh thin, white, unchanging when bruised; taste indistinctive; odor pleasant to somewhat nauseous.

Gills white, free, crowded to close, moderately broad toward the cap margin; edges floccose.

Stalk 80-120 mm long, 5-8(-10) mm thick at the apex; base bulbous; surface slightly floccose to slightly floccose-scaly, white, stuffed to hollow. Annulus membranous, fragile, sometimes torn, white, persistent, and apical on the stalk. Volva thick, membranous, firmly appressed to the bulbous base, free above, white, and splitting into lobes.

Spores white in deposit, globose to subglobose, (8-)8.5-11.0(-12.4) × 8.0-10.2(-11.0) μm, hyaline, amyloid. Basidia two-spored, clavate, 29-34 × 7.3-8.8 μm, hyaline. Pleurocystidia absent. Cheilocystidia absent, but cylindrical to sac-like cells of the partial veil sometimes occurring on gill edges; these cells 24-34 × 7-16 μm in size, hyaline. Gill trama divergent from a central strand; hyphae more or less interwoven, cylindrical to inflated, 3.5-20 μm wide; inflated hyphae common in the central strand, hyaline. Subhymenium more or less cellular; hyphae (6-)8-10 μm wide, hyaline. Cap cuticle filamentous, gelatinous; hyphae interwoven, radially arranged, cylindrical, 2.8-5.6 μm wide, hyaline to yellowish. Cap trama loosely interwoven; hyphae 2.8-30 μm wide, cylindrical to inflated, hyaline. Caulocystidia clavate, overlapping; apical cells 7-15 μm wide, arising from interwoven branched cylindrical hyphae 2.3-7.1 μm wide. Annulus consisting of hyaline hyphae and sphaerocysts; hyphae interwoven, cylindrical, 2.3-5.9 μm wide; sphaerocysts hyaline, globose to ellipsoidal or ovoid, 28-37.4 × 11.7-28 μm in size. Volva consisting of hyaline hyphae and sphaerocysts; hyphae interwoven, mainly cylindrical, 2.9-5.9(-10.5) μm wide; sphaerocysts scattered, subglobose to ellipsoidal or ovoid, 58.5-93.6 × 42-60.8 μm, single or in short chains. Oleiferous hyphae yellowish, refractive, thin-walled, 5-7 μm wide, scattered in the flesh and volva. Clamp connections absent.

Useful references. The following references contain information pertinent to *A. bisporigera* in North America.

- *A Field Guide to Western Mushrooms* (Smith 1975)—A color photograph and a good description and discussion are given.
- *Mushrooms in their Natural Habitats* (Smith 1949)—See under *A. verna* for treatment of *A. bisporigera*.
- *Mushrooms of North America* (Miller 1972)—A description and discussion are given, included as part of *A. verna* and *A. virosa*.
- *The Mushroom Hunter's Field Guide* (Smith 1974)—Consult for reference only.

Amanita ocreata Peck

Figs. 19, 35

Common names. None.

Distinguishing characteristics. Cap 50-120 mm broad, convex to plane when expanded; surface dry to more or less viscid, white to buff, often pinkish buff on the disc. Gills free, subdistant, white. Stalk 15-20 mm thick; base bulbous, white, often discoloring pale brown when bruised. Annulus apical on the stalk, membranous, white, usually collapsing in early stages. Universal veil leaving a white sac-like to cup-like volva around the stalk base. Spores amyloid, ovoid to more or less ellipsoidal, rarely subglobose, 9-14 × 7-10 μm in size. Potassium hydroxide staining the cap surface bright yellow.

Observations. *A. ocreata* is similar in appearance to the white species of *Amanita* in this group and has been confused with one or more of these, especially *A. verna*. *A. bisporigera* and *A. virosa* usually have globose to subglobose spores that are somewhat smaller than those of *A. ocreata*. In addition, *A. bisporigera* has predominantly two-spored basidia. *A verna* has somewhat smaller spores and a negative to pale yellow reaction with potassium hydroxide. The caps of *A. verna* and other white species in this group do not typically develop buff to pinkish buff colors.

Distribution and seasonal occurrence. To date *A. ocreata* has only been reported from California, from San Diego County to San Francisco County and inland to the foothills of the Sierra Nevada Range. It is likely to be found elsewhere and should be watched for, especially on the Pacific coast north of California. It fruits from January through April in California, depending on location and weather.

Habitat and habit. This species is terrestrial, solitary to gregarious, and is apparently associated with coast live oak (*Quercus agrifolia*).

Technical description. Cap 50-120 mm broad, convex to bluntly convex to subglobose when young, becoming broadly convex to plane; margin entire, decurved when young but becoming plane with age, not striate; surface more or less viscid, shiny, smooth, glabrous to somewhat silky-fibrillose, rarely with thin patches of buff to white universal veil remnants but without scales, and white to buff with pinkish buff on the disc with age. Flesh white, 5-15 mm thick on the disc, narrowing to the margin; taste mild to slightly metallic; odor mild, indistinctive.

Gills white, free or slightly attached by a narrow decurrent tooth, broad, somewhat ventricose, subdistant; edges entire, not floccose.

Stalk 80-200 mm long, 15-20 mm thick at the apex, and equal to tapering upward; base abruptly bulbous, subglobose to obovoid; surface finely fibrillose above the annulus, appressed silky-fibrillose below, white, staining pale brown when bruised; flesh solid, white, unchanging when exposed. Annulus apical on the stalk, sometimes collapsing and evanescent, striate above, floccose below, white. Volva sac-like, appressed at the base; margin conspicuously free, collapsing with age, white.

Spores white in deposit, 9-14 × 7-10 μm, ovoid to more or less ellipsoidal, rarely subglobose, smooth, hyaline, amyloid. Basidia two- or four-spored, clavate, 37-55 × 10-15 μm, hyaline. Pleurocystidia lacking. Cheilocystidia lacking, but sac-like to more or less clavate cells of the partial veil present on gill edges.

Gill trama divergent from a central strand; hyphae 2.9-29.2 μm wide, cylindrical to inflated, hyaline to yellowish. Subhymenium interwoven; hyphae cylindrical to slightly inflated, 3.5-5.7(-7.0) μm wide, hyaline. Cap cuticle composed of interwoven radially arranged cylindrical hyphae 2.2-11.0 μm wide; these hyphae hyaline to yellowish. Cap trama composed of hyaline loosely interwoven cylindrical to inflated hyphae 2.9-29.2 μm wide. Caulocystidia clavate, 90-150 × 9.5-30 μm, hyaline, overlapping. Annulus composed of interwoven cylindrical hyphae 2.9-8.8 μm wide, with clavate end cells 35-68 × 14-23 μm, single or in short chains. Universal veil composed of interwoven cylindrical to inflated hyphae 3.9-13.1 μm wide; sphaerocysts scattered. Volva composed of hyaline hyphae and sphaerocysts; hyphae 18-25 μm wide; sphaerocysts scattered, spherical to broadly ellipsoidal, 63-84(-207) × 25-50(-92) μm. Oleiferous hyphae scattered, greenish yellow, and refractive, present in the trama, cap surface, and volva. Clamp connections absent.

Useful references. There are no general North American references available on this species. Ammirati et al. (1977) provide a technical description and discussion.

Amanita phalloides (Fr.) Link

Figs. 20, 36

Common names. Death cap, destroying angel, death cup, poison amanita, deadly amanita, deadly agaric.

Distinguishing characteristics. Cap 50-150 mm broad, convex to plano-convex when expanded; surface viscid, becoming dry and often glistening with age; color variable, fuscous, some shade of olive, olive yellow, greenish, yellowish green or nearly white, at times with a brownish to grayish cast or with blackish streaks, usually darkest in the center. Gills close to subdistant, free, white or with age slightly yellowish. Stalk 10-35 mm thick, with a thick bulbous base; color white to grayish green. Partial veil leaving a membranous persistent white to grayish green annulus at the apex of the stalk. Universal veil leaving a prominent white cup-like volva around the stalk base. Spores globose to ellipsoidal, 7-12.8 × 5.5-10.2 μm in size, amyloid.

Observations. In its typical form *A. phalloides* is easily recognized and is well known in the literature for its death-dealing toxins. When, however, the cap color is pale to almost white, it may be confused with *Amanita verna*. Potassium hydroxide apparently does not stain the cap surface yellow. Concentrated sulfuric acid (H_2SO_4) stains the gills purple to lilac. This color reaction with sulfuric acid seems to be characteristic of *A. phalloides* and can be used to help separate pale forms of *A. phalloides* from *A. verna*.

Spore size is somewhat variable in *A. phalloides*. Specimens studied by Tanghe and Simons (1973) had smaller spores, (6.5-)7.4-8.8(-9.6) × (5.6-)6.1-7.3(-9.0) μm, than the range reported here. Similar differences in spore size have been reported in Europe by Romagnesi (1962) and Courtillot and Staron (1970).

During the last part of the 19th century and into the first quarter of the 20th century various species were reported under the name *A. phalloides*. One of these species, *A. brunnescens* Atk., was described as new and to date has not been shown to contain lethal amounts of amatoxins. *A. porphyria* (Alb. & Schw. ex Fr.) Secr. and *A. citrina* Schaeff. ex S. F. Gray [=*A. mappa* (Batsch ex Lasch) Quél.] were at times also confused with *A. phalloides*. These species, as well as *A. brunnescens*, have a less distinct volva, and in *A. brunnescens* the bulb often splits longitudinally. *A. citrina* and *A. porphyria* have not been shown to contain amatoxins. At times white species of *Amanita* have been reported as *A. phalloides*, but these are most likely the white deadly species, such as *A. virosa*.

Distribution and seasonal occurrence. So far *A. phalloides* has not been reported in Canada. It occurs in areas of the United States adjacent to Canada, namely the states of New York and Washington, and is to be expected in Canada. This species fruits in late summer and fall and into the winter, depending on location and weather.

Habitat and habit. This species is terrestrial, growing solitary to gregarious in humus. It is reported as occurring with several species of conifers as well as under hardwoods, especially oak (*Quercus*).

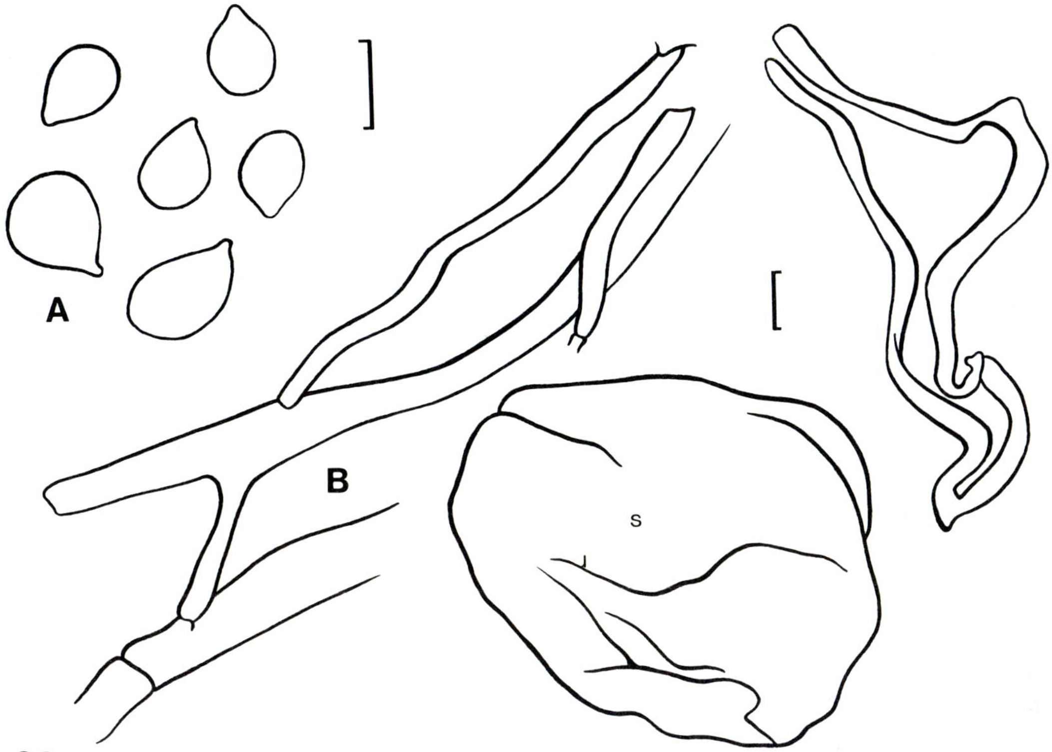

20

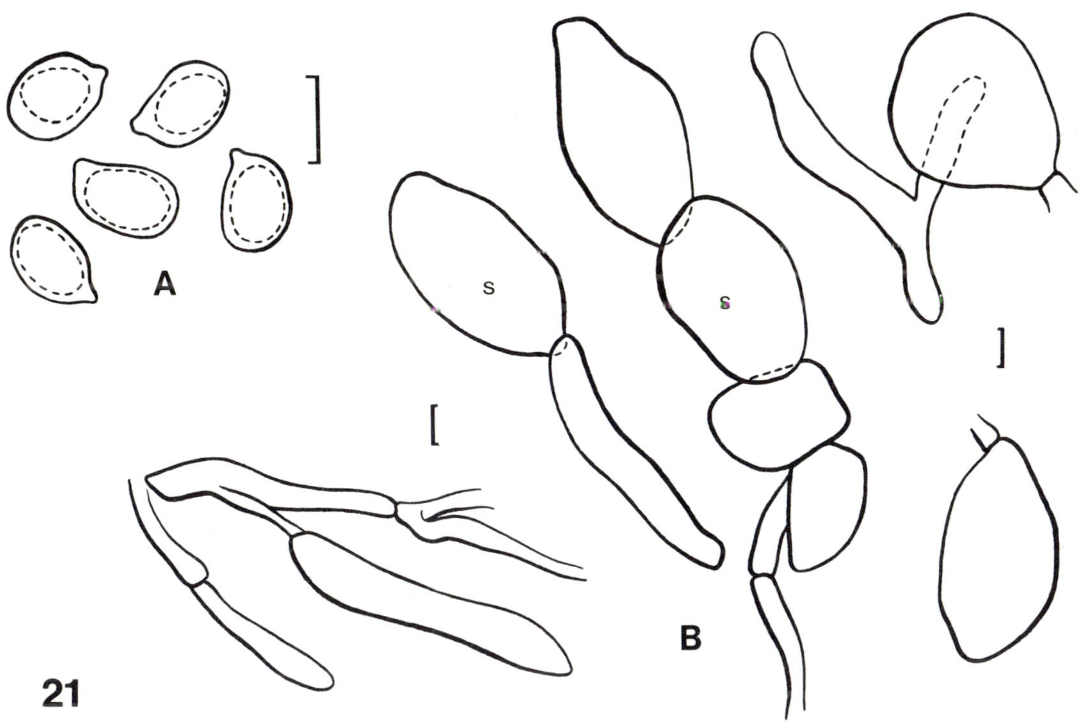

21

Fig. 20. *Amanita phalloides*: A, spores; B, hyphae and a sphaerocyst (*s*) of the volva.

Fig. 21. *Amanita verna*: A, spores; B, hyphae and sphaerocysts (*s*) of the volva.

Scale line 10 μm

Technical description. Cap (50-)70-150 mm broad, conic to obtusely convex expanding to plano-convex; margin not striate; surface viscid, becoming dry and glistening when old, not hygrophanous, smooth, sometimes with radiating innately fibrillose streaks, glabrous, rarely with one or two large patches of universal veil tissue; color fuscous, dark olive, pale olive, olive yellow, or greenish to yellowish green or nearly white, sometimes with a brownish or grayish cast, at times with blackish streaks, usually darkest on the disc. Flesh white or yellowish green below the cap surface; taste not recorded; odor more or less disagreeable.

Gills white during all stages of development or becoming very pale yellow with age, unchanging when bruised, free or when young attached by a slight decurrent tooth, close to subdistant, broad; edges similarly colored, entire becoming slightly eroded with age; several tiers of lamellulae present.

Stalk 50-150 mm long, (8-)10-20(-35) mm thick at the apex, more or less equal above a thick globose to pyriform or irregularly shaped abrupt bulb; surface dry to moist, glabrous to silky appressed-fibrillose, white during all stages of development, unchanging or becoming yellowish to brownish when bruised. Flesh soft, stuffed except toward the base where it is often hollow; flesh at base solid, white, unchanging when exposed. Annulus apical on the stalk, membranous, skirt-like, white to grayish green, persistent. Volva cup-like, prominent, membranous, appressed to the bulbous base of the stalk and free above, usually lobed, white.

Spores white in deposit, 7-12.8 × 5.5-10.2 μm, globose to subglobose or ellipsoidal, smooth, hyaline, amyloid. Basidia two- or four-spored, clavate, 35-48 (-60) × 9.5-11.0 μm. Pleurocystidia lacking. Cheilocystidia lacking, but clavate to sac-like cells of the partial veil present on the gill edges.

Gill trama divergent from a central strand; hyphae interwoven to parallel, cylindrical to inflated, 3.5-11 (-27.0) μm wide, hyaline. Subhymenium composed of interwoven cylindrical to inflated hyphae up to 20 μm wide. Cap cuticle filamentous, gelatinous; hyphae more or less radially arranged, cylindrical, 3-8(-10) μm wide, hyaline to slightly yellowish. Cap trama composed of loosely interwoven cylindrical hyphae 3-8 μm wide and inflated hyphae 5-18.7(-45) μm wide; both types of hyphae mostly hyaline. Caulocystidia abundant on the apex of the stalk, subcylindrical to clavate to sac-like, 12.4-49.6 × 10-25 μm, thin-walled, hyaline. Annulus composed of interwoven cylindrical hyphae; end cells clavate, 3.5-7.1(-17.5) μm wide; sphaerocysts abundant, 28.4-52.0 × 12-23 μm, spherical to pyriform or ellipsoidal, terminal or in short chains; all three structures hyaline. Universal veil hyphae from cap interwoven, mostly cylindrical, 3.5-5.9(-11.0) μm wide; sphaerocysts 10-30 × 9.4-15 μm. Volva composed of cylindrical hyphae 3.5-9.4(-20) μm wide; scattered sphaerocysts ellipsoidal to subglobose, 40-76.1 × 38-52 μm. Oleiferous hyphae refractive, 5.9-7.1 μm wide, present in the annulus, volva, and stalk. Clamp connections absent.

Useful references. The following references contain information pertinent to *A. phalloides* in North America.

- *A Field Guide to Western Mushrooms* (Smith 1975)—A color photograph and a description and discussion are presented.
- *Mushrooms of North America* (Miller 1972)—A description and discussion are given.
- *The Savory Wild Mushroom* (McKenny 1971)—A black-and-white photograph and a discussion are given.

Also see technical publications by Ammirati et al. (1977) and Tanghe and Simons (1973). An excellent color photograph is presented in *Scientific American* (Litten 1975).

Amanita verna (Bull. per Fr.) Roques

Fig. 21

Common names. Destroying angel, angel of death, white death cap, spring amanita.

Distinguishing characteristics. Cap 40-120 mm broad, convex to plane when expanded; surface viscid when fresh, becoming dry, white to creamy or at times yellowish on the center. Gills free, close to crowded, white. Stalk 8-20 mm thick, with a bulbous to ovoid base, and white; surface at times broken into scales. Partial veil leaving an often pendant membranous white annulus apical on the stalk. Universal veil forming a white cup-like volva around the stalk base. Spores broadly ellipsoidal to ellipsoidal, amyloid, mostly 10-11 × 6-7.5 μm in size.

Observations. The name *Amanita verna* has been used extensively in the North American literature on mushrooms and the species has been variously interpreted. It is basically a white mushroom that in general size and stature is similar to other deadly *Amanita* species, for example *A. virosa.* Here we have used ellipsoidal spores and the negative to pale (slightly yellowish) reaction with potassium hydroxide to distinguish it from *A. virosa* and *A. bisporigera.* The species is in need of thorough study in order to separate it clearly from the other deadly white species as well as from *A. phalloides.* Compare also *A. ocreata* and the whitish or pale forms of *A. phalloides.*

Distribution and seasonal occurrence. *A. verna* is widely distributed and has been reported in both the eastern and western United States. It is more common in the east than in the west. Because it has been confused with the other white, deadly *Amanita* species, however, it is difficult to establish its true distribution in Canada. Generally, it occurs in forested areas across the southern part of Canada. The species fruits in the spring or early summer on the Pacific coast, but in general, during the summer and fall.

Habitat and habit. *A. verna* is terrestrial, growing solitary to scattered or gregarious in hardwood and mixed forests. It has been reported in the west under Douglas-fir.

Technical description. Cap 40–120 mm broad, ovoid to nearly cylindrical when young, expanding to convex, sometimes becoming plane; margin decurved, entire, smooth; surface dry or viscid, glabrous to appressed-fibrillose, white to creamy or yellowish on the disc. Flesh moderately thick, white; taste not recorded; odor pleasant to somewhat nauseous.

Gills white, free, close to crowded, moderately broad; edges floccose; lamellulae often sharply squared near the stalk.

Stalk 60–150(–200) mm long, 8–20 mm thick at the apex, more or less equal with a bulbous to ovoid base; surface white, silky to silky-fibrillose above the annulus, and floccose to somewhat scaly below; flesh white, solid to fragile, soft. Annulus pendant, membranous, white, apical on the stalk. Volva cup-like, membranous, appressed to the base of the stalk but free in the upper part, splitting into lobes, white.

Spores white in deposit, (9.5–)10.2–11.0(–12.4) × 6.0–7.4(–8.0) μm, broadly ellipsoidal to ellipsoidal, flattened along the axis, hyaline, amyloid. Basidia four-spored, clavate to ventricose, 32.9–38.0 × 8.8–9.5 μm, hyaline, thin-walled. Pleurocystidia absent. Cheilocystidia absent, but sac-like cells of the partial veil present on gill edges 20–40 × 14–18 μm in size.

Gill trama divergent from a central strand; hyphae cylindrical to inflated, more or less interwoven, 9–13 (–16) μm wide, thin-walled, hyaline. Subhymenium cellular; hyphae inflated, 8–13 μm wide, hyaline. Cap cuticle gelatinous, filamentous; hyphae interwoven, more or less radially arranged, cylindrical, 3.6–9.0 μm wide, hyaline; scattered inflated hyphae near trama 13–16 μm wide, clavate to fusiform, terminal or between other cells. Cap trama loosely interwoven; hyphae cylindrical to inflated, 3.7–16 μm wide, hyaline. Caulocystidia clavate, overlapping, 108–144 × 18 μm; inflated tips of interwoven hyphae 3.6–10.8 μm wide. Annulus composed of hyaline interwoven cylindrical hyphae 2.9–7.3(–12.4) μm wide; sphaerocysts scattered, subglobose to ellipsoidal or more or less cylindrical, 27–42 × 14.6–25.6 μm, single and terminal, hyaline. Volva composed of interwoven cylindrical hyphae 3.6–9.0 μm wide; sphaerocysts abundant, 26–66(–144) × 18.0–43.4 μm, subglobose to ellipsoidal or fusiform or clavate-cylindrical, terminal or in short chains, hyaline. Oleiferous hyphae refractive, scattered in the surface and flesh of the cap and stalk and the annulus, 3.5–7.0 (–10) μm wide. Clamp connections absent.

Useful references. The following references contain information pertinent to *A. verna* in North America.

- *Mushrooms in their Natural Habitats* (Smith 1949)—A color photograph, description, and discussion are presented.
- *Mushrooms of North America* (Miller 1972)—The pertinent description and discussion are included as part of *A. virosa.*
- *The Mushroom Hunter's Field Guide* (Smith 1974)—Black-and-white photographs and a discussion are presented.
- *The Savory Wild Mushroom* (McKenny 1971)—This reference includes a black-and-white photograph and a description and discussion.

Amanita virosa (Fr.) Bertillon

Figs. 22, 37

Common names. Destroying angel, deadly amanita, angel of death, white death cap.

Distinguishing characteristics. Cap 30–140 mm broad, conic to convex becoming plane with age; surface viscid when fresh to dry; color basically white but with pale yellow to creamy discolorations developing in the center and sometimes overall with age. Gills free or nearly so, close to crowded, and white but at times becoming sordid to pale dingy yellow with age. Stalk 10–23 mm thick; base enlarged and usually bulbous; color white, becoming somewhat dingy with age. Partial veil leaving a membranous skirt-like white annulus apical on the stalk. Universal veil leaving a distinct cup-like volva around the stalk base; color white or somewhat yellowish at times. Spores globose to subglobose, mostly 8.8–11.7 × 8.4–10.2 μm in size. Potassium hydroxide staining the cap surface yellow.

Observations. *A. virosa* is the most common deadly white *Amanita* in southeastern Canada and southward into the eastern United States. It has caused many poisonings in this region.

Because *A. virosa* is very similar to *A. bisporigera* and *A. verna,* the three species cannot readily be separated on macroscopic characteristics. The yellow staining reaction of the cap surface when potassium hydroxide is applied to it and the globose to subglobose spores are helpful in separating *A. virosa* from *A. verna. A. bisporigera* has two-spored basidia whereas those of *A. virosa* are four-spored. A two-spored form of *A.*

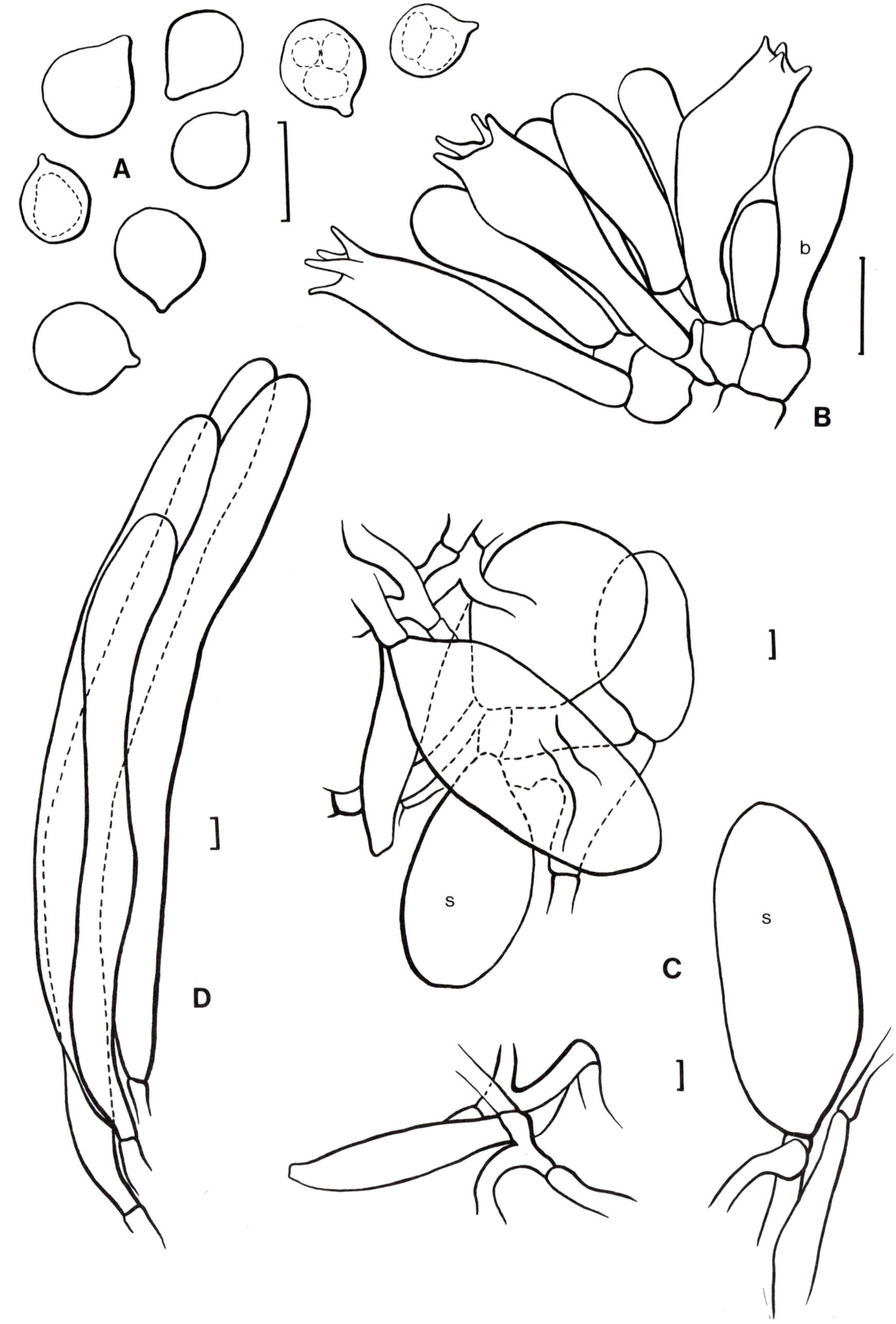

Fig. 22. *Amanita virosa*: A, spores; B, basidia and basidioles (*b*); C, hyphae and sphaerocysts (*s*) of the volva; D, caulocystidia.

Scale line 10 μm

virosa has been reported by A. H. Smith (Personal commun.), but in our studies, we have not encountered such a form. Spore size in *A. virosa* tends to vary as it does in *A. phalloides*. Some collections have spores that fit into the size ranges given above for length and width, while others favor the upper or lower part of these ranges. Tanghe and Hillhouse (1973) give a smaller spore size (7-9 × 6.8-8.8 μm) than we found in our collections.

Distribution and seasonal occurrence. *A. virosa* is common in eastern Canada, especially in the southern portion, but it also occurs northward into the boreal forest region, extending southward into the eastern United States. It occurs in the summer and fall and fruits into October when the weather remains warm. In some years it is common in wooded parks and lawns during August along the Lake Ontario shore.

Habitat and habit. The species is terrestrial, growing solitary to scattered or in groups in forested areas such as mixed woods and hardwood forests, and also in conifer woods. It is also found in lawns and parks where there are trees.

Technical description. Cap 30-140 mm broad, conic to convex, finally becoming plane when old; margin incurved, becoming decurved and faintly striate with age; surface glabrous, viscid to dry, not hygrophanous, white with pale yellow or creamy discolorations on the disc, sometimes dingy to yellowish when old, and staining yellow with potassium hydroxide. Flesh moderately thick on the disc, tapering toward the margin, white; taste not recorded; odor negligible to pungent and disagreeable.

Gills white to sordid or pale dingy yellow with age, free or reaching the stalk by a line, up to 10-20 mm broad, close to crowded, ventricose, narrowing toward the stalk; edges floccose.

Stalk 80-240 mm long, 10-23 mm thick at the apex, enlarged gradually toward the base; base ovoid to globose; surface glabrous to faintly floccose above the annulus and glabrous to fibrillose-scaly below the annulus, white to dingy, stuffed.

Annulus membranous, flaring to skirt-like, white becoming dingy to slightly yellowish with age, apical on the stalk. Volva cup-like, firmly appressed to the base of the stalk, free at the top, membranous, persistent, and white to yellowish where inserted in the soil.

Spores white in deposit, (7.7-)8.8-11.7 × (7.3-)8.4-10.2(-11.7) μm, globose to subglobose, hyaline, amyloid; apiculus blunt and 0.7-2.0 μm long. Basidia clavate, four-spored, 40.2-47.5 × 11.0-13.1 μm. Pleurocystidia lacking. Cheilocystidia indistinct, but gill edges lined by clavate to sac-like cells and interwoven hyphae from veil; cells 22-29 × 15-20 μm.

Gill trama divergent from a central strand, more or less tightly interwoven, moderately branched, cylindrical to inflated, 2.9-18.3 μm wide, hyaline. Subhymenium composed of hyaline short inflated compactly interwoven hyphae 9.4-10.5 μm wide. Cap cuticle thin, filamentous, gelatinous; hyphae interwoven, more or less radially arranged, cylindrical, 2.9-6.0 μm wide, moderately branched, hyaline. Cap trama loosely interwoven; hyphae cylindrical to inflated, 3.7-30.0 μm wide, hyaline. Caulocystidia clavate-cylindrical, terminal, 18.0-23.4 μm wide, parallel, overlapping to slightly deflected; internal hyphae parallel, branched, 4.4-9.0 μm wide. Annulus composed of interwoven branched cylindrical thin-walled hyphae 2.9-11 μm wide and inflated ellipsoidal to cylindrical hyaline cells occurring terminally or in short chains 25-36 × 11-20 μm long. Volva composed of interwoven cylindrical hyphae 3-12(-20) μm wide and some scattered inflated subglobose to cylindrical cells 90-120(-144) × 54-80 μm in size; cells often terminal, occurring singly or in short chains; hyphae and cells hyaline. Oleiferous hyphae refractive, irregular, 9-12 μm wide, scattered throughout the fruit body. Clamp connections absent.

Useful references. The following references are pertinent to *A. virosa* in North America.

- *Edible and Poisonous Mushrooms of Canada* (Groves 1979)—This book contains color and black-and-white photographs, a description, and a discussion.
- *Mushrooms of North America* (Miller 1972)—Presented are a color photograph and a description and discussion, combined with *A. bisporigera* and *A. verna*.

Conocybe Fayod — Bolbitiaceae

In general appearance, species of *Conocybe* are similar to several other genera of small brown mushrooms. Therefore, without the aid of a microscope and careful study, it is difficult to be sure of generic identification.

The distinguishing characteristics of *Conocybe* include usually small, slender fruit bodies; a conic to bell-shaped cap (convex in certain annulate species); a long, slender, and very fragile stalk; ascending-adnate gills, often colored rust brown from the spores; a persistent annulus present or absent; a rust brown to ocherous rust or ocher brown spore deposit; usually smooth spores, with an apical germ pore; a cellular cap cuticle as viewed in a section of the cap surface; and often the presence of characteristic flask-shaped or bowling-pin-shaped cheilocystidia.

The species are terrestrial, usually occurring in pastures, lawns, or other grassy areas, on compost piles, in terrariums and greenhouses, or on dung. They also occur in forested areas and may fruit on wood or in moss.

Species of *Galerina* might be confused with *Conocybe.* The easiest way to separate them is microscopically; *Galerina* typically has roughened spores that lack an apical germ pore, as well as a filamentous cap cuticle. In general, *Galerina* occurs with moss although a few species grow on wood and at least one species, *G. venenata,* a dangerously poisonous one, grows in lawns. Certain species of *Psilocybe,* in general appearance, look like *Conocybe. Psilocybe* species have darker spore deposits than *Conocybe,* namely some shade of chocolate to blackish brown, fuscous purple, or purple brown, and a filamentous cap cuticle.

Only one species of *Conocybe* occurring in Canada, *C. filaris,* is known to contain amatoxins. The description follows. See Chapter 12 for a discussion of the hallucinogenic species of this genus.

Conocybe filaris (Fr.) Kühner

Figs. 23, 38

Common names. None.

Distinguishing characteristics. Fruit body small; stalk 10–35 mm long. Cap more or less conic in early stages, becoming more convex to nearly plane, often with a broad umbo, and striate when moist; color reddish brown in the center and dark brown at the margin with reddish brown striations, brown to yellow brown when old. Stalk very pale yellowish or brownish at the apex, darker brown downward; base blackish brown; surface usually with whitish to grayish or pale brownish fibrils. Inner veil forming a conspicuous annulus near the apex of the stalk. Spores 7.3–8(–9.9) × 4.0–4.8(–5.4) μm. Cheilocystidia ventricose to fusiform-ventricose, with tapered apices.

Observations. *C. filaris* appears to be an uncommon species in most of North America but does occur frequently in the Pacific northwest. It has only occasionally appeared in the North American literature and is not described or illustrated in any of the recent popular mushroom books. The species has been placed in the genus *Pholiotina,* as *P. filaria* (Fr.) Singer, by some workers. The names *Pholiota rugosa* Peck and *Pholiotina rugosa* (Peck) Singer have also appeared in the North American literature for this species. Identification of this species is difficult; therefore, close attention to details of the species description is essential before a final identification is made. Compare other brown-spored genera, especially *Galerina,* to be sure of genus identification. The technical description given below is taken in part from the work of Kits van Waveren (1970).

Distribution and seasonal occurrence. The distribution of this species is poorly documented in North America. It is fairly common in the Pacific northwest. It fruits in the spring, summer, and fall.

Habitat and habit. This species grows in small or large groups, usually gregarious, rarely solitary. It is found in rich clay soil, often in gardens and parks, especially around greenhouses. It also occurs occasionally in moss, on mossy logs, along paths, and in sawdust and compost piles.

Technical description. Cap 5–25 mm broad, conic to obtusely conic in early stages, becoming more or less convex to nearly plane and then often with a distinct obtuse umbo; margin strongly striate when moist; surface smooth to finely wrinkled; color dark reddish brown in the center and along the furrows, dark brown between the furrows, and yellowish brown to pale brownish yellow near the margin; marginal veil elements absent. Flesh thick; odor and taste not distinctive.

Stalk 15–35 mm long, 0.75–2.0 mm wide, equal, sometimes slightly thickened at the base, firm, hollow, pruinose over the apex, coarsely fibrillose-striate to woolly below the annulus; color whitish to grayish over pale yellow to brown, increasingly brown toward the base, blackish brown at the base. Annulus conspicuous, median or lower on the stalk, ascending, horizontal, or descending, often detached and sliding down the stalk, whitish to pale brown, felt-like, coarsely striate-plicate.

Spores orange brown in deposit, 7.3–8.0(–9.9) × 4.0–4.8(–5.4) μm, ellipsoidal, smooth, thick-walled, with a distinct apical germ pore, rusty brown, nonamyloid. Basidia four-spored, short-clavate, 15–25 × 5.1–7.3 μm, hyaline. Pleurocystidia lacking. Cheilocystidia ventricose to fusiform-ventricose with tapered apices, 20–50 × 5.8–10.6 μm, hyaline, thin-walled.

Gill trama hyphae compactly interwoven to more or less parallel, cylindrical to inflated, 2.9–20.0 μm wide, hyaline to yellowish. Subhymenium composed of hyaline compactly interwoven cylindrical hyphae 2.9–5.1 μm wide. Cap cuticle composed of more or less globose-pedicellate closely packed cells forming a hymeniform layer 20–36 μm wide; cells 12–18 μm wide at the apex and 3.7–5.8 μm wide at the base, hyaline to brownish, with thin to thickened walls. Cap trama composed of yellowish loosely interwoven cylindrical to inflated hyphae 4.6–16.1 μm wide. Caulocystidia clavate to fusiform-ventricose, sometimes slightly capitate, 18–36 × 5.1–7.0 μm, hyaline. Stalk hyphae parallel to interwoven at the surface, 3.7–4.4(–11.0) μm wide, hyaline to brownish. Annulus hyphae interwoven, cylindrical, 3.7–8.8 μm wide, hyaline to brownish. Oleiferous hyphae not observed. Clamp connections present.

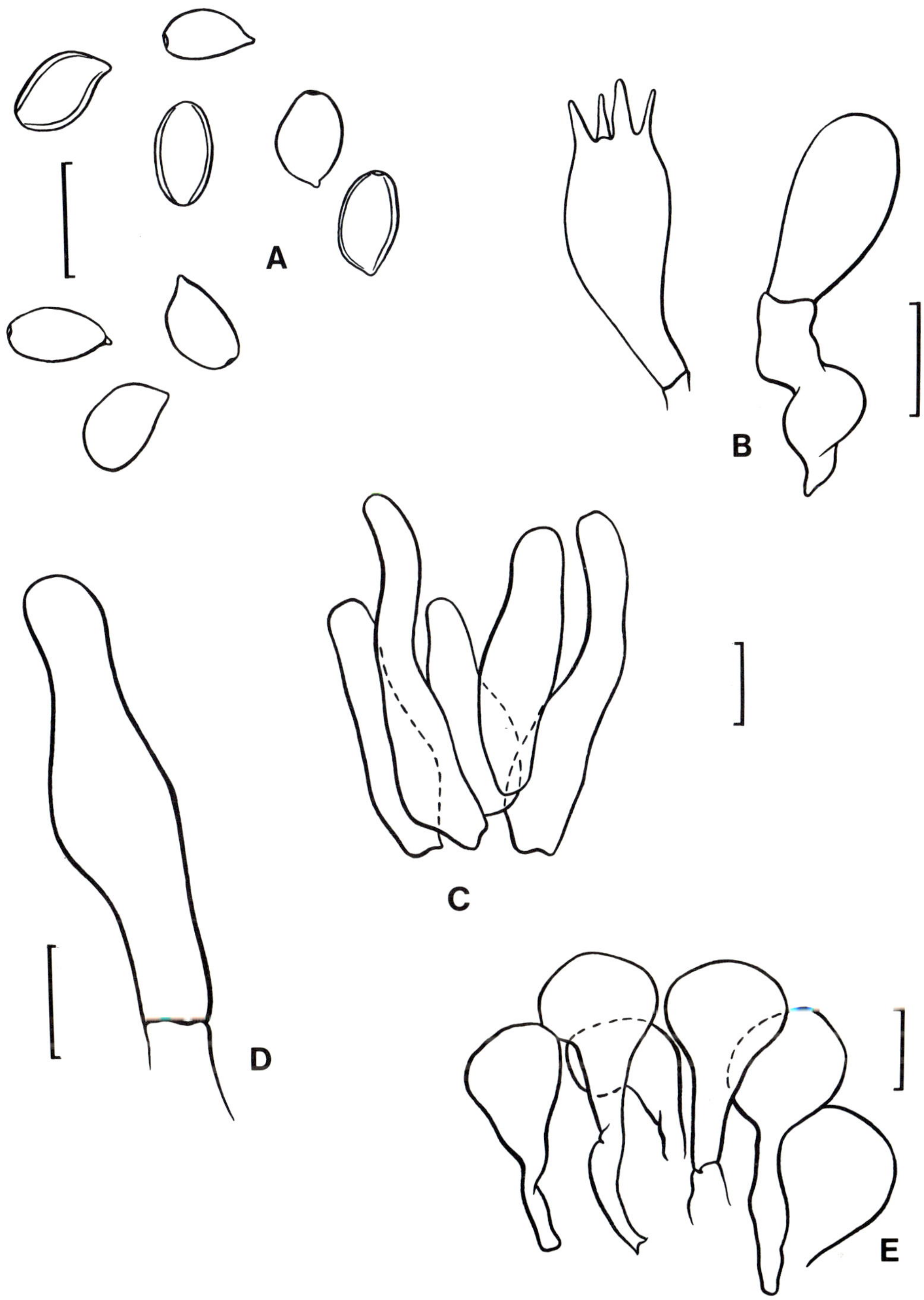

Fig. 23. *Conocybe filaris*: A, spores; B, basidium and basidiole; C, cheilocystidia; D, caulocystidium; E, cellular cuticle.

Scale line 10 μm

Useful references. There are few field guides or mushroom floras that describe and illustrate this species, but see Stamets (1978). Kits van Waveren (1970) and Smith (1934) provide further technical information.

Cortinarius Fr.

Cortinariaceae

Until the early 1950's no one suspected that the genus *Cortinarius* contained deadly species. *C. orellanus* was the first poisonous species to be elucidated. Grzymala (1965*a*, 1965*b*) reported numerous cases of poisoning, with several fatalities, in Poland. At least two other species, *C. speciosissimus* and *C. gentilis*, are reported as toxic. The toxin isolated from *C. orellanus* is called orellanin. The toxin in *C. gentilis* and *C. speciosissimus* may also be orellanin, but this has not been proven.

The interval between ingestion of *C. orellanus* and the onset of symptoms, the latent period, varies from three days to two weeks; the average delay is about eight days. Symptoms associated with poisoning by *C. orellanus* are varied. Often the initial symptoms are dryness and burning of the mouth; an intense thirst; stomach and intestinal disorders and pain, such as nausea, vomiting, abdominal pain, constipation, or even diarrhea; extreme heat or cold; headache; and often pain in the lumbar region. Eventually kidney and liver necrosis develops. Death commonly occurs in two or three weeks. The fatality rate is about 15%. For further information see works by Favre et al. (1976), Grzymala (1959, 1965*a*, 1965*b*), Hulmi et al. (1974), and Skirgiello et al. (1957).

Of the species involved, only *C. gentilis* is known to occur in North America. *C. gentilis, C. orellanus,* and *C. speciosissimus* are described below, even though the last two species have not yet been documented from North America, because it is likely all three occur on this continent. Moser (1969) presents a description and provides color plates of *C. speciosissimus.*

Cortinarius is the largest genus of mycorrhizal mushrooms in North America. The species are difficult to identify and the edibility of many species is unknown.

As with *Inocybe, Hebeloma,* and other brown-spored genera it is essential to learn to identify the genus. Although *Cortinarius* is a diverse genus and difficult to characterize, most people with a little practice can learn to identify it in the field. Some helpful guidelines follow.

The fruit bodies of *Cortinarius* species vary in size from very small to very large and robust; therefore, the size is not helpful in identifying the genus. The shape of the cap and stalk also varies. The cap surface may be smooth, silky, fibrillose, or scaly, and in several species it is glutinous to viscid. The presence or absence of a glutinous to viscid cap surface and features of the stalk are helpful in identifying some of the subgenera in *Cortinarius*.

The gills in *Cortinarius* are attached to the stalk and are often coated with spores when mature. The spores are usually rusty brown although some are cinnamon brown to duller brown. Often the spores can be found deposited on the stalk, on fibrils of the cortina, or on the cap of a second fruit body growing beneath a sporulating fruit body. The color of the young gills varies depending on the species and is an important character for species identification. The color may be pallid to some shade of brown, but frequently the gills are highly colored, such as yellow, green, orange, red, violet, or purple. The cap also varies in color and may be brightly colored. In some species the cap surface stains whereas in several others the color fades as moisture is lost from the cap surface, a condition described as hygrophanous.

In young specimens a well-developed cortina, formed from a fibrillose partial veil, is present. This veil usually becomes lost with age, but often a dense coating of fibrils can be found on the cap margin as a zone near the stalk apex or as longitudinal fibrils on the stalk surface.

The stalk may be equal, tapered, ventricose, clavate, or bulbous. Usually the stalk apex is not pruinose or scaly as it frequently is in *Hebeloma* and *Inocybe.* The stalk surface of some species of *Cortinarius* is coated with a glutinous to viscid layer.

The spores in *Cortinarius* are typically ornamented with small warts, wrinkles, or ridges. As in *Inocybe* and *Hebeloma* the spores lack an apical germ pore. Pleurocystidia are rarely found in the genus; however, they occur in *C. violaceus* (L. ex Fr.) S. F. Gray, a species central to the genus. Cheilocystidia are uncommon and when present are often not well differentiated. The cap cuticle is typically filamentous.

Three species of *Cortinarius* are described below. Most people find them difficult to identify and must consult a specialist in mushroom taxonomy before a final determination is made.

Cortinarius gentilis Fr. Figs. 24, 39

Common names. None.

Distinguishing characteristics. Cap conic to bell-shaped or more expanded and usually with a prominent umbo; color when moist brownish orange to yellowish brown or cinnamon brown, fading to paler brown and changing markedly when dry to yellow or somewhat golden yellow. Gills subdistant to distant, cinnamon to cinnamon brown or orangish brown. Stalk long, equal, or tapered downward, dark brown with one or more shiny yellow fibrillose zones about midstalk. Partial veil yellow. Spores broadly ellipsoidal to ovoid, warted, 8-9(-10) × 5.8-6.6 μm in size. Pleurocystidia or cheilocystidia absent, or cheilocystidia present as narrow cylindrical-clavate end cells.

Observations. The outstanding features of *C. gentilis* are the hygrophanous cap that changes color markedly on drying from a dark brown to strongly yellow, the distant gills, and the dark brown stalk with zones of yellow fibrils. The color change of the cap is so dramatic that faded specimens often look like a totally different species. The warty, broadly ellipsoidal to ovoid spores are characteristic of *C. gentilis* and its relatives.

Distribution and seasonal occurrence. *C. gentilis* has not commonly been reported in the North American literature. Based on herbarium specimens it appears to be widely distributed and most common in the western mountains. It usually fruits in the late summer and fall, depending on weather and location. At times it is locally abundant.

Habitat and habit. This species is terrestrial or humicolous, or it sometimes occurs on rotten conifer wood. It grows singly, scattered, gregarious, or caespitose and typically occurs in conifer forests.

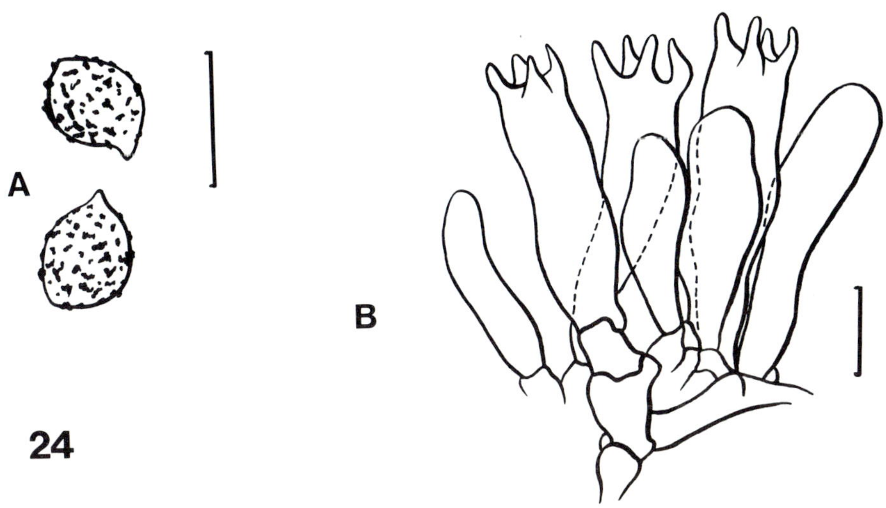

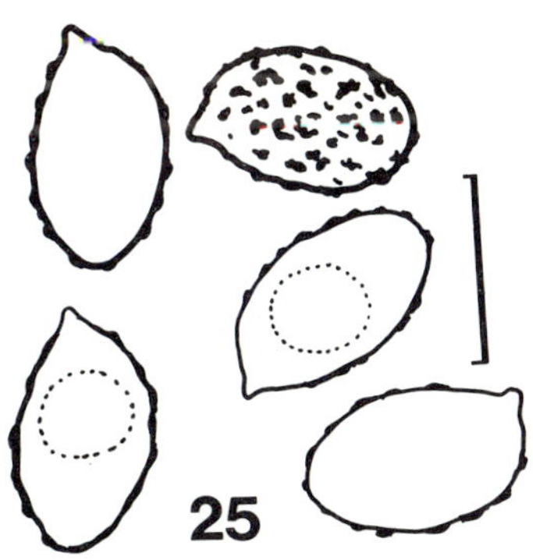

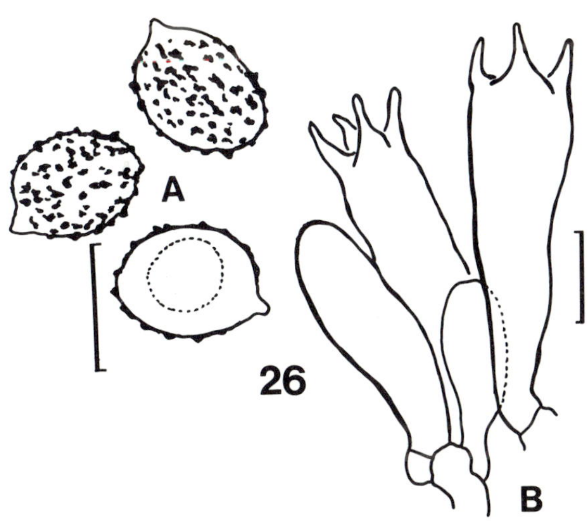

Fig. 24. *Cortinarius gentilis*: A, spores; B, basidia.
Fig. 25. *Cortinarius orellanus*: A, spores.

Fig. 26. *Cortinarius speciosissimus*: A, spores; B, basidia.
Scale line 10 μm

Technical description. Cap 30-50 mm broad, more or less conic in early stages, becoming bell-shaped to round-convex or more expanded to elevated with age, usually abruptly umbonate, sometimes obtuse with a small acute umbo or prominently obtuse; margin incurved in early stages and silky-cortinate or spotted by yellowish veil remnants; surface entirely glabrous when moist, silky-shiny when dry, hygrophanous, Sudan brown to cinnamon rufous when moist, changing markedly when dry and becoming wax yellow to amber yellow or paler. Flesh thick on the disc, narrow toward the margin, the same color as the cap, and hygrophanous; taste mild; odor negligible to slightly radish-like.

Gills cinnamon when young, becoming a darker brown with age, broadly adnate, 6-9 mm broad, subdistant in early stages, becoming finally very distant, thick, sometimes interveined; edges finely floccose.

Stalk 50-100 mm long, 3-7 mm thick, more or less straight, equal or tapering, stuffed then hollow, cartilaginous, naked at the apex, fibrillose-scaly elsewhere, having amber yellow bands or zones in the middle part, antique brown when moist; flesh uniformly colored. Cortina yellowish to yellow ocher.

Spores yellowish brown in deposit, 8.0-9.0(-10.0) × 5.8-6.6 μm, broadly ellipsoidal to ovoid, with the sides slightly unequal in profile, verruculose, with an indistinct plage, yellow brown, nonamyloid. Basidia four-spored, clavate, 29.2-40.0 × 8.8-11.0 μm, hyaline to yellowish brown, often more or less collapsed, thin-walled. Pleurocystidia and cheilocystidia absent, or gill edge with scattered to clustered, cylindrical clavate hyphal ends 3.7-4.4 μm wide, hyaline.

Gill trama compactly interwoven to parallel; hyphae more or less cylindrical, 4.6-8.2(-13.1) μm wide; walls thin to slightly thickened, encrusted, hyaline to yellowish brown. Subhymenium interwoven; hyphae more or less cylindrical, 4.4-5.1 μm wide, hyaline to yellowish. Cap cuticle filamentous; hyphae more or less radially arranged and parallel, cylindrical, 4.0-6.6(-16.2) μm wide, yellowish brown, thin-walled to slightly thick-walled, and spirally encrusted. Cap trama loosely interwoven; hyphae cylindrical to inflated, 15-25 μm wide, yellowish, thin-walled, finely encrusted. Caulocystidia lacking. Stalk hyphae parallel, cylindrical, 3.7-9.5 μm wide, thin-walled to slightly thickened, sometimes encrusted, hyaline to ocherous or yellow brown. Oleiferous hyphae 4.4-9.5 μm wide, refractive, greenish yellow, contorted, scattered in the cap trama and stalk. Clamp connections present.

Useful references. *C. gentilis* is not described or illustrated in any of the common North American mushroom books or field guides. It is, however, described by Kauffman (1932).

Cortinarius orellanus Fr.

Fig. 25

Common names. None.

Distinguishing characteristics. Cap convex and often broadly umbonate; surface finely tomentose to tomentose-scaly usually overall; color uniformly tawny ocher, orange brown, or reddish brown and fading to yellowish brown or tawny umber. Gills distant to subdistant, ocher yellow becoming rusty orange brown to bright cinnamon brown. Stalk long, equal to tapered downward, yellow to tawny yellow with lemon yellow at the apex, becoming more brownish with age, lacking veil fibrils on the surface. Spores ellipsoidal to almond-shaped, finely to strongly warty, 9.5-12.4 × 5.1-6.7(-7.3) μm in size. Cystidia lacking.

Observations. *C. orellanus* is distinguished by the orange brown to reddish brown cap, the distant gills, and the yellow to tawny yellow or golden yellow stalk. In general appearance it is similar to certain species of *Cortinarius* in the subgenus *Dermocybe* but it is not related to this group. There is some evidence that a related species, *C. fluorescens* Hook., contains orellanin (D. M. Simons, Personal commun.). It, too, is not known from North America.

Distribution and seasonal occurrence. *C. orellanus* occurs in Europe. Although it has not yet been documented from North America, it most likely occurs here and should fruit in the late summer and fall.

Habitat and habit. The species grows singly to caespitose and is found in mixed forests, often near pine stumps.

Technical description. Cap 30-85 mm broad, convex-expanded, often broadly umbonate; margin somewhat decurved; edges more or less recurved and sometimes tearing in old specimens; surface finely tomentose to appressed tomentose-scaly usually over the entire cap, glabrous in older specimens; color uniformly tawny ocher, orange brown, foxy orange, or reddish brown fading to yellowish brown or tawny umber, with the margin sometimes paler, Flesh thick on the disc, yellowish or tawny yellow, sometimes reddish brown; taste and odor strongly to faintly radish-like.

Gills ocher yellow, rusty orange brown, or bright cinnamon brown to tawny orange, paler when young, adnate, or shallowly emarginate, thick, fairly broad, more or less ventricose, distant to subdistant, sometimes interveined; edges even to uneven.

Stalk 30-90 mm long, 4-12 mm thick at the apex, equal or tapered downward, sometimes ventricose, firm; surface fibrillose-striate, without discernible veil remnants. Color yellow to tawny yellow; apex light lemon yellow, becoming deeper yellowish brown to dull light brown; base often ocher and paler; cortina pale yellowish to ocher.

Spores rusty yellow brown, 9.5-12.4 × 5.1-6.6(-7.3) µm, ellipsoidal to almond-shaped in side view, ellipsoidal to ovoid or subfusiform in face view; surface finely to strongly verrucose, rusty cinnamon brown. Basidia four-spored, clavate, 28-40 × 7.3-11.0 µm, hyaline to yellowish brown, often more or less collapsed and darker yellow brown. Pleurocystidia and cheilocystidia lacking.

Gill trama compactly interwoven; hyphae more or less parallel, 5.1-30.7 µm wide, cylindrical to inflated, thin- to slightly thick-walled and more or less encrusted, hyaline to yellowish. Subhymenium compactly interwoven, more or less collapsed; hyphae more or less cylindrical, 3.5-6.6 µm wide, hyaline to yellowish. Cap cuticle filamentous; hyphae interwoven, radially arranged and parallel, cylindrical, 7.0-16.7 µm wide, thin- to slightly thick-walled, encrusted, yellowish to ocherous; end cells upturned, scattered, cylindrical and 4-4.6 µm wide or subclavate and 8-10 µm wide, hyaline to ocherous. Cap trama loosely interwoven; hyphae cylindrical to more or less inflated, 11.7-23.4 µm wide, hyaline to yellowish. Caulocystidia lacking. Stalk hyphae parallel, more or less cylindrical, 3.5-11.7(-15.5) µm wide, thin-walled to slightly thickened, finely to coarsely encrusted, hyaline to yellowish. Oleiferous hyphae refractive, ocherous yellow, 3.5-4.6 µm wide, scattered in the stalk trama. Clamp connections present throughout the fruit body.

Useful references. This species is not covered in any of the common North American mushroom books or field guides.

Cortinarius speciosissimus Kühner & Romagnesi

Fig. 26

Common names. None.

Distinguishing characteristics. Cap conic-convex to convex and umbonate; surface more or less smooth in the center and elsewhere minutely fibrillose-scaly to almost tomentose; color tawny reddish or tawny date brown, with a pale brown, pale ocherous, or tawny buff margin, coated with yellowish veil fragments in early stages. Taste and odor strongly radish-like. Gills pale ocher and eventually a deep tawny reddish brown, subdistant to fairly crowded. Stalk long, equal to slightly thickened at the base or at the apex; surface silky-fibrillose, pale ocherous yellow to deeper ocher or tawny and becoming rusty especially near the midstalk, with yellow ring-like zones and patches on the middle and lower stalk. Spores broadly ellipsoidal to subglobose, fairly warty and punctate, 8.8-11.0(-12.0) × 6.6-8.0(-8.5) µm in size. Pleurocystidia lacking; cheilocystidia lacking or present as poorly differentiated clavate to cylindrical end cells.

Observations. *C. speciosissimus* has not been documented in North America and the above description is from European material. It is similar in appearance to *C. gentilis,* which is fairly common in the conifer forests of North America.

Distribution and seasonal occurrence. This species has not been reported from North America but undoubtedly occurs here. It should fruit during the late summer or fall, depending on location and weather.

Habitat and habit. *C. speciosissimus* is terrestrial. In Europe it occurs in conifer woods where *Vaccinium* is present and usually occurs under pine.

Technical description. Cap 25-80 mm broad, convex or conic-convex then expanding, acutely or obtusely umbonate; margin often slightly wavy, lobed and upturned when old, sometimes becoming slightly torn or incised, covered with veil fragments in early stages; surface soon more or less smooth on the disc, elsewhere minutely appressed fibrillose-scaly or almost tomentose, especially near the margin; color tawny reddish or tawny date brown, with paler brown, pale ocherous, or tawny buff margin. Flesh yellowish ocherous; taste and odor faintly to strongly radish-like.

Gills pale ocher, soon becoming bright ocher, gradually deepening to tawny ocher or deep tawny reddish brown, adnate, subdistant or fairly crowded, rather thick, veined on the sides and at the base; edges more or less even, then slightly and finely floccose.

Stalk 50-110 mm long, 5-14 mm thick, more or less equal or slightly thickened at the base or at the apex, stuffed then hollow; surface silky-fibrillose, striate, pale ocherous yellow soon tinged deeper ocher or tawny, and rusty, especially in the middle; base clavate-bulbous, 7-20 mm wide, white-tomentose; flesh yellowish to tawny rusty near the base. Cortina yellowish. Veil fragments forming a ring-like zone or patches on the lower part of the stalk.

Spores yellow brown in deposit, 8.8-11.0(-12.0) × 6.6-8.0(-8.5) µm, broadly ellipsoidal to subglobose, with the sides slightly unequal in side view, elliptic-ovate in face view, punctate, yellow brown, nonamyloid. Basidia four-spored, clavate, 36.5-44.0(-50.0) × 9.5-12.4(-15.0) µm, hyaline to yellowish or ocherous brown, often more or less collapsed. Pleurocystidia lacking. Cheilocystidia lacking or occurring as clusters of cylindrical to subclavate end cells 5-10 µm wide.

Gill trama interwoven; hyphae more or less parallel, cylindrical to inflated, 4.4-29.2 µm wide, hyaline to yellowish; walls of hyphae thin to slightly thickened, some encrusted. Subhymenium interwoven; hyphae cylindrical, 3.7-5.1 µm wide, hyaline to yellowish;

walls of hyphae thin to slightly thickened. Cap cuticle filamentous; hyphae more or less radially arranged and parallel, cylindrical and 4.6–12.9 μm wide in the upper portion, more or less inflated and 16–35.1 μm wide in the lower layer; hyphal walls thin to thickened and often encrusted, hyaline to brownish yellow. Cap trama loosely interwoven; hyphae cylindrical to inflated, 5.8–29.2 μm wide, hyaline. Caulocystidia lacking. Stalk hyphae parallel, more or less cylindrical, 4.6–11.7 μm wide, hyaline to yellowish; walls of hyphae slightly thickened and encrusted. Oleiferous hyphae greenish yellow, refractive, 3.5–7.0 μm wide, scattered in the cap trama and cuticle and the stalk. Clamp connections present.

Useful references. This species is not reported in the North American literature.

Galerina Earle — Cortinariaceae

Galerina is another genus of minute to small or, less commonly medium-sized, brown-spored mushrooms that should be totally avoided as food. Most *Galerina* species occur in moss beds or on mossy logs and are very small to minute mushrooms. Little is known about the edibility of these species and they are not generally collected for food. The larger *Galerina* species, which in comparison with other mushroom genera would be considered medium to fairly small in size, occur in lawns, perhaps growing from buried wood, on wood, or on humus. Some of these larger *Galerina* species are known to contain amatoxins. The species involved include *G. autumnalis, G. marginata,* and *G. venenata.*

General characteristics of the genus *Galerina* include comparatively small to medium-sized fruit bodies; a typically thin, glabrous cap, often hygrophanous, losing moisture readily and fading, usually yellow red to orange brown when fresh, often with radial striations on the margins; attached gills, pallid honey color in early stages, usually becoming some shade of rusty brown from spores; a slender, equal stalk, at times with a fibrillose annular zone, usually somewhat fragile and often more or less similar in color to the cap surface; ocherous yellow, rusty brown, or cinnamon brown spores in deposit, typically with a warted to wrinkled surface, some with a plage, less commonly smooth, rarely with an apical germ pore; and typically the presence of cheilocystidia, as well as pleurocystidia in several species.

Among small, brown-spored mushrooms such as *Galerina,* confusion with similar-appearing genera, such as *Conocybe* and some of the smaller *Pholiota* and *Cortinarius* species, is always a possibility. Species of *Conocybe* are so similar in appearance to *Galerina* species that the best way to separate them is by microscopic characteristics. *Conocybe* species typically have smooth spores with an apical germ pore, but most importantly, they have a cellular cap cuticle, made up of more or less isodiametric to inflated cells. *Galerina* species usually have roughened spores that lack an apical germ pore and a cap surface composed of filamentous hyphae, long in relation to their width. The smaller *Pholiota* species that occur on wood are similar in size and stature to the wood-inhabiting *Galerina* species. Again, microscopic differences are the best means for distinguishing the two. In *Pholiota* the spores are smooth and, in this group of species, they have a distinct apical germ pore (use 100× lens). Small *Cortinarius* species can usually be separated from *Galerina* by the absence of cheilocystidia. Keep in mind that the identification of small brown-spored genera is difficult, requiring some expertise. When a poisoning results from the ingestion of a brown-spored mushroom, have the identification verified by an expert. Several brown-spored species are extremely dangerous.

In North America there are three *Galerina* species that contain amatoxins: *G. autumnalis, G. marginata,* and *G. venenata.* These three are described here. A fourth species, *G. sulciceps* (Berk.) Boedijn (also called *Phaeomarasmius sulciceps* (Berk.) Scherffel and *Marasmius sulciceps* Berk.) is a tropical species, occurring in Java, Sumatra, and Ceylon. It grows on wood, and it is extremely toxic, causing death within 7–24 hours in some victims. The toxin or toxins until recently were unknown. Besl (1981), however, reported that *G. sulciceps* contains amatoxins. Smith and Singer (1964) give a description of this species.

Galerina autumnalis (Peck) Smith & Singer — Figs. 27, 40

Common name. Autumnal galerina.

Distinguishing characteristics. Cap 25–65 mm broad, usually more or less convex, slightly viscid, hygrophanous, yellow brown fading to buff when dry; margin with radial striations. Gills attached to the stalk, pale brown or pale tawny becoming darker brown to more ferruginous with age, and close, Stalk 3–8 mm thick, more or less equal, pale to light brown becoming darker brown from the base upward as the specimens age, usually with a thin narrow whitish zone near the stalk apex. Spores with a wrinkled surface, 8.8–11.0 × 5.1–6.6 μm in size. Conspicuous fusiform-ventricose pleurocystidia and cheilocystidia present. Species usually occurring in clusters, typically on wood.

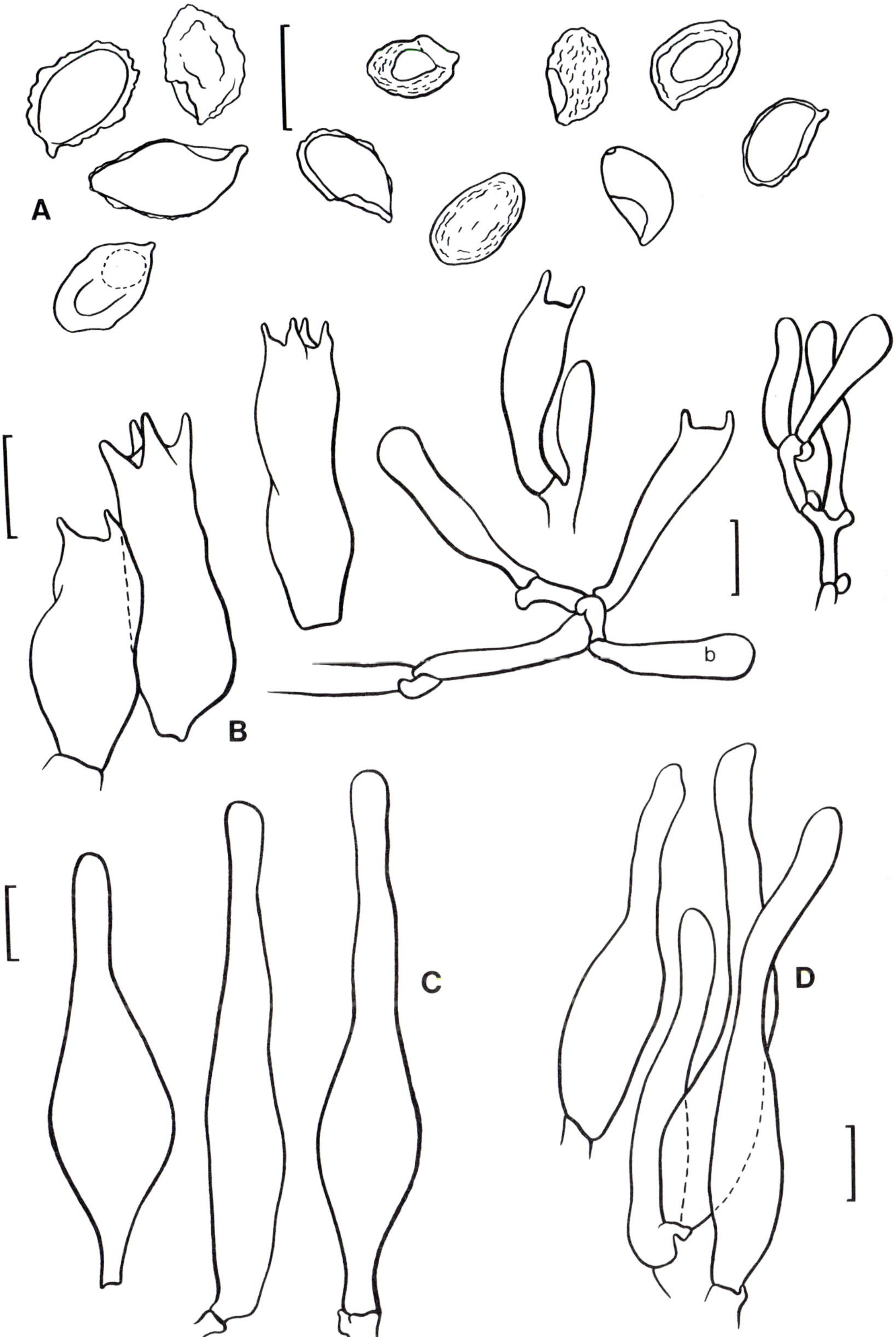

Fig. 27. *Galerina autumnalis*: A, spores with a plage; B, basidia and basidioles; C, cheilocystidia and pleurocystidia; D, caulocystidia.

Scale line 10 μm

Observations. *G. autumnalis* is fairly easy to recognize but because it grows on wood it is occasionally confused with species of *Pholiota* and in some instances even with the white-spored *Armillariella mellea*. This confusion is no doubt responsible for the majority of poisonings by this species and related *Galerina* species, because they are not intentionally collected for food. *Galerina autumnalis* has been called *Pholiota autumnalis* Peck in some books (Groves 1962), adding to the problem. Spore characteristics are the most reliable means for separating wood-inhabiting species of *Galerina* and *Pholiota*. The spore surface is wrinkled and there is no apical germ pore in *Galerina*; the spore surface is smooth and an apical germ pore is present in *Pholiota*.

Distribution and seasonal occurrence. *G. autumnalis* is widely distributed in forested areas. It fruits from spring into fall, but it is usually uncommon in hot, dry seasons.

Habitat and habit. This species is usually caespitose, although at times it grows scattered on decaying wood, usually on hardwood logs or stumps but occasionally on conifer wood. It is also occasionally terrestrial, growing from buried wood or woody debris.

Technical description. Cap 25–64 mm broad, convex expanding to plane, sometimes with a slightly obtuse umbo; margin translucent-striate when moist, opaque when dry; surface viscid when fresh, hygrophanous, strong yellowish brown to brown or brownish orange when moist, fading to light orange yellow or pale yellow with the disc sometimes remaining dark. Flesh relatively thick in the disc, tapering to the margin, watery brown fading to pale yellow or buff; taste mild or like fresh meal; odor slightly resembling fresh meal when flesh is crushed.

Gills colored similarly to the cap in early stages, darker and more ferruginous with age, bluntly adnate or with a slight decurrent tooth, soon seceding, moderately close, narrow to moderately broad (±5 mm), thin; edges more or less entire.

Stalk 30–90 mm long, 3–8 mm thick at the apex, equal to more or less clavate, straight to curved, central, hollow, moist to dry, somewhat striate, glabrous to pruinose above the annulus; lower stalk covered with a thin more or less evanescent layer of grayish appressed fibrils, pale brown in background or colored similarly to the cap, becoming darker brown; base usually with some white mycelia. Flesh cartilaginous, pale brown to darker brown. Annulus membranous or having a fibrillose zone, thin, often evanescent, apical on the stalk.

Spores dull brownish orange in deposit, 8.8–11.0 × 5.1–6.6 μm, ellipsoidal to broadly ellipsoidal, with the sides slightly unequal in side view, ovate in face view; surface wrinkled-verruculose, with a distinct plage, thick-walled, yellowish brown to brownish orange, nonamyloid. Basidia four-spored, short-clavate, 22–30 × 4.8–6.6 μm, hyaline. Pleurocystidia fusiform-ventricose with long tapered apices; these apices sometimes slightly enlarged or obtuse, 40–80 × 8.8–11.7 μm; necks 3.7–5.1 μm wide, hyaline, thin-walled. Cheilocystidia similar to pleurocystidia.

Gill trama interwoven; hyphae more or less parallel, cylindrical to fusiform, 3.5–10.5 μm wide, hyaline to yellowish. Subhymenium interwoven; hyphae cylindrical, 3.0–4.5(–5.1) μm wide, hyaline. Cap cuticle filamentous, gelatinous; hyphae compactly interwoven, more or less radially arranged, cylindrical, 2.3–5.1 μm wide, hyaline to yellowish, at times slightly encrusted. Cap trama loosely interwoven; hyphae 4.5–10.2 μm wide, hyaline to yellowish. Caulocystidia at the apex of the stalk, fusiform-ventricose with long tapered apices, 51–73 × 10.2–13.1 μm, hyaline. Stalk hyphae parallel to interwoven, cylindrical, 3.0–4.5 μm wide, hyaline to yellowish. Annulus composed of interwoven cylindrical hyphae 2.3–4.5 μm wide; cells dissociated. Oleiferous hyphae ocherous, refractive, 3.7–5.1 μm wide, scattered in the stalk. Clamp connections present.

Useful references. The following books contain information pertinent to *G. autumnalis* in North America.

- *A Field Guide to Western Mushrooms* (Smith 1975)—The discussion appears under *Pholiota mutabilis* (Fr.) Kummer.
- *Edible and Poisonous Mushrooms of Canada* (Groves 1979)—The discussion appears under *Pholiota marginata,* as both are listed as species of *Pholiota.*
- *Mushrooms of North America* (Miller 1972)—A color photograph and a description and discussion are presented.
- *The Mushroom Hunter's Field Guide* (Smith 1974)—A color photograph and a description and discussion are presented.
- *The Savory Wild Mushroom* (McKenny 1971)—A color photograph and a description and discussion are presented.

Galerina marginata (Fr.) Kühner

Fig. 28

Common names. None.

Distinguishing characteristics. Similar in appearance to *G. autumnalis.* Cap 15–40 mm broad, in general convex; surface moist but not viscid; margin striate, yellow brown fading to tan. Gills crowded, pale brown to darker brown or brownish orange. Stalk 2–9 mm thick, equal or slightly enlarged below, about the color of the cap but paler and with the lower portion darkening and becoming browner with age. A more or less thin annulus or annular zone present, near the stalk

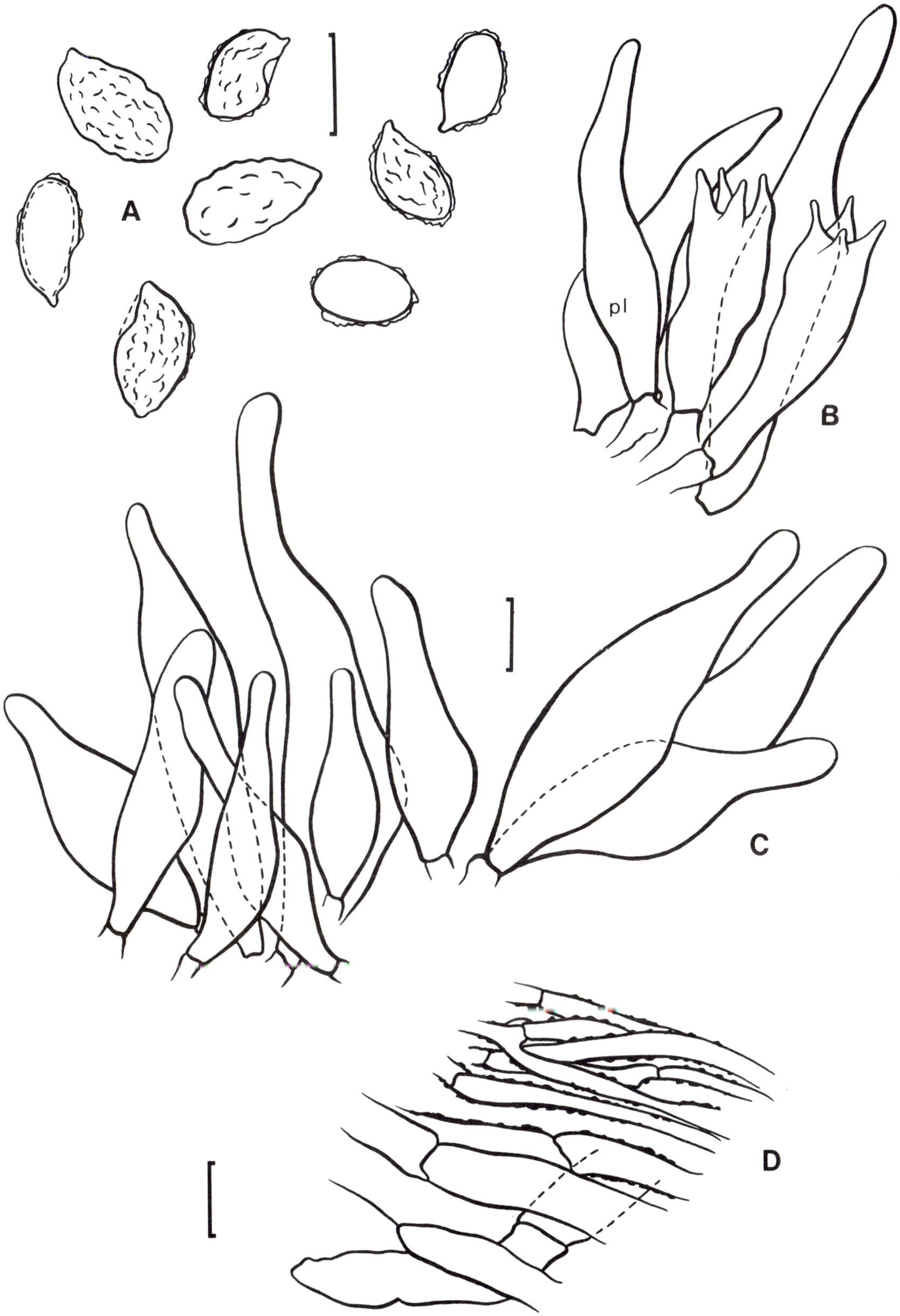

Fig. 28. *Galerina marginata*: A, spores; B, basidia and pleurocystidia (*pl*); C, cheilocystidia; D, filamentous cuticle with narrow hyphae at the cap surface.

Scale line 10 μm

apex. Spore surface wrinkled; spores 8.8-11.0 × 4.4-5.8 μm in size. Conspicuous ventricose to fusiform-ventricose pleurocystidia and cheilocystidia present.

Observations. In our experience, this species is difficult to separate from *G. autumnalis*. The distinct gelatinous layer on the cap of *G. autumnalis* is the best distinguishing characteristic, although at times this feature is difficult to demonstrate, especially after specimens are dried.

Distribution and seasonal occurrence. The species is widely distributed in forested areas and usually occurs in the summer and fall. It is commonly reported in Europe.

Habitat and habit. *G. marginata* is gregarious to caespitose, growing typically on or near the wood of conifers.

Technical description. Cap 17-40 mm broad, obtuse to convex, expanding to broadly convex, plane, or slightly umbonate; margin incurved in early stages, translucent-striate, moist but not viscid; surface glabrous, hygrophanous, pale to strongly yellowish brown fading to tan. Flesh thin, pliant, pale brownish yellow to nearly white; taste and odor slightly resembling fresh meal.

Gills pallid brown to brownish orange at maturity, broadly adnate to subdecurrent, crowded, narrow; edges even and whitish.

Stalk (20-)30-60 mm long, 2-9 mm thick at the apex, equal to slightly enlarged downward, straight to curved, central, becoming hollow, moist to dry, pruinose above the annulus; lower stalk covered with pallid fibrils that are more or less evanescent, paler than the cap, browner downward with a reddish brown to dark yellow brown base. Flesh more or less the same color as the surface, becoming reddish brown. Annulus more or less membranous to fibrillose, relatively thin, often evanescent, and located on the upper half of the stalk.

Spores brownish orange in deposit, 8.8-11.0 × 4.4-5.8 μm, elliptic with the sides slightly unequal in side view, ovate in face view, warty to finely wrinkled on the surface, with a distinct plage; wall thickened, brownish orange to rusty brown, nonamyloid. Basidia four-spored, clavate, 21.9-27.4 × 6.6-8.8 μm, hyaline. Pleurocystidia and cheilocystidia similar, ventricose to fusiform-ventricose; apices sometimes tapered, 26-58 μm long, 7.3-12.4(-16.0) μm wide at the base, 3.7-5.1 μm wide at the apex, hyaline, thin-walled.

Gill trama composed of interwoven to more or less parallel hyphae; these hyphae cylindrical to somewhat inflated, 3.7-8.3(-20.1) μm wide, hyaline to yellowish. Subhymenium composed of compactly interwoven cylindrical hyphae; these hyphae 2.9-3.7(-5.1) μm wide, yellowish brown. Cap cuticle filamentous; hyphae interwoven, more or less radially arranged, cylindrical, 3.5-5.1(-6.6) μm wide, hyaline to yellowish, more or less gelatinous in potassium hydroxide, some encrusted. Cap trama composed of loosely interwoven cylindrical to more or less inflated hyphae; these hyphae 3.7-7.3(-11.0) μm wide, yellowish, finely encrusted near the cap surface. Caulocystidia fusiform-ventricose with tapered apices, 37.0-51.0 × 10.6-11.0 μm, hyaline, thin-walled, clustered at the apex of the stalk. Stalk hyphae parallel, 3.5-6.6(-11.0) μm wide, narrow at the surface, hyaline to yellowish. Annulus composed of interwoven cylindrical hyphae 2.2-5.1 μm wide; these hyphae hyaline to yellowish. Oleiferous hyphae refractive, yellowish to ocherous, 3.5-5.1 μm wide, scattered in the cap trama and stalk. Clamp connections present.

Useful references. Some color photographs, although generally quite poor, are presented in *Edible and Poisonous Mushrooms of Canada* (Groves 1979), along with a description and discussion. The species is reported as *Pholiota marginata* (Batsch ex Fr.) Quél.

Galerina venenata Smith — Fig. 29

Common names. None.

Distinguishing characteristics. Cap 10-35 mm broad, more or less convex, glabrous, moist and not viscid, hygrophanous, cinnamon brown to pale bay brown or reddish cinnamon, fading to yellowish white or buff. Gills golden tawny becoming duller brown with age, subdistant. Stalk 3-5 mm thick, enlarged somewhat toward the base, brownish, similar in color to that of the cap. A band-like ring or zone present above midstalk. Taste of raw flesh typically disagreeable, leaving a burning sensation in the throat. Fungi apparently terrestrial, reported from lawns. Spore surface wrinkled; spores 9.1-11.0 × 5.1-6.6 μm in size. Fusiform-ventricose pleurocystidia and cheilocystidia present.

Observations. In general *G. venenata* resembles *G. autumnalis.* It is reported as terrestrial but may fruit from buried wood as related species are wood inhabiting. It is dangerous and has caused several cases of poisoning in the northwest United States, including one fatality in Washington State in December 1981. It can be confused with several of the hallucinogenic lawn-inhabiting mushrooms, for example species of *Psilocybe* or *Conocybe*. See sections on *Psilocybe* (Ch. 12) and *Conocybe* (Chs. 7 and 12) for characteristics of these genera. The technical description given below is adapted from the work of Smith and Singer (1964).

Distribution and seasonal occurrence. *G. venenata* has been reported from Oregon and Washington and can be

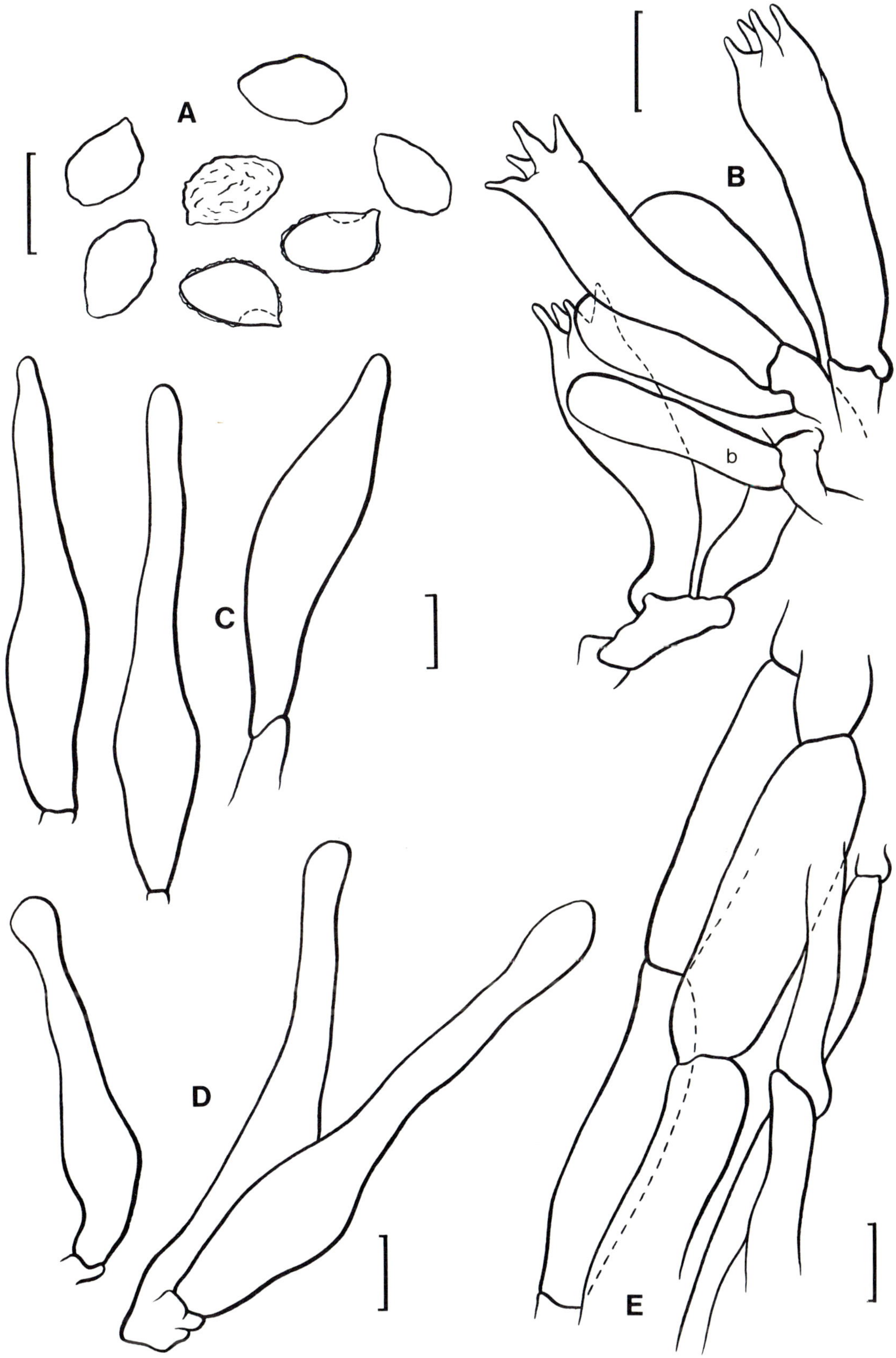

Fig. 29. *Galerina venenata*: A, spores; B, basidia and basidioles (*b*); C, pleurocystidia and cheilocystidia; D, caulocystidia; E, cap cuticle with narrower hyphae near the cap surface.

Scale line 10 μm

expected to occur along the west coast of Canada. It fruits in the fall and winter.

Habitat and habit. The species is gregarious in lawns. It is thought to grow from buried wood but this has not been verified.

Technical description. Cap 10–35 mm broad, broadly convex, expanding to plane or with a slightly uplifted arched margin, depressed on disc; margin crenate to lacerated; surface glabrous and moist, hygrophanous, near cinnamon brown to pale bay brown or reddish cinnamon, fading to very dingy yellowish white or pinkish buff. Flesh moderately thick, yellowish white; odor resembling fresh meal; taste also resembling fresh meal but slowly becoming bitter and disagreeable, leaving a burning sensation in the throat.

Gills golden tawny becoming dull cinnamon, broadly adnate, broad, subdistant; edges even.

Stalk 30–40 mm long, 3–5 mm thick at the apex, enlarged somewhat toward the base, terete or compressed, pruinose at the apex, glabrous below the annulus, brownish in color becoming white at the base, often with a growth of white mycelia at the base. Annulus very thin, membranous, quickly collapsing, evanescent, apical in position.

Spores (8–)9.1–11.0 × 5.1–6.6 μm, elliptic with the sides slightly unequal in side view, elliptic in face view, finely wrinkled and warty with a distinct plage, russet to ocherous yellow. Basidia four-spored, clavate, 20.1–29.2 × 7.3–9.1 μm, hyaline, thin-walled. Pleurocystidia and cheilocystidia fusiform-ventricose with long tapered apices, 41.1–70.2 × 9.4–14.3 μm, hyaline, thin-walled.

Gill trama composed of parallel to interwoven hyphae; these hyphae cylindrical to inflated, 5.1–14.6(–21.9) μm wide, yellow brown. Subhymenium composed of hyaline cylindrical interwoven hyphae 3.7–5.1 μm wide. Cap cuticle filamentous; hyphae interwoven, more or less radially arranged, cylindrical, 3.7–8.0 μm wide, hyaline to yellowish, at times slightly to conspicuously encrusted. Cap trama composed of loosely interwoven hyphae; these hyphae cylindrical to inflated, 3.7–18.3 μm wide, hyaline to yellowish, some finely encrusted near the cap cuticle. Caulocystidia fusiform-ventricose with long tapered apices, 51–73 × 10.2–11.0 μm, hyaline, thin-walled, clustered at the apex of the stalk. Stalk hyphae parallel, 3.7–21.9(–25) μm wide, hyaline to yellowish. Annulus composed of hyaline to yellowish interwoven cylindrical hyphae 3.7–5.1 μm wide. Oleiferous hyphae refractive, ocherous to gold, 3.0–7.0 μm wide, scattered in the trama and stalk. Clamp connections present.

Useful references. The following books contain information on *G. venenata* pertinent to North America.

- *A Field Guide to Western Mushrooms* (Smith 1975)—A color photograph is presented, along with a description and a discussion.
- *Edible and Poisonous Mushrooms of Canada* (Groves 1979)—A discussion is presented.
- *Mushrooms of North America* (Miller 1972)—A description and discussion are given.
- *The Savory Wild Mushroom* (McKenney 1971)—A discussion is presented.

Lepiota (Pers. ex Fr.) S. F. Gray — Lepiotaceae

There are several interpretations of the genus *Lepiota*. Modern authors have often used a narrow definition of *Lepiota* and in doing so have recognized several segregate genera including *Chlorophyllum*, *Leucocoprinus*, *Leucoagaricus*, and *Macrolepiota*. A broad definition of *Lepiota* is used here, however, which includes all the above genera. Species of *Lepiota* have in common distinctly free gills; a spore deposit that is typically white to pale cream, although it is green in one species; a partial veil that usually forms an annulus, although it is sometimes lost in mature specimens; and a more or less fleshy-fibrous stalk that is often somewhat clavate. Species of *Lepiota* vary from very tiny to large. Often the cap is convex, with a raised knob in the center. There is usually no universal veil and consequently no volva.

The structure of the cap cuticle varies but often is composed of filamentous hyphae. The spores are smooth and of various shapes; they sometimes have an apical germ pore and are nonamyloid, dextrinoid, or more rarely amyloid.

In general appearance certain species of *Amanita* resemble the genus *Lepiota*. The presence of a universal veil and consequently a volva in *Amanita* separates it from *Lepiota*. Species of *Armillaria* and *Cystoderma* also may resemble *Lepiota* in general appearance. These genera have species with attached gills. Species of *Agaricus* have the same appearance as some *Lepiota* species, but *Agaricus* can be easily distinguished by its blackish brown to dark brown spore deposit. Species of *Lepiota* that produce amatoxins are in general poorly known. Cases of poisoning have been reported from Israel, southern Europe, and Chile, and amatoxins have been detected in several species by thin-layer chromatography. To date the following species seem to be involved: *L. brunneoincarnata* Chodat & Martin, *L. brunneolilacea* Bon. & Boif, *L. clypeolarioides*, *L. fuscovinacea* Miller & Lange, *L. heimii* Locquin, *L. helveola sensu* Huijsman, *L. helveola sensu* Josserand, *L. locanensis* Espinosa, *L. pseudohelveola* Kühner ex Hora, *L. rufescens*, *L. subincarnata* Lange, *L. castanea* Quél., *L. felina* (Pers. ex Fr.)

Karst., *L. griseovirens* Maire, and *L. ochraceofulva* Orton. Some of these species have been reported from North America, but apparently are uncommon here; a number are tropical or subtropical species. There is one known case of poisoning in North America from this group of *Lepiota* species. Dr. D. E. Stuntz (Personal commun.) reported a mild case of poisoning by *L. helveola*. Selected species reported to have occurred in North America are described below. They are for the most part small species and difficult to identify. Microscopic examination and reference to technical literature is essential before a positive identification can be made.

Lepiota clypeolarioides Rea

Fig. 30

Common names. None.

Distinguishing characteristics. Cap tan with small reddish brown scales and fibrils. Gills white, becoming yellowish. Annulus narrow and similar in color to the cap. Stalk below annulus scaly and similar in color to the cap. Spores dextrinoid, 5.8–7.3 × 4–4.4 μm in size.

Observations. *L. clypeolarioides* appears to be uncommon in North America and has not been reported in any recent North American literature. We studied specimens from Nova Scotia and California that were collected by Dr. A. H. Smith in 1931 and 1937, respectively, and W. Sundberg (Personal commun.) reports its occurrence in Oregon and Washington.

Distribution and seasonal occurrence. Although uncommon, the species may be widely distributed. It fruits from late summer to early fall.

Habitat and habit. *L. clypeolarioides* has been reported growing under conifers such as spruce (*Picea*) in North America. Rea (1922) indicates that it grows in woods and along hedgerows.

Technical description. Cap 30–50 mm broad, convex, obtusely umbonate, then plane and depressed; surface appressed fibrillose-scaly, tan in color with reddish brown fibrils; flesh thick, white.

Gills white, becoming yellowish, free, crowded.

Stalk 75–100 mm long, 4–8 mm wide, gradually narrowed upward, smooth at the apex, scaly below the annulus, similar in color to the cap.

Spores white in deposit, 5.8–7.3 × 4.0–4.4 μm, although W. Sundberg (Personal commun.) gives a somewhat larger spore size, (5.5–)6–8.4 × 3.7–5.6 μm, ellipsoidal to ovoid, with the sides slightly unequal in side view, thick-walled, yellowish, dextrinoid. Basidia four-spored, clavate, 27.0 × 6.6–8.8 μm, thin-walled, hyaline. Pleurocystidia lacking. Cheilocystidia 18.3–27 × 6.6–8.8 μm, fusiform-ventricose to clavate, mucronate or with a more or less acute apex, thin-walled, hyaline.

Gill trama composed of hyaline cylindrical interwoven to parallel hyphae 3.5–13.1 μm wide. Subhymenium more or less cellular in places, compacted, interwoven; hyphae more or less cylindrical and short, 3.5–5.1 μm wide, hyaline. Cap cuticle filamentous; hyphae compactly interwoven to more or less radially arranged below, similar to a trichoderm at the surface; end cells pileocystidium-like, 23.4–120 × 9.4–15.5 μm, some thick-walled with walls more or less yellow brown, with rounded to more or less acute apices; hyphae otherwise cylindrical, 7.0–23.4 μm wide, thin- to slightly thick-walled and finely encrusted, hyaline to yellowish. Cap trama loosely interwoven; hyphae 2.9–16.7 μm wide, cylindrical to more or less inflated, hyaline, thin-walled. Caulocystidia clavate to fusiform, 4.5–13.1 μm wide, thin- to slightly thick-walled, hyaline to yellowish brown. Stalk hyphae parallel to interwoven, 3.5–14.3 μm, narrow at the surface, hyaline to yellowish. Annulus hyphae interwoven, 2.3–5.1(–8.1) μm wide, more or less cylindrical, hyaline; brown clavate-fusiform end cells present on outer annulus surface. Oleiferous hyphae not observed. Clamp connections present.

Useful references. There are no books or field guides pertinent to North America that contain information on this species.

Lepiota helveola Bres.

Fig. 31

Common names. None.

Distinguishing characteristics. Cap fibrillose to fibrillose-scaly on the margin, tomentose to finely squarrose in the center; color moderate brown to reddish brown or dark grayish reddish brown or paler (light brown, light reddish brown, light yellowish brown to brownish pink), typically paler-colored toward the margin. Gills white to creamy, not staining. Annulus, when present, more or less floccose and similar in color to scales on the stalk. Stalk fibrillose, tinted or shaded with colors of the cap, staining cinnamon brown when bruised or handled, with scales below the annulus. Spores dextrinoid, 5.8–8.0 × 3.0–4.4 μm in size.

Observations. *L. helveola* has rarely been reported in the North American literature and is a poorly known species. Collections studied for this treatment are from western North America.

Distribution and seasonal occurrence. *L. helveola* appears to be fairly common along the Pacific coast, but in general, its distribution is poorly known. It fruits during the fall.

Habitat and habit. The species grows solitary to scattered in soil and humus in mixed hardwood-coniferous forests and under conifers.

Technical description. Cap 15-30 mm broad, convex then expanded, more or less umbonate; surface scaly, madder brown in color. Flesh somewhat thick, white, becoming reddish when dry.

Gills creamy white, free, crowded, ventricose; edges fimbriate.

Stalk 20-40 mm long, 3-7 mm wide, equal, fibrillose-tomentose, similar in color to the cap. Annulus whitish, evanescent.

Spores white in deposit, 5.8-8.0 × 3.0-4.4 μm, elliptic to more or less kidney-shaped in side view, smooth, thick-walled, hyaline, dextrinoid, lacking an apical germ pore. Basidia four-spored, clavate, 22.0-27.7 × 6.6-8.0 μm, hyaline. Pleurocystidia lacking. Cheilocystidia clavate to sac-like, 18.3-27.7 × 7.3-13.1 μm, thin-walled, hyaline.

Gill trama interwoven to subparallel; hyphae cylindrical to inflated, 5.1-16.0 μm wide, hyaline. Subhymenium cellular; hyphae interwoven, more or less short-cylindrical, 5.1-8.8 μm wide, hyaline. Cap cuticle filamentous; hyphae radially arranged, in places forming a trichoderm of cylindrical-clavate erect end

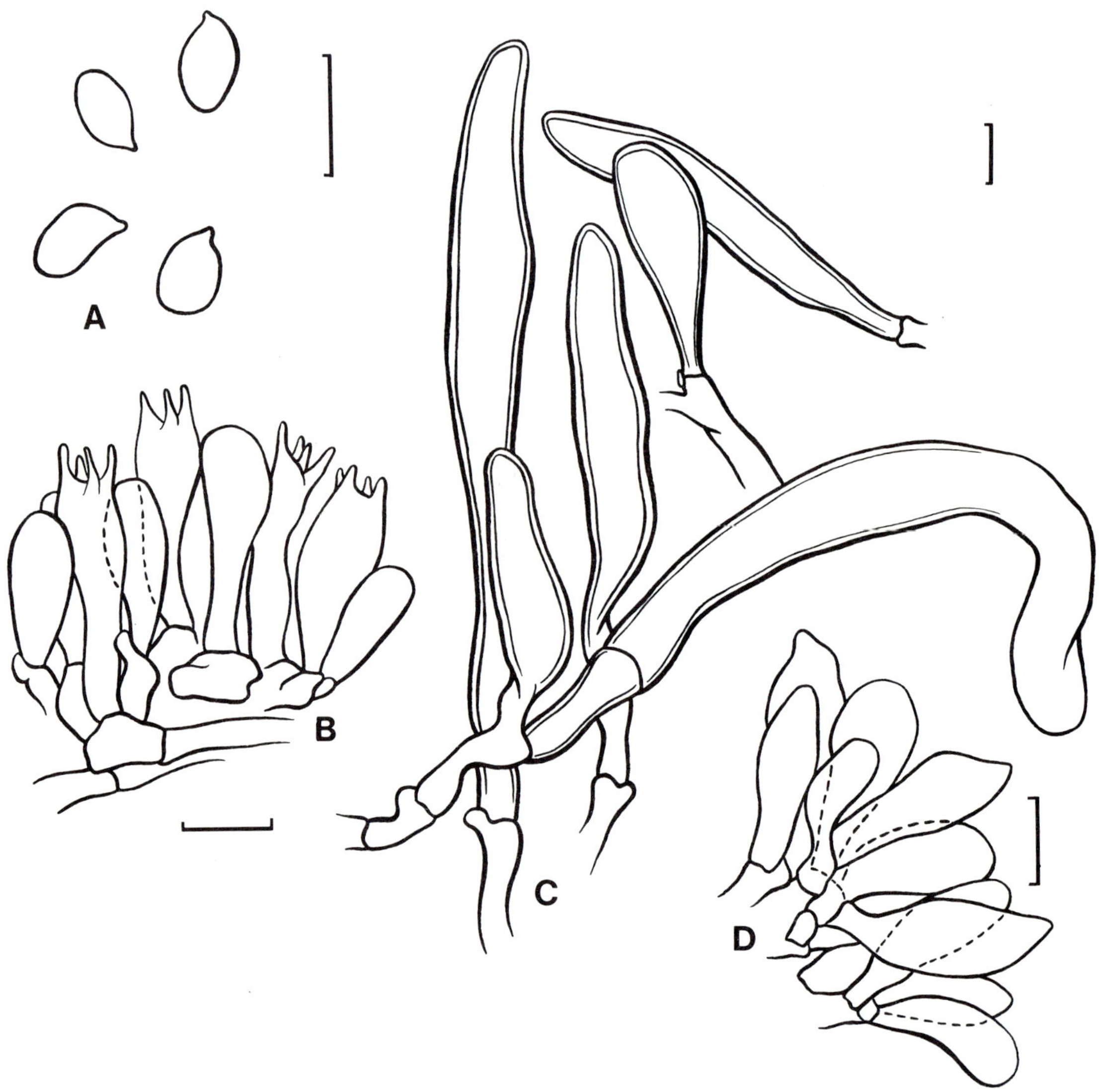

Fig. 30. *Lepiota clypeolarioides*: A, spores; B, basidia; C, pileocystidia and caulocystidia; D, cheilocystidia.

Scale line 10 μm

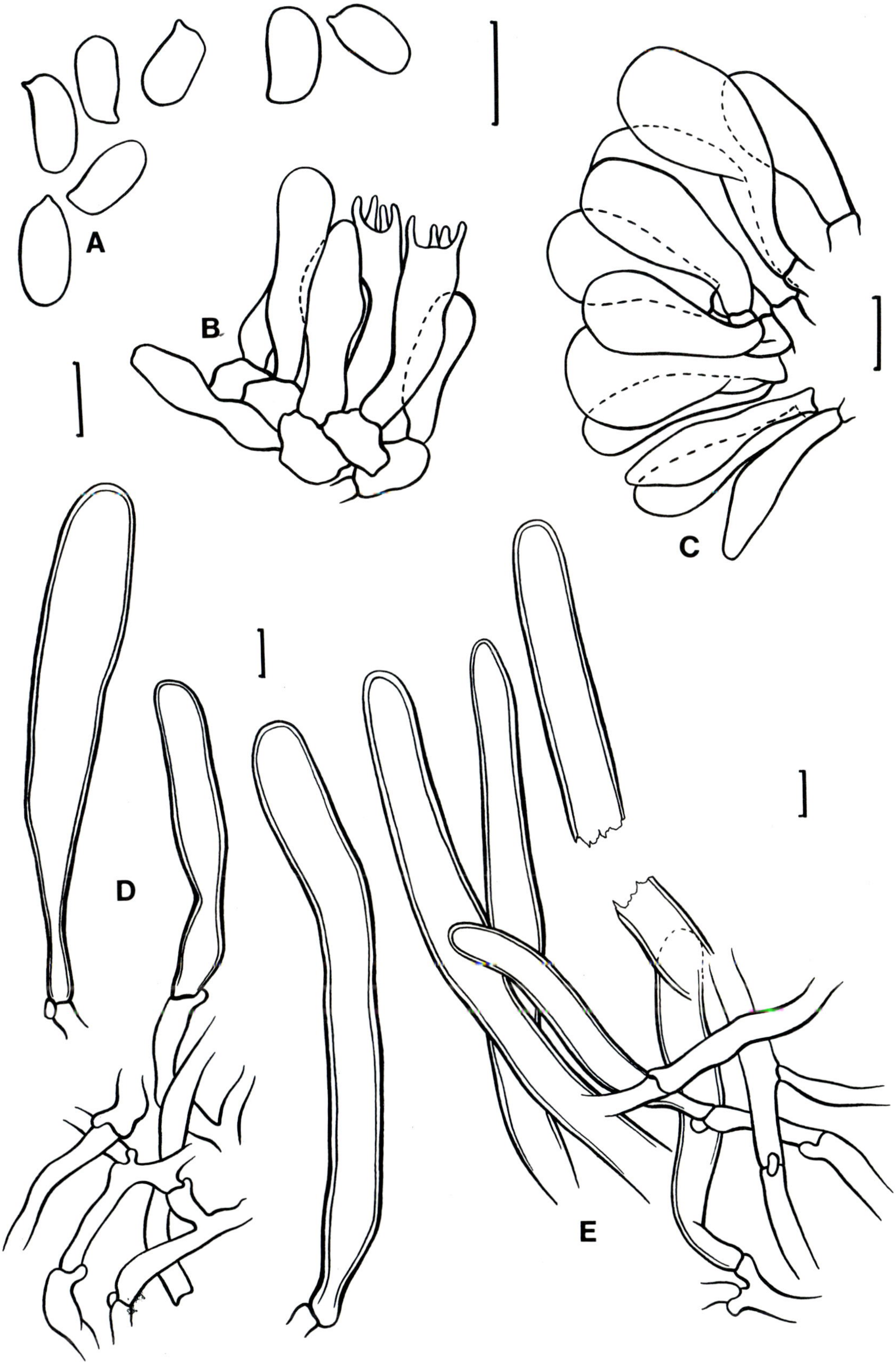

Fig. 31. *Lepiota helveola*: A, spores; B, basidia; C, cheilocystidia; D, caulocystidia; E, pileocystidia.

Scale line 10 μm

cells (pileocystidia); these cells 45-350 × 3.5-7.1 μm, thick-walled, hyaline to yellowish brown; hyphae otherwise more or less cylindrical, 7-13.1 μm wide, more or less thin-walled, hyaline to yellowish brown. Cap trama interwoven; hyphae more or less cylindrical, 3.5-9.4 μm wide, hyaline. Caulocystidia cylindrical to clavate, 46-200 × 10.5-16.7 μm, thick-walled, brownish. Stalk hyphae parallel to interwoven, more or less cylindrical, 3.5-10.0 μm wide, hyaline. Annulus interwoven; hyphae cylindrical, 3.0-5.1 μm wide, hyaline. Oleiferous hyphae refractive, greenish yellow, 5.1-7.3 μm, scattered. Clamp connections present.

Useful references. This species is not included in any of the popular field guides pertinent to North America. Sundberg (1967), however, should be consulted for a good technical description.

Lepiota rufescens (Berk. & Br.) Lange

Fig. 32

Common names. None.

Distinguishing characteristics. Cap appearing mealy to powdery or granular although at times appearing almost smooth with age; color whitish to sordid cream color or tinted pinkish cinnamon when fresh, staining reddish when bruised or with age. Gills white and often staining reddish on edges. Annulus whitish to creamy, powdery and soon lost. Stalk covered with powdery material and colored like the cap. Spores nonamyloid, 4.4-5.1(-6.6) × 2.5-2.9 μm in size.

Observations. *L. rufescens* was reported from Michigan by Helen V. Smith in 1954. The description here is based on material from the University of Michigan Herbarium.

Distribution and seasonal occurrence. The species appears to be uncommon but widely distributed. It fruits during the summer in the Great Lakes region and in the fall on the west coast.

Habitat and habit. *L. rufescens* is scattered in mixed woods and coniferous forests.

Technical description. Cap 15-25 mm broad, conical to broadly convex; surface covered by a dense coating of evanescent mealy pyramid-shaped scales, becoming evenly powdery or granular, at times smooth to finely fibrillose; margin appendiculate in early stages; color whitish, sordid creamy, or light pinkish cinnamon when fresh, becoming grayish and changing to reddish or dull ferruginous when bruised or touched. Flesh fairly thick, whitish, slowly turning reddish where cut; taste bitter; odor not distinctive.

Gills white, becoming sordid brownish when dried, often staining reddish on edges, free, almost touching the stalk, close, broad; three tiers of lamellulae present.

Stalk 20-40 mm long, 1.5-4 mm wide, thick, tubular, enlarged somewhat at the base, covered in early stages with powdery material (as is the cap), similar in color to the cap at the base, turning reddish or dingy when handled. Annulus evanescent, consisting of a zone of whitish to creamy powder.

Spores white in deposit, 4.4-5.1(-6.6) × 2.5-2.9 μm, ellipsoidal to broadly ellipsoidal, smooth, thin-walled,

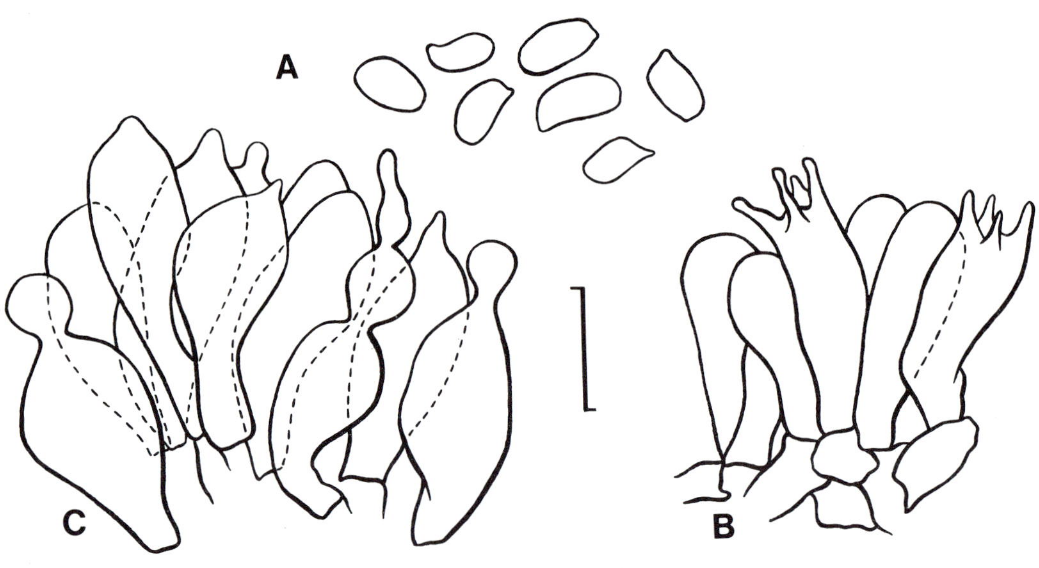

Fig. 32. *Lepiota rufescens*: A, spores; B, basidia; C, cheilocystidia.

Scale line 10 μm

hyaline, nonamyloid, not dextrinoid. Basidia four-spored, clavate, 12.4–14.6(–18.3) × 5.8–6.6(–7.3) µm, hyaline. Pleurocystidia lacking. Cheilocystidia clavate to clavate-mucronate or fusiform-ventricose, 21.9–29.2 × 7.3–9.5(–11.0) µm, hyaline; apices tapered and filiform to capitate.

Gill trama interwoven to parallel; hyphae cylindrical to inflated, 2.9–11.7(–18.3) µm wide, hyaline. Subhymenium filamentous to more or less cellular; hyphae interwoven, more or less short-cylindrical, 2.9–5.1 µm wide, hyaline to yellowish. Cap cuticle appearing more or less cellular, composed of ellipsoidal to subglobose or globose pedicellate cells 25.6–48.0 µm wide and cylindrical hyphae 2.9–5.1 µm wide; hyphae thin-walled, hyaline to yellowish. Cap trama loosely interwoven; hyphae cylindrical to inflated, 2.9–19.2 µm wide, hyaline. Caulocystidia in the form of clavate-cylindrical end cells 5.8–16.7 µm wide, recurved, rare. Stalk hyphae parallel in the core to interwoven on the surface, 3.5–16.0 µm wide, hyaline to yellowish; stalk surface below annulus covered by globose (15.5–35.1 µm wide) to subglobose-ellipsoidal cells (15.5–48.6(–51) × 13–29.2(–41) µm); these cells hyaline to yellowish, thin-walled to slightly thick-walled. Annulus interwoven, composed of hyphae 3.5–5.8 µm wide and scattered spherical cells 15–29.2 µm wide; all cells hyaline to yellowish brown. Oleiferous hyphae refractive, greenish yellow, 5.8–8.4 µm wide, scattered in cap trama and stalk. Clamp connections present.

Useful references. There are no books or field guides pertinent to North America that contain information on this species.

Naematoloma Karst.

Strophariaceae

Naematoloma fasciculare has been reported as poisonous by Herbich et al. (1966) in Austria, Narita (1957) in Japan, and Wasiljkow (1966) in Russia. Chilton (1972) also discusses this species. There are no reports of poisoning by this species in North America. The species typically has a bitter taste that usually does not disappear on cooking. Some nonbitter forms have been reported (Smith 1975), however, and there are reports that cooking destroys the bitter taste (McIlvaine and Macadam 1973). The onset of symptoms is delayed until nine hours after ingestion of the mushroom meal; liver and kidney damage eventually develop (Herbich et al. 1966). The toxin or toxins have not yet been reported in the literature.

Naematoloma fasciculare has been confused with the honey mushroom, *Armillariella mellea*, and inadvertently gathered for food along with the latter (Chilton 1972). Similar confusion of *Galerina* species with *Armillariella mellea* has also been noted.

The genus *Naematoloma* is rather difficult to define and to separate from the genera *Psilocybe* and *Stropharia*. Certain modern authors would place all *Naematoloma* species in *Psilocybe*; others would place the *Naematoloma* species in the genus *Hypholoma*. The genus *Naematoloma* is used in this treatment mainly because the species of concern, *N. fasciculare*, is most commonly found as a species of *Naematoloma* in North American literature.

The characteristics that help to define *Naematoloma*, as used here, include a purple brown to dull cinnamon brown spore deposit; a fleshy and often brightly colored cap, with yellow to orangish colors; attached gills, either adnexed to adnate or subdecurrent, usually shaded with spores when mature; a fleshy-fibrous stalk or sometimes a more slender and cartilaginous stalk; the presence of a partial veil of fibrils, leaving fibrils on the cap margin but no distinct annulus on the stalk when rupturing; a cap cuticle composed of filamentous hyphae; smooth spores with an apical germ pore; the presence of a characteristic type of cystidium, namely chrysocystidia, in the hymenium; a wood or woody debris habitat, but this species is also found in peat-like soil and mosses.

To separate *Naematoloma* from *Psilocybe* requires the use of microscopic characteristics, and even these are not always reliable. Usually the presence of chrysocystidia in the gills is an indication that the species is a *Naematoloma*. *Stropharia* has been traditionally separated from *Naematoloma* by the presence of an annulus in *Stropharia*. This distinction works most of the time.

Naematoloma fasciculare (Huds. ex Fr.) Karst.

Figs. 33, 41

Common names. Clustered woodlover, sulfur tuft, sulfur cap.

Distinguishing characteristics. Species often growing in clusters, on decaying logs or stumps on both conifers and hardwoods. Cap 10–80 mm broad, more or less convex, citron orange to orange yellow or orange ocherous, with the margin at times more olive yellow in color. Gills narrow and crowded, sulfur yellow to greenish yellow, usually with a distinct greenish to olive tint before spores darken gills. Flesh typically with a bitter taste. Stalk yellow to greenish yellow and becoming brown to rusty brown from the base upward. Partial veil thin, leaving a slight zone on the stalk at times. Spores purple brown in deposit, smooth, 5.1–8(–9.5) × 3.7–4.4 μm in size, with an apical germ pore. Chrysocystidia present.

Observations. *N. fasciculare* is an attractive, conspicuous species and a common element of the western mushroom flora. It also occurs in the east but is less common there. Although the taste is usually very bitter, mild forms have been reported.

Distribution and seasonal occurrence. The species occurs across North America. It is common in the west but rare to infrequent in the east. It fruits occasionally in the spring but is usually found in the summer and fall. It may also fruit in the winter, depending on location and weather conditions.

Habitat and habit. *N. fasciculare* usually occurs in clusters but is also gregarious to scattered. It grows on decaying logs, stumps, or wood chips, usually on conifer wood but also on hardwoods.

Technical description. Cap (10–)20–50(–80) mm broad, obtusely conic to bell-shaped, becoming umbonate tc nearly plane; margin incurved in early stages, usually more or less decurved in expanded caps; surface moist or somewhat lubricous when wet, glabrous or at times breaking into spot-like scales, occasionally slightly silky toward the margin; edge with veil fragments in early stages; color ocherous orange on the disc, light greenish yellow (citron yellow tinged green) to brilliant greenish yellow or tinged light olive brown along the margin, sometimes becoming amber brown on the disc with age and pale yellow or sordid olive on the margin. Flesh moderately thick, tapering gradually toward the margin, light greenish yellow, brilliant greenish yellow, or bright yellow, becoming sordid brown after cutting; taste typically very bitter, rarely mild; odor indiscernible.

Gills pale sulfur yellow, pale greenish yellow, light greenish yellow, or brilliant greenish yellow, becoming moderately greenish yellow and then more or less olive, adnate, seceding, crowded, narrow; several tiers of lamellulae present; edges even.

Stalk 50–120 mm long, 4–10 mm wide, narrowed downward, terete or compressed, hollow, often wavy to undulating and more or less contorted; apex pruinose to glabrous but unpolished; below appressed-fibrillose; color pale greenish yellow in early stages but soon becoming brownish to brownish orange or rusty brown, especially below. Flesh yellowish. Partial veil leaving a fibrillose evanescent zone. Universal veil lacking.

Spores purple brown in deposit, 5.1–8.0(–9.5) × 3.7–4.4 μm, ellipsoidal to slightly ovoid, slightly flattened on one side, smooth; apex truncate; a distinct apical germ pore present. Basidia four-spored, clavate, 17–22 × 5.1–6.6 μm, hyaline to yellow. Pleurocystidia ventricose to clavate-mucronate, 26–42 × 6–11 μm, thin-walled; contents hyaline to golden yellow and refractive; chrysocystidia with yellow refractive inclusion body. Cheilocystidia fusiform-ventricose, 20–30 × 6.6–9.0 μm, hyaline to yellowish, thin-walled.

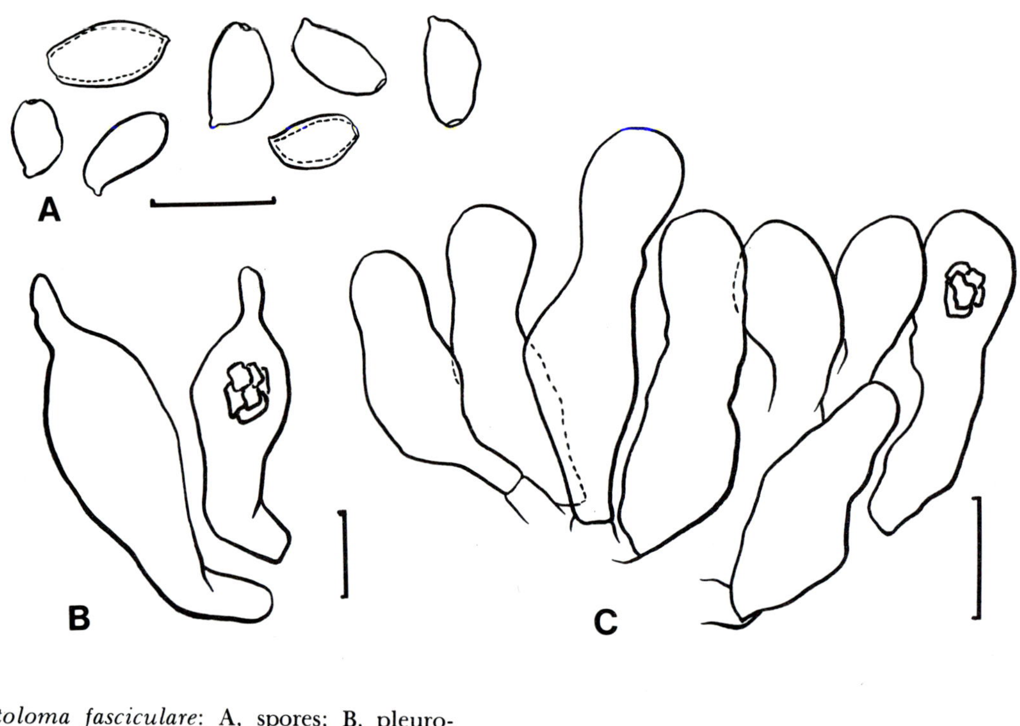

Fig. 33. *Naematoloma fasciculare*: A, spores; B, pleurocystidia; C, cheilocystidia. Note the chrysocystidia, containing crystals.

Scale line 10 μm

Gill trama interwoven to more or less parallel; hyphae cylindrical to inflated or fusiform, 3.5–25.6 μm, hyaline to yellowish. Subhymenium hyphae compacted, cylindrical to inflated, interwoven, 4.0–5.1 μm wide, hyaline to yellowish. Cap cuticle filamentous; hyphae interwoven, radially arranged; hyphae of the outermost layer of the cutis cylindrical, 1.8–5.1 μm wide, thin-walled, yellow brown; hyphae of the inner layer of the cutis interwoven to parallel, cylindrical to inflated, 15.1–30.0 μm wide, thin- to slightly thick-walled and encrusted, hyaline to yellow brown. Cap trama loosely interwoven; hyphae 6–50 μm wide, hyaline to yellowish. Caulocystidia broadly clavate to clavate-mucronate or fusiform-ventricose, clustered, 22–25 × 7.3–8.8 μm. Stalk hyphae parallel, more or less cylindrical, 3.7–10.2(–12.4) μm wide, hyaline to yellowish. Oleiferous hyphae refractive, yellowish to ocherous, contorted, 3.7–9.5 μm, scattered in the stalk and cap trama. Clamp connections present.

Useful references. The following books contain information on *N. fasciculare* pertinent to North America.

- *A Field Guide to Western Mushrooms* (Smith 1975)—A color photograph and a description and discussion are presented.
- *Edible and Poisonous Mushrooms of Canada* (Groves 1979)—A discussion is given.
- *Mushrooms in their Natural Habitats* (Smith 1949)—A description and discussion, as well as a color photograph on a reel, are presented.
- *Mushrooms of North America* (Miller 1972)—A color photograph and a description and discussion are given.
- *The Mushroom Hunter's Field Guide* (Smith 1974)—A black-and-white photograph and a description and discussion are presented.
- *The Savory Wild Mushroom* (McKenny 1971)—A black-and-white photograph and a description and discussion are given.

Literature

Abdel-Malak, S. Chemotaxonomic significance of alkaloids and cyclopeptides in *Amanita* species. University Microfilms, Ann Arbor, MI; 1974.

Ammirati, J. F.; Thiers, H. D.; Horgen, P. A. Amatoxin-containing mushrooms: *Amanita ocreata* and *A. phalloides* in California. Mycologia 69(6):1095–1108; 1977.

Bartelli, I. Don't pick poison (when gathering mushrooms for food in Michigan). Cooperative Extension Service, Michigan State University, Bulletin E-1080; 1977.

Becker, C. E.; Tong, T. G.; Boerner, U.; Roe, R. L.; Scott, R. A. T.; MacQuarrie, M. B.; Bartter, F. Diagnosis and treatment of *Amanita phalloides*-type mushroom poisoning—Use of thioctic acid. West. J. Med. 125:100–109; 1976.

Benedict, R. G. Mushroom toxins other than *Amanita*. Kadis, S., et al., eds. Microbiological toxins. Vol. 8. New York, NY: Academic Press; 1972.

Besl, H. Amatoxine im Gewächshaus: *Galerina sulciceps*, ein tropischer Giftpilz. Z. Mykol. 47:253–255; 1981.

Buck, R. Mycetism. New Engl. J. Med. 280:1363; 1969.

Chilton, W. S. Poisonous mushrooms. Pac. Search, March; 1972.

Coker, W. C. The amanitas of the eastern United States. J. Elisha Mitchell Sci. Soc. 33(1 and 2):1–88; 7 plates; 1917.

Courtillot, M.; Staron, T. *Amanita virosa* Fr. Précisions sur l'espèce, mise en évidence de sa toxine principle (virosine). Ann. Phytopathol. 2:561–584; 1970.

Faulstich, H.; Cochet-Meilhac, M. Amatoxins in edible mushrooms. FEBS (Fed. Eur. Biochem. Soc.) Lett. 64:73–75; 1976.

Favre, H.; Leski, M.; Christeler, P.; Vollenweider, E.; Chatelanat, F. Le *Cortinarius orellanus*: Un champignon toxique provoquant une insuffisance rénale aiguë retardée. Schweiz. Med. Wochenschr. 106:1097–1102; 1976.

Ford, W. W. The distribution of poisons in the amanitas. J. Pharmacol. Exp. Ther. 1:275–287; 1909.

Groves, J. W. Edible and poisonous mushrooms of Canada. Ottawa, Ont.: Agriculture Canada. Agric. Can. Publ. 1112; 1979 [first printed in 1962].

Grzymala, S. Zur toxischen Wirkung des orangefuchsigen Hautkopfes (*Dermocybe orellana* Fr.). Dtsch. Z. Gerichtl. Med. 49:91–99; 1959.

Grzymala, S. Etude clinique des intoxications par les champignons du genre *Cortinarius orellanus* Fr. Bull. Med. Leg. 8(1):6–70; 1965*a*.

Grzymala, S. Les recherches sur la frequence des intoxications par les champignons. Bull. Med. Leg. 8(2):200–210; 1965*b*.

Heim, R. Sur un cas d'empoisonement mortel cause au Mexique par l'*Amanita bisporigera* Atk. Rev. Mycol. 22:208–216; 1957.

Herbich, J.; Lohwag, K.; Rotter, R. Tödliche Vergiftung mit dem grünblattrigen Schwefelkopf. Arch. Toxicol. 21: 310–320; 1966.

Hulmi, S.; Sipponen, P.; Forsström, J.; Vilska, J. *Cortinarius* causes kidney damage. Duodecim 90:1044–1050; 1974 [in Finnish].

Kauffman, C. H. *Cortinarius* Fries. North Am. Flora 10:282–348; 1932.

Kits van Waveren, E. The genus *Conocybe* subgenus *Pholiotina* I. The European annulate species. Persoonia 6:119–165 (in part); 1970.

Lincoff, G.; Mitchel, D. H. Toxic and hallucinogenic mushroom poisoning—A handbook for physicians and mushroom hunters. New York, NY: Van Nostrand Reinhold Co.; 1977.

Litten, W. The most poisonous mushrooms. Sci. Am. 232 (Mar.): 90–101; 1975.

McIlvaine, C.; Macadam, R. K. One thousand American fungi. New York, NY: Dover Publ. Inc.; 1973 [first printed in 1902].

McKenny, M. The savory wild mushroom (revised and enlarged by Stuntz, D. E.). Seattle, WA: University of Washington Press; 1971.

Meixner, A. Amatoxin-Nachweis in Pilzen. Z. Mykol. 45(1):137–139; 1979.

Miller, O. K., Jr. Mushrooms of North America. New York, NY: E. P. Dutton & Co., Inc.; 1972.

Moser, M. *Cortinarius* Fr. Untergattung *Leprocybe* subgen. nov., die Rauh Kopfe. Z. Pilzkd. 35:213–248; 1969.

Narita, D. A case of poisoning by *Naematoloma fasciculare* (*Hypholoma fasciculare*). Nippon Kingakki Kaiho 2:6; 1957.

Rea, C. British basidiomycetes. Cambridge, England: Cambridge University Press; 1922.

Ridgway, A. Color standards and color nomenclature. Washington, D.C.: R. Ridgway; 1912.

Romagnesi, H. Petit atlas des champignons. Vol. II. Paris, France: Bordas; 1962.

Skirgiello, A.; Nespiak, A.; Grzymala, S. Erfahrungen mit *Dermocybe orellana* (Fr.) in Polen. Z. Pilzkd. 23:138–143; 1957.

Smith, A. H. New and unusual agarics from Michigan. Ann. Mycol. 32:478–479; 1934.

Smith, A. H. Mushrooms in their natural habitats. 2 Vols. Portland, OR: Sawyer's, Inc.; 1949.

Smith, A. H. The mushroom hunter's field guide. Ann Arbor, MI: University of Michigan Press; 1974 [first printed in 1963].

Smith, A. H. A field guide to western mushrooms. Ann Arbor, MI: University of Michigan Press; 1975.

Smith, A. H.; Singer, R. A monograph of the genus *Galerina* Earle. New York, NY: Hafner Publ. Co.; 1964.

Smith, H. V. A revision of Michigan species of *Lepiota*. Lloydia 17:307–328; 1954.

Stamets, P. *Psilocybe* mushrooms and their allies. Seattle, WA: Homestead Book Co.; 1978.

Sundberg, W. J. The family Lepiotaceae in California. Masters Thesis. San Francisco, CA: San Francisco State University; 1967.

Tanghe, L. J.; Hillhouse, E. Dependence of spore shape on maturity of carpophore in the *Phalloidae* section of *Amanita*. McIlvainea 1(2):1–8; 1973.

Tanghe, L. J.; Simons, P. M. *Amanita phalloides* in the eastern United States. Mycologia 65:99–108; 1973.

Vergeer, P. P. A potential problem with the Meixner test for amatoxins in mushrooms. Mycena News 29(10):79; 1979.

Wasiljkow, B. P. Die Vergiftungsfälle des büscheligen Schwefelkopfes, *Hypholoma fasciculare* (Fr.) Quél. Schweiz. Z. Pilzkd. 41:117–121; 1963.

Wieland, T. 1968. Poisonous principals of mushrooms in the genus *Amanita*. Science 159:946–952; 1968.

Fig. 34. *Amanita bisporigera.*

Fig. 35. *Amanita ocreata.*

Fig. 36. *Amanita phalloides.*

Fig. 37. *Amanita virosa.*

Fig. 38. *Conocybe filaris.*

Fig. 39. *Cortinarius gentilis.*

Fig. 40. *Galerina autumnalis.*

Fig. 41. *Naematoloma fasciculare.*

8

MONOMETHYLHYDRAZINE

Toxicology and Symptomatology

Morels (species of *Morchella*) and false morels (species of *Gyromitra*), of the class Discomycetes in the division Ascomycota, are among the most sought after fungi of the mushroom flora that abounds in the spring, both in Europe and in certain parts of North America. On both continents many severe, sometimes fatal poisonings have been caused by false morels, specifically by *Gyromitra esculenta*.

The toxin found in false morels is called gyromitrin. This compound is very unstable and is easily hydrolyzed at moderate cooking temperatures to toxic monomethylhydrazine. Gyromitrin does not occur in the free state in fresh fungus tissue but is produced by the hydrolysis of insoluble, nonvolatile precursors. Gyromitrin has been reported in *Gyromitra esculenta* and may also occur in some strains of *Gyromitra gigas* (in part *G. montana* Harmaja) and *Gyromitra fastigiata* (=*G. brunnea* Underw. and *G. underwoodii* Seaver) (Benedict 1972). Gyromitrin or its hydrazine derivatives have not been found in other species of *Gyromitra* or in species of the closely related genus *Discina*, but these species have not been systematically tested. Wells and Kempton (1968) reported a poisoning by *Gyromitra infula* (subsequently identified as *Gyromitra ambigua* on the basis of spore characteristics). The symptoms produced by poisoning caused by *Gyromitra ambigua* are similar to those produced by *Gyromitra esculenta* but the presence of gyromitrin in this species has not been documented. *Disciotis venosa* (Pers. ex Fr.) Bondatsev, *Gyromitra californica* (Phillips) Raitvir (=*Helvella californica* Phillips), *Gyromitra korfii* (Raitvir) Harmaja (=*G. fastigiata* (Krombh.) Rehm *sensu* McKnight), and *Gyromitra sphaerospora* (Peck) Sacc. have also been listed as suspected, dangerous, or not recommended.

Because of the uncertainty surrounding the identification of a number of species of *Gyromitra*, it is difficult to determine which ones, if any, are safe to eat. Until the species are better known and each is systematically checked for hydrazines, it is best to avoid the entire genus *Gyromitra*.

When you consider the large number of people who eat false morels, the incidence of poisoning is rather low. If you do decide to eat false morels, follow these suggestions carefully for preparing them. Parboil the specimens twice, and discard the cooking water each time. Rinse the pieces and cook them thoroughly. Monomethylhydrazine has a boiling point of 87.5°C and is readily decomposed in air. Most of the risk in eating these fungi comes from consuming them raw or from preparing them incorrectly. They can be dangerous when braised in a little water and consumed with the broth. Eat only small quantities, and never eat false morels several days in succession. When collecting specimens, avoid all that are soft, blackened, or dried. Use only young, firm fruit bodies. Never eat false morels raw.

Species of *Helvella* are also suspect and should be eaten with caution. Follow the same procedures for cooking them as those used for false morels, and never eat them raw. Other genera of the Discomycetes that are reported as poisonous include *Sarcosphaera, Verpa, Morchella, Peziza*, and *Disciotis*; they are especially dangerous when they are eaten raw but some people experience adverse reactions even when these fungi are cooked (Lincoff and Mitchel 1977). There are some reports (N. Smith-Weber, Personal commun.) that species of *Disciotis* are edible, which seems logical because the genus is related to *Morchella*. On the other hand, as the genus *Discina* is closely related to *Gyromitra*, some species of *Discina* may be toxic. *Sarcosphaera* and *Disciotis* are not treated further here. Smith and Smith (1973) give a key and descriptions for them, although in their publication, *Disciotis venosa* is placed in the genus *Peziza*. Further discussion of the other genera mentioned appears under *Gyromitra* in this chapter and in Chapter 13. See Chapters 2 and 3, "Ascomycota", for a description of the general features of this division.

Chemically, monomethylhydrazine is an extremely simple molecule. It contains two nitrogen atoms, hydrogen, and a single methyl (CH_3) group (List and Luft 1969). Monomethylhydrazine is often used as a propellant for rockets. In fact, people in the aerospace industry who come in contact with these hydrazine rocket fuels often exhibit the same kinds of symptoms as do individuals poisoned by *Gyromitra esculenta*.

Monomethylhydrazine acts directly on the central nervous system by interfering with the normal utilization and function of vitamin B_6. This vitamin is associated with several key cellular enzymes that are involved in amino acid metabolism. Furthermore, convincing evidence has been obtained from studies on mammals to suggest that exposure to monomethylhydrazine can increase the number of malignant tumors in experimental animals. It is unlikely, however, that there is any direct connection between mushroom poisoning and cancer.

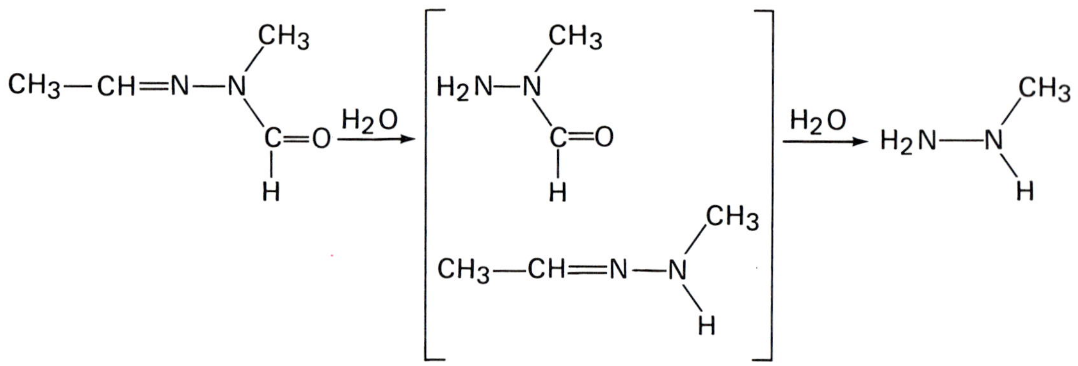

An interesting aspect of monomethylhydrazine poisoning is the very narrow margin between a dose that causes no symptoms and a dangerous or lethal dose (Back and Pinkerton 1967). Individual tolerance varies considerably. Some people may be poisoned by one meal of *Gyromitra esculenta,* whereas others may require two or three times as much to experience an adverse effect.

The first symptoms of monomethylhydrazine poisoning are nausea and vomiting, which start 7–10 hours after the fungus has been consumed. Often the victim feels bloated before the onset of the nausea and persistent vomiting. Victims often have watery or even bloody diarrhea. Muscle cramps and abdominal pain are also associated as symptoms. In severe cases, liver damage, jaundice, high fever, dizziness, loss of coordination, and convulsions are further symptoms. Death can occur 2–4 days after ingestion in very severe cases.

Pyridoxine hydrochloride, a preparation of one form of vitamin B_6, has been used as a treatment for poisoning by *Gyromitra.* This compound usually protects against the convulsive action of various hydrazine compounds. The handbook by Lincoff and Mitchel (1977) should be consulted for further details of treatment.

Symptom-producing Species

Morchellaceae
- *Disciotis*
  - *D. venosa* (suspected)

Helvellaceae
- *Gyromitra*
  - *G. ambigua*
  - *G.* (*Neogyromitra*) *californica* (suspected)
  - *G. caroliniana* (suspected)
  - *G. esculenta*
  - *G. fastigiata* (=*G. brunnea* and *G. underwoodii*) (suspected)
  - *G. korfii* (=*G. fastigiata sensu* McKnight) (suspected)
  - *G. gigas* (at least in part *G. montana*)
  - *G. infula*
  - *G. sphaerospora* (suspected)

Pezizaceae
- *Sarcosphaera*
  - *S. crassa* (suspected)

All the species listed occur in North America. The list is arranged alphabetically within families. Most of the species in this list occur or are likely to occur in Canada or in areas of the United States adjacent to Canada. The remainder of this chapter contains a description of the genus *Gyromitra,* followed by descriptions of those species known to contain the toxin, arranged alphabetically.

Descriptions of the Genus and Species

Gyromitra Fr.

Helvellaceae

Species of *Gyromitra*, commonly called the false morels, are widely distributed in North America. At certain times of the year, especially in the spring, some false morels, particularly *G. esculenta*, are abundant and comprise a large proportion of the flora of the fleshy fungi.

The genus *Gyromitra* belongs to the class Discomycetes of the Ascomycota. Although it has a cap and stalk like gill mushrooms and boletes of the Basidiomycota, *Gyromitra* is not related to these fungi. Other fleshy genera belonging to the Discomycetes include *Morchella* (morels), *Helvella, Verpa,* and others.

The apothecia of *Gyromitra* have a distinct stalk that may be grooved or even interiorly folded. The stalk is typically hollow and is sometimes narrow in relation to the width of the cap or thick to more massive than the cap. The cap varies in shape; it may be saddle-shaped, folded to convoluted and brain-like, or more or less irregularly lobed. The surface varies from smooth to strongly wrinkled but is never distinctly pitted as in the genus *Morchella*. The color of the cap surface is ocherous to yellow brown, reddish brown to dark brown, or bay red. The stalk is often white or tinted with the colors of the cap but it is usually not darker in color than the cap surface. The sterile undersurface of the cap is continuous with the stalk and is also paler than the cap surface. In *Gyromitra* and other stalked Discomycetes the spore-producing layer is situated on the cap surface, rather than on the undersurface of the cap as in gill mushrooms and boletes.

Species now placed in *Gyromitra* have been, and still are, placed in *Helvella* by some authors. The two genera are difficult to separate because the best features separating them require detailed study of the development of the fruit body and of the microscopic characteristics of the ascus. The best way to separate the genera in the field is by the color of the apothecium, especially the cap. The apothecium of *Helvella* is typically white to pallid or pale gray to black, whereas that of *Gyromitra* is bay to red brown, light brown, or tan to ocherous. *Helvella* species that have brown apothecia often are cup- or saucer-shaped but may be saddle-shaped. The cap of the apothecium of *Gyromitra* is not normally cup-like.

Morchella can be separated from both *Gyromitra* and *Helvella* by the distinct pits on the cap surface. *Verpa* species have a more or less conic cap that may have elongated pits or longitudinal ridges and folds or may be nearly smooth. They are like morels in shape and general appearance. For further discussion of *Helvella, Morchella,* and *Verpa* refer to Chapter 13. See Chapters 2 and 3, "Ascomycota", for a description of the general features of the division.

The key to selected species of *Gyromitra* that follows is adapted from a volume by Smith and Smith (1973). *G. californica, G. caroliniana,* and *G. korfii,* which are only suspected of containing the toxin, are not treated further here. Smith and Smith (1973) provide descriptions for *G. californica* and *G. caroliniana*. Harmaja (1973) discusses *G. korfii*.

Key to selected species

1. Cap typically saddle-shaped, either like a simple saddle or a compound structure 2

   Cap folded to convoluted or brain-like, occasionally irregularly saddle-shaped 4

2. Cap compound and more or less saddle-shaped. Stalk thick and composed of several interconnecting channels. Spores ornamented. Fungi growing on humus, fruiting in the spring .... ***G. fastigiata***

   Cap not compound. Stalk typically with a single channel. Spores smooth. Fungi fruiting in the late summer and fall, often occurring on rotting wood but also terrestrial 3

3. Spores mostly 16–22(–26) × 6.5–9.5(–10) μm in size, ellipsoidal to oblong-subfusiform. Fungi often occurring on rotting wood.................. ***G. infula***

   Spores mostly 22–28 × 8.0–9.5 μm in size, ellipsoidal to subfusiform. Fungi associated with northern coniferous forests, found on disturbed sites or bare soil............................. ***G. ambigua***

4. Cap at times more or less saddle-shaped, usually somewhat rounded to more or less flattened, and typically strongly wrinkled to folded or convoluted to brain-like. Stalk usually with a single channel, which can be obscured when the stalk is compressed. Spores smooth, 16.5–23.5 × 7–9.5 μm in size. Fungi typically fruiting in spring, or into summer in the mountains, associated with coniferous forests............................***G. esculenta***

   Cap usually folded to brain-like. Stalk thick and typically composed of several interior folds or channels, especially near the base. Spores smooth to faintly reticulate, (21–)24–36(–37.5) × (9–)10.5–16 μm in size; mature spores with a small apiculus at each end ±1 μm long .... ***G. gigas***

   see discussion under *G. gigas* for similar species

Gyromitra ambigua (Karst.) Harmaja

Figs. 42, 45

Common names. None.

Distinguishing characteristics. Cap 30–100 mm broad, saddle-shaped or at time irregularly lobed; surface smooth to irregularly wrinkled; color dark reddish brown or somewhat blackish when dried. Stalk 10–60 mm long, 7–20 mm thick, more or less equal or enlarged below, cylindrical; surface smooth, depressed, or folded, usually with a single channel that may be stuffed; color pale brownish and often violet-tinged. Ascospores ellipsoidal to subfusiform, smooth, 22.6–28.2 × 8.0–9.5 μm in size, eight per ascus. Fungi fruiting in the late summer and early fall.

Observations. Compare *G. ambigua* with *G. infula*. The technical description below is adapted in part from that of Harmaja (1969).

Distribution and seasonal occurrence. The distribution of *G. ambigua* is not well known. According to Harmaja (1969), it fruits from August to November in Europe. We have collected it in middle to late August in the boreal forest.

Habitat and habit. This species is solitary or occurs in groups, often near conifers; it is also found on disturbed sites, on bare soil, and along paths and roadsides (Harmaja 1969).

Technical description. Cap 30–100 mm broad, saddle-shaped to, more rarely, irregularly lobed. Surface of the hymenium smooth to irregularly wrinkled with age, dark reddish brown, often black when dry. Excipulum with indistinctive taste and odor.

Stalk 10–60 mm long, 7–20 mm thick; surface smooth, depressed to folded, somewhat tomentose, equal or expanded downward, hollow with a single chamber or stuffed; color pale brownish, more or less violet-tinged.

Ascospores white to yellowish, 22.6–28.2 × 8.0–9.5 μm, ellipsoidal to subfusiform, with narrowed apices, hyaline, biguttulate, blue in the reagent Cotton Blue, smooth; walls thickish. Asci eight-spored, cylindrical, 180–290 × 10.5–17.9(–20.0) μm, reddish brown in potassium hydroxide, darker in Melzer's reagent. Paraphyses clavate to cylindrical, thin-walled, 3.5–9.4 μm wide, sometimes encrusted; contents reddish.

Excipulum interwoven; hyphae cylindrical to somewhat inflated, (3.7–)9.4–14.3 μm wide, hyaline, sometimes with reddish brown contents. Stalk hyphae interwoven, cylindrical to somewhat inflated, 4.8–16.7 μm wide, hyaline; cuticle hyphae projected as erect hyphal ends. Oleiferous hyphae reddish brown, refractive, scattered in the stalk, 3.5–7.2 μm wide.

Useful references. Harmaja (1969) gives a description and discussion of European specimens. Wells and Kempton (1968) describe this species from Alaska under the name *G. infula*.

Gyromitra esculenta (Fr.) Fr.

Figs. 43, 46

Common names. False morel, beefsteak morel, edible morel, brain mushroom, lorel (lorchel), edible gyromitra, elephant ears.

Distinguishing characteristics. Cap 20–100 mm broad, rounded or sometimes more or less flattened or saddle-shaped; surface sometimes nearly smooth, but typically strongly wrinkled to folded into many convolutions and appearing more or less brain-like; color variable, sometimes yellowish to yellowish brown, more commonly light to dark reddish brown and becoming darker on drying. Stalk 20–150 mm long, 10–25 mm thick, cylindrical, often tapering upward and enlarged at the base; surface often longitudinally grooved, hollow within and typically with a single channel, which is obscured when the stalk is compressed; color white to pale brown or tinted with the color of the cap. Eight ascospores per ascus, ellipsoidal, smooth, 16.7–23.4 × 7.3–9.4 μm in size. Fungi fruiting in the spring, particularly in coniferous forests.

Observations. The early fruiting period, the rounded shape and brain-like configuration of the cap, and the more or less cylindrical stalk with a single channel are good field characteristics. The color of the cap varies but it, too, can be helpful in identification.

Distribution and seasonal occurrence. This species is widely distributed in temperate coniferous forests across North America and Europe. It typically fruits in the spring, but fruiting may occur in early summer in the western mountains, depending on elevation, slope exposure, rate of snow melt, and weather conditions. It often fruits at the same time as *Morchella*.

Habitat and habit. This species is sometimes solitary but is usually found scattered or in groups. It is terrestrial, typically found with conifers, especially pine (*Pinus*). However, it has also been reported with aspen (*Populus*) (Smith and Smith 1973).

Technical description. Cap 20–100 mm broad, irregularly subglobose, sometimes more or less flattened or depressed on top and tending to be saddle-shaped, irregularly three- to five-lobed. Surface of the hymenium nearly smooth in some but more often strongly wrinkled or folded, but not pitted; color variable, yellowish to yellowish brown or more commonly light to dark

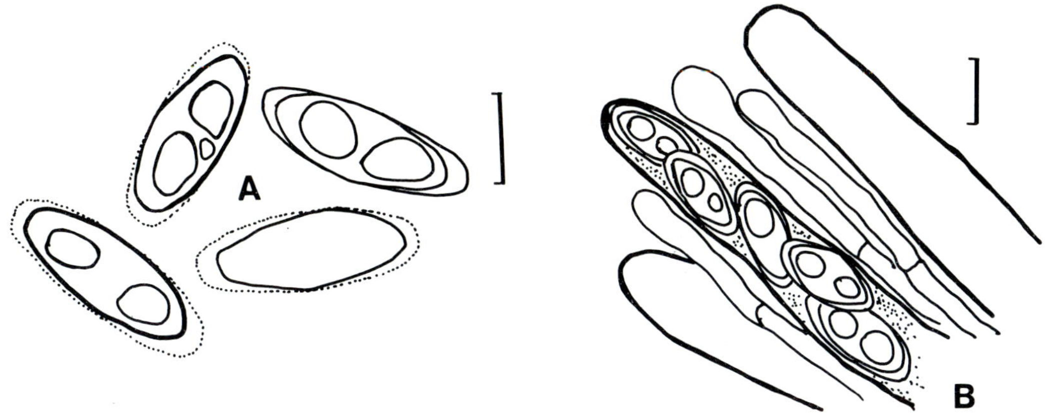

42

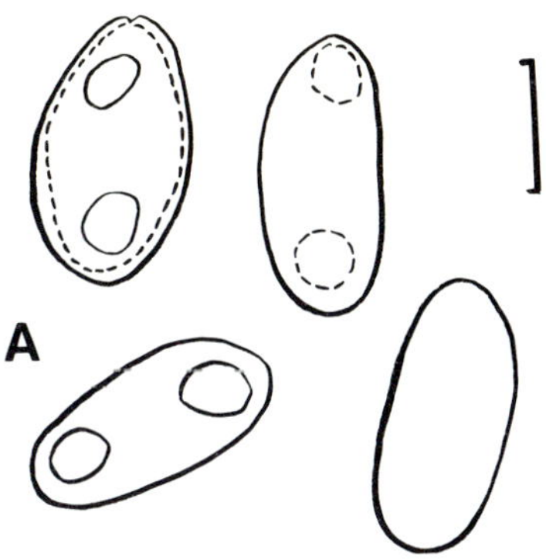

43

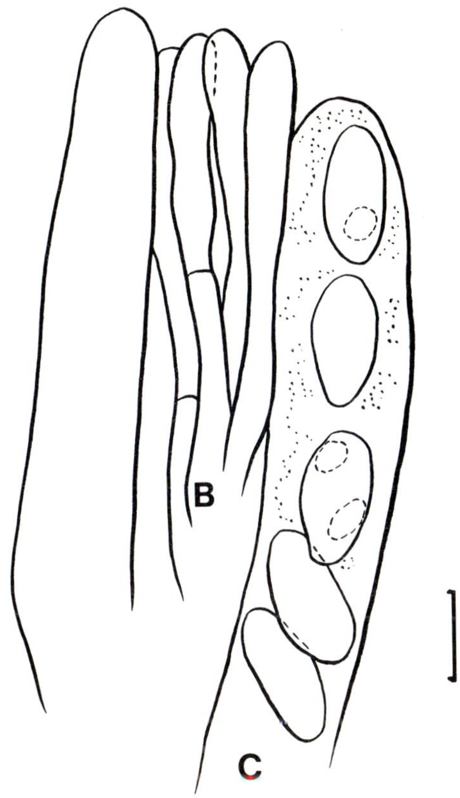

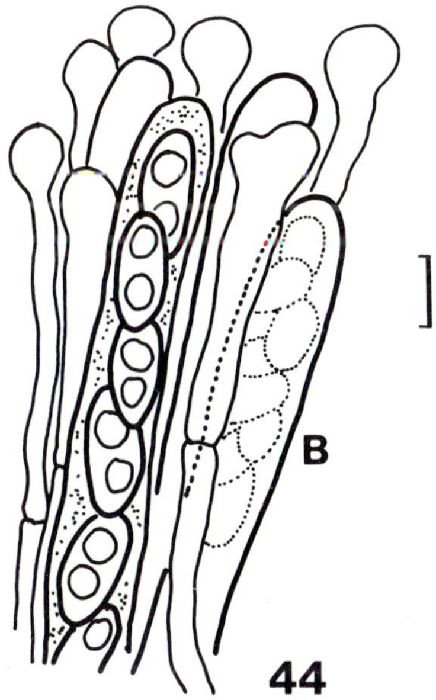

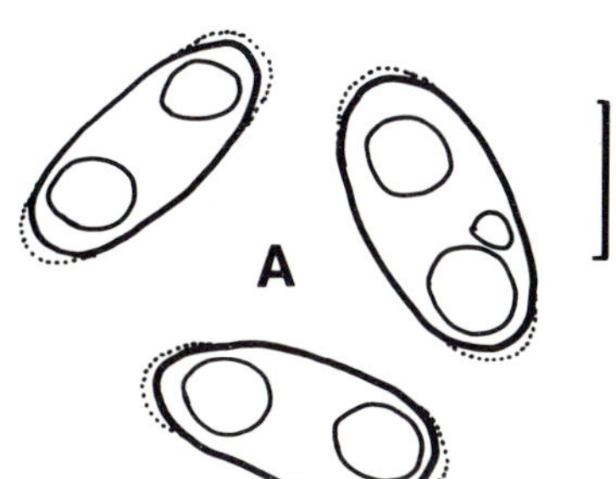

44

Fig. 42. *Gyromitra ambigua*: A, spores; B, asci and paraphyses.

Fig. 43. *Gyromitra esculenta*: A, spores; B, asci and paraphyses.

Fig. 44. *Gyromitra infula*: A, spores; B, asci and paraphyses.

Scale line 10 μm

reddish brown, usually darkening when dry. Excipulum whitish to pallid, thin; taste and odor mild.

Stalk 20–150 mm long, 10–25 mm thick, cylindrical but often tapering upward and often enlarged at the base to form a short irregular bulb, hollow, with a single channel or stuffed, terete to compressed and more or less wrinkled to sulcate, sometimes fused with the inner surface of the cap, glabrous to scurfy; color white to pale brown or tinted with colors of the cap. Flesh soft, white, hollow with a single channel that may be obscured when the stalk is compressed.

Ascospores 16.7–23.4 × 7.3–9.4 μm, ellipsoidal, smooth, biguttulate, thin- to slightly thick-walled, hyaline. Asci eight-spored, clavate-cylindrical, thin-walled, 180–250 × 14–16.2 μm; contents hyaline to reddish brown. Paraphyses cylindrical, septate, 5.8–6.6 μm wide, thin-walled; contents dull red to reddish brown.

Excipulum interwoven; hyphae 4.4–17.9 μm wide, cylindrical to inflated, hyaline. Stalk hyphae interwoven, 4.4–23.4 μm wide, hyaline to yellowish; hyphal ends erect, cylindrical-clavate, 4.4–12.9 μm wide. Oleiferous hyphae scattered, reddish, 4.4–9.4 μm wide.

Useful references. The following books contain information on *G. esculenta* pertinent to North America.

- *How to Know the Non-Gilled Fleshy Fungi* (Smith and Smith 1973)—A good key to *G. esculenta* and other species is included.
- *A Field Guide to Western Mushrooms* (Smith 1975)—A color photograph, a black-and-white photograph, and a description and discussion are presented.
- *Mushrooms in their Natural Habitats* (Smith 1949)—A description and discussion under the name *Helvella esculenta* (Pers.) Fr., as well as a color photograph on a reel, are presented.
- *Mushrooms of North America* (Miller 1972)—A color photograph and a description and discussion are given.
- *The Mushroom Hunter's Field Guide* (Smith 1974)—A color photograph, a black-and-white photograph, and a description and discussion are presented.
- *The Savory Wild Mushroom* (McKenny 1971)—A black-and-white photograph and a description and discussion are given.

Gyromitra fastigiata (Krombh.) Rehm

Fig. 47

Common names. Brown false morel, brown gyromitra.

Distinguishing characteristics. Cap 50–120 mm broad, somewhat saddle-shaped but often lobed and folded with lobes sometimes overlapping but not fused at their margins or with the stalk; surface more or less wrinkled, strong brown to moderate reddish brown. Stalk 60–90 mm long, 25–60 mm thick, usually expanded at the base and longitudinally ribbed; interior composed of several interconnecting channels; surface white. Ascospores 24–30 × 11.5–14 μm, ellipsoidal; surface coarsely reticulate-warty.

Observations. *G. fastigiata* is well discussed by McKnight (1973) under the name *G. brunnea* Underw. Another name for this species is *Helvella underwoodii* Seaver.

Distribution and seasonal occurrence. This species occurs in the Great Lakes region and in the southeastern and midwestern United States. It fruits in early to late spring.

Habitat and habit. This species is solitary or can be found in groups on soil, litter, or well-rotted wood in hardwood forests.

Technical description. Cap 50–120 mm broad, somewhat saddle-shaped but often lobed and folded; lobes not fused at their margins; margin thick, rounded, sterile, not fused with the stalk. Surface of the hymenium more or less wrinkled; color strong brown to moderate reddish brown, often with more reddish overtones; sterile undersurface rough and scurfy in early stages, light grayish yellowish brown to yellowish gray or whitish. Excipulum with indistinctive odor and taste.

Stalk 60–90 mm long, 25–60 mm thick, expanded at the base, sometimes with branches visible below the lowest lobes of the cap, coarsely and irregularly ribbed longitudinally; surface white, matted, fibrillose; interior composed of connecting channels that are hollow or stuffed.

Ascospores 24–30 × 11.5–14.0 μm, ellipsoidal, slightly flattened, thick-walled, hyaline with one to three guttulae, coarsely reticulate with small slender spines arising from the intersections; ornamentation 0.5–1.4 μm high. Asci eight-spored, cylindrical; apex broadly rounded, 18–25 × 400–525 μm; walls showing no reaction in Melzer's reagent; contents bright yellowish orange. Paraphyses 5–9 μm wide, with three to five septa, thin-walled, light yellowish brown, with granular contents.

Excipulum composed of slender interwoven cells.

Useful references. The following books contain information on *G. fastigiata* pertinent to North America.

- *Mushrooms of North America* (Miller 1972)—A color photograph and a description and discussion are given, under the name *G. brunnea.*
- *The Mushroom Hunter's Field Guide* (Smith 1974)—A black-and-white photograph and a description and discussion are presented, under the name *Helvella underwoodii.*

See also the technical description and discussion by McKnight (1973) and Smith and Smith (1973).

Gyromitra gigas (Krombh.) Quél.

Fig. 48

Common names. Snow mushroom, snow morel, giant helvella.

Distinguishing characteristics. Cap irregularly shaped but basically rounded to ellipsoidal, 50-180 mm across; surface strongly convoluted to folded, strong yellowish brown to strong brown or moderate brown. Stalk thick (30-150 mm) and fleshy, hollow with several channels; base even to expanded; color white or nearly so. Ascospores smooth to faintly reticulate, (21.4-)24.3-35.8 (-37.5) × (9-)10.7-15.8 μm in size.

Observations. Perhaps more has been written about *G. gigas* in recent years than about any of the other species of *Gyromitra*. The description of *G. gigas* given below is taken from McKnight (1971) and is similar to the one given by Smith (1949) and Smith and Smith (1973). Harmaja (1973) calls the American form of this species *G. montana* Harmaja and is still not convinced that the true *G. gigas* occurs in North America. There are other related species, such as *G korfii* (=*G. fastigiata sensu* McKnight); for more information consult the technical literature. Until the problem of identification is settled, we prefer to use the name *G. gigas* rather than *G. montana* for this species. Because some forms of *G. gigas* appear to contain gyromitrin, it is essential to be absolutely sure of species identification when studying toxins or poisonings in this group.

G. gigas sensu McKnight (1971) is similar in appearance to *G. korfii*. The two are best separated on the basis of spore shape; spores of *G. korfii* are more tapered toward the ends and in general are distinctly fusiform. The spores of *G. korfii* also have a well-developed apiculus at each end compared with the apiculi of *G. gigas*, which are shorter. *G. gigas* is found in western North America, often around melting snowbanks in the mountains. *G. korfii* occurs in the west but is most often found in the east.

Distribution and seasonal occurrence. *G. gigas* occurs in western North America and is most abundant in the mountains, especially the Rockies. It fruits in the spring and early summer, depending somewhat on the year and the location.

Habitat and habit. This species is solitary or can be found in groups, on soil in coniferous or mixed forests, often associated with melting snowbanks.

Technical description. Cap 50-180 mm across, irregularly shaped but roughly globose to ellipsoidal, strongly convolute. Surface of the hymenium flattened against the stalk and sometimes fused with it; color strong yellowish brown to strong brown or moderate brown. Excipulum white or nearly so; odor and taste indistinctive.

Stalk thick and fleshy, 20-140 mm long, 30-150 mm thick, even or expanded at the base, with rounded longitudinal ribs, and hollow with several connecting channels; surface white or nearly so.

Ascospores whitish in deposit, (21.4-)24.3-35.8 (-37.5) × (9-)10.7-15.8 μm, ellipsoidal, typically flattened on one side, smooth or very faintly roughened with incomplete reticulum, and with one to three guttulae; apiculi very short and truncate or more often broadly rounded or lacking, 0-1.1 μm long, hyaline. Asci eight-spored, cylindrical, 350-400 × 18-24 μm, strong orange yellow in Melzer's reagent. Paraphyses with two to four septa above the branches; terminal cells cylindrical-capitate, contorted, 4-12 μm wide, pale yellow in Melzer's reagent.

Excipulum consisting of compacted hyphae 7-13 μm wide.

Useful references. The following books contain information on *G. gigas* pertinent to North America.

- *How to Know the Non-Gilled Fleshy Fungi* (Smith and Smith 1973)—A technical description and discussion are included.
- *Edible and Poisonous Mushrooms of Canada* (Groves 1979)—*G. gigas* is discussed under *G. esculenta*.
- *Mushrooms of North America* (Miller 1972)—A color photograph and a description and discussion are given.
- *The Mushroom Hunter's Field Guide* (Smith 1974)—A color photograph, a black-and-white photograph, and a description and discussion are presented.
- *The Savory Wild Mushroom* (McKenny 1971)—A black-and-white photograph and a description and discussion are given.

An additional technical description and a discussion are given by McKnight (1971).

Gyromitra infula (Schaeff. ex Fr.) Quél.

Figs. 44, 49

Common names. Hooded false morel, hooded helvella, bay gyromitra.

Distinguishing characteristics. Cap 30-100 mm broad, typically saddle-shaped but sometimes indistinctly so or even more irregular in outline; surface smooth to distinctly wrinkled; color cinnamon, dull yellowish brown to dark reddish brown, or grayish reddish brown to dark brown. Stalk 10-60 mm long, 7-20 mm thick, equal or enlarged downward; surface smooth or irregularly depressed to folded, hollow with a single channel; color dingy brownish or at times nearly white, often with purplish brown to purplish or reddish purple tints. Fungi fruiting from the summer late into the fall, often occurring on rotting wood. Eight spores per ascus, ellipsoidal to oblong-subfusiform, smooth, 16.8-21.9(-26) × 6.4-9.5(-10) μm in size.

Observations. *G. infula* is best characterized by the saddle-shaped cinnamon to reddish brown cap, the stalk with a single hollow channel, and the habit of growth on rotten wood or soil rich in woody debris. It typically fruits after *G. esculenta*. *G. ambigua* has no doubt been confused with *G. infula* in North America. They are similar in color and stature, and are best distinguished by microscopic characteristics. The main difference seems to be in spore size. The spores of *G. ambigua* are larger (22.6-28.2 × 8-9.5 μm) than those of *G. infula*. There are other microscopic differences but they require careful study and considerable knowledge of the group. Harmaja (1969) gives a discussion of these two species in Europe. *G. ambigua* has not been generally reported from North America. We have, however, collected material in the Canadian boreal forest.

Distribution and seasonal occurrence. *G. infula* is widely distributed and common in North America. It occasionally occurs in the spring or early summer (McKenny 1971) but is typically a late summer and fall species. It is sometimes abundant in a particular location.

Habitat and habit. This species is solitary to scattered and typically grows on rotten wood, but also in humus or in woody debris.

Technical description. Cap 30-100 μm broad, typically saddle-shaped but sometimes indistinctly so, occasionally quite irregular in outline or lobed; surface of the hymenium smooth to distinctly wrinkled or somewhat convoluted; color cinnamon, dull yellowish brown to dark reddish brown, or grayish reddish brown to dark brown, becoming darker with age or on drying. Excipulum with a mild taste and indistinctive odor.

Stalk 10-60 mm long, 7-20 mm thick; surface smooth to irregularly depressed or folded but never strongly fluted, more or less tomentose, equal or expanded downward, hollow with a single channel or stuffed; color dingy brownish to nearly white, often with purplish brown to purplish or reddish purple tints.

Ascospores white to yellowish in deposit, 16.8-21.9 (-26.0) × 6.6-9.5(-10.0) μm, ellipsoidal to oblong-subfusiform with broad rounded apexes, smooth, thick-walled, hyaline, deep blue in Cotton Blue reagent. Asci eight-spored, 200-250 × 11.7-13.1 μm, cylindrical, reddish brown, but darker reddish brown in Melzer's reagent; apex obtuse-truncate. Paraphyses 3.5-5.8 μm wide, thin-walled, straight, septate with expanded apex 7-12.9 μm wide, sometimes encrusted; contents reddish brown.

Excipulum compactly interwoven; hyphae 3.7-13.1 μm wide, hyaline, thin-walled. Stalk interwoven; hyphae somewhat short, cylindrical, 3.5-16.7 μm wide, hyaline to brownish; cuticle hyphae interwoven, 9.4-11.7 μm, having clavate erect hyphal ends. Oleiferous hyphae refractive, 3.7-7.0 μm wide, red brown.

Useful references. The following books contain information on *G. infula* pertinent to North America.

- *A Field Guide to Western Mushrooms* (Smith 1975)—A color photograph, a black-and-white photograph, and a description and discussion are presented.
- *Edible and Poisonous Mushrooms of Canada* (Groves 1979)—A color photograph, a black-and-white photograph, and a description and discussion are included.
- *Mushrooms in their Natural Habitats* (Smith 1949)—A description and discussion, as well as a color photograph on a reel, are presented.
- *Mushrooms of North America* (Miller 1972)—A color photograph and a description and discussion are given.
- *The Mushroom Hunter's Field Guide* (Smith 1974)—A color photograph, a black-and-white photograph, and a description and discussion are presented.
- *The Savory Wild Mushroom* (McKenny 1971)—A black-and-white photograph and a description and discussion are given.

Literature

Back, K. C.; Pinkerton, M. L. Toxicology and pathology of repeated doses of monomethylhydrazine in monkeys. Aerospace Medical Research Laboratory, Wright-Paterson Air Force Base, OH. AMRL-TR 66-199; 1967.

Benedict, R. G. Mushroom toxins other than *Amanita*. Kadis, S., et al., eds. Microbiological toxins. Vol. 8. New York, NY: Academic Press; 1972.

Groves, J. W. Edible and poisonous mushrooms of Canada. Ottawa, Ont.: Agriculture Canada. Agric. Can. Publ. 1112; 1979 [first printed in 1962].

Harmaja, H. A neglected species, *Gyromitra ambigua* (Karst.) Harmaja n. comb., and *G. infula* s. str. in Fennoscandia. Karstenia 9:13-19; 1969.

Harmaja, H. Amendments of the limits of the genera *Gyromitra* and *Pseudorhizina*, with the description of a new species, *Gyromitra montana*. Karstenia 13:48-58; 1973.

Lincoff, G.; Mitchel, D. H. Toxic and hallucinogenic mushroom poisoning—A handbook for physicians and mushroom hunters. New York, NY: Van Nostrand Reinhold Co.; 1977.

List, P. H.; Luft, P. Nachweis und Gehaltsbestimmung von Gyromitrin im frischen Lorcheln. Arch. Pharmacol. 303:143-146; 1969.

McKenny, M. The savory wild mushroom (revised and enlarged by Stuntz, D. E.). Seattle, WA: University of Washington Press; 1971.

McKnight, K. H. On two species of false morels (*Gyromitra*) in Utah. Great Basin Nat. 31:35–47; 1971.

McKnight, K. H. Two misunderstood species of *Gyromitra* (false morels) in North America. Mich. Bot. 13:147–162; 1973.

Miller, O. K., Jr. Mushrooms of North America. New York, NY: E. P. Dutton & Co., Inc.; 1972.

Smith, A. H. Mushrooms in their natural habitats. 2 Vols. Portland, OR: Sawyer's, Inc.; 1949.

Smith, A. H. The mushroom hunter's field guide. Ann Arbor, MI: University of Michigan Press; 1974 [first printed in 1963].

Smith, A. H. A field guide to western mushrooms. Ann Arbor, MI: University of Michigan Press; 1975.

Smith, H. V.; Smith, A. H. How to know the non-gilled fleshy fungi. Dubuque, IA: William C. Brown Co.; 1973.

Wells, V. L.; Kempton, P. E. Studies on the fleshy fungi of Alaska II. Mycologia 60:888–901; 1968.

Fig. 45. *Gyromitra ambigua.*

Fig. 46. *Gyromitra esculenta.*

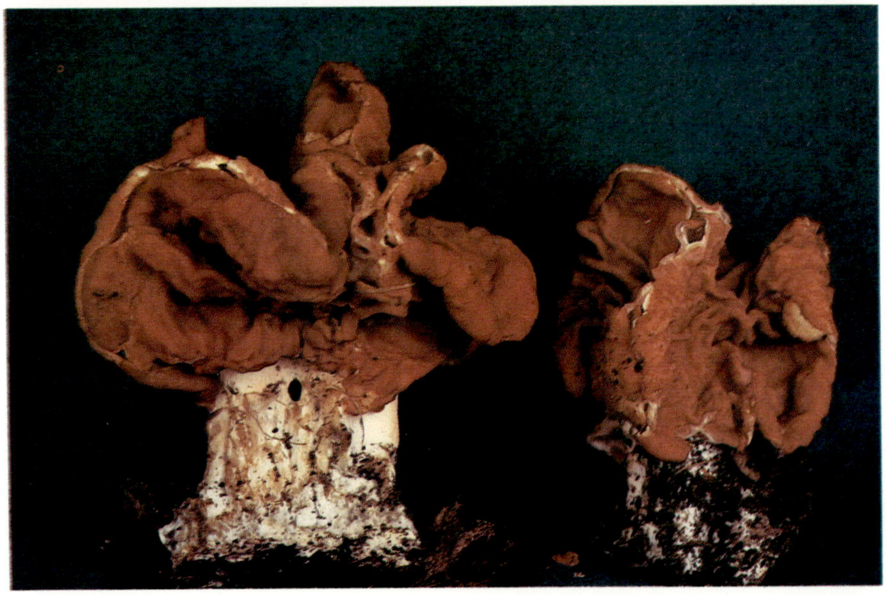

Fig. 47. *Gyromitra fastigiata.*

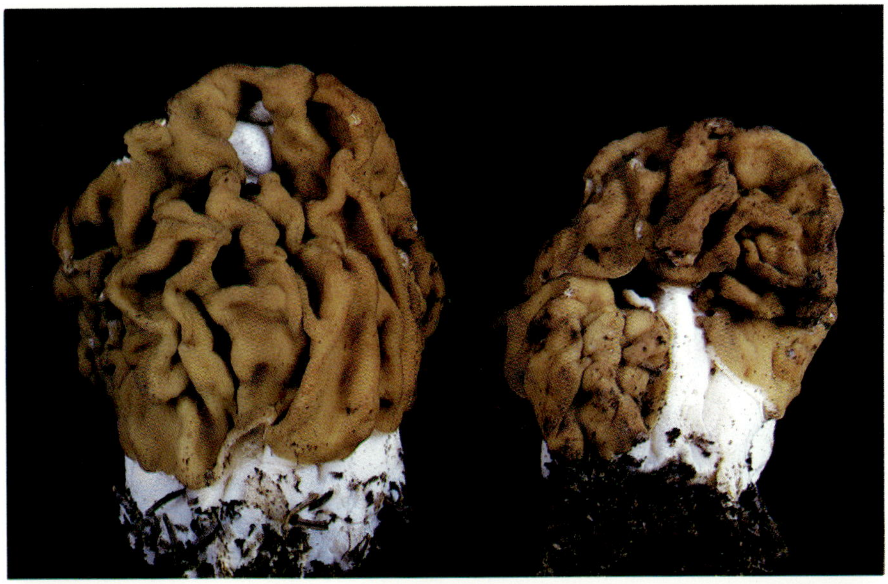

Fig. 48. *Gyromitra gigas.*

Fig. 49. *Gyromitra infula.*

9

COPRINE

Toxicology and Symptomatology

A few mushrooms, mainly in the genera *Coprinus* and *Clitocybe,* when consumed with alcohol cause a reaction very similar to that produced by the drug Antabuse (disulfiram), which is given to alcoholics. The reaction is attributed to the toxin coprine (N^5-[1-hydroxycyclopropyl]glutamine). Coprine is an amino acid, a derivative of glutamine, one of the common amino acids found in cellular proteins (Hatfield and Schaumberg 1975, 1978; Lindberg et al. 1975). *Coprinus atramentarius* is the most widely studied coprine-containing species in the genus *Coprinus* producing this reaction. Hatfield and Schaumberg (1978) list *Coprinus atramentarius, C. quadrifidus* Peck, *C. insignis* Peck, and *C. variegatus* Peck as coprine-containing species. They state that *Coprinus comatus* (Müller ex Fr.) S. F. Gray and *C. micaceus* (Bull. ex Fr.) Fr. do not contain coprine. Cochran and Cochran (1978*a*, 1978*b*) report a similar reaction with *Clitocybe clavipes.*

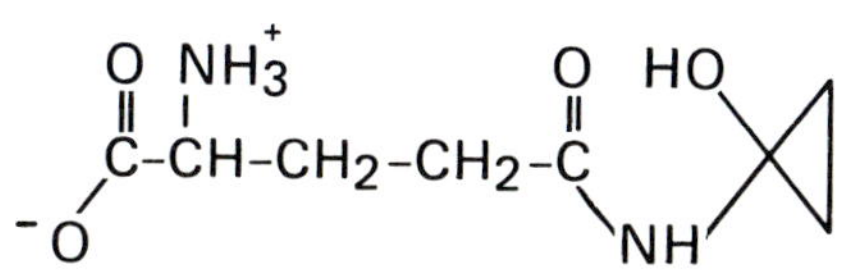

coprine

Other mushroom-alcohol reactions that do not exhibit disulfiram-type symptoms have been reported. Groves (1964) noted an adverse reaction with species of *Morchella* and alcohol, and Shaffer (1965) with *Pholiota squarrosa* and alcohol. *Boletus luridus* and *Laetiporus (Polyporus) sulphureus* have also been associated with alcohol-related poisonings. These mushrooms, however, apparently do not contain coprine and are dealt with further in Chapter 13.

The symptoms of coprine poisoning are similar to acetaldehyde poisoning and result from the blockage of the activity of an enzyme called aldehyde dehydrogenase. Coprine itself likely does not affect the enzyme but is metabolized in the body to a compound that directly acts on the enzyme (Lincoff and Mitchel 1977).

The symptoms of coprine poisoning begin with a slight rise in blood pressure accompanied by marked flushing, especially in the upper half of the body, face, and eyes. The heartbeat may rise to as much as 140 beats a minute, and the victim may experience a pounding headache. In addition, the victim may experience pericardial pain and rapid breathing. After approximately 15 minutes of the above kinds of reactions, blood pressure drops and the victim feels weak, dizzy, and nauseated. Vomiting usually occurs and occasionally the victim actually faints.

The reaction of humans to coprine poisoning is dependent on the level of alcohol in the blood. One drink taken several hours before a meal of coprine-containing mushrooms would probably not cause the above symptoms, whereas three or four drinks taken several hours after the mushrooms are eaten would set off a reaction. Some reports suggest that *C. atramentarius* eaten raw does not cause the characteristic reaction. If the specimens eaten were not simply low in coprine, this finding would suggest that coprine is formed during the cooking process.

The coprine-alcohol reaction usually does not require medical treatment. In severe cases, however, raising the feet and administering oxygen and possibly an antihypotensive prove effective. Lincoff and Mitchel (1977) give further details.

Symptom-producing Species

Coprinaceae
- *Coprinus*
  - *C. atramentarius*
  - *C. insignis*
  - *C. variegatus* (=*C. quadrifidus* and *C. ebulbosus*)

Tricholomataceae
- *Clitocybe*
  - *C. clavipes*

The species listed above produce a disulfiram-type reaction in humans when ingested with alcoholic beverages. All the species listed occur in North America. The list is arranged alphabetically within families. Most of the species in this list occur or are likely to occur in Canada or in Areas of the United States adjacent to Canada. The remainder of this chapter contains descriptions in alphabetic order of the genera in this group.

A species description for *Clitocybe clavipes* is included in this chapter. Of the three species of *Coprinus* listed, only *C. atramentarius* is formally described here, with brief reference therein to *C. variegatus* and *C. insignis.*

Description of Genera and Species

Clitocybe (Fr.) Quél. — Tricholomataceae

The genus *Clitocybe* is widely distributed in Canada and the United States. It is a moderately large genus and in certain modern treatments includes the genus *Lepista*. In general, the genus *Clitocybe* can be distinguished by decurrent to subdecurrent gills; a usually white spore deposit, although the color sometimes ranges from buff or cream color to pale yellowish or pale pinkish buff; often fibrous fruit bodies varying in size from small to large, more or less fleshy, and usually with a centrally attached stalk; a cap with the margin enrolled in the early stages and often with the center depressed at maturity; no partial or universal veil; normally smooth-walled spores, although the walls in some species are slightly roughened or wrinkled; and a cap cuticle typically composed of filamentous hyphae.

Several small genera have been separated from *Clitocybe*. Two of these, *Omphalotus* and *Hygrophoropsis*, are fairly common and at least one, *Omphalotus*, contains poisonous species. *Omphalotus* is treated in Chapter 13 and in poison cases should be closely compared with *Clitocybe*. *Hygrophoropsis aurantiaca* (Wulfen ex Fr.) Maire, also sometimes classified in *Clitocybe* and *Cantharellus*, has been reported as poisonous, but other reports list it as edible, so its status is not yet resolved. The cap, stalk, and gills of this species are usually some shade of orange, but the color at times is pale yellow or strongly brownish. The gills are decurrent, crowded, and repeatedly forked. The orange color and the crowded forked gills help to distinguish it from *Clitocybe*. It often fruits on or near decayed wood.

Other genera that can be confused with *Clitocybe* are *Cantharellus*, *Collybia*, *Laccaria*, *Lactarius*, *Leucopaxillus*, *Lyophyllum*, *Hygrophorus*, and *Tricholoma*. *Cantharellus* typically has blunt-edged, more or less fold-like to ridge-like gills rather than the distinct sharp-edged, blade-like gills of a typical mushroom. *Laccaria* differs in having rather thick and somewhat waxy-looking gills that are often pinkish to purplish in color. The spores are often spiny or roughened with fine points. *Hygrophorus* has many species with decurrent gills; however, the rather thick, waxy gills and long, narrow basidia of this genus separate it from *Clitocybe*. Certain *Leucopaxillus* species have the same stature as *Clitocybe*. All *Leucopaxillus* species have amyloid spores and usually exhibit amyloid ornamentation on the spore surface. *Tricholoma* species are best separated from *Clitocybe* by the notched to adnexed or sinuate gill attachment. *Collybia* is generally separated from *Clitocybe* by the more cartilaginous stalk and adnate to adnexed gill attachment. *Lactarius* species with decurrent gills can be separated from *Clitocybe* by their brittle consistency, the presence of latex in the fruit body, and the amyloid ornamentation on the spores. When *Lyophyllum* has decurrent gills, it is difficult to separate from *Clitocybe*. Check for the presence of siderophilous granules in the basidia of *Lyophyllum* to separate it from *Clitocybe*. As the above white-spored genera may be difficult to separate by the simple comparisons described here, consult carefully the general and technical keys in Chapter 5 when making an identification.

Clitocybe clavipes (Fr.) Kummer — Figs. 50, 52

Common name. Clavate-stalked clitocybe.

Distinguishing characteristics. Cap 20–90 mm broad, plane with the margin narrowly incurved; center slightly raised, more or less depressed with age; surface often moist, glabrous to matted-fibrillose; color brownish to grayish brown to olive brown, darkest in the center, usually paler with age. Gills decurrent, close to subdistant, narrow, often forked and interveined; color white, then creamy to pale yellowish. Odor usually fragrant. Stalk 6–12 mm at the apex; base clavate to somewhat bulbous, 10–35 mm thick; surface streaked with sordid olive buff to pale grayish fibrils, usually whitish beneath the streaks, generally darkest at the base; base densely tomentose. Spores white in deposit, ellipsoidal, smooth, 6.6–9.5(–10.0) × 3.7–5.1 μm in size.

Observations. *C. clavipes* is usually considered edible, except when consumed with alcohol. It is a widely distributed and often common species. The fruit bodies do not decay as readily as those of other mushrooms, often remaining in the woods for a considerable time. Although *C. clavipes* is usually easy to recognize, it does vary in color and texture of the cap, depending on age, moisture, and weather. In some forms the cap color is decidedly grayish and the overall color is generally darker on drying. The margin of the cap is often pale or whitish and the center is usually darker. In wet weather the flesh becomes watery and flaccid; in dry weather, the flesh may become quite brittle. The gills sometimes tend to be subdistant and somewhat thick, thus resembling a *Hygrophorus*. The stalk typically has a clavate base but is often somewhat bulbous or at times

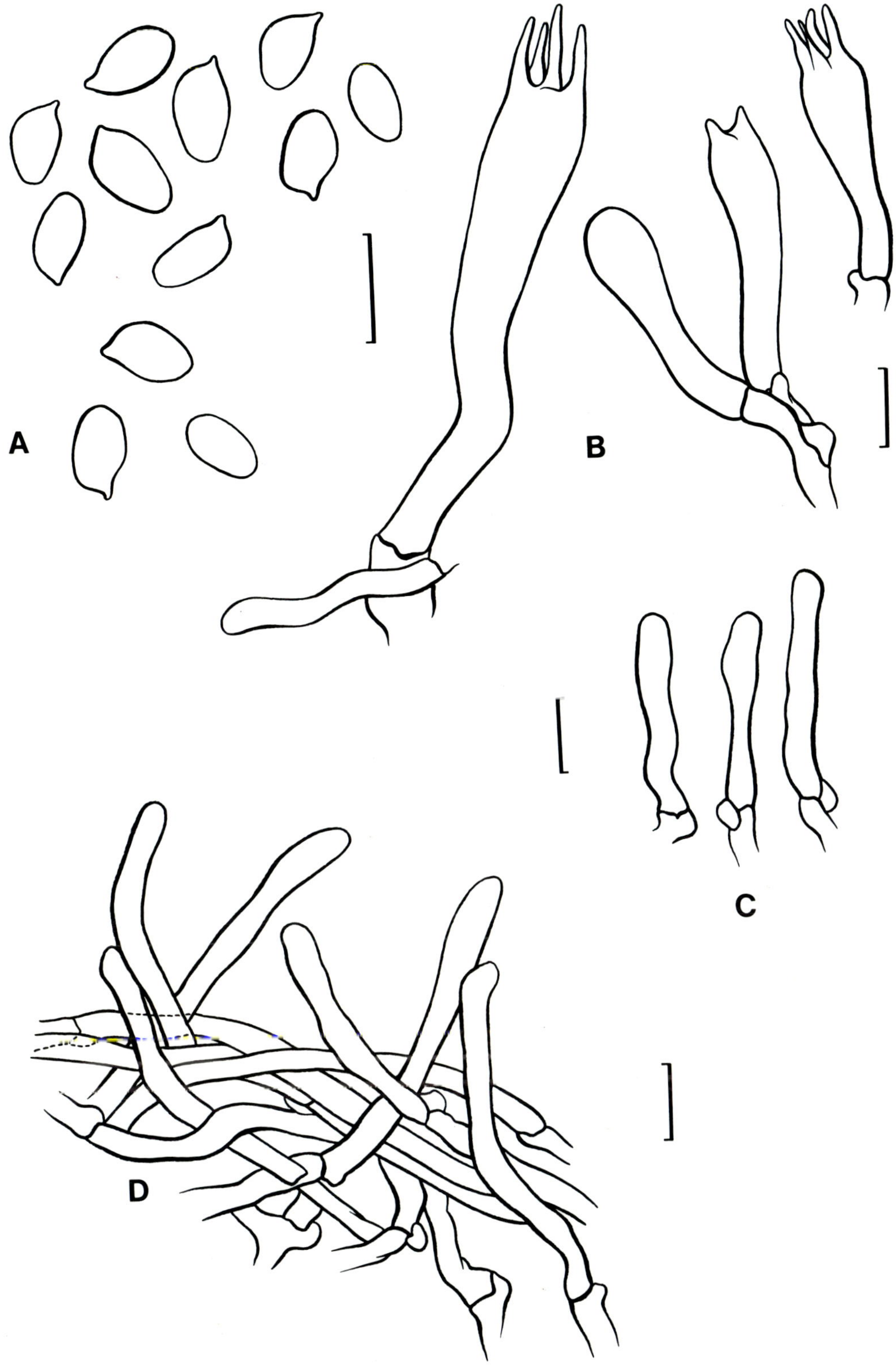

Fig. 50. *Clitocybe clavipes*: A, spores; B, basidia; C, cheilocystidia; D, cap cuticle.

Scale line 10 μm

nearly equal. *C. nebularis* is sometimes confused with *C. clavipes*. *C. nebularis*, however, has a larger cap that is grayish to drab; close to crowded whitish gills; and a disagreeable odor and taste (see description in Ch. 13). *C. avellaneialba* Murrill is also similar to *C. clavipes*. *C. avellaneialba*, however, occurs in the west, usually on or near rotting wood; typically it has a dark brown to olive brown cap when fresh, broadly fusiform spores, and no distinctive odor or taste.

Distribution and seasonal occurrence. *C. clavipes* occurs quite commonly across Canada and the United States. It fruits in the summer and fall, as late as November, depending on weather and location.

Habitat and habit. This species is terrestrial, growing solitary, scattered, or gregarious, rarely subcaespitose, and sometimes occurring in arcs. It typically occurs under conifers but is also found in mixed woods or, less commonly, under hardwoods.

Technical description. Cap 20-90 mm broad, plane in early stages, slightly umbonate at times, shallowly depressed or broadly umbilicate on the disc with age; margin narrowly incurved, rarely enrolled, soon elevated to broadly wavy, sulcate at times; surface moist but not truly hygrophanous, glabrous to matted-fibrillose, nearly scurfy on the disc; color brownish or grayish brown to olive brown, darkest on the disc, paler with age, darker and gray on drying. Flesh thick, watery in wet weather, whitish; odor fruity; taste mild.

Gills white then pale yellowish to creamy, decurrent, close to subdistant, narrow, often forked and interveined; edges even in early stages, broadly wavy with age.

Stalk 35-60 mm long, 6-12 mm thick at the apex, 10-35 mm thick at the base, clavate, somewhat bulbous, rarely equal, solid-stuffed with the interior usually soft and white, often curved and compressed below; base densely tomentose; surface fibrillose-streaked, with sordid olive buff or pale grayish over a whitish background color and darkest at the base.

Spores white in deposit, 6.6-9.5(-10.0) × 3.7-5.1 µm, ellipsoidal, slightly inequilateral in profile, smooth, hyaline, nonamyloid. Basidia four-spored, narrowly clavate, 25-44 × 5.5-8.0 µm, hyaline. Pleurocystidia and cheilocystidia somewhat indistinct, cylindrical, 20-30 × 3.7-4.4 µm, hyaline.

Gill trama interwoven, more or less parallel; hyphae more or less cylindrical, 3.5-13.1 µm wide, hyaline. Subhymenium hyphae interwoven, cylindrical, 3.7-4.4 µm wide, hyaline. Cap cuticle filamentous; hyphae interwoven, more or less radially arranged, cylindrical to clavate, 2.9-5.1 µm wide; hyphal ends projecting, hyaline to brownish gray, thin-walled, slightly encrusted. Cap trama loosely interwoven; hyphae cylindrical, 5.0-12.9 µm wide, hyaline. Caulocystidia narrowly clavate with recurved hyphal ends, 16 × 3.5-5.1 µm, hyaline; stalk hyphae 3.5-23 µm wide, cylindrical to inflated, parallel to interwoven, branched, hyaline; hyphae on surface and from base more interwoven and narrower than those in the flesh. Oleiferous hyphae refractive, yellowish, 5.1-7.3 µm wide, not septate, scattered. Clamp connections present.

Useful references. The following books contain information on *C. clavipes* pertinent to North America.

- *Edible and Poisonous Mushrooms of Canada* (Groves 1979)—A color photograph, a black-and-white photograph, and a description and discussion are provided.
- *Mushrooms of North America* (Miller 1972)—A color photograph and a description and discussion are given.

Bigelow (1965) provides a technical description and a key to this and related species.

Coprinus (Pers. ex Fr.) S. F. Gray

Coprinaceae

The genus *Coprinus* is widely distributed in North America. Its species occupy a variety of habitats and vary considerably in size and stature. They all have a dark brown to blackish brown or black spore deposit and are commonly called inky caps because the gills and often the flesh of the cap dissolve, by a process of autolysis, into an inky fluid at maturity. The gills in *Coprinus* are fairly thin and crowded and their faces are parallel rather than at an angle as in most mushrooms.

Some of the common species of *Coprinus* are easy to recognize in the field, but often microscopic characteristics are needed for species identification. The gills often produce specialized cells, paraphyses and cystidia, which are used in identification. The spores are smooth or ornamented, typically with thick walls and a lens-shaped apical germ pore.

The general appearance and the color of the spore deposit of certain *Coprinus* species resemble those of species of *Panaeolus* or *Psathyrella*. Neither of these genera, however, have deliquescent gills or flesh. The genus *Pseudocoprinus* is also similar to *Coprinus* but the gills are not deliquescent. Species of *Pseudocoprinus* are classified in *Coprinus* in this treatment.

A number of the larger *Coprinus* species are edible and some, such as *C. comatus*, are highly sought for food. In general, species of *Coprinus* are considered edible if the young specimens are gathered and eaten before any breakdown or digestion of the specimens occurs. *C. insignis* Peck from Europe is reported as poisonous (Smith 1975), although there are no reports of poisoning by this species in North America. However, *C. insignis* and *C. variegatus* contain coprine (Hatfield and Schaumberg 1978).

C. atramentarius, a species that is edible and popular as food, often causes poisoning when consumed with alcohol. There are reports of an alcohol reaction with other species of *Coprinus,* such as *C. comatus* and *C. micaceus,* but these poisonings have not been documented. Hatfield and Schaumberg (1978) state that *C. comatus* and *C. micaceus* do not contain coprine. Lincoff and Mitchel (1977) list *C. fuscescens* Fr. as suspected of containing coprine, although we have no information on this species in North America.

Coprinus atramentarius (Bull. ex Fr.) Fr.

Figs. 51, 53

Common names. Inky cap, inky coprinus, gray inkcap.

Distinguishing characteristics. Cap 20–60(–80) mm broad, up to 60 mm high, ovoid to conic or bell-shaped; margin flattened against the stalk when young; surface dry and glabrous or with a silvery luster in early stages; margin typically with distinct radial lines or striations; center sometimes with very small scales; color light grayish brown to sordid brownish in the center, gray toward the margin, becoming blackish from the margin inward as the cap tissue and gills begin to break down. Gills very crowded, free from the stalk or slightly attached; color in early stages whitish to very pallid grayish, becoming more grayish or somewhat brownish but finally black and dissolving into an inky fluid. Stalk 40–150 mm long, 6–12 mm thick, equal or enlarged somewhat below; surface dry and more or less silky-fibrillose; color white, with veil leaving a scurfy-fibrillose to minutely scaly grayish brown to brownish coating on the base of the stalk; more or less distinct basal annular zone present. Spores black in deposit, smooth, ovoid, with an apical germ pore, thick-walled, 7.3–10.2(–12) × 5.1–5.8 μm in size.

Observations. *C. atramentarius* is often called the "true" inky cap. It is widely distributed in North America and is sometimes abundant, occurring in dense clusters. Its availability, together with the firm, fleshy texture of the young fruit bodies, makes it attractive as an apparently edible species.

There are other coprine-containing species of *Coprinus* that are similar in appearance to *C. atramentarius,* including *C. insignis* Peck and *C. variegatus* Peck. According to Patrick (1979) *C. variegatus* is the same species as *C. quadrifidus* and *C. ebulbosus* (Peck) Peck, but *C. variegatus* is the correct name for the species. Patrick (1979) characterizes the *Coprinus variegatus* group by its large size, stature similar to *C. atramentarius,* evanescent annulus, and a dense, matted-fibrillose veil that becomes torn and separated into conspicuous appressed-tomentose patches on the surface of the cap. He further states that it most frequently fruits in the spring and early summer. *C. variegatus* is shown in Fig. 54 here. Groves (1979) reports it as *C. quadrifidus. C. insignis* is best separated from *C. atramentarius* by its roughened spores. Smith (1975) provides a color photograph, a brief description, and a discussion of this species.

Distribution and seasonal occurrence. *C. atramentarius* is a cosmopolitan mushroom that occurs throughout North America. It fruits in the spring, summer, and fall, depending on weather and location. It tends to favor cool, wet weather and fruits most commonly in the fall (Patrick 1979).

Habitat and habit. This species is caespitose and often occurs in large clusters where woody or organic material has accumulated. It often fruits around stumps and has been reported in gardens, lawns, city dumps, along roadsides, around old buildings, and in wooded areas around decaying trees.

Technical description. Cap 20–60(–80) mm broad, up to 60 mm high, ovoid to conic in early stages, conic to bell-shaped when expanded; margin flattened against the stalk when young, longitudinally wrinkled to striate; surface dry; disc smooth or broken up into very minute scales, glabrous or with a silvery luster in early stages; disc light grayish brown to sordid brownish, gray toward the blackish margin. Flesh soft, pallid watery brownish gray, soon deliquescent; taste and odor indistinctive.

Gills whitish to very pallid grayish in early stages, becoming brownish to grayish and finally black, deliquescing into an inky fluid, free or slightly attached at the apex of the stalk, crowded, up to 10 mm broad; edges even.

Stalk 40–150 mm long, 6–12 mm thick, equal or enlarged toward the base, central, hollow, dry, more or less silky-fibrillose above, often splitting into segments; color white. Flesh white. Veil leaving a more or less scurfy-fibrillose to minutely scaly coating of brownish to grayish brown fibrils up to the basal annulus. True universal veil lacking.

Spores black in deposit, 7.3–10.2(–12) × 5.1–5.8 μm, ovoid, smooth, with an apical germ pore; apex truncate; wall thickened, blackish. Basidia four-spored, clavate, 19–25 × 6.6–7.3 μm, hyaline. Pleurocystidia cylindrical to ventricose, 100–120 × 30–40 μm, hyaline, thin-walled. Brachybasidioles vesiculate, 11–16.8 × 9–12.4 μm, thin-walled, collapsing. Cheilocystidia clavate to ventricose, 60–80 × 29–35 μm, hyaline, thin-walled.

Gill trama hyphae more or less parallel, somewhat cylindrical, 3.5–5.1 μm wide, hyaline. Subhymenium more or less cellular, interwoven; hyphae 5.8–11.0 μm wide, hyaline. Cap cuticle more or less cellular to filamentous; hyphae interwoven, more or less radially arranged, 3–7.3(–23.0) μm wide, cylindrical to barrel-shaped, hyaline. Cap trama loosely interwoven; hyphae

Fig. 51. *Coprinus atramentarius*: A, spores; B, basidia and brachybasidioles (*b*); C, pleurocystidia; D, cheilocystidium.

Scale line 10 μm

cylindrical to inflated, 3–24 µm wide, hyaline. Caulocystidia lacking; stalk hyphae parallel to more interwoven at the surface, more or less cylindrical, (3–)8–11 µm wide, hyaline. Oleiferous hyphae refractive, greenish, 3.7–7.3 µm wide, thin-walled, scattered in the cap and gill trama and the stalk. Clamp connections present, more obvious in the hymenium and stalk.

Useful references. The following books contain information on *C. atramentarius* pertinent to North America.

- *A Field Guide to Western Mushrooms* (Smith 1975)—A color photograph, a black-and-white photograph, and a description and discussion are presented.
- *Edible and Poisonous Mushrooms of Canada* (Groves 1979)—A black-and-white photograph and a description and discussion are presented.
- *Mushrooms in their Natural Habitats* (Smith 1949)—A description and discussion, as well as a color photograph on a reel, are presented.
- *Mushrooms of North America* (Miller 1972)—A color photograph and a description and discussion are given.
- *The Mushroom Hunter's Field Guide* (Smith 1974)—A color photograph, a black-and-white photograph, and a description and discussion are presented.
- *The Savory Wild Mushroom* (McKenny 1971)—A color photograph, a black-and-white photograph, and a description and discussion are given.

Literature

Bigelow, H. E. The genus *Clitocybe* in North America: section *Clitocybe*. Lloydia 28(2):139–180; 1965.

Cochran, K. W.; Cochran, M. W. *Clitocybe clavipes*: Antabuse-like reaction to alcohol. Mycologia 70:1124–1126; 1978*a*.

Cochran, K. W.; Cochran, M. W. Interaction of *Clitocybe clavipes* and alcohol. McIlvainea 3(2):32–35; 1978*b*.

Groves, J. W. Poisoning by morels when taken with alcohol. Mycologia 56:779–780; 1964.

Groves, J. W. Edible and poisonous mushrooms of Canada. Ottawa, Ont.: Agriculture Canada; 1979. Agric. Can. Publ. 1112 [first printed in 1962].

Hatfield, G. M.; Schaumberg, J. P. Isolation and structural studies of coprine, the disulfiram-like constituent of *Coprinus atramentarius*. Lloydia 38(6):489–496; 1975.

Hatfield, G.; Schaumberg, J. The disulfuram-like effect of *Coprinus atramentarius* and related mushrooms. Rumack, B. H.; Salzman, E., eds. Mushroom poisoning: diagnosis and treatment. West Palm Beach, FL: CRC (Chemical Rubber Company) Press; 1978.

Lincoff, G.; Mitchel, D. H. Toxic and hallucinogenic mushroom poisoning—A handbook for physicians and mushroom hunters. New York, NY: Van Nostrand Reinhold Co.; 1977.

Lindberg, P.; Bergman, R.; Wickberg, B. Isolation and structure of coprine, a novel physiologically active cyclopropane derivative from *Coprinus atramentarius* and its synthesis via 1-aminocyclopropanol. J. Chem. Soc. Chem. Commun. 1957:946–947; 1975.

McKenny, M. The savory wild mushroom (revised and enlarged by Stuntz, D. E.). Seattle, WA: University of Washington Press; 1971.

Miller, O. K., Jr. Mushrooms of North America. New York, NY: E. P. Dutton & Co., Inc.; 1972.

Patrick, W. W. Comparative morphology and taxonomic disposition of *ebulbosus, quadrifidus,* and *variegatus* in the genus *Coprinus* (Agaricales). Mycotaxon 10(1):142–154; 1979.

Shaffer, R. L. Poisoning by *Pholiota squarrosa*. Mycologia 57:318–319; 1965.

Smith, A. H. Mushrooms in their natural habitats. 2 Vols. Portland, OR: Sawyer's, Inc.; 1949.

Smith, A. H. The mushroom hunter's field guide. Ann Arbor, MI: University of Michigan Press; 1974 [first printed in 1963].

Smith, A. H. A field guide to western mushrooms. Ann Arbor, MI: University of Michigan Press; 1975.

Fig. 52. *Clitocybe clavipes.*

Fig. 53. *Coprinus atramentarius.*

Fig. 54. *Coprinus variegatus.*

10

MUSCARINE

Toxicology and Symptomatology

Muscarine is a toxin found in various families of mushrooms, which affects the parasympathetic nerve endings of the autonomic nervous system. Muscarine was the first mushroom toxin to be isolated from and named after a poisonous mushroom. Remarkably, *Amanita muscaria,* for which the toxin was named, contains very little muscarine. In fact, it is now widely accepted that *A. muscaria* normally does not contain enough muscarine to cause serious human poisoning. Many species of *Inocybe* and certain species of *Clitocybe* are rich in muscarine.

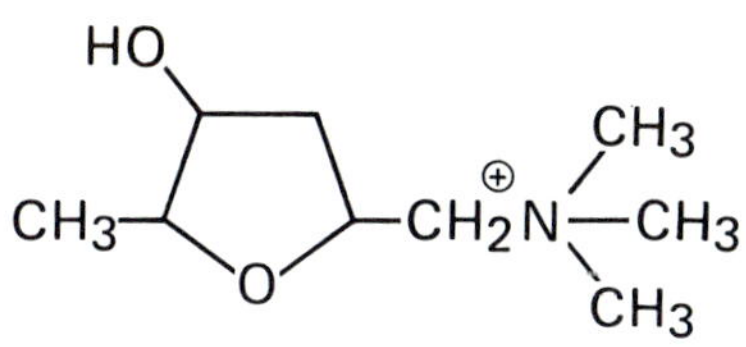

muscarine

Muscarine is a relatively small and simple molecule chemically, containing nine carbon atoms in addition to hydrogen, oxygen, and nitrogen. The muscarine content of mushrooms has been measured by various bioassay methods, most of which give erroneously high determinations because of other substances that interfere with measurements. A good analytical procedure for isolating and measuring muscarine has recently been developed. The procedure utilizes gas chromatography and is probably the most efficient method for measuring muscarine in mushroom tissues (Eugster and Schleusener 1969). A dose of about 40 mg of muscarine is life threatening in humans. Therefore, depending on the species of mushroom and the concentration of toxin it contains, as little as 50–100 g or as much as 5 kg of muscarine-containing mushrooms would have to be ingested to be life threatening in humans. Unfortunately, accurate data on levels of toxins in many of the known muscarine-containing mushrooms are scarce, especially the fluctuations associated with seasonal and environmental variation.

Biologically, muscarine combines with the acetylcholine receptor sites, commonly called cholinergic receptors, of specific effector cells innervated by the autonomic nervous system. The cells are mostly associated with the parasympathetic division of the autonomic nervous system. This part of the nervous system exerts control over involuntary muscles and glandular secretions, over which we have no conscious control.

The symptoms of muscarine poisoning begin about 15–20 minutes after the mushrooms are eaten. The victim feels nauseated and vomiting is often accompanied by diarrhea. Vomiting the undigested mushroom tissue from the stomach often removes much of the poison from the victim's system. Some people incorrectly believe that cooking the mushrooms destroys the muscarine. In fact, muscarine remains stable even after prolonged heating. Besides nausea, vomiting, and diarrhea, victims often show constriction of the muscular region at the back of the mouth, a painful urge to urinate, profuse sweating and salivation, constriction of the pupil of the eye associated with blurred vision, excessive secretion of tears, difficulty in breathing as a result of constriction of the bronchial region or because of obstruction of air passages by profuse secretion of mucus, decreased blood pressure, and occasionally pulmonary edema, slowing of the heart rate, and cardiac insufficiency. The combination of perspiration, salivation, and watering eyes does not occur in any other kind of mushroom poisoning. No direct effects on the central nervous system, such as delirium, confusion, or convulsions, are associated with muscarine poisoning. In severe poisonings during the later stages, the victim may go into a coma because of lack of sufficient oxygen reaching the brain, as a result of the cardiac effects of muscarine.

The alkaloid atropine completely blocks all the effects of muscarine on the peripheral autonomic nervous system. When atropine is carefully administered according to the size and condition of the victim, it is the antidote of choice. Physicians should refer to the work of Lincoff and Mitchel (1977) for information on the medical treatment of muscarine poisoning.

Symptom-producing Species

Amanitaceae
- *Amanita*
  - *A. gemmata* (suspected)
  - *A. muscaria* (usually not physiologically active)
  - *A. pantherina* (usually not physiologically active)
  - *A. parcivolvata* (suspected)

Boletaceae
- *Boletus*
  - *B. calopus*
  - *B. luridus*
  - *B. pulcherrimus* (=*Boletus eastwoodiae*) (suspected)
  - *B. satanas* (suspected)

Cortinariaceae
- *Hebeloma*
  - *H. crustuliniforme* (suspected)
- *Inocybe* (most species)
  - *I. fastigiata*
  - *I. geophylla*
  - *I. lacera*
  - *I. mixtilis*
  - *I. napipes*
  - *I. patouillardi*
  - *I. pudica*
  - *I. sororia*
  - *I. subdestricta*

Tricholomataceae
- *Clitocybe*
  - *C. dealbata* (=*C. sudorifica*)
  - *C. dilatata* (=*C. cerussata* var. *difformis*)
  - *C. morbifera*
  - *C. nebularis* (suspected)
  - *C. rivulosa*
  - *C. truncicola* (suspected)
- *Hygrophoropsis*
  - *H. (Clitocybe) aurantiaca* (doubtful)
- *Mycena*
  - *M. pura*
- *Omphalotus*
  - *O. illudens* (suspected)
  - *O. olearius*
  - *O. olivascens* (suspected)
  - *O. subilludens* (suspected)

Russulaceae
- *Russula*
  - *R. emetica* (suspected)

Most of the species listed above contain or are suspected to contain sufficient muscarine to cause a poisoning reaction when ingested by humans. Two of the species of *Amanita* listed, *A. muscaria* and *A. pantherina,* although containing muscarine, normally do so in quantities so minute as to be physiologically insignificant. These two species, as well as *A. gemmata,* are described in Chapter 11. Of the other genera listed, only *Clitocybe* and *Inocybe* are treated further in this chapter. See Chapter 13 for descriptions of the genus *Boletus* and its important species, *Clitocybe nebularis, Hebeloma crustuliniforme, Omphalotus illudens* and its relatives, and *Russula emetica.* The species listed under *Clitocybe* and *Inocybe,* except for *Clitocybe nebularis* and *Clitocybe truncicola* (Peck) Sacc. whose toxicities have not been verified, are described here in alphabetic order.

Descriptions of Genera and Species

Clitocybe (Fr.) Quél. — Tricholomataceae

The genus *Clitocybe* is widely distributed in Canada and the United States. It is a moderately large genus and in certain modern treatments includes the genus *Lepista.* In general, the genus *Clitocybe* can be distinguished by decurrent to subdecurrent gills; a usually white spore deposit, although the color sometimes ranges from buff, cream color, pale yellowish buff, or pale pinkish buff; often fibrous fruit bodies varying in size from small to large, more or less fleshy, and usually with a centrally attached stalk; a cap with the margin enrolled in the early stages and often with the center depressed at maturity; no partial or universal veil; normally smooth-walled spores, although the walls in some species are slightly roughened or wrinkled; and a cap cuticle typically composed of filamentous hyphae.

Several small genera have been separated from *Clitocybe.* Two of these, *Omphalotus* and *Hygrophoropsis,* are fairly common and at least one, *Omphalotus,* contains poisonous species. *Omphalotus* is treated in Chapter 13 and in poison cases should be closely compared with *Clitocybe. Hygrophoropsis aurantiaca* (Wulfen ex Fr.) Maire, also sometimes classified in *Clitocybe* and *Cantharellus,* has been reported as poisonous, but other reports list it as edible, so its status is not yet resolved. The cap, stalk, and gills of this species are usually some shade of orange, but the color at times is pale yellow or strongly brownish. The gills are decurrent, crowded, and repeatedly forked. The orange color and the crowded forked gills help to distinguish it from *Clitocybe.* It often fruits on or near decayed wood.

Other genera that can be confused with *Clitocybe* are *Cantharellus, Collybia, Laccaria, Lactarius, Leucopaxillus, Lyophyllum, Hygrophorus,* and *Tricholoma. Cantharellus* typically has blunt-edged, more or less fold-like to ridge-like gills rather than the distinct sharp-edged, blade-like gills of a typical mushroom. *Laccaria* differs in having rather thick and somewhat waxy-looking gills that are often pinkish to purplish in color. The spores are often spiny or roughened with fine points. *Hygrophorus* has many species with decurrent gills; however, the rather thick, waxy gills and long, narrow basidia of this genus separate it from *Clitocybe.* Certain *Leucopaxillus* species have the same stature as *Clitocybe.* All *Leucopaxillus* species have amyloid spores and usually exhibit amyloid ornamentation on the spore surface. *Tricholoma* species are best separated from *Clitocybe* by the notched to adnexed or sinuate gill attachment. *Collybia* is generally separated from *Clitocybe* by the more cartilaginous stalk and adnate to adnexed gill attachment. *Lactarius* species with decurrent gills can be separated from *Clitocybe* by their brittle consistency, the presence of latex in the fruit body, and the amyloid ornamentation on the spores. When *Lyophyllum* has decurrent gills, it is difficult to separate from *Clitocybe.* Check for the presence of siderophilous granules in the basidia of *Lyophyllum* to separate it from *Clitocybe.* As the above white-spored genera may be difficult to separate by the simple comparisons described here, consult carefully the general and technical keys in Chapter 5 when making an identification.

At least four species of *Clitocybe, C. dealbata* (=*C. sudorifica*), *C. dilatata* (=*C. cerussata* var. *difformis*), *C. morbifera,* and *C. rivulosa,* are reported as containing muscarine or as capable of producing muscarine-like symptoms when ingested by man. These are described or discussed below. Although *Omphalotus* species also contain muscarine, the genus is discussed more appropriately in Chapter 13.

Clitocybe dealbata (Fr.) Kummer

Figs. 55, 68

Common name. Sweat-producing clitocybe.

Distinguishing characteristics. Cap 15–55 mm broad, convex to plane, becoming slightly to distinctly depressed; margin incurved and narrowly enrolled in early stages; surface moist and dull, glabrous to silky; color white to whitish buff, sometimes tinged flesh color. Odor and taste indistinctive. Gills decurrent, close to crowded, narrow; color white to whitish, darkening with age to pale dingy yellow. Stalk 10–30 mm long, 3–5 mm thick, equal or with the base slightly enlarged; surface fibrillose to silky; color whitish to grayish buff. Spores white in deposit, smooth, ellipsoidal, 4.4–5.1 × 2.6–3.0 μm in size.

Observations. *C. dealbata* has at least two interpretations. As described here, it is the same species as *C. sudorifica* Peck. *C. dealbata sensu* Moser is a different but related mushroom, characterized by an opaque, pubescent to appressed-silky, fibrillose, white to grayish white cap; an odor and taste like fresh meal; close to crowded, narrow, dull white gills; a more or less equal to downward tapered, white to watery gray stalk; and subglobose to ellipsoidal spores, 4–6.6 × 2.9–4.4 μm in size. The two are similar in appearance; taste, odor, and spore size are the best characters for separating them. The mushroom we call *C. dealbata* is probably quite widespread, occurring across Canada wherever favorable habitats are found. *C. dealbata sensu* Moser, however, appears to be rare (H. E. Bigelow, Personal commun.).

C. morbifera and *C. rivulosa* are related to *C. dealbata* and descriptions of these species should be consulted before attempting to identify a species in this group. All contain muscarine, and it is therefore important to learn the general features that distinguish this group of mushrooms. They all are relatively small species, pale colored, with close, fairly narrow, and usually decurrent gills, often occurring in grassy areas, especially lawns. Because they are similar in size, stature, and color to the edible *Marasmius oreades* (Bolt. ex Fr.) Fr., they may be gathered by mistake by collectors seeking the edible mushroom for food. *M. oreades* is characterized by a white spore deposit; a glabrous cap, often reddish to dull brown in color but sometimes pale buff to pallid, like *C. dealbata;* pallid to white gills, deeply notched to almost free, broad, and not very close together; and a pallid stalk, smooth above, and woolly over the lower third or half.

Distribution and seasonal occurrence. *C. dealbata* is widely distributed in North America and no doubt occurs across Canada. It is reported most commonly in Eastern Canada and the United States. It fruits mainly in the fall until the cold weather prevents fruiting, but it also can be found during the summer. It may also occur in the winter, when the weather remains mild.

Habitat and habit. This species is terrestrial, often occurring in lawns, fields, meadows, and open areas; it has also been reported on leaves and along roadsides. It is scattered to numerous, sometimes growing in arcs, but occasionally solitary. It often occurs with other mushrooms such as *Marasmius oreades.*

Technical description. Cap 15–55 mm broad, convex in early stages with the margin incurved and narrowly enrolled, but becoming plane and slightly to distinctly depressed, with the margin lobed and wavy; surface moist, not striate, dull, glabrous, finely matted-fibrillose under the hand lens, sometimes obscurely zoned but not rivulose; color white to whitish buff or flesh-toned. Flesh thin, pliant, whitish; taste and odor indistinctive.

Gills white to whitish in early stages darkening with age to pale dingy yellow, becoming darker than the cap, decurrent, close to crowded, thin, narrow to moderately broad, 2–4 mm wide.

Stalk 10–30 mm long, 3–5 mm wide at the apex, equal or with the base slightly enlarged, often curved, at times compressed, stuffed becoming hollow; surface fibrillose to covered by a dense silky down, whitish to

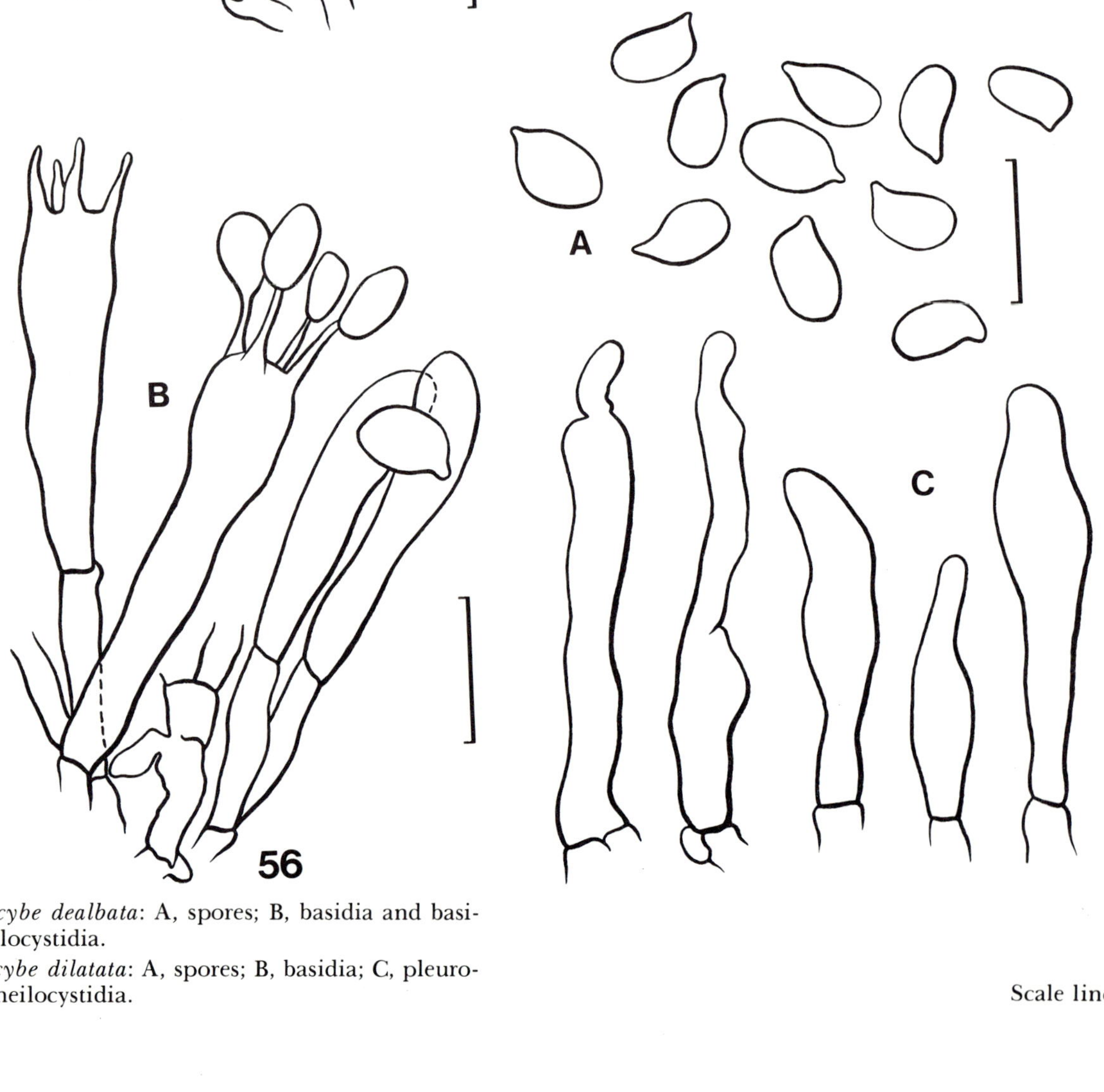

Fig. 55. *Clitocybe dealbata*: A, spores; B, basidia and basidioles; C, cheilocystidia.

Fig. 56. *Clitocybe dilatata*: A, spores; B, basidia; C, pleurocystidia and cheilocystidia.

Scale line 10 μm

grayish buff when the coating is appressed. Partial or universal veil lacking.

Spores white or rarely cream in deposit, 4.4–5.1 × 2.6–3.0 μm, ellipsoidal, inequilateral in side view, smooth, hyaline, nonamyloid. Basidia four-spored, clavate, 14.6–23.0 × 4.4–5.1 μm, hyaline. Pleurocystidia and cheilocystidia, when present, inconspicuous, cylindrical to more or less clavate, 16.1–30.0 × 4.4–5.1 μm, hyaline. Gill trama interwoven; hyphae more or less parallel, 2.2–5.1(–7.3) μm wide, hyaline. Subhymenium hyphae branched, cylindrical, 2.2–3.7 μm wide, hyaline. Cap cuticle filamentous; hyphae interwoven, parallel or radially arranged, cylindrical, 2.2–3.7 μm wide, hyaline. Cap trama loosely interwoven; hyphae cylindrical, 5.1–10.6 μm wide, hyaline. Caulocystidia cylindrical to fusiform with tapered wavy tips, 21.9–30.0 × 3.0–4.4 μm, hyaline. Stalk hyphae 3.0–5.1 μm wide, parallel in the core; surface hyphae interwoven, 2.9–3.7 μm wide. Oleiferous hyphae refractive, yellowish, 2.2–3.7(–5.1) μm wide, scattered. Clamp connections present throughout the fruit body.

Useful references. The following books contain information on *C. dealbata* pertinent to North America.

- *Edible and Poisonous Mushrooms of Canada* (Groves 1979)—A black-and-white photograph and a description and discussion are presented.
- *Mushrooms of North America* (Miller 1972)—A color photograph and a description and discussion are given.
- *The Savory Wild Mushroom* (McKenny 1971)—A black-and-white photograph, a color photograph, and a description and discussion are given.

Clitocybe dilatata Pers. ex Karst.

Figs. 56, 69

Common names. None.

Distinguishing characteristics. Cap 15–150 mm broad, with the margin enrolled and remaining so well into maturity; shape convex, expanding to plane with a raised center, becoming shallowly depressed in the center with age; color grayish when young, becoming whitish to watery buff as the cap expands, in dry weather chalky white. Gills adnate to short-decurrent and tending to become more distinctly decurrent with age, close to crowded, narrow to broad; color pale buff to pale pinkish buff. Stalk 50–120 mm long, 7–28 mm thick, central to somewhat eccentric, equal to somewhat enlarged toward the base; color about the same as the cap surface. Spore deposit white; spores smooth, ellipsoidal, 4.5–6(–6.5) × 3–3.5 μm in size. Fungi gregarious to caespitose in open areas, occurring on bare soil or in low vegetation, often on road shoulders.

Observations. This species is found in many references under the name *C. cerussata* var. *difformis* (Fr.) Bres. In general appearance *C. dilatata* is similar to *Lyophyllum decastes* (Fr. ex Fr.) Singer, but *L. decastes* usually has a more brownish cap, white gills and stalk, gills that do not tend to be decurrent, spores that are larger (5–7 × 5–6 μm in size) and broadly ellipsoidal to globose, and siderophilous granules in the basidia.

There are other caespitose *Clitocybe* and *Lyophyllum* species that resemble *C. dilatata*. A number of them are difficult to identify and are easily confused with the similar *C. dilatata*. An expert in this group should be consulted to verify identifications. The technical description given below follows that of Bigelow (1965).

Distribution and seasonal occurrence. *C. dilatata* occurs in western North America along the Pacific coast and inland into the Yukon Territory. This species fruits primarily in midsummer and fall, and occasionally in the spring, depending on location and weather. It appears to fruit commonly after a cool, wet period.

Habitat and habit. This species is usually caespitose, terrestrial, and gregarious. It usually occurs in large clusters in open or bare soil, or in low vegetation, often along the edges of roads. The caps are often deformed because they are crowded together in dense clusters.

Technical description. Cap 15–150 mm broad, convex in early stages with the margin decurved, incurved and enrolled when very young, finally becoming plane with the margin broadly decurved to nearly horizontal, remaining narrowly enrolled for some time, becoming very wavy and sinuate with age, occasionally elevated, not striate; disc soon becoming rounded or umbonate, broadly and shallowly depressed with age; surface not hygrophanous, finely matted-fibrillose under a hand magnifying glass, densely covered with silky down near the margin and radiate-fibrillose on the disc; color grayish when young, becoming whitish with translucent watery buff areas with expansion in wet weather, chalky white in dry weather, usually deformed from mutual pressure of adjacent caps. Flesh thick on the disc, gray in early stages becoming watery whitish in wet weather, whitish in dry weather, firm; odor indiscernible; taste somewhat sour and disagreeable.

Gills adnate to short-decurrent, with age moderately decurrent, close to crowded, thin, narrow to broad, up to 10 mm wide, forked at times, not interveined; color buff to pale pinkish buff; edges even to wavy.

Stalk 50–120 mm long, 7–28 mm thick, central or somewhat eccentric, solid or sometimes open, fibrous; surface fibrillose-scurfy to fibrillose-striate, usually compressed, often curved, equal or with the base somewhat enlarged; bases sometimes connate; color same as that of the cap surface, although the base can become sordid-stained with handling. Partial and universal veils lacking.

Spores white in deposit, 4.5-6(-6.5) × 3-3.5 μm, ellipsoidal, smooth, nonamyloid. Basidia two- or four-spored, 20-26 × 4-6.5 μm, basically clavate. Pleurocystidia and cheilocystidia not well-differentiated, cylindrical to clavate or fusiform.

Gill trama regular to interwoven; hyphae cylindrical, 1.5-6(-10) μm wide, hyaline. Cap cuticle thin, somewhat gelatinous in potassium hydroxide; hyphae 2-4 μm wide, cylindrical, hyaline. Cap trama thick; hyphae cylindrical to somewhat inflated, 2-9(-15) μm wide. Oleiferous hyphae present. Clamp connections present, most common in the cap trama.

Useful references. The following books contain information on *C. dilatata* pertinent to North America.

- *Mushrooms in their Natural Habitats* (Smith 1949)—A description and discussion, as well as a color photograph on a reel, are presented under the name *Clitocybe cerussata* var. *difformis.*
- *Mushrooms of North America* (Miller 1972)—A color photograph and a description and discussion are given.

Bigelow (1965) gives an excellent treatment of *C. dilatata* and related species.

Clitocybe morbifera Peck

Fig. 57

Common names. There are no common names for *C. morbifera,* although those applied to *C. dealbata* (=*C. sudorifica*) have probably been used.

Distinguishing characteristics. In general much like *C. dealbata.* Cap 10-35 mm broad, convex to plane or sometimes depressed; margin striate; color in early stages grayish brown, fading to grayish white or white. Odor resembling fresh meal. Gills close, narrow, adnate to short-decurrent; color white to pallid or faintly creamy. Stalk 4-8 mm thick, equal or tapered upward; color whitish gray to pallid, sometimes with a dingy brown base. Spores white in deposit, smooth, ellipsoidal, 4.4-5.8(-6.6) × 2.9-3.7 μm in size.

Observations. *C. morbifera* is closely related to *C. dealbata* and is best separated from it by its hygrophanous cap. The grayish brown cap and the slightly colored stalk of *C. morbifera* also help to distinguish it from *C. dealbata.*

Distribution and seasonal occurrence. According to H. E. Bigelow (Personal commun.), *C. morbifera* is rare in North America. Its distribution is not well known and its seasonal occurrence is probably similar to that of *C. dealbata.*

Habitat and habit. This species occurs in lawns and grassy areas; its habit is similar to that of *C. dealbata.*

Technical description. Cap 10-35 mm wide, convex in early stages, becoming plane, depressed at times, hygrophanous, striate at the margin, glabrous; color grayish brown, fading to white or grayish white. Flesh thin, fragile, white; taste disagreeable, alkaline-bitter; odor resembling fresh meal.

Gill white to pallid with a pale creamy tint, adnate to short-decurrent, seceding, close, narrow.

Stalk 15-25 mm long, 4-8 mm wide at the apex, equal or tapered upward, stuffed then hollow, more or less compressed; surface sparsely silky; color whitish gray to pallid, with a dingy brown base. Partial and universal veils lacking.

Spores white in deposit, 4.4-5.8(-6.6) × 2.9-3.7μm, ellipsoidal, slightly inequilateral, hyaline, nonamyloid. Basidia four-spored, narrowly clavate, 21-25 × 4.4-5.1 μm, hyaline. Pleurocystidia and cheilocystidia, when present, narrowly cylindrical or narrowly fusiform-ventricose with tapered wavy apices, 16.8-25.6 × 2.2-3.7 μm, hyaline. Gill trama hyphae more or less parallel, cylindrical to inflated, 5.1-7.3(-14.6) μm wide, hyaline. Subhymenium tightly interwoven; hyphae more or less cylindrical, 2.2-2.9 μm wide, hyaline. Cap cuticle compactly interwoven; hyphae filamentous, cylindrical, 2.2-3.7 μm wide, hyaline. Cap trama loosely interwoven; hyphae 3.7-15 μm wide, hyaline. Caulocystidia consisting of recurved hyphal ends, cylindrical to fusiform, 2.2-4.0 μm wide; stalk hyphae more or less parallel, 2.5-8.2 μm wide. Oleiferous hyphae not observed. Clamp connections present throughout the fruit body.

Useful references. There are no North American references available on this species.

Clitocybe rivulosa (Fr.) Kummer

Fig. 58

Common names. None.

Distinguishing characteristics. Cap 10-55 mm broad, convex to plane, sometimes somewhat depressed in the center; color whitish hoary over watery brown areas, fading to light buff or white; surface sometimes with concentric ridges, marked with branching lines much like a branching river system. Odor and taste slightly resembling fresh meal. Gills dingy white to pale buff, adnate to more or less decurrent, narrow, close. Stalk 25-35 mm long, 3-10 mm thick; surface covered with a silky down; color the same as the cap, fading to white. Spores white in deposit, smooth, ellipsoidal, mostly 4.5-6 × 3-4 μm in size.

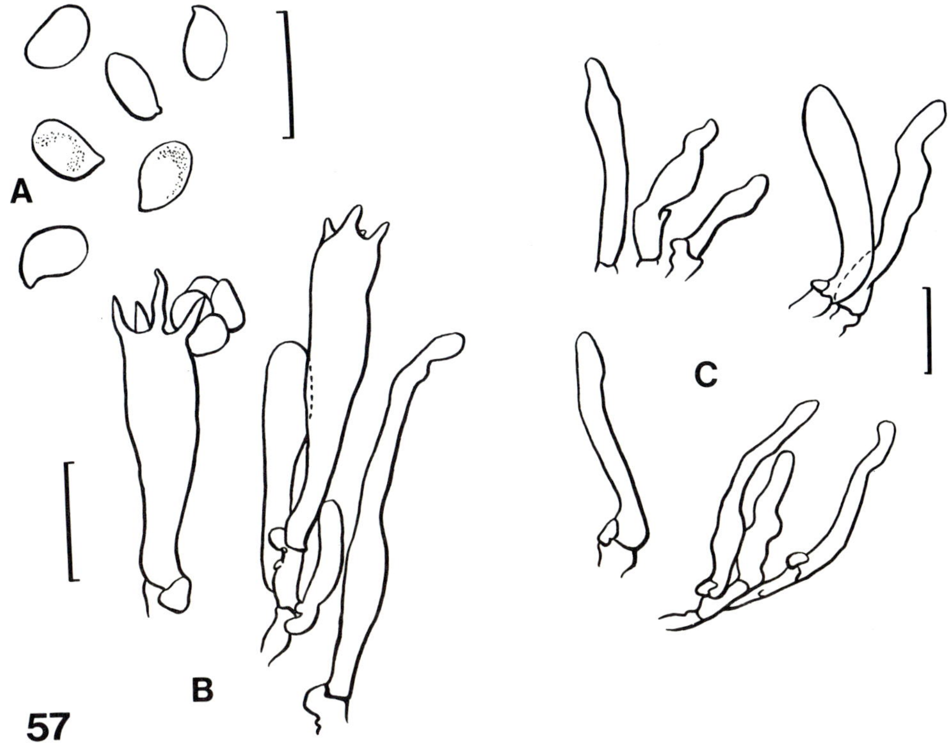

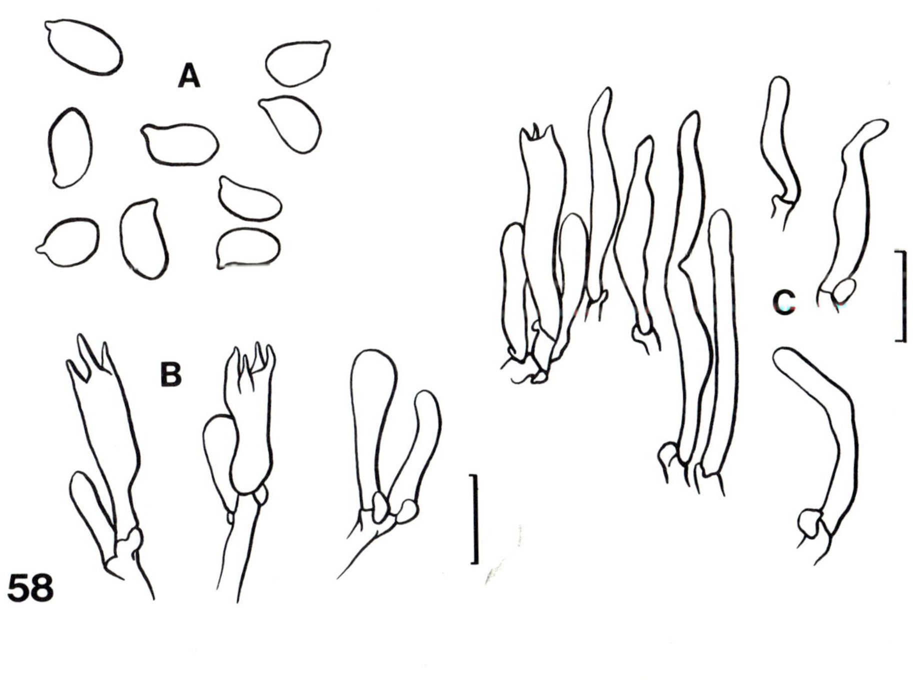

Fig. 57. *Clitocybe morbifera*: A, spores; B, basidia; C, pleurocystidia and cheilocystidia.

Fig. 58. *Clitocybe rivulosa*: A, spores; B, basidia; C, pleurocystidia and cheilocystidia.

Scale line 10 μm

Observations. Compare *C. dealbata* and *C. morbifera.* The rivulose cap surface is a key characteristic in identifying this species. *C. rivulosa* var. *angustifolia* Kauffman occurs on pine needles and has been reported from Michigan and Washington State.

Distribution and seasonal occurrence. This species has been reported from the eastern United States west to the Great Lakes and in Eastern Canada. It fruits in the fall.

Technical description. Cap 10-55 mm broad, broadly convex to plane; margin narrowly enrolled, not striate, finally somewhat upturned and broadly wavy; disc sometimes depressed; surface uneven, sometimes with concentric ridges or rivulose (especially in those with larger caps); color whitish hoary over watery brown areas, light buff or white when faded or in dry weather. Flesh thin, pale watery brown when moist, white when dry; odor and taste slightly resembling fresh meal, with a faintly pungent aftertaste.

Gills dingy white to pale buff, adnate to short-decurrent, finally moderately decurrent, narrow, rarely forked, close to nearly crowded, interveined, with faces sometimes veined.

Stalk 25-35 mm long, 3-10 mm wide at the apex, equal or enlarged at the apex or the base, often curved, solid; surface covered by a thin usually appressed silky down, same color as the cap, fading to white. Partial and universal veils lacking.

Spores white in deposit, (3.0-)4.5-5.8(-6.6) × 2.9-3.7(-4.5) μm, ellipsoidal, smooth, nonamyloid, hyaline. Basidia four-spored, clavate, 15.0-19.0 × 4.4-5.1 μm, hyaline. Pleurocystidia rare when present, cylindrical to narrowly clavate, fusiform-ventricose with wavy tapered apices, 21.9-37.0 × 3.7-4.4 μm. Cheilocystidia similar to pleurocystidia.

Gill trama interwoven to more or less parallel; hyphae 3.7-9.4 μm wide, hyaline. Subhymenium hyphae cylindrical, 2.2-3.0 μm wide, hyaline. Cap cuticle filamentous, compactly interwoven; hyphae cylindrical, 2.2-5.1 μm wide, hyaline. Cap trama loosely interwoven; hyphae more or less cylindrical, 3.7-7.0 μm wide, hyaline. Caulocystidia cylindrical to somewhat clavate, comprising recurved hyphal ends 3.5-4.6 μm wide. Oleiferous hyphae refractive, greenish yellow, 2.2-4.4 μm wide, scattered in the trama and cap cuticle. Clamp connections present.

Useful reference. The following book contains information on *C. rivulosa* pertinent to North America.

- *Edible and Poisonous Mushrooms of Canada* (Groves 1979)—The species is discussed under *C. dealbata.*

Inocybe (Fr.) Fr. — Cortinariaceae

Because many species of *Inocybe* contain muscarine, learn how to recognize the genus and how it differs from other brown-spored genera. Most of the species are medium sized, with a cap 15-50 mm broad and a stalk 20-80 mm long by 2-8 mm wide; a few species are large and fleshy. The cap in many species is some shade of brown; a few are white, lilac, or blackish. Usually the cap surface is dry, fibrillose to somewhat scaly, and often radially cracked. A few species have a silky, sticky to viscid cap surface. The cap is typically conic to bell-shaped in the early stages, sometimes becoming convex at maturity.

An inconspicuous partial veil of fibrils, the cortina, is often present but usually does not leave a distinct annulus or a conspicuous layer of fibrils on the cap margin. At times a slight fibrillose zone may be present near the stalk apex. There is usually no universal veil.

The gills are attached to the stalk and usually have grayish brown to dull brown or umber-colored faces with contrasting light-colored edges. The clay to dark brown color of the mature gills is imparted by the similarly colored spores. The stalk is often more or less equal in thickness but may be enlarged at the base. Often the stalk apex is scurfy to powdery.

Most species of *Inocybe* are terrestrial but occasionally they occur on very rotten wood. They occur in the woods or near trees or shrubs and most species are considered to be associated with the roots of higher plants. They often occur in large numbers, and they are sometimes common along the edges of dirt roads in certain regions, for example the boreal forest.

Microscopic study is usually indispensible for distinguishing the various species of this genus. The spores are typically ellipsoidal, but some are irregular to angular in outline. The spore surface is often smooth but some are nodulose, warty, or spiny. Cheilocystidia are present in all species of *Inocybe,* and many species have pleurocystidia, often in the form of metuloids. The cap cuticle is composed of filamentous, more or less radially arranged hyphae.

Four common genera, *Cortinarius, Hebeloma, Galerina,* and *Pholiota,* may be confused with *Inocybe.* Species of *Cortinarius* often have a rust brown spore deposit rather than the clay to dark brown color of *Inocybe,* and the spore surface is covered with small warts or very small points or dots. The spore surface may also be finely wrinkled. Cheilocystidia and pleurocystidia are usually lacking. Species of *Pholiota* usually grow on wood rather than in soil like the terrestrial

Inocybe, and those species of *Pholiota* that are terrestrial have dry, fibrillose to fibrillose-scaly stalks and are usually caespitose, a habit of growth rarely found in *Inocybe.* The spores of *Pholiota* species typically have an apical germ pore, in contrast with the spores of *Inocybe.* Most species of *Galerina* are small and slender and frequently grow in moss beds. The larger species of *Galerina* often occur on wood; their caps are moist to viscid, normally hygrophanous, and not fibrillose. They usually have a roughened spore surface, with a distinct plage. *Hebeloma* is perhaps most easily confused with *Inocybe* because both genera have dull brown spore deposits and similarly colored gill faces. *Hebeloma* species, however, are usually larger and fleshier than *Inocybe* species. The cap surface of *Hebeloma* is viscid when fresh and the spore surface is punctate-roughened to finely warty or wrinkled.

Inocybe contains many species, and their identification is difficult. The several species that are described and discussed in this chapter have significant amounts of muscarine and are relatively common in North America. Muscarine poisoning from the ingestion of *Inocybe* is not common in Canada. However, in August 1977 two people were poisoned by a mixture of *Inocybe* species collected near Bancroft, Ont. The patients were treated at the Oshawa General Hospital, Oshawa, Ont.

Inocybe fastigiata (Schaeff. ex Fr.) Quél.

Figs. 59, 70

Common names. None.

Distinguishing characteristics. Cap prominently umbonate; surface radially fibrillose and rimose; margin often becoming split; color dull yellowish to yellowish brownish gray or dark yellowish brown, sometimes paler, with the center usually darker than the margin. Gills close to crowded; color whitish to pallid in early stages, becoming dull brown. Stalk more or less equal, white or tinged with cap colors. Spores ellipsoidal to somewhat bean-shaped, smooth, 6.6–10.0 × 4.4–5.8 µm in size. Pleurocystidia lacking.

Observations. *I. fastigiata* is one of the most common species of *Inocybe* found and is encountered by most mushroom collectors. It is a variable species, with several varieties differentiated on the basis of size, color, and presence or lack of odor. It should be closely compared with *I. sororia,* which is somewhat paler and has a distinct odor of green corn.

Distribution and seasonal occurrence. This species is widely distributed in North America. It fruits throughout the mushroom season, depending on the location, but appears to be most common from June through October.

Habitat and habit. *I. fastigiata* is found in moist woods, in coniferous and mixed hardwood-coniferous forests. It grows solitary to scattered and is sometimes gregarious.

Technical description. Cap 20–70 mm wide, conic in early stages, becoming conic-bell-shaped and sometimes ovoid-bell-shaped, later more or less expanded, usually more or less acute, prominently umbonate; margin becoming split and bent backward; surface dry, innately radially fibrillose, striped, long-rimose; color dull yellow ocher to rich yellowish fuscous, sometimes soot brown on the umbo, paler on the margin. Flesh white; taste not recorded; odor lacking to spermatic or strong and disagreeable.

Gills whitish in early stages, soon tinged olive or gray, darker with age, adnexed, becoming sinuate-free, narrowed posteriorly, not broad, more or less ventricose, close to crowded.

Stalk 40–80 mm long, 4–10 mm wide, equal or tapering upward, sometimes twisted, obscurely striate, more or less fibrillose or scurfy, white to slightly fuscous or tinged with cap colors; flesh solid.

Spores yellowish brown in deposit, 6.6–10.2 × 4.4–5.8 µm, elliptic to reniform or slightly inequilateral in side view, ovate to elliptic in face view, smooth, thin-walled, yellowish brown, nonamyloid. Basidia four-spored, clavate, 27.0–30.0 × 8.0–8.8 µm, hyaline. Pleurocystidia lacking. Cheilocystidia sac-like to obovoid or broadly clavate, 25.6–62 × 8.0–21.9 µm, thin-walled, hyaline.

Gill trama hyphae more or less parallel, cylindrical to more or less inflated, 3.0–25.6 µm wide, hyaline to yellowish brown. Subhymenium hyphae interwoven, cylindrical, 3.7–4.8 µm wide, hyaline. Cap cuticle filamentous; hyphae parallel, radially arranged, cylindrical, 3.5–12.9 µm wide; walls thin to thickened and encrusted, brownish. Cap trama loosely interwoven; hyphae cylindrical to inflated, 12–30.0 µm wide, hyaline or brownish near the surface. Caulocystidia sac-like to broadly cylindrical-clavate, 19.7–44.0 × 7.3–14.6 µm, hyaline, thin-walled; stalk hyphae parallel to interwoven on the surface, 3.7–14.6 µm wide, hyaline to yellowish. Oleiferous hyphae refractive, yellowish to ocherous, 3.7–6.6 µm, scattered in the stalk. Clamp connections present.

Useful references. The following references contain information pertinent to *I. fastigiata* in North America.

- *Edible and Poisonous Mushrooms of Canada* (Groves 1979)—A color photograph and a description and discussion are presented.
- *Mushrooms of North America* (Miller 1972)—A description and discussion are given. The description is not particularly accurate, for example, *I. fastigiata* does not exhibit the odor of green corn that Miller ascribes to it.

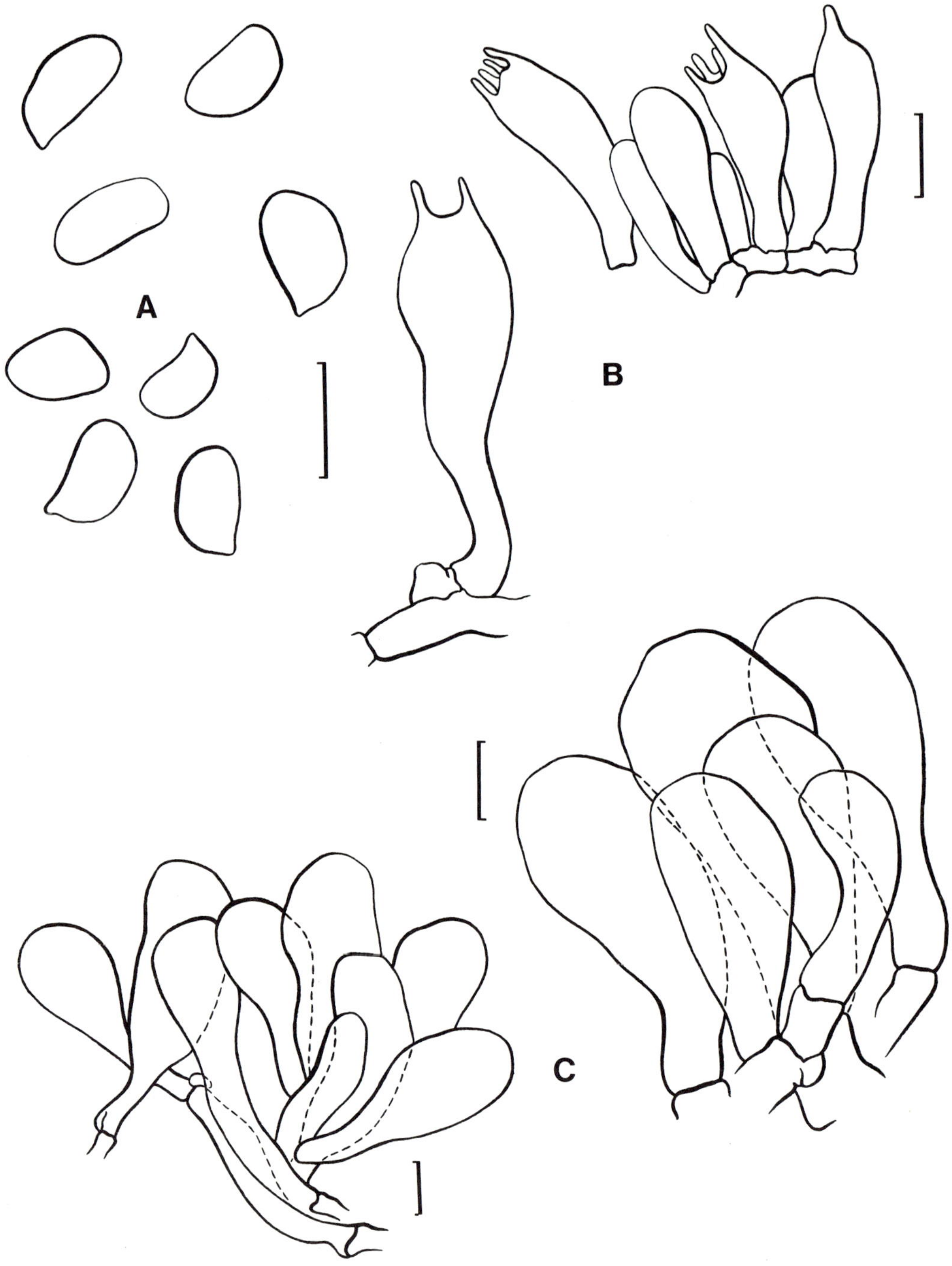

Fig. 59. *Inocybe fastigiata*: A, spores; B, basidia; C, cheilocystidia.

Scale line 10 μm

Inocybe geophylla (Sow. ex Fr.) Kummer

Figs. 60, 71

Common name. White inocybe.

Distinguishing characteristics. Cap conic to bell-shaped with a sharp pointed center, white; surface silky, lubricous when wet, but often dry. Gills whitish in early stages, then yellow brown to pale clay color. Stalk equal to clavate, narrow, white to somewhat grayish. Spores more or less ellipsoidal to ovoid, smooth, 6.6-10.2(-11.1) × 4.4-5.8 μm in size. Pleurocystidia common, thick-walled, with the apex encrusted.

Observations. This small white species is widely distributed and easily recognized. Besides the typical white form, it also occurs in a lilac form, often referred to as *I. geophylla* var. *lilacina* Peck or *I. lilacina* (Boud.) Kauffman. In the lilac form, the cap, gills, and stalk are tinted with lilac to light vinaceous purple or vinaceous lilac; however, the cap may fade to a dingy cream color with age. Both forms of this species may occur in the same area. *I. pudica* Kühner is also a white species but it soon develops pink to red stains. *I. geophylla* does not stain with age or on handling.

Distribution and seasonal occurrence. This species is fairly common and widely distributed. It fruits from spring to the end of the fall mushroom season, depending on the area.

Habitat and habit. *I. geophylla* is terrestrial, found in mixed woods and coniferous and hardwood forests, and also in lawns where trees occur. It is scattered to gregarious, usually occurring in groups.

Technical description. Cap 15-30 mm broad, in early stages more or less conic to bell-shaped becoming expanded-bell-shaped to almost plane with a more or less small persistent umbo; surface lubricous when wet, often dry, not hygrophanous, silky smooth to fibrillose-silky, sometimes splitting radially on the margin; color white to whitish, sometimes becoming dingy with age. Flesh thin, white; taste indiscernible or slightly disagreeable; odor disagreeable to nauseous.

Gills whitish to grayish in early stages, pale to moderately yellow brown in later stages, adnate to adnexed, close, moderately broad.

Stalk 20-60 mm long, 2-4 mm wide, equal or somewhat clavate below, central, stuffed, silky-fibrillose, dry or slightly sticky near the base when moist, white to grayish white. Flesh whitish. Partial veil leaving a faint fibrillose ring.

Spores dull yellowish brown in deposit, 6.6-10.2(-11.1) × 4.4-5.8 μm, ellipsoidal, slightly inequilateral in side view, more or less elliptic to ovate in face view, thin-walled, smooth, yellowish brown. Basidia four-spored, clavate, 21.9-25.6 × 8.8 μm, hyaline to yellowish. Pleurocystidia fusiform-ventricose to obovate, 46.8-73.0 × 15.0-17.8 μm, thick-walled; apex encrusted, hyaline. Cheilocystidia broadly fusiform-ventricose to obovoid, thick-walled, 46-64.5 × 15-17.8 μm, and apically encrusted, but also sac-like to broadly clavate, 15-25 × 7.3-11.0 μm, thin-walled, and hyaline. Cap cuticle filamentous; hyphae more or less parallel, radially arranged, cylindrical, 4.0-7.3 μm wide, hyaline to yellowish; hyphae just below the cuticle compacted, 3.7-16.1 μm wide, with walls thin to slightly thickened and finely encrusted, hyaline to yellowish. Cap trama loosely interwoven; hyphae cylindrical to inflated, 5.1-22.3 μm wide, hyaline to yellowish. Caulocystidia sac-like to clavate, 20-23 × 12.9-16.7 μm, hyaline to yellowish, thin- to thick-walled and clavate to fusiform-lanceolate, 52-81.9 × 12.9-16.7 μm, hyaline, slightly thickened at the apex, encrusted; stalk hyphae parallel to interwoven, 3.5-9.4 μm wide, hyaline. Partial veil interwoven; hyphae 3.5-10.5 μm wide, hyaline. Oleiferous hyphae not observed. Clamp connections present.

Useful references. The following references contain information pertinent to *I. geophylla* in North America.

- *Edible and Poisonous Mushrooms of Canada* (Groves 1979)—A color photograph and a description and discussion are presented.
- *Mushrooms in their Natural Habitats* (Smith 1949)—A description and discussion and a color photograph on a reel are presented.
- *Mushrooms of North America* (Miller 1972)—A color photograph and a description and discussion are given.
- *The Savory Wild Mushroom* (McKenny 1971)—A discussion is given.

Inocybe lacera (Fr.) Kummer

Figs. 61, 72

Common names. None.

Distinguishing characteristics. Cap more or less convex when expanded; surface fibrillose to scaly, becoming lacerate-rimose; color variable, yellowish brown or olive brown to deep brown, becoming darker and somewhat more reddish brown with age. Gills whitish to grayish brown or sometimes darker, broad and ventricose. Stalk more or less equal, fibrillose-streaked, pale above and brownish below, becoming more or less entirely brown to almost black at the base. Spores ellipsoidal to cylindrical and smooth, 8-14.6 × 4.4-5.1 μm in size. Pleurocystidia encrusted, thick-walled.

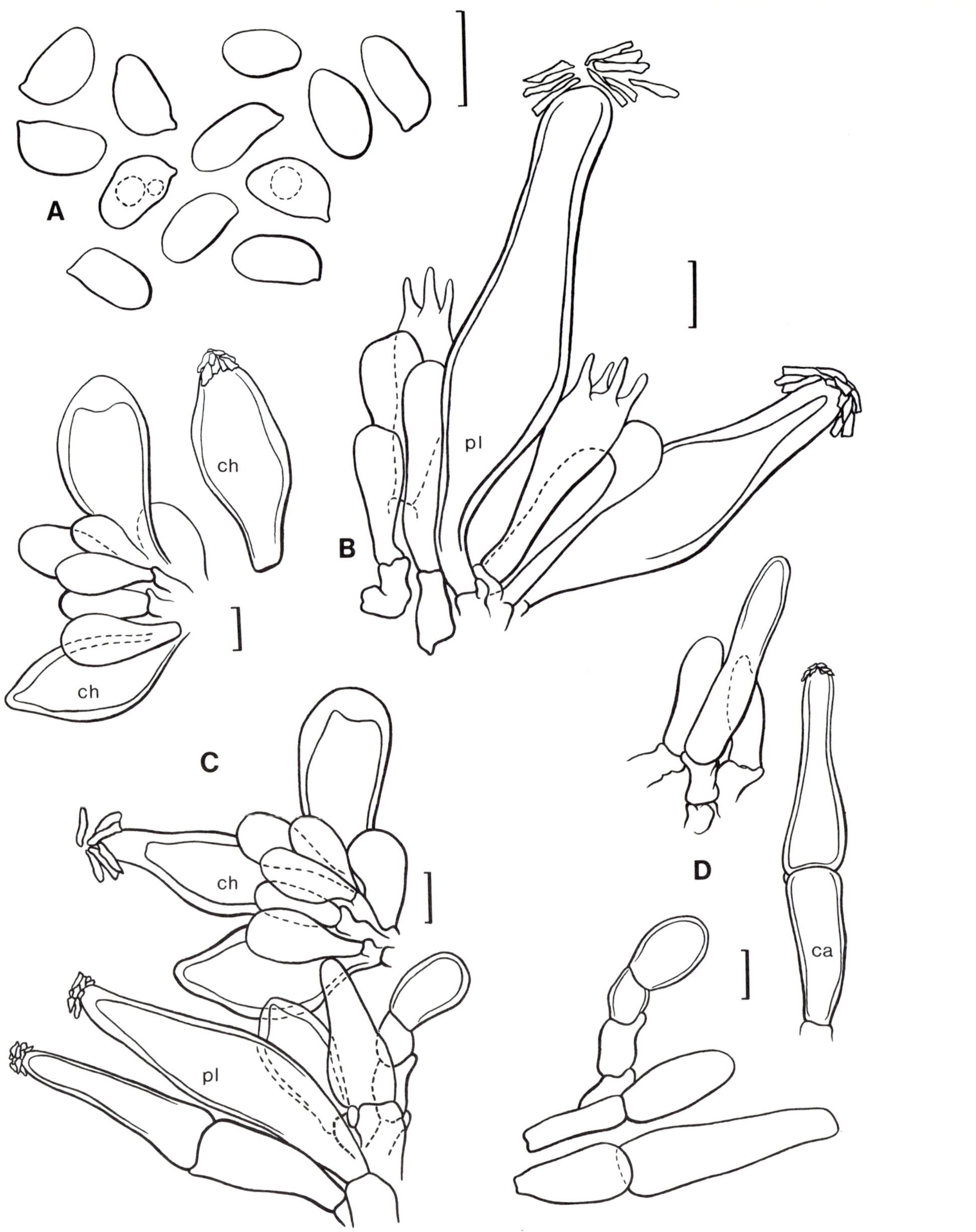

Fig. 60. *Inocybe geophylla*: A, spores; B, basidia and pleurocystidia (*pl*); C, cheilocystidia (*ch*); D, caulocystidia (*ca*).

Scale line 10 μm

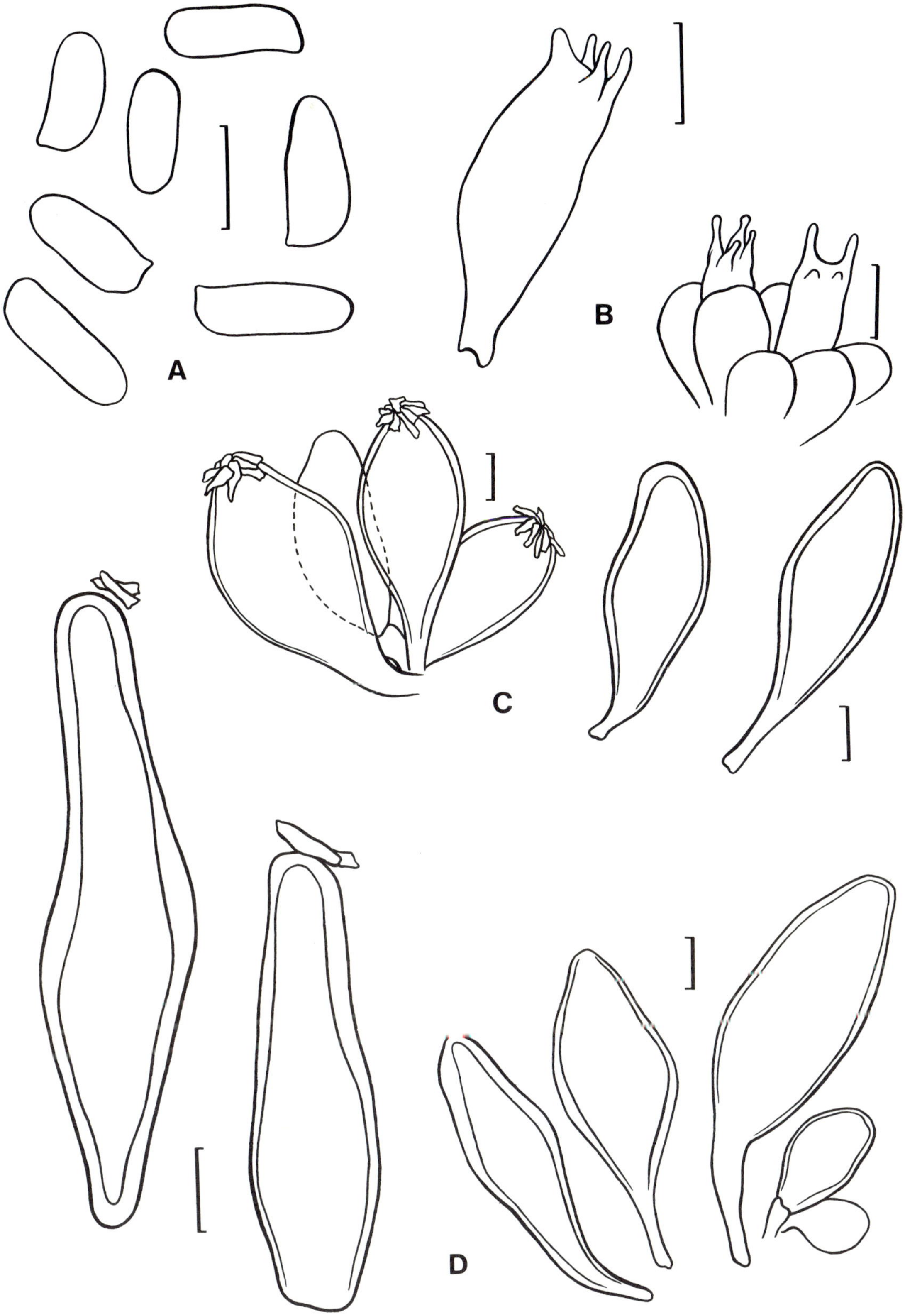

Fig. 61. *Inocybe lacera*: A, spores; B, basidia; C, cheilocystidia; D, pleurocystidia.

Scale line 10 μm

Observations. *I. lacera* occurs in various forms. It has only occasionally been reported in North American literature and there are no keys or descriptions of the various forms available in the literature.

Distribution and seasonal occurrence. This species is widely distributed in Canada and other parts of North America. It often occurs early in the mushroom season and it fruits throughout the summer and fall.

Habitat and habit. *I. lacera* is terrestrial, found in hardwood, coniferous, and mixed forests and commonly along roadsides. It also occurs in sphagnum bogs and in lawns where trees and shrubs grow. It grows scattered to gregarious, usually occurring in groups.

Technical description. Cap 10-65 mm broad, in early stages more or less convex to more or less bell-shaped, becoming more convex, dry with age; surface not hygrophanous, fibrillose to fibrillose-scaly becoming scaly to lacerate-rimose; margin sometimes more or less fibrillose, becoming glabrous; color moderate olive brown to deep yellowish brown or moderate brown to deep brown when young, varying through dark grayish red brown to brownish orange ocherous with age or sometimes shaded reddish. Flesh thin, whitish; taste mild; odor slight to nauseous.

Gills whitish to pallid in early stages, becoming dark grayish brown to grayish light brown or sometimes darker, adnate to adnexed, close, broad, ventricose.

Stalk 20-50 mm long, 3-6 mm wide, equal to nearly equal, central, solid, dry, fibrillose-streaked; color whitish pallid above, brownish below, becoming dark and finally more or less entirely brown, often almost black at the base. Flesh whitish. Partial veil evanescent; annular zone lacking.

Spores yellow brown in deposit, 8.0-14.6 × 4.4-5.1 μm, ellipsoidal to cylindrical, inequilateral in side view, thin-walled, smooth, yellow brown, nonamyloid. Basidia four-spored, clavate, 21.9-27.0 × 8.8-9.5 μm, hyaline to yellowish. Pleurocystidia and cheilocystidia similar, fusiform-ventricose; apices obtuse to acute, encrusted, 51.3-70.2 × 14.3-23.4 μm, thin-walled to thick-walled, hyaline to yellowish; sac-like clavate cheilocystidia also present, 16.7-35.1 × 11.7-14.3 μm, thin-walled, hyaline.

Gill trama parallel; hyphae cylindrical to inflated, 3.7-7.0(-20.4) μm wide, hyaline. Subhymenium hyphae cylindrical, interwoven, 4.4-5.8 μm wide, hyaline. Cap cuticle filamentous; hyphae parallel, radially arranged, agglutinated into fascicles, with some hyphal ends recurved; hyphae more or less cylindrical, 3.7-11.0 μm wide (up to 18.3 μm wide in the area just below the cuticle); hyphal walls slightly thickened, encrusted, brown to rusty brown. Cap trama loosely interwoven; hyphae cylindrical to inflated, 3.5-21.9 μm wide, yellow brown. Caulocystidia lacking; stalk hyphae parallel to interwoven at the surface, 3.5-12.0 μm wide, hyaline to yellowish. Oleiferous hyphae refractive, yellowish, 3.5-4.4 μm wide, scattered in the stalk. Clamp connections present.

Useful references. There are no books or field guides pertinent to North America that contain information on this species.

Inocybe mixtilis (Britzelm.) Sacc.

Fig. 62

Common names. None.

Distinguishing characteristics. Cap viscid when moist, silky when dry, umbonate; center brownish orange; margin pale yellowish, tan, light brown or tinted brownish orange yellow. Gills whitish to pale in early stages, becoming grayish brown with white edges. Stalk bulbous at the base, white but sometimes slightly yellowish. Spores angular to nodulose, 7.3-9.5(-10.2) × 5.1-6.6 μm in size. Pleurocystidia thick-walled, with the apices encrusted.

Observations. *I. mixtilis* contains a considerable amount of muscarine. It is rarely reported in North American literature, but it appears to be a fairly widespread species. In Kauffman's treatment of *Inocybe* (Kauffman 1924), it was described under the name *I. trechispora* (Berk.) Karst. Microscopically it can be easily separated from species such as *I. geophylla* and *I. fastigiata* because of its angular to nodulose spores. The angular to nodulose spores and thick-walled cystidia are characteristics it shares with *I. decepientoides* Peck and certain other species. *I. decepientoides,* however, which is fairly common in Canada (occurring in forested areas as well as in lawns under trees and shrubs), has a more or less scaly cap that is umber in color, umbrinous gills when mature, and a slightly tapered stalk that is pallid above and often dark reddish brown below.

Distribution and seasonal occurrence. *I. mixtilis* has been reported from various parts of Canada and the United States and appears to be widely distributed. Reports indicate that it fruits in the summer and fall.

Habitat and habit. *I. mixtilis* is terrestrial, occurring scattered to gregarious in various habitats in forested areas. This species is reported from low, moist areas as well as from open, sandy, aspen woods.

Technical description. Cap 20-25 mm broad, more or less conic to bell-shaped-convex, becoming expanded to plane, umbonate; surface viscid, becoming silky, glabrous, and not hygrophanous when dry; disc brownish orange; margin pale yellowish, tan, light brown, or tinted brownish orange yellow. Flesh solid, white, and unchanging; taste not recorded; odor more or less spermatic or slightly radish-like.

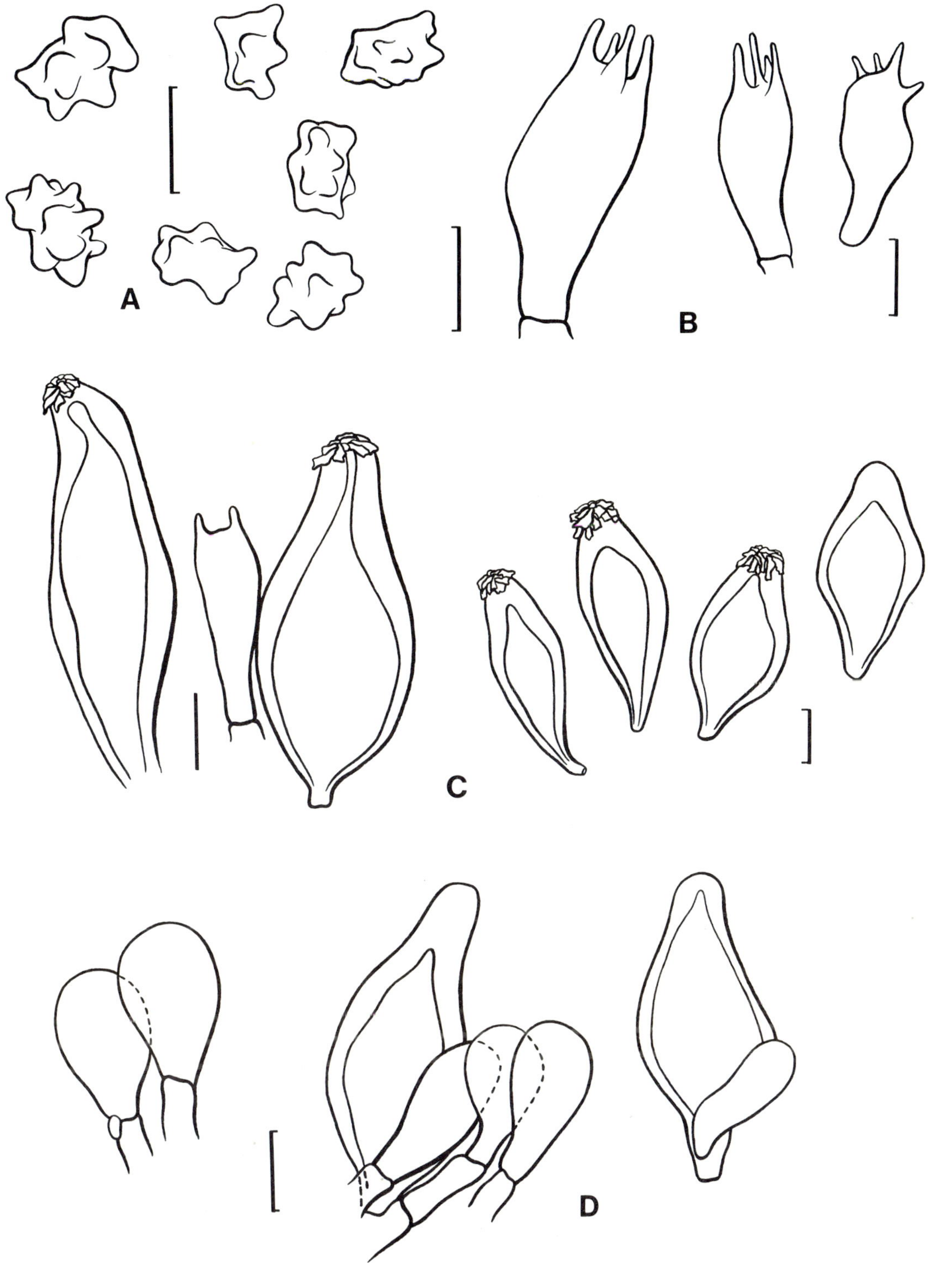

Fig. 62. *Inocybe mixtilis*: A, spores; B, basidia; C, pleurocystidia and cheilocystidia; D, caulocystidia.

Scale line 10 μm

Gills white to whitish in early stages but grayish brown at maturity, sinuate-adnexed, close, ventricose, moderately broad; edges minutely fringed.

Stalk 20-50 mm long, 3-5 mm wide, tapered slightly upward, bulbous at the base, central, solid, pruinose above, innately striate below, white but sometimes becoming slightly yellowish. Flesh whitish. Annulus and universal veil lacking.

Spores yellow brown in deposit, 7.3-9.5(-10.2) × 5.1-6.6 μm, ellipsoidal in general outline, angular, nodulose, yellow brown. Basidia four-spored, clavate, 24.1-25.6 × 8.8-9.5 μm. Pleurocystidia and cheilocystidia fusiform-ventricose to pear-shaped and sac-like; apices encrusted, 35.0-46.8 × 15.2-22.0 μm, thick-walled.

Gill trama parallel; hyphae cylindrical to inflated, 5.1-35 μm wide, hyaline to yellowish. Subhymenium hyphae interwoven, cylindrical, 2.2-3.7 μm wide, hyaline. Cap cuticle filamentous; hyphae parallel, radially arranged, cylindrical, 3.7-6.6 μm wide, hyaline to pale yellow, with walls thin to slightly thickened and finely encrusted; hyphae just below the surface cylindrical to somewhat inflated, 8.0-11.0 μm wide, yellow brown, encrusted. Cap trama loosely interwoven; hyphae cylindrical to inflated, 5.8-30.0 μm wide, hyaline. Caulocystidia broadly fusiform-ventricose to pear-shaped and sac-like, 20-42 × 13.1-18.7(-23.4) μm, thin- to thick-walled, some with encrusted apices, hyaline, clustered at the apex and base of the stalk; stalk hyphae parallel, cylindrical to inflated, 5.0-23.4 μm wide, hyaline. Oleiferous hyphae not recorded. Clamp connections present.

Useful references. There are no books or field guides pertinent to North America that contain information on this species.

Inocybe napipes Lange

Figs. 63, 73

Common name. Turnip bulb inocybe.

Distinguishing characteristics. Cap surface silky-fibrillose, somewhat radially cracked in dry weather; color uniformly umbrinous to brown, at times with a superficial silvery coating. Gills pale brown. Stalk with a bulbous turnip-shaped base; color pallid above to pale brown below, darker with age. Spore angular, nodulose. Pleurocystidia thick-walled, encrusted.

Observations. *I. napipes* has only occasionally been reported in the North American literature and was originally described from Europe. It has several features in common with *I. mixtilis,* but it is generally more brownish and has an evenly umbrinous to brown cap surface. *I. napipes* contains large amounts of muscarine and is a potentially dangerous mushroom.

Distribution and seasonal occurrence. *I. napipes* appears to be widespread in North America and has been reported from the west coast, the Great Lakes region, and Nova Scotia. It fruits in the spring, summer, and fall.

Habitat and habit. This species is terrestrial, often found in coniferous forests. It has also been reported in sandy soil in aspen forests. It grows singly, scattered, or gregariously.

Technical description. Cap 10-25 mm broad, convex, becoming expanded and umbonate; surface appressed silky-fibrillose, shining, tending to become rimose toward the margin in dry weather, uniformly umbrinous, tawny olive, or snuff brown. Flesh white to pallid, unchanging, thin, firm; odor unpleasant, either resembling chestnut catkins or seeming somewhat rancid.

Gills white, becoming pale brown to dull grayish brown, adnexed to narrowly adnate, subventricose, close; edges similar in color to the faces.

Stalk 40-60 mm long, 3-5 mm thick, terete; base with a turnip-shaped bulb and a rounded margin; surface smooth, satiny, glabrous or nearly so, slightly fibrillose-flocculose at the apex, pallid above and pale brown below, becoming yellowish tan or cinnamon brown above and darker below with age. Annulus and universal veil lacking.

Spores yellow brown in deposit, 7.3-10.6 × 5.8-6.6 μm, ellipsoidal in outline, angular, nodulose, yellow brown. Basidia four-spored, short-clavate, 20-25.6 × 7.3-8.8 μm, hyaline. Pleurocystidia broadly fusiform-ventricose to clavate, with acute, mucronate, or tapered apices, 48.0-70.2 × 13.1-23.4 μm, pedicellate, thick-walled, apically encrusted, hyaline. Cheilocystidia similar to pleurocystidia, fusiform-ventricose, 35-75.0 × 14.0-23.0 μm, thick-walled, hyaline.

Gill trama hyphae more or less parallel, cylindrical to inflated, 3.5-11.7(-29.2) μm wide, hyaline to yellowish. Subhymenium interwoven, cylindrical; hyphae 3.0-5.1 μm wide, hyaline. Cap cuticle filamentous; hyphae parallel to radially arranged, more or less cylindrical, 4.7-12.9 μm wide, thin- to thick-walled and encrusted. Cap trama loosely interwoven; hyphae cylindrical to inflated, 5.1-26 μm wide, hyaline to yellowish brown. Caulocystidia clavate, lanceolate to sac-like, (30-)58.5-70.0 × 8.0-13.7 μm, thin-walled, hyaline; stalk hyphae parallel in the core to interwoven on the surface, more or less cylindrical, 3.5-9.4 μm wide, hyaline. Oleiferous hyphae refractive, ocherous, 4.6-7.3 μm wide, scattered in the cap trama and stalk. Clamp connections present.

Useful reference. The following book contains information on *I. napipes* that is pertinent to North America.

- *The Savory Wild Mushroom* (McKenny 1971)—A black-and-white photograph, a color photograph, and a description and discussion are given.

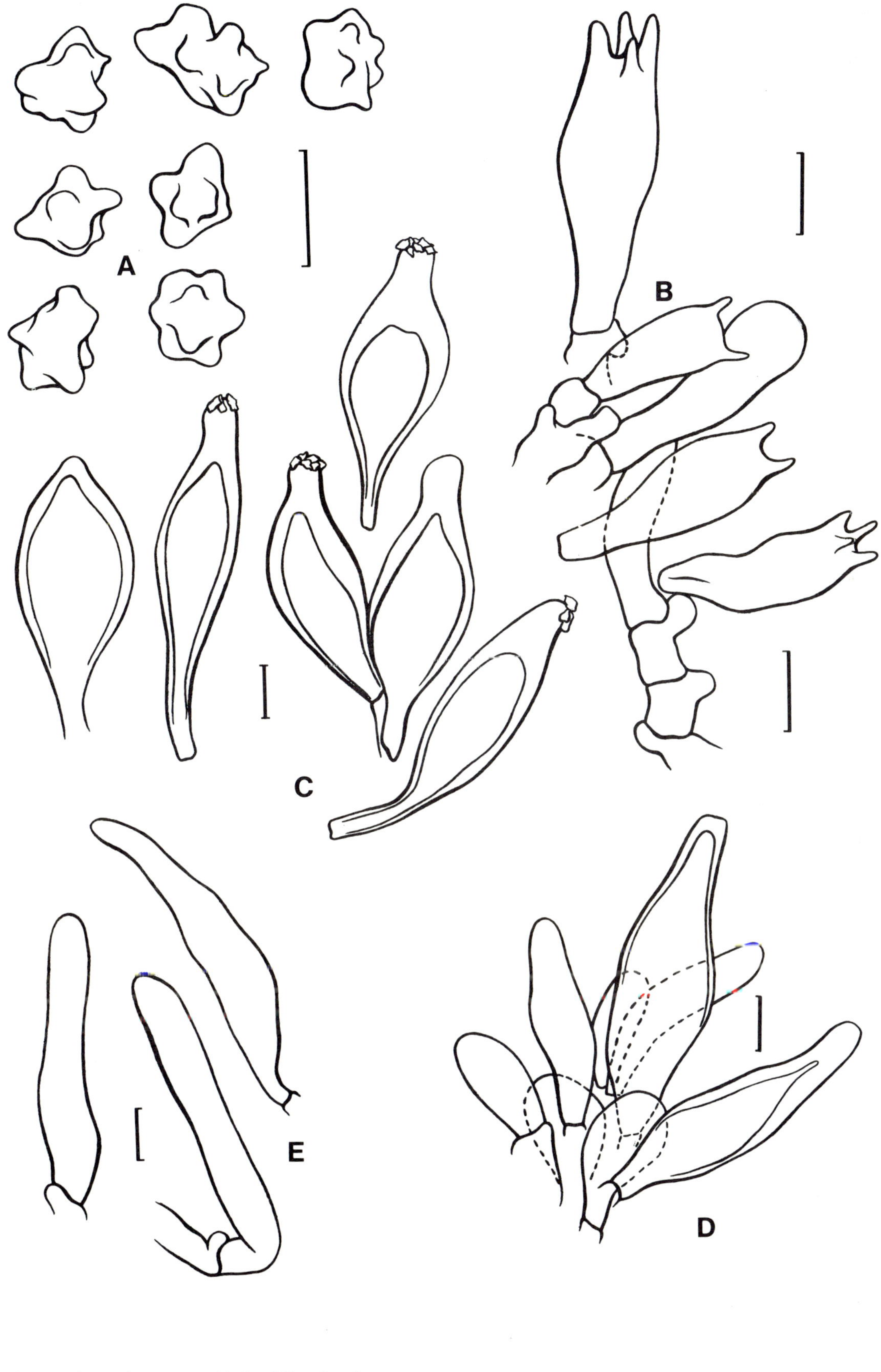

Fig. 63. *Inocybe napipes*: A, spores; B, basidia; C, pleurocystidia; D, cheilocystidia; E, caulocystidia.

Scale line 10 μm

Inocybe patouillardi Bres.

Fig. 64

Common names. Patouillard's inocybe, red-staining inocybe.

Distinguishing characteristics. Fruit body large, with the stalk up to 25 mm thick. Cap slightly lubricous when moist, silky to slightly fibrillose-streaked on the margin; color more or less cinnamon, slowly turning reddish wherever touched or bruised. Gills white, becoming dingy brownish and staining reddish where bruised. Stalk equal, long and thick, pure white, somewhat silky-streaked, staining reddish where bruised; base becoming flushed buff to dingy yellowish at times. Spores ellipsoidal and smooth, 11.7–14.6 × 5.8–7.3 μm in size. Pleurocystidia lacking.

Observations. *I. patouillardi* is a distinctive species and is widely distributed in Europe, where it has caused a number of poisonings. It has rarely been reported in the

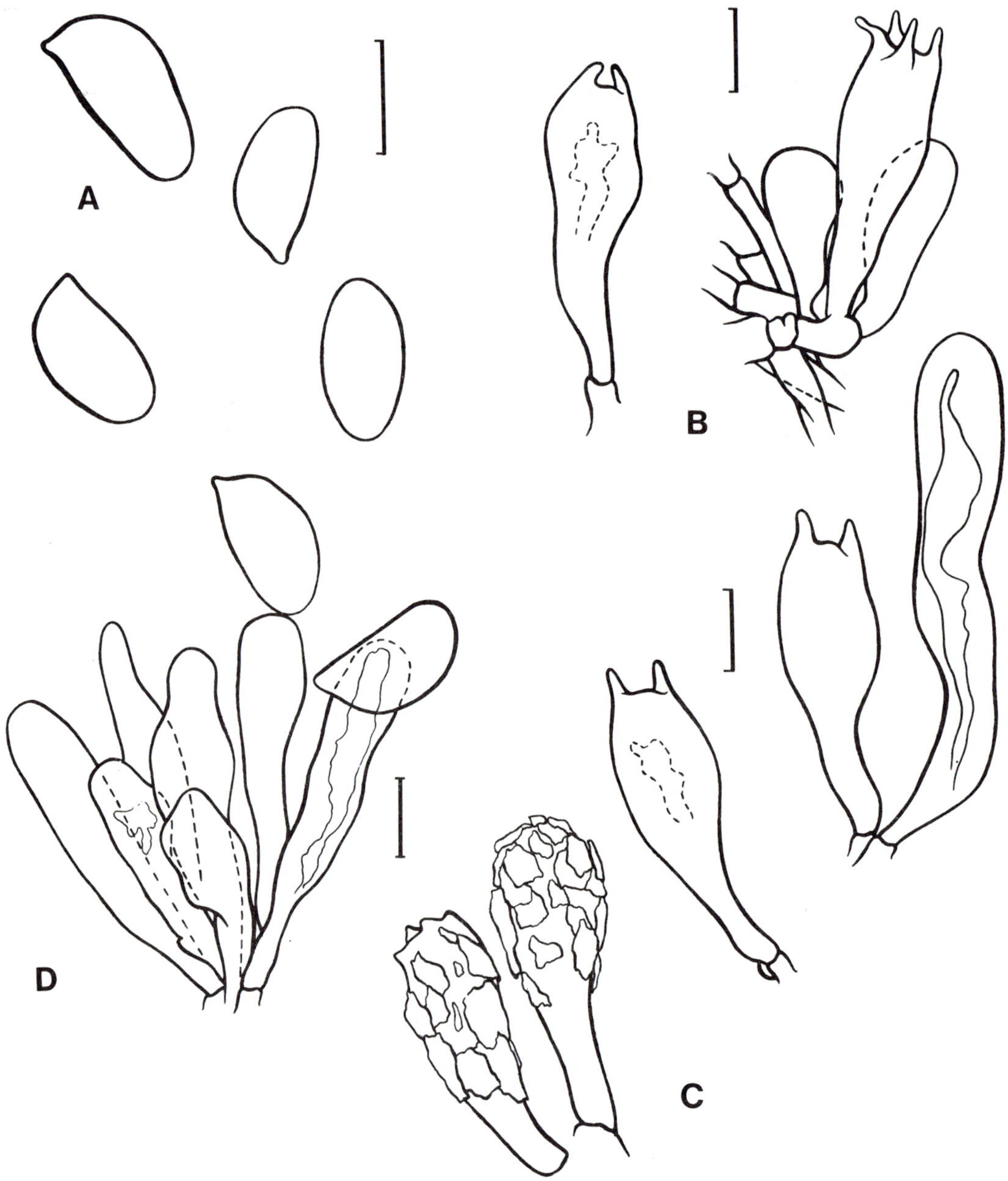

Fig. 64. *Inocybe patouillardi*: A, spores; B, basidia; C, encrusted basidioles; D, cheilocystidia.

Scale line 10 μm

North American literature but is included here because it is potentially dangerous. Its larger size makes it more attractive for foragers collecting edible mushrooms than most other species of *Inocybe*.

Distribution and seasonal occurrence. *I. patouillardi* appears to be uncommon in North America. Collections in North America have been made during the summer.

Habitat and habit. This species is terrestrial, found in mixed woods. It is reported from hardwood forests in Europe. It occurs singly or in groups.

Technical description. Cap 40–70 mm broad, persistently bell-shaped; margin incurved, lobed, and wavy; surface slightly lubricous when moist, smooth, innately silky, and unbroken, but becoming a little fibrillose-streaked toward the margin; color cinnamon buff to cinnamon, slowly turning reddish (carrot red to vinaceous) wherever touched or bruised. Flesh firm, white, slowly turning red where exposed; taste not recorded; odor faintly aromatic, like nasturtium.

Gills almost pure white in early stages, becoming dingy brownish, staining red where bruised, adnexed, broadly rounded, nearly ventricose, not broad, 4.5 mm wide, rather close.

Stalk 55–75 mm long, 9–25 mm thick, terete, equal, solid; surface glabrous, satiny, shining, faintly longitudinally silky-streaked, pure white, staining red where bruised, becoming more or less flushed with dingy ocher or buff at the base. Flesh white, slowly turning reddish where cut. Annulus and universal veil lacking.

Spores yellow brown in deposit, 11.7–14.6 × 5.8–7.3 μm, elliptic and inequilateral or almond-shaped in side view, elliptic in face view, smooth, thin-walled, yellow brown. Basidia four-spored, clavate, 35.1–41.2 × 10.5–11.7 μm, hyaline to yellowish; contents sometimes with crystalline inclusions. Pleurocystidia lacking. Cheilocystidia narrowly sac-like, clavate or cylindrical, or fusiform-ventricose, 52.0–70.2 × 9.4–11.7 μm, hyaline, thin-walled; contents crystalline.

Gill trama compactly interwoven to parallel; hyphae cylindrical to somewhat inflated, 3.5–14.6 μm wide, hyaline; contents crystalline. Subhymenium interwoven; hyphae cylindrical, 3.7–5.1 μm wide, hyaline to yellowish. Cap cuticle filamentous; hyphae parallel, radially arranged, agglutinated by crystalline material, more or less cylindrical, 3.7–8.0(–18.3) μm wide, hyaline to yellowish, thin-walled to finely encrusted. Cap trama loosely interwoven; hyphae cylindrical to inflated, 3.7–22.0 μm wide, hyaline to yellowish. Caulocystidia clavate to cylindrical, 36.5–66 × 11.0–12.4 μm, hyaline to yellowish, single to clustered on the apex of the stalk; stalk hyphae parallel, more or less cylindrical, 3.7–13.5 μm wide, hyaline to yellowish. Oleiferous hyphae refractive, ocherous to gold-colored, 3.7–8.8 μm wide, scattered in the cap trama. Clamp connections present.

Useful references. There are no field guides pertinent to North America that contain information on *I. patouillardi*. However, Pilát and Ušák (1954) give a description and discussion and provide a color photograph of this species.

Inocybe pudica Kühner

Figs. 65, 74

Common name. Blushing inocybe.

Distinguishing characteristics. Cap obtusely umbonate; surface lubricous when moist, silky-smooth and shining when dry; color pure white or slight ocherous to brownish in the center, staining pinkish orange to salmon orange or entirely rufous with age or on bruising. Odor strongly spermatic. Gills whitish in early stages, becoming more or less cinnamon brown and flushed salmon orange like the cap; edges usually white. Stalk equal, with the base at times slightly bulbous; color white, becoming flushed like the cap, often intensely so. Spores ellipsoidal and smooth, 8–10.2 × 4.4–5.1 μm in size. Pleurocystidia thick-walled and encrusted.

Observations. This very distinctive species is similar in general appearance to *I. geophylla* and was considered a variety of it, var. *lateritia* Weinm., by earlier authors. *I. geophylla* has a sharply pointed, pure white cap and a stalk that never stains red. The red stains in *I. pudica* are usually pronounced but at times may only occur as a tint of red along the edge of the cap. Microscopically *I. pudica* is similar to *I. geophylla*. The technical description given below is adapted from that of Stuntz (1947).

Distribution and seasonal occurrence. *I. pudica* is common on the west coast, but there are no reports of it in the literature from the eastern United States or Eastern Canada. It fruits during the fall and winter, depending on weather and location.

Habitat and habit. *I. pudica* is terrestrial. It occurs in conifer woods and is reported as common under pole-size Douglas-fir. It is found scattered to gregarious.

Technical description. Cap 10–60 mm broad, usually averaging 30–40 mm, bell-shaped, becoming expanded and obtusely umbonate; surface lubricous when moist, soon dry, silky-smooth, shining; color in early stages pure white, but becoming slightly ocherous to nearly tawny olive at the center, with the whole cap or parts of it slowly flushed with salmon orange, bittersweet pink, apricot orange, or flesh ocher, sometimes at later stages entirely rufous, bittersweet orange, or flame scarlet. Flesh 4–6 mm at the center, white, usually slowly

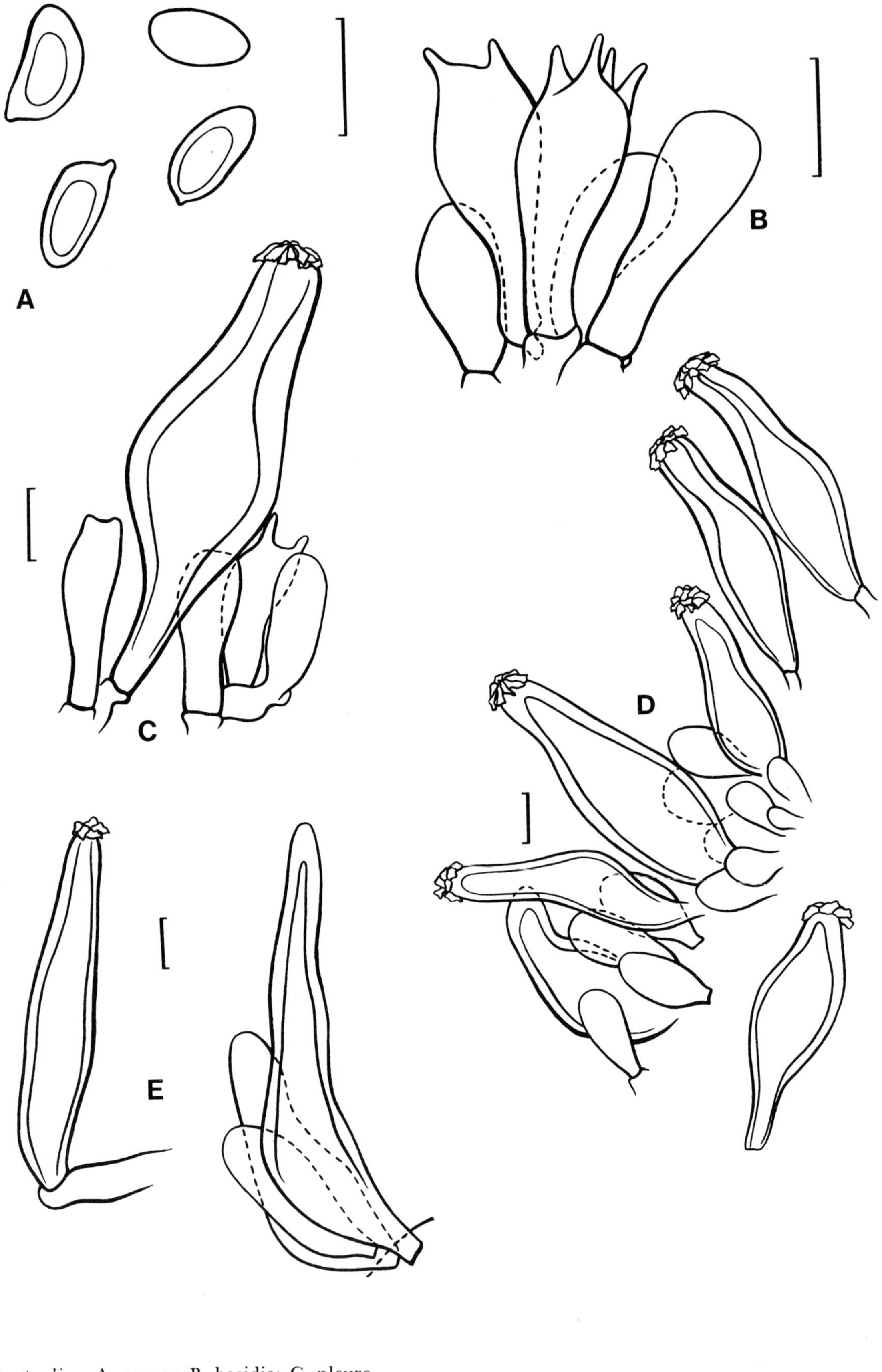

Fig. 65. *Inocybe pudica*: A, spores; B, basidia; C, pleurocystidia; D, cheilocystidia; E, caulocystidia.

Scale line 10 μm

becoming reddish where exposed; taste not recorded; odor strong, spermatic.

Gills adnexed to adnate, more or less broadly rounded at both ends, ventricose, broad, with the edges convex and white-fimbriate; color in early stages usually white, but sometimes tinted salmon orange, becoming cinnamon brown to snuff brown and flushed like the cap.

Stalk 50-100 mm long, 4-10 mm thick, cylindrical, equal; base not or only slightly rounded-bulbous; apex solid or sometimes hollow, densely white pruinose; flesh white; surface glabrous, satiny, lubricous below when moist, white, becoming flushed like the cap with salmon orange or darker, often intensely so. Annulus and universal veil lacking.

Spores yellow brown in deposit, 8.0-10.2 × 4.4-5.1 μm, ellipsoidal, slightly inequilateral, smooth, yellow brown. Basidia four-spored, clavate, 19.9-25.6 × 8.0-9.5 μm, hyaline to yellowish. Pleurocystidia fusiform-ventricose, 48.0-58.0 × 16.1-17.9 μm; apices 6.6-8.8 μm wide, thick-walled, encrusted, yellowish. Cheilocystidia similar to pleurocystidia or clavate to obovoid and sac-like, thin-walled.

Gill trama parallel; hyphae cylindrical to inflated, 5.8-23.0 μm wide, hyaline or yellowish near the hymenium. Subhymenium interwoven; hyphae cylindrical, 3.7-5.1 μm wide, hyaline to yellowish. Cap cuticle filamentous; hyphae interwoven, more or less radially arranged, more or less cylindrical, 3.5-8.2 μm wide, hyaline. Cap trama loosely interwoven; hyphae cylindrical to inflated, 3.5-29.2 μm wide, hyaline or yellowish near the cuticle, thin-walled to slightly thickened and encrusted near the cuticle. Two types of caulocystidia present: thick-walled, fusiform-ventricose to lanceolate, 54.8-70.2(-87.9) × 15.5-19.9 μm, and encrusted at the apex; or thin-walled, sac-like and clavate to obovoid, 23-29.2 × 11.7-15.5 μm, and hyaline. Stalk hyphae parallel to interwoven on the surface, more or less cylindrical, 3.5-14.3 μm wide, hyaline. Oleiferous hyphae refractive, greenish yellow, 3.5-14.3 μm wide, scattered. Clamp connections present.

Useful references. The following books contain information on *I. pudica* that is pertinent to North America.

- *A Field Guide to Western Mushrooms* (Smith 1975)—A color photograph and a description and discussion are presented.
- *The Savory Wild Mushroom* (McKenny 1971)—A black-and-white photograph and a description and discussion are given.

Inocybe sororia Kauffman

Figs. 66, 75

Common names. None.

Distinguishing characteristics. Cap prominently umbonate; surface dry and silky-fibrillose, becoming radially cracked; color creamy to more or less pale straw color or stronger yellow, somewhat darker with age. Odor strongly resembling green corn. Gills whitish to yellowish, becoming olive buff to stronger yellow; edges white, minutely fringed. Stalk more or less equal, with a somewhat bulbous base; color whitish, becoming dingy. Spores ellipsoidal, smooth, 11-14.6 (-15.3) × 5.8-6.6 μm in size. Pleurocystidia lacking.

Observations. The pale straw-colored cap and the distinct odor of green corn make *I. sororia* fairly easy to identify in the field. It is similar in some characteristics to *I. fastigiata*; however, *I. fastigiata* has a spermatic odor and usually a darker cap surface than *I. sororia.* Both species are fairly common and can be found growing together in the same area.

Distribution and seasonal occurrence. The species is widely distributed across Canada and the United States and at times is fairly common. It fruits during the summer, fall, and winter, depending on weather and location.

Habitat and habit. *I. sororia* is terrestrial, scattered to gregarious, and often found under conifers.

Technical description. Cap 20-50(-70) mm wide, more or less conic-bell-shaped, becoming expanded and prominently umbonate; umbo more or less conic or acute; margin split and recurved with age; surface dry, innately silky-fibrillose, at later stages long-rimose; color in early stages creamy to straw color or mustard yellow, sordid and darker with age, with a darker umbo. Flesh pallid or yellowish, thin; odor strongly resembling green corn; taste not recorded.

Gills whitish or yellowish, becoming olive buff to mustard yellow or ocherous, attenuate-adnate, narrow, close to crowded; edges white and minutely fringed.

Stalk 30-70 mm long, 2-5 mm wide, equal or tapering upward; base more or less bulbous, in early stages silky-cortinate, becoming glabrous, innately fibrillose, solid, pruinose at the apex, whitish, becoming dingy. Annulus and universal veil lacking.

Spores yellow brown in deposit, 11.0-14.6(-15.3) × 5.8-6.6 μm, ellipsoidal, inequilateral in side view, elliptic in face view, smooth, yellow brown. Basidia four-spored, clavate, 29.0-35.1 × 11.0-11.7 μm, hyaline. Pleurocystidia lacking. Cheilocystidia clavate to fusiform-ventricose; apices more or less capitate to tapered, 36.5-76.2 × 11.0-15.5(-17.9) μm, thin-walled, hyaline.

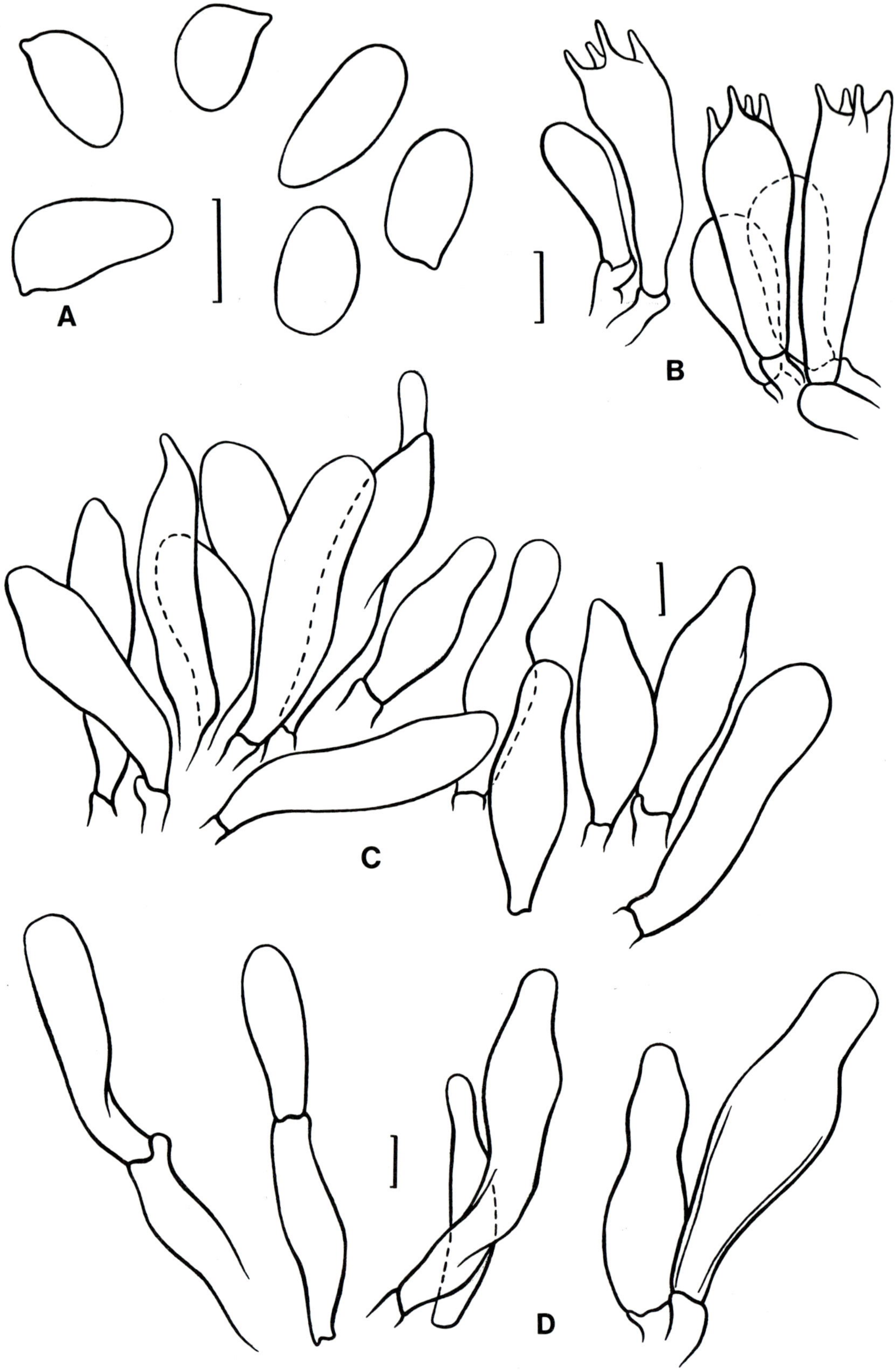

Fig. 66. *Inocybe sororia*: A, spores; B, basidia; C, cheilocystidia; D, caulocystidia.

Scale line 10 μm

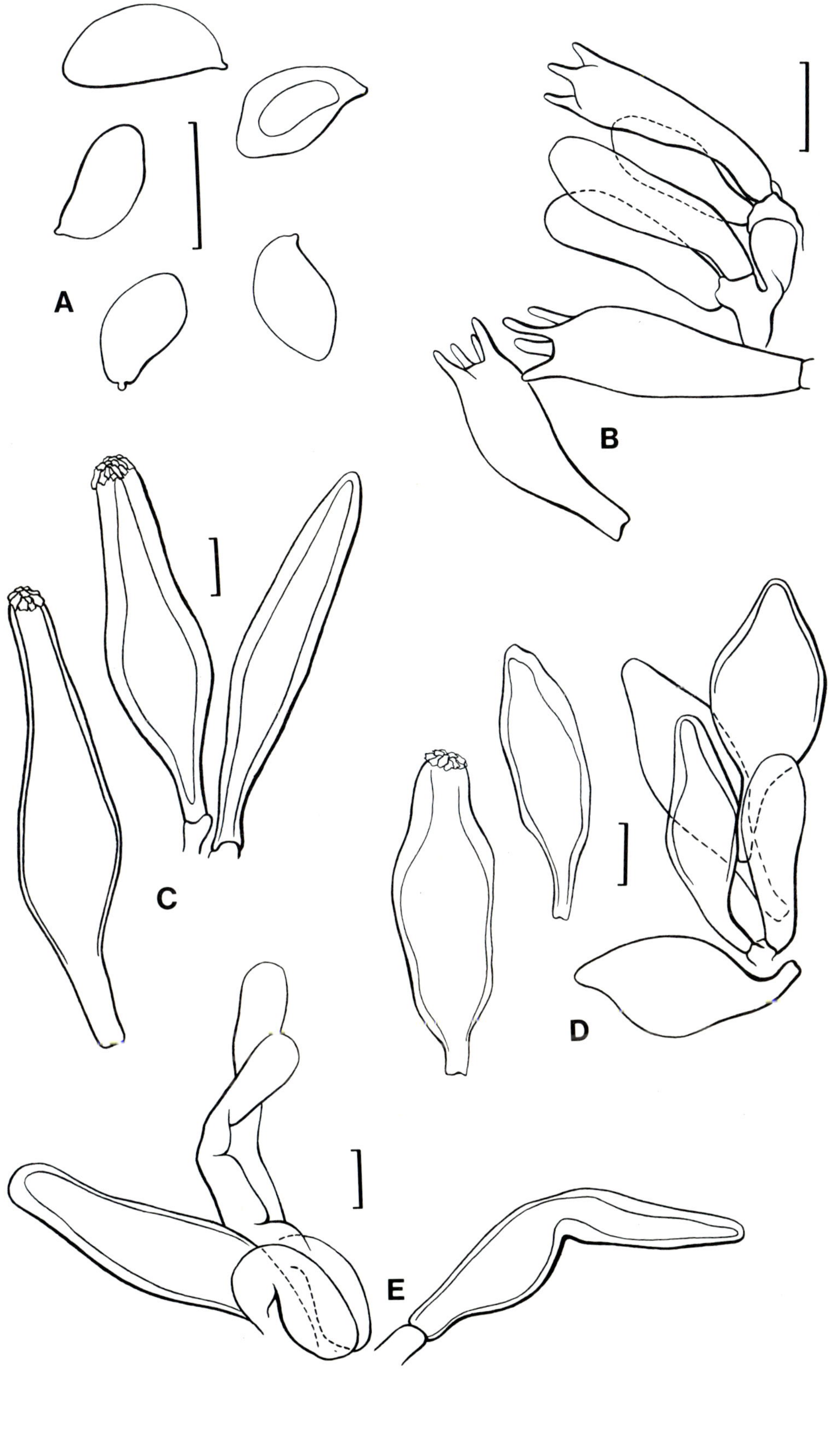

Fig. 67. *Inocybe subdestricta*: A, spores; B, basidia; C, pleurocystidia; D, cheilocystidia; E, caulocystidia.

Scale line 10 μm

Gill trama parallel; hyphae cylindrical to inflated, 3.7-5.1(-17.3) μm wide, hyaline to yellowish. Subhymenium hyphae interwoven, more or less cylindrical, 3.7-5.1 μm wide, hyaline. Cap cuticle filamentous; hyphae parallel, more or less radially arranged, more or less cylindrical, 3.7-8.8(-18.3) μm wide, narrow on the surface, hyaline to yellowish; walls thin to slightly thickened and finely encrusted. Cap trama loosely interwoven; hyphae cylindrical to inflated, 7-20 μm wide, hyaline. Caulocystidia clavate to cylindrical or fusiform-ventricose; apex mucronate to nearly capitate or tapered, 41.2-86.9 × 11.7-14.3 μm, hyaline, occurring in clusters at the apex of the stalk; stalk hyphae parallel to interwoven, more or less cylindrical, 3.5-16.7 μm wide, hyaline to yellowish. Oleiferous hyphae refractive, yellowish, 3.5-7.0 μm wide, scattered in the cap trama and stalk. Clamp connections present.

Useful reference. The following book contains information on *I. sororia* pertinent to North America.

- *Mushrooms in their Natural Habitats* (Smith 1949)—A color photograph on a reel and a description and discussion are presented.

Inocybe subdestricta Kauffman

Figs. 67, 76

Common names. None.

Distinguishing characteristics. Cap surface appressed-fibrillose; color brown on the margin to umbrinous in the center. Odor spermatic. Gills pallid, becoming umbrinous with white edges. Stalk equal, with the base rounded to thickened; surface powdery at the apex and fibrillose downward; color pale pinkish cinnamon. Spores ellipsoidal to almond-shaped, smooth, 8.8-11.0(-12.0) × 5.1-5.8 μm in size. Pleurocystidia thick-walled.

Observations. *I. subdestricta* was originally described by C. H. Kauffman from material in Michigan. It has not commonly appeared in North American literature and adequate descriptions can only be found in technical literature.

Distribution and seasonal occurrence. *I. subdestricta* seems widely distributed and is probably very common. It has been collected in British Columbia and as far east as Nova Scotia. It fruits in the late spring, summer, and fall.

Habitat and habit. The species is terrestrial and is found singly to scattered under conifers and hardwoods.

Technical description. Cap 20-25 mm broad, bell-shaped, becoming expanded and umbonate; surface appressed-fibrillose, more or less shining, at later stages lacerated-scaly to rimose, or both; color umbrinous to dark chestnut at the center, tawny olive to dark brown or rufous brown toward the margin. Flesh firm, pallid, unchanging when bruised. Odor spermatic to nauseous; taste not recorded.

Gills pallid in early stages, becoming dingy umbrinous, adnate to sinuate-adnexed, ventricose, medium broad, moderately close, with white edges.

Stalk 25-30 mm long, 4-5 mm wide, terete, equal; base rounded and slightly thickened, solid; surface pruinose at the apex, slightly to densely fibrillose downward; color pale pinkish cinnamon or tinged rufous.

Spores yellowish brown in deposit, 8.8-11.0(-12.0) × 5.1-5.8 μm, elliptic and slightly inequilateral to almond-shaped in side view, ovate in face view, smooth, thin-walled, yellowish brown. Basidia four-spored, clavate, 23.4-26.4 × 8.0-8.4 μm, hyaline. Pleurocystidia fusiform to fusiform-ventricose, 58.5-76.2 × 12.9-18.0 μm, thick-walled, apically encrusted, hyaline. Cheilocystidia clavate and sac-like to fusiform-ventricose or fusiform-lanceolate, (20-)41-58.0 × (10-)12.9-14.3 μm, hyaline, thin-walled to thick-walled (like pleurocystidia), sometimes apically encrusted.

Gill trama compactly interwoven to parallel; hyphae cylindrical to somewhat inflated, 3.0-15.0 μm wide, hyaline or yellowish near the subhymenium. Subhymenium interwoven; hyphae short-cylindrical, 3.7-5.8 μm wide, hyaline to yellowish. Cap cuticle filamentous; hyphae parallel to radially interwoven, more or less cylindrical, 3.7-7.3(-14.6) μm, thin- to thick-walled, finely encrusted, brownish. Cap trama loosely interwoven; hyphae cylindrical to inflated, 3.7-20 μm wide, hyaline. Caulocystidia fusiform-ventricose to lanceolate, (20-)35-70.2 × 13-16.7 μm, thin- or thick-walled, yellowish; stalk hyphae parallel in the core to interwoven on the surface, more or less cylindrical, 3.5-12.9 μm wide, hyaline to yellowish. Oleiferous hyphae refractive, greenish yellow, 3.5-8.0 μm wide, scattered in the stalk and cap trama. Clamp connections present.

Useful references. *I. subdestricta* has not appeared in any of the common mushroom guides. Technical descriptions are provided by Kauffman (1924) and Grund and Stuntz (1968).

Literature

Bigelow, H. E. The genus *Clitocybe* in North America. Sect. *Clitocybe*. Lloydia 28(2):139-180; 1965.

Eugster, C. H.; Schleusener, E. Stereomere Muscarine kommen in der Natur vor. Gas-chromatographische Trennung der Norbasen. Helv. Chim. Acta 52:708-715; 1969.

Groves, J. W. Edible and poisonous mushrooms of Canada. Ottawa, Ont.: Agriculture Canada. Agric. Can. Publ. 1112; 1979 [first printed in 1962].

Grund, D. W.; Stuntz, D. E. Nova Scotian inocybes. I. Mycologia 60(2):406-425; 1968.

Kauffman, C. H. *Inocybe*. N. Am. flora 10:227-260; 1924.

Lincoff, G.; Mitchel, D. H. Toxic and hallucinogenic mushroom poisoning—A handbook for physicians and mushroom hunters. New York, NY: Van Nostrand Reinhold Co.; 1977.

McKenny, M. The savory wild mushroom (revised and enlarged by Stuntz, D. E.). Seattle, WA: University of Washington Press; 1971.

Miller, O. K., Jr. Mushrooms of North America. New York, NY: E. P. Dutton & Co., Inc.; 1972.

Pilát, A.; Ušák, O. A handbook of mushrooms. London, England: Spring Books; 1951.

Smith, A. H. Mushrooms in their natural habitats. 2 Vols. Portland, OR: Sawyer's, Inc.; 1949.

Smith, A. H. A field guide to western mushrooms. Ann Arbor, MI: University of Michigan Press; 1975.

Stuntz, D. E. Studies in the genus *Inocybe*. I. New and noteworthy species from Washington. Mycologia 39:21-55; 1947.

Fig. 68. *Clitocybe dealbata.*

Fig. 69. *Clitocybe dilatata.*

Fig. 70. *Inocybe fastigiata.*

Fig. 71. *Inocybe geophylla.*

Fig. 72. *Inocybe lacera.*

Fig. 73. *Inocybe napipes.*

Fig. 74. *Inocybe pudica.*

Fig. 75. *Inocybe sororia.*

Fig. 76. *Inocybe subdestricta.*

11

IBOTENIC ACID — MUSCIMOL

Toxicology and Symptomatology

The three major toxins found in *Amanita muscaria* and its relatives are the isoxazole substances muscimol, ibotenic acid, and muscazone. The concentrations of these substances are variable, depending on the specific specimen, environment, location, and time of year. For example, as much as 10 times the amount of isoxazoles has been reported in specimens collected in the summer, compared with those collected in the fall. Ibotenic acid is relatively unstable and is easily converted to muscimol. Muscimol has been reported to comprise 0.03–0.1% of the fresh weight of the tissue. Although this amount may seem small, it becomes significant when you consider that 75–80% of the fresh weight of a mushroom is water. Besides the three toxins already mentioned, muscarine (Chapter 10) is present in very small amounts (Simons 1971, Hatfield and Brady 1975, Chilton and Ott 1976). Although other chemical compounds have been reported as being present and active as toxins in this group (Hardin and Arena 1974), the above four are the ones, at present, that require the most attention. Of course, the complex symptoms observed may indeed involve chemical substances that have as yet been unidentified.

The red form of *A. muscaria* has been reported to contain the highest levels of the isoxazole compounds. Comparative studies on the colored forms of *A. muscaria*, however, do not substantiate these claims (Benedict 1966, Chilton and Ott 1976). Although one of the pigments in the cap surface, musca-aurin I, has ibotenic acid bound to the chromophore, this pigment is present in only a very small amount. The belief that the toxicity can be reduced by removing the outer layer of the mushroom cap is therefore unfounded.

All the naturally occurring isoxazole derivatives are biologically active and are reported to interfere with the normal utilization of some important amino acids. These toxins affect the central nervous system, although everyone does not experience the same symptoms of toxicity. The psychoactive compounds of *A. muscaria* are not metabolized by the body, but rather, are passed into the urine. Some authors believe that ibotenic acid and muscimol can resemble the hallucinatory drug LSD (Hatfield and Brady 1975) in their influence on neural transmission.

The symptoms from ingesting mushrooms of this group appear within 20–90 minutes, depending on the amount ingested. Victims seldom vomit after ingesting *A. muscaria*. Muscimol is known to be an effective agent for suppressing vomiting. People who have eaten mushrooms from this group may appear inebriated. Derangement of the senses, manic behavior, delirium, and perceiving very small objects as enormous ones have often been experienced. Sometimes the victim has an intense desire for exaggerated physical activity. Often the victim experiences a deep, death-like sleep from which he or she is extremely difficult to wake. Some have visions during this period.

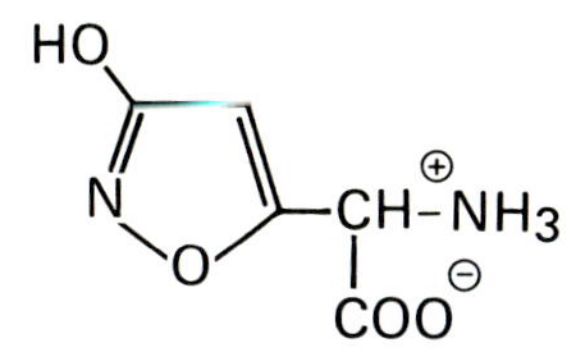

ibotenic acid

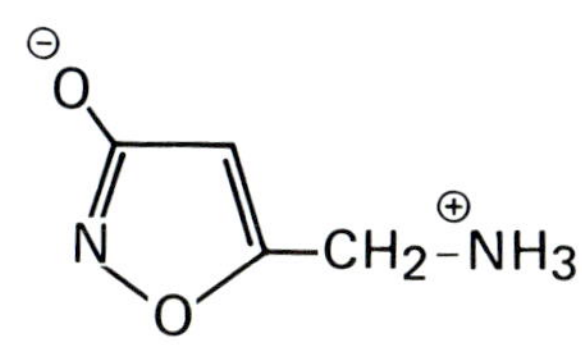

muscimol

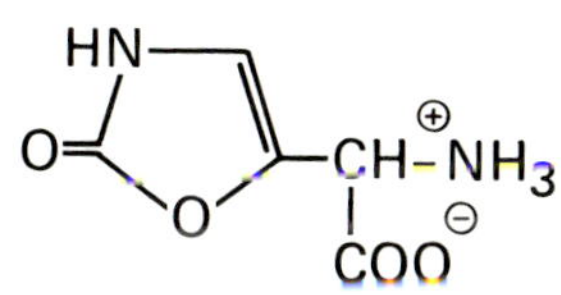

muscazone

The severity of the above symptoms depends on the amounts of the toxins ingested. When considerable mushroom material containing the toxins is ingested, incoherence, confusion, and great excitement, with visual and acoustic hallucinations, may be experienced. A patient disturbed in this state may display attacks of rage. The victim may also display spontaneous muscle twitchings and spasms, sometimes violent and erratic. In some reported cases convulsions have occurred during the unconscious period. Often, the victim cannot remember anything that occurred during the period of intoxication. Although the symptoms generally disappear in 12 hours, the victim may not feel really well for several days. Death from ingesting mushrooms from this group is rare; however, fatalities have been reported from muscimol poisoning when large quantities were eaten or the victim was suffering from some other unrelated organic illness.

Clinical data available on temperature, pupil size, heart rate, and blood pressure are highly variable, probably because of varying amounts of other toxins, such as muscarine and other cholinomimetics, contained in the mushrooms. Individual sensitivity to the toxins in these mushrooms also varies, as well as do seasonal and even environmental factors, which affect the amounts of toxins present in these mushrooms.

No safe or effective drug has been developed for treatment of poisoning from mushrooms in this group. People suspected of ingesting these mushrooms should be given prompt medical attention. Animal studies have suggested that diazepam or phenobarbital may potentiate the effects of muscimol. Atropine should not normally be administered (Lincoff and Mitchel 1977).

Symptom-producing Species

Amanitaceae
- *Amanita*
  - *A. cothurnata*
  - *A. crenulata* (suspected)
  - *A. frostiana* (suspected)
  - *A. gemmata*
  - *A. muscaria* (including several varieties)
  - *A. pantherina*

Amanita crenulata may be poisonous, but its toxicity has not yet been documented. *Amanita frostiana* is still suspect but there is no proof that it contains the toxins found in *Amanita muscaria*. *Amanita smithiana* Bas (=*A. solitaria* (Bull. per Fr.) Mérat *sensu* Stuntz) is reported to contain toxic compounds (Benedict 1972, Chilton and Ott 1976). In Japan, *Amanita strobiliformis* (Paulet ex Vittad.) Bertillon is reported to contain ibotenic acid (Benedict 1972) but this substance has not been found in specimens from North America (Chilton and Ott 1976). Bas (1969) gives a description and discussion of *Amanita smithiana, Amanita solitaria,* and *Amanita strobiliformis. Tricholoma muscarium* Kawakami ex Hongo is also reported to contain ibotenic acid. It is reported as edible but intoxicating by Imazeki and Hongo (1957). Miller (1972) also reports *Amanita cokeri* (Gilbert & Kühner) Gilbert, *Panaeolus campanulatus,* and *Panaeolus retirugis* as containing pantherin (=muscimol) or related toxins. There is no chemical evidence for muscimol in *Panaeolus* (S. Chilton, Personal commun.). Descriptions are included in this chapter for the *Amanita* species that appear on the list.

Descriptions of the Genus and Species

Amanita Pers. ex S. F. Gray

Amanitaceae

Anyone eating mushrooms or dealing with poisonous mushrooms must learn to recognize the genus *Amanita*. Although some species of *Amanita* are edible, many of them are poisonous, and species such as *A. phalloides* are among the most poisonous mushrooms known.

There are four principal morphological characteristics that are shared by most *Amanita* species: the universal veil, the annulus, free or slightly attached gills, and a white spore deposit.

The universal veil completely covers the young mushroom in the button stage and breaks when the stalk elongates and the cap enlarges and expands (Fig. 1 F–H, A). The universal veil breaks in various ways depending on its structure and consistency. When the universal veil is membranous and tough, it often breaks by a slit across the top of the cap, allowing the cap to become free.

In such cases the universal veil is usually left as a cup, the volva, around the stalk base. This type of development is characteristic of *A. phalloides* and its relatives. In some species the universal veil is membranous but more fragile and breaks in an irregular pattern as the mushroom enlarges. In these cases a large piece of universal veil material is often left on the cap, with the remainder forming a cup-like volva around the stalk base. Development of this type is found in *A. calyptroderma* Atk. & Ballen, an edible species, occurring on the west coast of North America. In other species of *Amanita* the universal veil is more fragile and not membranous. For example, the universal veil of *A. muscaria* typically breaks into small patches or warts that are variously arranged over the cap surface. The remainder of the universal veil, the volva, is associated with the lower stalk surface and the apex of the bulbous stalk base, where it forms ring-like patches (Fig. 85). In *A. pantherina* the universal veil also breaks into small patches or warts on the cap surface, but the volva typically occurs as a collar-like rim on the apex of the stalk base, indicating where the universal veil broke (Fig. 89). Certain *Amanita* species have a universal veil that crumbles easily into small pieces that range from mealy to more or less powdery. In these species the cap surface has granules or small patches of soft tissue that are easily weathered away, so that the cap of mature specimens lacks any remains of universal veil. The volva in these species usually occurs as loosely attached granules or small patches on the lower stalk and at its base. The volva usually also becomes lost or weathered away with age and at times can be found on the ground around the stalk base. In these species, particularly, both young and mature specimens must be studied. In many *Amanita* species the volva is buried in the soil and litter and may not be evident unless the entire mushroom, including the stalk base, is carefully removed from the soil.

The annulus is formed from the partial veil, a more or less thin membranous layer of tissue that extends from near the apex or middle of the stalk to the edge of the cap (Fig. 1 H and A). In the button stage and other young stages the partial veil covers the young, developing gills. When the mushroom matures, the partial veil typically breaks along the cap edge, leaving a thin, skirt-like ring on the upper part of the stalk. Sometimes the partial veil breaks away from the stalk rather than from the cap edge, leaving a fringe of tissue hanging from the cap edge. Some *Amanita* species do not have a partial veil and therefore have no annulus. These species have traditionally been placed in a separate genus, *Amanitopsis* or *Vaginata*. Now, they are usually placed in the genus *Amanita*.

The gills are often free in *Amanita* but in some species, such as *A. silvicola* Kauffman, they are attached or just reach the stalk (Fig. 7 A and B).

The spore deposit in *Amanita* is typically white. The gills are also usually white or pale colored, but because the spore color is not always the same color as the gills, be sure to take a spore deposit.

Amanita species also have in common a divergent gill trama and a unique stalk trama composed of very large, clavate hyphae and narrow, branching, longitudinally arranged, interwoven hyphae. The divergent gill trama is often used as a key characteristic (Fig. 14C).

There are several species of *Amanita*, closely related to *Amanita muscaria*, which contain ibotenic acid and muscimol; however, *A. muscaria* and *A. pantherina* are the two species most commonly involved in this type of mushroom poisoning.

In summary, the diagnostic characteristics of this group of *Amanita* species are a white spore deposit; nonamyloid spores; free to slightly attached white gills; a well-developed annulus, except in forms that do not have an annulus at all and specimens in which the partial veil forms a fringe on the cap edge rather than an annulus; a universal veil that breaks into warts or small patches on the cap surface and leaves zones and patches on the lower stalk surface or a distinct, collar-like rim on the apex of the stalk base; and a striate cap margin, although the striations may be faintly developed in the early stages or in some forms. A general key that separates *A. muscaria* and its relatives follows. Jenkins (1974, 1977, 1978) provides a technical treatment of this group, complete with keys.

Key to common species

1. Cap white to whitish, cream color, buff, or shaded slightly with yellow, yellow brown, or cinnamon pink tones, not truly yellow, orange, red, or brown 2

   Cap some shade of true yellow, orange, red, or brown 4

2. Stalk slender, with an abruptly bulbous base; volva forming a close-fitting enrolled collar on the apex of the bulb. Cap white to whitish or slightly cream color or with the center shaded somewhat yellow to yellow brown .... ***A. cothurnata***

   compare with pale form of *A. muscaria*

   Stalk with zones or patches, or both, on the lower surface, on and above the apex of the bulb; or volva delicate, leaving easily removed warts and patches on the cap surface and a slight white to pallid granular covering on the apex of the stalk bulb. Cap white to buff, slightly pink, grayish, or tinged yellow 3

3. Cap white to buff or slightly cinnamon pink, usually with ascending rings of volval tissue on the lower stalk surface, and sometimes with a slight rim on the apex of the bulb. Annulus typically persistent .... ***A. muscaria* var. *alba***

   Cap whitish to grayish or tinged yellowish. Cap surface with thin delicate whitish warts or patches that are easily removed and with a volva forming an easily removed white to pallid, mealy covering on the apex of the stalk bulb; annulus often lost in mature specimens .... ***A. crenulata***

4. Fruit body small, 2–7 cm high. Cap bright orange to yellow; stalk white to yellow. Annulus usually present and yellowish; volva yellow, thin, and easily removed. Spores globose to subglobose .... ***A. frostiana***

   Not with the above combination of characteristics 5

5. Cap usually dark brown to pale brown or yellowish brown, occasionally paler or more yellowish, with white to creamy white warts or small patches that are scattered or more or less concentrically arranged on the cap surface. Volva a more or less distinct collar-like roll on the apex of the bulbous stalk base and usually with zones and patches just above the basal bulb .... ***A. pantherina***

   compare with *A. gemmata*

   Cap distinctly yellow, orange, or red or a combination of these colors 6

6. Cap crimson to dark red overall or with the margin orange red .... ***A. muscaria* var. *muscaria***

   Cap some shade of yellow or orange or sometimes somewhat reddish orange 7

7. Cap light yellow or maize yellow to orange, sometimes reddish orange in the center. Volva forming

distinct zones or patches on the lower stalk surface .... ***A. muscaria* var. *formosa***
when the universal veil is tinted yellowish, it may be ***A. muscaria* var. *flavivolvata***

Cap creamy yellow, pale yellow, or occasionally pale yellowish beige or champagne color. Volva forming creamy to tan and more or less easily removable patches on the apex of the stalk bulb and just above the bulb................. ***A. gemmata***

Amanita cothurnata Atk.

Figs. 77, 83

Common name. Booted amanita.

Distinguishing characteristics. Cap pure white to cream color or slightly yellow to yellow brown in the center. Universal veil remnants white, randomly arranged, occurring as warts or patches on the cap surface, easily removed or weathered away. Stalk slender, white, abruptly bulbous; volva forming a close-fitting enrolled collar on the apex of the bulb. Spores nonamyloid, 8.8-11.7(-12.4) × 6.0-8.0 (-8.8) μm in size.

Observations. In its typical form *A. cothurnata* is a fairly distinct species. Its slender habit and white color together with the collar-like volva distinguish it from the pale form of *A. muscaria*. Its relationship seems to lie with *A. pantherina*, and some workers consider it a variety of this species, namely *A. pantherina* var. *multisquamosa* (Peck) Jenkins.

Distribution and seasonal occurrence. This species is found primarily in eastern North America and rarely in the western coastal region. It fruits in summer and early fall.

Habitat and habit. *A. cothurnata* is terrestrial and grows solitary to gregarious in hardwood forests or under conifers.

Technical description. Cap 30-90 mm broad, hemispherical to convex-expanding to plane, occasionally depressed; margin faintly to strongly striate; surface viscid, glabrous, covered with numerous white floccose warts and angular to diffuse patches, eroded when old, randomly arranged; cap color white to yellowish white, sometimes pale yellow to tan on the disc. Flesh white; odor and taste indiscernible.

Gills white, free, crowded, broadest at the margin; edges floccose.

Stalk 35-130 mm long, 3-12 mm wide at the apex, cylindrical to tapering upward, slightly expanded at the apex; base bulbous, subglobose to ovoid; surface minutely to strongly floccose to fibrillose, white to creamy white, hollow. Annulus apical, floccose-membranous, often pendant or flaring, ragged, white with a yellowish white edge, persistent. Volva white to whitish, membranous, appressed to the stalk base, never breaking into rings; margin free, extending up to 10 mm up the stalk before forming a distinct roll or collar, with threads of tissue connecting the volval margin to the stalk.

Spores white in deposit, 8.8-11.7(-12.4) × 6.0-8.0(-8.8) μm, subglobose to broadly ellipsoidal, adaxially flattened, thin-walled, smooth, hyaline, nonamyloid. Basidia two- (rarely) or four-spored, clavate to subcylindrical, 33-62 × 4-11 μm, thin-walled, hyaline. Pleurocystidia and cheilocystidia lacking; sterile cells present on gill edges, cylindrical to sac-like, 20-25 × 14-18 μm, hyaline.

Gill trama bilateral, divergent from a central strand; hyphae moderately branched, cylindrical to inflated, 2.6-16 μm wide; hyphae in central strand narrower and typically cylindrical, hyaline. Subhymenium hyphae cylindrical to slightly inflated, branched, 6.6-10 μm wide, hyaline. Cap cuticle filamentous, gelatinous; hyphae radially arranged, more or less cylindrical, 2.3-6.6(-10) μm wide, hyaline to yellowish. Cap trama loosely interwoven; hyphae cylindrical to inflated, 5.5-27 μm wide, hyaline. Caulocystidia clavate, 80-120(-385) × 11-16(-40) μm, overlapping, hyaline. Annulus interwoven; hyphae mainly cylindrical, 3-5(-8) μm wide, hyaline; sphaerocysts also present, scattered, ovoid to ellipsoidal or globose (30-)47-77 × (20-)23.4-70.2 μm, hyaline, thin-walled. Universal veil interwoven; hyphae cylindrical to inflated, 3-20 μm wide, branched, hyaline; sphaerocysts also present, (15-)35-65(-81) × 11-26(-42) μm, narrowly ellipsoidal, pear-shaped or subglobose to globose, terminal or in short chains, hyaline. Volva interwoven; hyphae mainly cylindrical or inflated, 2.6-25 μm wide; sphaerocysts also present, globose to ovoid or ellipsoidal, 25-50 × 20-40 μm, thin-walled, hyaline, scattered. Oleiferous hyphae refractive, thin-walled, grayish yellow, 5-7.1 μm wide, scattered in the cuticle and the volva. Clamp connections present in the tissue of the cuticle, trama, volva, and stalk.

Useful references. The following books contain information on *A. cothurnata* pertinent to North America.

- *Edible and Poisonous Mushrooms of Canada* (Groves 1979)—A discussion is presented.
- *Mushrooms in their Natural Habitats* (Smith 1949)—A color photograph on a reel and a description and discussion are presented.
- *Mushrooms of North America* (Miller 1972)—A discussion is presented.

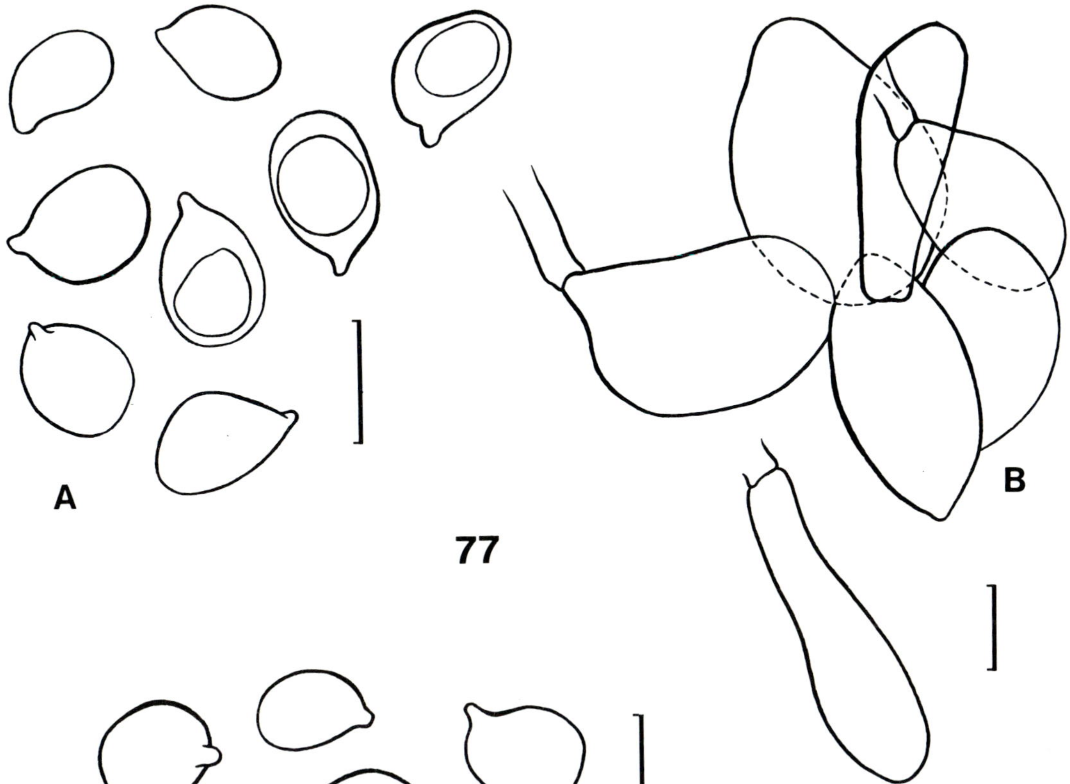

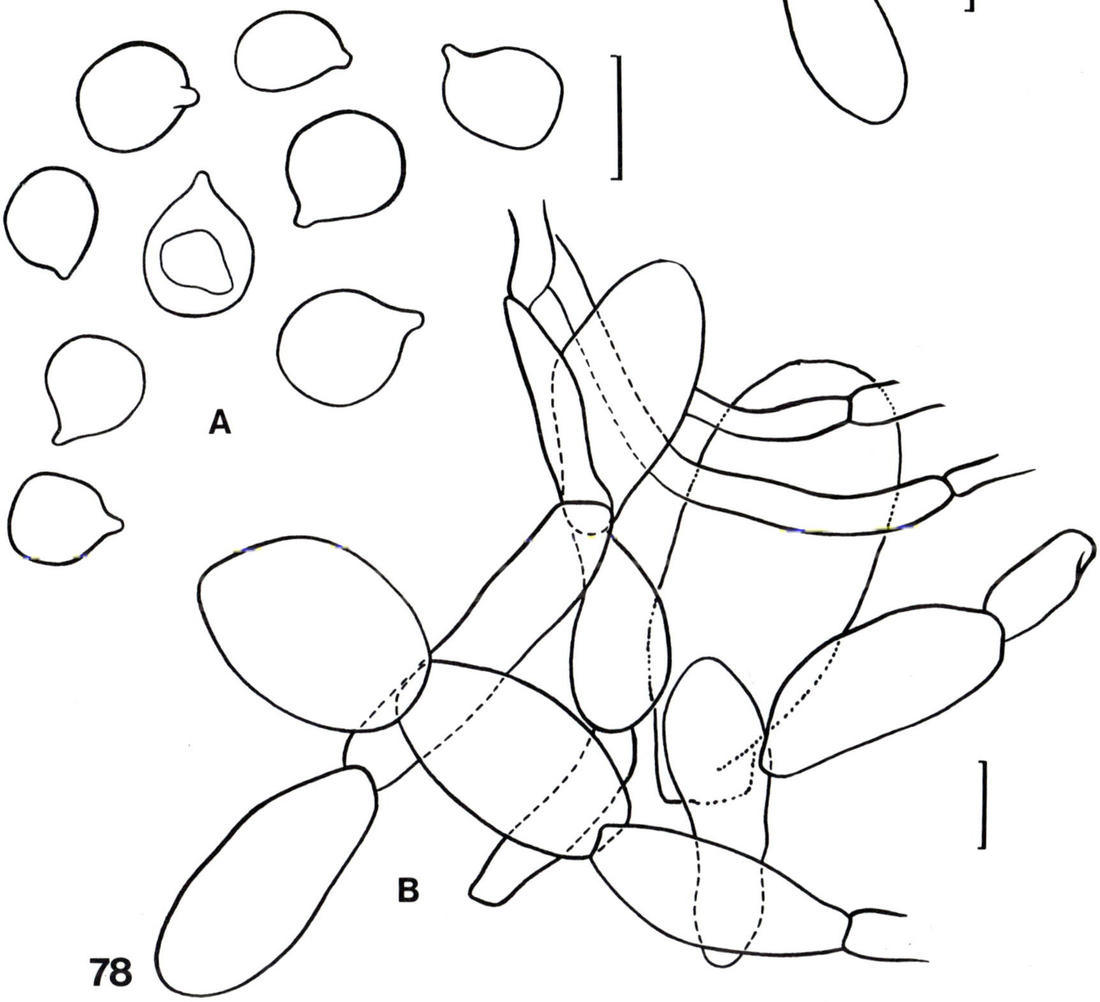

Fig. 77. *Amanita cothurnata*: A, spores; B, volval elements.

Fig. 78. *Amanita crenulata*: A, spores; B, volval elements.

Scale line 10 μm

Amanita crenulata Peck

Fig. 78

Common names. None.

Distinguishing characteristics. Cap whitish to grayish or tinged yellowish. Universal veil forming delicate, thin, whitish warts or patches on the cap surface, with these randomly scattered or in a more or less continuous layer and easily removed. Stalk base bulbous; volva typically forming a slight white to pallid mealy covering on the apex of the bulb; volval material easily removed and often lacking in mature specimens. Annulus usually only found in young specimens, rarely present on mature specimens. Spores nonamyloid, (8.0–)9.5–10.2 × 7.0–8.0 μm in size.

Observations. This species is commonly mistaken for *A. gemmata*, but it can be distinguished by its mealy volva, which is usually lost at maturity, and by its globose to subglobose spores. It has not been commonly reported in North America. In the original description, Peck (1900) mentioned that this species was eaten without causing harm. However, Buck (1965) reported a poisoning by *A. crenulata* that was characterized by a brief period of low blood pressure.

Distribution and seasonal occurrence. This species is found in eastern North America; it has been reported from Massachusetts and the surrounding area. It apparently fruits in late summer, usually not until September, and fall.

Habitat and habit. *A. crenulata* is terrestrial, growing solitary to gregarious in mixed woods.

Technical description. Cap 25–55 mm wide, ovoid, expanding to convex or nearly plane; margin striate; surface glabrous, viscid, covered with thin white floccose warts or patches of universal veil tissue that are easily removed and are scattered randomly; cap color white to grayish white, tinged yellowish. Flesh white; odor and taste not recorded.

Gills white, close, free or decurrent by a tooth, broadest near the margin; edges floccose.

Stalk 25–70 × 4–10 mm, equal or tapering upward, swollen at the base; surface floccose above the annulus, smooth to fibrillose below, white, stuffed or hollow. Annulus apical, floccose-membranous, evanescent, white. Volva floccose-mealy, evident at the apex of the bulb, appressed to eroding and easily removed.

Spores white in deposit, (8.0–)9.5–10.2 × 7.0–8.0 μm, globose to broadly ellipsoidal, adaxially flattened, thin-walled, hyaline, nonamyloid. Basidia two- or four-spored, 37–48 × 8.4–9.0 μm, clavate to subcylindrical, thin-walled, hyaline.

Gill trama divergent from a central strand; hyphae interwoven, cylindrical to inflated, 3–10 μm wide, hyaline; hyphae of the central strand narrower and more cylindrical. Subhymenium hyphae cylindrical, 4.5–7.0 μm wide, hyaline. Cap cuticle filamentous, gelatinous; hyphae interwoven, cylindrical, 2–7 μm wide, more or less radially arranged, hyaline to yellowish. Cap trama interwoven; hyphae cylindrical to inflated, 3.7–28 μm wide, hyaline. Caulocystidia clavate, 100–300 × (15–)20–50 μm, overlapping; terminal cells hyaline; stalk hyphae interwoven, cylindrical, 3–8 μm wide, hyaline. Annulus interwoven; hyphae mostly cylindrical to inflated, especially the terminal cells, 3–38 μm wide, hyaline. Universal veil warts composed of loosely interwoven cylindrical hyphae 3–7 μm wide; sphaerocysts also present, subclavate to narrowly ellipsoidal, ellipsoidal, or ovoid, 30–70 × 20–50 μm, terminal or in short chains, hyaline. Volva interwoven; hyphae cylindrical, 3–7 μm wide, hyaline; sphaerocysts also present, abundant, clavate to broadly ellipsoidal or subglobose, 25–95 × 20–30 μm, hyaline, thin-walled, scattered, single or in short chains. Oleiferous hyphae scattered in the cuticle, 3.5–7.0 μm, greenish yellow. Clamp connections lacking to rare.

Useful references. There is no reference to this species in recent popular mushroom floras or handbooks. Workers interested in this species should consult the works of Jenkins (1974, 1977, 1978). Kreiger (1967) gives a description and brief discussion of this species.

Amanita frostiana (Peck) Sacc.

Fig. 79

Common names. None.

Distinguishing characteristics. Fruit body usually small to medium, more or less slender. Cap margin usually strongly striate. Cap bright orange, at times fading on the margin to yellow or, rarely, whitish. Universal veil remnants usually yellow, occasionally white, forming warts and patches that are usually randomly scattered but often more dense toward the center and relatively easy to remove. Stalk base bulbous. Volva membranous, white to yellowish, with a free collar; in addition, lower stalk surface and apex of bulb coated with a yellow cottony covering. Annulus usually yellow. Spores globose to subglobose, nonamyloid, 7.0–10.2(–11) × 7.0–10.2 μm in size.

Observations. This species is easily mistaken for smaller forms of *A. muscaria* and a very similar-appearing species, *A. flavoconia* Atk. It is easily separated from *A. muscaria* by the globose spores and the yellow, more or less cottony volval remains on the cap, lower stalk surface, and apex of the bulbous stalk base.

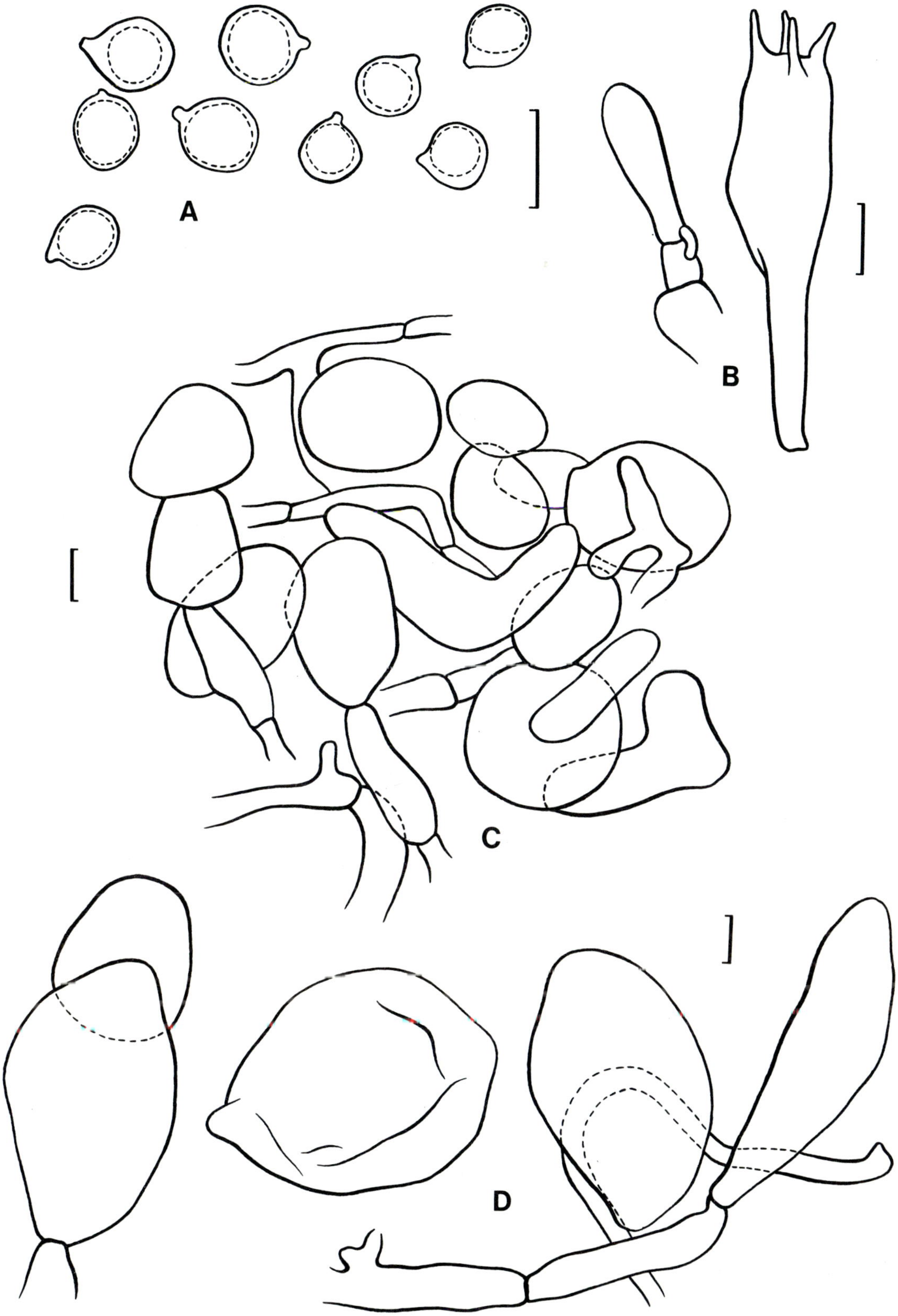

Fig. 79. *Amanita frostiana*: A, spores; B, basidium; C, volval elements; D, universal veil.

Scale line 10 μm

A. flavoconia has ellipsoidal, amyloid spores and the cap margin is not striate or at most only faintly so. Microscopic examination is the easiest and surest way of separating it from *A. frostiana*.

Distribution and seasonal occurrence. The species is found in southeastern Canada and into the southeastern United States from July to September. It often appears after *A. flavoconia* begins to fruit. It apparently is not a common species.

Habitat and habit. *A. frostiana* is terrestrial, found solitary to gregarious in mixed woods.

Technical description. Cap 20-60 mm wide, convex to plane, slightly depressed; margin conspicuously striate; surface viscid, glabrous, bearing randomly scattered floccose yellow or whitish yellow warts and yellow floccose patches that are easily removed; cap color deep orange or reddish orange on the disc, fading to yellow on the margin. Flesh thin, white to yellowish white; odor and taste not recorded.

Gills white, tinged creamy yellow, free, close, broad; edges white-floccose to yellow.

Stalk 47 mm long, 4-9 mm wide at the apex, tapering upward, slightly expanded at the apex; base bulbous, ovoid to globose; surface fibrillose to floccose above the annulus; stalk hollow to stuffed, white or pale yellow. Annulus apical, membranous to floccose, pendant, yellow, fragile, evanescent. Volva membranous, appressed to the bulb, white to yellowish over the bulb; collar free.

Spores white in deposit, 7.0-10.2(-11.0) × 7.0-10.2 μm, globose to subglobose, hyaline, nonamyloid. Basidia four-spored, clavate, 45-55(-64) × 5-11 μm, hyaline. Pleurocystidia and cheilocystidia lacking; saclike to clavate remnants of the partial veil present on the gill edges.

Gill trama divergent from a central strand; hyphae more or less interwoven, cylindrical to inflated, 2.3-17.2 μm wide, hyaline; hyphae of the central strand cylindrical and narrow. Subhymenium cellular; hyphae inflated, 9-10.2 μm wide, hyaline. Cap cuticle filamentous; hyphae interwoven, more or less radially arranged, cylindrical, 2.5-9.5 μm wide, hyaline to yellow. Cap trama loosely interwoven; hyphae cylindrical to inflated, 3.5-11.9(-14) μm wide, hyaline. Caulocystidia clavate, 23-35 μm wide, composed of overlapping terminal cells formed from branched interwoven cylindrical hyphae 3.5-8.1 μm wide, hyaline. Annulus interwoven; hyphae mostly cylindrical, 3.5-8.1 μm wide; sphaerocysts also present, scattered, ovoid to subcylindrical, (27-)58-70(-160) × (14-)20-55(-70.2) μm, terminal or in short chains, hyaline. Universal veil present as warts on the cap surface, loosely interwoven; hyphae cylindrical, 3-8 μm wide, hyaline; sphaerocysts also present, globose to ellipsoidal or clavate-fusiform, (55-)70-160 × 38-55(-70) μm, hyaline. Volva loosely interwoven; some hyphae cylindrical, 3.5-9.2 μm wide; most sphaerocysts globose to ellipsoidal, 27-70.2 × (25-)35-47(-70.2) μm, hyaline, thin-walled. Oleiferous hyphae refractive, thin-walled, yellowish, scattered in the cap cuticle, annulus, and stalk. Clamp connections rare, but when present occurring in the annulus and hymenium.

Useful reference. The following book provides information on *A. frostiana* pertinent to North America.

- *Edible and Poisonous Mushrooms of Canada* (Groves 1979)—A poor color photograph and a description and discussion are presented.

Amanita gemmata (Fr.) Bertillon in Dechambre

Figs. 80, 84

Common names. None.

Distinguishing characteristics. Fruit body small to medium and usually slender. Cap pale yellow to golden yellow, usually somewhat darker in the center, paler or occasionally whitish on the margin. Universal veil remnants white to dirty white or pale cream color, randomly arranged, occurring as warts and patches or forming a more or less continuous covering over the cap surface, usually more dense in the center. Stalk base bulbous, with more or less easily removable white to creamy tan patches of volval material on the upper bulb or sometimes on the stalk above the bulb, but less commonly present as irregular zones. Spores nonamyloid, (8.8-)9.5-11.7(-12.4) × 7.3-8.8(-10.2) μm in size.

Observations. *A. gemmata* is one of the better known species in *Amanita*, but it has often been mistaken for other species. It is variable and because of the yellow to whitish colors of the cap, it has been mistaken for *A. crenulata* and other similar-appearing species. It may hybridize with *A. pantherina* and produce fruit bodies with intermediate coloration and toxicity. *A. russuloides* Peck and *A. junquillea* Quél. are considered to be the same as *A. gemmata*. *A. velatipes* Atk. is a species similar in appearance but is probably related to *A. pantherina*. Jenkins (1974, 1977, 1978) discusses these species. Page (1975) reported a poisoning that he attributed to *A. gemmata*, with chiefly cardiovascular manifestations.

Distribution and seasonal occurrence. This species occurs throughout most of Canada and the United States. It fruits from spring through late fall.

Habitat and habit. *A. gemmata* is terrestrial, occurring solitary to gregarious under conifer and deciduous broad-leaved trees.

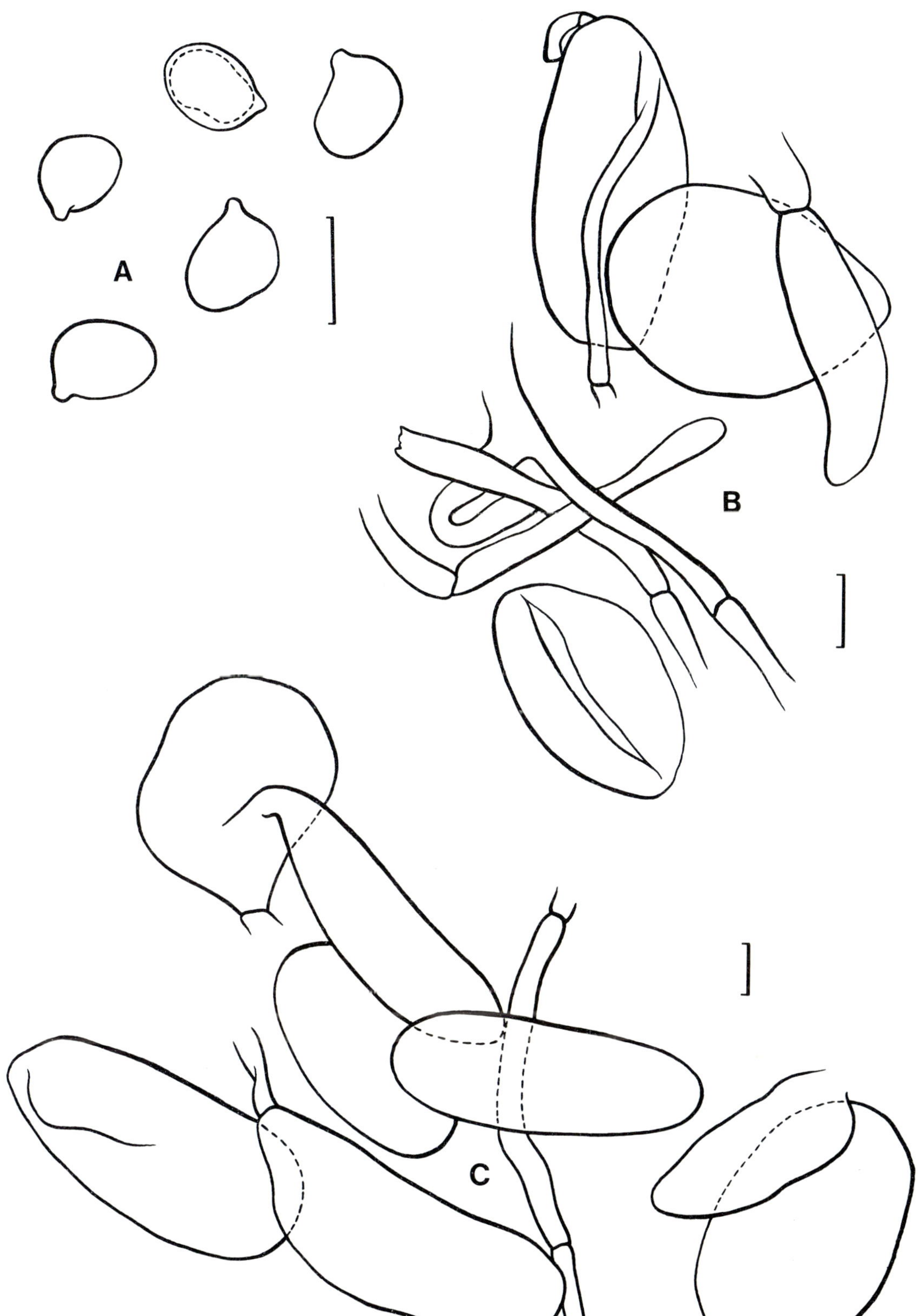

Fig. 80. *Amanita gemmata*: A, spores; B, volval elements; C, elements of the universal veil from the cap surface.

Scale line 10 μm

Technical description. Cap 25–120 mm broad, hemispherical to convex, becoming expanded to flat, occasionally depressed in the center; margin conspicuously striate, slightly recurved with age; surface smooth, viscid, bearing many to few small warts or floccose patches that are white to creamy white and randomly arranged but more dense in the center; color dull yellow when young, fading to dingy yellow or buff on the disc and pale yellow to creamy white on the margin. Flesh fairly thick on the disc, thinner toward the margin, white to yellowish white under the surface; odor indistinctive; taste not recorded.

Gills free, close, broadest toward the margin and tapering to the point of attachment, white to creamy white; edges floccose, with several tiers of shorter lamellulae.

Stalk 40–120 mm long, 5–19 mm thick at the apex, equal to tapering upward; base bulbous, subglobose to ovoid or rarely somewhat rooting; apex slightly floccose and slightly pruinose-striate, more or less glabrous to appressed floccose toward the base, white to pale creamy tan. Flesh stuffed to hollow, white. Annulus apical to median on the stalk, pendant, membranous; upper surface floccose-striate, white with yellowish white tints, fragile, evanescent. Volva appressed to the bulb, white to creamy tan; margin free in early stages, disappearing to irregularly torn.

Spores white in deposit, (8.8–)9.5–11.7(–12.4) × 7.3–8.8(–10.2) μm, subglobose to broadly elliptic and adaxially flattened in side view, smooth, hyaline, nonamyloid. Basidia two- (rarely) to four-spored, clavate, 30–40 × 8–11 μm, thin-walled, hyaline. Pleurocystidia and cheilocystidia lacking; gill edges lined with clavate to sac-like cells from the annulus.

Gill trama divergent from a central strand; hyphae cylindrical to inflated, 2.2–9 μm wide, thin-walled, hyaline to yellowish; hyphae in the central strand narrower and typically cylindrical. Subhymenium interwoven; hyphae branched, cylindrical to slightly inflated, 6–9 μm wide, hyaline. Cap cuticle filamentous, interwoven, radially arranged; hyphae cylindrical, 2.7–4 μm wide, thin-walled, hyaline to yellowish, gelatinizing in potassium hydroxide. Cap trama interwoven; hyphae cylindrical to more or less inflated, 3.7–14.6 μm wide, thin-walled, branched, hyaline to yellowish. Caulocystidia clavate to cylindrical, 50–60 × 9.4–10.2 μm, thin-walled, hyaline, abundant on the stalk apex. Annulus interwoven; hyphae cylindrical, 3–9 μm wide; sphaerocysts also present, clavate to ellipsoidal, 29–55 × 30–70 μm, terminal, hyaline. Universal veil present as warts on the cap surface, loosely interwoven; hyphae cylindrical to inflated, 3.5–8 μm wide, thin-walled, hyaline; sphaerocysts also present, 58.5–70.2 × 17.5–40 μm, ellipsoidal, hyaline. Volva interwoven; hyphae cylindrical, 4.4–7.3 μm wide, hyaline; sphaerocysts also present, ellipsoidal to subglobose, 35–70 × 20–35 μm, hyaline. Oleiferous hyphae refractive, 3.7–6 μm wide, scattered, most abundant in the region just below the cap cuticle. Clamp connections rare, present in the annulus, gill trama, subhymenium, and cap trama.

Useful references. The following books provide information on *A. gemmata* pertinent to North America.

- *Edible and Poisonous Mushrooms of Canada* (Groves 1979)—A poor color photograph and a description and discussion are presented.
- *Mushrooms in their Natural Habitats* (Smith 1949)—A color photograph on a reel and a description and discussion (as *A. junquillea*) are presented.
- *Mushrooms of North America* (Miller 1972)—A description and a poor discussion are given.

Amanita muscaria (L. per Fr.) Hook.

Figs. 81, 85-87

Common names. Fly agaric, fly mushroom, scarlet fly cup, fly-blown agaric, fly poison amanita, fly amanita, bug agaric, soma, and false orange.

Distinguishing characteristics. Cap dark red to crimson red overall or margin orange red (var. *muscaria*); light yellow to maize yellow or orange (var. *formosa* (Gonnermann & Robenhorst) Vesley); or whitish to cream or shaded cinnamon pink (var. *alba* Peck). Universal veil breaking into white or whitish warts or small patches that are more or less randomly arranged over the cap surface. Volva forming two or more somewhat ascending zones or patches on or above the apex of the stalk bulb, sometimes with a slight collar-like rim developing on the bulb apex. Annulus typically present, white and membranous. Spores nonamyloid, (9–)11–13.1(–14.6) × (6.6–)7.3–8.8(–9.1) μm in size.

Observations. *A. muscaria* comes in several color forms. The three most common forms, each called a separate variety, are listed in the previous section; a brown to reddish brown form, var. *regalis* (Fr.) Sacc., apparently is uncommon in North America. The caps are sometimes uniformly colored, but often the margin is paler than the corresponding central part of the cap. The caps in all these forms have a white covering of universal veil tissue resembling cottage cheese, which is randomly distributed over the surface or in more or less concentric zones. Specimens with a yellowish universal veil are sometimes called *A. muscaria* var. *flavivolvata* (Singer) Jenkins. Old or weathered specimens may lack universal veil elements on the cap surface. The volva characteristically forms ascending zones of tissue on the lower portion of the stalk or in some specimens it may be in the form of a slight collar-like rim on the apex of the bulb. The universal veil never forms a cup- or sheath-

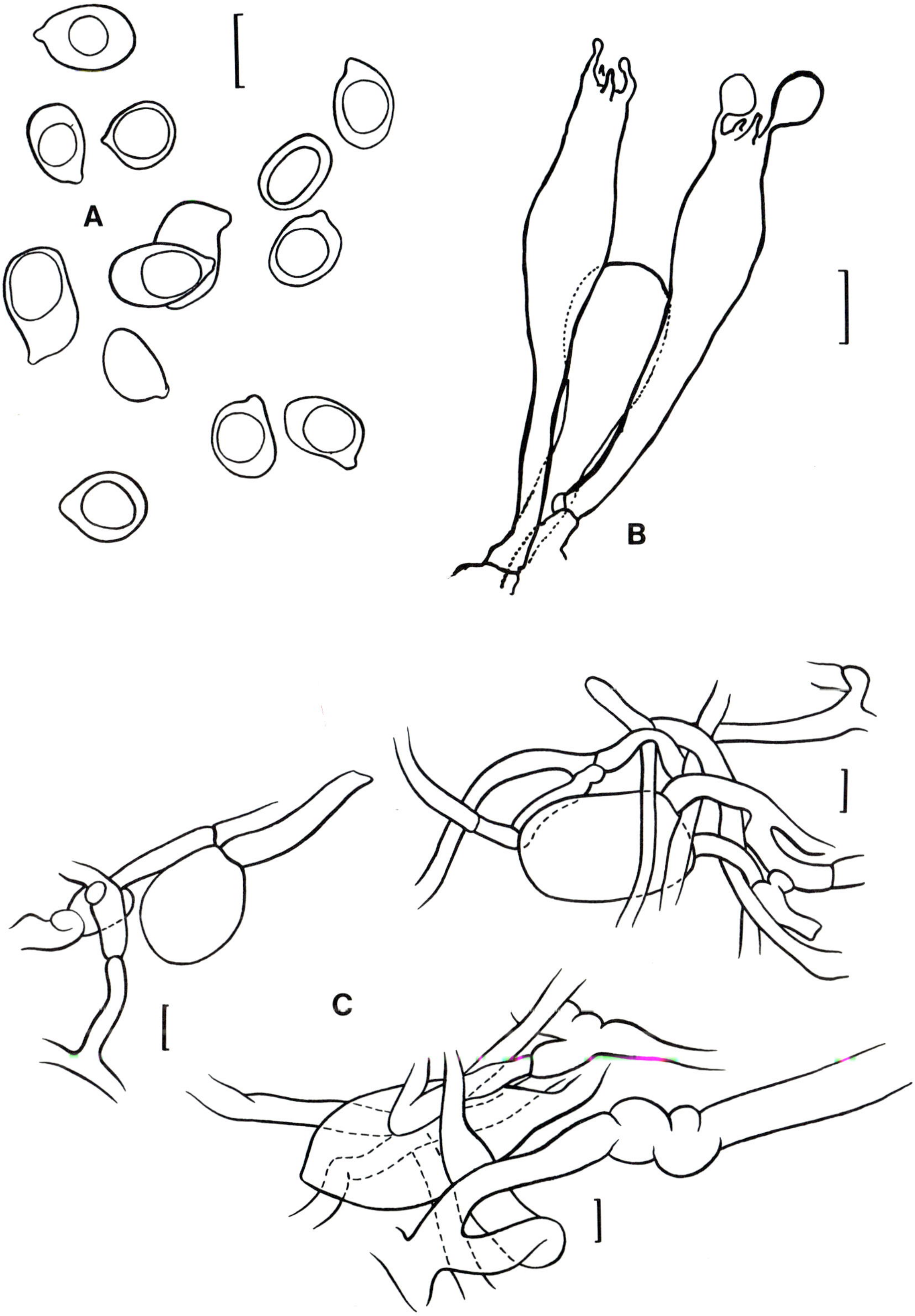

Fig. 81. *Amanita muscaria*: A, spores; B, basidia; C, elements of the universal veil.

Scale line 10 μm

like volva around the stalk base. All forms of *A. muscaria* have a partial veil and usually an annulus; however, occasionally the partial veil breaks in such a way that the tissue that normally forms the annulus remains attached as a fringe on the cap edge. This species should be carefully compared with *A. frostiana*, *A. gemmata*, and *A. pantherina*.

Distribution and seasonal occurrence. *A. muscaria* is widespread and common in Canada and in the United States. The dark red form commonly illustrated in European floras occurs primarily in western North America, with only occasional reports from eastern North America. The yellow to orange form is common in eastern North America and also occurs in the west. The whitish to buff form of *A. muscaria* var. *alba* occurs in the west, but is more often found in eastern North America and can be very abundant in the Great Lakes region. In the Great Lakes region and eastern North America, *A. muscaria* fruits in June and July and then appears again from middle to late August until October. In the western mountains it is a summer or early fall species and along the Pacific coast, typically a fall to winter species.

Habitat and habit. *A. muscaria* is terrestrial, usually found scattered to gregarious; it occurs sometimes in fairy rings and, less commonly, solitary. It is found in forested areas, on edges of pastures, in open areas, and along roadsides. It is usually associated with conifers, but can be common in areas containing aspen and beech.

Technical description. Cap 70-240 mm broad, hemispherical when young, expanding to convex or nearly plane; margin faintly striate to strongly striate or tuberculate-striate; surface viscid, shiny when dry, smooth, red, reddish orange to orange, yellowish, or rarely whitish to buff or brown to reddish brown, darkest at the center, lighter toward the margin, adorned with more or less persistent white to yellowish pyramid-shaped warts or small patches, with the edge at times appendiculate. Flesh thick, white; odor indistinctive; taste not recorded.

Gills white, free, up to 15 mm wide, not strongly ventricose, with edges even to floccose.

Stalk 80-220 mm long, 30-50 mm wide, equal to tapering upward; base bulbous, stuffed becoming hollow; surface silky-fibrillose to slightly floccose above the annulus and fibrillose-scaly below; color white to pallid to yellowish below the annulus. Flesh white. Annulus median to apical, pendant, membranous, white; margin frayed, sometimes with yellowish floccose patches. Volva inconspicuous, white to yellowish white or more yellowish, appressed to the base of the stalk, broken into several concentric scaly rings on the upper part of the bulb and the lower part of the stalk, rarely forming a slight collar-like rim on the apex of the bulbous base.

Spores white in deposit, (9-)11-13.1(-14.6) × (6.6-)7.3-8.8(-9.1) μm, broadly ellipsoidal, adaxially flattened; walls thin, smooth, hyaline, nonamyloid. Basidia four-spored, clavate, 49-62 × 7-10.2 μm, hyaline. Pleurocystidia and cheilocystidia absent; sterile cells on the gill edges 20-28 × 8-12 μm in size, sac-like to clavate, hyaline.

Gill trama divergent from a central strand; hyphae 3.5-25 μm wide, cylindrical to inflated, but mainly cylindrical and narrow in the central strand. Subhymenium hyphae 4.6-10.2 μm wide, cylindrical to slightly inflated, hyaline. Cap cuticle filamentous, gelatinous; hyphae interwoven, more or less radially arranged, more or less cylindrical, 2.2-7(-12) μm wide, hyaline to yellow. Cap trama interwoven; hyphae 3.7-8.8(-17.9) μm wide, cylindrical to more or less inflated, hyaline. Caulocystidia clavate, 10-40 × 170-320 μm, overlapping or recurved, parallel, hyaline. Stalk hyphae more or less parallel, cylindrical, 2.2-9.8 μm wide, hyaline. Annulus interwoven; hyphae predominantly cylindrical, 2.3-9.5 μm wide, hyaline; sphaerocysts also present, narrowly ellipsoidal to clavate and 60-120 × 9.5-14.2 μm in size or globose and 35-75 μm wide, scattered, single or in short terminal chains, hyaline. Universal veil (from cap surface) interwoven; hyphae cylindrical, 2.8-7.2 μm wide, hyaline; sphaerocysts also present, ellipsoidal to globose, 29.1-63 × 29.1-44.8 μm in size, single or in short chains, hyaline. Volva interwoven; hyphae cylindrical, 3-10 μm wide, hyaline; sphaerocysts also present, globose to broadly ellipsoidal, (25-)50-72 × 25-50 μm in size, single or in short chains, hyaline. Clamp connections present in the hymenium, volva, annulus, and stalk. Oleiferous hyphae refractive, yellowish, scattered in the cuticle, stalk, and volva.

Useful references. The following books contain information on *A. muscaria* pertinent to North America.

- *A Field Guide to Western Mushrooms* (Smith 1975)—A color photograph and a description and discussion are presented.
- *Edible and Poisonous Mushrooms of Canada* (Groves 1979)—A color photograph and a description and discussion are given.
- *Mushrooms in their Natural Habitats* (Smith 1949)—A color photograph on a reel and a description and discussion are presented.
- *Mushrooms of North America* (Miller 1972)—A color photograph and a description and discussion are given.
- *Mushroom Pocket Field Guide* (Bigelow 1974)—A color photograph and a description and discussion are presented.
- *The Mushroom Hunter's Field Guide* (Smith 1974)—A color photograph, a black-and-white photograph, and a description and discussion are given.
- *The Savory Wild Mushroom* (McKenny 1971)—A color photograph, a black-and-white photograph, and a description and discussion are given.

Amanita pantherina (DC. per Fr.) Krombh.

Figs. 82, 88, 89

Common names. Panther, panther amanita.

Distinguishing characteristics. Cap dark brown to pale brown or yellowish brown, usually paler on the margin, occasionally paler or more yellowish overall. White to creamy white or pallid warts or patches of universal veil tissue, or both, scattered or in concentric zones on the cap. Stalk with a basal bulb that has a more or less distinct collar of volval tissue around its apex; volva usually forming zones or patches on the lower stalk surface above the bulbous base. Spores nonamyloid, (8-)9.5-11.7(-14.6) × (6.5-)7.3-8.0(-10.2) μm in size.

Observations. *A. pantherina* in its typical form, that is with a dark brown to tan cap, is a distinctive mushroom. However, the cap color can be quite variable, either paler or in some instances, more yellowish. Therefore, this species should be closely compared with *A. gemmata* and *A. cothurnata* as well as with yellowish or pale forms of *A. muscaria*. Based on variation in color and toxicity, there is some indication that *A. gemmata* and *A. pantherina* hybridize in nature. Bleached or faded specimens of *A. pantherina* may appear very similar to *A. gemmata* and often are practically impossible to distinguish.

Distribution and seasonal occurrence. *A. pantherina* is common in the mountains and coastal region of western North America; it is less commonly found in eastern North America. It usually fruits in the fall, but it also occurs in the spring and summer in western North America.

Habitat and habit. The species is terrestrial, occurring solitary to gregarious, usually under conifers and occasionally under hardwoods.

Technical description. Cap 30-160 mm wide, hemispherical to convex, becoming flat and occasionally depressed slightly in the center; margin faintly to strongly striate, appendiculate; surface viscid, glabrous, covered with soft white floccose patches, scales, or pyramid-shaped warts randomly arranged but usually more dense in the center; cap color blackish brown to olivaceous brown, fading to pale grayish or yellowish brown. Flesh white, yellowish under the cuticle, up to 1.5 cm thick at the center, tapering toward the margin; odor indiscernible; taste not observed.

Gills white to whitish cream, free to attached along a line, crowded, broadest at the margin; edges floccose to finely scalloped; lamellulae in one or two tiers, truncate at the stalk.

Stalk 40-180 mm long, 8-25 mm thick, equal to tapering upward; apex slightly expanded, stuffed to hollow, fibrous-fleshy; surface smooth to silky or lightly floccose above the annulus and lacerate or fibrillose-scaly below, white to pallid in color; bulb subglobose to ovoid, 15-40 mm wide. Annulus apical to median, membranous, pendant, persistent, white to yellowish white; edges often torn. Volva membranous, white, adherent to the bulb, torn into one or two rings on the lower stalk; collar free or enrolled.

Spores white in deposit, (8-)9.5-11.7(-14.6) × (6.5-)7.3-8.0(-10.2) μm, broadly ellipsoidal and adaxially flattened in side view, thin-walled, hyaline, nonamyloid. Basidia four-spored, clavate, 40.8-62 × 4-12 μm. Pleurocystidia and cheilocystidia absent; clavate to sac-like sterile cells present along the gill edges.

Gill trama divergent from a central strand; hyphae interwoven, cylindrical to inflated, 4-25 μm wide, hyaline. Subhymenium hyphae cylindrical to slightly inflated, branched, 6-8 μm wide, hyaline. Cap cuticle filamentous; hyphae interwoven, radially arranged, cylindrical, 1.8-3.5 mm, hyaline to yellowish, gelatinizing in potassium hydroxide. Cap trama loosely interwoven; hyphae cylindrical to inflated, 18 μm wide, hyaline. Caulocystidia terminal, clavate, 32-320 × 10-30 μm, hyaline, parallel, and overlapping; stalk hyphae otherwise interwoven, cylindrical, 2.9-3.5 μm wide, hyaline. Annulus interwoven; hyphae cylindrical, 3.5-4.7 μm wide, hyaline; sphaerocysts also present, scattered, subglobose to ellipsoidal or clavate, 16-60 × 9.5-20 μm, hyaline. Universal veil interwoven; hyphae cylindrical, 3.7-14.0 μm wide; sphaerocysts also present, globose to subglobose or ellipsoidal to clavate, 58-95(-160) × 37.0-76.0 μm, single or in loose chains, hyaline. Clamp connections rare.

Useful references. The following books provide information on *A. pantherina* pertinent to North America.

- *A Field Guide to Western Mushrooms* (Smith 1975)—A color photograph and a description and discussion are presented.
- *Edible and Poisonous Mushrooms of Canada* (Groves 1979)—A description and discussion are given.
- *Mushrooms in their Natural Habitats* (Smith 1949)—A color photograph on a reel and a description and discussion are given.
- *Mushrooms of North America* (Miller 1972)—A color photograph of the yellow form and a description and discussion are given.
- *The Mushroom Hunter's Field Guide* (Smith 1974)—A black-and-white photograph and a description and discussion are given.
- *The Savory Wild Mushroom* (McKenny 1971)—A color photograph, a black-and-white photograph, and a description and discussion are given.

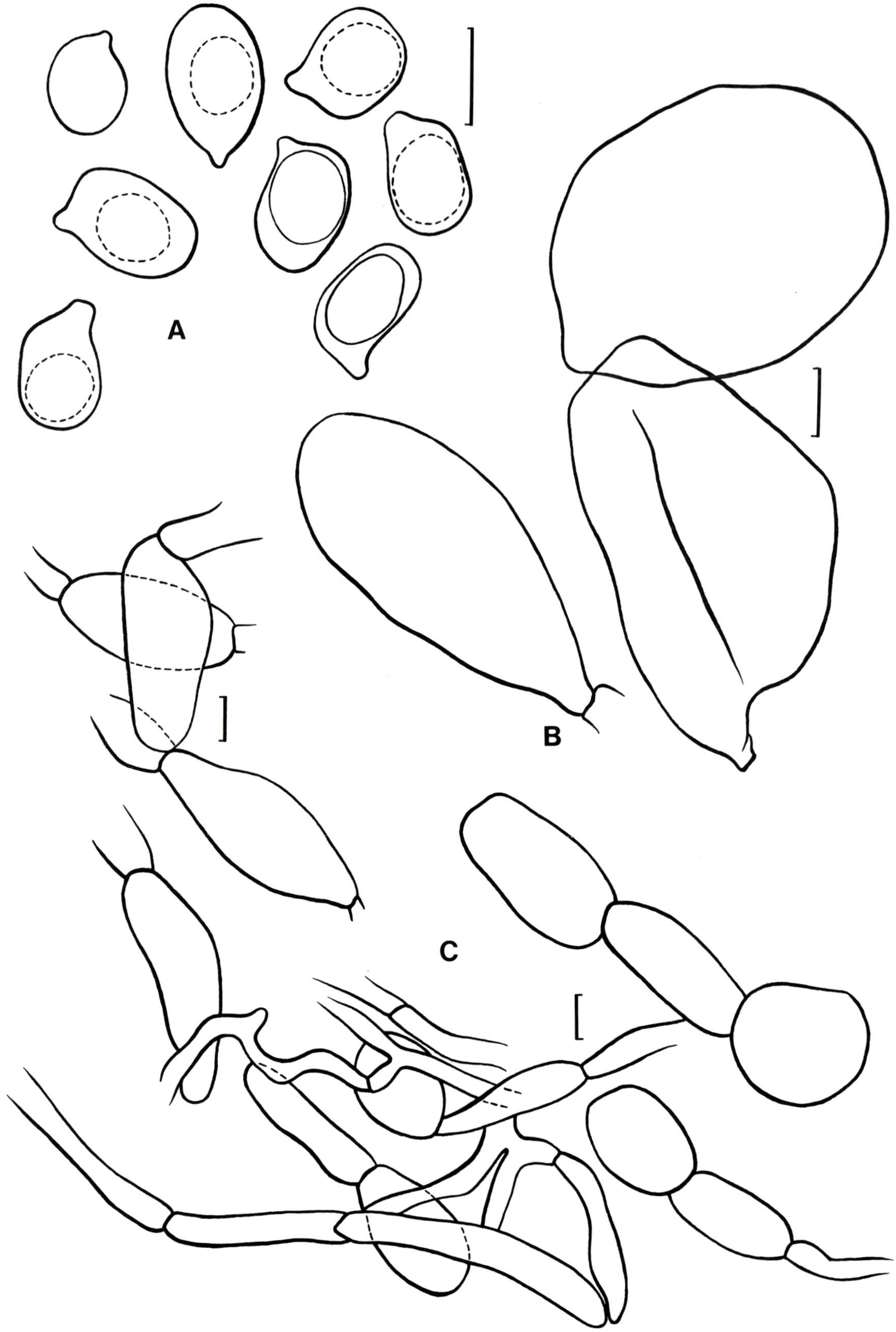

Fig. 82. *Amanita pantherina*: A, spores; B, sphaerocysts of the cap warts; C, volval elements.

Scale line 10 μm

Literature

Bas, C. Morphology and subdivision of *Amanita* and a monograph of its section *Lepidella*. Persoonia 5(4):285–579; 1969.

Benedict, R. G. Chemotaxonomic significance of isoxazole derivatives in *Amanita* species. Lloydia 29:333–342; 1966.

Benedict, R. G. Mushroom toxins other than *Amanita*. Kadis, S., et al., eds. Microbiological toxins. Vol 8. New York, NY: Academic Press; 1972.

Bigelow, H. E. Mushroom pocket field guide. New York, NY: MacMillan; 1974.

Buck, R. W. Poisoning by *Amanita crenulata*. N. Engl. J. Med. 272:475–476; 1965.

Chilton, W. S.; Ott, J. Toxic metabolites of *Amanita pantherina*, *A. cothurnata*, *A. muscaria*, and other *Amanita* species. Lloydia 39:150–157; 1976.

Groves, J. W. Edible and poisonous mushrooms of Canada. Ottawa, Ont.: Agriculture Canada. Agric. Can. Publ. 1112; 1979 [first printed in 1962].

Hardin, J. W.; Arena, J. M. Human poisoning from native and cultivated plants. Durham, NC: Duke University Press; 1974.

Hatfield, G. M.; Brady, L. R. Toxins of higher fungi. Lloydia 38:36–55; 1975.

Imazeki, R.; Hongo, T. Fungi of Japan. Vol. I. Osaka, Japan: Hoikusha; 1957.

Jenkins, D. A floristic study of *Amanita* section *Amanita* in North America. Ph.D. Thesis. Knoxville, TN: University of Tennessee; 1974.

Jenkins, D. A taxonomic and nomenclatural study of the genus *Amanita* section *Amanita* in North America. Bibl. Mycol. 57:1–126; 1977.

Jenkins, D. A study of *Amanita* types I. Taxa described by C. H. Peck. Mycotaxon 7(1):23–44; 1978.

Krieger, L. C. C. The mushroom handbook. New York, NY: Dover Publ. Inc.; 1967 [first printed in 1936].

Lincoff, G.; Mitchel, D. H. Toxic and hallucinogenic mushroom poisoning—A handbook for physicians and mushroom hunters. New York, NY: Van Nostrand Reinhold Co.; 1977.

McKenny, M. The savory wild mushroom (revised and enlarged by Stuntz, D. E.). Seattle, WA: University of Washington Press; 1971.

Miller, O. K., Jr. Mushrooms of North America. New York, NY: E. P. Dutton & Co., Inc.; 1972.

Page, L. B. Poisoning due to *Amanita gemmata*. Boston Mycol. Club Bull. 3:12; 1975.

Peck, C. H. New species of fungi. Torr. Bot. Club Bull. 27:15; 1900.

Simons, D. M. The mushroom toxins. Del. Med. J. 42:177–187; 1971.

Smith, A. H. Mushrooms in their natural habitats. 2 Vols. Portland, OR: Sawyer's, Inc.; 1949.

Smith, A. H. A field guide to western mushrooms. Ann Arbor, MI: University of Michigan Press; 1975.

Fig. 83. *Amanita cothurnata.*

Fig. 84. *Amanita gemmata.*

Fig. 85. *Amanita muscaria* var. *muscaria.*

Fig. 86. *Amanita muscaria* var. *formosa.*

Fig. 87. *Amanita muscaria* var. *alba.*

Fig. 88. *Amanita pantherina.*

Fig. 89. *Amanita pantherina.*

12

HALLUCINOGENS

Toxicology and Symptomatology

Several mushroom genera have species known to be hallucinogenic. The toxins implicated are psilocybin and psilocin, as well as related compounds such as baeocystin and norbaeocystin. Most hallucinogenic species are in the genera *Psilocybe* and *Panaeolus*, but certain species of *Conocybe*, *Gymnopilus*, *Pluteus* (Saupe 1981), and perhaps *Stropharia* also contain these compounds.

Psilocybe and *Panaeolus* species, because of their hallucinogenic properties, have gained considerable attention since the 1950's, and in the 1970's they became widely known as magic mushrooms. In the coastal region of the Pacific northwest, they are now considered as a source of recreational drugs by some people. Species of *Psilocybe* and *Panaeolus*, and possibly those of other genera, have a long and intriguing history as intoxicants used in religious rites of the native people of the western hemisphere, especially in Mexico and Central America.

The compounds found in these mushrooms are indole alkaloids, specifically hydroxyltryptamine derivatives. Many of the species that contain psilocybin and psilocin stain blue when injured or with age. There is some indication that the blue staining reaction is related to the presence of these compounds.

Psilocybin is the phosphate ester of psilocin, chemically 3-[2-(dimethylamino)ethyl]-1*H*-indol-4-ol. Psilocybin and psilocin are similar in many respects to one of the essential amino acids, tryptophan, found in cellular proteins. The compounds also have similarities with a group of higher plant hormones known as auxins. Readers interested in the detection of psilocybin should refer to articles by Brown et al. (1972), Aaron et al. (1973), Zabik and Maickel (1974), Cashman and Thornton (1975), Folen (1975), and Masoud (1975).

The psilocybin present in hallucinogenic mushrooms is rapidly hydrolyzed to psilocin, the true pharmacologic agent, when the mushrooms are ingested. Hallucinogens achieve their effects because they are chemically similar to neurotransmitters interacting with the brain. Psilocin resembles serotonin in its effect on nerve transmission.

The hallucinogenic symptoms of psilocin poisoning have been the topic of many historical accounts and of numerous popular books and articles. The symptoms listed below are taken from the work of Hollister et al. (1960) and are descriptive of the psilocybin effect in man.

From 0 to 30 minutes, the victim feels dizzy, giddy; nausea, abdominal discomfort; weakness, muscle aches and twitches, shivering; anxiety, restlessness; and numbness of lips.

From 30 to 60 minutes, the symptoms include visual effects (blurring, brighter colors, sharper outlines, longer afterimages, visual patterns with the eyes closed); increased hearing; yawning, tearing, sweating, facial flushing; decreased concentration and attention, slow thinking, feelings of unreality, depersonalization, dreaminess; and incoordination and tremulous speech.

From 60 to 120 minutes, experiences include increased visual effects (colored patterns and shapes, mostly with the eyes closed); wave motion of viewed surfaces; impaired distance perception; euphoria, rumination, increased perception; and slowed passage of time.

From 120 to 140 minutes, there is a waning and nearly complete resolution of the above effects. The victim usually returns to normal between 4 and 12 hours after ingesting these mushrooms. Later effects may include headache, fatigue, and a contemplative state, or less commonly, uncontrollable laughter, a tingling of the skin, synesthesia, difficulty in breathing, decreased appetite, and transient sexual feelings.

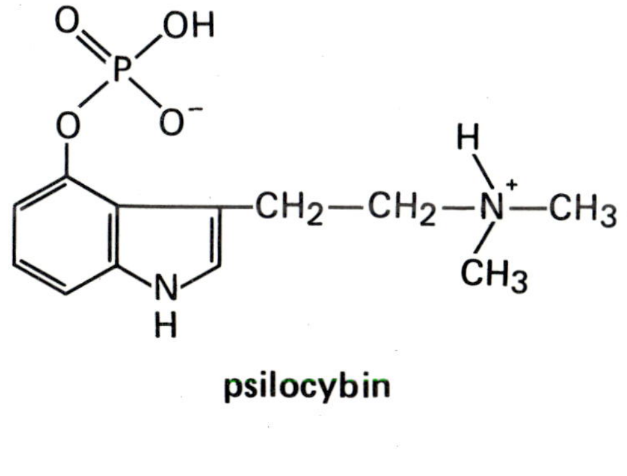

psilocybin

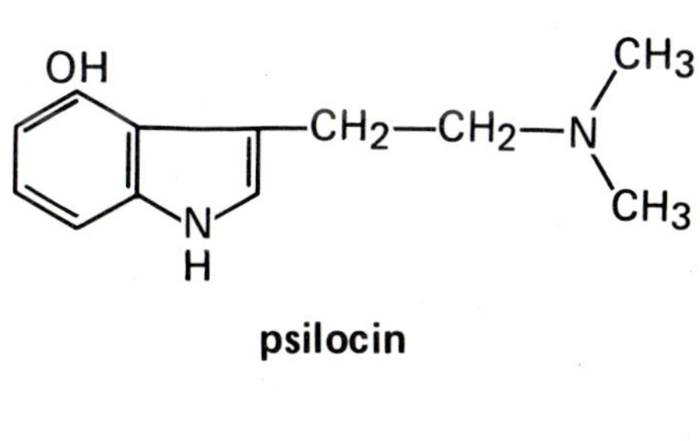

psilocin

Needless to say, the hallucinogenic and euphoric sensations produced by eating these mushrooms are highly individualistic. Detailed descriptions are available in the literature cited at the end of this chapter. The intensity of intoxication depends on how the mushrooms are prepared and eaten: fresh and raw, dried and raw, cooked, dried and smoked, or even as a liquid alcoholic extract. The effects of the mushrooms are most pronounced when the stomach is otherwise empty. The effects of the toxins begin within 15 to 30 minutes after ingestion and are said to last from 4 to 12 hours.

At present in Canada, possession of the hallucinogenic compounds psilocybin and psilocin is restricted by law. However, eating hallucinogenic mushrooms may also be dangerous. Many individuals have been stricken with serious side effects after eating these mushrooms. Symptoms such as severe dysphoria, which is a chronic feeling of illness or discontent, vomiting, prostration, and occasionally paralysis have been reported. These mushrooms can be quite dangerous when ingested by young children. Elevated body temperatures and convulsions have been reported as well as one fatality, that of a 6-year-old child. No specific treatment can be recommended for psilocybin poisoning in humans. Lincoff and Mitchel (1977) provide further information.

Potentially Hallucinogenic Species

- Bolbitiaceae
  - *Conocybe*
    - *C. cyanopus*
    - *C. smithii*
- Coprinaceae
  - *Panaeolus*
    - *P. campanulatus* (depending on the species concept followed)
    - *P. castaneifolius*
    - *P. (Copelandia) cyanescens*
    - *P. fimicola*
    - *P. foenisecii*
    - *P. sphinctrinus*
    - *P. subbalteatus*
- Cortinariaceae
  - *Gymnopilus*
    - *G. aeruginosus*
    - *G. luteus*
    - *G. spectabilis* (certain specimens)
    - *G. validipes*
    - *G. viridans*
- Pluteaceae
  - *Pluteus*
    - *P. salicinus*
- Strophariaceae
  - *Psilocybe*
    - *P. baeocystis*
    - *P. caerulescens*
    - *P. caerulipes*
    - *P. cubensis*
    - *P. cyanescens*
    - *P. cyanofibrillosa*
    - *P. pelliculosa*
    - *P. semilanceata*
    - *P. strictipes*
    - *P. stuntzii*
  - *Stropharia*
    - *S. semiglobata* (suspected)

This list contains many of the species found in North America that are known to contain psilocybin, psilocin, or related compounds, arranged by family. Lincoff and Mitchel (1977) list several other species suspected of containing these toxins. *Stropharia semiglobata* is included although it is not clear to date whether it contains any hallucinogenic toxins. Several other species of *Psilocybe* are also known to contain hallucinogens but are not included here because they are primarily subtropical to tropical.

The remainder of this chapter contains descriptions of the genera in this group, arranged alphabetically. Descriptions of the species known to be of importance in Canada are included in alphabetic order for each genus.

Descriptions of Genera and Species

Conocybe Fayod — Bolbitiaceae

In general appearance species of *Conocybe* are similar to several other genera of small brown mushrooms. Therefore, without the aid of a microscope and careful study, it is difficult to be sure of generic identification.

The distinguishing characteristics of *Conocybe* include usually small, slender fruit bodies; a conic to bell-shaped cap (convex in certain annulate species); a long, slender, and very fragile stalk; ascending–adnate gills, often colored rust brown from the spores; a persistent annulus present or absent in some species; a rust brown to ocherous rust or ocher brown spore deposit; usually smooth spores, with an apical germ pore; a cellular cap cuticle as viewed in a section of the cap surface; and often characteristic flask-shaped or bowling-pin-shaped cheilocystidia.

The species usually occur in pastures, lawns, or other grassy areas, on compost piles, in terrariums and greenhouses, or on dung. They occasionally occur in forested areas, not commonly on wood and less commonly on moss.

Species of *Galerina* might be confused with *Conocybe*. The easiest way to separate them is microscopically; *Galerina* typically has roughened spores that lack an apical germ pore, as well as a filamentous cap cuticle. Species of *Galerina* generally occur with moss although a few species grow on wood and at least one, *G. venenata*, a dangerously poisonous species, grows in lawns. Certain species of *Psilocybe*, in general appearance, look like *Conocybe*. *Psilocybe* species have darker spore deposits than *Conocybe*, namely some shade of chocolate to blackish brown, fuscous purple, or purple brown, and the cap cuticle is composed of filamentous hyphae.

Described below are two species of *Conocybe* that have been shown to contain the hallucinogenic drug psilocybin. Like several species of *Psilocybe* that contain psilocybin, psilocin, or related compounds, these species of *Conocybe* stain blue or blue green on the stalk, especially on the base, when the flesh is injured or with age.

Conocybe cyanopus (Atk.) Kühner

Fig. 90

Commons names. None.

Distinguishing characteristics. Cap 7-25 mm broad, more or less semiglobate, only faintly striate; color rusty brown to darker brown. Gills not crowded; color rusty brown with white flocculose edges. Stalk thin; color white, discoloring grayish to brownish at times; base staining deep blue green. Annulus absent. Spores smooth, with a distinct apical germ pore, and 6.5-9.5(-11.0) × 4.8-5.1(-5.8) μm in size. Cheilocystidia ventricose, with acute to capitate apices. Fungi occurring in lawns.

Observations. *C. cyanopus* is not common but has been reported from both eastern and western North America. According to Watling (Benedict et al. 1967) *C. cyanopus* differs from *C. smithii* in the absence of a cinnamon flush in the gills, wider spores, larger size and sturdier stature of the fruit bodies, greater number of pileocystidia, narrower and slightly different-shaped cheilocystidia (rarely more or less capitate), white silky stalk, and darker color of the cap which is hardly, if at all, striate. No cases of poisoning have been reported by *C. cyanopus*. However, it is likely to be collected along with hallucinogenic species of *Psilocybe*, where they grow together in lawns or in other grassy areas. It is too small to be of value as food.

Distribution and seasonal occurrence. This species is reported from the eastern and western United States and probably occurs throughout North America, although it is not commonly reported. It appears to fruit during the summer and fall, depending on location and weather.

Habitat and habit. *C. cyanopus* grows scattered to gregarious in lawns and grassy areas.

Technical description. Cap 7-25 mm broad, convex, almost globose, slightly expanding, obscurely umbonate; margin faintly striate, covered with a few filamentous veil fragments; surface smooth to slightly wrinkled over the disc; color rusty brown, snuff brown to dark brown, or almost umber. Flesh thin, whitish to brownish; taste and odor indiscernible.

Gills rusty brown, attached in early stages, not crowded; edges white, flocculose.

Stalk 20-40 mm long, 1-1.5 mm thick, central, silky-striate, white becoming grayish then brownish streaked at the apex when handled or at maturity, deep blue green at the base. Flesh white, slightly darker at the base, becoming brownish. Annulus lacking; partial veil fragments present on the cap margin. Universal veil lacking.

Spores rusty brown in deposit, 6.5-9.5(-11.0) × 4.8-5.1(-5.8) μm, ellipsoidal to slightly inequilateral, smooth, thick-walled; apex truncate, with a distinct but small apical germ pore, rusty to cinnamon brown, nonamyloid. Basidia four-spored, short-clavate, 18.3-25.9 × 7.3-8.8 μm, hyaline. Pleurocystidia absent. Cheilocystidia ventricose; apices acute to capitate, 14.6-38.7 × 6.6-14.6 × 5.8 (at apex) μm, hyaline, thin-walled.

Gill trama compactly interwoven; hyphae cylindrical to inflated, 3.5-18.0 μm wide, hyaline to yellowish. Subhymenium cellular; hyphae 5.8-7.1 μm wide, often collapsed, hyaline. Cap cuticle in the form of a cellular to hymeniform layer composed of globose-pedicellate cells that are 33-48(-55) × 12.8-18.3 × 3.0-4.4 (at base) μm in size; bases of cells brown, slightly thick-walled; hyphae just below the cuticle interwoven, 4.0-6.0 μm wide, brown; pileocystidia filiform or cylindrical, scattered, 33-48 × 4.0-5.1 μm, yellowish to brown, slightly thick-walled. Cap trama loosely interwoven; hyphae cylindrical to more or less inflated, 5.1-14.6 μm wide, with walls thin to irregularly thickened and encrusted, hyaline to yellowish brown. Caulocystidia ventricose to fusiform-ventricose with tapered or capitate apices, 18-29.2 × 11-14 × 4.4-5.8 (at apex) μm, hyaline, single to clustered; stalk hyphae parallel in the core to interwoven on the surface, 4.4-15.0 μm wide, cylindrical to more or less inflated, hyaline. Oleiferous hyphae not observed. Clamp connections present.

Useful references. Benedict et al. (1967) and Stamets (1978) provide additional descriptions and drawings.

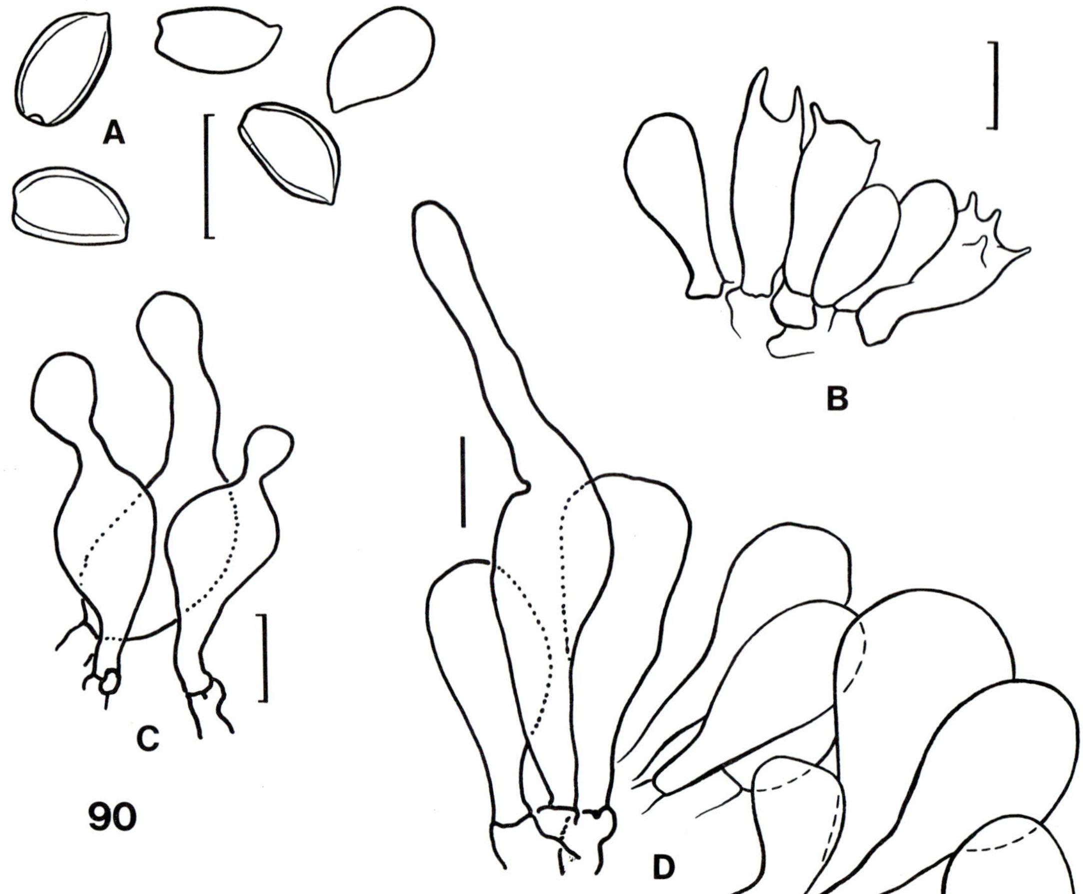

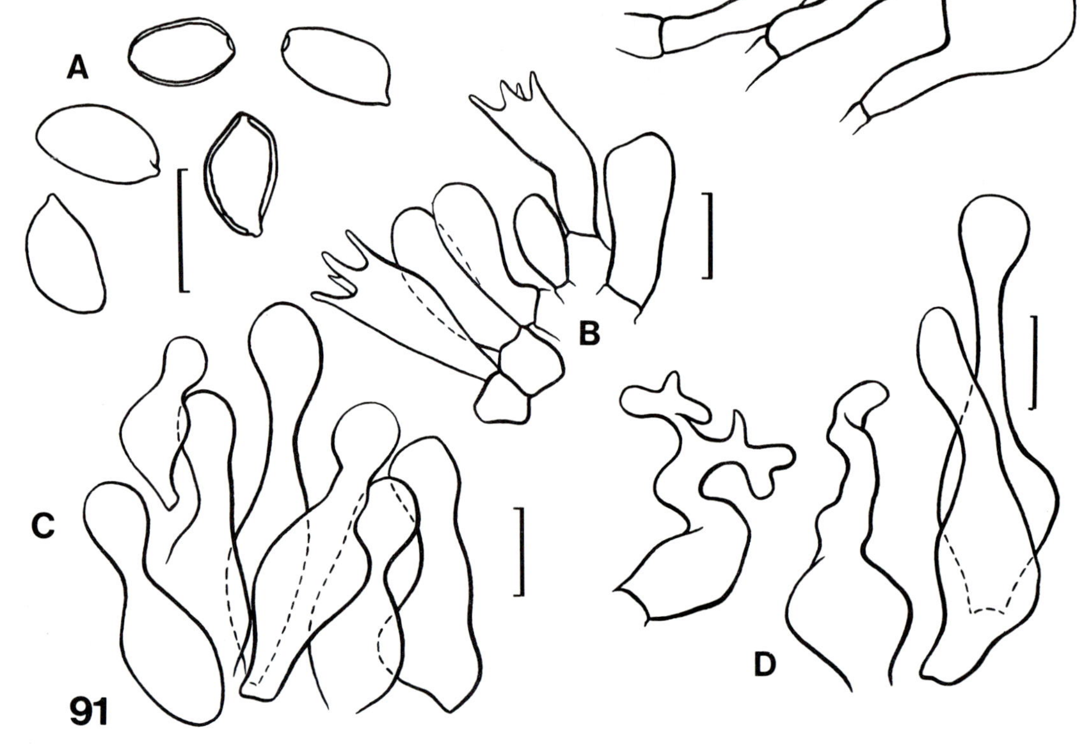

Fig. 90. *Conocybe cyanopus*: A, spores; B, basidia; C, cheilocystidia; D, cuticle and pileocystidia.

Fig. 91. *Conocybe smithii*: A, spores; B, basidia; C, cheilocystidia; D, caulocystidia.

Scale line 10 μm

Conocybe smithii Watling

Figs. 91, 105

Common names. None.

Distinguishing characteristics. Cap 3-13 mm broad, more or less conic, distinctly striate; color ocherous tawny to tawny cinnamon. Gills ocherous buff, becoming cinnamon rust from spores. Stalk very thin, pure white or slightly yellowish; base tinged greenish gray or grayish blue. Annulus absent. Spores smooth, with an apical germ pore, 6.5-9.5 × 4.4-5.1(-5.8) μm in size. Cheilocystidia fusiform-ventricose, with capitate apices.

Observations. *C. smithii* is related to *C. cyanopus* but seems to be somewhat more common. There are no known reports of poisoning by it. See comments under *C. cyanopus*.

Distribution and seasonal occurrence. The species seems to be widely distributed but is less common in western North America than in the northcentral and northeastern United States and adjacent areas of Canada. It fruits during the late spring, the summer, and perhaps the early fall.

Habitat and habit. *C. smithii* grows scattered and is found in various habitats including swampy areas, cool bogs, mossy logs, moist soil, or under dense cover and is often associated with mosses. It also grows in lawns (D. Malloch, Personal commun.).

Technical description. Cap 3-13 mm broad, obtusely conic, expanding at the margin to become distinctly umbonate, hygrophanous, glistening, striate for two-thirds of the margin; color ocherous tawny to tawny cinnamon, darker on the disc, fading to pinkish buff or dingy tan on drying. Flesh watery white; taste and odor indiscernible.

Gills ocherous buff in early stages becoming brownish to cinnamon rust, adnate, soon seceding, subdistant to crowded, narrow to moderately broad, edges more or less the same color as the faces to minutely flocculose and whitish.

Stalk 10-70 mm long, 0.75-1.0 mm thick, slightly swollen at the base, up to 1.5 mm wide, central, fragile, smooth to faintly pubescent at the apex, fibrillose becoming glabrous at the base, pure white, thin, watery, and with age faintly yellowish to ocherous pallid; base tinged greenish gray or grayish blue, becoming yellowish with age. Flesh watery white. Annulus or universal veil lacking.

Spores rusty brown in deposit, 6.5-9.5 × 4.4-5.1(-5.8) μm, ellipsoidal, slightly inequilateral in side view, smooth, thick-walled, slightly truncate, with a small apical germ pore, rusty brown, nonamyloid. Basidia four-spored, short-clavate, 18.3-20.0 × 7.3-8.0 μm, hyaline. Pleurocystidia absent. Cheilocystidia fusiform-ventricose with capitate apices, 20-32.9 × 9.5-10.6(-15.0) × 6.2-8.8 (at apex) μm, hyaline, thin-walled.

Gill trama compactly interwoven to parallel; hyphae cylindrical to inflated, 2.9-18.6 μm wide, hyaline. Subhymenium cellular; hyphae inflated, 3.7-7.2 μm wide, hyaline. Cap cuticle in the form of a cellular to hymeniform layer composed of globose-pedicellate cells that are 29.2-36.5 × 11.0-14.6 × 3.7-5.1 (at base) μm and hyaline to yellowish; bases of cells brown, thick-walled, encrusted; pileocystidia tapered fusiform-ventricose to filiform-cylindrical, 3.7-5.1 μm wide, rare, hyaline to yellowish; hyphae just below the cuticle compactly interwoven, more or less cylindrical, 3.7-7.3 μm wide, thin-walled, some spirally encrusted, hyaline to brownish. Cap trama loosely interwoven; hyphae cylindrical to more or less inflated, 3.8-9.9 μm wide, hyaline. Caulocystidia fusiform-ventricose with capitate to nearly capitate or filiform apices, 36-44 × 11.0-12.0 × 3.7-7.0 (at apex) μm, hyaline, clustered; stalk hyphae parallel to interwoven, more or less cylindrical, 3.7-10.2 μm wide, hyaline to yellowish. Oleiferous hyphae not observed. Clamp connections present.

Useful references. Benedict et al. (1967) and Stamets (1978) provide technical descriptions and illustrations.

Gymnopilus Karst.

Cortinariaceae

Most species of *Gymnopilus* are medium sized, although some are small and a few such as *G. spectabilis* and relatives are very large. The fruit bodies typically occur on wood, usually decayed logs or stumps, or grow from buried wood; a few are apparently terrestrial. In general the species are brightly colored, often orange brown to red brown, although some are more yellow to blue or green. The gills are attached and are usually some shade of yellow in the early stages, becoming rusty orange brown to rusty orange from similarly colored spores. The stalk usually appears centrally attached but may be slightly eccentric; its color is usually similar to that of the cap. A partial veil is present in some species and may leave a distinct annulus. The flesh often has a bitter taste. *Gymnopilus* spores are roughened, with a wrinkled to warted surface, and lack an apical germ pore. However, the ornamentation is sometimes so small that the spores appear smooth under the microscope, even when magnified at 1000×; nevertheless they are always somewhat roughened, never smooth. The cap cuticle is filamentous, and cheilocystidia are typically present. For a comparison of similar genera, see *Cortinarius* (Ch. 7); *Inocybe* (Ch. 10); and *Hebeloma, Phaeolepiota,* and *Pholiota* (Ch. 13).

The literature contains scattered reports of poisoning by the larger species of *Gymnopilus,* for example by *G. spectabilis* (Walters 1965, Buck 1967) and by *G. validipes* (Hatfield et al. 1978). The symptoms include visual hallucinations and generally do not appear to be particularly severe. Recent research (Hatfield et al. 1978) shows that at least some specimens of *G. aeruginosus, G. luteus, G. spectabilis, G. viridans,* and *G. validipes* contain psilocybin. This report is the first documentation of psilocybin in *Gymnopilus.*

The descriptions below are taken directly from Hesler (1969); only *G. spectabilis* is illustrated by a color plate.

Gymnopilus aeruginosus (Peck) Singer

Fig. 92

Common names. None.

Distinguishing characteristics. Cap 20–50 (rarely 60–230) mm broad, convex, dry; color variable depending on the age of the fruit body, bluish gray green to bluish green and sometimes mottled with green and yellow, at times with pinkish to reddish patches, becoming pinkish buff or brownish to drab when older; surface often breaking into patches or scales that become tawny to blackish. Flesh white, tinged greenish; taste bitter. Gills adnexed to adnate or somewhat decurrent, especially when young; color creamy buff to pale yellowish orange. Stalk up to 120 mm long and up to 40 mm thick, about the same color as the cap. Annulus lacking, but sometimes a slight apical zone is present from fibrils of the partial veil. Cheilocystidia present. Spores ellipsoidal, roughened, 6–8.5(–9) × (3.5–)4–4.5 μm in size, dextrinoid.

Observations. *G. aeruginosus* is distinguished by dextrinoid spores, a green to yellow cap, and white flesh with greenish tints. The lack of a distinct annulus helps to separate it from *G. spectabilis* and relatives.

Distribution and seasonal occurrence. The species occurs across the United States and Canada. It fruits from May to November, depending on weather and location.

Habitat and habit. *G. aeruginosus* grows on the wood of conifer and hardwood trees, on sawdust, timbers, logs, or stumps. It is caespitose in habit.

Technical description. Cap 20–50 (more rarely 60–230) mm broad, convex; margin even; color in early stages dull bluish gray green or blue green to variegated green and yellow, sometimes with patches of salmon or livid red, becoming warm buff to pinkish buff or at times brown to drab, especially when dried; surface dry, fibrillose-scaly, becoming rimose to sometimes extensively divided into areas by little cracks; each division, when present, with two to eight cushion-like to fibrillose tawny or blackish scales. Flesh pallid or whitish, tinged greenish or dull bluish green, when drying becoming yellowish to vinaceous; odor mild; taste bitter.

Gills adnexed to adnate, often decurrent by a line in early stages, frequently seceding; color cream buff to pale ocherous orange, broad to medium broad, crowded or close; lamellulae numerous; edges even to slightly roughened.

Stalk (30–)50–120 mm long, (4–)10–15(–40) mm thick, similar in color to the cap, appressed-fibrillose or glabrous, dry, sometimes striate, solid, becoming more or less hollow, equal, sometimes connate, with three to eight stalks joined at the base. Partial veil composed of yellowish fibrils, evanescent, leaving an apical evanescent zone.

Spores orange rufous, ferruginous, xanthine orange, or cinnamon rufous in deposit, 6–8.5(–9) × (3.5–)4–4.5 μm, inequilateral in side view, elliptic in face view, verruculose, without an apical germ pore, dextrinoid. Basidia four-spored, 24–28 × 5–7 μm. Pleurocystidia 23–35 × 5–7 μm, ventricose, rare. Cheilocystidia 20–38 × (3–)5–7(–9) μm, flask-shaped to ventricose, capitate to nearly capitate to more rarely not capitate, sometimes extending up the sides for a short distance.

Gill trama interwoven; hyphae 5–12(–25) μm wide. Subhymenium not distinctive. Cap cuticle filamentous; hyphae repent or occurring in tufts, brownish, sometimes encrusted. Cap trama interwoven. Caulocystidia 34–60 × 3–7 μm, cylindric-clavate, sac-like, or more rarely in clusters. Clamp connections present.

Useful references. There are no general books or field guides pertinent to North America that contain information on *G. aeruginosus.*

Gymnopilus luteus (Peck) Hesler

Common names. None.

Distinguishing characteristics. Cap 50–100 mm broad, convex; surface dry, fibrillose to silky, at times somewhat scaly; color buff yellow to slightly more yellowish. Flesh pale yellow; taste bitter. Gills adnexed, pale yellow becoming dark ferruginous, close. Stalk 40–75 mm long, 6–16 mm thick, similar in color to the cap to more ferruginous yellow when handled. Partial veil leaving a more or less distinct fibrillose to rather membranous annulus. Cheilocystidia present. Spores ellipsoidal, roughened, 6–9 × 4–5(–5.5) μm in size. Caulocystidia absent.

Observations. *G. luteus* seems closely related to *G. spectabilis* (Hesler 1969). According to Hesler's discussion, *G. luteus* has adnexed gills that are more narrow than those of *G. spectabilis*. The hyphae are also distinctly radially arranged in the cap trama in *G. luteus*. The absence of caulocystidia is used to separate *G. luteus* from *G. spectabilis* and related species in the key provided by Hesler (1969).

Distribution and seasonal occurrence. This species has been reported from the eastern United States. It fruits from August to September.

Habitat and habit. *G. luteus* grows on decaying wood, logs, and trunks of trees.

Technical description. Cap 50–100 mm broad, convex; margin incurved and slightly surpassing the gills; surface dry, appressed floccose-fibrillose or silky, sometimes minutely floccose-squamulose toward the center; color buff yellow, ranging from warm buff to antimony yellow. Flesh firm, fleshy, pale yellow; odor pleasant; taste bitter.

Gills adnexed, rounded behind, pale yellow becoming dark ferruginous with age, thin, close, moderately narrow.

Stalk 40–75 mm long, 6–16 mm thick, similar in color to the cap, becoming ferruginous yellowish on handling, fibrillose, solid, firm, thickened above or below. Partial veil forming a fibrillose to nearly membranous annulus.

Spores 6–9 × 4–5(–5.5) μm, ellipsoidal, slightly inequilateral, verruculose, without an apical germ pore, ferruginous, dextrinoid. Basidia four-spored, 22–25 × 5–6 μm. Pleurocystidia 26–31 × 4–7 μm, ventricose, scattered, inconspicuous. Cheilocystidia 24–32 × 4–6 μm, ventricose to flask-shaped, capitate to nearly capitate.

Gill trama composed of nearly parallel hyphae 3–5 μm wide. Subhymenium not with a distinctive structure. Cap cuticle composed of a zone of dark brown repent filamentous hyphae. Cap trama composed of radial hyphae. Caulocystidia lacking. Clamp connections present.

Useful references. There are no general books or field guides pertinent to North America that contain information on *G. luteus*.

Gymnopilus spectabilis (Fr.) Smith

Figs. 93, 106

Common names. None.

Distinguishing characteristics. Cap large, up to 180 mm broad, usually broadly convex; surface dry, usually with small scales on the margin; color bright, usually some shade of orange ranging from buff yellow to tawny orange. Taste typically bitter. Gills close to crowded; color pale yellow to orange yellow, becoming rusty with age. Stalk thick, 8–10(–30) mm at the apex, clavate or tapered downward, colored as the cap. Partial veil at times leaving a nearly membranous or fibrillose zone near the stalk apex. Spores roughened, under the microscope appearing dark rusty brown to orange brown in potassium hydroxide, 8–10.2 × (4.5–) 6.6–7.3 μm in size.

Observations. *G. spectabilis* is distinguished by the large size and bright colors; however, there are other large species of *Gymnopilus*, such as *G. luteus*, that are similar in appearance, and these should be checked carefully. In older literature *G. spectabilis* was sometimes placed in the genus *Pholiota*.

Distribution and seasonal occurrence. This species is widely distributed in North America. It fruits from summer throughout the winter, depending on location and weather.

Habitat and habit. *G. spectabilis* grows singly at times but often caespitose, on stumps and logs or on dead or living trees. It is at times terrestrial, probably growing from buried wood. It occurs on both hardwoods and conifers.

Technical description. Cap (50–)80–180 mm broad, convex-expanding to broadly convex or nearly plane, striped when moist; surface smooth, slightly silky-fibrillose from remnants of a thin veil, at times becoming slightly lacerated and with minute scales except over the disc; color buff yellow to near brilliant orange, strong orange, or tawny orange at maturity. Flesh relatively thin except on the disc, firm, in early stages pale yellowish becoming brilliant yellow with age; taste bitter; odor slight.

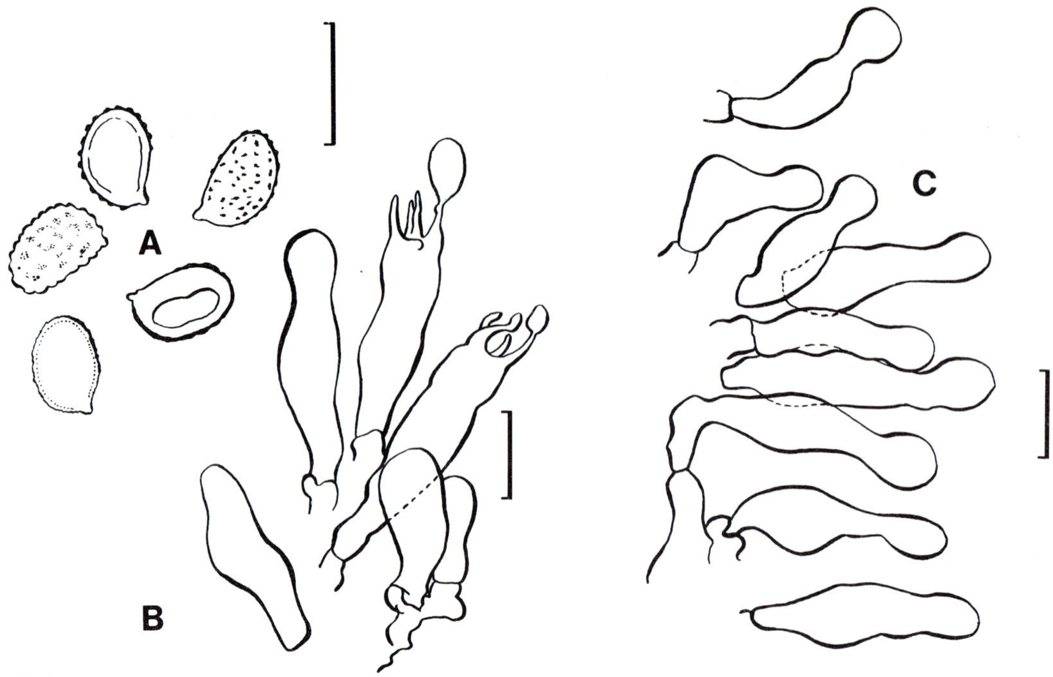

92

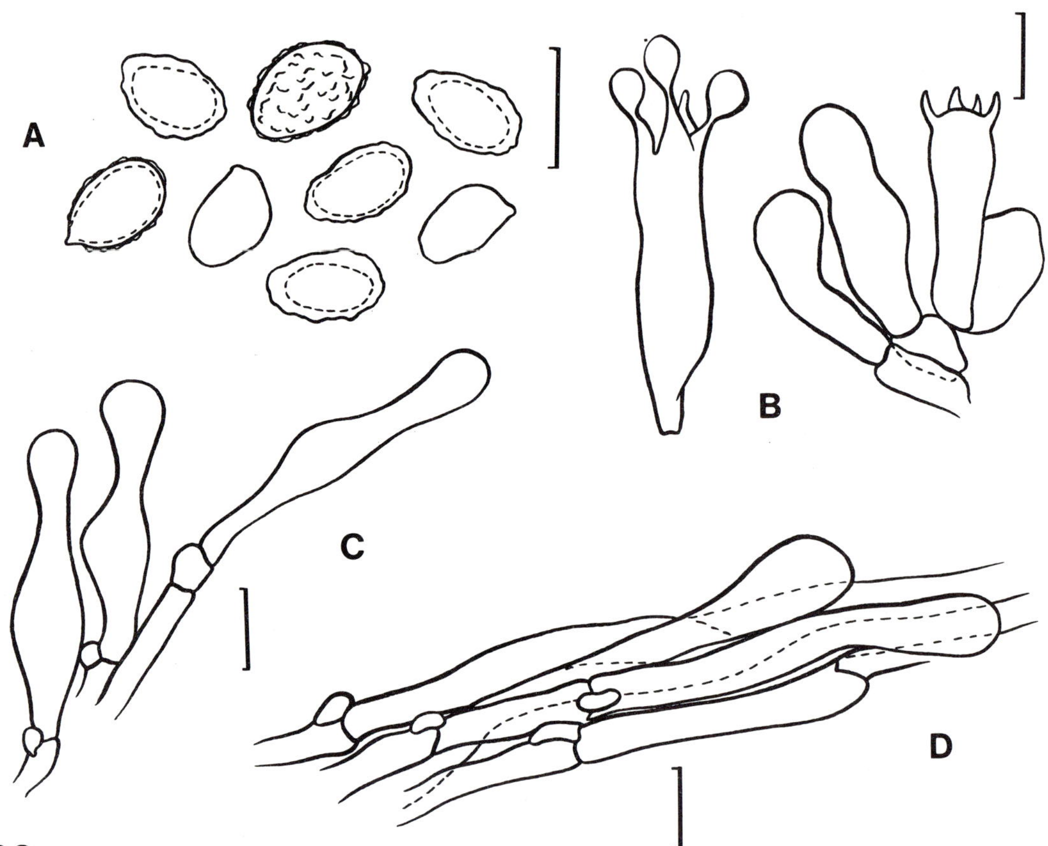

93

Fig. 92. *Gymnopilus aeruginosus*: A, spores; B, basidia; C, cheilocystidia.

Fig. 93. *Gymnopilus spectabilis*: A, spores; B, basidia; C, cheilocystidia; D, caulocystidia appressed to the stalk.

Scale line 10 μm

Gills pale yellow to deep orange yellow or rusty when mature, adnate to slightly decurrent, becoming emarginate-hooked with age or with descending lines on the stalk apex, close to crowded, rather narrow to moderately broad; edges finely scalloped.

Stalk 30-200 mm long, 8-10(-30) mm wide at the apex, clavate to tapered at the base, sometimes ventricose or nearly equal, central, solid, slightly fibrillose from veil remnants; color pale yellow to more or less similar in color to the cap. Flesh compact, hard, pale yellow. Partial veil thin, nearly membranous to fibrillose, leaving an indistinct more or less apical annulus or a slightly fibrillose pale yellow zone. Universal veil absent.

Spores strong orange to vivid orange or dull rusty brown in deposit, 8.0-10.2 × (4.5-)6.6-7.3 μm, elliptic to more or less ovate to slightly almond-shaped in side view, ovate to elliptic in face view, verruculose to finely wrinkled, dark rusty brown to orange brown, dextrinoid. Basidia four-spored, clavate, 22-36.5 × 6.6-8.8 μm, hyaline to brownish. Pleurocystidia inconspicuous, fusiform-ventricose, 25-29.2 × 5.8-6.4 μm, hyaline. Cheilocystidia fusiform-ventricose with capitate apices, 23-38.7(-43) × 5.8-7.3(-8.0) × 5.8-8.0 (at apices) μm, hyaline to yellowish brown.

Gill trama interwoven to more or less parallel; hyphae cylindrical to inflated, 3.3-18.3 μm wide, hyaline to yellow with pigment diffusing out in potassium hydroxide. Subhymenium interwoven; hyphae cylindrical, 3.3-5.1 μm wide, hyaline to yellow. Cap cuticle filamentous, interwoven, more or less radially arranged; hyphae cylindrical, 2.2-5.8(-7.2) μm wide, hyaline to brownish, at times encrusted, forming fascicles where scales occur. Cap trama interwoven; hyphae cylindrical to inflated, 3.7-20 μm wide, hyaline to yellowish. Caulocystidia ventricose to clavate; apices nearly capitate to capitate, 11-18.3(-35) × 5.1-7.3 μm with apices 6.6-7.3 μm wide, hyaline to yellowish, clustered, occurring above the annulus. Stalk hyphae parallel at the core to interwoven on the surface, 3.7-6.6(-9.5) μm wide, hyaline to yellowish. Annulus interwoven; hyphae cylindrical, 3.3-6.2 μm wide, hyaline to yellowish, thin-walled, sometimes encrusted. Oleiferous hyphae yellowish, refractive, with granular contents, 3.7-11.0 μm wide, scattered in the trama, cap cuticle, and stalk. Clamp connections present.

Useful references. The following books contain information on *G. spectabilis* pertinent to North America.

- *Edible and Poisonous Mushrooms of Canada* (Groves 1979)—A color photograph and a description and discussion of this species, listed as *Pholiota spectabilis* (Fr.) Gill., are given.
- *Mushrooms in their Natural Habitats* (Smith 1949)—A color photograph on a reel, a description and discussion, and a photograph of the western form are presented.
- *Mushrooms of North America* (Miller 1972)—A color photograph and a description and discussion are given.

For an additional technical description see *North American Species of Gymnopilus* (Hesler 1969).

Gymnopilus validipes (Peck) Hesler

Common names. None.

Distinguishing characteristics. Cap 75-150 mm broad, convex to plane; surface scaly; color ocherous. Flesh white to yellowish; taste mild. Gills adnate to decurrent with a tooth, yellowish white becoming cinnamon, close. Stalk 100-130 mm long, 25-50 mm thick, yellowish white. Annulus present. Cheilocystidia present. Spores ellipsoidal, ornamented, 7.5-10 × 4.5-5(-5.5) μm in size.

Observations. *G. validipes* is distinguished from *G. spectabilis* and its relatives by the ocherous cap, yellowish white stalk, and mild taste. It is not well known.

Distribution and seasonal occurrence. According to Hesler (1969), this species is known from New York State. It fruits in September.

Habitat and habit. *G. validipes* was originally found fruiting in a small excavation near a farmhouse. There are few other reports of it in the literature.

Technical descriptions. Cap 75-150 mm broad, convex becoming nearly plane, ocherous, squamulose or floccose-squamulose. Flesh white, tinged with yellow next to the gills; taste mild.

Gills adnate or decurrent by a tooth, yellowish white becoming cinnamon, narrow, close, thin.

Stalk 100-130 mm long, 25-50 mm thick, yellowish white, striate at the top by the decurrent gills, firm, solid, white within, more or less annulate from remnants of the partial veil.

Spores 7.5-10 × 4.5-5(-5.5) μm, usually elliptic but more rarely nearly ovate in face view, slightly inequilateral in side view, verrucose, ferruginous; only a few dextrinoid when initially tested, but all becoming dextrinoid in about an hour. Basidia two- and four-spored, 23-28 × 5-7 μm. Pleurocystidia 23-28 × 5-7 μm, ventricose, inconspicuous. Cheilocystidia 23-30 × 3-7 μm, flask-shaped, capitate or nearly capitate.

Gill trama composed of nearly parallel hyphae 5-9 μm wide. Subhymenium not distinctive in structure. Cap cuticle composed of repent filamentous hyphae, with bundles of brown hyphae toward the disc. Cap trama composed of radial hyphae. Caulocystidia and pileocystidia lacking. Clamp connections present.

Useful references. There are no general books or field guides pertinent to North America that contain information on *G. validipes*.

Gymnopilus viridans Murrill

Common names. None.

Distinguishing characteristics. Cap up to 122 mm broad, convex with a raised or depressed center; surface dry to moist, appressed-fibrillose or breaking into patch-like scales; color ocherous to apricot or ocherous orange becoming greenish-spotted or discolored where handled. Gills close to crowded, adnate to hooked; color dingy yellowish brown to ferruginous when mature. Stalk up to 62 mm long, 20 mm thick, enlarged below; color pale yellowish to yellow with darker streaks. Cheilocystidia present. Spores ellipsoidal, minutely verruculose, not dextrinoid, (6–)7–8.5 × 4.5–5 μm in size.

Observations. *G. viridans* is poorly known. Hesler (1969) states that the green stains suggest a relationship to *G. aeruginosus*. According to Hesler *G. aeruginosus* is distinct from *G. viridans* by its greenish cap in early stages with scales that sometimes become blackish, yellow gills that become yellowish orange, dextrinoid spores, and the presence of pleurocystidia.

Distribution and seasonal occurrence. This species grows in the Pacific northwest, specifically in Washington around Seattle, and has also been reported from southern Ontario. It fruits from October to November.

Habitat and habit. *G. viridans* occurs caespitose on coniferous logs (Washington), and gregarious to caespitose on logs of *Betula papyrifera* (Ontario).

Technical description. Cap 50–122 mm broad, convex to obtusely convex; disc with a large umbo to shallowly depressed; surface moist to dry, more or less glabrous to innately fibrillose or appressed-fibrillose, at times smooth but usually breaking into appressed-fibrillose patches or slightly raised patch-like scales; edge opaque, enrolled to incurved, in places coated with brownish fibrils from the veil; margin ocherous warm buff to antimony yellow or in places ocherous buff, orange buff, apricot yellow, or, where moist, cadmium yellow, developing grayish to almost tea green discolorations or becoming green-spotted; scales sometimes a light bay color similar to the disc, mainly cadmium yellow to near ocherous orange and usually watery moist (darker colors seemingly associated with moist areas). Flesh solid, firm, up to 15 mm on the disc and gradually thinner at the edge, pale to light yellow, more yellowish over the gills and toward the edge and paler toward the surface of the disc, never white, though at times very pale, sometimes mottled over the gills, usually with a dull watery zone (olive yellow) where the gills meet the cap flesh; taste bitter; odor fragrant.

Gills up to 8 mm wide, close to crowded, emarginate with a decurrent tooth to adnate or more or less hooked, not ventricose, tapering from the base to the margin; edges somewhat uneven and more or less yellowish to yellow, staining or discoloring more or less Sudan brown to raw sienna with age; faces deep buff, dingy yellow brown, or slightly more brownish to ferruginous when mature.

Stalk 40–62 mm long; apex 10–20 mm thick; base slightly enlarged; surface scurfy and bright yellow to dull pale yellow at the apex, but dull pale yellowish streaked with Sudan brown to argus brown or Mars yellow fibrils below; base and lower stalk similarly colored, although yellowish ground color becoming less pronounced with age; fibrillose annular zone of collapsed fibrils present near the apex; outer layer tough; interior stuffed with a softer pith, not becoming truly hollow but sometimes with a narrowly hollowed center above or with short hollowed areas (in part caused by larvae); flesh pale to light yellow with some brighter yellow to watery yellow streaks toward the center, yellowish to yellow near the surface or darker-colored as the surface.

Spores rich argus brown to rich Sudan brown in deposit, (6–)7–8.5 × (4–)4.5–5 μm, ellipsoidal, unequal in side view, minutely verruculose, without an apical germ pore, not dextrinoid. Basidia four-spored, 20–24 × 6–7 μm. Pleurocystidia absent. Cheilocystidia 20–26 × 5–7 μm, fusiform-ventricose; neck filiform, capitate, or nearly capitate.

Gill trama nearly parallel; hyphae 4–10 μm broad. Subhymenium not distinctive in structure. Cap cuticle composed of repent filamentous brown hyphae that are more or less encrusted. Cap trama composed of interwoven hyphae. Pileocystidia lacking. Caulocystidia 35–43 × 4–7 μm, similar to the terminal elements of the surface hyphae, cylindrical to flask-shaped; apices at times nearly capitate. Clamp connections present.

Useful references. There are no books or field guides pertinent to North America that contain information on *G. viridans*.

Panaeolus (Fr.) Quél. — Coprinaceae

The genus *Panaeolus* is cosmopolitan, with several temperate as well as tropical to subtropical species. Many species of *Panaeolus* occur on dung, but a large number are terrestrial and occur on soil, especially in grassy areas such as pastures. The distinguishing characteristics of *Panaeolus* include a bell-shaped or less commonly a conic or convex cap that usually does not expand much with age, is viscid to moist when fresh, and is glabrous or has fragments of white to whitish partial veil tissue on the margin; a cellular cap cuticle, composed of more or less spherical to isodiametric cells as viewed under the microscope; attached gills, mottled or variegated by patches of mature, black spores; the presence of cheilocystidia; a slender stalk, usually long in comparison with the diameter of the cap, more or less fragile; the presence of a partial veil, sometimes as an annulus or as fragments along the margin of the cap; a usually black to purple black or, less commonly, a dark grayish brown fuscous spore deposit; smooth spores most commonly, but warted spores found in some species, and with an apical germ pore. Apparently most of the hallucinogenic species of *Panaeolus* do not stain blue on the stalk; *P. subbalteatus* does, however, and it also develops grayish blue to greenish gray colors on the cap when bruised.

Panaeolus can be confused with certain dark-spored genera such as *Coprinus* or *Psathyrella*. *Coprinus* species can usually be separated by their deliquescent gills. *Psathyrella* is somewhat more difficult to separate from *Panaeolus*. The easiest way to separate these genera is by mounting the spores in concentrated sulfuric acid (H_2SO_4). The spores of *Psathyrella* usually fade or discolor when mounted in H_2SO_4, as compared with their color in water; the spores of *Panaeolus* typically retain their color in H_2SO_4. *Coprinus* spores usually also fade or discolor when mounted in H_2SO_4. Both *Coprinus* and *Pathyrella* lack the mottled gills found in *Panaeolus*. *Psathyrella* species are usually more fragile than species of *Panaeolus*.

Because of their dark spore deposits species of *Psilocybe, Stropharia,* and *Naematoloma* might be confused with *Panaeolus*. The cap cuticle in each of these genera is filamentous, however, composed of cylindrical hyphae, not cellular as in *Panaeolus*.

There is some disagreement over the classification of certain species of *Panaeolus*. For example, *Panaeolus foenisecii* is sometimes placed in the genus *Panaeolina* because it has warty spores that are a dark vinaceous brown to dark blackish brown in deposit. Other authors have placed this species in *Psathyrella*.

Several species of *Panaeolus,* including several tropical and subtropical species, have been shown to contain psilocin or psilocybin (Guzmán and Vergeer 1978). In the north temperate region of North America, especially in the Pacific northwest, *P. campanulatus sensu* Hora, *P. castaneifolius, P. foenisecii, P. sphinctrinus,* and *P. subbalteatus* have been used to some degree for their hallucinogenic properties. Ola'h (1969) reported at least 10 species that contain psilocin or psilocybin; of these at least *P. castaneifolius, P. fimicola* (Fr.) Gill., *P. foenisecii, P. sphinctrinus* (Fr.) Quel., and *P. subbalteatus* have been reported in Canada and adjacent areas of the United States. Stamets (1978) also treats a number of psilocybin-containing species of *Panaeolus*. Ola'h (1969) did not detect hallucinogens in *P. campanulatus;* however, according to Guzmán et al. (1976), this species is used as a drug mushroom on the Pacific coast.

These conflicting reports probably arise because of the various concepts of *P. campanulatus* that are currently in existence. Possibly, however, various chemical races of this species occur, some lacking the hallucinogens and others containing one or more of these compounds. This possibility cannot be thoroughly studied, though, until the concept of *P. campanulatus* is clear. As Singer (1975) suggests, the name *campanulatus* should be abandoned if a correct concept of the species cannot be determined.

Because the concept of *P. campanulatus* and certain other species remains unclear, the correct descriptions for these species are difficult to determine. Therefore, reports of poisonings may be unreliable. Presented here, however, are some of the more commonly encountered species, using the concepts currently accepted.

Panaeolus campanulatus (Fr.) Quél. — Fig. 107

Common name. Bell-shaped panaeolus.

Distinguishing characteristics. Cap 10–50 mm broad, bell-shaped to somewhat conic; surface glabrous or with white veil fragments along the margin; color reddish brown to grayish brown, grayish cinnamon buff or paler with age, with reddish brown tints persisting on the center. Gills subdistant; color gray spotted with black; edges white, at times with droplets. Stalk long and slender, 30–140 mm × 1.5–5 mm; color grayish red to reddish brown. Partial veil not forming an annulus. Spores smooth, thick-walled, with a conspicuous apical germ pore, ovoid to lemon-shaped, 13–18 × 6–8.5 μm in size.

Observations. Because the concept of *P. campanulatus* is unclear, it is variously described in the literature. As described here, the species exhibits reddish brown to pinkish brown colors in the cap and is somewhat larger than *P. sphinctrinus*. The spores of *P. campanulatus* are also significantly smaller than those of *P. sphinctrinus*. However, the spore sizes reported for *P. campanulatus* vary, depending on the author, and this discrepancy, coupled with the general confusion over the concept of *P. campanulatus,* makes it impossible to determine whether spore size is significant in separating these species.

Ola'h (1969), for example, describes the spores of *P. campanulatus* as 15–18.5 × 10–12.5 μm in size. He agrees, in general, that *P. campanulatus* is somewhat larger than *P. sphinctrinus* and that it has a somewhat darker cap. However, he indicates that the cap of *P. campanulatus* varies in color and may be pale, like that of *P. sphinctrinus. P. papilionaceus* (Bull. ex Fr.) Quél., another species reported as hallucinogenic by at least some authors, may be the same as *P. campanulatus.*

There are various reports on the toxicity of *P. campanulatus.* It is, in general, reported as being hallucinogenic and traces of psilocybin have been reported in it (Guzmân et al. 1976). Ola'h (1969) says that *P. campanulatus* does not contain psilocybin or psilocin, and Stamets (1978) says it is not known to be active in North America.

Distribution and seasonal occurrence. This species is widespread in North America. It fruits during the spring, summer, or fall, depending on location and weather.

Habitat and habit. *P. campanulatus* grows singly to gregarious on horse and cow dung and on well-manured soil.

Technical description. Cap (10-)20-50 mm broad, conic, bell-shaped, or obtusely conic and sometimes mammillate; margin incurved in early stages; surface somewhat viscid in early stages, soon becoming dry, not hygrophanous, even in early stages, splitting or cracking with age but not finely wrinkled or just scarcely so, glabrous or when young appendiculate with white veil fragments that form a denticulate margin; color reddish brown to grayish brown when young, paler (grayish cinnamon buff or paler) with age. Flesh grayish white; taste indistinctive; odor indiscernible or sometimes, when cut, like brown sugar.

Gills ascending-adnate, seceding with age, subdistant, unequal, moderately broad, ventricose, gray in early stages, spotted with black at maturity, with one to three tiers of lamellulae; edges sometimes white-fimbriate, often guttulate.

Stalk 30-140 mm long, 1.5-3.5(-5) mm thick, cylindrical, straight, nearly equal, stiff, hollow or fistulous, densely grayish pruinoise overall, often longitudinally striate at the apex, smooth or grooved below; color grayish red to reddish brown. Annulus and universal veil absent.

Spores black in deposit, 13-18 × 8-11(-12.5) × 6-8.5 μm, ovoid to lemon-shaped, smooth, thick-walled, with an apical germ pore resulting in a truncate appearance, blackish to blackish brown, nonamyloid. Pleurocystidia lacking. Two types of cheilocystidia present: cylindrical filiform or narrowly fusiform-ventricose hairs, 18-26 × 5.8 μm, thin-walled, and hyaline; and clavate cells, 32-40(-55) × 8-12 μm, thin-walled, and hyaline. Cap cuticle a multilayered palisade of two types of cells: pear-shaped or subglobose cells, 10-24 × 15-28 μm and hyaline; and filamentous pileocystidia, which may be swollen at the base or apex, sometimes with dark exudations, 36-62 × 4-6 μm.

Useful references. *P. campanulatus* is reported in several mushroom guides. The concept of this species often differs from publication to publication and may not agree entirely with the concept of *P. campanulatus* given here. The following books contain information on this species pertinent to North America.

- *Mushrooms in their Natural Habitats* (Smith 1949)—A description and discussion, as well as a color photograph on a reel, are presented.
- *Mushrooms of North America* (Miller 1972)—A description and discussion are given.
- *The Savory Wild Mushroom* (McKenny 1971)—A black-and-white photograph and a description and discussion are given.

Ola'h (1969) should also be consulted by workers interested in this species.

Panaeolus foenisecii (Fr.) Kühner

Figs. 94, 108

Common name. Haymaker's panaeolus.

Distinguishing characteristics. Cap 10-40 mm broad, conic to convex; color some shade of dark brown when moist and fresh, fading quickly to pale tan or pallid. Gills moderately close to subdistant; color pallid to brownish pallid but some darker brown to deep vinaceous brown, becoming with age somewhat mottled with patches of spores; edges of gills often pale or whitish. Stalk 1.5-3.5 mm thick; color pallid above and brownish to dark brown below. Partial veil absent. Spores mostly 11-18.3 × 7-9.5 μm in size, usually dark vinaceous brown to dark blackish brown in deposit; surface roughened; apical germ pore present.

Observations. *P. foenisecii* is a common species in lawns and other grassy areas. It often occurs with other lawn mushrooms such as *Marasmius oreades* (Bolt. ex Fr.) Fr. and *Conocybe lacteus* (Lange) Métrod.

P. foenisecii has been classified in various genera including *Psilocybe, Panaeolina,* and recently *Psathyrella. Panaeolus castaneifolius* (Murrill) Smith is a similar species with a thicker stalk (4-6 mm thick) and a rather strong odor and unpleasant taste. The species would appear to be difficult to distinguish from *P. foenisecii.* According to Smith (1972) *P. castaneifolius* occurs in the northeastern United States and in Quebec. See his work for a description of this species. Stamets (1978) reports that *P. castaneifolius* grows across North America and latently contains psilocybin.

There are various reports on the edibility of *P. foenisecii.* Although it is reported as edible by several authors the species is not often recommended. Certain populations contain low levels of psilocybin and psilocin (Ola'h 1969, Robbers et al. 1969). Miller (1972) reported the poisoning of a four-year-old boy by *P. foenisecii.* The child was comatose for a short time. A

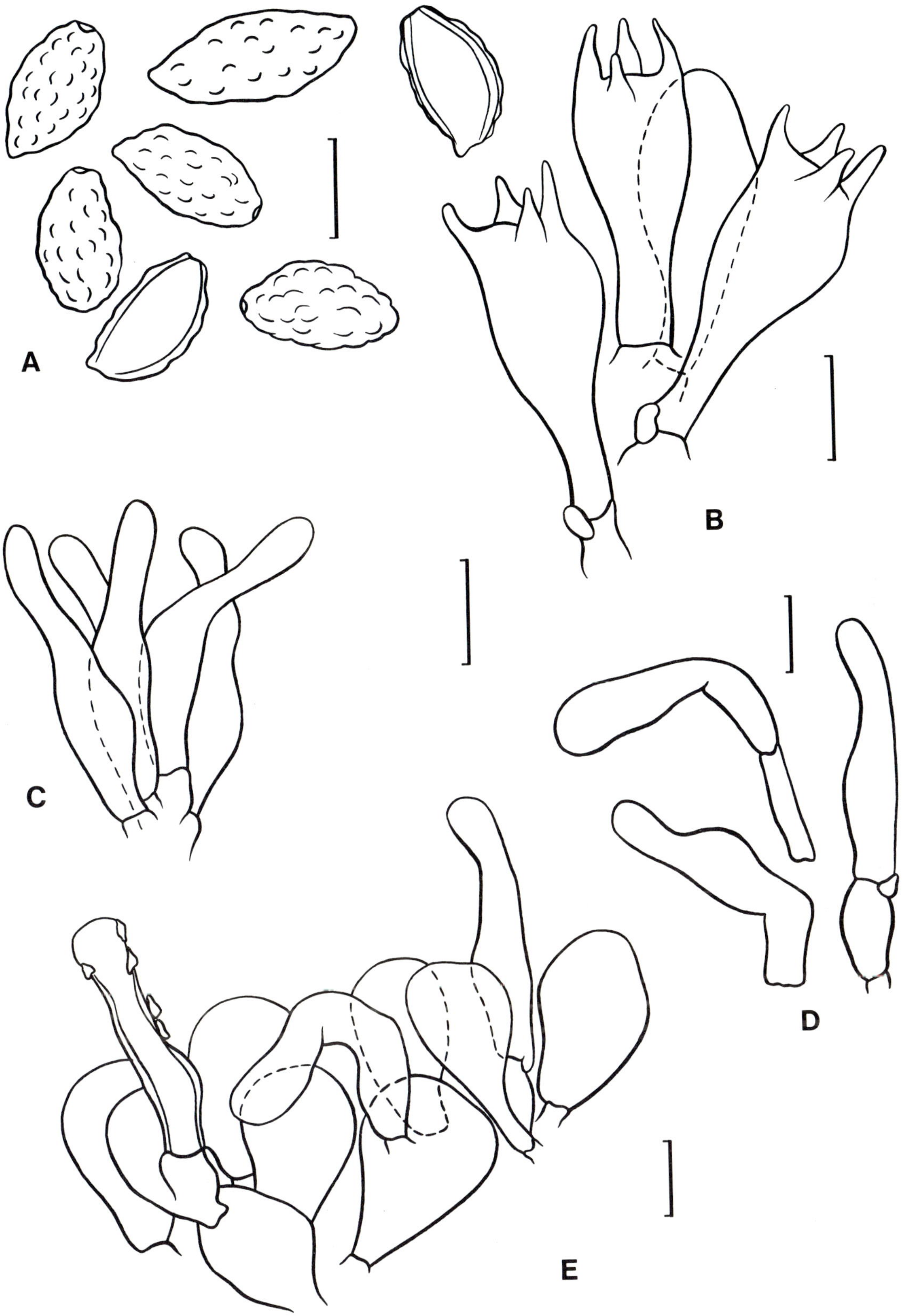

Fig. 94. *Panaeolus foenisecii*: A, spores; B, basidia; C, cheilocystidia; D, caulocystidia; E, cuticle and pileocystidia.

Scale line 10 μm

second case of poisoning, also involving a young child, was reported by Miller (Personal commun.).

Distribution and seasonal occurrence. This cosmopolitan species occurs throughout North America. It fruits in the spring, summer, and fall but is most commonly reported from late spring to early summer. The species may appear in the summer on well-watered lawns.

Habitat and habit. *P. foenisecii* grows scattered to gregarious, usually not singly, and often among other lawn mushrooms. It occurs in lawns, fields, meadows, and other grassy places.

Technical description. Cap 10-40 mm broad, obtusely conic to convex, becoming broadly umbonate to plane; surface moist, strongly hygrophanous, glabrous; margin at times faintly striate when moist, at times with the cuticle variously disrupted from weathering; veil fragments lacking; color, when moist, dull cinnamon brown, dark brown to grayish red brown, dull grayish brown, or dingy grayish vinaceous brown, at times with a purplish cast when fresh, fading to brownish pink avellaneous, dingy buff, or pallid. Flesh thin, fragile, watery brown fading to pinkish brown and finally pallid; taste acidulous; odor fungoid.

Gills pallid to brownish pallid, becoming darker, deep reddish brown (chocolate brown) to dull dark grayish red, or deep vinaceous brown, mottled with age, rounded-adnate, soon seceding, moderately close to subdistant, broad and ventricose; edges even, whitish.

Stalk 40-100 mm long, 1.5-3.5 mm thick, equal, central, tubular, faintly appressed fibrillose-striate, pruinose at the apex, twisted-striate at times, slightly pubescent at the base; color pallid to brownish gray, paler above, dingy brown at the base; flesh fragile. Partial and universal veils lacking.

Spores dark vinaceous brown to dark blackish brown in deposit, thick-walled, 11.0-18.3(-20.1) × 7.0-9.5 μm, elliptic to slightly inequilateral in side view, ovate in face view; apex truncate, snout-like, with an apical germ pore; surface dark brown, nonamyloid, roughened with irregular low warts. Basidia four-spored, clavate, 24-32.9 × 8-12.8 μm, hyaline; contents granular. Pleurocystidia lacking. Cheilocystidia narrowly fusiform-ventricose with obtuse to tapered apices, 28-42 × 6.6-12.4 μm, hyaline, thin-walled.

Gill trama more or less parallel; hyphae 3.5-11.0 μm wide, more or less cylindrical, hyaline. Subhymenium cellular; hyphae interwoven, 5.8-11.0 μm broad, hyaline. Cap cuticle cellular; cells sac-like and pedicellate to pear-shaped or subglobose, 11-20 μm wide, hyaline; hyphae just below the cuticle cylindrical to inflated, 3.5-11.0 μm wide, interwoven, encrusted, brownish, giving rise to cylindrical, nearly clavate to fusiform-capitate pileocystidia which project above the cuticle; these cells scattered, 25-36.5 × 4.4-5.8 μm, thin-walled to slightly thick-walled, hyaline to brown, more or less encrusted. Cap trama loosely interwoven; hyphae cylindrical to inflated, 3.5-20.0 μm, thin-walled to slightly encrusted, hyaline to brownish. Caulocystidia filiform, zigzagged to fusiform-ventricose with undulate tapered necks and obtuse apices, 28-40 × 5.4-7.3(-11.0) μm, hyaline. Stalk hyphae parallel, more or less cylindrical, 3.5-7.3 μm wide, hyaline to pale brown. Oleiferous hyphae not observed. Clamp connections present.

Useful references. The following books contain information on *P. foenisecii* pertinent to North America.

- *Edible and Poisonous Mushrooms of Canada* (Groves 1979)—A black-and-white photograph and a description and discussion are given.
- *Mushrooms in their Natural Habitats* (Smith 1949)—A description and discussion, as well as a color photograph on a reel, are presented.
- *Mushrooms of North America* (Miller 1972)—A color photograph and a description and discussion are given.
- *The Savory Wild Mushroom* (McKenny 1971)—A black-and-white photograph and a description and discussion are given.

Smith (1972) provides a good technical description of this species.

Panaeolus sphinctrinus (Fr.) Quél.

Figs. 95, 109

Common names. None.

Distinguishing characteristics. Cap 15-30 mm broad, hemispherical to ovoid-bell-shaped; margin usually with whitish partial veil fragments, especially when young; color brownish gray to olive gray or grayish green, at times with blackish or reddish tints in the center of the cap. Gills subdistant; color grayish to grayish black, mottled blackish with age; edges often whitish. Stalk 2-3 mm thick, equal; color grayish brown above and brownish red purple to reddish brown toward the base. Annulus typically lacking. Spores smooth with a distinct apical germ pore, 14.5-18 × 10.5-12.6 μm in size.

Observations. *P. sphinctrinus* and *P. campanulatus* are closely related and difficult to distinguish. There has been much confusion in the literature concerning both species and no doubt the two have been confused in poison cases as well. The work by Ola'h (1969) separates these species on the basis of cap color and size of fruit body. In general, *P. campanulatus* is larger than *P. sphinctrinus.* The cap, when fresh, is reddish brown or pinkish brown in *P. campanulatus* and grayish olive to grayish green or brownish gray in *P. sphinctrinus.* The color of the cap is not a consistent difference, however, because certain forms of *P. campanulatus* are reported as pale when fresh, resembling the grayish white color

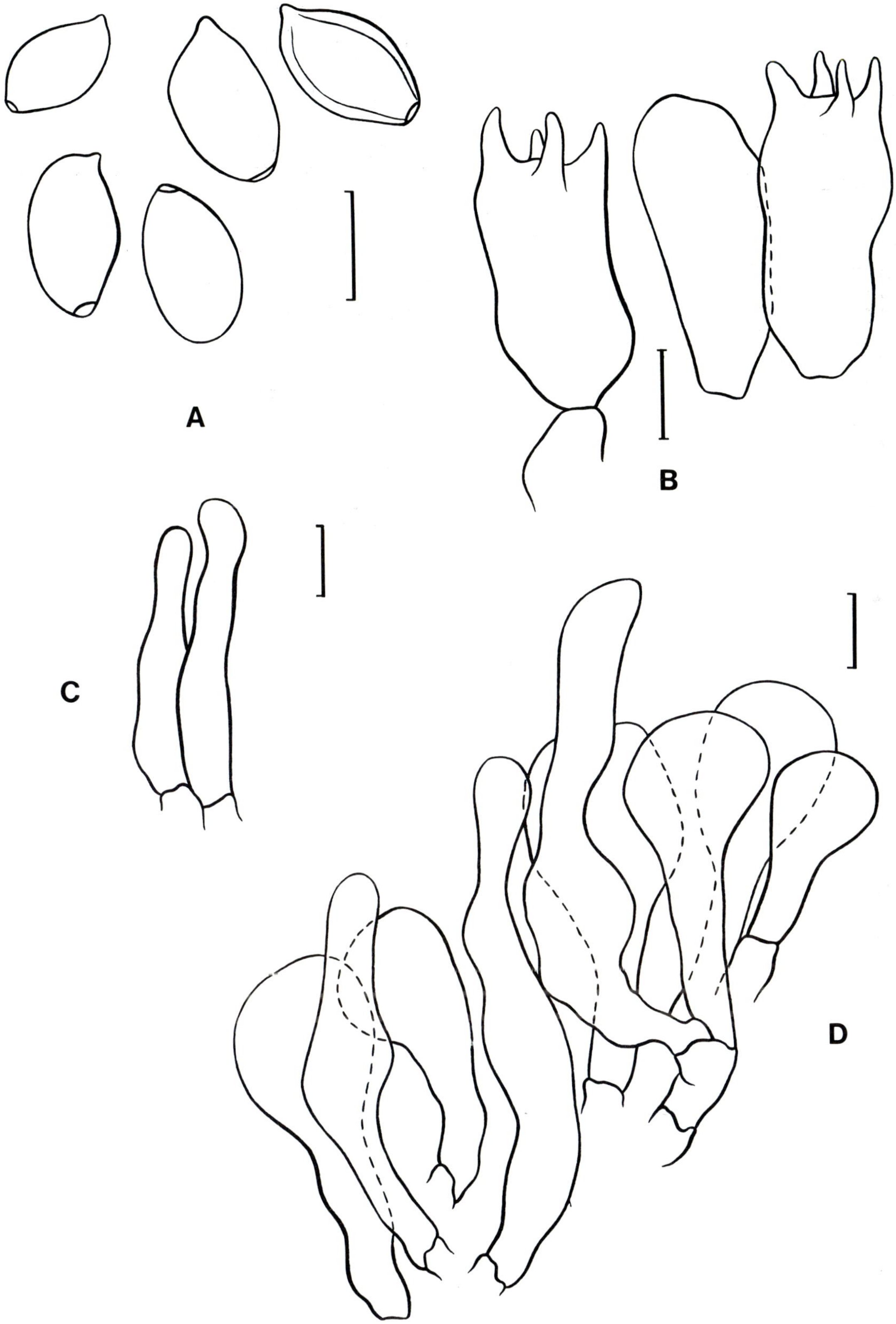

Fig. 95. *Panaeolus sphinctrinus*: A, spores; B, basidia; C, cheilocystidia; D, cuticle and pileocystidia.

Scale line 10 μm

of *P. sphinctrinus*. Tints of olive or green on the cap may be a better color characteristic of *P. sphinctrinus*.

Guzmān (1972) and Stamets (1978) provide additional descriptions of *P. sphinctrinus*. Guzmān (1972) gives a description of *P. sphinctrinus* var. *minor* Singer.

Ola'h (1969) reports psilocybin and psilocin in *P. sphinctrinus*. According to Guzmān et al. (1976) *P. sphinctrinus* is used as a recreational drug mushroom in the Pacific northwest, although some users find it devoid of hallucinogenic activity. Perhaps there are chemical races of this species, some containing hallucinogens, as well as of *P. campanulatus*. Another species, *P. retirugis* (Fr.) Quél., may also be hallucinogenic; it is similar in appearance to *P. campanulatus* and *P. sphinctrinus* (Stamets 1978). It is not treated here.

Distribution and seasonal occurrence. This species is widely distributed in North America. It fruits in the spring, summer, and fall.

Habitat and habit. *P. sphinctrinus* occurs singly or in groups. It occurs on horse and cow dung and in pastures and meadows.

Technical description. Cap 15-25(-30) mm broad, 12-17(-20) mm high, hemispherical, ovoid-bell-shaped, rarely umbonate; margin more or less incurved, appendiculate with whitish fragments of the veil, especially when young; surface smooth or when dry more or less areolate, sometimes more or less hygrophanous; color brownish gray to olive gray, grayish green, or grayish white, sometimes blackish, sometimes tinted reddish on the disc. Flesh thin, colored like the surface; taste and odor indistinctive.

Gills adnate to ascending, seceding, subdistant, broad, grayish to grayish black, becoming mottled blackish; edges whitish to whitish gray, floccose.

Stalk 40-80(-100) mm long, 2-3 mm thick, equal, slightly enlarged at the base, straight, hollow, striate at the apex, with a grayish to grayish olive pruinose coating; ground color grayish brown above and brownish red purple to reddish brown toward the base. Flesh more or less the same color as the surface. Annulus and universal veil lacking.

Spores blackish in deposit, 14.5-18 × 10.5-12.6 μm, lemon-shaped, smooth, thick-walled, truncate at the apex with a distinct apical germ pore, brownish black in water, more brownish in potassium hydroxide, nonamyloid. Pleurocystidia lacking. Cheilocystidia 15-25(-50) × 4-6(-19) μm, numerous, variable in shape, cylindrical to clavate, zigzagged, rounded at the apex, hyaline. Cap cuticle a multilayered palisade of two types of cells: subglobose to ovoid or pear-shaped cells which are hyaline; and cylindrical somewhat zigzagged pileocystidia which are cylindrical-clavate to clavate, or enlarged at the apex or base and sometimes more or less pedicellate, with the wall often thickened and colored brownish black.

Useful references. The following books contain information on *P. sphinctrinus* pertinent to North America.

- *Edible and Poisonous Mushrooms of Canada* (Groves 1979)—A color photograph and a discussion of *P. sphinctrinus*, considered to be the same as *P. campanulatus* Fr. *sensu* Kauffman in this reference, are presented.
- *Mushrooms of North America* (Miller 1972)—*P. sphinctrinus* is considered to be the same as *P. campanulatus* in this reference.

A description of *P. sphinctrinus* is presented in works by Ola'h (1969), Stamets (1978), and Guzmān (1972).

Panaeolus subbalteatus (Berk. & Br.) Sacc.

Figs. 96, 100

Common names. None.

Distinguishing characteristics. Cap 8-50 mm broad, hemispherical to conic or somewhat bell-shaped-convex, expanding to almost flat with a small umbo; color deep reddish brown to deep brown when fresh, sometimes dark chestnut at the center, fading to clay brownish or bright brownish yellow but usually retaining some tint of rufous brown or copper color, sometimes becoming grayish blue to greenish gray at the margin when bruised. Gills close to nearly close; color flesh-toned to pinkish when young, finally dull reddish brown and more or less blackish mottled. Stalk 2-8 mm thick, more or less equal; base somewhat enlarged; color reddish brown to blackish when fresh, fading to pale grayish red brown or clay color, turning blue where handled. Annulus lacking. Spores smooth, with a distinct apical germ pore, (10-)11-14.6(-15.3) × 7.3-9.5 μm in size.

Observations. *P. subbalteatus* occurs in Europe and North America. Watling (1977) gives a good description and discussion of this species as it occurs around Edinburgh, Scotland. Our description, at least in part, is taken from this paper. The species is also described in works by Ola'h (1969), Stamets (1978), and Guzmān (1972).

Guzmān et al. (1976) report this species as common in the Pacific northwest, where it is used as a recreational drug mushroom.

The bluing reaction in the cap or stalk of *P. subbalteatus* was reported by Ola'h (1969) and Watling (1977). It is difficult to determine how consistent the reaction is in this species, mainly because there are few reports of it in the North American literature.

P. venenosus Murrill and *P. rufus* Overholts are apparently the same as *P. subbalteatus*. Murrill (1916) reports four severe poisonings by *P. venenosus*.

Distribution and seasonal occurrence. This species is widespread in North America. It fruits in the late spring, summer, and fall.

Habitat and habit. *P. subbalteatus* is gregarious, occurring in large clusters. It grows on manured soil, cultivated gardens, compost beds, and dung and has also been reported on beds of cultivated mushrooms.

Technical description. Cap 8-50 mm broad, hemispherical, conic, semiglobate, obtusely conic to bell-shaped-convex, expanding to almost flat with a small umbo; margin sometimes recurved, not striate; surface moist, smooth becoming finely wrinkled, cracking with age, lacking veil fibrils; color, when moist, deep reddish brown to deep brown on the margin, dark chestnut on the disc, drying clay brownish to light brownish yellow with a darker marginal area, usually retaining rufous brown or copper throughout, becoming slightly grayish blue to greenish slate on the margin where bruised. Flesh thick on the disc, tapering to the margin, grayish to reddish brown; taste slightly mealy; odor indistinctive.

Gills flesh-colored or pinkish when young, darkening to grayish red, brownish, or dull reddish brown, becoming more or less blackish mottled to olivaceous black, slightly to broadly adnate, close to nearly close; edges whitish to somewhat minutely fringed.

Stalk 35-90 mm long, 2-8 mm wide, more or less equal or narrowed downward, more or less enlarged where inserted in the substratum, hollow, central, straight to curved, longitudinally striate, covered with a fine white bloom or white tomentose below; ground color reddish brown to blackish when moist, fading to pale grayish red brown or clay color, turning blue where handled; flesh of stalk similar to cap flesh. Annulus and universal veil lacking.

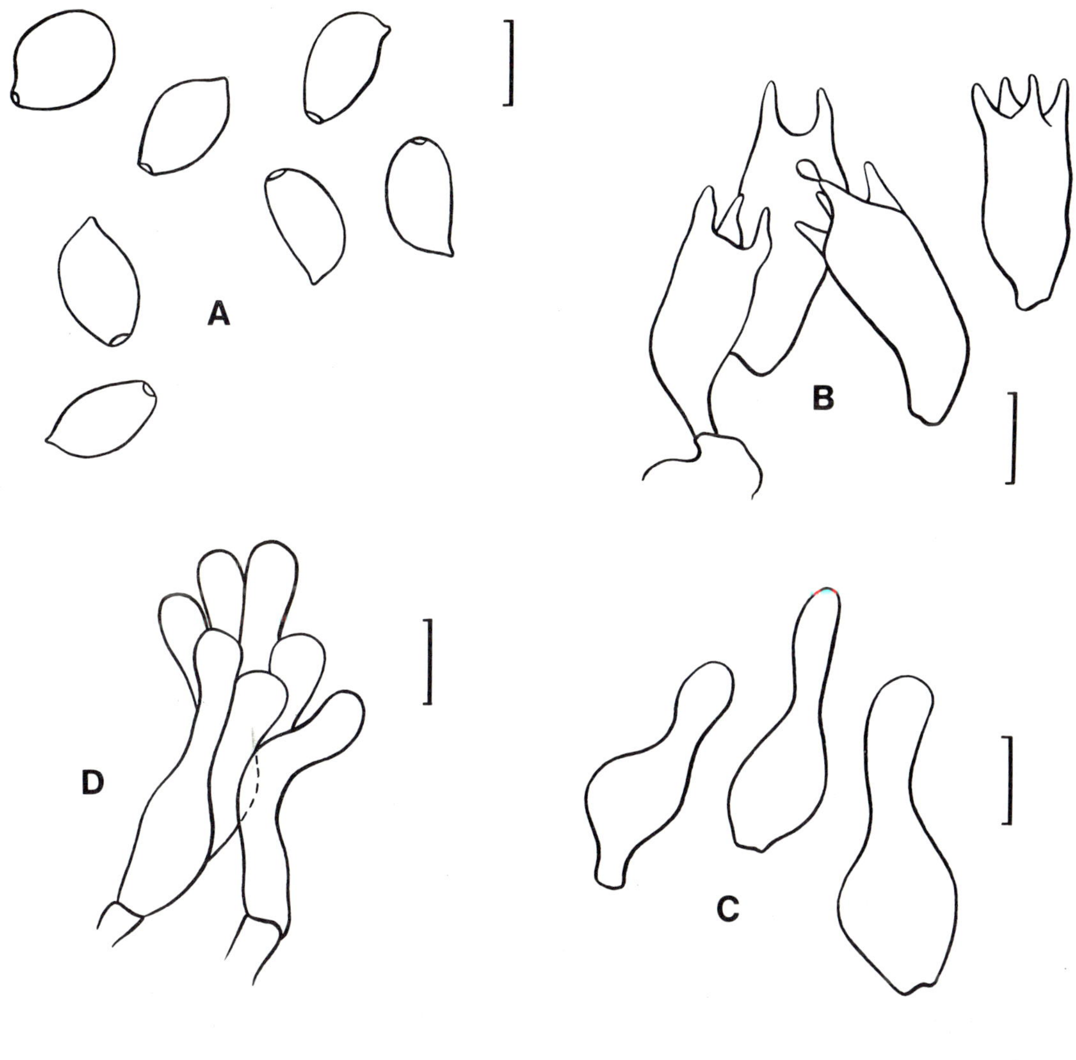

Fig. 96. *Panaeolus subbalteatus*: A, spores; B, basidia; C, cheilocystidia; D, caulocystidia.

Scale line 10 μm

Spores blackish brown to dull blackish in deposit, (10–)11.0–14.6(–15.3) × 7.3–9.5 μm, ellipsoidal, slightly inequilateral in side view, ovate in face view, thick-walled, smooth, brownish to blackish, nonamyloid; apex truncate, with a distinct apical germ pore. Basidia two- and four-spored, short-clavate, 21.9–29.2 × 8.5–10.2 μm, hyaline. Pleurocystidia lacking. Cheilocystidia variable in shape, broadly fusiform-ventricose with elongate to flask-shaped necks, 21.9–40.0 × 5.1–13.9(–16.1) μm, hyaline, thin-walled; apices capitate to nearly capitate or clavate.

Gill trama interwoven to parallel; hyphae cylindrical to inflated, 3.7–14.6 μm wide, thin-walled to slightly thick-walled and finely encrusted, hyaline to brownish in potassium hydroxide. Subhymenium nearly cellular to cellular; hyphae 3.7–9.0 μm wide, hyaline to brownish. Cap cuticle cellular, composed of a compact layer of vesiculose to globose-pedicellate cells; these cells 15–23 μm wide, hyaline to slightly brownish below, bearing scattered amorphous crystalline encrustations; cells just below the cuticle 7–11 μm wide, globose, lacking a pedicel and intermixed with interwoven cylindrical hyphae 3.5–7.3 μm wide; pileocystidia filiform to clavate or pear-shaped, 17–26 × 3.5–10.2 μm, hyaline to yellowish brown, with granular contents. Cap trama interwoven; hyphae inflated to vesiculose, 10–20 μm wide, hyaline. Caulocystidia filiform to narrowly fusiform-ventricose and capitate, 22–36.5 × 5.1–7.3(–9.5) μm, hyaline, clustered at the apex of the stalk. Stalk hyphae parallel, cylindrical to inflated, 3.7–16.1 μm wide, hyaline. Oleiferous hyphae not observed. Clamp connections present.

Useful references. Guzmán (1972), Ola'h (1969), Stamets (1978), and Watling (1977) provide additional descriptions.

Pluteus Fr. — Pluteaceae

The Pluteaceae (=Volvariaceae) contains three genera in North America, *Pluteus, Chamaeota*, and *Volvariella*. The three genera have in common free gills, a pinkish brown to vinaceous brown or more pinkish to vinaceous spore deposit, smooth spores that are even in outline, and a convergent gill trama. The genus *Pluteus* lacks an annulus and volva, *Chamaeota* has only an annulus, and *Volvariella* has only a volva. Confusion with members of the family Rhodophyllaceae is a strong possibility. Some mushrooms in the Rhodophyllaceae have pink spores, attached gills, spores that are angular in outline, and a regular to irregular gill trama. See *Entoloma* (Ch. 13).

Pluteus and *Volvariella* are the two genera most commonly encountered. *Pluteus* is particularly common in most years on wood or woody debris. Both genera are known to contain edible species such as *Pluteus cervinus* (Schaeff. ex Fr.) Quél. and *Volvariella volvacea* (Bull. ex Fr.) Singer. Saupe (1981) isolated psilocybin and psilocin from *Pluteus salicinus*, which adds it to the list of potentially psychoactive mushrooms. To date there are no reports that *P. salicinus* has been ingested for its hallucinogenic properties. It is described below.

Pluteus salicinus (Pers. ex Fr.) Kummer

Common names. None.

Distinguishing characteristics. Cap 30–50 mm broad, obtusely rounded, becoming convex to plane or slightly umbonate; surface dry, minutely scurfy at the center to minutely scaly on the margin; color blackish to bluish fuscous brown to fuscous, becoming drab to bluish gray with age; taste unpleasant. Gills free, crowded; color white, becoming pinkish with age. Stalk 40–70 mm long, 2.5–6 mm thick above, equal or nearly so; color whitish in early stages, developing bluish to greenish stains below with age or when injured. Spores 7–9 × 5–6 μm, even in outline, smooth. Pleurocystidia with apical horn-like projections. Fruit bodies occurring on hardwood logs.

Observations. *P. salicinus* is not commonly reported in the general mushroom literature. Smith and Smith (1979) report var. *salicinus*, with an opaque cap margin, and var. *achloes* Singer, with a striate cap margin when moist. Both have pleurocystidia with apical, horn-like projections. The greenish to blue stains, especially around the stalk base, are the best field characteristics for identifying *P. salicinus* and related species. The blue to greenish staining in *P. salicinus* is correlated with the presence of psilocybin and psilocin in this species (Saupe 1981). *P. cyanopus* Quél. is no doubt related to *P. salicinus* as it also has bluish to greenish stains on the stalk base.

Distribution and seasonal occurrence. *P. salicinus* seems to occur mainly east of the Great Plains and fruits in the summer and fall seasons. It apparently is not a common species.

Habitat and habit. *P. salicinus* occurs solitary to scattered on hardwood logs.

Technical description. Cap 30-50 mm broad, obtuse when young, expanding to shallowly depressed or more often becoming convex to plane or with the margin at times uplifted, more or less umbonate; surface dry, minutely scurfy over the disc and often minutely scaly toward the margin, not hygrophanous; color dark gray to brown to bluish fuscous brown or blackish when fresh, becoming with age more drab or grayish brown to bluish gray. Flesh gray to watery; odor disagreeable; taste unpleasant.

Gills crowded, moderately broad, free and fairly remote, whitish in the early stages, becoming pinkish with age, grayish on the edges where bruised.

Stalk 40-70 mm long, (2.5-)3-4(-6) mm thick at the apex, equal or nearly so, glabrous, hollow, moderately fragile, whitish overall in early stages, developing bluish to greenish stains below and on the base with age or when injured.

Spores 7-9 × 5-6 μm, ellipsoidal, smooth, nearly hyaline in potassium hydroxide, nonamyloid. Basidia four-spored, 30-35 × 7-8 μm, clavate, hyaline with granular contents. Pleurocystidia abundant, 50-70 × 11-18 μm, fusiform-ventricose with a crown of three to five horns around the apex; wall not greatly thickened but rigid. Cheilocystidia similar to pleurocystidia or clavate to vesiculose, 20-30 × 9-15 μm. Gill trama convergent. Cap cuticle composed of radially arranged hyphae with cystidium-like end cells 100 × 9-20 μm in size and containing grayish brown pigment in potassium hydroxide. Cap trama composed of loosely interwoven hyphae. Clamp connections present.

Useful references. *P. salicinus* is not described or illustrated in any of the references used here. Groves (1975) briefly discusses it under *P. cervinus*.

Psilocybe (Fr.) Quél. — Strophariaceae

Psilocybe is a moderately large genus, with a variety of species. The genus is poorly known in North America and there is no comprehensive treatment of the genus available for Canada or the United States. Recently the hallucinogenic species of *Psilocybe* have received considerable attention. The species and their distributions are thus becoming better known, and several new species have been described.

Psilocybe is a somewhat difficult genus to recognize because of similarities in general appearance with certain species of other genera. Characteristics that are helpful in recognizing *Psilocybe* include a typically conic to bell-shaped cap, less commonly convex, often with a viscid surface; usually adnate gills that are typically blackish from spores when mature, not mottled as in *Panaeolus*; a central stalk, often long and slender; the presence of a partial veil in some species, at times leaving an annulus; the development of greenish to blue green stains on the stalk and sometimes on the cap upon handling; a purple brown to dark purple violaceous or chocolate brown spore deposit; typically smooth spores with an apical germ pore; and a more or less interwoven to radially arranged cap cuticle composed of filamentous, cylindrical hyphae.

Some species of *Psilocybe* growing in lawns or on moss beds generally resemble certain species of *Panaeolus* or *Conocybe*. In these other two genera, however, the cap cuticle is cellular, composed of more or less spherical to isodiametric cells when viewed under the microscope. Also, the spore deposit of *Conocybe* is not blackish but typically rust brown, and the gills are never blackish.

Psilocybe is also difficult to separate from the genera *Stropharia* and *Naematoloma* (=*Hypholoma*), and all three are no doubt closely related. According to some authors the relationship is close enough to combine these genera into the one genus, *Psilocybe*.

In this treatment, however, we have followed the traditional classification, using the three genera *Psilocybe*, *Naematoloma*, and *Stropharia*. To separate these genera requires close attention to both macroscopic and microscopic characteristics. Some helpful comparisons that can be used as a guide in identifying these genera follow. When the fruit body has a distinct annulus, it is probably a *Stropharia* species, but check *Psilocybe* also. When an annulus is lacking and the gills have specialized gloeocystidia called chrysocystidia, the specimen is probably a *Naematoloma*. When a distinct annulus and chrysocystidia are lacking, the stalk is viscid, and the mushroom occurs on dung or manured soil, the specimen is probably a *Stropharia* species (but annulate species of *Psilocybe* also occur on dung). When an annulus or chrysocystidia are not present, check *Psilocybe* first. When an annulus and chrysocystidia are present, check *Stropharia*. When the cap or the stalk stains blue or blue green, the specimen is most likely a *Psilocybe*.

Certain species of *Pholiota* and *Psathyrella* are also similar in appearance to *Psilocybe*. Species of *Psathyrella*, however, usually have a cellular cap cuticle. *Pholiota* species typically have a distinctly brown spore deposit (not blackish or violet to purple tinted), and they frequently occur on wood.

The species of concern here are those that are hallucinogenic, containing psilocybin or psilocin or closely related compounds. They are characterized by the greenish to blue green stains that develop on the fruit body, especially at the stalk base. The stains are usually present in these hallucinogenic mushrooms, though some stain slowly. Guzmán and Vergeer (1978) list a total of 73 hallucinogenic species in the genus *Psilocybe*; many of these are tropical or subtropical in distribution. Several species occur in the United States and Canada. Some of these, such as *P. cubensis* and *P. caerulescens,* are southern in distribution, occurring in Florida and other areas of the southern United States. The hallucinogenic species reported from Canada and the adjacent northern areas of the United States include: *P. baeocystis, P. caerulipes, P. cyanescens, P. liniformans* var. *americana* Guzmán & Stamets (Stamets et al. 1981), *P. moelleri* Guzmán, *P. merdaria* (Fr.) Ricken, *P. pelliculosa, P. quebecensis* Ola'h & Heim, *P. semilanceata, P. silvatica* (Peck) Singer & Smith, *P. strictipes,* and *P. stuntzii.* Of these species *P. baeocystis, P. cyanescens, P. cyanofibrillosa* Guzmán & Stamets (Stamets et al. 1980), *P. moelleri, P. pelliculosa,* and *P. stuntzii* seem to be western in distribution, reported primarily from the Pacific coast. *P. caerulipes, P. quebecensis,* and *P. silvatica* are mainly eastern in distribution; however, *P. silvatica* has been reported from the west by Stamets (1978). *P. cyanescens* occurs on the Pacific coast and was originally described from England. It is probably widespread in North America, but to date the reports are few. Both *P. semilanceata* and *P. strictipes* have been reported from eastern and western North America, and *P. semilanceata* seems to be widely distributed and fairly common.

Not all the species discussed above are described here. We have included only some of the more common species. Singer and Smith (1978) provide a description of *P. silvatica,* and Singer (1969) describes *P. moelleri,* although he reported it as *P. merdaria* var. *macrospora* Singer, *P. quebecensis* is described by Ola'h and Heim (1967). Also, Stamets (1978) treats a large number of psilocybin-containing species in the Strophariaceae and in *Panaeolus.* A key to selected species of psilocybin-containing species of *Psilocybe* is presented below, adapted in part from the work of Stamets (1978).

Key to selected species

1. Partial veil present, typically leaving a fairly well-formed persistent membranous annulus on the stalk. Fungi growing on woody debris or in lawns or fields, in the Pacific northwest..... ***P. stuntzii***

   Partial veil present but not typically leaving a persistent well-formed annulus, though at times with an annular zone remaining. Substrate and distribution various.. 2

2. Stalk extremely viscid to glutinous below the annular zone in fresh moist specimens. Partial veil and cap glutinous. Fungi growing in dung or well-manured ground.. ***Stropharia semiglobata***

   Not with the above combination of characters.. 3

3. Cap normally conic, conic-bell-shaped to bell-shaped, and usually not expanding further at maturity (occasionally expanding to more or less convex).. 4

   Cap normally expanding to broadly convex to plane at maturity.. 8

4. Cap sharply conic, with a distinct umbo. Fruit bodies often collected in grassy areas on or near dung. Spores 11.7-14.6 × 7-8 μm...................... .... ***P. semilanceata***

   Not with all the above features 5

5. Cap with a thin gelatinous cuticle. Fruit bodies typically growing on woody debris. Spores 6-9.5(-11.5) × 4.5-5 μm .... see ***P. silvatica*** not described here; Singer and Smith (1958) describes this species

   Not as above. Spores typically longer and wider.. 6

6. Cap margin incurved when young, more or less like a *Collybia* species.................................... 7

   Cap margin straight when young, like a *Mycena* species, and remaining so. Fruit bodies usually growing in association with woody debris. Spores 9.5-11.7 × 5.8-6.6 μm in size........ ***P. pelliculosa***

7. Stalk long in relation to cap diameter (ratio of cap diameter to stalk length ±0.30). Fruit bodies often occurring on woody debris. Spores 9-12.4(-17.5) × 6.6-7.3(-8.0) μm in size.................. ***P. strictipes***

   Stalk shorter in relation to cap diameter (ratio of cap diameter to stalk length ±0.60). Habitat various. Spores 8.8-14.6 × 6.6-7.3 μm in size..... .... ***P. baeocystis***

8. Partial veil thinly membranous and at maturity present as a more or less persistent annular zone on the stalk............................ check ***P. stuntzii***

   Partial veil composed of fibrils (not truly membranous) and leaving a slight nonpersistent zone of fibrils on the stem, if at all......................... 9

9. Fruit bodies growing in decayed hardwood substratum, particularly birch (*Betula*) and maple (*Acer*), in midwestern and eastern North America. Spores 7.3-10.2 × 4.4-5.1 μm in size.................. .... ***P. caerulipes***

   Not with the above combination of characteristics 10

10. Fungi reported from Quebec. Fruit bodies usually occurring on wood, both hardwood and conifer. Stalk yellowish brown overall. Spores 9-11 × 6-8 μm in size.................................. *P. quebecensis* not described here; Ola'h and Heim (1967) provide a description

    Fungi mainly reported from western North America, especially the Pacific northwest..... 11

11. Cap margin not incurved when young. Stalk white overall, but turning blue when handled. Fruit bodies growing on woody substrata or in grassy areas. Spores 9-15 × 5-8 μm in size........ .... ***P. cyanescens***

    Cap margin incurved when young. Habit more or less like a *Collybia* species............................ 12

12. Stalk long in relation to cap diameter (ratio of cap diameter to stalk length ±0.30) .... check ***P. strictipes***

Stalk shorter in relation to cap diameter (ratio of cap diameter to stalk ±0.60) .... check ***P. baeocystis***

Psilocybe baeocystis Singer & Smith

Figs. 97, 111

Common names. None.

Distinguishing characteristics. Cap 15-55 mm broad, conic, expanding to convex or plane; surface viscid, only faintly striate on the margin when young; color olive brown to buffy brown or yellowish brown, often with a greenish zone around the margin, staining greenish when touched or bruised. Gills nearly close; color grayish to purplish dark cinnamon, becoming more grayish purple with age. Stalk 2-3 mm thick, equal; color white, sometimes yellowish above, and becoming brownish on the base. Partial veil present but inconspicuous. Spores grayish pale violet to blackish in deposit, thick-walled, with a broad apical germ pore, 8.8-14.6 × 6.6-7.3 μm in size.

Observations. *P. baeocystis* seems to be endemic to southwest Canada and northwest United States. It is probably used as a source of recreational drugs throughout those areas. *P. strictipes* is similar in appearance to *P. baeocystis,* and the two should be closely compared. A new species, *P. cyanofibrillosa,* is somewhat similar to *P. baeocystis* (Stamets et al. 1978).

Distribution and seasonal occurrence. This species has been reported from British Columbia and southward into California. It fruits in the fall and into the winter, depending on location and weather.

Habitat and habit. *P. baeocystis* occurs singly, scattered, or in groups. It grows on peat moss, on wood mulch, and in grass, probably in both lawn and pasture grasses.

Technical description. Cap 14-54 mm broad, conic when young, expanding to convex or plane although sometimes only slightly so, broadly umbonate; margin incurved, faintly striate when young; surface viscid, smooth, hygrophanous; color dark grayish yellow brown or olive brown to moderate yellowish brown or buffy brown, finally with a greenish zone around the margin, becoming greenish when touched or bruised, strong copper brown in the center when dry. Flesh similar in color to the cap or paler; taste and odor indistinctive.

Gills dark brownish moderate orange or dark cinnamon, or gray with purplish tints, becoming grayish pale violet, nearly heliotrope gray, adnate to hooked, nearly close; edges whitish.

Stalk 50-70 mm long, 2-3 mm thick, equal to nearly equal, central, stuffed with loose fibrils, with fine white fibrils over the surface; apex yellowish, becoming white to brownish at the base; flesh brownish. Partial veil inconspicuous.

Spores grayish pale violet to blackish in deposit, 8.8-14.6 × 6.6-7.3 μm in size, mostly 11-12 × 6.3-7 μm, ellipsoidal, thick-walled, with a broad apical germ pore, deep yellow brown to olivaceous brown. Basidia four-spored, clavate, 22-32.9 × 6.6-7.3 μm, hyaline, thin-walled. Pleurocystidia rare to absent, like cheilocystidia. Cheilocystidia flask-shaped to fusiform-ventricose, with filiform tapered apices, 21-30 × 6.6-8.8 μm, hyaline, thin-walled.

Gill trama compactly interwoven; hyphae more or less parallel, cylindrical to inflated, 5.1-18.3 μm wide, hyaline to brownish. Subhymenium hyphae cylindrical to nearly cellular, 3.5-7.0 μm, brownish. Cap cuticle gelatinous, filamentous; hyphae interwoven to more or less radially arranged, cylindrical, 3.9-5.8 μm wide; hyphae just beneath the cuticle more or less parallel, more or less cylindrical, 7.0-9.0 μm wide, yellowish, thin-walled. Cap trama loosely interwoven; hyphae cylindrical to inflated, 3.5-30.0 μm wide, yellowish brown. Caulocystidia flask-shaped with filiform apices, 25-32.9 × 8.8-11.0 μm, like cheilocystidia, hyaline; stalk hyphae parallel, cylindrical to inflated, 5.0-15.0(-23.4) μm wide, hyaline to yellowish. Partial veil interwoven; hyphae cylindrical, 2.3-3.5 μm wide, hyaline. Oleiferous hyphae refractive, greenish yellow, 3.5-5.8 μm wide, scattered in the stalk. Clamp connections present.

Useful references. *P. baeocystis* is not illustrated or described in any of the popular mushroom guides referred to in this treatment. It is discussed and described by Lincoff and Mitchel (1977) and Stamets (1978) and was originally described by Singer and Smith (1958).

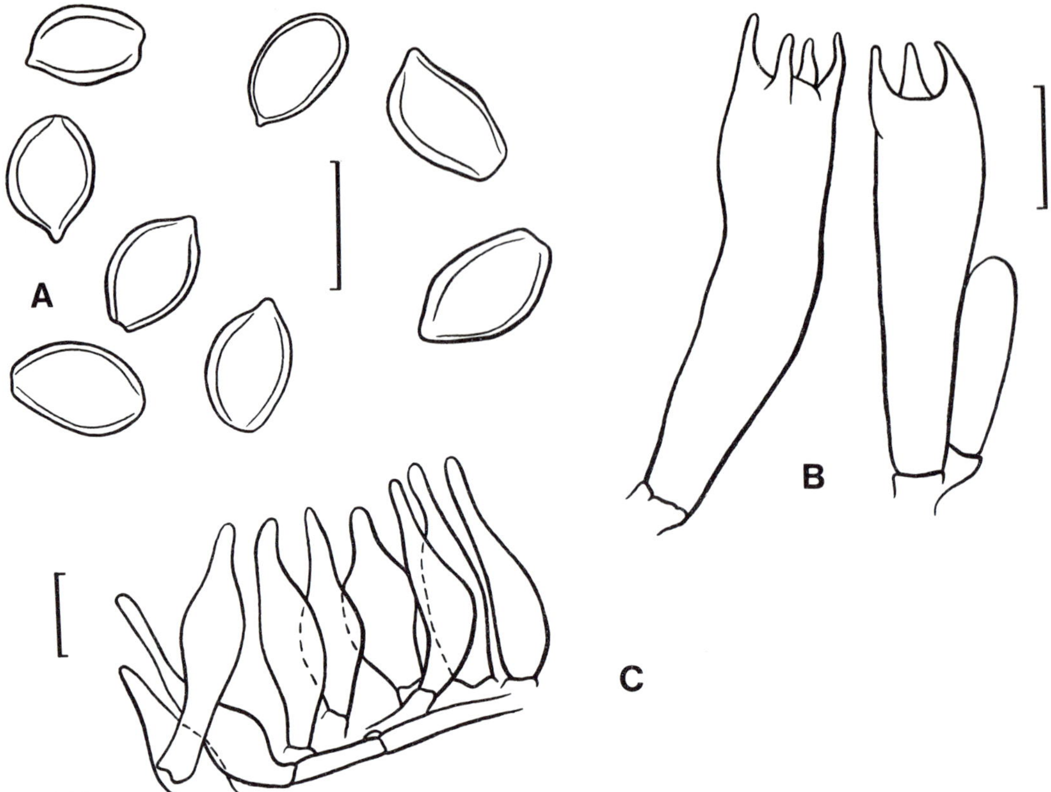

97

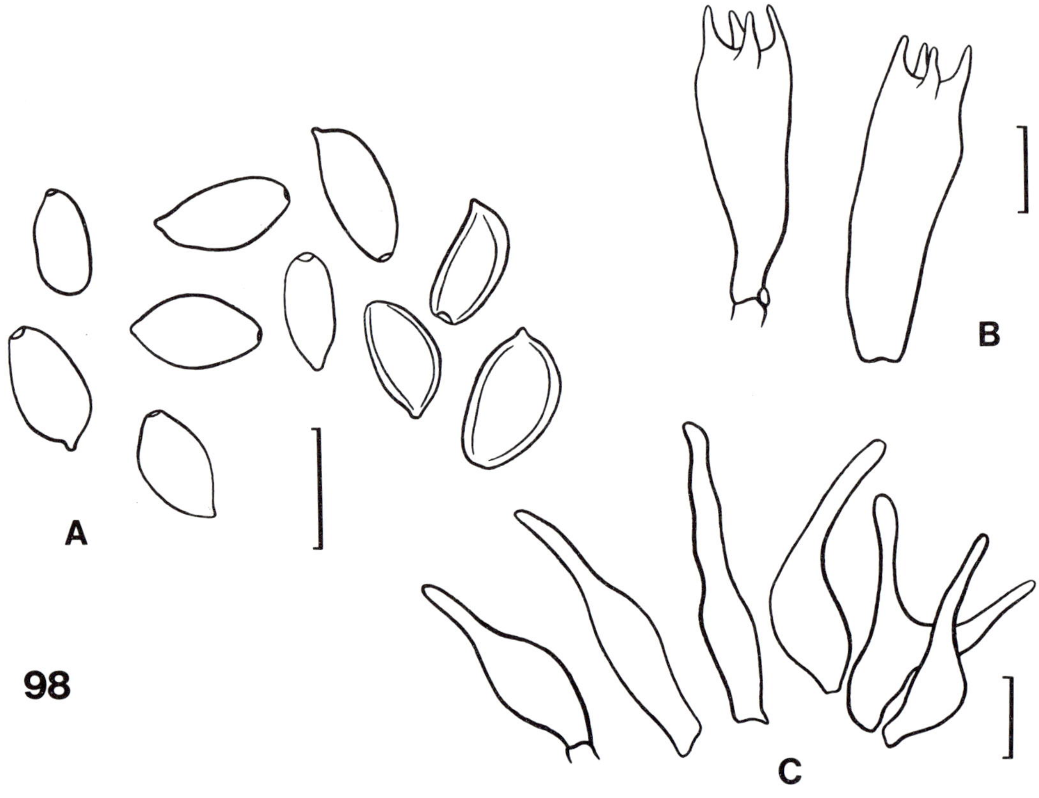

98

Fig. 97. *Psilocybe baeocystis*: A, spores; B, basidia; C, cheilocystidia.

Fig. 98. *Psilocybe caerulipes*: A, spores; B, basidia; C, cheilocystidia.

Scale line 10 μm

Psilocybe caerulipes (Peck) Sacc.

Fig. 98

Common names. None.

Distinguishing characteristics. Cap 10-35 mm broad, obtusely conic, becoming convex to plane, viscid, striate on the margin when fresh; color moderate brown to cinnamon brown, fading to cinnamon buff or ocherous buff. Gills close to crowded; color sordid brown to cinnamon, cinnamon brown, or rusty cinnamon. Stalk 2-3 mm wide, equal or slightly enlarged below; color whitish to buff, becoming brownish below; flesh bluish when injured. Spores dark purple brown in deposit, thick-walled, with a distinct apical germ pore, 7.3-10.2 × 4.4-5.1 μm in size.

Observations. *P. caerulipes* has only been reported from eastern North America. Singer and Smith (1958) report that the blue to greenish stains on the stalk may not develop until a few hours after the flesh has been injured. It is not known how commonly this species is used as a drug plant.

Distribution and seasonal occurrence. This species occurs in Canada and the United States west to the midwest and Great Lakes region, and into the southeastern United States. It fruits during the summer and less commonly in the fall. The species does not seem to be particularly common.

Habitat and habit. *P. caerulipes* occurs singly or gregariously, or is caespitose. It grows on debris and logs of hardwood trees. It seems to be most common on the wood of beech (*Fagus*) and maple (*Acer*).

Technical description. Cap 10-35 mm broad, when young obtusely conic to convex with an incurved margin, becoming broadly convex to plane or retaining

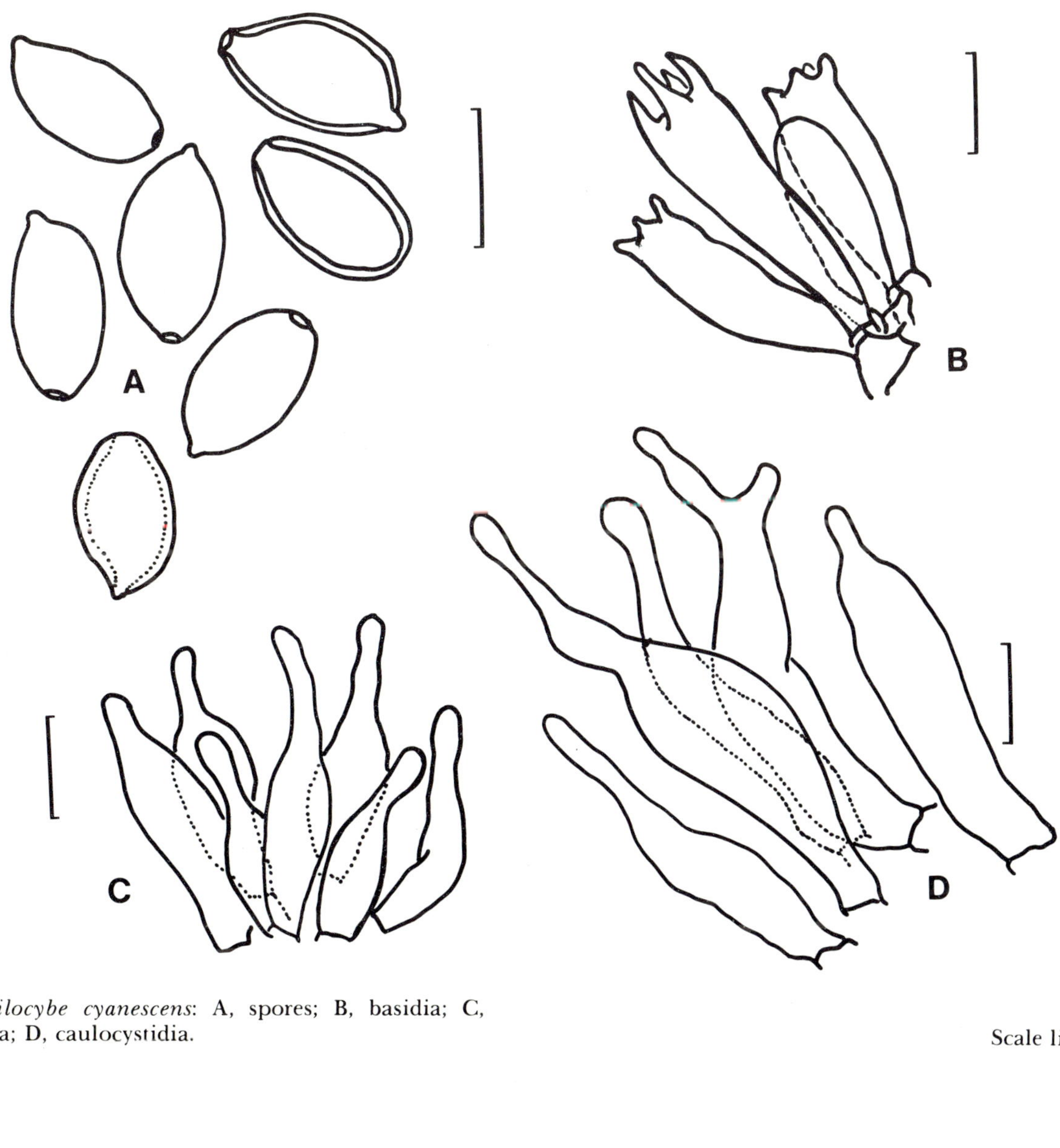

Fig. 99. *Psilocybe cyanescens*: A, spores; B, basidia; C, cheilocystidia; D, caulocystidia.

Scale line 10 μm

a slight umbo, at times quite irregular; surface viscid, soon becoming dry and shiny, hygrophanous, translucent-striate when moist, glabrous; color watery moderate brown, such as cinnamon brown to dingy sayal brown, to dingy ocherous buff or cinnamon buff, becoming sordid brownish with age, often with greenish stains along the margin or overall. Flesh thin, pliant, bluish when injured; taste like fresh meal; odor indiscernible.

Gills sordid brown, becoming cinnamon to cinnamon brown or rusty cinnamon, adnate or arcuate-adnate or with a decurrent tooth, close to crowded, narrow; edges white, minutely fringed.

Stalk 30-60 mm long, 2-3 mm thick, equal or slightly enlarged downward, central, stuffed with a pith, becoming hollow; surface pruinose near the apex, downward-appressed white to grayish fibrillose, whitish to buff in early stages, becoming dingy brown on the lower part with age, pallid to bluish when dried; basal mycelia white; flesh tough, becoming bluish where injured. Partial veil forming a thin evanescent whitish fibrillose zone apically on the stalk; no distinct annulus.

Spores very dark purple brown in deposit, 7.3-10.2 × 4.4-5.1 μm, ellipsoidal, flattened on one side, smooth, thick-walled, with a truncate apex and a distinct apical germ pore, yellow brown. Basidia two- and four-spored, short-clavate, 21-30 × 7.3-8.0 μm, hyaline. Pleurocystidia lacking, or present near the gill edges and similar to cheilocystidia. Cheilocystidia flask-shaped to narrowly fusiform-lanceolate; apex tapered and filiform, rarely divided into two short branches, (18-)25.0-32.0 × 4.0-7.3 μm, hyaline.

Gill trama interwoven to more or less parallel; hyphae cylindrical to inflated, 5.1-18.3 μm wide, hyaline to yellowish. Subhymenium interwoven; hyphae cylindrical, 3.7-5.1 μm wide, hyaline to yellowish. Cap cuticle gelatinous, filamentous; hyphae more or less radially arranged to loosely interwoven, cylindrical, 3.0-5.1 μm wide, hyaline; hyphae of the layer just below the cuticle compactly interwoven, more or less cylindrical, 5.1-9.0(-11.0) μm, yellowish brown; walls thin to slightly thickened, spirally encrusted. Cap trama interwoven; hyphae cylindrical to inflated, 3.5-22.0 μm wide, hyaline. Caulocystidia flask-shaped to fusiform-ventricose with tapered filiform apices, (20-)37-40 × 7.3-8.0 μm, hyaline; stalk hyphae parallel to interwoven, more or less cylindrical, 3.7-5.1(-11.0) μm wide, hyaline. Oleiferous hyphae not recorded. Clamp connections present.

Useful reference. The following book contains information on *P. caerulipes* pertinent to North America.

- *Mushrooms of North America* (Miller 1972)—A description and discussion are given.

Psilocybe cyanescens Wakefield

Figs. 99, 112

Common names. None.

Distinguishing characteristics. Cap usually 20-40 mm broad, obtusely conic in early stages, expanding to convex or nearly plane; margin often undulating in expanded caps; surface viscid, striate when moist; color chestnut brown when fresh, fading to yellowish, staining bluish when bruised. Gills close to subdistant; color cinnamon brown, becoming smokey brown. Stalk usually 2.5-5 mm thick; color whitish, staining bluish when bruised. Partial veil in the form of a cortina, composed of white fibrils, quickly disappearing, leaving a slight annular zone at times. Spores fuscous in deposit, thick-walled, with an apical germ pore, 9-15 × 5-8 μm in size. Fungi particularly common in areas enriched with woody material, such as mulch.

Observations. Stamets (1978) reports *P. cyanescens* as very potent, containing psilocin or psilocybin or both. The brown, viscid cap and the presence of a more or less well-developed veil in young specimens are helpful characteristics in identifying this species. The more or less convex cap with an undulating to wavy margin also helps to distinguish *P. cyanescens*. The technical description below is adapted in part from the work of Singer and Smith (1958).

Distribution and seasonal occurrence. This species was originally described from England. Singer and Smith (1958) suggest that it may have been introduced into Europe. It occurs from California northward along the coast into British Columbia. It fruits in the fall and early winter.

Habitat and habit. *P. cyanescens* grows caespitose, gregarious, or scattered in humus, especially in areas rich in woody debris, such as wood chips or decayed coniferous wood. It is often found under Douglas-fir (*Pseudotsuga taxifolia*) or cedar (*Thuja*) and in heavily mulched rhododendron beds (Stamets 1978).

Technical description. Cap 20-40(-75) mm broad, obtusely conic to conic-convex, expanding to convex or nearly plane; margin often undulating or wavy in expanded caps, striate to translucent-striate when moist; surface smooth, viscid, with a separable gelatinous cuticle, hygrophanous; color when fresh and moist chestnut brown, fading to yellowish brown or yellowish to ocherous, staining bluish when touched. Flesh more or less the same color as the cap surface, staining bluish to bluish green when bruised.

Gills adnate to adnate-decurrent, close to subdistant, up to 5 mm broad; color cinnamon to cinnamon brown, becoming deep smokey brown as spores mature; edges paler.

Stalk 60-80(-100) mm long, 2.5-5(-7 mm) thick, fairly rigid; base somewhat enlarged, often curved near the base; color whitish, changing to bluish when bruised or on drying, with rhizomorphs present at the stalk base. Partial veil a cortina, more or less well-developed on young specimens, evanescent, sometimes leaving an obscure annular zone.

Spores fuscous in deposit, 9–12 × 5–8 μm, slightly broader in face view than in side view, ellipsoidal, more ovate in face view, thick-walled, smooth, with a distinct broad apical germ pore, yellow brown. Basidia four-spored, 16–26 × 7.2–9.3 μm, clavate, hyaline. Pleurocystidia rare, 21–28 × 5.5–8.8 μm, basidiole-like but mucronate, hyaline. Cheilocystidia 12–26 × 5–8 μm, fusiform-ventricose, with the neck 2–4 μm wide, projecting from the hymenium; apex obtuse. Cap cuticle composed of an epicutis and a subcutis; hyphae of the epicutis loosely interwoven, embedded in a gelatinous matrix; hyphae of the subcutis yellowish, not conspicuously encrusted; pileocystidia lacking. Cap trama interwoven; hyphae cylindrical to inflated. Clamp connections present.

Useful references. There are no general books or field guides pertinent to North America that contain information on this species.

A description of *P. cyanescens* was given by Singer and Smith (1958). Stamets (1978) also provides a description and discussion, and a color photograph of this species.

Psilocybe pelliculosa (Smith) Singer & Smith

Figs. 100, 113

Common name. Liberty caps.

Distinguishing characteristics. Cap 8–15(–30) mm broad, conic to obtusely bell-shaped, with a straight margin; surface viscid and striate when moist and fresh; color dark dingy yellow brown to light olive brown or more olive, fading with age to pinkish buff, often developing grayish green tones with age. Gills close; color dull cinnamon brown to darker blackish brown, with pale edges. Stalk approximately 2 mm thick, enlarged somewhat at the base; color pallid to grayish or brownish, dark yellowish brown below with age. Partial veil rudimentary or lacking; hence annulus lacking. Spores purplish brown in deposit, smooth, 9.5–11.7 × 5.8–6.6 μm in size, with an apical germ pore. Fungi growing on rich humus and woody debris.

Observations. *P. pelliculosa* is a common species in the Pacific northwest and is used there as a drug plant. In general appearance it is similar to *P. semilanceata*, which, according to Guzmán et al. (1976), has larger spores than *P. pelliculosa*. *P. semilanceata* var. *microspora* Singer is apparently the same as *P. pelliculosa*.

Distribution and seasonal occurrence. This species occurs primarily in southwestern Canada and southward along the Pacific coast; we have a single report of it from Lake Timagami, Ont. It fruits from the fall into the winter, depending on location and weather.

Habitat and habit. *P. pelliculosa* is scattered, gregarious, or caespitose and is found in humus or debris, often in or near conifer forests.

Technical description. Cap 8–15(–30) mm broad, at least half as high, obtusely conic with a straight margin when young, remaining so or becoming more broadly conic to bell-shaped, never fully expanding; margin translucent, striate when moist; surface viscid, hygrophanous, glabrous, smooth; color near deep yellowish brown, dark dingy yellow brown to more olive or light olive brown, usually fading to pale orange yellow or dull pale pinkish buff, often developing greenish gray tints with age. Flesh thin, pliant, bluish where injured; odor indiscernible to slightly musty; taste not recorded.

Gills dull cinnamon brown to blackish brown, adnate, markedly ascending, eventually seceding, close, rather narrow to moderately broad; edges pallid.

Stalk 60–80 mm long, 1.5–2.0 mm thick, equal above, enlarged toward the base, pruinose at the apex, appressed silky-fibrillose below, pallid to grayish or brownish, very dark yellowish brown below with age. Flesh rather tough, bluish to greenish where injured. Partial veil rudimentary or lacking; annulus lacking.

Spores purplish brown in deposit, 9.5–11.7 × 5.8–6.6 μm, ellipsoidal, more or less inequilateral in profile, elliptic in face view, thick-walled, smooth, with a distinct apical germ pore, yellowish brown. Basidia four-spored, clavate, 20–25.6 × 6.6–8.0 μm, hyaline. Pleurocystidia absent to rare, like cheilocystidia. Cheilocystidia flask-shaped to narrowly fusiform-lanceolate with tapered apices, occasionally divided into two branches, 22–25.2 × 5.8–6.6 μm, hyaline.

Gill trama parallel; hyphae more or less cylindrical, 3.7–13.1 μm wide, hyaline. Subhymenium hyphae interwoven, cylindrical, 3.0–4.4 μm wide, hyaline. Cap cuticle gelatinous, filamentous; hyphae interwoven, more or less radially arranged, cylindrical, 2.2–3.0(–5.1) μm wide, hyaline; layer just below the cuticle compactly interwoven; hyphae cylindrical, 3.5–8.0 μm wide, brownish, thin-walled to slightly thickened, encrusted. Cap trama loosely interwoven; hyphae cylindrical to somewhat inflated, 2.5–18.0(–25) μm, hyaline. Caulocystidia flask-shaped to fusiform-lanceolate, 25–32 × 6.6–8.0 μm, hyaline; apices acute to obtuse or filiform; stalk hyphae more or less parallel, cylindrical, 3.7–8.0 μm wide, hyaline to yellowish. Oleiferous hyphae not observed. Clamp connections present.

Useful reference. The following book contains information on *P. pelliculosa* pertinent to North America.

- *A Field Guide to Western Mushrooms* (Smith 1975)—A color photograph and a description and discussion are presented.

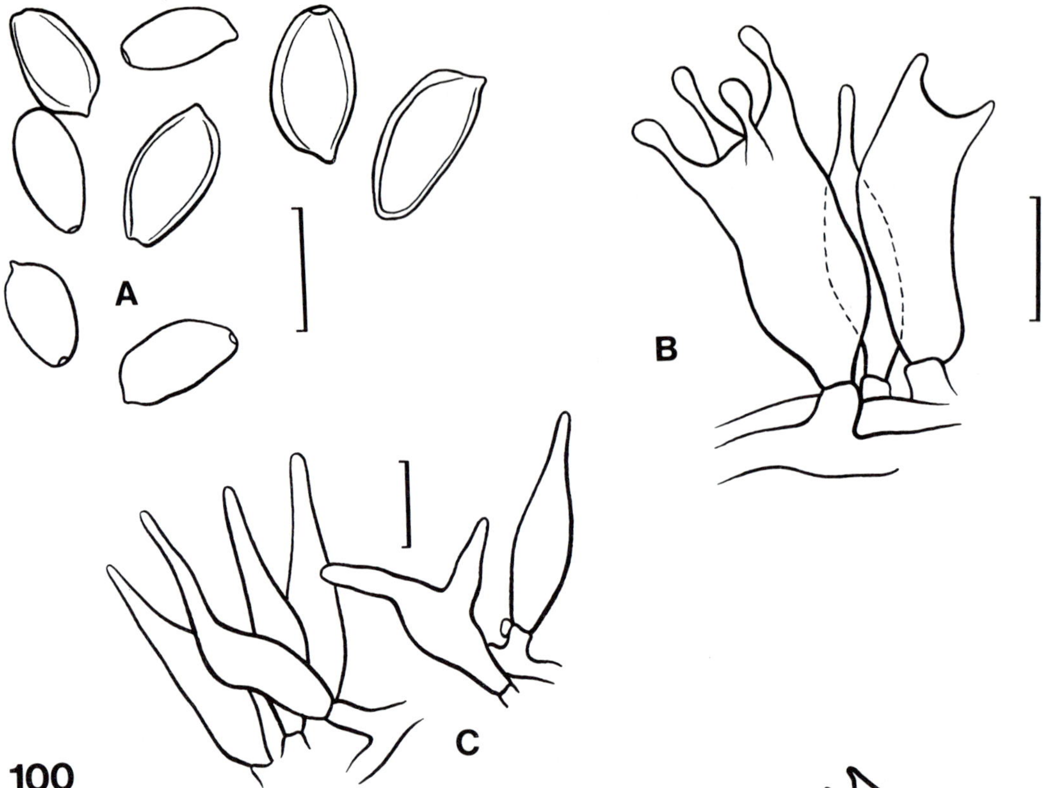

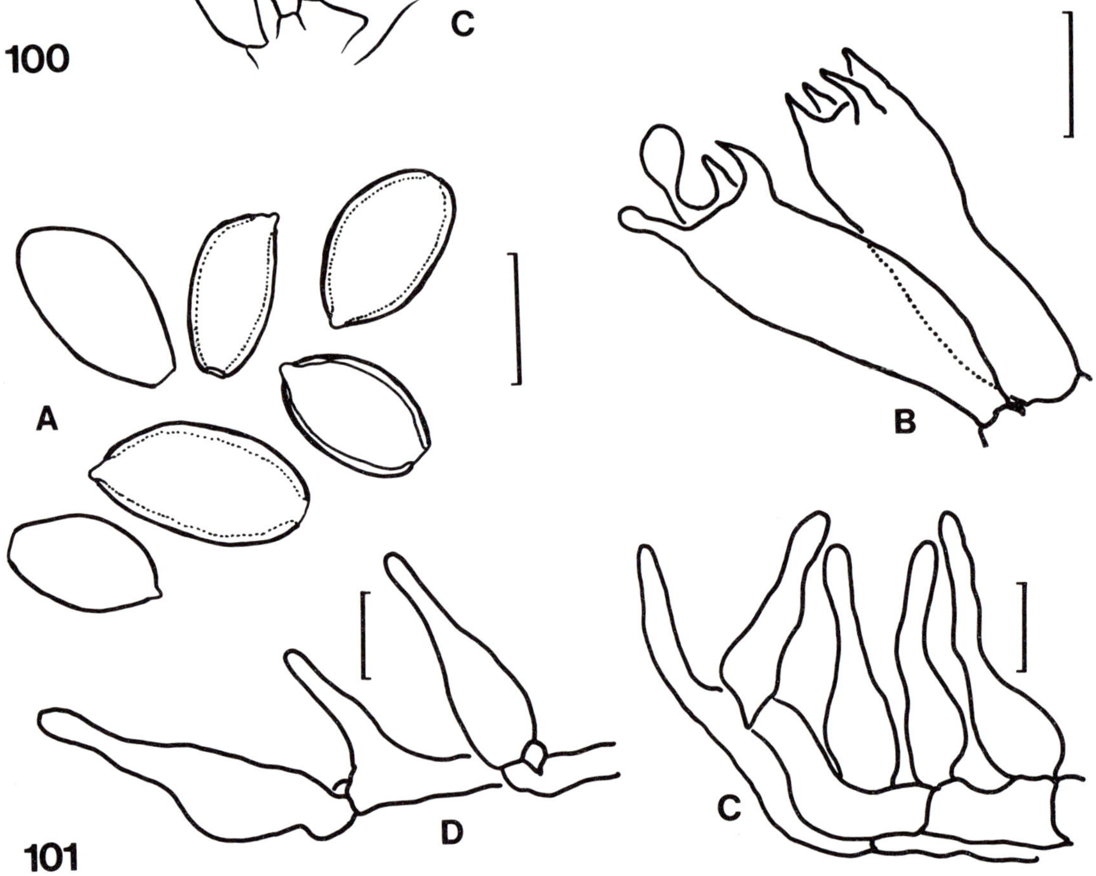

Fig. 100. *Psilocybe pelliculosa*: A, spores; B, basidia; C, cheilocystidia.

Fig. 101. *Psilocybe semilanceata*: A, spores; B, basidia; C, cheilocystidia; D, caulocystidia.

Scale line 10 μm

Psilocybe semilanceata (Fr. ex Secr.) Kummer

Figs. 101, 114

Common name. Liberty caps.

Distinguishing characteristics. Cap 8-14(-30) mm broad, sharply conic with a distinct central point, rarely expanding; surface more or less sticky when moist, sometimes with veil fragments on the margin when young; color olive umber to light brownish olive, fading to clay brown or pale buff, often flushed with grayish green at maturity. Gills crowded; color dark reddish brown to purplish black with white edges. Stalk 50-70(-110) mm long, 4-6 mm thick, equal or slightly enlarged at the base; color yellowish pallid to buff, frequently bluish to greenish on the base where injured. Spores purple brown in deposit, smooth, 11.7-14.6 × 7-8.8 μm in size, with an apical germ pore.

Observations. This species is widespread in North America and has been reported from Eastern Canada and the Pacific coast. Both the typical form and *P. semilanceata* var. *caerulescens* Cooke were described from Europe. The variety *caerulescens* markedly turns blue when the flesh is injured, but it is not considered distinct from the typical form, probably only exhibiting an extreme reaction to handling. *P. pelliculosa* is similar in appearance to *P. semilanceata* but has smaller spores. Both species are commonly collected and used as hallucinogens in British Columbia and south along the Pacific coast.

Distribution and seasonal occurrence. *P. semilanceata* is reported from Eastern and Western Canada. It is reported to be most common along the Pacific coast but is probably widespread across Canada and the United States. This species fruits in the fall and into the winter, depending on location and weather.

Habitat and habit. This species grows scattered to gregarious. It is terrestrial, occurring in grassy areas such as lawns, meadows, and pastures, sometimes on or near dung, along roadsides, along the edge of forested areas, and in open woods.

Technical description. Cap 8-14(-30) mm broad, sharply conic with a distinct acute umbo, rarely becoming expanded; margin incurved, often fluted and wrinkled, ornamented with fibrillose veil remnants when young; surface smooth, somewhat viscid to slightly sticky when wet, drying shiny, hygrophanous; color raw umber or light brownish olive to pale buff to clay brown, soon flushed grayish green at maturity. Flesh thin at the margin, thick on the disc, white to pallid; taste and odor not recorded.

Gills dark reddish brown to purplish black, adnate to adnexed, crowded; edges white.

Stalk 50-70(-110) mm long, 4-6 mm thick, equal to slightly enlarged near the base, smooth, tough, minutely pruinose at the apex, yellowish pallid to buff when moist, often bluish or greenish at the base when injured.

Spores purple brown in deposit, 11.7-14.6×7.0-8.8 μm, ellipsoidal, smooth, thick-walled with an apical germ pore, yellowish brown. Basidia four-spored, clavate, 20-30 × 7.3-8.8 μm, hyaline. Pleurocystidia lacking. Cheilocystidia flask-shaped to fusiform-ventricose with tapered filiform apices, 25.0-29.2 × 6.6-8.8 μm, hyaline, thin-walled.

Gill trama interwoven; hyphae more or less parallel, more or less cylindrical, 5.1-18.6 μm wide, hyaline to brownish. Subhymenium hyphae cylindrical, 3.7-5.1 μm wide, hyaline to brownish. Cap cuticle gelatinous, filamentous; hyphae interwoven, more or less radially arranged, cylindrical, 3.0-4.4 μm wide, hyaline to yellowish; zone just below the cuticle compactly interwoven; hyphae cylindrical, 5.8-8.0 μm wide, yellowish brown. Cap trama loosely interwoven; hyphae cylindrical to somewhat inflated, 8.2-18.7 μm wide, yellowish; contents granular; surface finely encrusted. Caulocystidia flask-shaped with tapered filiform apices, 25.6-32.9 × 7.3-8.8 μm, like cheilocystidia, hyaline, present on the stalk apex; stalk hyphae parallel, more or less cylindrical, 4.5-11.8 μm wide, hyaline to yellowish. Oleiferous not observed. Clamp connections present.

Useful references. Although this species is fairly common in North America, it is not described or illustrated in any of the general mushroom guides. It has been described and illustrated, however, in several books dealing with hallucinogenic fungi (Haard and Haard 1976, Ott 1976, Stamets 1978).

Psilocybe strictipes Singer & Smith

Fig. 102

Common names. None.

Distinguishing characteristics. Cap 20-40 mm broad, bell-shaped to convex or plane; margin viscid and striate when moist; color dull yellowish brown to light olive brown, fading to cinnamon buff or pallid on the margin, often becoming dingy overall. Gills close; color pallid, becoming a dark chocolate. Stalk 100-130 mm long, 2-3 mm thick, equal or slightly enlarged at the base; color pallid and staining brownish where fibrils are removed, turning bluish when bruised. Partial veil sometimes leaving zones near the stalk apex, but no true annulus is formed. Spores purplish in deposit, smooth, thick-walled with an apical germ pore, 9-12.4(-17.5) × 6.6-7.3(-8.0) μm in size.

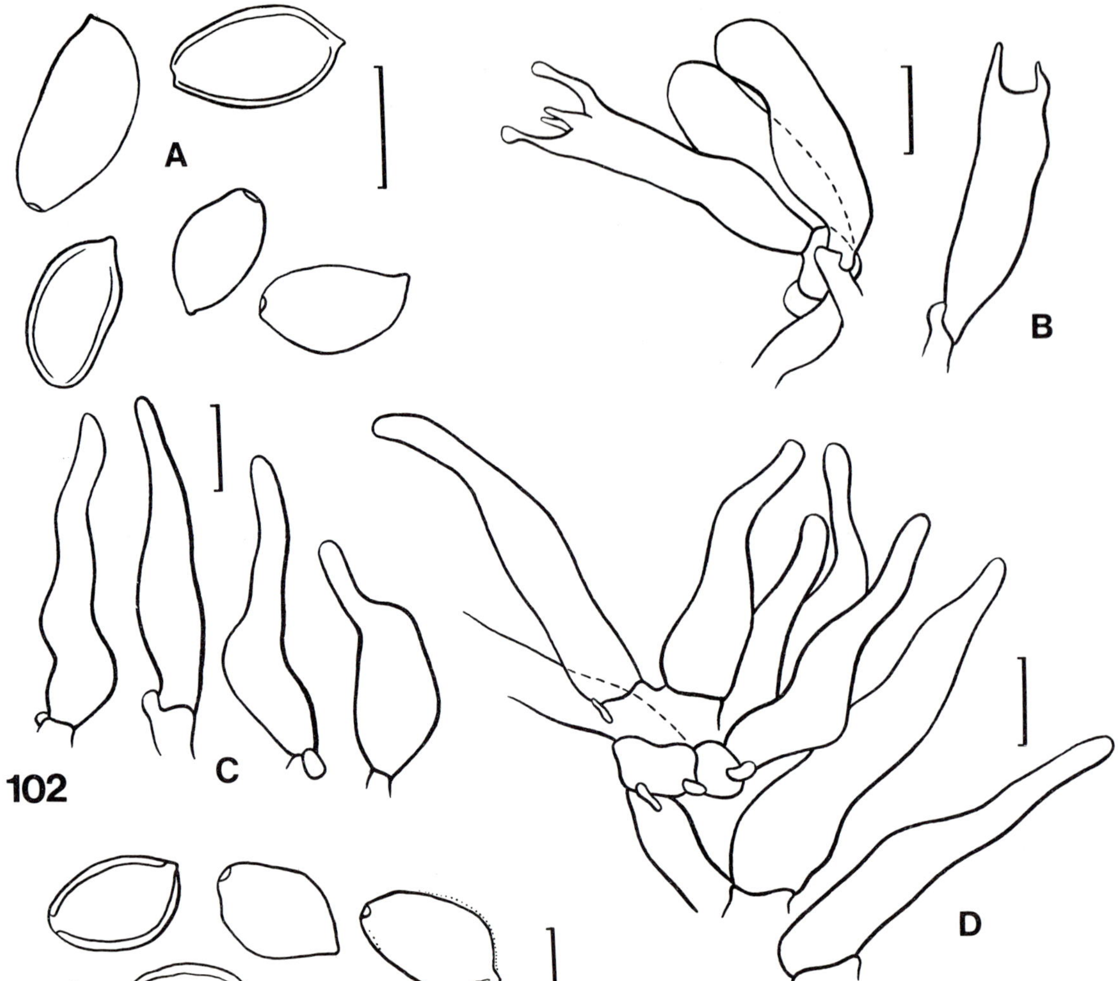

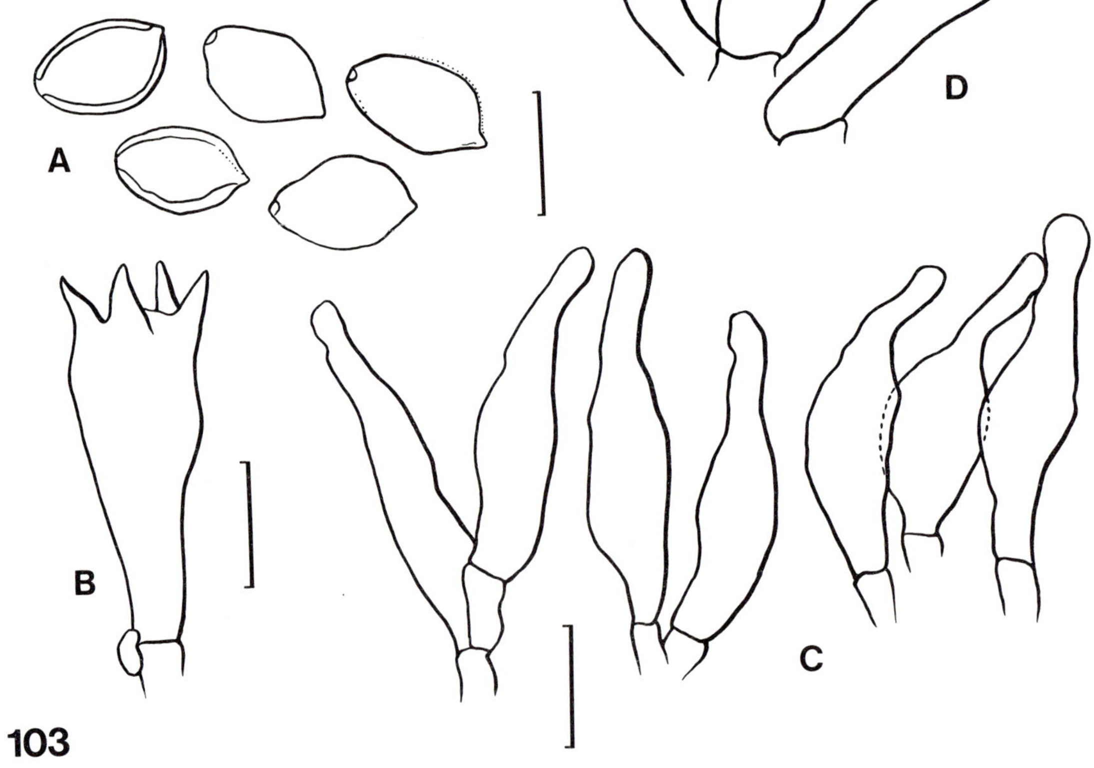

Fig. 102. *Psilocybe strictipes*: A, spores; B, basidia; C, cheilocystidia; D, caulocystidia.

Fig. 103. *Psilocybe stuntzii*: A, spores; B, basidium; C, cheilocystidia.

Scale line 10 μm

Observations. This species has been reported in Canada from British Columbia (Guzmán et al. 1976). It was originally described from Oregon and occurs in Washington as well. *P. strictipes* is another *Psilocybe* species that seems to occur primarily in the Pacific coast region. We have no records of *P. strictipes* from eastern North America.

Distribution and seasonal occurrence. *P. strictipes* has been reported on the west coast from British Columbia south to Oregon and fruits in the fall and winter.

Habitat and habit. This species is gregarious to subcaespitose and is found on decaying conifer debris, on soil that is high in woody material, on twigs and leaves, in coniferous and mixed woods, and on the edges of fields.

Technical description. Cap 20-40 mm broad, bell-shaped to convex when young, becoming broadly convex to plane or with an elevated wavy margin with age; margin finely striate when moist; surface viscid, hygrophanous; color dull yellowish brown to light olive brown, fading to cinnamon buff on the disc, pallid on the margin, becoming dingy overall, staining greenish to bluish green where bruised. Flesh the same color as the surface; taste mild; odor indiscernible.

Gills pallid becoming dark chocolate, bluntly adnate to depressed adnate, close, not crowded, narrow.

Stalk 100-130 mm long, 2-3 mm wide, equal or slightly enlarged at the apex, straight, central, stuffed, pallid from appressed fibrils on the surface, fibrillose over the lower stalk, strigose at the base, staining brownish where the fibrils are removed, turning bluish when bruised. Flesh cartilaginous, dingy brownish with a pallid brown pith. Partial veil often leaving zones near the apex of the stalk; distinct annulus lacking.

Spores purplish in deposit, 9-12.4(-17.5) × 6.6-7.3(-8.0) μm, elliptic to inequilateral in side view, oblong in face view, thick-walled, smooth; apex truncate, with an apical germ pore, yellowish brown. Basidia four-spored, clavate, 25.6-29.2 × 7.3-8.8 μm, hyaline. Pleurocystidia none or rare near the gill edges, like cheilocystidia. Cheilocystidia rounded to fusiform-ventricose or fusiform, with filiform tapered apices, 29.2-35.0 × 7.3-9.5 μm, hyaline, thin-walled; apices 2.2 μm wide.

Gill trama interwoven; hyphae more or less parallel, more or less cylindrical, 3.7-18.3 μm, hyaline. Subhymenium interwoven; hyphae cylindrical, 3.7-5.1 μm wide, hyaline to yellow brown, thin-walled. Cap cuticle gelatinous, filamentous; hyphae interwoven, more or less radially arranged, cylindrical, 3.0-3.7 μm wide, hyaline; hyphae just below the cuticle compactly interwoven, more or less parallel, cylindrical, 3.7-7.3 μm wide, hyaline to yellow brown to brown, thin-walled to thick-walled and encrusted. Cap trama loosely interwoven; hyphae cylindrical to somewhat inflated, 7.3-18.6 μm wide, hyaline. Caulocystidia narrowly flask-shaped to fusiform-lanceolate with tapered zigzagged apices, 30-44 × 7.3-9.1 μm, hyaline, thin-walled, clustered on the stalk apex; stalk hyphae parallel, more or less cylindrical, 3.7-12.2 μm wide, hyaline to yellowish. Oleiferous hyphae not recorded. Clamp connections present.

Useful references. This species has not been described in any of the mushroom field guides used here.

Psilocybe stuntzii Guzmán & Ott

Figs. 103, 115

Common names. Stuntz's psilocybe, Stuntz's blue legs.

Distinguishing characteristics. Cap 5-35 mm broad, usually bell-shaped to convex, slightly viscid to lubricous; color fulvous brown fading to ocherous or pale ocher. Gills brownish violet, chocolate brown to blackish violet, with whitish edges. Stalk 1.5-5 mm thick, equal; color whitish to ocherous. Annulus membranous, thin; partial veil remnants also often on the cap margin of young fruit bodies. Spores purple violet in deposit, smooth, with an apical germ pore, (8.2-)9.3-10.4(-12.6) × 6-7.1(-7.7) μm in size.

Observations. *P. stuntzii* is a common species in the Pacific northwest, often used as a hallucinogen in the Seattle-Olympia area. It is a very distinct species, especially when annulate specimens are seen, and is related to *P. venenata* (Imai) Imazeki & Hongo from Japan and *P. subaeruginascens* Hohnel from Java. *P. stuntzii* appears to have been incorrectly identified as *P. cyanescens* Wakefield by Weil (1975) and Haard and Haard (1976). Harris (1976) called it *P. pugetensis* Guzman, a name with no official status. Both psilocybin and psilocin have been detected in *P. stuntzii*. The technical description below is taken from that of Guzmán and Ott (1976); specimens were provided by G. Guzmán.

Distribution and seasonal occurrence. This species has been collected in British Columbia and Washington state. It fruits from August to December, depending on location and weather.

Habitat and habit. This species grows solitary or in dense clusters, in landscaped areas covered with mulch and in lawns and fields.

Technical description. Cap 5-35 mm broad, convex to bell-shaped, or somewhat nipple-shaped, sometimes becoming shallowly depressed at the disc; surface glabrous, slightly viscid or lubricous, fulvous brown fading to ocherous or pale ocher, hygrophanous, smooth or sometimes striate at the margin; margin with traces of a silky white veil adhering in young stages after the annulus is formed. Flesh white throughout, tough in the stalk, staining blue; odor and taste slightly resembling fresh meal.

Gills sinuate-adnate to short-decurrent, narrow, brownish violet, chocolate brown to blackish violet, with whitish edges.

Stalk 35-75 × 1.5-5.5 mm, central, equal, zigzagged, glabrous to slightly fibrillose, dry, whitish to ocherous, staining blue where injured, becoming hollow. Annulus membranous to scaly-membranous, thin, smooth below, slightly striate above.

Spores deep violaceous to dark purple violaceous in deposit, (8.2-)9.3-10.4(-12.6) × 6-7.1(-7.7) × 5.5-6.6 μm, smooth, elliptic in profile, more or less rhombiform in face view; apex truncate from an apical germ pore, dingy yellow brown. Basidia four-spored, 16.5-25 × 5.5-7.7 μm, subcylindrical and slightly constricted at the middle. Pleurocystidia lacking. Cheilocystidia 22-27.5 × 4.4-6.6 μm, fusiform-lanceolate or fusiform-flask-shaped, with an elongate zigzagged neck 1.6-2.2 μm in diameter, hyaline, forming a sterile band on the gill edge. Subhymenium thin, hyaline, with yellowish brown wall pigment in places; segments irregularly arranged.

Gill trama parallel; hyphae elongate-cylindrical to inflated, hyaline. Cap cuticle thin, filamentous; hyphae more or less gelatinized, 1.6-5 μm wide, hyaline to yellowish; hyphae of subcutis 5 μm or more broad, brownish to brownish red. Clamp connections present.

Useful references. *P. stuntzii* was published in 1976, but to date it has not appeared in any of the mushroom guides used here. Lincoff and Mitchel (1977) and Stamets (1978) provide a description and discussion of this species.

Stropharia (Fr.) Quél.

Strophariaceae

Stropharia is a fairly large genus and there is no adequate treatment of it for Canada or the United States. The species are generally widely distributed and there are several different kinds. To date only a few have been reported as poisonous. Certain recent treatments place *Stropharia* species in the genus *Psilocybe.*

Stropharia in general appearance resembles certain species of *Psilocybe, Naematoloma,* and *Pholiota.* All three should be carefully compared before making a final generic determination.

The characteristics that help to define *Stropharia* include a lilac to sooty lilac or a purplish smoky to purplish brown spore deposit; the presence of a partial veil, often leaving a distinct annulus or an annular zone, and sometimes leaving veil material on the cap margin; a moist to viscid cap, often becoming more or less convex with age; attached gills, either adnexed to adnate, usually not decurrent, not mottled as in *Panaeolus*; variable habitat, usually terrestrial and humicolous, sometimes with fruit bodies occurring on wood or woody debris or sawdust, less commonly on dung; smooth spores with an apical germ pore; the presence of chrysocystidia, a special type of gloeocystidium, in the gills; and an interwoven cap cuticle composed of filamentous cylindrical hyphae. The interwoven cap

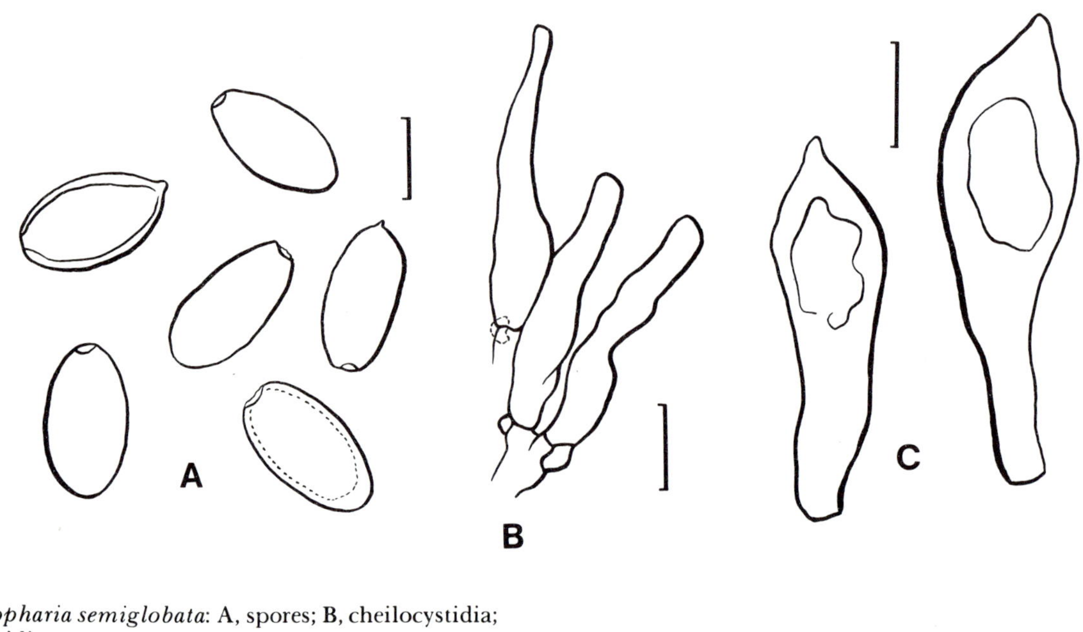

Fig. 104. *Stropharia semiglobata*: A, spores; B, cheilocystidia; C, pleurocystidia.

Scale line 10 μm

cuticle separates *Stropharia* from *Panaeolus* and most *Psathyrella* species; these other genera typically have a cellular cap cuticle as viewed under the microscope.

For a comparison of *Stropharia* with *Naematoloma* and *Psilocybe*, see the discussions presented under those genera. Certain *Pholiota* species are similar in appearance to some species of *Stropharia*. Usually the two genera can be separated on spore deposit color; the spores of *Pholiota* are typically dull brown to dark brown in deposit, at times rusty brown to yellow brown.

Stropharia semiglobata (Batsch ex Fr.) Quél.

Figs. 104, 116

Common names. Hemispherical (hemispheric) stropharia, halfglobe mushroom.

Distinguishing characteristics. Cap 10–40 mm broad, hemispherical to convex, viscid; color evenly bright pale yellow, light or straw yellow, becoming dingy or developing an olive tint at times, sometimes fading to whitish. Gills almost distant; color grayish in early stages, becoming purplish brown to dark purple brown. Stalk 2–5 mm thick, with an equal to bulbous base, and viscid below the annular zone when fresh; color white to buff or shaded with cap colors. Annulus sometimes present but often evanescent, or partial veil delicate and sometimes not leaving an annulus. Fungi usually associated with dung or manured fields. Spores smooth, with an apical germ pore, very large, 16.1–19.0 × 8.8–11.0 μm in size.

Observations. The outstanding characteristics of this species include its hemispherical, yellow, viscid cap, its viscid stalk that typically does not have a well formed annulus, and its occurrence on dung. This species, as well as *S. coronilla* and *S. aeruginosa* (Curt. ex Fr.) Quél., has been reported as poisonous, but there is no convincing evidence that *S. aeruginosa* is poisonous in North America. *S. semiglobata* has been classified as a *Panaeolus* by some researchers.

Distribution and seasonal occurrence. *S. semiglobata* is widely distributed in North America and occurs across Canada. It probably fruits throughout the year, depending on location and weather, and has been reported in the spring, summer, and fall from most areas.

Habitat and habit. This species occurs solitary or in groups, on dung, in manured soil, on lawns, and often in livestock corrals.

Technical description. Cap 10–40 mm broad, obtuse to hemispherical, becoming convex to nearly plane; surface glabrous, viscid; color evenly bright pale yellow, light yellow, or straw yellow in early stages, fading to dull yellow or dingy yellow with an olive tint, sometimes fading to whitish. Flesh thick on the disc, thin on the margin, soft and watery, buff to very pale yellowish; taste bitterish or indistinctive; odor indistinctive.

Gills grayish when young, becoming sordid purplish brown to dark purple brown with age, adnate with a slight decurrent tooth, almost distant, very broad, 6–8 mm, becoming ventricose; edges white, fimbriate.

Stalk 30–120 mm long, 2–5 mm wide, equal above, bulbous at the base, central, stuffed becoming hollow, silky-hairy above the annulus, viscid on the lower portion, white to buff or almost the same color as the cap surface. Flesh white with yellowish pith, at times sordid yellowish at the base. Annulus very delicate, often forming an apical evanescent fibrillose zone, white. Universal veil lacking.

Spores dark purple brown in deposit, 16.1–19.0 × 8.8–11.0 μm, ellipsoidal, smooth, thick-walled, truncate at the apex with a small apical germ pore, yellow brown. Basidia four-spored, clavate, 33–40 × 13–14.6 μm, hyaline. Pleurocystidia ventricose, thin-walled; apex obtuse to mucronate; contents yellow, crystalline in potassium hydroxide. Cheilocystidia filiform-cylindrical to fusiform with tapered to nearly capitate apices, 30–44 × 3.7–4.1 μm, hyaline, thin-walled.

Gill trama interwoven to parallel; hyphae cylindrical, 3.7–6.6 μm wide, hyaline. Subhymenium hyphae interwoven, cylindrical, 3.7–4.5 μm wide, hyaline. Cap cuticle gelatinous, filamentous; hyphae interwoven, cylindrical, 2.2–3.7 μm wide, hyaline to yellowish. Cap trama loosely interwoven; hyphae more or less cylindrical, 4.0–12.6 μm wide, hyaline to yellowish. Caulocystidia filiform-cylindrical, 30–40 × 2.2–3.7 μm, hyaline; stalk hyphae parallel in the core to interwoven on the surface, cylindrical, 2.7–8.8 μm wide, hyaline. Annulus composed of interwoven hyaline to yellowish cylindrical hyphae 3.7–5.1 μm wide. Oleiferous hyphae greenish yellow, refractive, 3.7–5.1 μm wide, scattered in the cap cuticle and trama. Clamp connections present.

Useful references. The following books contain information on *S. semiglobata* pertinent to North America.

- *Edible and Poisonous Mushrooms of Canada* (Groves 1979)—A color photograph, as well as a description and discussion, is included.
- *Mushrooms in their Natural Habits* (Smith 1949)—A description and discussion, as well as a color photograph on a reel, are presented.
- *Mushrooms of North America* (Miller 1972)—A description and discussion are given.

Literature

Aaron, J. J.; Sanders, B.; Winefordner, J. D. Analytical study of some important hallucinogens by a combined fluorimetric and phosphorimetric method. Clin. Chem. Acta 45:375-386; 1973.

Benedict, R. G.; Brady, L. R.; Smith, A. H.; Tyler, V. E., Jr. Occurrence of psilocybin and psilocin in certain *Conocybe* and *Psilocybe* and *Psilocybe* species. Lloydia 25:156-159; 1962.

Brown, J. K.; Shapazian, L.; Griffin, G. D. A rapid screening procedure for some "street drugs" by thin-layer chromatography. J. Chromatogr. 64:129-133; 1972.

Buck, R. W. Psychedelic effect of *Pholiota spectabilis.* New Engl. J. Med. 267:391; 1967.

Cashman, P. J.; Thornton, J. I. A specific microcrystalline test for indolamine derivatives. Microchem. J. 20:511-518; 1975.

Folen, V. A. X-ray powder diffraction data for some drugs, excipients, and adulterants in illicit samples. J. Forensic Sci. 20:348-372; 1975.

Groves, J. W. Edible and poisonous mushrooms of Canada. Ottawa, Ont.: Agriculture Canada. Agric. Can. Publ. 1112; 1979 [first printed in 1962].

Guzmân, G. Los Especies Conocidas del Genero *Panaeolus* en Mexico. Bol. Soc. Mex. Micol. 6:17-53; 1972.

Guzmân, G.; Ott, J. Description and chemical analysis of a new species of hallucinogenic *Psilocybe* for the Pacific northwest. Mycologia 68:1261-1267; 1976.

Guzmân, G.; Ott, J.; Boydston, J.; Pollock, S. H. Psychotropic mycoflora of Washington, Idaho, Oregon, California and British Columbia. Mycologia 68:1267-1272; 1976.

Guzmân, G.; Vergeer, P. P. Index of taxa in the genus *Psilocybe.* Mycotaxon 6(3):464-476; 1978.

Haard, R.; Haard, K. Poisonous and hallucinogenic mushrooms. Seattle, WA: Cloudburst Press; 1976.

Harris, B. Growing wild mushrooms. Berkeley, CA: Wingbow Press; 1976.

Hesler, L. R. North American species of *Gymnopilus.* New York, NY: Hafner Publ. Co.; 1969.

Hollister, L. E.; Prusmack, J. J.; Paulsen, J. A.; Rosenquist, N. Comparison of three psychotropic drugs (psilocybin, JB-329 and IT-290) in volunteer subjects. J. Nerv. Ment. Dis. 131:428-434; 1960.

Lincoff, G.; Mitchel, D. H. Toxic and hallucinogenic mushroom poisoning—A handbook for physicians and mushroom hunters. New York, NY: Van Nostrand Reinhold Co.; 1977.

Masoud, A. N. Systematic identification of drugs of abuse: I. Spot tests. J. Pharm. Sci. 64:841-844; 1975.

McKenny, M. The savory wild mushroom (revised and enlarged by Stuntz, D. E.). Seattle, WA: University of Washington Press; 1971.

Miller, O. K., Jr. Mushrooms of North America. New York, NY: E. P. Dutton & Co., Inc.; 1972.

Murrill, W. A. A very dangerous mushroom. Mycologia 8:186-187; 1976.

Ola'h, G. M. Le genre *Panaeolus.* Mus. Natl. Hist. Nat., Paris; 1969.

Ola'h, G. M.; Heim, R. Une nouvelle espèce nord-américaine de *Psilocybe* hallucinogène: *Psilocybe quebecensis.* C.R. Hebd. Sci. Acad. Sci., Sér. D264:1601-1604; 1967.

Ott, J. Recreational use of hallucinogenic mushrooms in the United States. Salzman, E., ed. Mushroom poisoning. Cleveland, OH: CRC (Chemical Rubber Co.) Press; 1976.

Robbers, J.; Tyler, V. E.; Ola'h, G. M. Additional evidence supporting the occurrence of psilocybin in *Panaeolus foenisecii.* Lloydia 32:399-400; 1969.

Saupe, S. G. Occurrence of psilocybin/psilosin in *Pluteus salicinus* (Pluteaceae). Mycologia 73:781-784; 1981.

Singer, R.; Smith, A. H. Mycological investigations on teonanacatl, the Mexican hallucinogenic mushroom: Part II. A taxonomic monograph of *Psilocybe,* section *Caerulescentes.* Mycologia 50:262-303; 1958.

Singer, R. Mycoflora australis. Nova Hedwigia 29:1-405; 1969.

Singer, R. The agaricales in modern taxonomy. 3d rev. ed. Lehre, Germany: J. Cramer; 1975.

Smith, A. H. Mushrooms in their natural habitats. 2 Vols. Portland, OR: Sawyer's, Inc.; 1949.

Smith, A. H. The North American species of *Psathyrella.* Mem. N.Y. Bot. Gard. 24:1-633; 1972.

Smith, A. H. A field guide to western mushrooms. Ann Arbor, MI: University of Michigan Press; 1975.

Smith, A. H.; Smith, H. V. How to know the gilled mushrooms. Dubuque, IA: William C. Brown Co.; 1973.

Stamets, P. Psilocybe mushrooms and their allies. Seattle, WA: Homestead Book Co.; 1978.

Stamets, D. E.; Beug, M. W.; Bigwood, J. E.; Guzmân, G. A new species and a new variety of *Psilocybe* from North America. Mycotaxon 11(2):476-484; 1980.

Walters, M. B. *Pholiota spectabilis,* a hallucinogenic fungus. Mycologia 57:837; 1965.

Watling, R. A *Panaeolus* poisoning in Scotland. Mycopathologia 61(3):187-190; 1977.

Weil, A. The natural mind. Boston, MA: Houghton Mifflin; 1972.

Zabik, J. E.; Maickel, R. P. 1974. Relevance of street drug analyses in the forensic laboratory to clinical toxicology of drug abuse. Drug Addict 4:203-217; 1974.

Fig. 105. *Conocybe smithii.*

Fig. 106. *Gymnopilus spectabilis.*

Fig. 107. *Panaeolus campanulatus.*

Fig. 108. *Panaeolus foenisecii.*

Fig. 109. *Panaeolus sphinctrinus.*

Fig. 110. *Panaeolus subbalteatus.*

Fig. 111. *Psilocybe baeocystis.*

Fig. 112. *Psilocybe cyanescens.*

Fig. 113. *Psilocybe pelliculosa.*

Fig. 114. *Psilocybe semilanceata.*

Fig. 115. *Psilocybe stuntzii.*

Fig. 116. *Stropharia semiglobata.*

13

GASTROINTESTINAL IRRITANTS

Toxicology and Symptomatology

Many fungi cause gastrointestinal irritation or disturbance. Some are consistent offenders, whereas others seem to affect only certain individuals. Species that are harmless to some but harmful to others, causing acute illness, are considered partial offenders by Lincoff and Mitchel (1977). Some fungi are toxic when eaten raw or improperly prepared. Others cause problems when eaten in large quantities. For some species the toxins have been isolated and identified; for others they remain unknown. Long cooking, parboiling, or special methods of preservation such as packing in salt apparently destroy or inactivate some toxins or irritants. However, some fungi are toxic regardless of the manner of preparation.

Hatfield and Brady (1975) suggest that gastrointestinal disturbances and hypersensitive responses are probably the most common adverse effects arising from the ingestion of mushrooms and other fungi. Symptoms normally include nausea with vomiting or diarrhea or both, often accompanied by abdominal pain. Poisonings range from mild to severe; rarely, a few species have been reported to have caused human deaths.

Symptoms usually appear within 15 minutes to 4 hours, although with certain species such as *Gomphus floccosus* they have a delay of 8–14 hours. They are typically transient, and often terminate spontaneously or soon after the ingested material is removed from the intestinal tract (Hatfield and Brady 1975, Lincoff and Mitchel 1977). Certain poisonous species such as *Tricholoma pardinum* and *Entoloma lividum* cause such severe poisoning that recovery may take several days. *Entoloma lividum* has been reported as also causing liver damage (Tyler 1963).

Symptom-producing Species

BASIDIOMYCOTA

AGARICALES

Agaricaceae
- *Agaricus*
  - *A. albolutescens*
  - *A. arvensis*
  - *A. californicus*
  - *A. glaber*
  - *A. hondensis*
  - *A. meleagris*
  - *A. placomyces*
  - *A. silvaticus*
  - *A. subrufescentoides*
  - *A. sylvicola*
  - *A. xanthodermus*

Amanitaceae
- *Amanita* spp.

Boletaceae
- *Boletus*
  - *B. bicolor*
  - *B. calopus* var. *frustosus*
  - *B. erythropus*
  - *B. huronensis*
  - *B. luridus*
  - *B. miniato-olivaceus*
  - *B. pulcherrimus* (=*B. eastwoodiae*)
  - *B. satanas*
  - *B. sensibilis* (=*B. miniato-olivaceus* var. *sensibilis*)
  - *B. subvelutipes* (suspected)
- *Tylopilus*
  - *T. felleus*

Cortinariaceae
- *Hebeloma*
  - *H. crustuliniforme*
  - *H. fastibile*
  - *H. mesophaeum*
  - *H. sinapizans*
- *Phaeolepiota*
  - *P. aurea*
- *Pholiota*
  - *P. abietis* (perhaps =*P. limonella*)
  - *P. hiemalis*
  - *P. highlandensis*
  - *P. squarrosa*

Lepiotaceae
- *Lepiota*
  - *L. molybdites*
  - *L. naucina*

Paxillaceae
- *Paxillus*
  - *P. involutus*

Rhodophyllaceae (=Entolomataceae) (rhodophylloid mushrooms)
- *Entoloma*
  - *E. albidum*
  - *E. lividum*
  - *E. nidorosum*
  - *E. rhodopolium*

Nolanea
N. verna

Russulaceae
Lactarius
L. chrysorheus
L. deceptivus
L. piperatus var. *glaucescens*
L. piperatus var. *piperatus*
L. repraesentaneus
L. rufus
L. scrobiculatus var. *canadensis*
L. scrobiculatus var. *scrobiculatus*
L. subvellereus var. *subdistans*
L. torminosus
L. uvidus
L. vellereus
Russula
R. emetica

Strophariaceae
Stropharia
S. coronilla

Tricholomataceae
Armillariella
A. mellea (commonly eaten but not tolerated by some)
Clitocybe
C. irina
C. nebularis
Collybia
C. dryophila
Mycena
M. pura
Omphalotus
O. illudens
O. olearius
O. olivascens
O. subilludens
Tricholoma
T. inamoenum
T. pardinum
T. pessundatum
T. venenatum
Tricholomopsis
T. platyphylla

APHYLLOPHORALES

Cantharellaceae
Gomphus
G. floccosus

Clavariaceae
Ramaria
R. flavobrunnescens
R. formosa (suspected)
R. gelatinosa var. *gelatinosa*
R. gelatinosa var. *oregonensis*
R. pallida

Polyporaceae (polypores)
Albatrellus
A. dispansus
Laetiporus
L. (Polyporus) sulphureus

SCLERODERMATALES

Sclerodermataceae
Scleroderma
S. cepa
S. citrinum

ASCOMYCOTA

PEZIZALES

Helvellaceae
Helvella
H. lacunosa

Morchellaceae (commonly eaten but not tolerated by some)
Morchella
M. angusticeps
M. conica
M. esculenta
Verpa
V. bohemica

Pezizaceae
Sarcosphaera
S. crassa

Only those species that are the most consistently implicated in gastrointestinal disturbances are listed above, alphabetically, according to order, family, and genus, with the Basidiomycota preceding the Ascomycota. Most of the species listed here occur or are likely to occur in Canada or in areas of the United States adjacent to Canada. Descriptions of all the genera follow in alphabetic order, with general discussions for the polypores and the rhodophylloid mushrooms appearing under their common names. Descriptions of species known to be consistent offenders and of importance in Canada are included for each genus, where warranted. The discussions for each genus provide explanations concerning the relative importance of species.

Descriptions of Genera and Species

Agaricus L. ex Fr.

Agaricaceae

Most people who are interested in mushrooms for food are acquainted with the genus *Agaricus*. It includes the common cultivated mushroom, *Agaricus bisporus* (Lange) Möller & Schaeff. (=*Agaricus brunnescens* Peck), which is sold commercially, and the meadow mushroom, *Agaricus campestris* Fr., which is collected for food in Europe and North America.

Agaricus is characterized by medium to large, typically stout and fleshy fruit bodies, although a few species in the *Agaricus deminutivus* Peck group are small; a fibrillose to scaly cap surface in some species, but glabrous in others; usually close to crowded gills, narrow to moderately broad, and free or nearly so, although the gills approach the stalk very closely and at times appear to be nearly attached, white to pinkish to grayish white in the early stages, and becoming dark purplish brown to blackish brown or chocolate brown from the spores at maturity; a usually fairly thick and fleshy stalk; typically no universal veil and the presence of a partial veil that often leaves a distinct annulus, although the annulus may be lost in some species; smooth, often ellipsoidal spores, usually without an apical germ pore but, when present, usually indistinct, and blackish brown to chocolate brown to reddish brown or purplish brown in deposit; and a cap cuticle composed of filamentous hyphae.

Because of the free gills and the annulus, species of *Agaricus* may be confused with certain species of *Amanita* or *Lepiota*. *Amanita* species have a pale, usually white spore deposit, a universal veil, and a divergent gill trama. The larger species of *Lepiota* most closely resemble *Agaricus*. The color of the spore deposit of *Lepiota* is usually pale, often white, except for that of the poisonous *Lepiota molybdites*, which is green. Some of the larger species of *Stropharia* are also similar in appearance to *Agaricus*. Refer to the description of *Stropharia* later in this chapter, alphabetically listed. The genus *Psalliota* is used in place of *Agaricus* by some authors.

Many deaths have no doubt occurred because collectors are not aware of the deadly species that look like the edible species they are seeking (Ch. 6, Table 2). This problem is especially important for collectors of *Agaricus* because often the species sought for food are white and similar in stature to the deadly, white or pale-colored *Amanita* species. When the collector searching for a white species of *Agaricus* encounters instead a deadly white *Amanita* and does not extract it from the ground carefully to check for a volva, an initial mistake in identification can easily be made. Unless the collector determines the color of the spore deposit, white in *Amanita* and dark brown in *Agaricus*, the error is compounded and a fatal poisoning may result.

When collecting *Agaricus* and other white or pale-colored mushrooms such as *Lepiota*, one simply cannot be too careful. Remove each specimen from the substrate carefully and check for the volva, representing the remains of a universal veil. Spore deposits must then be taken as a second precaution. These rules are especially important when collecting in a wooded area or among scattered trees, because *Amanita* species are typically associated with the roots of trees.

Although the genus *Agaricus* contains a number of good, edible species, it also contains several other species that have been reported as poisonous. Because the identification of *Agaricus* species is difficult and the names in many instances have been incorrectly applied, the number and identity of species that are poisonous are difficult to determine.

In the literature several species of *Agaricus* are reported to have caused gastrointestinal upset. *Agaricus bisporus* (=*A. brunnescens*), for example, the common cultivated mushroom, can be safely eaten in quantity by most people, yet it is known to cause acute but usually brief digestive upset in people who are sensitive to it. People sensitive to *A. bisporus* may be sensitive to other kinds of mushrooms as well. Species commonly listed as poisonous include *A. albolutescens*, *A. arvensis*, *A. glaber*, *A. meleagris*, *A. placomyces*, *A. sylvicola* (also spelled *silvicola*), *A. silvaticus*, and *A. xanthodermus*. Besides these, Kerrigan (1978) lists *A. californicus* Peck, *A. hondensis*, and *A. subrufescentoides* Murrill as poisonous. He states that species with a strong odor of creosote, phenol, or ink are in general poisonous. He adds that those that stain yellow when cut or bruised should also be avoided. He also suggests that poisonous species turn bright yellow when caustic soda is applied to the fruit body.

The species most often reported as causing gastrointestinal upset fall into two main groups, the *Agaricus sylvicola* group and the *Agaricus placomyces* group. Discussions of these two groups below are followed by a general description of a third group, which is centered around *Agaricus xanthodermus*. Descriptions of selected species follow these discussions in alphabetic order.

The *Agaricus sylvicola* group, which includes *A. arvensis*, contains species that are colored white overall, at least when young, and stain yellow when bruised or broken. The cap surface is fibrillose (innate to appressed) or fibrillose-scaly, and the fibrils and scales are not colored. The stalk usually has a conspicuous, persistent annulus, and the mushrooms are typically large and often robust. Some forms of these species produce an anise- or almond-like odor. The species in this group also often stain yellow when touched with a drop of potassium hydroxide, made up in a 5% aqueous solution.

Species in the *Agaricus sylvicola* group, including *A. arvensis,* do not seem to be as consistently reported as being poisonous as those in the *A. placomyces* group. In fact, they are fairly commonly eaten. However, they cannot be tolerated by some people and produce adverse reactions when eaten. Although we can reasonably assume that these reports of poisonings represent examples of simple mushroom intolerance among certain individuals, these species are so widespread in North America that it is also possible that certain forms or races contain toxic compounds. The presence of toxins, however, has not been documented.

Species in the *Agaricus placomyces* group are difficult to identify because of the uncertainty surrounding their correct concepts. We have included in this group *A. hondensis, A. placomyces,* and *A. silvaticus. A. meleagris* is discussed under *A. placomyces.*

Characteristics that aid in identifying the *Agaricus placomyces* group as treated here include a more or less dense to somewhat sparse coating of fibrils or scales on the cap surface, which are colored brown, brownish orange, reddish brown, vinaceous brown, grayish brown, grayish vinaceous, grayish, or in some instances blackish; flesh staining yellowish when cut, then becoming dark pink, dingy vinaceous, or reddish brown (change may occur slowly), or staining directly pinkish to reddish brown; a well-developed annulus, usually with more or less floccose patches on the undersurface; and at times the presence of an odor of creosote, not detectable in all specimens.

Two other *Agaricus* species reported as poisonous are *A. albolutescens* Zeller and *A. xanthodermus* Genevier. *A. xanthodermus* was reported from western North America (McKenny 1971; B. Isaacs, Personal commun.). *A. albolutescens* was described from the Pacific northwest. Both species are white to whitish, at least when young, and quickly stain an intense yellow when bruised. *A. albolutescens,* when fresh, has a white cap that is soon stained brilliant yellow to orange yellow; usually the entire cap surface becomes orange yellow.

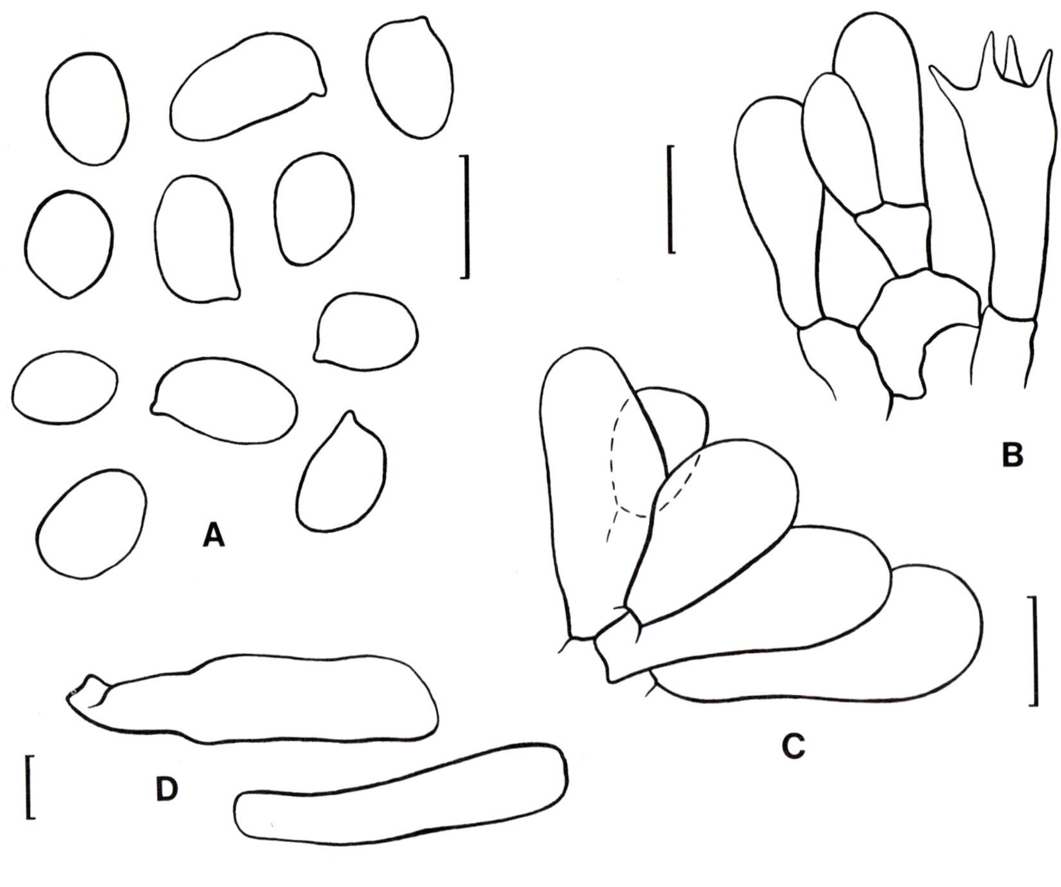

Fig. 117. *Agaricus arvensis*: A, spores; B. basidia; C, cheilocystidia; D, annulus elements.

Scale line 10 μm

The stalk surface, as well as the cap flesh, also stains yellow. The gills are described as pale grayish pink in the early stages, becoming dark grayish brown. The annulus is thick and membranous, with patches on the undersurface. It often has an odor of anise or almonds.

Agaricus xanthodermus (Fig. 177) is similar in appearance to *A. albolutescens,* but it also can be mistaken for *A. sylvicola* or *A. arvensis.* It is generally white in the early stages and stains yellow on all parts when bruised but apparently does not become as stained or discolored as *A. albolutescens. A. xanthodermus* is reported to have an unpleasant odor, either fetid or resembling creosote.

Agaricus arvensis Schaeff. ex Secr.

Figs. 117, 178

Common names. Field mushroom, plowed-land mushroom, prairie mushroom, horse mushroom.

Distinguishing characteristics. Cap usually fleshy and large, 40-200 mm broad, ovoid in early stages, becoming convex to plane when expanded; surface dry, sometimes somewhat scaly but usually appressed to innately fibrillose; color white overall or creamy to yellowish or yellowish brown in the center (usually with age). Surface and flesh of cap and stalk staining yellowish when cut or bruised. Gills free and fairly remote from the stalk, close, rather broad; color white or faintly grayish in early stages, slowly changing to dull pinkish vinaceous brown and finally dark blackish brown. Stalk 50-200 mm long, 10-20 mm thick, sometimes equal but usually somewhat clavate at the base; surface glabrous and silky; color white, yellowish-tinged with age or staining yellow when bruised. Annulus membranous, thick with cottony patches on the undersurface, white or tinged yellowish with age, persistent, sometimes also with remnants of the partial veil on the cap margin. Spores purplish brown to blackish brown in deposit, smooth, ellipsoidal to ovoid, thick-walled, lacking an apical germ pore, (6.6-)7.3-10.6(-11.7) × (4.4-)5.0-6.0 μm in size.

Observations. The large size, the white color overall when young, and the yellow stains help to distinguish this species. It is typically more robust and fleshy than *A. sylvicola* and has longer spores. Compare with the description of *A. sylvicola.*

A. arvensis var. *palustris* Smith has spores that are typically more ovoid than those of var. *arvensis;* also, var. *palustris* occurs in low wooded areas, whereas var. *arvensis* typically grows in meadow, lawns, or pastures. In southern Ontario a sylvan form is fairly common in low areas under hardwood trees or in mixed woods. Apparently both varieties have caused illness (Smith 1949; D. M. Simons, Personal commun.; D. Malloch, Personal commun.). *A. arvensis* may have an odor of anise, but this characteristic does not seem to be consistent.

Distribution and seasonal occurrence. *A. arvensis* has been reported in eastern and western North America and probably occurs across Canada, wherever a suitable habitat is available. McKenny (1971) states that this species, once common in the Puget Sound region, is now seldom found. It fruits in the summer and fall.

Habitat and habit. This species is terrestrial, gregarious, or scattered to solitary. It occurs in pastures, meadows, and lawns, as well as in wooded areas under hardwood trees and in mixed woods.

Technical description. Cap 40-200 mm broad, when young ovoid to subcylindrical, expanding to convex or plane; surface dry, not hygrophanous, opaque, innately fibrillose, sometimes appressed fibrillose-scaly; edge of margin often decorated with white veil remnants; color white overall or creamy to yellowish or yellowish brown on the disc, staining yellow when bruised. Flesh thick, up to 10 mm, firm, white, when exposed slowly tinged yellowish or staining yellowish near the surface when bruised; taste mild or very slightly resembling almonds; odor indiscernible or resembling anise.

Gills free and remote from the stalk, close, rather broad, 8-12 mm, nearly equal, with more or less even edges, white or faintly grayish in early stages, slowly changing to dull pinkish vinaceous brown, finally becoming dark blackish brown.

Stalk 50-200 mm long, 10-30 mm thick, equal above a slightly clavate base, central, stuffed becoming hollow, dry, glabrous and silky above and below the annulus, white, staining yellowish when bruised or becoming yellowish-tinged with age. Flesh white, when exposed slowly tinged yellowish. Annulus membranous, double, apical on the stalk, smooth and silky on the upper surface, with cottony patches beneath, white or tinged yellowish with age. Universal veil absent.

Spores purplish brown to blackish brown in deposit, (6.6-)7.3-10.6(-11.7) × (4.4-)5.0-6.0 μm, ellipsoidal to ovoid, adaxially flattened in side view, thick-walled, smooth, nonamyloid, dark grayish brown, lacking an apical germ pore. Basidia four-spored, short-clavate, 21.9-24.8 × 7.3-8.0 μm, hyaline to brownish. Pleurocystidia absent. Cheilocystidia sac-like to clavate, 14.6-26.0 × 8.8-13.1(-15.0) μm, hyaline to brownish.

Gill trama interwoven; hyphae branched, more or less cylindrical, 2.9-11.0 μm wide, hyaline to brownish. Subhymenium filamentous to nearly cellular, interwoven; hyphae more or less cylindrical, 4.4-6.2 μm wide, hyaline. Cap cuticle filamentous; hyphae interwoven to more or less radially arranged, cylindrical, 3.7-11 μm wide, hyaline to brownish yellow, thin-walled. Cap trama loosely interwoven; hyphae cylindrical to slightly inflated, 7.3-18.3 μm wide, hyaline. Caulocystidia lacking. Stalk hyphae parallel, more or less cylindrical, 3.7-13.1 μm wide, hyaline to yellowish. Annulus interwoven; upper layer compact; hyphae of the upper layer cylindrical, 3.7-8.0 μm wide, hyaline to yellowish; lower layer floccose; hyphae of the lower layer 5-15 μm wide, separating into cylindrical segments 15-40 μm long, hyaline to yellowish. Oleiferous hyphae refractive, greenish yellow, 5.8-7.3 μm wide, scattered in the cap trama and cuticle, and gill trama. Clamp connections not seen.

Useful reference. The following book contains information on *A. arvensis* pertinent to North America.

- *Mushrooms of North America* (Miller 1972)—A color photograph, as well as a description and discussion, are included.

Smith (1940) provides a description of *A. arvensis* var. *palustris*.

Agaricus hondensis Murrill

Figs. 118, 179

Common name. Felt-ringed agaricus.

Distinguishing characteristics. Cap 80-150 mm broad, hemispherical to convex in early stages, becoming flattened at maturity; surface dry; color variable, typically grayish vinaceous, pale fawn, pale lilac brown to vinaceous brown, or grayish reddish brown, with appressed fibrils and scales over a white to ivory ground color; fibrils and scales darkening with age; surface generally darker overall with age. Flesh staining yellowish, then becoming sordid dark pink to dingy vinaceous; odor indistinctive or resembling creosote. Gills free, close to crowded, narrow; color pink to pale grayish lilac, becoming more or less deep reddish brown. Stalk 80-140 mm long, 10-20 mm thick, usually with an abrupt bulbous base; color white, gradually darkening to grayish red. Annulus membranous, moderately thick; upper surface white; lower surface typically with loose dark pink to grayish vinaceous fibrillose patches or sometimes merely white-fibrillose. Spores purplish brown to chocolate brown in deposit, broadly ellipsoidal, smooth, thick-walled, without an apical germ pore, (5.1-)5.8-7.3(-8.8) × 3.7-4.4 μm in size.

Observations. This species is somewhat difficult to define and descriptions in the literature are sometimes contradictory. The colored fibrils on the cap surface usually darken with age, but some forms have a very pale cap when they are young. With age, the cap surface usually becomes dull grayish reddish brown to vinaceous brown because of a darkening of both the background color and the fibrils. Consequently, *A. hondensis* (and especially its mature fruit bodies) has often been confused with *A. silvaticus* and probably *A. placomyces*.

Some specimens of *A. hondensis* have a strong odor of creosote, which becomes more evident when the mushroom is cooked; the taste becomes unpleasant, soapy and metallic, with a slight taste of creosote (McKenny 1971). A case of poisoning was reported for this species by Dr. H. D. Thiers (Personal commun.).

A. glaber Zeller is considered to be the same species as *A. hondensis*. It has also been reported as poisonous (Zeller 1938). Smith (1975) lists *A. hillii* Murrill, *A. mcmurphyi* Murrill, and *A. bivelatoides* Murrill as synonyms for *A. hondensis*. He also states that *A. hondensis* has probably been confused with *A. silvaticus* in the past.

Distribution and seasonal occurrence. This species occurs along the Pacific coast from California to British Columbia. It fruits in the fall.

Habitat and habit. This species is scattered to gregarious and occurs under conifers or in mixed hardwood-coniferous forests.

Technical description. Cap 80-150 mm broad, hemispherical to convex when young, with the disc soon becoming flattened, expanding to broadly convex or plane, with the disc sometimes appearing slightly depressed as the margin becomes elevated, sometimes slightly umbonate; surface dry, not hygrophanous, opaque, innately fibrillose, at times with the fibrils aggregated into fascicles near the margin; surface with flat appressed fibrils or scales or appearing glabrous to the naked eye; background color nearly white to ivory in early stages, with fibrils and scales grayish vinaceous, pale fawn, or pale lilac brown, usually darkest on the disc; fibrils and scales gradually darkening to dull grayish reddish brown or vinaceous brown; surface darker overall with age. Flesh white, usually staining yellowish and then pinkish to sordid dark pink or dingy vinaceous where bruised; taste indistinctive; odor indistinctive or resembling creosote.

Gills free and rather remote from the apex of the stalk, close to crowded, narrow, equal, pink to pale grayish lilac in early stages, becoming more or less deep reddish brown at maturity; edges even.

Stalk 80-140 mm long, 10-20 mm thick at the apex, with a rather abrupt bulbous base, central, dry, innately silky above and below the annulus, white, gradually darkening to grayish red. Flesh similar in color to that of the cap. Annulus membranous, moderately thick, apical on the stalk; upper surface white and striate; lower surface typically with loose dark pink to grayish vinaceous cottony fibrillose patches, or merely loosely white-fibrillose. Universal veil lacking.

Spores purplish brown to chocolate brown in deposit, (5.1-)5.8-7.3(-8.8) × 3.7-4.4 μm, broadly ellipsoidal, slightly inequilateral in side view, smooth,

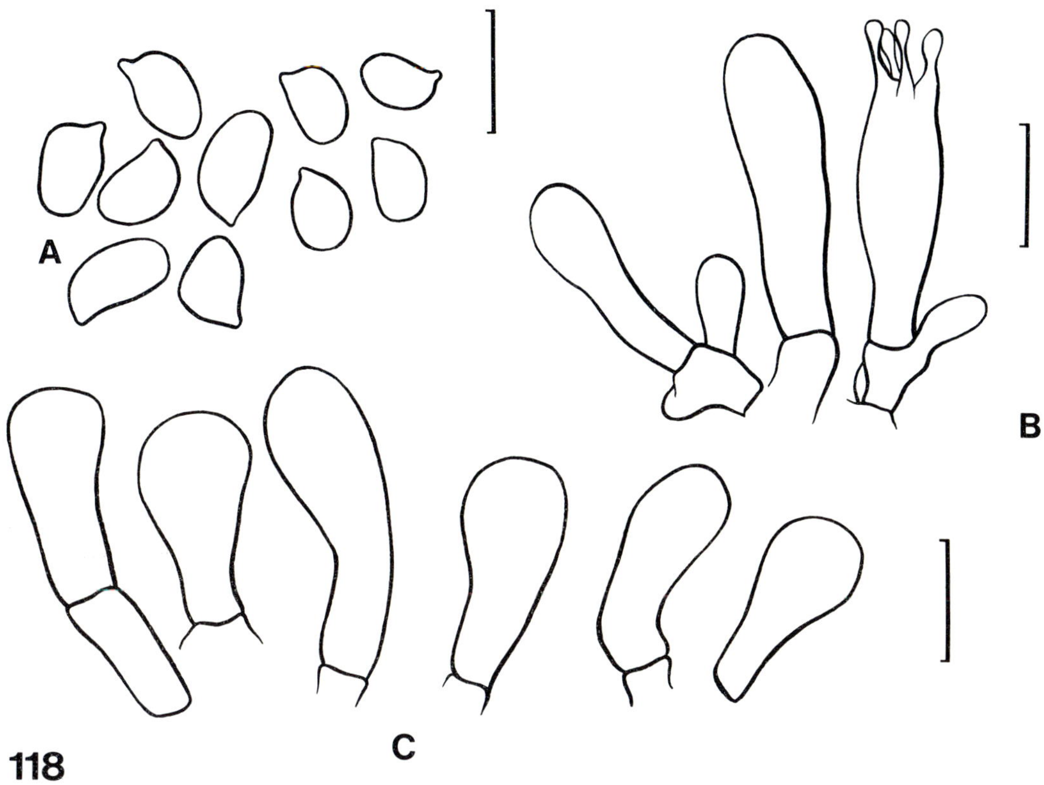

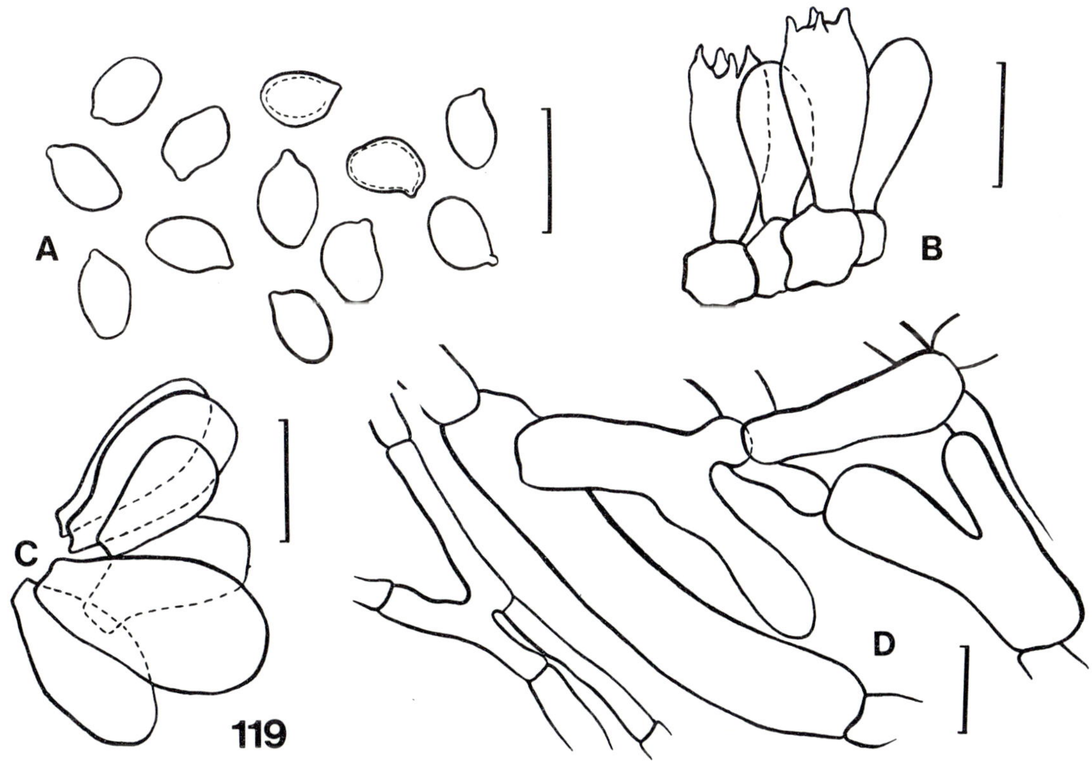

Fig. 118. *Agaricus hondensis*: A, spores; B, basidia; C, cheilocystidia.

Fig. 119. *Agaricus placomyces*: A, spores; B, basidia; C, cheilocystidia; D, annulus elements.

Scale line 10 μm

thick-walled, without an apical germ pore, dull brown. Basidia four-spored, clavate, 20–21.3 × 5.8–7.0 μm, hyaline. Pleurocystidia lacking. Cheilocystidia sac-like to clavate, 18.3–25.6 × 7.3–11.0 μm, hyaline to pale yellowish brown.

Gill trama tightly interwoven to more or less parallel; hyphae more or less cylindrical, 3.5–11.7 μm wide, hyaline. Subhymenium filamentous to subcellular; hyphae cylindrical, 3.7–8.7 μm wide, hyaline. Cap cuticle filamentous; hyphae interwoven, more or less radially arranged, parallel in fibrils, cylindrical, 4–11.0 μm wide; contents hyaline to brown, especially in the fibrils. Cap trama loosely interwoven; hyphae cylindrical to inflated, (3.7–)5–15.0(–18.3) μm wide, hyaline or with yellowish brown walls, especially near the cap cuticle. Caulocystidia lacking; stalk hyphae parallel to interwoven, cylindrical, 5–11 μm wide, hyaline. Annulus loosely interwoven; hyphae branched, cylindrical, 3.7–10.0 μm wide, hyaline to brownish. Oleiferous hyphae refractive, yellowish to pale brownish, 3.7–7.3 μm, scattered in the cap cuticle, gill trama, and stalk. Clamp connections not seen.

Useful references. The following books contain information on *A. hondensis* pertinent to North America.

- *A Field Guide to Western Mushrooms* (Smith 1975)—A color photograph and a description and discussion are presented.
- *Mushrooms in their Natural Habitats* (Smith 1949)—A description and discussion, as well as a color photograph on a reel, are presented.
- *The Savory Wild Mushroom* (McKenny 1971)—A black-and-white photograph and a description and discussion are given.

Agaricus placomyces Peck

Figs. 119, 180

Common names. Flat-capped agaricus, flat-top mushroom.

Distinguishing characteristics. Cap 40–150 mm broad, rounded to convex, typically becoming flattened at maturity; surface dry, covered with a more or less thin coating of grayish to grayish brown or blackish fibrils that usually form small scales or patches over the surface. Flesh of cap and stalk white in early stages, becoming pinkish with age, staining either pinkish or yellowish, becoming pinkish to reddish brown when bruised. Odor indiscernible or disagreeable to pungent, when strong reminiscent of creosote. Gills free, crowded, narrow; color pinkish in early stages, usually pale pink to bright pink but sometimes with the pinkish color not well-developed, becoming dark reddish brown when mature. Stalk 30–150 mm long, 9–15 mm thick at the apex, enlarged below, and usually with a somewhat flattened base; color white, becoming discolored sordid pinkish brown, sometimes staining like the flesh when bruised. Annulus membranous, white on the upper surface, white to brownish pink and with soft floccose patches on the lower surface; underside of the unbroken partial veil often with yellow to brown droplets. Spores dark brown to chocolate brown in deposit, broadly ellipsoidal, smooth, thick-walled, without an apical germ pore, (4.5–)5.1–6.6(–7.3) × 3.7–4.4 μm in size.

Observations. The descriptions of *A. placomyces* presented in the literature are sometimes contradictory, making reports of poisonings by this species and its relatives difficult to interpret.

As conceptualized here, *A. placomyces* is an *Agaricus* species with a coating of small, grayish to grayish brown scales over the cap surface. The whitish background color is usually apparent, at least on the cap margin. The fibrils and scales are never reddish tinted as in *A. silvaticus*. Smith (1975) and B. Isaacs (Personal commun.) agree that this species has yellowish to brown droplets on the underside of the unbroken partial veil. The odor apparently resembles creosote in some forms; in others the odor is reported to be indiscernible.

Several different *Agaricus* species have probably been called *A. placomyces*. Specimens with gray brown, flattened scales on the cap surface, bright pink gills when young, and an odor of creosote may be *A. meleagris* Schaeff. (McKenny 1971), often growing in clusters under deciduous trees or conifers (see Fig. 181). *A. meleagris* is also believed to have several variants; one that occurs in the Great Lakes region has a blackish, squamulose cap and bright pink gills when young and stains intensely yellow on the stalk (Isaacs, Personal commun.). These forms lack the yellow to brown droplets on the underside of the unbroken partial veil. The European *A. meleagris* can be characterized by a whitish cap with a dark sooty brown center and with sooty brown to dark grayish brown scales on the margin; pale gills in early stages, soon developing to a rich pink, becoming blackish brown at maturity; a stalk with a bulbous base, colored white, but staining intensely and immediately bright lemon yellow when bruised, with the spot or stained area changing to purplish brown; a thick, white annulus, staining like the stalk when touched, with yellowish brown tooth-like patches on the lower surface; a disagreeable smell, like ink, sweat, or moldy straw; flesh quickly staining lemon yellow when bruised, later turning purplish brown; spores 4–5 × 3 μm in size. This information is taken from a description by F. H. Möller (1950).

Because of the variation in *A. placomyces* and related species in North America, many years of intense field studies will be required before it is possible to distinguish the species and to determine where they occur and their correct names. *A. meleagris*, in one or more forms, occurs in North America. It seems to be related to *A. placomyces*, but they do not appear to be the same species; both have been reported in the literature as poisonous. Specimens smelling of creosote are reputed to cause illness but those lacking this odor have also made people sick. It is therefore best to avoid eating all species in this group.

Distribution and seasonal occurrence. *A. placomyces* may be widely distributed in North America but its exact distribution is unclear. It occurs in eastern North America but may be absent in the west. Smith (1975) states that it is found in the Great Lakes region and eastward. He has seen no authentic specimens from the west. This species fruits in the late summer and fall.

Habitat and habit. *A. placomyces* grows scattered to gregarious. It occurs in low hardwood forests or mixed woods.

Technical description. Cap 40-100(-150) mm broad, obtuse to convex or with a slightly flattened disc when young, becoming broadly convex to plane or with a broad low umbo with age; surface dry, not hygrophanous, opaque, when young covered by a more or less thin coating of fibrils that are grayish to dark grayish brown, light grayish brown, brownish gray to light brown, or very dark gray to blackish; these fibrils usually form small scales or are aggregated into bundles; disc darker than marginal area; white background color usually showing over the marginal area, sometimes becoming pinkish in wet weather. Flesh thin to moderately thick, white or pinkish with age (particularly in wet weather), staining either pinkish or yellowish then pinkish to reddish or reddish brown where bruised; taste mild, but often disagreeable in old caps; odor indiscernible or somewhat pungent-disagreeable when the flesh is crushed, strongly reminiscent of creosote.

Gills free, approximate to the apex of the stalk or becoming remote, crowded, narrow, pale dull pink when young, developing to a bright pink or sometimes developing scarcely any pinkish tints, finally becoming dark reddish brown at maturity.

Stalk (30-)70-150 mm long, 9-15 mm thick at the apex, slightly enlarged downward, usually with a slightly flattened base, central, stuffed, dry, smooth and silky when young; white rhizomorphs usually numerous; color white, but often discoloring to sordid pinkish brown, sometimes staining yellowish then pinkish to reddish where bruised. Flesh solid, white, sometimes staining the same color as the flesh of the cap. Annulus membranous, flabby, more or less apical on the stalk, silky; upper surface white; lower surface white to more or less brownish pink, with soft floccose patches, at least along the margin, often with yellow to brown droplets on the underside of the unbroken partial veil. Universal veil absent.

Spores dark brown to chocolate brown in deposit, (4.5-)5.1-6.6(-7.3) × 3.7-4.4 μm, broadly ellipsoidal, slightly inequilateral in side view, smooth, thick-walled, without an apical germ pore, blackish brown. Basidia four-spored, clavate, 13.9-16.8 × 6.6-7.3 μm, hyaline. Pleurocystidia absent. Cheilocystidia sac-like to clavate, 14.6-25.6 × 8.8-16.1 μm, thin-walled, hyaline.

Gill trama interwoven; hyphae branched, cylindrical to inflated, 3.5-15.5(-17.8) μm wide, hyaline. Subhymenium nearly cellular to cellular; hyphae compactly interwoven, 4.4-7.3 μm wide, hyaline. Cap cuticle filamentous; hyphae interwoven, more or less radially arranged, parallel in the fibrils, more or less cylindrical, 3.7-11.0 μm wide, hyaline to brown, especially in the fibrils. Cap trama loosely interwoven; hyphae branched, cylindrical to inflated, 3.7-16.8(-20.4) μm wide, hyaline. Caulocystidia lacking; stalk hyphae parallel, more or less cylindrical, 4.0-15.5 μm wide, hyaline. Annulus interwoven; hyphae in upper surface compact, more or less radially arranged, 3.5-9.5(-15.0) μm wide; hyphae on lower surface separating into cylindrical fragments 4-7(-13.0) μm wide. Oleiferous hyphae greenish yellow, refractive, not septate, contorted, 3.7-7.3 μm, scattered in the gill trama. Clamp connections not seen.

Useful references. The descriptions cited below for *A. placomyces* no doubt include at least *A. meleagris* and perhaps other species. They may therefore not be in complete agreement with each other or with the description presented above.

- *A Field Guide to Western Mushrooms* (Smith 1975)—A color photograph and a description and discussion are presented.
- *Edible and Poisonous Mushrooms of Canada* (Groves 1979)—A color photograph and a description and discussion are presented.
- *Mushrooms in their Natural Habitats* (Smith 1949)—A description and discussion, as well as a color photograph on a reel, are presented.
- *Mushrooms of North America* (Miller 1972)—A color photograph and a description and discussion are given.
- *The Mushroom Hunter's Field Guide* (Smith 1974)—A black-and-white photograph and a description and discussion are presented.
- *The Savory Wild Mushroom* (McKenny 1971)—A color photograph, a black-and-white photograph, and a description and discussion are given.

Agaricus silvaticus Schaeff. ex Vittad. Figs. 120,182

Common names. Woods psalliota, sylvan agaricus.

Distinguishing characteristics. Cap 40–150 mm broad, hemispherical to convex, expanding to plane, sometimes more or less ovoid in early stages with a flattened center; surface dry, covered by a more or less dense to sparse fibrillose coating that usually breaks into patches or scales; color of fibrils and scales deep brown to brownish orange or dark rusty brown to reddish brown. Flesh white, slowly staining reddish brown when cut. Odor usually indiscernible. Gills free, crowded, narrow to moderately broad; color whitish to pale pink, becoming light grayish red to grayish reddish brown. Stalk 60–110 mm long; apex 10–20 mm thick; base enlarged and usually bulbous; color white in early stages, but soon becoming tinged pale pinkish brown to dark dingy brown. Annulus whitish or at times tinged pinkish below; lower surface with floccose scales or patches. Spores dull brown to chocolate brown in deposit, smooth, thick-walled, without an apical germ pore, (5.8–)6.6–7.3(–8) × 3.0–3.7 μm in size.

Observations. The name *A. silvaticus* has no doubt been applied to several *Agaricus* species in North America. The description given here allows for a wide range of variation and can be applied to all related species. In some areas *A. silvaticus* has been confused with *A. hondensis* and with *A. placomyces*. The dark rusty brown to reddish brown scales and fibrils on the cap surface and the lack of yellow stains or discoloration on the surface and flesh of *A. silvaticus* are helpful in separating it from the other two species. *A. silvaticus* apparently lacks any special odor, which might also help to distinguish it from other forms. Europeans have reported that the flesh of *A. silvaticus* turns cherry red when cut or bruised, whereas the North American *A. silvaticus* turns reddish brown. Some North American investigators therefore claim that what we call *A. silvaticus* is in fact another species. This question cannot be settled here.

Distribution and seasonal occurrence. *A. silvaticus* appears to be widely distributed in North America. It fruits in spring, summer, and fall but is mostly abundant in the fall.

Habitat and habit. This species is terrestrial, growing scattered to gregarious. It occurs in forested areas under conifers or hardwoods.

Technical description. Cap 40–120(–150) mm broad, nearly ovoid with a somewhat flattened disc in the button stage, sometimes nearly hemispherical, expanding to convex to broadly convex and finally becoming more or less plane; margin sometimes raised with age; edge of margin slightly incurved or straight; surface dry, not hygrophanous, opaque, covered by a more or less dense to sparse fibrillose coating that usually breaks up into appressed patches or scales; fibrils and scales deep brown to brownish orange or dark rusty brown or reddish brown, sometimes becoming brownish grayish reddish orange with age; coating sometimes very thin and present only as colored fibrils near or over the disc.

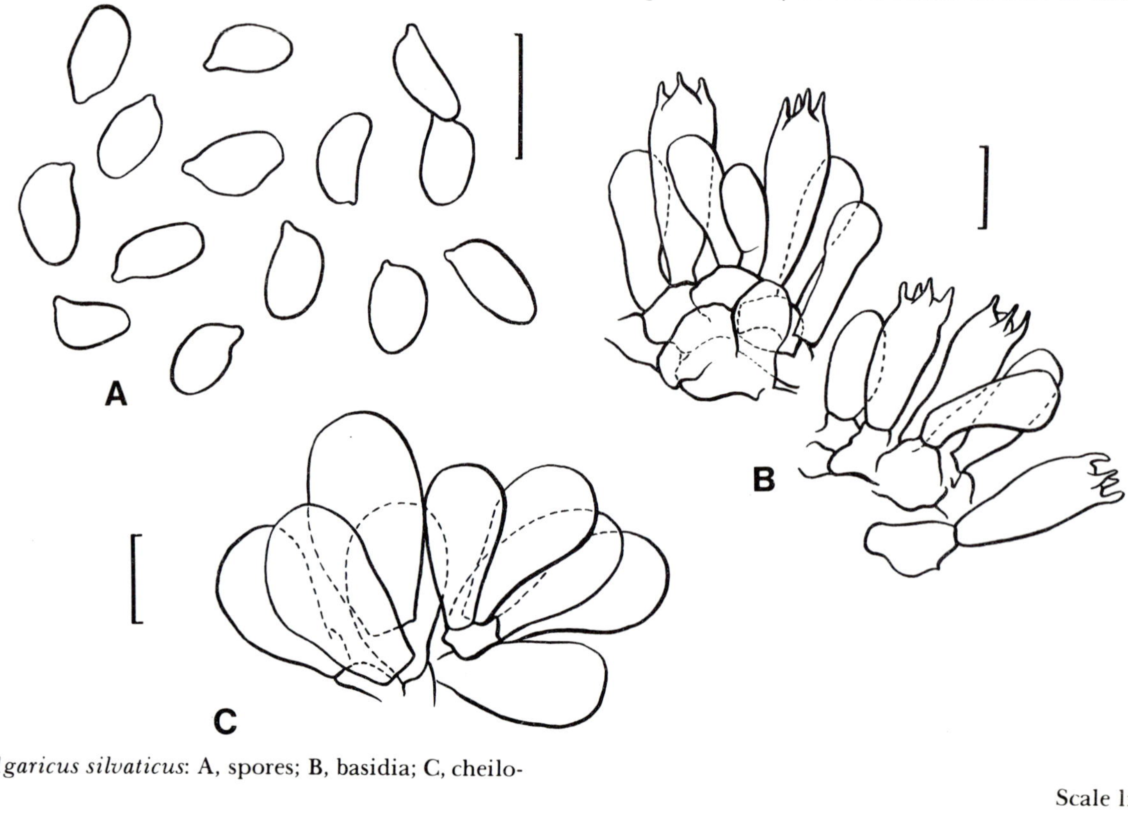

Fig. 120. *Agaricus silvaticus*: A, spores; B, basidia; C, cheilocystidia.

Scale line 10 μm

Flesh moderately thick, firm, rather moist, whitish to pallid, when cut slowly becoming reddish brown; taste indiscernible or slightly resembling almonds; odor usually indiscernible.

Gills free but sometimes almost reaching the stalk apex, crowded, narrow to moderately broad; edges even; color whitish to pale pink, becoming light grayish red when near maturity, finally turning grayish reddish brown.

Stalk 60-110 mm long, 10-20 mm thick, enlarged downward, central, hollow; base somewhat bulbous or sometimes with a flaring bulb that is flattened below; surface appressed-silky above and below the annulus; color whitish in early stages but soon becoming tinged pale pinkish brown and finally darker dingy brown. Flesh the same color as the cap flesh. Annulus apical on the stalk, whitish but with dark pink-tinged floccose scales or patches on the lower surface, sometimes merely white-fibrillose. Universal veil lacking.

Spores dull brown to chocolate brown in deposit, (5.8-)6.6-7.3(-8.0) × 3.0-3.7 μm, broadly ellipsoidal to ovoid, slightly inequilateral in side view, smooth, thick-walled, without an apical germ pore, dull brown. Basidia four-spored, clavate, 19.7-24.5 × 5.8-6.6 μm, hyaline. Pleurocystidia lacking. Cheilocystidia sac-like to clavate, 12-16.7(-18.0) × 7-8.8(-10) μm, hyaline.

Gill trama interwoven; hyphae cylindrical to inflated, 3.5-16.7 μm wide, hyaline. Subhymenium cellular, compact; hyphae 5.8-8.2 μm wide, hyaline. Cap cuticle filamentous; hyphae more or less radially arranged, parallel in the fibrils, cylindrical, 4-6.6 μm wide, hyaline or with brownish contents in fibrils. Cap trama loosely interwoven; hyphae cylindrical to inflated, 3.5-16.7 μm wide, branched, hyaline. Caulocystidia lacking; stalk hyphae more or less cylindrical, 3.5-17.8 μm wide, parallel, with some terminal cells deflected, hyaline. Annulus interwoven; hyphae cylindrical, 3.0-7.0 μm wide, hyaline, branched. Oleiferous hyphae refractive, greenish yellow to ocherous, 3.5-9.4(-11.7) μm wide, scattered in the gill trama and stalk. Clamp connections not seen.

Useful references. The following books contain information on *A. silvaticus* pertinent to North America. The descriptions may not be in complete agreement with the one presented here. They do, however, give a clear indication of what the *A. silvaticus* group is like.

- *Mushrooms of North America* (Miller 1972)—A color photograph and a description and discussion are given.
- *The Mushroom Hunter's Field Guide* (Smith 1974)—A color photograph, a black-and-white photograph, and a description and discussion are presented.
- *The Savory Wild Mushroom* (McKenny 1971)—A black-and-white photograph and a description and discussion are given.

Agaricus sylvicola (Vittad.) Peck — Figs. 121, 183

Common names. Woodland mushroom, wood-loving mushroom, sylvan mushroom.

Distinguishing characteristics. Cap 40-200 mm broad, hemispherical to convex, becoming more or less flattened with age; surface usually appressed-silky to fibrillose; color white to creamy white, becoming more or less yellowish over the disc. Surface and sometimes the flesh of the cap and stalk staining yellow when bruised. Gills free, crowded, narrow to moderately broad; color white to whitish when young, soon changing to pale pink or grayish pink, although the pink color may be faint to almost lacking, finally becoming darkish grayish reddish brown. Stalk 80-150 mm long, 8-25 mm thick at the apex, enlarged below, sometimes with a flaring abruptly bulbous base; surface silky above the annulus and fibrillose below; color white or with yellowish to tan or pinkish brown discolorations. Annulus membranous, persistent, thick; color white, with floccose yellowish patches of soft tissue on the lower surface. Spores reddish brown to chocolate brown in deposit, ellipsoidal, smooth, thick-walled, lacking an apical germ pore, 5.5-6.6(-7.3) × 3.7-4.4 μm in size.

Observations. Carefully compare *A. sylvicola* with *A. arvensis.* They are difficult, sometimes impossible, to distinguish using general appearance and color as criteria. They are best separated by spore size.

Some forms of *A. sylvicola* have a disagreeable to pungent odor of bitter almonds, whereas others have no discernible odor. The same variation occurs with *A. arvensis.*

A. sylvicola possibly has two other names, *A. abruptibulbus* Peck and *A. fabaceous* Berk. *A. abruptibulbus* has been used by several authors for the form with a distinctly flattened, bulbous stalk base; *A. fabaceous* has been rarely used but is perhaps the oldest name for North American species in this group. The naming has not been clearly determined and therefore *A. sylvicola* is retained here.

Distribution and seasonal occurrence. *A. sylvicola* occurs in several forms and is widely distributed. It fruits mainly from midsummer to fall. It has also been reported in June from eastern North America and the Pacific coast during the winter, depending on weather and location.

Habitat and habit. This species is rarely solitary, usually scattered to gregarious under hardwoods and conifers.

Technical description. Cap (40-)50-120(-200) mm broad, hemispherical to convex in early stages or with the disc somewhat flattened, expanding to broadly convex to plane with age; surface dry, not hygrophanous, opaque, appressed-silky to fibrillose or with the fibrils aggregated into obscure scales with age; scales occasionally somewhat recurved, white to creamy white, but usually becoming more or less yellowish over the disc, often staining yellowish where bruised; outer margin sometimes pale reddish brown with age. Flesh thick, soft, white, unchanging or changing to yellowish where bruised; taste more or less resembling almonds; odor pleasant to faintly pungent, but not distinctive.

Gills free and remote from the stalk apex, crowded, narrow to moderately broad, white to whitish in very young stages, soon pale pink to grayish pink, although the pink color of the immature gills is sometimes bright or sometimes almost lacking, finally becoming dark grayish reddish brown.

Stalk 80-150 mm long (8-)10-20(-25) mm thick at the apex, slightly enlarged downward, with or without a flaring abruptly bulbous flattened base, central, stuffed, soon becoming hollow, dry, silky above the annulus, more or less appressed-fibrillose below the annulus or with the fibrils aggregated into obscure patches, sometimes becoming glabrous, white or with yellowish to tan or sordid pinkish brown discoloration, sometimes staining yellowish when bruised. Flesh white, sometimes staining yellowish. Annulus membranous, typically double, median, white or with floccose yellowish patches of soft tissue on the lower surface. Universal veil lacking.

Spores reddish brown to chocolate brown in deposit, 5.5-6.6 (-7.3) × 3.7-4.4(-5.1) μm, ellipsoidal, slightly inequilateral in side view, smooth, blackish brown, thick-walled, lacking an apical germ pore or with the pore indistinct. Basidia four-spored, short-clavate, 20.0-50.0 × 7.3-8.2 μm, hyaline. Pleurocystidia lacking. Cheilocystidia sac-like to broadly clavate, 13-22 × 8-10 μm, hyaline.

Gill trama interwoven; hyphae branched, cylindrical, 2.2-12 μm wide, hyaline, thin-walled. Subhymenium nearly cellular to cellular, compactly interwoven; cells (6.6-)7.3-9.5 μm wide, hyaline. Cap cuticle filamentous; hyphae interwoven, radially arranged (especially the fibrils), cylindrical, 4.7-5.8 (-7.0) μm wide, yellow to greenish yellow. Cap trama loosely interwoven; hyphae cylindrical to more or less inflated, 3.5-12(-18) μm wide, hyaline to yellowish. Caulocystidia lacking. Stalk hyphae more or less parallel, cylindrical, 3.5-11.7 μm wide, yellowish. Annulus interwoven; hyphae cylindrical to inflated, 2.9-12(-25) μm wide, separating into cylindrical to ellipsoidal fragments. Oleiferous hyphae refractive, not septate, contorted, 3.5-9.5 μm wide, granular, yellowish green, scattered in the gill trama, stalk, and cap cuticle. Clamp connections absent.

Useful references. The following books contain information on *A. sylvicola* pertinent to North America.

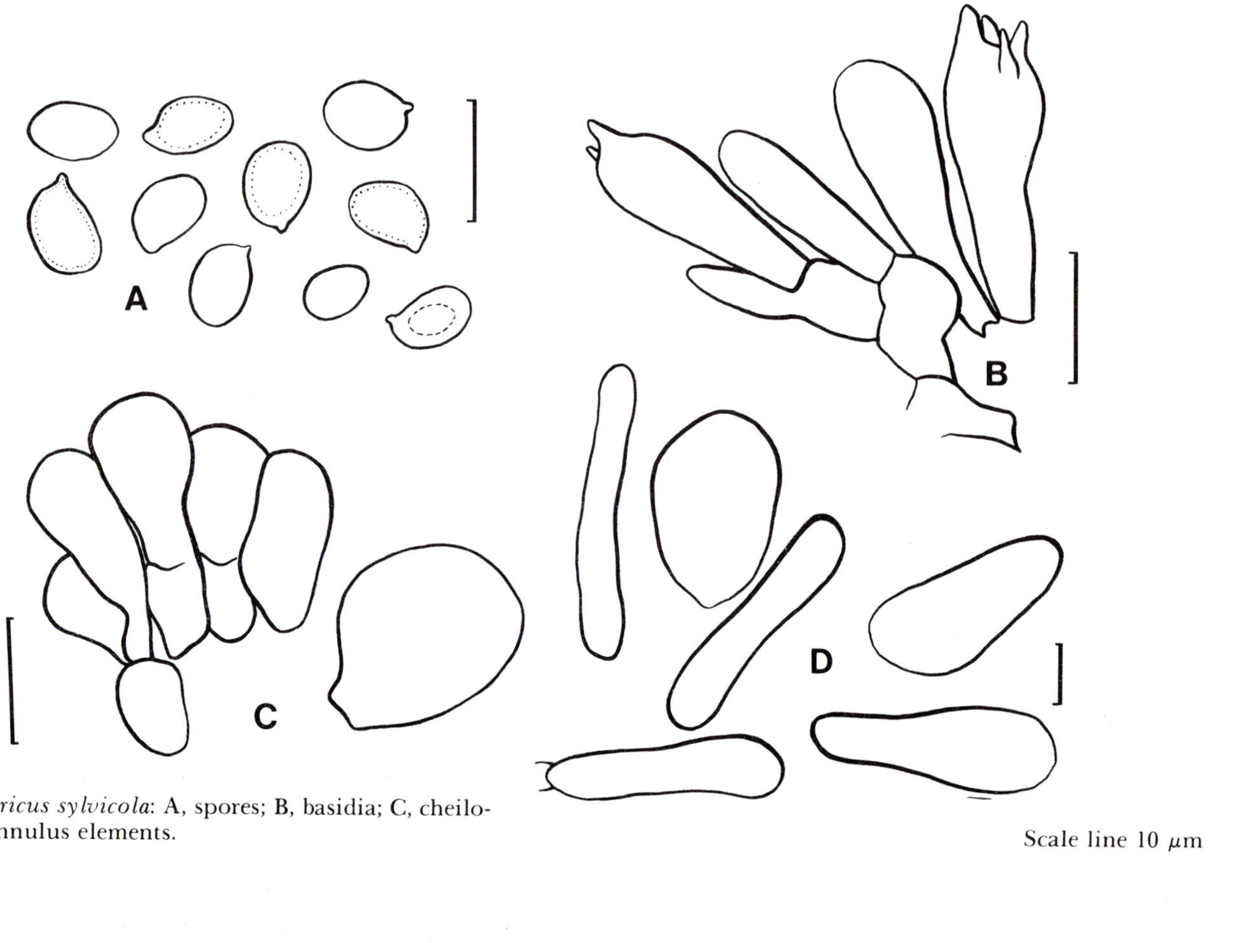

Fig. 121. *Agaricus sylvicola*: A, spores; B, basidia; C, cheilocystidia; D, annulus elements.

Scale line 10 μm

- *A Field Guide to Western Mushrooms* (Smith 1975)—A color photograph and a description and discussion are presented.
- *Edible and Poisonous Mushrooms of Canada* (Groves 1979)—A color photograph and a description and discussion are presented.
- *The Mushroom Hunter's Field Guide* (Smith 1974)—A black-and-white photograph and a description and discussion are presented.
- *The Savory Wild Mushroom* (McKenny 1971)—A black-and-white photograph and a description and discussion are given.

Amanita Pers. ex S. F. Gray

Amanitaceae

In the literature on poisonous mushrooms the genus *Amanita* has received the most attention. *A. phalloides* and its relatives and the *A. muscaria* - *A. pantherina* group have no doubt received the greatest amount of this attention. Chapters 7 and 11, respectively, provide a treatment of these species, as well as a detailed description of the genus.

Although a few species of *Amanita* are considered to be nonpoisonous, for example *A. caesarea* (Fr.) Schw., *A. calyptroderma* Atk. & Ballen, and, when thoroughly cooked, *A. rubescens* (Pers. ex Fr.) S. F. Gray and *A. vaginata* (Fr.) Vittad., the best policy is to avoid the entire genus. More than 25 species are listed as poisonous and many others, because of their relationship to known toxic species, can be presumed to be poisonous.

Lincoff and Mitchel (1977) list *A. citrina* and *A. porphyria* (Alb. & Schw. ex Fr.) Secr. as species containing toxic and psychotropic alkaloids but report no poison cases caused by these species in North America. *A. brunnescens* Atk., which is similar in appearance to *A. rubescens*, has been listed in several mushroom guides as poisonous. Early in this century it was confused with *A. phalloides* but it is not related to that species. There are no confirmed reports of poisoning by *A. brunnescens*. Abdel-Malak (1974) reported that *A. brunnescens*, as well as *A. flavoconia* Atk., *A. rubescens*, and several other species of *Amanita* contain cyclopeptides and alkaloids, but the toxins were not identified and the quantities were not determined. *A. flavorubescens* Atk. (=*A. flavorubens* Berk. & Mont.) has been listed as poisonous by Miller (1972), but there are no reports of toxins in this species and no cases of poisoning. An interesting account of *A. flavorubens* (=*A. flavorubescens*) is presented by Haines (1977).

Several species in the section *Lepidella* of *Amanita* have been suspected of being poisonous. *A. chlorinosma* (Austin) Lloyd and *A. cokeri* (Gilbert & Kühner) Gilbert, as well as *A. strobiliformis* (Paulet ex Vittad.) Bertillon, are species commonly listed as poisonous.

Most of these suspected species have a chlorine-like odor. They also have in common ellipsoidal, amyloid spores; pale white to grayish or pallid to buff caps covered with pyramidal warts; a large, rooting stalk base; and a universal veil that leaves a powdery coating or wart-like scales or both on the stalk surface, especially on the bulbous base. The confusion surrounding the identification of species in this group in North America makes it almost impossible to determine which species are toxic. *A. strobiliformis* and *A. smithiana* Bas (=*A. solitaria* (Bull. per Fr.) Mérat *sensu* Stuntz) have been listed as containing toxins (Lincoff and Mitchel 1977, Chilton and Ott 1976, Benedict et al. 1966). Specimens of *A. strobiliformis* from Japan contain ibotenic acid (Benedict 1972). No poisonings by *A. smithiana* have been reported in North America. According to Bas (1969), the only worker to do a modern taxonomic treatment of the section *Lepidella, A. solitaria* and *A. strobiliformis* do not occur in North America.

In summary, some species of the section *Lepidella* in North America are reported to be poisonous, and isoxazole derivatives or related compounds have been isolated from some other species from North America that were not identified with certainty. Only a critical taxonomic treatment, including extensive field work, can solve the taxonomic problems surrounding the North American species in the section *Lepidella*. This work must be done before the toxicity of these species can be clarified.

In addition to *A. muscaria, A. pantherina*, and *A. gemmata* (Ch. 11), Tyler (1963) listed *A parcivolvata* (Peck) Gilbert and *A. flavivolva* Murrill as probable muscarine-containing species.

A number of other *Amanita* species have been listed as poisonous or as suspected of being poisonous, but there is little or no information on their toxic components and there are no documented cases of poisoning. These include *A. spreta, A. volvata* (Peck) Martin, *A. agglutinata, A. flavoconia*, and *A. frostiana*. They should all be avoided, but they should not continue to be listed as poisonous, deadly, or otherwise, without evidence.

Armillariella Karst.

Tricholomataceae

Armillariella is a small genus. At least two species, *Armillariella mellea* and *Armillariella tabescens* (Scop. ex Fr.) Singer, occur in North America; of the two, *A. mellea* is the more widely distributed and the more common. The two species are similar in appearance; however, *A. mellea* has a partial veil that leaves an annulus on the stalk, whereas *A. tabescens* lacks an annulus. Many authors still place *Armillariella mellea* in the genus *Armillaria*, and *Armillariella tabescens* has been placed in the genera *Clitocybe* and *Monadelphus*.

Species of *Armillariella* often grow in clusters. They are frequently found fruiting on woody substrates but occur in various habitats and may be saprobic, parasitic, or symbiotic.

Both *Armillariella mellea* and *Armillariella tabescens* generally appear similar to species of *Clitocybe*. The gills are attached, at times decurrent, especially in *A. tabescens*, but may be subdecurrent to slightly sinuate. The cap surface is fibrillose to scaly, at times viscid to gelatinous; the cap cuticle is composed of filamentous hyphae. The spore deposit is white to light cream color; the spores are smooth and nonamyloid. The stalk is central and fibrous-fleshy; an annulus may be present. Clamp connections are uncommon to absent, but some can be found, for example, on the cells of the hymenium. Black rhizomorphs, resembling black shoelaces, are produced by *A. mellea* and *A. tabescens*.

The genus *Armillaria* can be separated from *Armillariella* by the amyloid spores and the generally larger and fleshier fruit bodies. The gill trama in *Armillaria* is divergent in young, developing specimens, whereas in *Armillariella* it is slightly divergent to more or less parallel.

Armillariella mellea is one of the most commonly eaten species of wild mushroom. However, it has a rather tough to coarse consistency, especially the stalks, and has sometimes caused gastrointestinal upset (North American Poison Mushroom Research Center). Poisoning of this type may be the result of improperly cooking the specimens or eating them raw. It is generally recommended that only the caps be eaten and the stalks discarded (McKenny 1971). Certain strains may conceivably be somewhat toxic but this possibility has not been proven. More likely, certain individuals simply cannot tolerate this species, especially when it is eaten in large quantities. We do not consider *A. mellea* to be a strictly poisonous species, but collectors should be aware that it can cause gastrointestinal upset.

Collectors have been known to confuse *Armillariella mellea* with the very poisonous *Galerina autumnalis* (Ch. 7). The two species can be found growing together on the same log, so take care when collecting *A. mellea* for food not to include the poisonous *G. autumnalis*.

Armillariella mellea (Vahl ex Fr.) Karst.

Figs. 122, 184

Common names. Honey mushroom, stump mushroom.

Distinguishing characteristics. Cap 30–120 mm broad, often convex with a somewhat raised center when mature; surface slightly sticky to viscid or even glutinous in wet weather, often with small dark scales, especially in the center; color variable, often some shade of tan to dark brown, but some forms strongly yellow, that is honey yellow to ocher yellow. Gills close to subdistant; color whitish, white to creamy, often with brownish stains or discolorations with age, attached, adnate to subdecurrent. Stalk 40–150 mm long, 6–20 mm thick; base often enlarged and club-shaped, but the stalk may be more or less equal or tapered toward the base; color whitish to tan or at times yellowish, often with brownish stains. Annulus usually distinct and more or less persistent. Black rhizomorphs resembling shoelaces associated with the fruit bodies. Fruit bodies growing in clusters, often on wood but at times on the ground, from buried wood. Spores white to creamy yellowish in deposit, smooth, nonamyloid, broadly ellipsoidal, (7.1–)8.2–10.2(–11) × 5.1–6.6(–7.0) μm in size.

Observations. *Armillariella mellea* is often called *Armillaria mellea* Fr. It has several forms, perhaps some of these representing distinct species or varieties (Smith 1974, Singer 1975, Anderson and Ullrich 1979). Some forms are difficult to identify, especially by the amateur collector. Smith (1974) described two forms: one is strongly yellowish, has a creamy yellowish spore deposit, grows in clusters in oak-hickory forests in late August to early September, and fruits in clusters from buried wood; the other is dark brown to almost blackish on the disc and grayish brown toward the margin, is distinctly scaly especially in the center, has a white spore deposit, grows in clusters on wood of hardwoods and conifers, and typically fruits in the fall. The second form appears to be the most common and occurs across North America.

Distribution and seasonal occurrence. *A. mellea* is found across North America, occurring in several

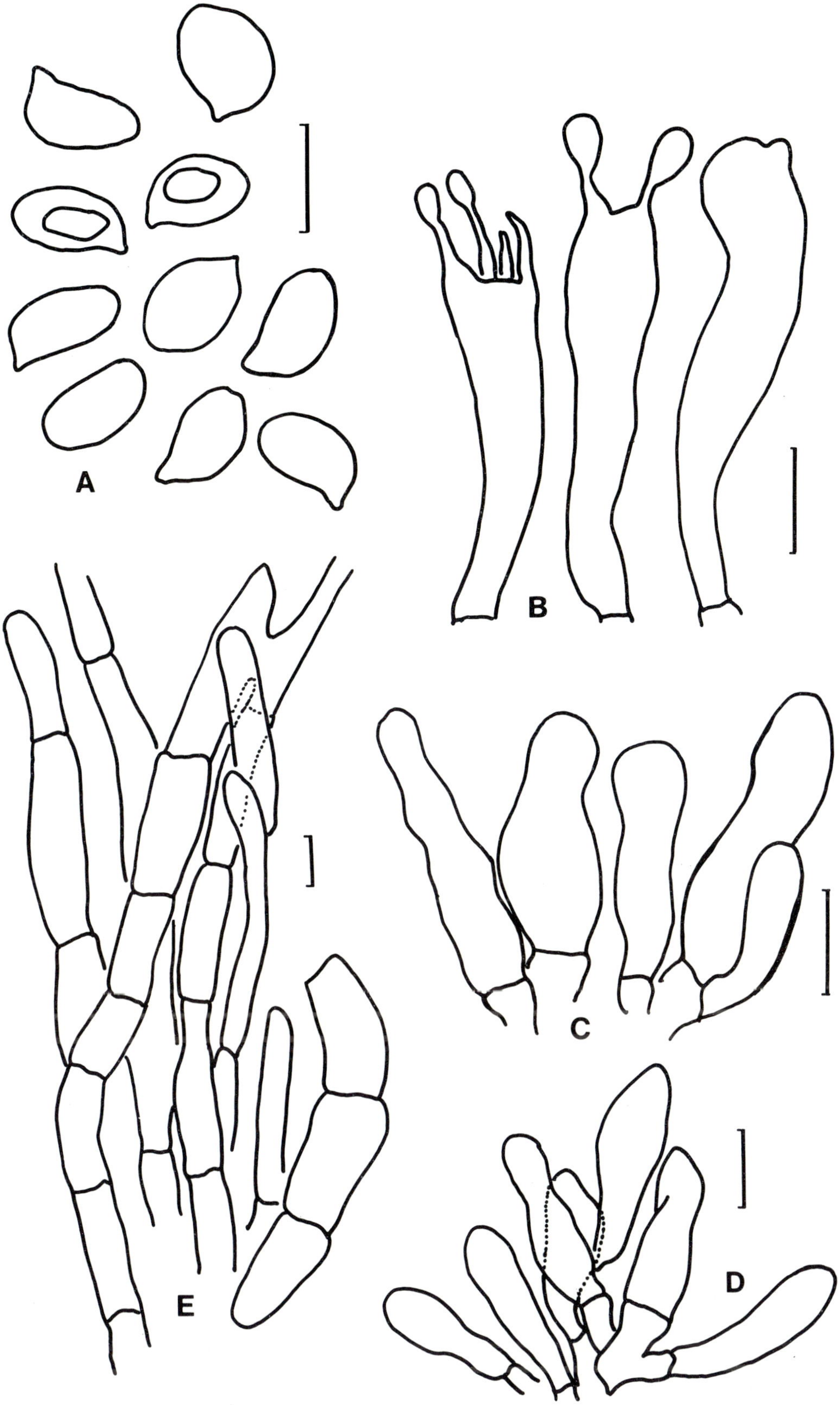

Fig. 122. *Armillariella mellea*: A, spores; B, basidia; C, cheilocystidia; D, caulocystidia; E, annulus elements.

Scale line 10 μm

forms. It fruits from the middle or late summer into the late fall.

Habitat and habit. This species grows in small clusters on wood of conifers and hardwoods, sometimes around stumps or the bases of living trees. It also may be terrestrial, growing from roots or buried wood.

Technical description. Cap 30-120 mm broad, in early stages more or less hemispherical, becoming convex or at times even more expanded, sometimes umbonate to nearly umbonate; margin enrolled in early stages, becoming incurved to decurved or with age somewhat uplifted, often striate, at times faintly so; surface slightly sticky to viscid or even glutinous in wet weather, nearly glabrous to minutely scaly, often with conspicuous small scales on the disc; color variable, honey yellow to ocher yellow, yellow brown, yellow buff, or tan, sometimes grayish brown to dark brown or almost blackish on the disc, sometimes more pinkish brown to rusty tinged, often darkest in the center. Flesh thin on the margin and thicker on the disc; color usually white, but at times with brownish to tan or rusty discolorations; odor mild to somewhat unpleasant; taste mild or slightly acrid and unpleasant.

Gills adnate to subdecurrent, close to subdistant, moderately broad; color white to pale cream, becoming somewhat tan with age, developing rusty brown to pinkish brown stains with age; surface often powdered with whitish to white spores.

Stalk 40-150 mm long, 6-20 mm thick at the apex, equal to enlarged below to a clavate base; stalk in some forms narrowed downward or tapered toward the base; surface finely fibrillose to fibrillose-scaly; color variable, usually somewhat paler than the cap, yellowish to white, buff, or tan, discoloring or staining brown to rusty brown below the annulus, usually pale to whitish above the annulus, stuffed then hollow; flesh fibrous-fleshy. Annulus more or less persistent to evanescent, fibrillose-membranous to cottony, apical on the stalk; color white to brownish or yellowish tinged, typically dry.

Spores white, creamy yellowish or cream color in deposit, (7.1-)8.2-10.2(-11.0) × 5.1-6.6(-7.0) μm, broadly ellipsoidal, inequilateral in side view, smooth, thin-walled, hyaline, nonamyloid. Basidia four-spored, rarely two-spored, clavate, 29.0-32.9 × 7.3-8.0 μm, hyaline. Pleurocystidia lacking. Cheilocystidia sac-like to fusiform-ventricose and nearly capitate, 23.0-27.0 × 6.6-7.3 μm, hyaline.

Gill trama hyphae slightly divergent to more or less parallel, cylindrical to inflated, (3.7-)8.8-15.0 μm wide, hyaline. Subhymenium filamentous; hyphae interwoven, 3.7-5.1 μm wide. Cap cuticle filamentous; hyphae interwoven, cylindrical, parallel in scales, 7.0-12.4 μm wide, constricted at the septa, thin-walled; walls and contents brownish. Cap trama loosely interwoven; hyphae inflated, 15.0-18.0(-22) μm wide, thin-walled, hyaline. Caulocystidia subclavate to fusiform-ventricose with nearly capitate tips, 18-23(-30) × 7.1-8.2(-9.4) μm, hyaline, located above the annulus; stalk hyphae interwoven to parallel, cylindrical to inflated, 4.5-14.5(-20) μm wide. Annulus interwoven; hyphae branched, thin-walled, 4.6-11.0 μm, cylindrical. Oleiferous hyphae refractive, greenish yellow, 7.0-12.0 μm, scattered in the cap cuticle, cap and gill trama, and stalk. Clamp connections rare to absent (only in some parts of the fruit bodies).

Useful references. The following books contain information on *A. mellea* pertinent to North America. Some of the references list this species in the genus *Armillaria*, for example, *Armillaria mellea*.

- *A Field Guide to Western Mushrooms* (Smith 1975)—A color photograph and a description and discussion are presented.
- *Edible and Poisonous Mushrooms of Canada* (Groves 1979)—A color photograph and a description and discussion are given.
- *Mushrooms of North America* (Miller 1972)—A color photograph and a description and discussion are given.
- *The Mushroom Hunter's Field Guide* (Smith 1974)—A color photograph, a black-and-white photograph, and a description and discussion are given.
- *The Savory Wild Mushroom* (McKenny 1971)—A black-and-white photograph and a description and discussion are given.

Boletus Dill. ex Fr.

Boletaceae

The genus *Boletus* belongs to the family Boletaceae. The Boletaceae, commonly called the boletes, and the gill mushrooms, comprising several families, together make up the order of fleshy fungi called the Agaricales.

Boletes resemble gill mushrooms in general stature and appearance. Boletes are fleshy like gill mushrooms, and the fruit bodies readily decay following maturity, or sometimes, through the action of insect larvae, while they are maturing. A cap and a distinct stalk are characteristic of all boletes. The undersurface of the cap is covered by a layer of tubes opening by small to coarse pores. This layer is often easily separated from the cap tissue and typically has the consistency of a soft sponge. In contrast, gill mushrooms have radiating, knife-like gills on the undersurface of the cap.

The spores in boletes are produced inside the tubes and fall from the pores as they are released from the basidia. When the cap of a bolete is separated from the stalk and placed on white paper, the spores fall into minute to small piles, each representing a pore. The color of the spore deposit varies considerably from genus to genus and is useful in separating bolete genera. The spore deposits for boletes in general range from pale yellow to cinnamon, olive to olive brown, dark yellow brown, vinaceous brown, vinaceous red, umber, fuscous, chocolate, or even purple.

Several boletes lack both a partial veil and a universal veil; however, some species, for example in the genus *Suillus*, have a veil that leaves an annulus on the stalk or veil remmants on the cap margin. *Fuscoboletinus* species also often have a veil that leaves an annular zone. *Pulveroboletus* has a bright yellow, delicate veil over the cap and stalk. Species of *Boletus* typically lack a veil, as do species of *Tylopilus* and the majority of *Leccinum* species.

A large family of fungi, the Polyporaceae, commonly called the polypores, is sometimes confused with the boletes. In the polypores the spore-bearing surface usually consists of minute to coarse tubes with pore-like openings, although sometimes the undersurface of the cap has gills; in general appearance their pores resemble those of the boletes. The polypores differ from the boletes by the typically leathery or corky or woody fruit bodies that do not decay readily. Although there are some almost fleshy species of polypores, they often do not have a distinct cap and stalk, as do the boletes. Although polypores can give a spore deposit, as do the boletes and gill mushrooms, spore deposits are often difficult or impossible to obtain from them. The spore deposit is usually white, but in some genera, it is brown.

The boletes are generally highly prized by food foragers. Many species are edible and certain species, such as *Boletus edulis* Bull. ex Fr., are of excellent flavor, texture, and quality.

As in all groups of fleshy fungi, there are, however, some poisonous species among the boletes. These are found primarily in the genus *Boletus*. The poisonous species are often those that have red to orange red tube pores and flesh that stains blue; however, some of the species with yellow tube pores and exhibiting a blue staining reaction have also caused gastrointestinal upset.

Some boletes are better classified as inedible rather than poisonous. These are the bitter-tasting species found especially in the genus *Tylopilus*. A description of the genus and of *Tylopilus felleus* appears later in this chapter. Consult the alphabetical listing. There are also several species listed as edible but of poor quality. They are not classified as poisonous and are not treated here.

The following species of *Boletus* have been reported as poisonous. Some of these poisonings are based on European reports, as well as on reports from North America. The most common of these species are described here in detail.

Boletus species with red to orange red tube pores and exhibiting a blue-staining reaction when cut or bruised are dangerous. These species, including *B. erythropus*, *B. luridus*, *B. pulcherrimus* (=*B. eastwoodiae*), *B. satanas*, and *B. subvelutipes*, are in the *Luridi* group (subsection *Luridi*). All should be avoided (Smith and Thiers 1971). The main characteristic of this group is the color of the young tube pores; they are orange to orange red, red brown, red, or dark brown. The majority of the species in this group also stain blue to blue green when cut or bruised.

B. calopus var. *frustosus* has been reported as poisonous by Miller (1972). This species belongs to a species group that has yellow tube mouths when young, usually stains blue when injured (both the flesh and the tubes), and often has a bitter taste. There is little information on the edibility of this group in general, and Miller does not explain why he lists *B. calopus* var. *frustosus* as poisonous. Eugster (1968) and Matzinger et al. (1972), however, have found traces of muscarine and its stereoisomers in this species.

K. Harrison (Personal commun.) recently reported poisoning by *B. huronensis* Smith & Thiers. Smith and Thiers (1971) and Grund and Harrison (1976) provide a description of this species.

The responsible toxins and toxic properties of poisonous species of *Boletus* are not well known. Krieger (1967) suggested that the symptoms produced in a poisoning by *B. sensibilis* were characteristic of muscarine poisoning. Investigating the levels of muscarine in various boletes, Eugster and associates did not find pharmacologically active amounts in *Boletus satanas* (Eugster 1968, Matzinger et al. 1972) but did find traces of muscarine and its stereoisomers in a variety of mushrooms, including *B. calopus* and *B. luridus*.

Boletus erythropus (Fr.) Krombh.

Figs. 123, 185

Common name. Red-footed bolete.

Distinguishing characteristics. Cap 60–150 mm broad, convex to plane; surface dry, usually smooth; color in early stages yellow brown to brownish moderate orange to grayish reddish orange to moderate reddish brown; center becoming deeper brown to more reddish with maturity, at times with the margin bright yellow and the center yellow brown to reddish. Flesh of cap pale greenish yellow to light greenish yellow, turning blue quickly when cut or exposed (the blue color eventually fading to yellowish gray). Tubes greenish yellow to somewhat olive, with minute pores; pore color red to reddish orange or more reddish brown, staining blue when injured. Stalk 60–120 mm long, 10–60 mm thick, usually more or less clavate or at times equal; surface dry, not reticulate; ground color yellow to orange yellow, with brown to reddish brown on the base, typically overlaid with a dark red to reddish orange pruinose to granular coating; base usually with yellow hair; surface turning blue when injured. Flesh of stalk similar to that of the cap except for reddish colors in the base, also staining blue. Spores olive brown in deposit, subcylindrical to subfusiform, smooth, 12.4–14.6(–16.1) × 4.4–5.8 μm in size.

Observations. *B. erythropus* is difficult to separate from related species such as *B. subvelutipes* Peck (cover photograph). The concepts of these species in this publication are based on those of Smith and Thiers (1971).

The yellow brown color of the young caps and the yellow hairs on the base of the stalk are helpful characteristics for identifying *B. erythropus,* but some specimens apparently lack the basal hairs (Thiers 1965). *B. subvelutipes* has a more yellow to orange yellow cap margin than *B. erythropus* when young, but it varies from yellow to cinnamon or reddish, with the darker colors toward the center. The yellow to orange yellow cap color of young specimens is helpful in separating it from *B. erythropus* but with age the colors may be similar. The red hairs at the base of the stalk in *B. subvelutipes* are probably the most helpful of the field characters for separating it from *B. erythropus.* Because these two species are easily mistaken for one another and there are other related species that are similar, it is difficult to correctly identify them. Consult an expert for species identification. The important thing is to be able to recognize this group, represented by *B. erythropus* and *B. subvelutipes,* as potentially poisonous.

Distribution and seasonal occurrence. *B. erythropus* and related species occur in Europe and North America. *B. erythropus* has been reported from eastern and western North America, but its actual distribution is unclear. It fruits in the summer and fall, depending on weather and location.

Habitat and habit. This species is solitary to gregarious, growing under hardwood trees. It has been reported

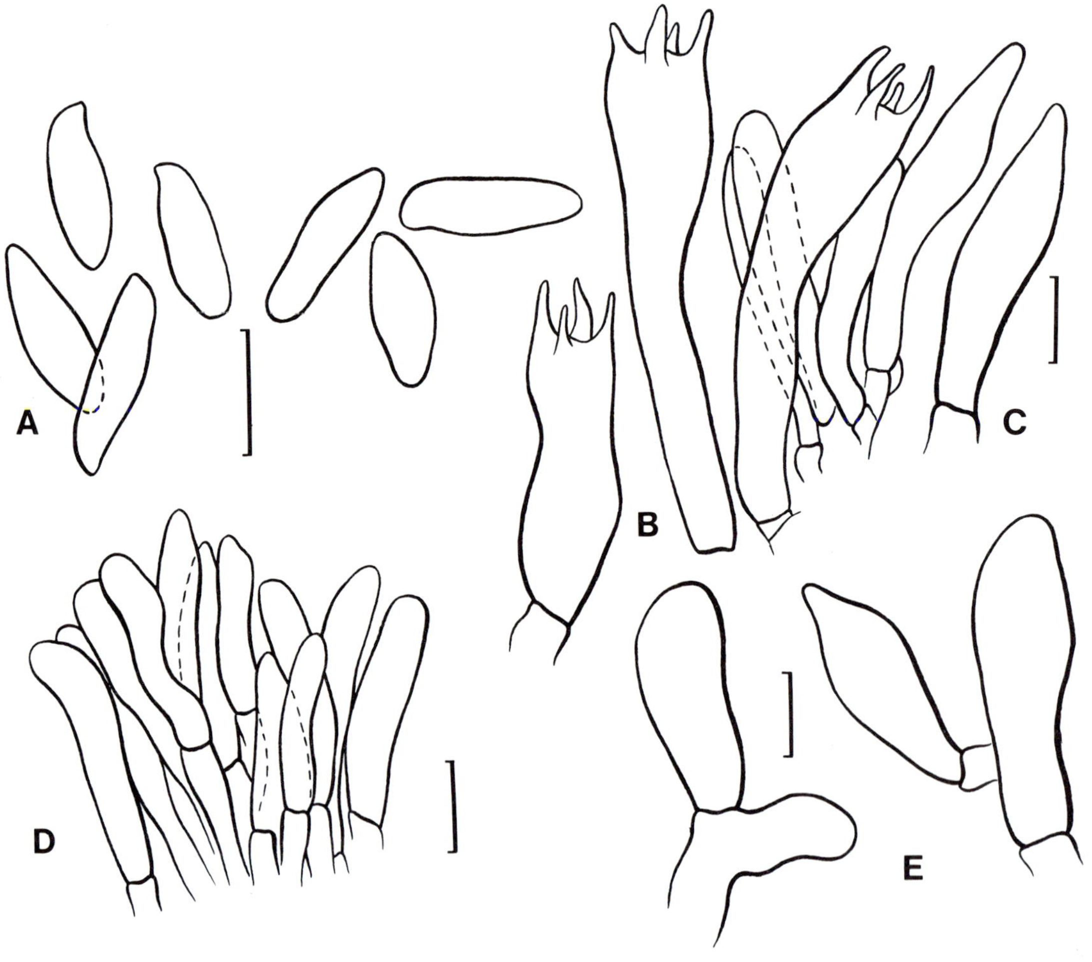

Fig. 123. *Boletus erythropus*: A, spores; B, basidia; C, pleurocystidia; D, cheilocystidia; E, caulocystidia.

Scale line 10 μm

from oak (*Quercus*) and beech-maple (*Fagus-Acer*) woods.

Technical description. Cap 60–150 mm broad, convex when young, becoming broadly convex to plano-convex to plane, occasionally umbonate or shallowly depressed with age; margin incurved to broadly decurved; surface dry, even, smooth and occasionally coarsely wrinkled to reticulate on the disc, more or less tomentose to occasionally appressed-fibrillose when young, often glabrous with age; color yellow brown to brownish moderate orange to moderate reddish brown or grayish reddish orange in early stages, becoming deep brown or deep red on the disc and strong brownish red toward the margin with age, sometimes bright yellow on the margin and yellow brown to reddish on the disc, usually darkening when bruised. Flesh 1–3 cm thick, firm, compact, pale greenish yellow to light greenish yellow, turning blue quickly when first exposed, then often fading to yellowish gray; taste mild or pungent to unpleasant; odor indiscernible or pungent to unpleasant.

Tubes deeply and broadly to narrowly depressed around the stalk, 10–20 mm long, brilliant greenish yellow, light greenish yellow, pale greenish yellow, or moderate olive, turning blue when bruised; pores small to minute, 0.5–1 mm broad, angular, dark red, moderate red, dark reddish orange, or moderate reddish brown, sometimes yellowish toward the margin and orange near the stalk, turning blue when bruised.

Stalk 60–120 mm long, 10–30(–60) mm thick at maturity, clavate to nearly clavate to occasionally equal, sometimes tapering upward, solid; surface dry, glabrous to more commonly granulose to obscurely pruinose or fibrillose, not reticulate; ground color lemon yellow to light yellow to near light orange yellow, with a dark red to deep reddish orange powdery bloom; surface turning blue where handled; base sometimes with yellow mycelia or with whitish rhizomorphs attached to the base; basal portion often moderate brown, moderate reddish brown, or dark reddish brown. Flesh pale greenish yellow to light greenish yellow or yellow throughout or more commonly reddish to rusty red in the base, turning blue instantly when exposed and often fading to grayish olive.

Spores olive brown in deposit, 12.4–14.6(–16.1) × 4.4–5.8 μm, subcylindrical to subfusiform, slightly inequilateral in side view, ocherous, nonamyloid. Basidia four-spored, clavate, 35–47 × 8–10.5 μm, hyaline to yellowish. Pleurocystidia scattered, fusiform-ventricose to nearly clavate, 25.6–36.5(–47) × 5.8–7.3(–13.1) μm, hyaline to yellowish. Cheilocystidia clavate to subfusiform, 31.2–32.0 × 4.4–8.8 μm, yellowish.

Tube trama divergent from a central strand; hyphae cylindrical, 3.5–7(–8.8) μm wide, hyaline to yellowish. Cap cuticle filamentous; hyphae compactly interwoven, with erect cylindrical to subfusiform hyphal ends, cylindrical, 3.5–4.7 μm wide, thin-walled to slightly thick-walled and finely encrusted, hyaline to yellow or ocherous. Cap trama loosely interwoven; hyphae branched, more or less cylindrical, 3.5–9.4(–11.7) μm wide, hyaline to ocherous. Caulocystidia fusiform-ventricose to ventricose, or broadly clavate with acute apices, 24.8–36.5 × 8.0–12.8 μm, hyaline to yellowish; stalk hyphae interwoven, more or less parallel, cylindrical, 3.5–8.2 μm wide, hyaline to yellowish; tomentum at base of stalk interwoven; hyphae cylindrical, 2.3–3.5 μm wide, hyaline to yellow. Oleiferous hyphae refractive, thin-walled, contorted, not septate, 3.7–4.4 μm wide, greenish yellow, scattered in the tube trama, cap cuticle, and stalk. Clamp connections not observed.

Useful references. There are no good references for *B. erythropus* among the general mushroom guides. However, the following book briefly deals with this species.

- *Edible and Poisonous Mushrooms of Canada* (Groves 1979)—The description of *B. subvelutipes* seems to include some features of *B. erythropus.*

There are at least four good technical books on the boletes of North America that should be consulted for identification of species: Snell and Dick (1970), Smith and Thiers (1971), Thiers (1975), and Grund and Harrison (1976). They may not all agree on the concept of *B. erythropus*, because it has been the subject of much controversy among students of *Boletus.* Smith and Smith (1973) also provide useful information for identifying boletes.

Boletus luridus Fr.

Fig. 124

Common names. None.

Distinguishing characteristics. Cap 50–120 mm broad; shape more or less convex at all stages; surface dry, at times with matted fibrils or fibrillose scales in the center; color variable, yellow to olive yellow flushed with pinkish red, or light orange yellow at the margin and darker orange yellow to brownish orange yellow or reddish on the disc. Flesh of cap yellowish to reddish, instantly staining blue when injured. Tubes greenish yellow, with small dark red to reddish brown pores; pores usually fading to orange yellow with age, staining greenish blue where injured. Stalk 60–150 mm long, 10–30 mm thick, equal to clavate with an enlarged base; surface dry; color variable, strong yellow overall or tinted orange red to blood red above with yellow below or at times yellow above and reddish below, reticulate (a raised net-like pattern) at the apex or nearly to the base. Flesh of stalk yellow or somewhat reddish below, staining blue to violaceous on exposure. Spores olive to olive brown in deposit, ellipsoidal to fusiform, smooth, 11–16.8 × 5.1–6.6 μm in size.

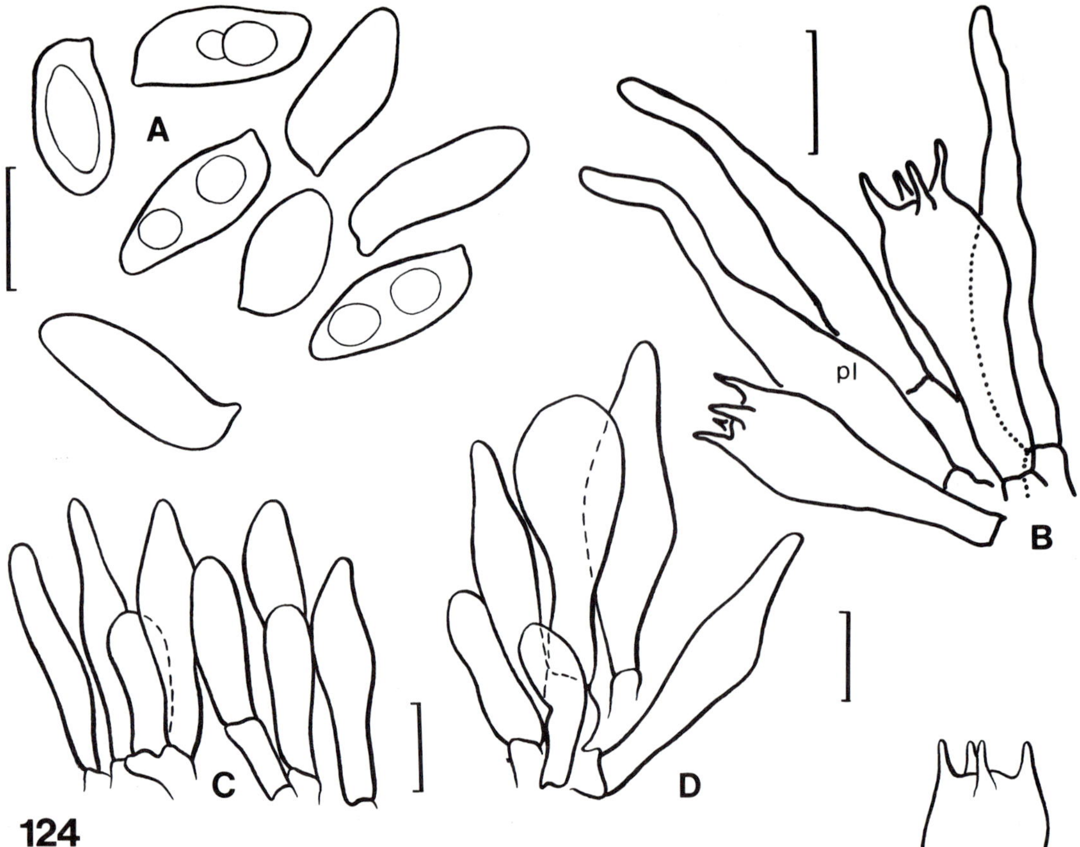

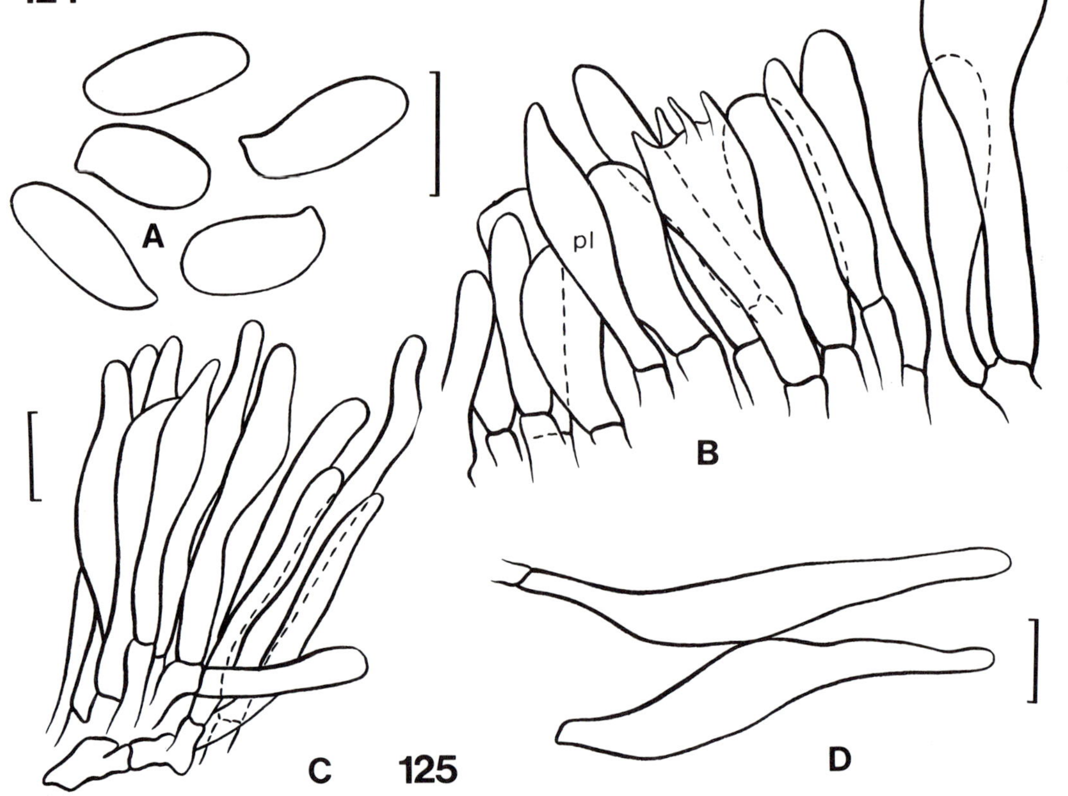

Fig. 124. *Boletus luridus*: A, spores; B, basidia and pleurocystidia (*pl*); C, cheilocystidia; D, caulocystidia.

Fig. 125. *Boletus pulcherrimus*: A, spores; B, basidia and pleurocystidia (*pl*); C, cheilocystidia; D, caulocystidia.

Scale line 10 μm

Observations. *B. luridus* occurs in Europe and North America. Some authors report it as edible but R. Singer, according to Smith and Thiers (1971), found it to be poisonous. In general appearance it is somewhat similar to *B. erythropus;* the distinctly reticulate stalk of *B. luridus* distinguishes it from *B. erythropus.*

Distribution and seasonal occurrence. *B. luridus* has been reported from the central area of the United States and eastward, and in Eastern Canada. It fruits in the summer and fall and at times is common.

Habitat and habit. The species is solitary to gregarious in hardwood forests; it seems to be associated with oak (*Quercus*).

Technical description. Cap 50-120 mm broad, convex becoming broadly convex to shallowly depressed at maturity; margin incurved in early stages; edge sterile, but neither crenate nor in segments; surface dry and unpolished to somewhat shiny, matted-fibrillose or with scattered minute fibrillose scales on the disc; color variable, yellow to olive yellow flushed with pinkish red, or sometimes light orange yellow on the margin and near brownish moderate orange yellow to dark orange yellow or reddish on the disc, soon staining greenish blue where handled. Flesh up to 3 cm thick, floccose, yellowish to reddish, with a red line above the tubes when freshly cut, instantly turning deep blue when injured; taste mild; odor indiscernible.

Tubes free or at least deeply depressed, rarely decurrent, 10-20 mm deep, moderate greenish yellow, but quickly staining greenish blue when injured; pores round, small, occurring at a density of about three pores per millimetre, dark red or brownish red to strong reddish brown, but soon fading to orange red, staining greenish blue when bruised.

Stalk 60-150 mm long, 10-30 mm thick, equal to clavate; base up to 40 mm thick, often flared at the apex, solid; surface dry, reticulate at the apex or nearly to the base, pruinose to scurfy, strong yellow overall or tinged orange red to blood red above and yellow below or sometimes yellow above and reddish below. Flesh bright yellow or streaked reddish in the base, soon staining blue to violaceous.

Spores olive to olive brown in deposit, 11.0-16.8 × 5.1-6.6 μm, ellipsoidal to fusiform, ventricose and somewhat inequilateral in side view, narrowly ovoid in face view, smooth, yellowish, nonamyloid. Basidia four-spored, clavate, 29.2-36.5 × 11.0-12.4 μm, hyaline to yellowish. Pleurocystidia fusiform, ventricose with long tapered necks, 33-48 × 7.3-13.5 μm, hyaline. Cheilocystidia similar to pleurocystidia or narrower, yellowish.

Tube trama divergent from a central strand of interwoven hyphae; hyphae cylindrical, 3.7-9.4(-11.0) μm wide, hyaline to yellowish, gelatinous. Subhymenium interwoven; hyphae short-cylindrical, 3.5-4.6 μm wide, hyaline. Cap cuticle filamentous, compactly interwoven; hyphae cylindrical, 3.7-5.8 μm wide; hyphal ends erect and aggregated in fascicles, thin- to thick-walled; contents and walls yellow to ocherous, sometimes encrusted. Cap trama loosely interwoven; hyphae branched, cylindrical to somewhat inflated, 3.7-8.8(-13.1) μm wide, hyaline to yellow. Caulocystidia narrowly fusiform-ventricose, 34.1-53.0 × 8.8-14.6 μm, hyaline to yellow; stalk hyphae compactly interwoven to more or less parallel, more or less cylindrical, 3.5-12.9 μm wide. Oleiferous hyphae refractive, not septate, contorted, ocherous, 3.5-5.1(-8.8) μm wide, scattered in the tube trama, stalk, and cap cuticle. Clamp connections not seen.

Useful references. The following books contain information pertinent to *B. luridus* in North America.

- *Mushrooms of North America* (Miller 1972)—A description and discussion are presented.
- *The Mushroom Hunter's Field Guide* (Smith 1974)—A black-and-white photograph and a description and discussion are given.

Consult, also, the technical references listed in the previous description for *B. erythropus.*

Boletus pulcherrimus Thiers & Halling

Figs. 125, 186

Common name. Alice Eastwood's boletus.

Distinguishing characteristics. Cap 80-200 mm broad, convex to plane; surface moist to dry and tomentose to fibrillose or somewhat scaly; color in early stages grayish yellow to reddish brown, often with reddish tones or blushes near the margin, usually unchanging with age or becoming somewhat more grayish. Flesh of cap greenish yellow to yellow, quickly changing to dark blue when bruised. Tubes pale to light greenish yellow, turning blue immediately when exposed; tube pores angular and dark red when young, fading to dark reddish brown, turning blue when injured. Stalk 80-150 mm long, 20-50 mm thick, with an enlarged clavate to bulbous base up to 95 mm thick; surface dry and conspicuously reticulate over the entire stalk or only on the upper two-thirds; ground color dark reddish orange to more orange or orange yellow, with dark red to dark reddish orange reticulations; base staining dark brown. Flesh of stalk yellow; flesh and surface of stalk staining blue when cut or exposed. Spores brown in deposit, more or less ellipsoidal to subfusiform or ventricose, smooth, 12.4-14.6(-16) × 5.1-5.8(-6.6) μm in size.

Observations. *B. pulcherrimus* has been known for years as *B. eastwoodiae* (Murrill) Sacc. & Trott. Thiers and Halling (1976) provide an explanation for the name change. It is one of the largest boletes in the western mushroom flora. According to Thiers and Halling (1976), the characteristics separating it from *B. satanas* and other red-pored boletes are the dark reddish brown, tomentose to fibrillose cap; the dark red pores; and the clavate to nearly bulbous stalk, colored pale reddish with dark red reticulations. This species is poisonous, causing more or less severe gastrointestinal upset.

Distribution and seasonal occurrence. *B. pulcherrimus* occurs on the Pacific coast from California to Washington and is to be expected in British Columbia. It fruits in the fall.

Habitat and habit. This species is solitary to gregarious and is found in mixed woods.

Technical description. Cap 80–200 mm broad, convex when young, becoming broadly convex to plane with age; margin entire and incurved, becoming decurved with age; surface moist to dry, glabrous to nearly tomentose, sometimes appressed-fibrillose with age, somewhat areolate or fibrillose-scaly when very old; color when young grayish yellow to reddish brown, often with distinct reddish tones or blushes near the margin, unchanging with age or with the tips of the fibrils becoming dark gray. Flesh 20–40 mm thick, brilliant greenish yellow to light yellow, quickly changing to dark blue when bruised; taste and odor indistinctive.

Tubes adnate to adnexed when young, becoming depressed, 5–10 mm long; color pale greenish yellow to light greenish yellow, turning blue immediately when exposed; pores up to 1 mm wide, angular, dark red when young, fading to dark reddish brown, sometimes yellowish toward the margin of the cap, turning blue immediately.

Stalk 80–150 mm long, 20–50 mm wide at the apex, up to 95 mm thick at the base, clavate to bulbous, rarely equal, solid; surface dry, conspicuously reticulate over the entire stalk or only the upper two-thirds; ground color dark reddish orange, moderately orange, or light orange yellow, with dark red to dark reddish orange reticulations; base staining dark brown where handled, turning blue when exposed. Flesh yellow, turning blue when exposed.

Spores brown in deposit, 12.4–14.6(–16.0) × 5.1–5.8(–6.6) μm, nearly ellipsoidal to subfusiform or ventricose, with a slight plage in side view, smooth, thin- to slightly thick-walled, ocherous, nonamyloid. Basidia four-spored, clavate, 36–48 × 7.3–13.1 μm, thin-walled, hyaline. Pleurocystidia fusiform-ventricose to nearly clavate, slightly protruding, 32–52 × 8–12 μm, hyaline. Cheilocystidia narrowly clavate to cylindrical, 22–44 × 3.7–5.8 μm, hyaline to yellowish.

Tube trama divergent from a distinct central strand of interwoven hyphae; hyphae more or less cylindrical, 3.7–10.6 μm wide, hyaline to ocherous. Subhymenium filamentous; hyphae branched, cylindrical, 3.7–4.4 μm wide. Cap cuticle filamentous, interwoven; hyphae cylindrical, 3.5–7.0(–11.7) μm wide; walls thin to slightly thickened, yellow to ocherous, encrusted; hyphae in fibrils parallel, more or less radially arranged, cylindrical, 5.0–11.7 μm wide, ocherous to brown, thick-walled, encrusted. Cap trama loosely interwoven; hyphae more or less cylindrical, 3.5–10.5 μm wide, hyaline to yellowish. Caulocystidia clavate to fusiform-ventricose, 29–55 × 6.6–18.3 μm, arranged in a close palisade, hyaline to yellowish; stalk hyphae interwoven, cylindrical, 3.5–7.3 μm wide, hyaline to yellowish. Oleiferous hyphae refractive, not septate, irregularly shaped, greenish yellow to ocherous, 3.7–10.6 μm wide, scattered in the tube trama, stalk, and cap cuticle. Clamp connections not seen.

Useful references. The following books contain information on *B. pulcherrimus*, listed as *B. eastwoodiae*.

- *Mushrooms of North America* (Miller 1972)—A color photograph and a description and discussion are presented.
- *The Mushroom Hunter's Field Guide* (Smith 1974)—An excellent color photograph, a black-and-white photograph, and a description and discussion are given.
- *The Savory Wild Mushroom* (McKenny 1971)—A black-and-white photograph and a description and discussion are given.

Thiers (1975) and Thiers and Halling (1976) contain good descriptions and discussions of this species.

Boletus satanas Lenz

Figs. 126, 187

Common name. Satan's bolete.

Distinguishing characteristics. Cap 100–200 mm broad at maturity, convex to somewhat flattened at maturity, usually remaining somewhat convex; surface dry, at times fibrillose to fibrillose-scaly; surface at the center often cracked; color olive buff to yellowish gray or buff brown with pinkish to rose tones when young, becoming predominantly pinkish by maturity. Flesh of cap thick; color in cap and stalk olive buff to deeper green, quickly turning blue when cut. Tubes usually some shade of greenish yellow, with very small round blood red pores staining blue when bruised. Stalk 60–120 mm long, 35–70 mm thick at the apex; base conspicuously bulbous, 90–140 mm thick; surface dry; ground color the same as the cap surface, with pink to vinaceous reticulations. Spores olive brown in deposit, more or less ventricose, smooth, 11–13.1(–16.7) × 4.8–5.1(–7.0) μm in size.

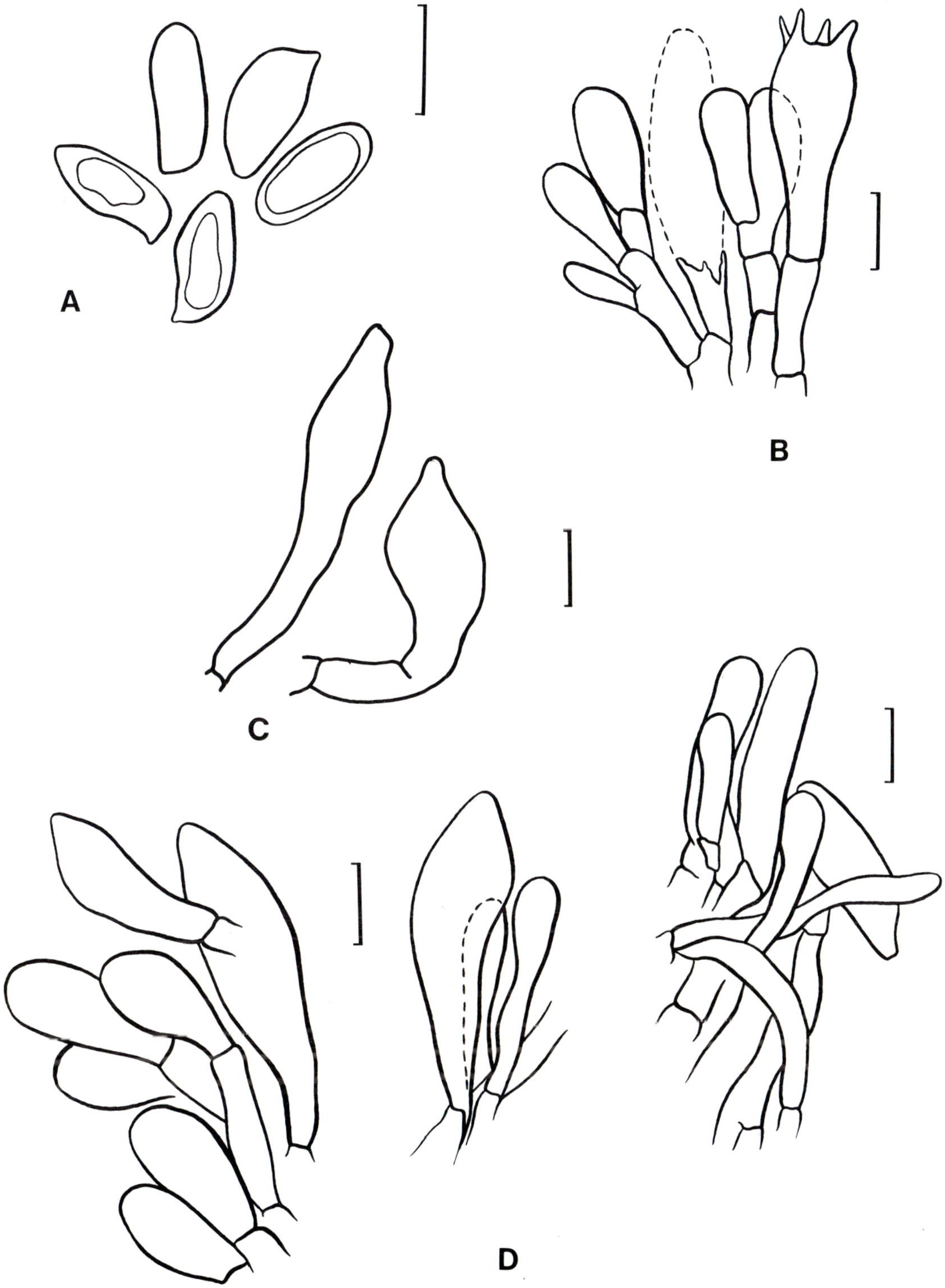

Fig. 126. *Boletus satanas*: A, spores; B, basidia; C, pleurocystidia; D, cheilocystidia.

Scale line 10 μm

Observations. The rose to livid pink color of the very large bulbous stalk base, the pale gray to pale buff cap that develops pinkish tones as it matures, and the small red pores on the undersurface of the cap help to distinguish this species. In early stages the stalk base is about as broad as the cap. The massive size of this bolete is helpful in its identification. *B. satanas* occurs in Europe and North America. It is usually considered poisonous and should be avoided.

Distribution and seasonal occurrence. *B. satanas*, also incorrectly spelled *B. satanus*, is apparently not widely distributed in North America. It has been reported from the west coast of California where it is at times common, and also from the southeastern United States. However, the latter report may not be authentic. This species fruits in fall, often the late fall, and in winter in California. It should be expected along the Pacific coast into British Columbia.

Habitat and habit. This species is solitary to scattered and occurs mostly under oak (*Quercus*). In Europe it grows under beech (*Fagus*).

Technical description. Cap 100-200 mm broad, convex to plano-convex with age; margin entire to eroded, incurved, becoming decurved with age; surface dry, unpolished, glabrous to obscurely fibrillose, occasionally fibrillose-scaly, becoming areolate on the disc or obscurely scaly on the disc and finely tomentose to glabrous toward the margin; color olive buff to yellowish gray or buff brown with pinkish to rose tones when young, with pink shades predominanting in older caps. Flesh 20-50 mm thick, olive buff to deeper green, readily turning blue when cut; taste and odor indistinctive.

Tubes 10-20 mm long, arcuate-decurrent to subdecurrent, becoming deeply to shallowly depressed with age, pale greenish yellow to sea green, blue when bruised; pores with a density of two per millimetre, very small, round, blood red, turning blue when bruised.

Stalk 60-120 mm long, 35-70 mm broad at the apex, conspicuously and abruptly bulbous; base 90-140 mm wide, solid; surface dry, reticulate, glabrous to obscurely tomentose over the nonreticulate portion, sometimes streaked and appearing appressed-fibrillose, similar in color to the cap, with pink to vinaceous reticulations; flesh similar in color to the flesh of the cap, turning blue when cut.

Spores olive brown in deposit, 11.0-13.1(-16.7) × 4.8-5.1(-7.0) μm, ellipsoidal in face view, ventricose with a plage in side view, smooth; walls thin to slightly thickened, ocherous, nonamyloid. Basidia four-spored, clavate, 23.4-29.2 × 9.5-10.5 μm, hyaline. Pleurocystidia clavate to fusiform-ventricose, 19.1-29.2(-35.2) × 7.0-8.2 μm wide, hyaline to yellowish. Cheilocystidia cylindrical to clavate, 18.3-24.1 × 5.1-7.3 μm wide, ocherous to reddish.

Tube trama divergent from a central strand; hyphae cylindrical, 2.3-7.0(-9.4) μm wide; walls thin to encrusted, yellowish, gelatinizing. Subhymenium filamentous; hyphae interwoven, cylindrical, 3.5-4.6 μm wide. Cap cuticle filamentous; hyphae interwoven, branched, more or less cylindrical, 3.7-7.3(-9.5) μm wide, with cylindrical to clavate apical cells; walls hyaline to ocherous, thin to encrusted. Cap trama loosely interwoven; hyphae more or less cylindrical, 4.4-9.4(-11.0) μm wide, hyaline to yellowish. Caulocystidia erect, clavate to subfusiform, 16.7-36.5 × 8.2-9.5(-13.1) μm, hyaline or with yellowish contents; stalk hyphae interwoven, cylindrical, 3.5-5.8(-13.1) μm, hyaline to yellowish. Oleiferous hyphae refractive, yellowish green, 3.5-5.8(-7.0) μm wide, more or less gnarled, scattered in the tube trama, stalk, and subcutis. Clamp connections not seen.

Useful references. The following books contain information on *B. satanas* pertinent to North America.

- *A Field Guide to Western Mushrooms* (Smith 1975)—A color photograph and a description and discussion are presented.
- *Mushrooms of North America* (Miller 1972)—A color photograph and a description and discussion are given.

Consult, also, publications by Thiers (1965, 1975) for further technical information.

Boletus sensibilis Peck

Figs. 127, 188

Common names. None.

Distinguishing characteristics. Cap 60-150(-300) mm broad, convex becoming broadly convex to nearly plane at maturity; surface dry; color grayish red to brick red or reddish brown in early stages, usually fading to dull cinnamon with age. Flesh of cap and stalk pale yellow to yellow, turning blue on exposure when cut or bruised. Tubes bright yellow, with round bright yellow pores; pores and tubes instantly staining blue when injured. Stalk 80-120 mm long, (6-)10-30 mm thick at the apex, equal or enlarged downward; surface at the apex reticulate; color brilliant yellow to dull red below. Spores olive brown in deposit, more or less ventricose, 10.6-13.1 × 3.7-4.4 μm in size.

Observations. *B. sensibilis* is in a group of species called the Stirpes *sensibilis*, which often have been incorrectly identified in the past and are still difficult to identify.

The entire group should be avoided. This species has also been called *B. miniato-olivaceus* var. *sensibilis* Peck.

B. sensibilis is closely related to *B. miniato-olivaceus* Frost and *B. bicolor* Peck. Workers interested in this group should consult works by Smith and Thiers (1971) and Grund and Harrison (1976), and other technical books and papers on *Boletus*. All three species are suspected to have caused poisonings of the gastrointestinal type. *B. bicolor* is reported as edible with a good flavor by some authors.

Distribution and seasonal occurrence. *B. sensibilis* is known from the Great Lakes region and eastward. It fruits from early summer into fall, depending on the weather. It is more common in wet seasons (Smith and Thiers 1971).

Habitat and habit. This species is scattered to gregarious, or at times solitary in hardwood forests, often where beech (*Fagus*) occurs.

Technical description. Cap 60-150(-300) mm broad, convex, expanding to broadly convex or nearly plane; margin even to slightly lobed, blunt; tubes exceeding the margin; surface dry, pruinose; color grayish red to brick red or reddish brown, slowly fading to dull cinnamon, quickly turning yellow with a drop of potassium hydroxide. Flesh thick, pale yellow, instantly turning blue when cut; odor and taste mild.

Tubes 8-15 mm long, adnate, slightly depressed at the apex of the stalk; color bright yellow, instantly turning blue when bruised; one or two pores per millimetre, round, bright yellow, and instantly turning blue when bruised with bruises slowly turning brownish to reddish.

Stalk 80-120 mm long, (6-)10-30 mm wide at the apex, solid, equal, or enlarged downward; surface reticulate at the apex, slightly so downward, tomentose at the base; color brilliant yellow to dull red below; flesh yellow, soon turning blue when cut.

Spores olive brown in deposit, 10.6-13.1 × 3.7-4.4 μm, oblong-elliptic to cylindrical in face view, ventricose and slightly inequilateral in side view from the presence of a plage, smooth, thin-walled to slightly thick-walled, yellowish, grayish to yellow in Melzer's reagent. Basidia four-spored, clavate, 20-30.7 × 7.3-9.5 μm, hyaline, thin-walled. Pleurocystidia fusiform to ventricose, with long tapered necks, 22.0-55.0 × 9.5-11.0 μm; necks 2.9-3.7 μm wide, hyaline, thin-walled. Cheilocystidia fusiform-ventricose, 25-35 × 5-8 μm, hyaline.

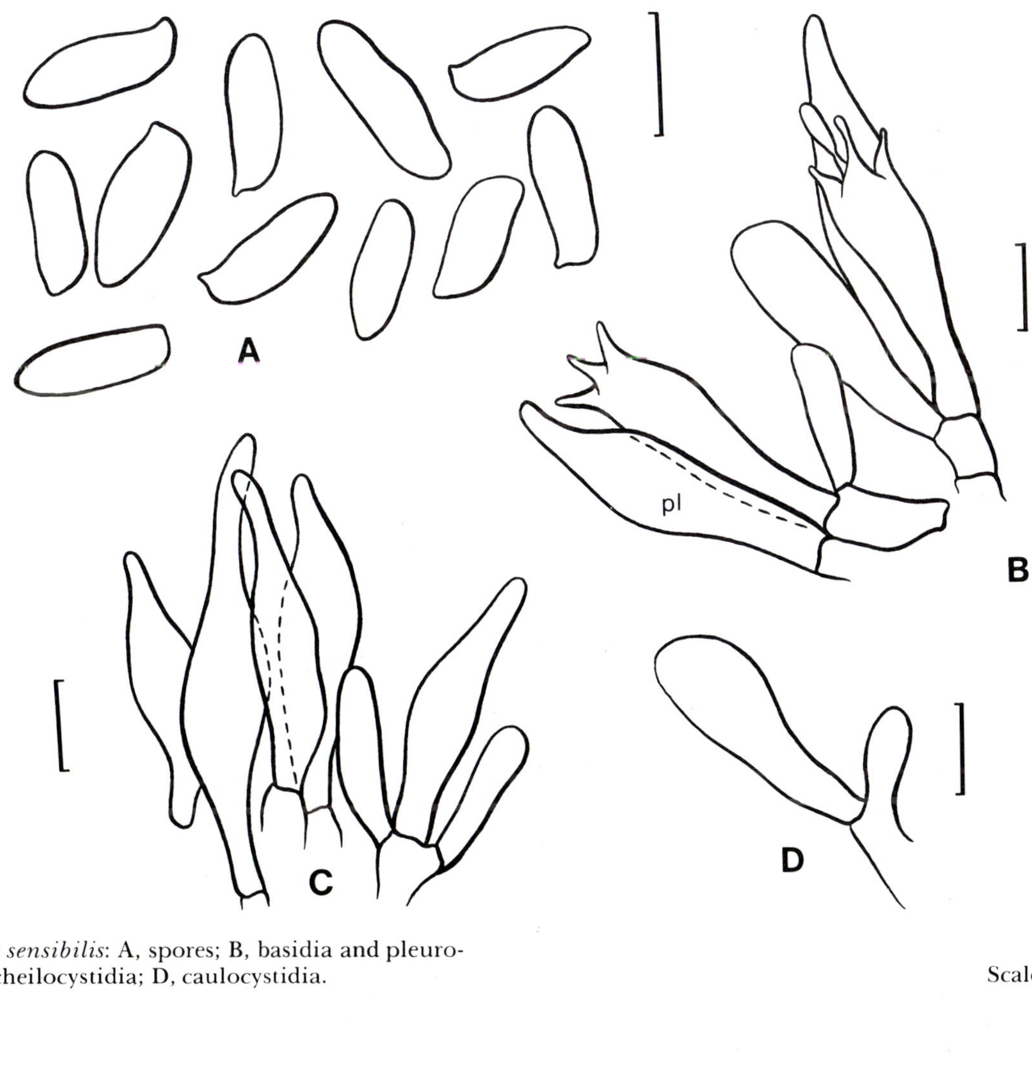

Fig. 127. *Boletus sensibilis*: A, spores; B, basidia and pleurocystidia (*pl*); C, cheilocystidia; D, caulocystidia.

Scale line 10 μm

Tube trama parallel to divergent from a central strand; hyphae more or less cylindrical, 3.7-12.4 μm wide, hyaline to yellowish. Subhymenium filamentous; hyphae cylindrical to slightly inflated, 3.7-5.1 μm wide, thin-walled. Cap cuticle filamentous, compactly interwoven; hyphae cylindrical, 3.7-5.8 μm; contents yellow; walls thin to slightly thickened and encrusted; hyphal ends obtuse to tapered. Cap trama interwoven; hyphae more or less cylindrical, 3.5-10.5 μm wide, branched, hyaline. Caulocystidia clavate to fusiform-ventricose, 15.3-32.9 × 5.1-10.6 μm, hyaline, thin- to slightly thick-walled; stalk hyphae interwoven to parallel, more or less cylindrical, 3.5-8.2 (-10.5) μm wide, hyaline. Oleiferous hyphae scattered, 4.4-5.8 μm wide. Clamp connections not seen.

Useful references. Workers interested in *B. sensibilis* and related species should consult the technical works of Snell and Dick (1970), Smith and Thiers (1971), and Grund and Harrison (1976).

Clitocybe (Fr.) Quél. — Tricholomataceae

The genus *Clitocybe* is widely distributed in Canada and the United States. It is a moderately large genus and in certain modern treatments includes the genus *Lepista.* In general, the genus *Clitocybe* can be distinguished by decurrent to subdecurrent gills; a usually white spore deposit, although the color sometimes ranges from buff or cream color to pale yellowish or pale pinkish buff; often fibrous fruit bodies, varying in size, more or less fleshy, and usually with a centrally attached stalk; a cap with the margin enrolled in the early stages and with the center depressed at maturity; no partial or universal veil; often smooth-walled spores, although the walls in some species are slightly roughened or wrinkled; and the cap cuticle typically composed of filamentous hyphae.

Several small genera have been separated from *Clitocybe.* Two of these, *Omphalotus* and *Hygrophoropsis,* are fairly common and at least one, *Omphalotus,* contains poisonous species. *Omphalotus* is treated later in this chapter, alphabetically listed, and in poison cases should be closely compared with *Clitocybe. Hygrophoropsis aurantiaca* (Wulfen ex Fr.) Maire, also classified in *Clitocybe* and *Cantharellus,* has been reported as poisonous, but other reports list it as edible, so its status is not yet resolved. The cap, stalk, and gills of this species are usually some shade of orange, but the color at times is pale yellow or strongly brownish. The gills are decurrent, crowded, and repeatedly forked. The orange color and the crowded, forked gills help to distinguish it from *Clitocybe.* It often fruits on or near decayed wood.

Other genera that can be confused with *Clitocybe* are *Cantharellus, Laccaria, Hygrophorus, Leucopaxillus, Tricholoma,* and *Collybia. Cantharellus* typically has blunt-edged, more or less fold-like to ridge-like gills rather than the distinct sharp-edged, blade-like gills of a typical mushroom. *Laccaria* differs in having rather thick and somewhat waxy-looking gills that are often pinkish to purplish in color. The spores are often spiny or roughened with fine points. *Hygrophorus* has many species with decurrent gills; however, the rather thick, waxy gills and long, narrow basidia of this genus separate it from *Clitocybe.*

Certain *Leucopaxillus* species have the same stature as *Clitocybe* species. All *Leucopaxillus* species have amyloid ornamentation, in the form of roughenings, on the spores. *Tricholoma* species are best separated from *Clitocybe* by the notched to adnexed or sinuate gill attachment. *Collybia* is generally separated from *Clitocybe* by the more cartilaginous stalk and adnate to adnexed gill attachment.

Besides the species of *Clitocybe* containing muscarine, treated in Chapter 10, and *Clitocybe clavipes* which gives a disulfuram-like reaction when eaten with alcohol, treated in Chapter 9, there are at least two other species that have caused gastrointestinal upset, *Clitocybe irina,* and *Clitocybe nebularis.* Both are considered edible but are known to cause problems with some people. *C. nebularis* is not often sought after for food because its skunkcabbage-like odor and taste make it unappealing. Consult also the treatment of *Omphalotus* later in this chapter, alphabetically listed, because in some publications its species are placed in the genus *Clitocybe.*

Clitocybe irina (Fr.) Bigelow & Smith — Figs. 128, 189

Common names. None.

Distinguishing characteristics. Cap 20-130 mm broad, convex to plane when expanded, usually with the margin enrolled in early stages and the center usually somewhat raised; surface nearly viscid to dry; color usually whitish to pallid, soon becoming pale dingy buff or more or less brownish pink to pale tan, often dingy brownish when old and water-soaked; surface at times pitted or water-spotted. Odor variable, pungent to fragrant or resembling green corn, at times faint. Gills crowded; color whitish in early stages, soon

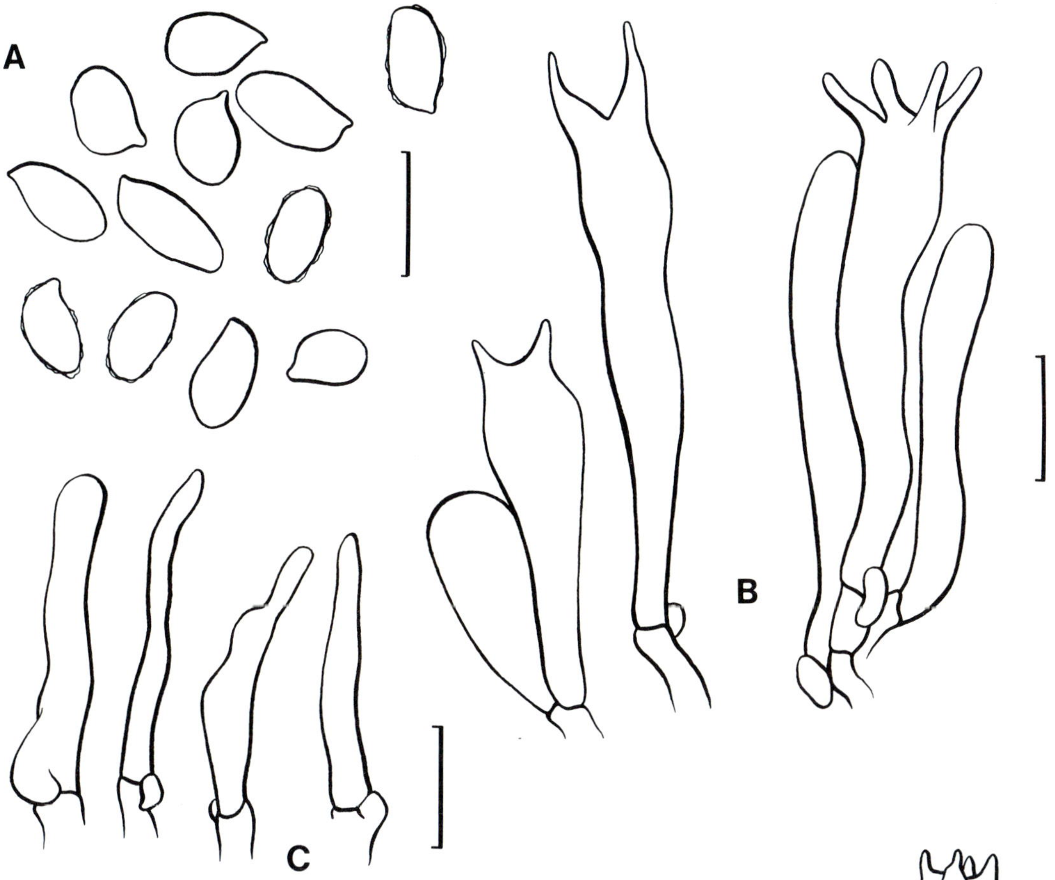

128

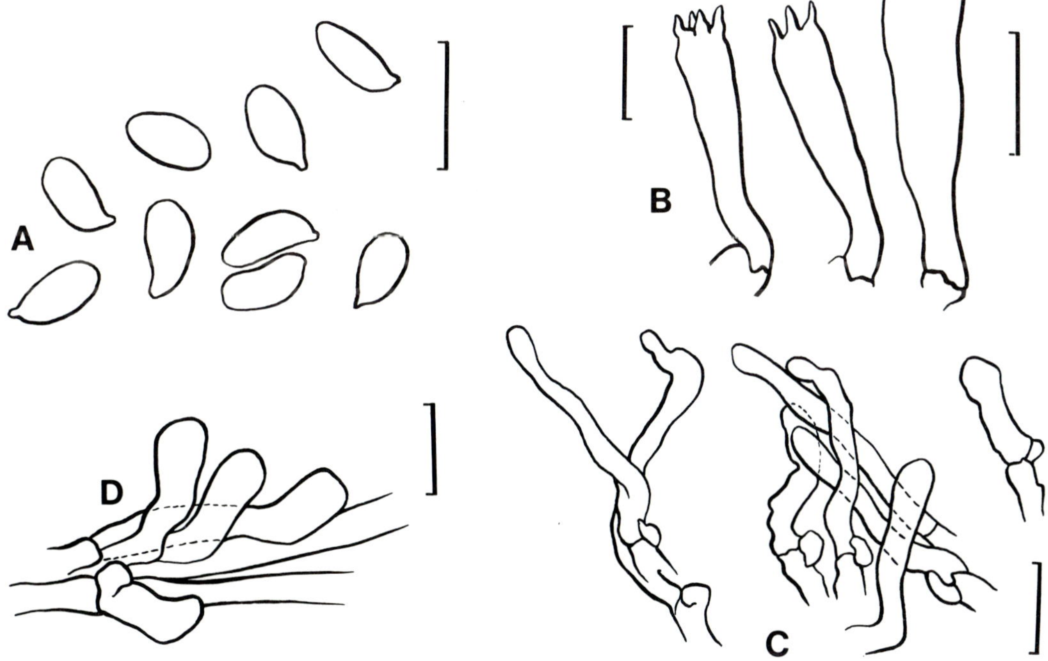

129

Fig. 128. *Clitocybe irina*: A, spores; B, basidia; C, cheilocystidia.

Fig. 129. *Clitocybe nebularis*: A, spores; B, basidia; C, cheilocystidia; D, caulocystidia.

Scale line 10 μm

becoming pale avellaneous to pale pinkish, pale brownish pink, or dingy buff, broadly adnate to subdecurrent, rarely truly decurrent. Stalk 10–25 mm thick at the apex, equal or enlarged at the base, clavate to bulbous; color pallid in early stages, becoming a sordid pinkish brown with age. Spores pale buff to pale pinkish buff in deposit, ellipsoidal; surface smooth to slightly roughened, 7.3–10.6 × 3.3–5.1 μm in size.

Observations. *C. irina* is a widely distributed species that is often quite common, at times forming large fairy rings or arcs containing hundreds of fruit bodies. *C. irina* is edible for most people and is commonly eaten. It apparently cannot be tolerated by some individuals, because it causes gastrointestinal upset. Smith (1962) reports that of 14 people who ate this species at the same gathering, two suffered gastrointestinal upset consisting of stomach cramps, nausea, and vomiting. It should be eaten with caution when trying it for the first time.

Its large size and occurrence in large numbers make it attractive as an edible mushroom. Difficulty in distinguishing *C. irina* from other large species of *Clitocybe* is highly probable, especially when the cap color of *C. irina* is whitish or pale.

C. robusta Peck (=*C. alba* (Bolt.) Singer) is one species likely to be mistaken for *C. irina,* especially because the two species can occur in the same habitat during the same season. *C. robusta* is fleshier; has a disagreeable odor and taste, a pale yellowish spore deposit, and smooth spores; and is paler overall than *C. irina. C. phyllophila* (Fr.) Quél. may also occur with one or both of these two species. It is very similar in color to *C. irina* and also has a pinkish spore deposit, but the spores are smooth and smaller (4–5 × 2.5–3.5 μm) than those of *C. irina.*

C. irina has also been called *Lepista irina* (Fr.) Bigelow, *Tricholoma irinum* (Fr.) Kummer, and *Rhodopaxillus irinus* (Fr.) Métrod.

Distribution and seasonal occurrence. *C. irina* is found across Canada, north into the Yukon Territory. It appears to be most common in the Great Lakes region and eastward. It usually fruits in the fall, often most commonly in October, but it has been collected in the summer and, rarely, as early as April.

Habitat and habit. *C. irina* is terrestrial, usually gregarious to caespitose. It is often found in arcs or fairy rings under conifers, in mixed woods, and in hardwood forests.

Technical description. Cap 20–130 mm broad, obtusely umbonate to convex with an enrolled margin, expanding to plane or nearly so, often retaining a low broad umbo, with the margin sometimes lobed or wavy; surface more or less viscid to the touch, soon dry, not hygrophanous, smooth, sometimes pitted or with water-soaked spots; margin white-cottony and occasionally sulcate-striate from pressure against the gills; color whitish when young or dry, usually soon becoming pale dingy buff (pale cinnamon pink, pale pinkish buff, or near vinaceous buff) or more or less brownish pink to pale tan, often becoming dingy brownish when old and water-soaked. Flesh thick, soft, whitish to pinkish or watery brownish pink; taste mild; odor sharply pungent, fragrant, or resembling green corn, but sometimes faint.

Gills whitish when young, soon becoming pale avellaneous to pale brownish pink, usually darker than the color of the spores, broadly adnate to subdecurrent or rarely decurrent, crowded, narrow to moderately broad; edges even to uneven.

Stalk 40–80 mm long, 10–25 mm thick at the apex, equal to clavate or sometimes rather bulbous; base up to 40 mm thick, central, solid; surface more or less fibrillose to slightly scabrous, sometimes scaly from the torn cuticle, becoming longitudinally striate; base with many mycelia binding the substratum; color pallid in early stages, becoming sordid pinkish brown with age. Flesh watery pinkish brown to whitish. Partial veil and universal veil lacking.

Spores pale creamy incarnate to pale pinkish buff or pale buff in deposit, 7.3–10.6 × 3.3–5.1 μm, ellipsoidal, smooth to finely warted, hyaline, nonamyloid. Basidia two- and four-spored, clavate, 23–44 × 6.6–7.3 μm, hyaline. Pleurocystidia not differentiated. Cheilocystidia, when present, filiform to narrowly fusiform, 18–42 × 2.2–3.7 μm, hyaline.

Gill trama interwoven. Cap cuticle filamentous; hyphae tightly interwoven, more or less radially arranged, cylindrical, 2.5–3.7 μm wide, hyaline, with some more or less erect hyphal tips. Cap trama loosely interwoven; hyphae more or less cylindrical, 3.7–10.2 μm wide, hyaline. Caulocystidia lacking; stalk hyphae more or less parallel to interwoven on the surface, cylindrical, 3.5–9.5 μm wide. Oleiferous hyphae refractive, greenish yellow, thin-walled, 3.3–7.3 μm wide, scattered in the gill trama and stalk. Clamp connections present.

Useful references. The following books provide information on *C. irina* pertinent to North America.

- *Edible and Poisonous Mushrooms of Canada* (Groves 1979)—A poor color photograph and a description and discussion are presented.
- *Mushrooms of North America* (Miller 1972)—A color photograph and a description and discussion are given.
- *The Mushroom Hunter's Field Guide* (Smith 1974)—A color photograph and a description and discussion are given.

Clitocybe nebularis (Fr.) Kummer

Figs. 129, 190

Common name. Graycap.

Distinguishing characteristics. Cap up to 150 mm broad, usually more or less plane with the margin somewhat decurved and the center slightly depressed; surface dry, radially streaked from fibrils; color grayish to pale brownish gray or somewhat paler. Flesh with a disagreeable odor and taste resembling skunk cabbage. Gills close, thin, forked, adnate to short- or moderate-decurrent; color whitish to buff or pale cream color. Stalk 25-40 mm thick at the apex, often enlarged toward the base; surface streaked with grayish brown fibrils, the same color as the cap surface, over a whitish ground color. Spore deposit typically pale yellowish; spores smooth, ellipsoidal, 5.8-8.0 × 3.7-4.4 μm.

Observations. *C. nebularis* at times is fairly common in the west but is rare in the east. The spore deposit color is reported as white in some publications, although it is typically a pale yellowish.

C. nebularis is considered edible, but it does not agree with everyone, so it should be eaten with caution. It has both a disagreeable odor, resembling skunk cabbage, and an equally unpleasant taste, and thus is not generally collected for food. It is reported to have caused poisonings in Europe (Alder 1950, 1956, 1959, 1961).

Distribution and seasonal occurrence. It appears to be widely distributed but rare in eastern North America and common to abundant in the west. It fruits mostly in the fall but occasionally is reported also in the spring. In warmer areas of the Pacific coast it may fruit into the winter.

Habitat and habit. The species is terrestrial; it grows solitary, scattered, or gregarious, on humus under conifers or in mixed woods. It is reported to be common in cedar-hemlock-alder (*Thuja-Tsuga-Alnus*) areas.

Technical description. Cap 90-150 mm broad, usually plane with the margin somewhat decurved and the disc shallowly depressed, convex with the margin incurved at times; disc nearly umbonate at times; margin sulcate-striate or split; surface not hygrophanous, conspicuously radiate-fibrillose; color grayish to pale drab or pale brownish gray, at times nearly pallid; flesh thick and tough, white; odor and taste disagreeable, resembling skunk cabbage.

Gills whitish to buff or pale cream color, adnate in early stages, soon becoming short-decurrent to moderately decurrent, thin, close, forked, not interveined, broad, arched.

Stalk 80-100 mm long, 25-40 mm thick at the apex; base often enlarged and tomentose, eccentric at times, straight to curved, occasionally compressed, solid, fleshy-fibrous; surface fibrillose-striate; fibrils near buffy brown over a silky whitish ground color, becoming sordid with handling. No partial or universal veil.

Spores pale creamy yellowish or yellowish white in mass, 5.8-8.0 × 3.7-4.4 μm, elliptic and slightly inequilateral in side view, smooth, hyaline, nonamyloid. Basidia four-spored, clavate, hyaline, 21.9-32.9 × 5.8-7.3 μm. Pleurocystidia not differentiated. Cheilocystidia, when present, narrowly clavate to filiform, 18-44 × 2.2-3.0 μm, hyaline.

Gill trama interwoven; hyphae more or less cylindrical, 4-9.5(-14.2) μm wide, hyaline. Subhymenium filamentous; hyphae compactly interwoven, cylindrical, 2.2-2.9 μm wide, hyaline. Cap cuticle filamentous; hyphae interwoven, more or less radially arranged, cylindrical, 2.2-3.7 μm wide, hyaline to yellowish. Cap trama loosely interwoven; hyphae more or less cylindrical, 5.1-14.6(-22.3) μm wide, hyaline. Caulocystidia cylindrical to clavate, single or in clusters, 10-30 × 5.1-6.6 μm, hyaline to yellowish; tips of interwoven hyphae 3.0-4.4 μm wide, covering a core of parallel hyphae 3.5-8.8 μm wide; tomentum at the stalk base interwoven; hyphae cylindrical, 2.3-6.6 μm wide, thin-walled to slightly thick-walled and encrusted. Oleiferous hyphae refractive, thin-walled, not septate, 3.5-6.0 μm wide, scattered in the gill trama. Clamp connections present.

Useful references. The following books provide information on *C. nebularis* pertinent to North America.

- *A Field Guide to Western Mushrooms* (Smith 1975)—A color photograph and a description and discussion are presented.
- *Mushrooms of North America* (Miller 1972)—A color photograph and a description and discussion are presented.
- *The Savory Wild Mushroom* (McKenny 1971)—A black-and-white photograph and a description and discussion are given.

Collybia (Fr.) Quél.

Tricholomataceae

Collybia is a moderately large genus of white-spored mushrooms. The species range from fairly large to minute, but they are generally small. The caps are convex when young, often becoming somewhat flattened with age. When young, the margin of the cap is enrolled to incurved rather than straight. The gills are attached, usually adnexed to adnate but not decurrent. The stalk is usually fairly thin, brittle to fibrous, and often with a soft center. There is no partial or universal veil. The fruit bodies do not usually revive when moistened, as is more characteristic of *Marasmius*. The spore deposit in *Collybia* is usually white to light cream color, or in some cases pinkish buff. The spores are smooth and nonamyloid. The cap cuticle is filamentous.

Most species of *Collybia* are small and not collected for food. In general they are considered nontoxic. *Collybia dryophila*, a species considered to be edible and choice by many authorities, has caused a poisoning in the Great Lakes region (L. Gillman, Personal commun.). Ms. Gillman was one of several victims. Several caps were fried in butter and symptoms developed about 15 minutes after eating them. Her symptoms began with a feeling of queasiness in the stomach, followed by rhythmical constriction of the throat muscles and several instances of vomiting during the first hour. An hour after onset of the initial symptoms, the face became numb, especially around the mouth and nose. She then felt dizzy when walking and experienced the room spinning when she lay down. Objects seemed a little more colorful and larger than usual, and sounds seemed louder.

This species is so widely distributed and variable that certain races or forms may possibly be poisonous, at least to some individuals. *Collybia dryophila* and its relatives need critical study before the edibility of this group can be accurately determined. Until then, caution should be used when eating *C. dryophila* and related species.

Collybia dryophila (Fr.) Quél.

Figs. 130, 191

Common names. Nut-brown collybia, oak-loving collybia.

Distinguishing characteristics. Cap 25–70 mm broad, convex to plane, with the center slightly uplifted at times; margin enrolled, but at times becoming elevated and wavy; surface dry to moist; color brownish to tan, yellowish brown, yellowish honey color, or somewhat more ocherous. Gills usually white to pallid in color, rarely yellow, attached (adnate to adnexed), crowded,

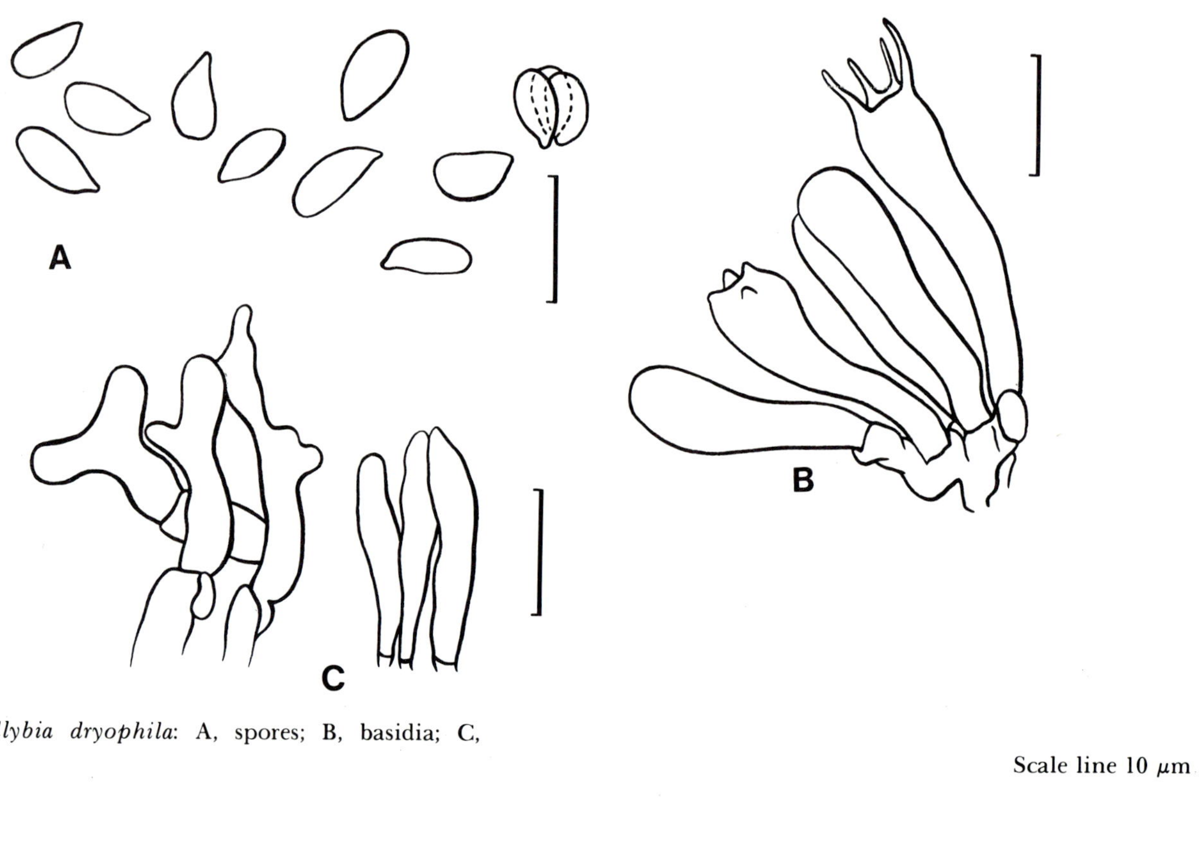

Fig. 130. *Collybia dryophila*: A, spores; B, basidia; C, cheilocystidia.

Scale line 10 μm

narrow. Stalk 30–80 mm long, 2–6 mm thick, equal or slightly enlarged below, glabrous; base somewhat spongy-thickened, often with numerous white rhizomorphs; color white to pallid or yellowish or tinted with the same color as the cap surface. Spores white in deposit, narrowly ellipsoidal in shape, smooth, nonamyloid, 5.1–7.0(–8.0) × 3.0–3.7 μm in size.

Observations. *C. dryophila* is widely distributed in North America. It occurs in a variety of habitats and in some years can be found throughout the mushroom season. The color of the cap is variable; usually it is some shade of yellow brown and often fades somewhat with age, especially on the margin. There is also a yellow-gilled form of *C. dryophila.* The stalk and cap of *C. dryophila* at times produce irregular masses of fungoid tissue that often obliterate the original mushroom. According to Smith (1974, photograph), this type of cancer-like growth is more frequent toward the end of the mushroom season. Ginns and Sunhede (1978) state that these tumors and galls are caused by species of *Christiansenia* (Corticiaceae).

C. dryophila is most commonly mistaken for *C. butyracea* (Bull. ex Fr.) Kummer. According to R. Halling (Personal commun.), *C. butyracea* has a vinaceous brown, greasy to lubricous, hygrophanous cap; nearly close to subdistant, broad, white gills with eroded edges; a clavate, striate, white to vinaceous-tinged stalk; and a pinkish buff to cream pink spore deposit. The spores are dextrinoid and the cap cuticle is composed of repent, filamentous, radially arranged hyphae. The cap cuticle of *C. dryophila* is composed of repent, short-celled, often branched hyphae that are not radially oriented, and it does not have dextrinoid spores.

Distribution and seasonal occurrence. This species is often common and widely distributed across Canada and other areas of North America. It fruits in the spring, summer, and fall. It is usually most abundant in the fall and persists until late in the season.

Habitat and habit. *C. dryophila* is terrestrial, often found on humus in hardwood, mixed, and coniferous forests. It grows scattered or gregarious, or occurs in clusters.

Technical description. Cap 25–70 mm broad, broadly convex to nearly plane, sometimes slightly umbonate; margin enrolled, but often becoming elevated and wavy; surface moist to dry, appearing nearly hygrophanous, opaque, slightly pubescent or glabrous; color variable, dark grayish reddish brown to dark brown or tan, often with yellowish tones, often becoming strong yellowish brown over the disc and more or less ocherous to yellowish honey color along the margin, sometimes developing a pale grayish moderate orange cast, at times one color overall. Flesh thin, pliant, pallid to whitish, often watery-textured; taste mild; odor more or less pleasant.

Gills whitish to pallid, although yellow in one variant, adnate to adnexed, crowded to very crowded, narrow; edges even.

Stalk 30–80 mm long, 2–6 mm thick, equal or nearly so or enlarged at the base, terete or compressed, central, hollow, glabrous, dull and unpolished when young, white to pallid or yellowish or tinted with the same colors as the surface of the cap; base somewhat spongy, thickened, with numerous white rhizomorphs; flesh cartilaginous; partial veil and universal veil lacking.

Spores white in deposit, 5.1–7.0(–8.0) × 3.0–3.7 μm, narrowly ellipsoidal, smooth, hyaline, nonamyloid. Basidia four-spored, clavate, 19–25 × 5.8–6.6 μm, hyaline. Pleurocystidia lacking. Cheilocystidia inconspicuous, filamentous to irregular-shaped, 17–25 × 3.7–5.8 μm, hyaline.

Gill trama interwoven to parallel; hyphae cylindrical, 3.7–8.8 μm wide, hyaline. Subhymenium filamentous, compactly interwoven; hyphae more or less cylindrical, 3.0–3.7 μm wide, hyaline. Cap cuticle filamentous, compactly interwoven, frequently branched, sometimes nodulose; hyphae cylindrical, 3.5–8.0 μm wide, yellowish to brownish, thin-walled. Cap trama loosely interwoven; hyphae more or less cylindrical, 3.5–11 μm wide, hyaline. Caulocystidia lacking; stalk hyphae interwoven to more or less parallel, cylindrical, 3.0–5.8 μm, hyaline. Oleiferous hyphae refractive, greenish yellow, 3.7–6.6 μm wide, scattered in the gill trama and cap cuticle. Clamp connections present.

Useful references. The following books provide information on *C. dryophila* pertinent to North America.

- *Edible and Poisonous Mushrooms of Canada* (Groves 1979)—A poor color photograph and a description and discussion are presented.
- *Mushrooms in their Natural Habitats* (Smith 1949)—A color photograph on a reel and a description and discussion are given.
- *Mushrooms of North America* (Miller 1972)—An excellent color photograph and a description and discussion are given.
- *The Mushroom Hunter's Field Guide* (Smith 1974)—A color photograph, a black-and-white photograph, and a description and discussion are given.
- *The Savory Wild Mushroom* (McKenny 1971)—A black-and-white photograph and a description and discussion are given.

Gomphus S. F. Gray

Cantharellaceae

The genus *Gomphus* belongs to the Cantharellaceae, a family of fungi commonly called the chanterelles. Besides the genus *Gomphus* this family contains the genus *Cantharellus,* known to most mushroom collectors, as well as *Craterellus* and sometimes *Polyozellus.* The species under consideration here, *Gomphus floccosus,* is classified as *Cantharellus floccosus* Schw. by some workers (Smith and Smith 1973).

The chanterelles are generally mushroom-like in stature, with a stalk and cap, but sometimes they have a more vase-like appearance. The undersurface of the cap, instead of having distinct, more or less thin gills with knifeblade-like edges, is nearly smooth to wrinkled and covered with low ridges or folds that often branch and interconnect, or it has thickened, blunt-edged gills that are often forked.

The genus *Gomphus* contains species that produce medium- to large-sized fruit bodies that have a fibrous to fleshy consistency. The spore deposit is ocherous to yellowish and the spores have a roughened or wrinkled surface. Often the fruit bodies are vase-like or trumpet-shaped, with a depressed center. The stalk is usually stout and is solid or hollow below the cap surface.

Cantharellus is somewhat difficult to separate from *Gomphus* and therefore several workers classify all the species in *Cantharellus. Cantharellus* as defined in this treatment includes those species that have typically medium to somewhat small fruit bodies; blunt gill-like ridges, wrinkles, or veins on the undersurface of the cap, or an essentially smooth undersurface; a solid or hollow stalk; thin-fleshed fruit bodies that are firm or fleshy; a white to buff, yellowish, or pinkish spore deposit; and smooth spores. *Cantharellus cibarius* Fr., a highly prized edible mushroom, is an example of this group.

Craterellus, another chanterelle, produces fruit bodies with a more or less thin to membranous consistency. The color of the fruit bodies is usually dark to sordid and some shade of gray, black, or brown. The undersurface of the cap is smooth, wrinkled, or covered with shallow vein-like ridges. The spores are smooth and white, cream, or buff in deposit. This genus is easily distinguished from *Gomphus* but is similar in appearance to certain species of *Cantharellus. Craterellus* and *Cantharellus* can be separated microscopically by checking the fruit bodies for clamp connections at the crosswalls of the hyphae. Species of *Craterellus* lack clamp connections; clamp connections are present in *Cantharellus.* The genus *Gomphus* has some species with clamp connections and some that lack them.

The genus *Polyozellus* is similar in appearance to *Gomphus* but is actually not related to it. The fruit bodies of *Polyozellus* usually occur in compound clusters, fused together as a group or in large masses; have a fleshy, soft to somewhat brittle consistency; are typically dull purplish to violaceous or bluish fuscous, becoming more blackish with age; have veins on the undersurface of the cap that are often interconnected in a net-like pattern or that form pore-like depressions; and have a white spore deposit, with spores that are angular and covered with coarse warts. The species common in North America is *Polyozellus multiplex* (Underw.) Murrill.

In general, the chantarelles are considered edible, although not all species have been tested. *Gomphus floccosus,* however, although edible for some people, has caused gastrointestinal upset in certain individuals and in laboratory animals (Smith 1949, Tyler 1963). A good description of the symptoms is not available, but their onset is delayed 8–14 hours after ingestion and the digestive upset usually includes diarrhea. *Gomphus floccosus* and *Gomphus kauffmanii* (Smith) Corner contain a considerable amount of α-tetradecylcitric acid (norcaperatic acid), which may be the toxic agent (Henry and Sullivan 1969, Carrano and Malone 1967, Miyata et al. 1966). We have no reports of poisoning by *Gomphus kauffmanii* but based on its close relationship with *Gomphus floccosus,* it should be avoided.

Poisoning by green chanterelles, *Cantharellus cibarius* infected with *Hypomyces luteovirens* (Fr.) Tul. (= *H. viridis* (Alb. & Schw. ex Fr.) Karst.), has been reported by Alder (1961). There are no similar reports from North America and we are still investigating the above case (J. F. Ammirati).

Gomphus floccosus (Schw.) Singer

Figs. 131, 192

Common names. Woolly chanterelle, scaly chanterelle.

Distinguishing characteristics. Cap 50–150 mm broad, stump-shaped in early stages, with the margin spreading and the center breaking up into scales and becoming hollowed, so that the mature fruit bodies are vase-like to funnel-shaped or trumpet-shaped; color reddish orange to orange, orange yellow, or even more yellowish, with the scales and the center usually more darkly colored than the margin. Undersurface of cap having decurrent low blunt ridges or folds that are forked to irregular in pattern and also often exhibiting shallow pore-like areas; color whitish to pale buff or yellowish. Stalk short and not sharply distinct from the cap, with the entire fruit body ranging from 80 to 200

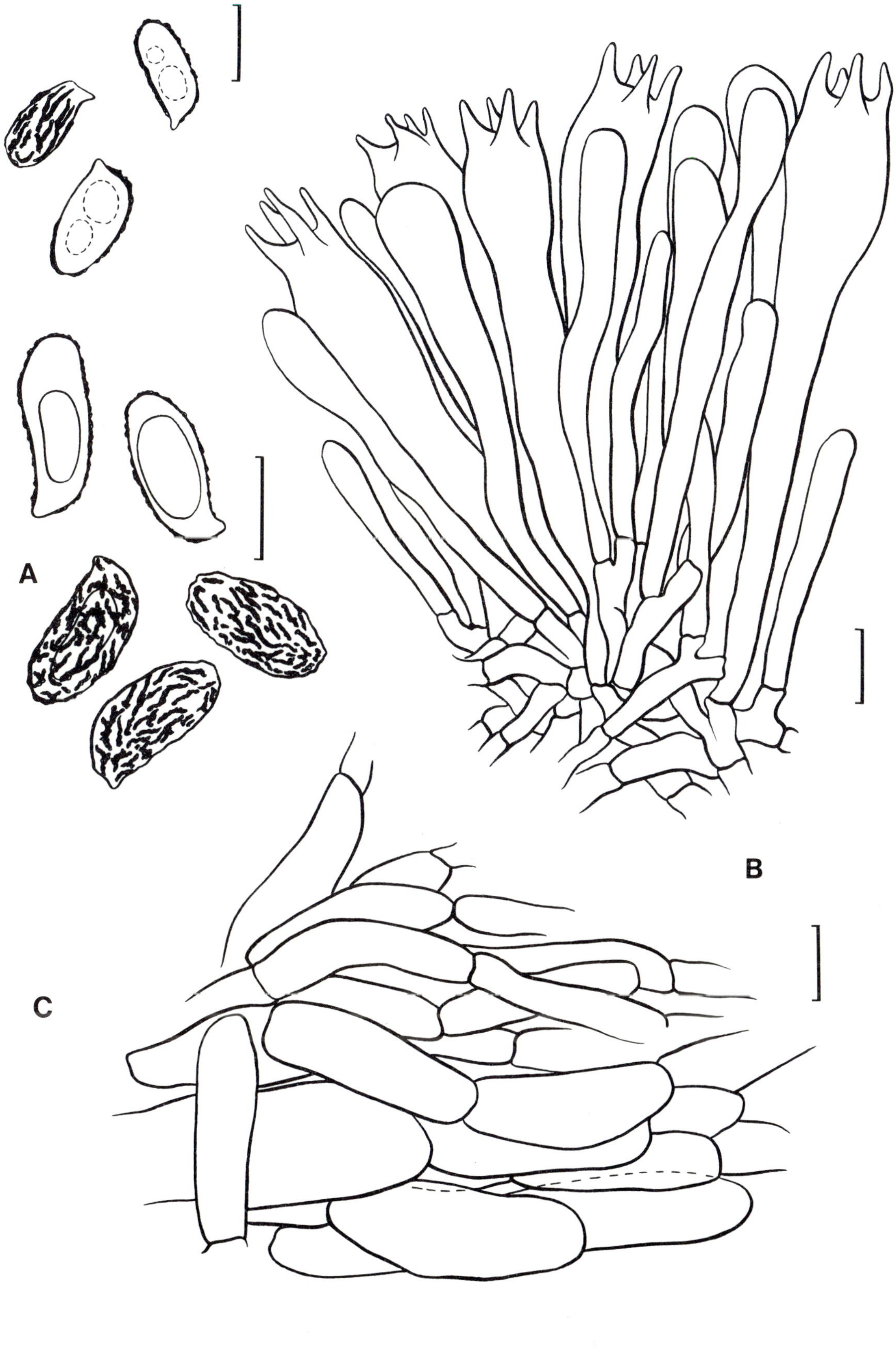

Fig. 131. *Gomphus floccosus*: A, spores; B, basidia; C, cap cuticle.

Scale line 10 μm

mm high, tapered toward the base, solid in early stages, becoming hollow; color whitish to pallid; spores ocherous in deposit, ellipsoidal, roughened to somewhat wrinkled on the surface, 12.4-16.8 × 5.8-7.3 μm in size.

Observations. *G. floccosus* is a widely distributed species. It varies in size and stature, configuration of the spore-producing surface, extent to which the stalk is hollowed, and color of the fruit bodies, as well as other features. Several forms and at least two subspecies have therefore been described (Smith and Morse 1947, Petersen 1971).

G. kauffmanii is closely related to *G. floccosus.* It is generally larger than *G. floccosus;* the cap is scalier and the scales are brown to clay color, yellowish brown, or ocherous tawny; the undersurface of the cap is covered with interconnecting folds and shallow pores that are yellow when young, becoming pinkish buff to pallid; the fruit bodies, especially the coarse scales, are very brittle. It has been reported from both eastern and western North America.

Distribution and seasonal occurrence. *G. floccosus* is found across North America wherever a suitable habitat is found. It fruits mainly in the late summer and fall.

Habitat and habit. *G. floccosus* is solitary, gregarious, or caespitose. It is associated with conifers and occurs from timberline to sea level.

Technical description. Cap 50-150 mm broad, truncate when young, soon becoming hollow in the center from the breaking up of tissue into floccose scales; margin finally spreading, giving the cap a funnel-shaped to trumpet-shaped appearance; surface dry, covered with flattened-appressed to recurved scales; color light orange to yellow or paler, fading to dingy yellow or pale yellow orange, with yellow, orange, or reddish orange scales. Flesh moderately thick, becoming thin on older caps, fibrous, white to pallid, unchanging; taste and odor indistinctive.

Hymenophore consisting of thick low blunt ridges or folds, forked or irregular in pattern, sometimes nearly pore-like, whitish to pale buff or more yellowish, decurrent almost to the base of the stalk.

Stalk short, not sharply distinct from the cap, with the entire fruit body 80-200 mm high, usually tapered toward the base, unpolished, solid in early stages, becoming hollow; surface whitish to pallid. Partial veil and universal veil absent.

Spores ocherous in deposit, 12.4-16.8 × 5.8-7.3 μm, ellipsoidal, with a plage visible in side view, narrowly elliptic in face view, verruculose, yellowish, with blue ornamentation in a solution of Cotton Blue and lactophenol, nonamyloid. Basidia four-spored, narrowly clavate, 70-94 × 10.5-13.1 μm, hyaline, thin-walled; contents refractive, crystalline. Pleurocystidia and cheilocystidia lacking. Paraphyses cylindrical, 73-82 × 3.5-5.8(-7.3) μm, hyaline.

Trama of hymenophore interwoven; hyphae cylindrical, 3.5-8.2(-9.4) μm wide, hyaline. Subhymenium filamentous, interwoven; hyphae cylindrical, 3.5-5.8 μm wide, hyaline; contents crystalline; surface finely encrusted. Cap cuticle filamentous, interwoven; hyphae cylindrical or more inflated, 3.5-7.0(-23.4) μm wide, more or less radially arranged in scales, hyaline to yellowish. Cap trama loosely interwoven; hyphae cylindrical or more inflated, 4.6-12.9(-20.0) μm wide, more compactly arranged near the cuticle, hyaline to yellow; contents granular. Caulocystidia lacking; stalk hyphae parallel in the core, interwoven on the surface, 3.5-9.4 μm wide, hyaline to yellowish, thin-walled to slightly thickened. Oleiferous hyphae refractive, yellowish, granular, thin-walled, not septate, contorted, 2.3-5.8(-7.3) μm wide, scattered in the trama of the hymenophore and stalk. Clamp connections lacking.

Useful references. The following books contain information on *G. floccosus* pertinent to North America.

- *A Field Guide to Western Mushrooms* (Smith 1975)—A color photograph (as *Cantharellus floccosus* and also showing *G. kauffmanii*) and a description and discussion are presented.
- *Edible and Poisonous Mushrooms of Canada* (Groves 1979)—A color photograph and a description and discussion are given.
- *Mushrooms in their Natural Habitats* (Smith 1949)—A color photograph on a reel and a description and discussion are given.
- *Mushrooms of North America* (Miller 1972)—A color photograph and a description and discussion are given.
- *The Mushroom Hunter's Field Guide* (Smith 1974)—A black-and-white photograph and a description and discussion are given.
- *The Savory Wild Mushroom* (McKenny 1971)—A black-and-white photograph and a description and discussion are given.

Hebeloma (Fr.) Quél. — Cortinariaceae

Hebeloma is poorly known in North America and the species are difficult to identify. Some species, such as *Hebeloma crustuliniforme,* are reported as poisonous, causing gastric disturbances; others such as *Hebeloma mesophaeum* (Pers.) Quél. are suspected of being poisonous. Avoid all species of *Hebeloma.*

As with *Inocybe* and other difficult and common brown-spored genera, it is best to learn to identify the genus *Hebeloma*. *Hebeloma* is a terrestrial genus, usually associated with the roots of higher plants. Its species are most abundant in forested areas, especially in the late summer and fall, and some fruit in arcs or fairy rings. They also occur in lawns where trees are present.

The fruit bodies are medium to large and tend to be fleshy. The cap surface is often some shade of pale brownish buff to brown, usually smooth and viscid to sticky when fresh. The gills are attached to the stalk and are usually some shade of dull brown with white edges. The apex of the stalk is often pruinose to scabrous, as in *Inocybe*. Some species have a partial veil of fibrils, which is usually not well developed and often can only be seen on young specimens. The spore deposit is usually some shade of brown, often a dull brown, or clay color. However, some species have a brick red to vinaceous brown spore deposit.

The spores of most species of *Hebeloma* have a roughened, punctate, finely warted, or wrinkled surface, but at times the ornamentation is minute, and some spores appear smooth even at a magnification of 1000×. Cheilocystidia are characteristic of *Hebeloma* species. Pleurocystidia are typically absent.

Hebeloma species are most likely to be confused with species of *Cortinarius* or *Inocybe*. Species of *Cortinarius* usually have a more or less rusty brown spore deposit. The spores typically coat the gills at maturity and often can be found deposited on the stalk or fibrils of the partial veil. *Hebeloma* species have dull brown gills at maturity, often with distinctly white edges. Cystidia are not common in *Cortinarius*; both cheilocystidia and pleurocystidia are usually lacking. *Hebeloma* typically has cheilocystidia.

Inocybe species are generally smaller than *Hebeloma* species and the cap often has a more pointed umbo. The cap surface of *Inocybe* species is usually dry and silky to fibrillose or scaly and often radially cracked, rarely silky and viscid to sticky. *Hebeloma* species typically have a smooth, viscid to sticky cap when fresh. The spore deposits of *Hebeloma* and *Inocybe* are somewhat similar but usually the spore deposits of *Inocybe* are a duller to darker brown than those of *Inocybe*. *Inocybe* spores are smooth or angular to nodulose, not punctate, finely warted, or wrinkled as in *Hebeloma*.

The species of *Hebeloma* that occur in North America are poorly known. There is no taxonomic treatment for the genus and little is known about their edibility. Price (1927) reported a case of poisoning by *Hebeloma crustuliniforme*, the only known case of poisoning by *Hebeloma* in North America. Other species reported as poisonous in North American field guides and handbooks on mushrooms include *Hebeloma fastibile* (Pers. ex Fr.) Kummer, *Hebeloma mesophaeum*, and *Hebeloma sinapizans*, but there are no documented cases of poisoning for North America.

Two species of *Hebeloma*, *H. crustuliniforme* and *H. sinapizans*, are described and discussed below. *H. crustuliniforme* is known to be poisonous. The edibility of the *H. sinapizans* is unknown, but it should not be eaten.

Hebeloma crustuliniforme (Bull. ex Saint Amans) Quél.

Figs. 132, 193

Common name. Poison pie.

Distinguishing characteristics. Fruit body of medium side. Cap surface slimy to viscid when fresh; color pale cream to pale crusty brown or pale tan. Flesh, when crushed, with an odor of radish. Gill faces white in early stages becoming dull brown; edges white, often beaded with drops of liquid. Partial veil absent. Stalk 6–10 mm at the apex, usually with a bulb at the base, white and granular near the apex. Spores ellipsoidal to almond-shaped, with the surface nearly smooth to minutely warted, 8.8–11.3(–12.4) × 5.1–6.6 μm in size.

Observations. *H. crustuliniforme* is a fairly common and somewhat variable species. The pale-colored slimy caps, white stem, dull brown gills frequently beaded with drops of liquid, and radish odor help to distinguish it in the field. The concept of *H. crustuliniforme* presented here generally agrees with the concept of the species as interpreted by most North American workers.

Distribution and seasonal occurrence. Because of the difficulty in identification of *Hebeloma* species, the exact range of this species is unknown; it appears to be widely distributed. It is primarily a late summer and fall species, but it occasionally fruits earlier in the year, depending on location and weather.

Habitat and habit. *H. crustuliniforme* is terrestrial and grows scattered to gregarious. It often occurs under conifers but has been reported from mixed woods and hardwood forests as well.

Technical description. Cap 30–70(–90) mm broad, convex with an enrolled margin in early stages, gradually expanding to broadly convex to plane; margin spreading or uplifted with age, opaque and even, remaining white-pruinose for some time; surface of cap viscid, appearing as if varnished; color very pale in early stages, pallid to near pale yellow or pale cream color, gradually becoming more or less brownish, pale crusty brown to pale alutaceous, at times with orange yellow tints on the disc and margin; disc occasionally darker brown to reddish brown. Flesh thick on the disc, tapering abruptly to the margin, firm, white; taste resembling radish or spice, or disagreeable, or bitter; odor usually like radish.

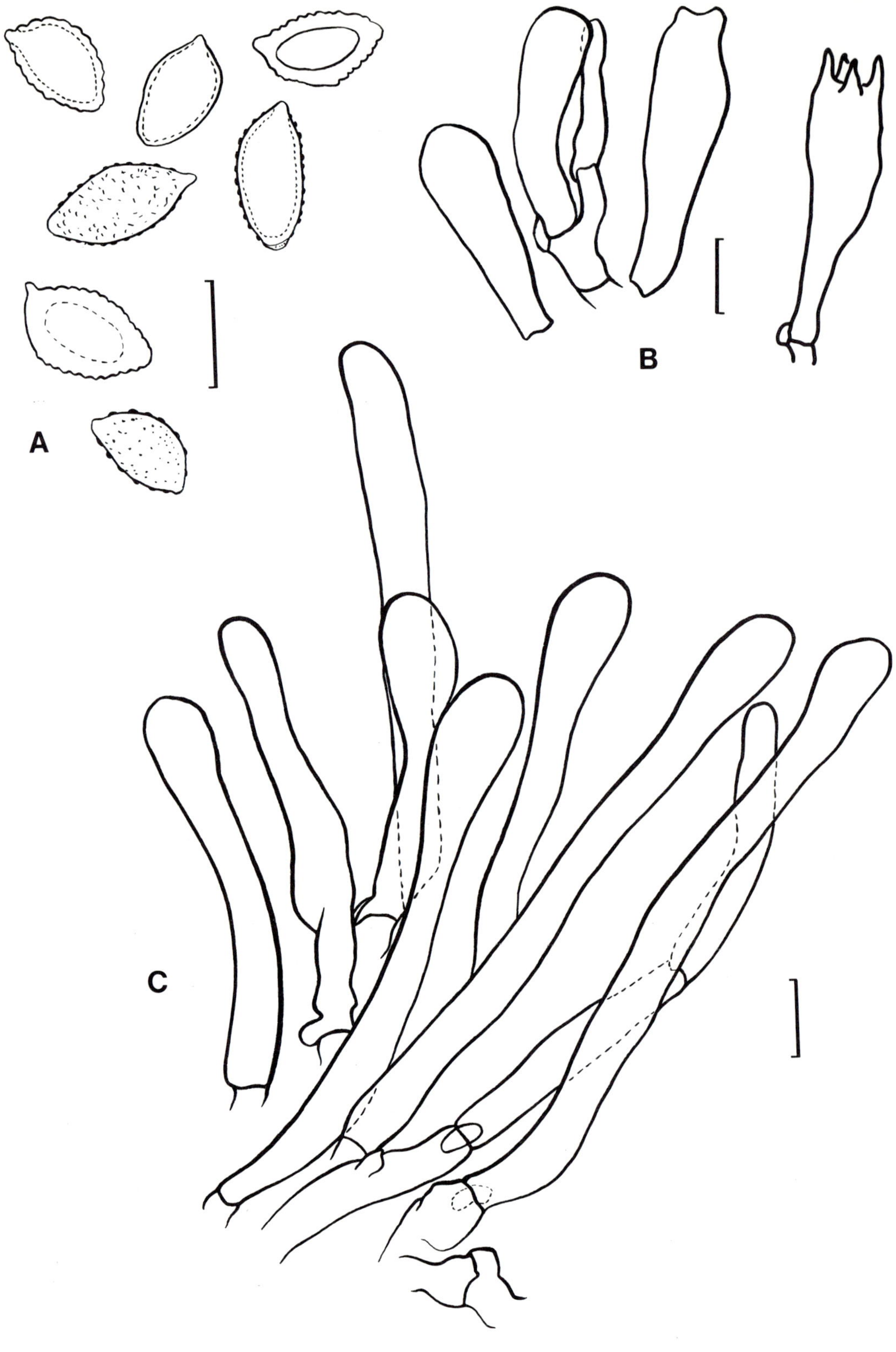

Fig. 132. *Hebeloma crustuliniforme*: A, spores; B, basidia; C, cheilocystidia.

Scale line 10 μm

Gills white to pallid or grayish when young, soon becoming pale grayish brown, finally turning dull brown to yellow brown, bluntly adnate becoming adnexed, close to crowded, narrow, becoming moderately broad; edges white, crenulate, and often beaded with drops of moisture.

Stalk 30-80 mm long, 6-10 mm thick, equal or abruptly enlarged at the base, central, solid; apex scurfy, fibrillose downward or sometimes lacerate with age; base surrounded by copious mycelia; color white, sometimes light yellowish to tan below. Flesh white. Partial and universal veils lacking.

Spores yellow brown to dull brown in deposit, 8.8-11.3(-12.4) × 5.1-6.6 μm, elliptic to almond-shaped and inequilateral in side view, elliptic in face view, nearly smooth to finely warted, pale brownish orange. Basidia four-spored, clavate, 30-35 × 10.2-11.9 μm, hyaline. Pleurocystidia lacking. Cheilocystidia cylindrical to narrowly clavate, 40-70 × 6-8 μm, thin-walled, hyaline.

Gill trama hyphae parallel, cylindrical, 4-6 μm wide, hyaline. Subhymenium filamentous; hyphae interwoven, more or less cylindrical, 3.0-3.5 μm wide. Cap cuticle gelatinous, filamentous; hyphae compactly interwoven, radially arranged, cylindrical, 3.5-4.6 μm wide, with some hyphal ends more or less upturned, hyaline to yellowish; walls thin to slightly thickened and encrusted. Cap trama loosely interwoven; hyphae cylindrical to inflated, 3.5-23.0 μm wide, hyaline. Caulocystidia clavate to filamentous, 40-60(-80) × 5.5-7.3 μm, hyaline, clustered on the stalk apex; hyphae of stalk parallel to more or less interwoven, more or less cylindrical, 3.5-16.0 μm wide, hyaline. Oleiferous hyphae refractive, greenish yellow, 4.4-6.6 μm wide, scattered in the gill trama and stalk. Clamp connections present.

Useful references. The following books contain information on *H. crustuliniforme* pertinent to North America.

- *A Field Guide to Western Mushrooms* (Smith 1975)—A color photograph and a description and discussion are presented.
- *Mushrooms in their Natural Habitat* (Smith 1949)—A color photograph on a reel and a description and discussion are given.
- *Mushrooms of North America* (Miller 1972)—A color photograph and a description and discussion are given.
- *The Savory Wild Mushroom* (McKenny 1971)—A black-and-white photograph and a description and discussion are given.

Hebeloma sinapizans (Paulet ex Fr.) Gill.

Figs. 133, 194

Common name. Because of incorrect identification as *H. crustuliniforme, H. sinapizans* has probably also been called poison pie.

Distinguishing characteristics. Fruit body usually large and robust, with a stalk 10-30 mm thick. Cap surface slightly viscid to sticky when fresh, cinnamon buff to some shade of medium brown, with the margin usually paler than the center. Odor like radish. Gills broad; color pallid to gray brown; edges often beaded with drops of liquid. Stalk stout, nearly equal; surface with distinct scales. Partial veil lacking. Spores ellipsoidal to almond-shaped, with a roughened wrinkled to warty surface 11-14.6(-15.3) × 7.3-8.8 μm in size. Cheilocystidia present.

Observations. *H. sinapizans* resembles *Cortinarius* species in size and general stature. However, the well-developed, fibrillose partial veil comprising the cortina, which is characteristic of the genus *Cortinarius,* is lacking in *H. sinapizans,* even in the button stage. Definite information on the edibility of *H. sinapizans* is not available. Its large size and tendency to fruit in large numbers make it attractive as an edible mushroom. It should be avoided, however, as should all species of *Hebeloma.*

Distribution and seasonal occurrence. The distribution of *H. sinapizans* is not well known, but based on herbarium collections it appears to be widely distributed across Canada and the United States. It fruits mainly in the late summer and fall, depending on weather and location. In some years it is uncommon, but it can be abundant in a particular location, especially during a wet, warm fall.

Habitat and habit. This species is terrestrial and grows scattered to gregarious-caespitose, often in arcs or fairy rings. It occurs in hardwood and conifer-hardwood forests. Smith (1975) reported it to be common in the west, under birch (*Betula*) and balsam (*Abies*).

Technical description. Cap 60-150 mm broad, convex, obtuse, becoming expanded to plano-convex; margin enrolled in early stages, even; surface viscid, glabrous to appressed-fibrillose; color pinkish buff to cinnamon buff or dull medium brown, with the margin usually paler than the disc. Flesh thick, whitish; taste bitter or radish-like; odor radish-like.

Gills pallid, becoming pale brownish to gray brown, adnexed, broad, close; edges beaded with drops of moisture, minutely fringed.

Stalk 25-120 mm long, 10-30 mm wide, stout, nearly equal, stuffed then hollow, floccose-scaly or scurfy at the apex, fibrillose-scaly below, white at the apex, pallid to brownish toward the base.

Spores clay brown in deposit, 11-14.6(-15.3) × 7.3-8.8 μm, ellipsoidal to almond-shaped, inequi-

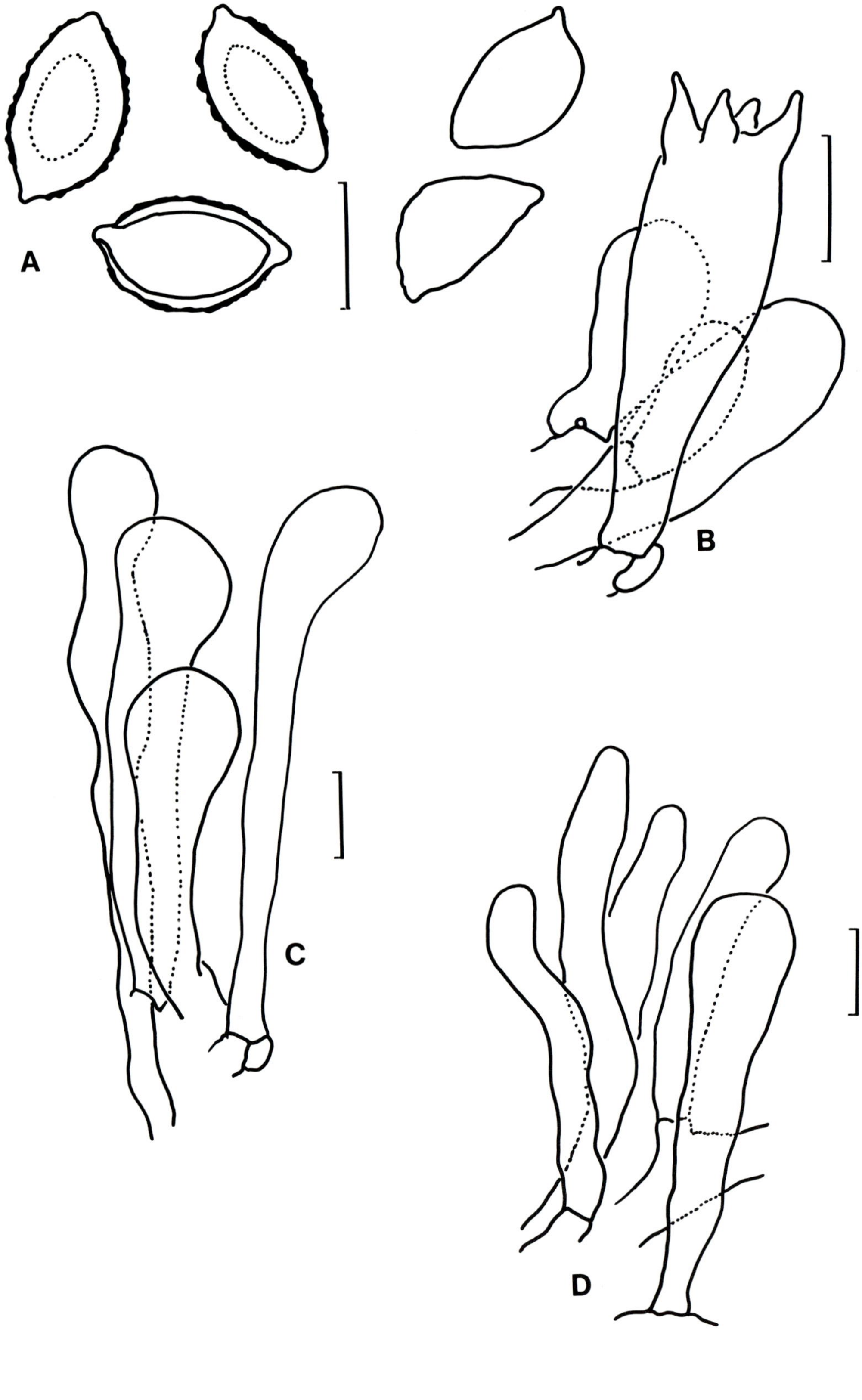

Fig. 133. *Hebeloma sinapizans*: A, spores; B, basidia; C, cheilocystidia; D, caulocystidia.

Scale line 10 μm

lateral in side view, elliptic in face view, thick-walled, ornamented, finely wrinkled to warty, yellow brown, nonamyloid. Basidia four-spored, clavate, 29.2-36.5 × 8.8-11.0 μm, thin-walled; contents granular, hyaline. Pleurocystidia lacking. Cheilocystidia clavate, 25-62 × 5.1-7.3(-10.2) μm, hyaline, thin-walled.

Gill trama parallel; hyphae cylindrical, 3.7-11.0 μm wide, hyaline. Subhymenium filamentous, interwoven; hyphae cylindrical, 3.7-4.4 μm wide, hyaline. Cap cuticle gelatinous, filamentous; hyphae compactly interwoven, radially arranged, cylindrical, 3.7-5.8(-8.0) μm wide, thin-walled, hyaline to yellowish, with scattered hyphae containing brown granules. Cap trama loosely interwoven; hyphae cylindrical to more or less inflated, 3.7-14.6 μm wide, hyaline, thin-walled. Caulocystidia cylindrical to clavate, 36-44 × 4.4-8.8 μm, hyaline, clustered at the stalk apex; stalk hyphae parallel, more or less cylindrical, 3.7-12.4 μm wide, hyaline. Oleiferous hyphae refractive, greenish yellow, 3.7-11.0 μm wide, scattered in the gill trama, stalk, and cap cuticle. Clamp connections present.

Useful references. The following books contain information on *H. sinapizans* pertinent to North America.

- *A Field Guide to Western Mushrooms* (Smith 1975)—A color photograph and a description and discussion are presented.
- *Edible and Poisonous Mushrooms of Canada* (Groves 1979)—A black-and-white photograph and a description and discussion are given.

Helvella L. ex Fr. — Helvellaceae

The genus *Helvella* is similar in appearance to *Gyromitra* (Ch. 8), and some workers place all the species in the genus *Helvella*. *Helvella*, as used in this treatment, however, excludes species of *Gyromitra*. These two genera belong to the family Helvellaceae, which is classified in the major division Ascomycota. See Chapters 2 and 3, "Ascomycota", for a discussion of the main features of this division.

The apothecia of *Helvella* are typically stalked and have a cap that is variously shaped, depending on the species. The cap may be cup- to saucer-shaped or saddle-shaded to somewhat irregularly lobed. The stalk may be short or long in relation to the width of the cap. The stalk surface is smooth in some species and longitudinally ribbed or ridged in others. In certain species, such as *Helvella lacunosa*, the stalk is deeply furrowed, with sharp ridges on the surface and longitudinal chambers in the interior. The apothecia in *Helvella* are often more or less grayish but vary from white to pallid to pale gray to black, or less commonly brownish. The brownish species can usually be separated from *Gyromitra* by their size and the shape of the cap. The brownish species of *Helvella* usually are smaller than *Gyromitra* species and often have a cup- to saucer-shaped cap. In *Gyromitra* the cap is not normally cup-shaped. In general, species of *Gyromitra* are larger and more robust than *Helvella* species. Compare also the genera *Morchella* and *Verpa* later in this chapter, alphabetically listed, and see the general and technical keys to genera of the Ascomycota in Chapter 5.

Species of *Helvella* to our knowledge have not caused cases of severe poisoning in North America. McKenny (1971) suggests that only fresh young specimens be collected for eating, thus indicating that older specimens may be poisonous. Smith (1975) and McKenny (1971) have suggested that *Helvella* species are poisonous when they are eaten raw.

In this publication we have described only *Helvella lacunosa*, a common and widely distributed species. Color illustrations are also included for *Helvella crispa* Fr. (Fig. 195) and *Helvella elastica* Bull. ex Fr. (Fig. 196), which are mentioned under *Helvella lacunosa* (Fig. 197). Other color illustrations of these species, as well as other species of *Helvella*, can be found in the references listed under *Helvella lacunosa*. A good technical treatment of *Helvella* for the Great Lakes region can be found in work by Weber (1972), and a key with descriptions to several species is provided by Smith and Smith (1973).

Helvella lacunosa Fr. — Fig. 197

Common name. Elfin saddle.

Distinguishing characteristics. Cap 15-60 mm broad, often saddle-shaped, but sometimes indistinctly so and irregularly convex to miter-shaped; surface smooth to wrinkled; color pale gray to medium gray (occasionally grayish brown), or black, less commonly white. Stalk 40-100 mm long, 15-20 mm thick, deeply furrowed with sharp longitudinal ridges or ribs on the surface and with longitudinal chambers in the interior; color white or tinted with colors of the cap. Apothecia occurring in late summer and fall, and into the winter where the weather remains above freezing. Eight spores per ascus, ellipsoidal, with a smooth surface, becoming somewhat warty with age, (16.5-)17.5-21(-22) × (10.5-)12-14 μm in size.

Observations. *H. lacunosa* is one of the most common species of *Helvella* in North America. The main field characters for identification are the grayish, pale gray to blackish, or at times grayish brown color of the cap and the conspicuous furrows and ridges of the stalk. *H. crispa* (Fig. 195) is similar in appearance to *H. lacunosa*, but the cap is white to buff or creamy white and the stalk is whitish to white. *H. crispa* is also villous on the undersurface of the apothecium and the cap margin is enrolled in the early stages. These features help to separate *H. crispa* from the rare white forms of *H. lacunosa*. *H. elastica* (Fig. 196) has a saddle-shaped cap that is nearly smooth and usually pale brown to buff. The stalk is long and slender, 50-100 mm long, 2-8 mm thick, smooth, and white. The technical description given below is taken from that of Weber (1972).

Distribution and seasonal occurrence. *H. lacunosa* occurs across North America. It usually fruits in late summer and fall but also in winter in areas where the temperature does not drop below freezing. *H. crispa* and *H. elastica* are sometimes found growing in the same area as *H. lacunosa*.

Habitat and habit. *H. lacunosa* grows scattered to densely gregarious. It is terrestrial or grows on decaying wood, under conifers and hardwood trees. At times this species is common in conifer plantations.

Technical description. Apothecium 15-60 mm broad, irregularly convex to miter-shaped with the sides appressed to the stalk, or saddle-shaped and not compressed; margin variable in curvature, never curving over the hymenium, attached to the stalk in places; surface of hymenium even to coarsely wrinkled or with small protuberances; color pale gray to medium gray, or black, occasionally grayish brown, rarely white; undersurface sterile, rarely white, usually some shade of gray or black, glabrous, ribbed, with ribs extending to the stalk apex.

Stalk 40-130 mm long, 10-30 mm thick, even or tapered toward the apex; ridges with rounded and cord-like or double-edged ribs branching and forming a network; color dingy white to gray or black, darkest on the ribs, often paler near the base, glabrous, lacunary internally and externally, more or less longitudinally chambered at some point in development.

Ascospores (16.5-)17.5-21(-22) × (10.5-)12-14 μm, ellipsoidal, smooth or occasionally verruculose to finely wrinkled with age, each containing one oil droplet. Hymenium 250-380 μm thick. Paraphyses clavate to nearly capitate with the widest portion 4-10 μm wide, hyaline to light brown.

Ectal excipulum of apothecium 50-130 μm thick, hyaline to grayish, with a palisade of clavate to rounded hyphal end cells forming the outer edge. Medullary excipulum 80-300(-600) μm thick; hyphae interwoven, 3-10(-20) μm wide, hyaline. Ectal excipulum of stalk 60-100(-400) μm thick, hyaline to pale grayish brown, with a palisade of clavate to rounded hyphal end cells forming the outer edge. Medullary excipulum interwoven; hyphae 3-6 μm wide, hyaline, at times refractive.

Useful references. The following books contain information on *H. lacunosa* pertinent to North America.

- *A Field Guide to Western Mushrooms* (Smith 1975)—A color photograph and a description and discussion are presented.
- *Mushrooms of North America* (Miller 1972)—A color photograph and a description and discussion are given.
- *The Mushroom Hunter's Field Guide* (Smith 1974)—A black-and-white photograph and a description and discussion are given.
- *The Savory Wild Mushroom* (McKenny 1971)—A black-and-white photograph and a description and discussion are given.

Lactarius Pers. ex S. F. Gray — Russulaceae

The main characteristics of the genus *Lactarius* include the presence of a latex that exudes from injured tissues of the fruit body; the often large size of the fruit bodies; the brittle consistency of the flesh caused by the presence of sphaerocysts; adnate gills in the early stages, often becoming decurrent; a whitish to yellowish spore deposit; strongly amyloid ornamentation on the spores; a mild to acrid flavor of the flesh and latex; and a terrestrial habit, with fruit bodies associated with the roots of trees.

Lactarius is distinguished from *Russula* by the presence of latex, but in many other respects the two genera are alike and are grouped together in the family Russulaceae. Both genera possess spores with amyloid ornamentation, characteristic pleurocystidia, characteristic vesiculose cells called sphaerocysts that are bound together by filamentous connective hyphae, and a cap that is often more or less funnel-shaped.

Macroscopic characteristics used to distinguish the species of *Lactarius* generally include color of the latex (including its discoloration, if any), features of the cap surface, and taste of the flesh and latex. Important microscopic characteristics include structure of the cap cuticle, the shape and size of the spores, and size and features of the spore ornamentation.

Species of *Lactarius* are found throughout the United States and Canada. Most fruit during the summer and fall. They often comprise a large portion of the summer mushroom flora. The genus as represented in North America contains about 300 species.

Species of *Lactarius* are frequently collected for food, especially *L. deliciosus* (L. ex Fr.) S. F. Gray and its relatives. Besides the so-called excellent edible species in this genus, several others have been listed as poisonous, suspected, or not recommended. The flesh or latex of many of these dangerous species have a hot, acrid taste. The compound responsible for the acrid taste of *L. vellereus* var. *vellereus* and *L. piperatus* var. *piperatus* (=*L. pergamenus* (Fr.) Fr.) is an azulene dialdehyde called isovelleral (List and Hackenberg 1973, Magnusson et al. 1972). This compound is very reactive and is converted to nonacrid products by heat, oxidation, or alcohol.

Benedict (1972) listed the following species of *Lactarius* as toxic, indicating that many of them are toxic only when eaten raw: *L. piperatus* var. *glaucescens*, *L. torminosus*, *L. helvus* Fr., *L. rufus*, and *L. uvidus*. Pilát (1951) listed *L. helvus* as mildly poisonous when eaten in large quantities. Lincoff and Mitchel (1977) include the above species as reported as or suspected of being poisonous and list in addition *L. chrysorheus*, *L. repraesentaneus*, *L. scrobiculatus* var. *scrobiculatus*, and *L. scrobiculatus* var. *canadensis*.

The above-mentioned species are those most commonly labeled as poisonous. They usually cause gastrointestinal upset, characterized by nausea, abdominal pain, vomiting, and diarrhea. The only known fatality in North America resulted from the ingestion of *L. piperatus* var. *glaucescens* (=*L. glaucescens* Crossland) (Charles 1942). The victim was a 5-year-old child who had a heart defect. It is believed that a small piece of mushroom contributed to his death (Lincoff and Mitchel 1977).

With the exception of *L. helvus* these species are presented in the following pages. The presence of *L. helvus* in North America apparently has not been verified; the species that has been called *L. helvus* in North America is actually *L. aquifluus* Peck (Smith 1975). *L. helvus* has an acrid taste and white latex, whereas *L. aquifluus* has a watery latex and a pronounced fragrant odor that persists for years in dried collections. Smith (1975) provides a full description and discussion, as well as a color photograph of *L. aquifluus*. Described in the pages following are a few fairly common, acrid species: *L. vellereus* var. *vellereus*, *L. subvellereus* var. *subdistans*, *L. piperatus* var. *piperatus*, and *L. deceptivus*. *L. theiogalus* Fr. has been reported as the cause of one poisoning in Ottawa (S. Thomson, Personal commun.). Apparently it was mistaken by the collector for *L. deliciosus*. The specimens from this poisoning are not available for study. *L. theiogalus* has been confused in North America with other species such as *L. chrysorheus*; therefore, without specimens, it is impossible to verify which species was involved. It should be considered as suspect. See the treatment of *Lactarius* by Hesler and Smith (1979) for a description of *L. theiogalus*.

Key to species

1. Latex quickly changing color upon exposure to air and upon injury to tissue or gill........................ 2
   Latex color unchanging or at times drying or staining tissues another color after a period of time.. 5

2. Latex changing to sulfur yellow upon exposure to air and staining tissues and gills yellow to clay color upon injury.. 3
   Latex changing to lilac violet upon exposure to air and staining tissues and gills purplish upon injury.. 4

3. Cap various shades of yellow to pale olive buff; margin bearded. Gills bruising yellowish to clay color. Stalk spotted. Taste of flesh acrid. Cap cuticle composed of hyphae crowded in bundles .... ***L. scrobiculatus* var. *canadensis***
   see also *L. scrobiculatus* var. *scrobiculatus*
   Cap pale buff to cinnamon, pruinose becoming glabrous; margin striate. Gills not spotting or staining ocherous. Stalk not spotted. Taste of flesh immediately acrid. Cap cuticle in the form of ixocutis.. ***L. chrysorheus***

4. Cap pale golden yellow; margin bearded. Stalk whitish to more or less the same color as the cap and spotted. Taste of latex mild to slightly acrid; taste of flesh finally bitterish to slightly acrid or both. Spores not reticulate... ***L. repraesentaneus***
   Cap brownish gray lilac; margin pruinose. Stalk yellow, not spotted. Taste of latex bitter; taste of flesh mild to bitter. Spores partially reticulate... .... ***L. uvidus***

5. Latex drying glaucous green and staining white paper yellow. Cap dry, smooth, white; $FeSO_4$-$7H_2O$ color reaction on cap flesh vinaceous cinnamon to vinaceous red. Gills crowded, narrow, forked...... ***L. piperatus* var. *glaucescens***
   Latex color unchanging or staining tissues a color other than green.. 6

6. Cap basically white, at times with yellowish brown surface hues.. 7
   Cap more highly colored than above.................. 8

7. Cap glabrous, smooth, dry. Flesh with mild odor. Latex copious, white, unchanging, acrid. Gills white, crowded, forked. Stalk white, dry, pruinose............................ ***L. piperatus* var. *piperatus***
   Cap tomentose to pubescent; margin enrolled. Flesh with odor often finally pungent............. 9

8. Cap red brown, dry; margin striate to ribbed. Gills grayish. Flesh with mild odor.............. ***L. rufus***
   Cap pinkish yellow to pinkish orange, with colors in concentric rings; margin bearded. Gills whitish, becoming vinaceous to cream colored.

Flesh with slight to pungent odor .... *L. torminosus*

9. Cap woolly, tomentose; margin cottony, enrolled. Flesh with pungent odor. Latex white, unchanging, staining tissue brownish. Gills close to subdistant. Spores 5-9 μm wide. Cap cuticle not distinct from cap trama; hyphae thin-walled .... *L. deceptivus*

   Cap velvety, dry, finely tomentose; margin enrolled in early stages. Latex white and unchanging, or becoming creamy yellow. Flesh and gills staining brownish. Cap cuticle distinct from cap trama; hyphae thick-walled 10

10. Spores narrow, 4.0-6.6 μm wide. Gills subdistant to distant. Flesh with pungent odor. Latex stinging-acid .... *L. subvellereus* var. *subdistans*

   Spores broad, 7.5-9 μm wide. Gills close to subdistant. Flesh with mild odor in early stages, and with acrid taste. Latex with mild taste .... *L. vellereus* var. *vellereus*

Lactarius chrysorheus Fr.

Figs. 134, 198

Common name. Yellow-juiced lactarius.

Distinguishing characteristics. Cap pale-colored, buff to cinnamon. Gills not becoming brown or spotted vinaceous. Latex white, becoming sulfur yellow upon exposure to air. Taste immediately acrid. Spores yellowish in deposit, broadly ellipsoidal, 6-9.5(-11.7) × 5.5-6.6(-7) μm in size.

Observations. *L. chrysorheus* is often confused with *L. theiogalus* and with the pale, unstained fruit bodies of *L. vinaceorufescens* Smith. *L. theiogalus* is smaller and more reddish brown or tawny red than *L. chrysorheus*. *L. vinaceorufescens* is separated from *L. chrysorheus* by the darker-colored cap, pallid to buff, becoming pinkish cinnamon to pink to vinaceous rose, and finally changing to dark vinaceous red; the gills spotting and staining pink to sordid reddish brown; and the acrid taste developing only after the latex turns yellow.

Kauffman (1971) and others reported *L. chrysorheus* as poisonous. It is advisable never to eat species of *Lactarius* that have a latex which turns bright sulfur yellow upon exposure to air.

Distribution and seasonal occurrence. This species is common and widely distributed, occurring in the northern, eastern, and western United States and southern Canada. It fruits from July to October.

Habitat and habit. *L. chrysorheus* grows subcaespitose or in groups. It occurs on the ground under oak (*Quercus*), spruce (*Picea*), and pine (*Pinus*) in deciduous, coniferous, and mixed woods. *L. vinaceorufescens* is more frequent in mixed hardwood-conifer stands on low ground, whereas *L. chrysorheus* often grows under oak (*Quercus*) on sandy soil.

Technical description. Cap 30-100 mm broad, nearly conic to convex becoming plane, soon slightly depressed over the disc, expanding to broadly funnel-shaped; surface of cap glabrous, moist to lubricous or merely almost viscid, soon becoming dry and whitish-glaucous, with watery spots occurring more or less in zones, or with strong concentric zones, or sometimes without distinct zones, nearly hygrophanous; margin thin, incurved, at times moist and striate, faintly pruinose becoming glabrous; color pallid to pale yellowish, pinkish cinnamon, or orangish pink buff, with darker spots, fading to whitish. Flesh thin, firm, and whitish but soon becoming yellow where cut; latex copious, white, usually changing to sulfur yellow, but color change sometimes delayed; taste strongly acrid; odor slight.

Gills adnate to short-decurrent, close, forked near the stalk at times, pallid to pinkish cinnamon to pale orange buff or slightly darker, not discoloring or spotting to vinaceous or brown; lamellulae numerous, occurring in tiers.

Stalk 30-80 mm long, 10-20 mm thick, equal or nearly so; surface nearly viscid to glabrous; base more or less strigose; color often uneven, becoming flushed orange buff with age. Flesh thick, whitish within, soon becoming yellowish to sulfur yellow where cut, stuffed becoming hollow.

Spores pale yellow in deposit, 6.0-9.5(-11.7) × 5.5-6.6(-7.0) μm, broadly elliptic in side view, hyaline; ornamentation amyloid, partially reticulate with separate warts more or less 0.5-1.0 μm high; plage indistinct. Basidia 37-45 × 8.0-10.5 μm, four-spored, clavate; contents granular, hyaline. Pleurocystidia of two types: macrocystidia, 45-80 × 6.0-12.4 μm, clavate to nearly cylindrical or fusiform, tapering to an obtuse or acute point, with refractive yellowish to hyaline contents; and pseudocystidia, filamentose, abundant, with tapering obtuse apices, and with refractive yellowish to hyaline contents. Cheilocystidia 37-45 × 6-8 μm, fusiform, acute to mucronate, similar to macrocystidia.

Gill trama interwoven; hyphae 3.5-10.5(-15.0) μm wide, hyaline. Subhymenium filamentous; hyphae interwoven, 3.5-4.4 μm wide, hyaline. Cap cuticle in the form of a thin ixocutis, filamentous, gelatinous in the outer layer; hyphae interwoven, 2.3-6.0 μm wide, hyaline, septate, with hyphal ends somewhat recurved. Cap trama heteromerous, interwoven; hyphae 3.5-14.3 μm wide, hyaline; sphaerocysts 23.0-48.0 × 29.2-35.0 μm, subglobose to ellipsoidal, hyaline, occurring in rosettes. Stalk cuticle consisting of nondifferentiated protruding caulocystidia, with scattered recurved hy-

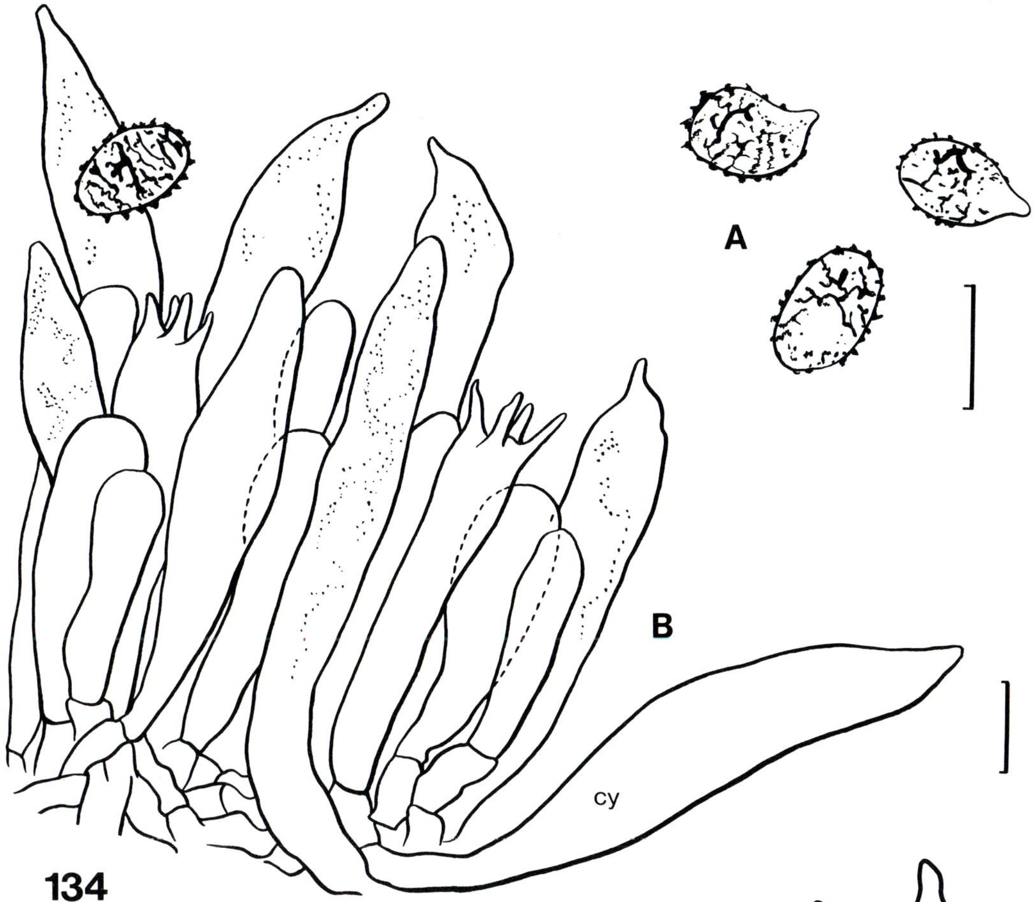

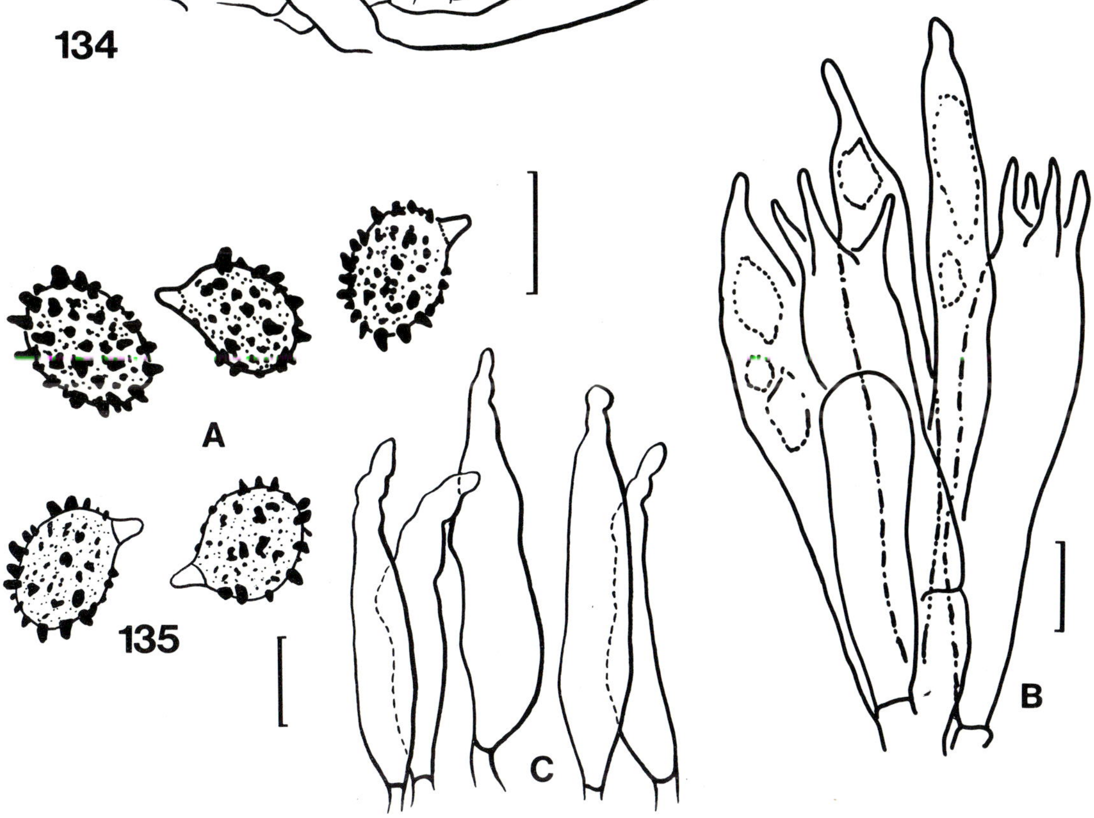

Fig. 134. *Lactarius chrysorheus*: A, spores; B, basidia and macrocystidia (*cy*).

Fig. 135. *Lactarius deceptivus*: A, spores; B, basidia and macrocystidia; C, cheilocystidia.

Scale line 10 μm

phal ends; surface hyphae interwoven, 3-9 μm wide, septate, cylindrical, thin-walled, hyaline. Lactiferous hyphae 3.5-9.4 μm wide, refractive, yellowish, with granular contents, abundant in the trama and stalk. Clamp connections absent.

Useful references. The following books contain information on *L. chrysorheus* pertinent to North America. These references may be for *L. vinaceorufescens*, especially the ones by Groves (1979) and Miller (1972).

- *Edible and Poisonous Mushrooms of Canada* (Groves 1979)—A color photograph and a description and discussion are given.
- *Mushrooms of North America* (Miller 1972)—A description and discussion are given.
- *The Mushroom Hunter's Field Guide* (Smith (1974)—A black-and-white photograph, a color photograph, and a description and discussion are given.

Lactarius deceptivus Peck

Figs. 135, 199

Common names. Deceptive lactarius, deceiving lactarius.

Distinguishing characteristics. Cap white, with yellow brown surface hues; surface woolly, with a cottony roll of tissue on the margin. Latex and flesh with an acrid taste; odor pungent. Spores white in deposit, subglobose to broadly ellipsoidal, 6-12.1 × 5-9 μm in size.

Observations. *L. deceptivus* is an extremely common species in North America. The color of the gills tends to vary, especially with age. It ranges from white to creamy in the early stages, sometimes becoming ocherous to tan or brownish, but occasionally pinkish.

Kauffman (1971) reported that Peck had eaten the species in quantity and found it to be fair in quality. Groves (1979) says the acrid taste disappears on cooking and reported it edible, but he noted the danger of confusing the species with *L. vellereus*, which has been reported as poisonous.

Distribution and seasonal occurrence. This species is common and widely distributed, occurring in northeastern to southeastern North America, west to the central United States, and adjacent southern Canada. It fruits from June to October.

Habitat and habit. *L. deceptivus* is gregarious and terrestrial. It occurs in mixed woods containing hemlock (*Tsuga*), in coniferous woods under white pine (*Pinus strobus*), under oak (*Quercus*), along the edges of bogs, and under aspen (*Populus*).

Technical description. Cap 35-240 mm broad, convex, becoming depressed, finally becoming funnel-shaped; margin enrolled, cottony, concealing the young gills in early stages; surface dry, glabrous in early stages but dull and unpolished, soon becoming torn into patches of fibrils or appearing squamulose, sometimes rimose, soft to the touch; ground color chalk white, staining brownish, often tan with yellowish or tan areas, or leather color overall; disc darker yellow to ocherous; margin pale yellow to cream. Flesh 3-15 mm thick, white, firm, not bruising; latex white, unchanging, staining the flesh brownish, copious; taste of flesh and latex acrid, at times delayed, becoming extreme to anesthetizing in the throat; odor mild in early stages, becoming strong and pungent.

Gills 5-65 mm long, 1-7 mm deep, adnate-decurrent, close to subdistant, moderately broad, somewhat forked; edges often beaded with hyaline drops; color white to ivory yellow in early stages, becoming cream to ocherous, avellaneous to dingy tan where bruised, at times staining dull pink from latex.

Stalk 18-90 mm long, 10-40 mm thick, short, hard, equal or tapered downward; surface dry, fibrillose to pubescent to squamulose, similar in color to the cap, white becoming stained brownish, possibly with a ring of robin's-egg blue at the apex. Flesh thick, firm, white, stuffed, becoming hollow.

Spores white to pale ocherous buff in deposit, 6.0-12.1 × 5-9 μm, subglobose to broadly elliptic in side view, hyaline; ornamentation amyloid, spiny-warty; prominences 0.5-1.5 μm high, not reticulate, in the form of isolated warts and rough particles; apiculus prominent; plage hyaline to diffusely amyloid. Basidia 32.0-58.8 × 5.5-13.1 μm, with prominent sterigmata 5.8-20.2 μm long, two- to four-spored, clavate, hyaline. Pleurocystidia in the form of macrocystidia, 47.7-96.0 × 6.6-10.5 μm, subcylindrical to nearly clavate, at times ventricose; apices tapered and obtuse to acute, often multiconstricted, yellowish to hyaline, abundant, projecting from the hymenium; contents granular. Cheilocystidia 32.0-58.0 × 3.7-7.0 μm, fusiform-ventricose to subcylindrical, with tapered and obtuse apices, at times multiconstricted, yellowish to hyaline, embedded or projecting from the hymenium.

Gill trama interwoven; hyphae 3.7-8.8 μm wide, hyaline. Subhymenium filamentous; hyphae interwoven, 3.5-11.7 μm wide, cylindrical to inflated, septate, thick-walled, hyaline. Cap cuticle in the form of a trichoderm, interwoven loosely to form the tomentum; hyphae 3.0-15.3 μm wide, cylindrical, blunt-ended, septate, thin-walled, hyaline to pale yellowish brown. Cap trama heteromerous, interwoven; hyphae 3.5-12.4 μm wide, cylindrical, septate, hyaline; sphaerocysts 23.4-35.1 × 8.8-19.1 μm, subglobose to ellipsoidal, inflated, hyaline. Stalk cuticle in the form of a trichoderm; surface hyphae uplifted, bending; terminal elements forming the caulocystidia. Caulocystidia hair-like to nearly ventricose, with long filiform hyphal end cells, thick-walled, unicellular to many-partitioned, 100-550 × 3.0-7.3 µm, pale yellow with internal dextrinoid material in Melzer's reagent, hyaline; surface hyphae interwoven, 3.5-4.6 µm wide, thin-walled, hyaline. Lactiferous hyphae 3.5-9.4 µm wide, yellowish, granular, refractive, scattered in the trama, cuticle, and stalk. Clamp connections absent.

Useful references. The following books contain information on *L. deceptivus* pertinent to North America.

- *Edible and Poisonous Mushrooms of Canada* (Groves 1979)—A black-and-white photograph and a description and discussion are given.
- *The Mushroom Hunter's Field Guide* (Smith 1974)—A black-and-white photograph and a description and discussion are given.

Lactarius piperatus **var.** ***glaucescens*** **(Crossland) Hesler & Smith**

Common names. None.

Distinguishing characteristics. Cap surface dry, smooth, basically white. Gills crowded, narrow, forked. Latex colored glaucous green upon drying, staining white paper yellow. Spores white to yellowish in deposit, subglobose to broadly ellipsoidal, 6.5-9.0 × 5.5-6.6 µm in size.

Observations. *L. piperatus* var. *glaucescens* is also known in North America as *L. glaucescens* Crossland. It is often confused with *L. vellereus*, which differs by its woolly cap, the distant nonforking gills, and the latex that stains cut or bruised surfaces brown. Microscopically the low ornamentation of the spores and the prominent cellular layer of the cap cuticle are distinguishing characteristics of *L. piperatus* var. *glaucescens*.

Distribution and seasonal occurrence. This variety is widely distributed and occurs in eastern North America from the southeastern to the northeastern states, as well as in southeastern Canada, but it is infrequently found in the central states. It fruits from summer through fall.

Habitat and habit. *L. piperatus* var. *glaucescens* grows scattered to gregarious on soil or moss in hardwood forests.

Technical description. Cap 40-108 mm broad, convex when young, with an enrolled margin and depressed disc, becoming plano-convex or retaining the incurved margin and the depressed disc; surface dry, dull, minutely velvety to glabrous, sometimes areolately cracked, not hygrophanous; color white to buff, becoming stained or spotted with yellowish buff or dingy yellow brown. Flesh 5-10 mm thick at the disc, hard; color pale cream, unchanging when cut; latex pale cream, unchanging but at times becoming dingy pale green when dried, staining white paper yellow; taste strongly acrid; odor mild to pungent.

Gills very narrow, 1-2 mm wide, arcuate-decurrent, crowded, often forked; color dull cream; edges entire; lamellulae numerous, varying in length.

Stalk 30-100 mm long, 10-22 mm thick, usually tapering to the base, sometimes nearly equal, straight or curved, terete; surface minutely pruinose to glabrous, similar in color to the cap surface or paler. Flesh solid; color whitish, staining glaucous green where latex dries.

Spores white to yellowish in deposit, 6.5-9.0 × 5.5-6.6 µm, subglobose to broadly elliptic and unequal in side view, hyaline; ornamentation amyloid, finely warted, with prominences of up to 0.2 µm high, not reticulate; plage indistinct. Basidia 30-45 × 7.5-9.0 µm, four-spored, narrowly clavate, hyaline. Pleurocystidia and cheilocystidia present as macrocystidia and pseudocystidia, 37-60 × 7.5-10.5 µm, cylindrical to ventricose to nearly clavate, with tapering obtuse apices, and with granular hyaline contents.

Gill trama interwoven to cellular; hyphae 3.7-8.0 µm wide, hyaline. Subhymenium filamentous, compactly interwoven; hyphae 3.7-5.1 µm wide, hyaline. Cap cuticle not differentiated from the trama, composed of interwoven filamentous hyphae and sphaerocysts; sphaerocysts globose to ellipsoidal, pedicellate, 11.7-13.1 µm wide, surmounted by a turf of interwoven hyphae 2.3-5.1 µm wide and scattered pileocystidia 46-70 × 3.5-5.1 µm in size; pileocystidia clavate to nearly clavate or cylindrical, granular, hyaline. Cap trama heteromerous, filamentous; hyphae interwoven, 3.7-10.2 µm wide, hyaline; sphaerocysts 29.2-36.5 × 18.3-23.4 µm, globose to ellipsoidal, hyaline, occurring in rosettes. Stalk cuticle in the form of a thick zone of vertical hyphae 3-14 µm wide; hyphae loosely arranged, giving rise at the surface to caulocystidia; caulocystidia 55-88 × 5.8-7.3 µm, hyaline, clavate to subcylindrical, with granular contents. Lactiferous hyphae 5.1-8.2 µm wide, granular, hyaline to yellowish, scattered in the trama, stalk, and gills. Clamp connections absent.

Useful references. There are no books or field guides pertinent to North America that contain information on this variety.

Lactarius piperatus (Fr.) S. F. Gray var. *piperatus*

Figs. 136, 200

Common names. Pepper cap, peppery lactarius.

Distinguishing characteristics. Cap dull white, dry, glabrous. Latex copious, white, unchanging, tasting quickly and strongly acrid. Gills white, becoming dull yellow with age, crowded, forked. Stalk white, dry, pruinose. Spores white in deposit, subglobose to ellipsoidal, 4.5-7.6(-8.5) × 5-5.8(-6.5) μm in size.

Observations. *L. piperatus* var. *piperatus* is the typical variety of *L. piperatus*. Hesler and Smith (1979) believe this variety is synonymous with *L. pergamenus* (Fr.) Fr. The latex never stains or dries green.

Fries (1821) stated that the variety is edible. Others have reported that when this variety is cooked, the intensely acrid taste disappears and the flavor is good. The variety is often confused with *L. vellereus*, which is distinguished by its woolly cap, distant nonforking gills, and white latex that stains bruised or cut surfaces brown.

Distribution and seasonal occurrence. This variety is common throughout eastern North America but apparently occurs infrequently in Eastern Canada. It fruits from summer through fall.

Habitat and habit. *L. piperatus* var. *piperatus* grows scattered to gregarious on soil and grass in hardwood forests.

Technical description. Cap 30-200 mm broad, convex, becoming plane to depressed, finally funnel-shaped; margin enrolled in early stages, becoming elevated, even, smooth, or variously uneven to wrinkled; surface dry, without zones, glabrous to unpolished; color white, finally often becoming stained dingy tan or becoming creamy white overall, or both. Flesh white, unchanging and not staining, or becoming at times discolored slightly to yellowish; latex copious, milky white, not changing; taste quickly and strongly acrid; odor mild.

Gills narrow, 1-2 mm wide, adnate-subdecurrent, crowded, forked, thin, white, becoming pale cream; edges even.

Stalk 20-80 mm long, 10-30 mm thick, equal, at times tapering downward; surface dry, pruinose; color white; flesh firm, solid.

Spores white in deposit, 4.5-7.6(-8.5) × 5.0-5.8(-6.5) μm, subglobose to elliptic in side view, hyaline; ornamentation amyloid, finely warted with faint connecting lines, incompletely reticulate; prominences more or less 0.2 μm high; plage indistinct. Basidia 36.5-48.0 × 6-8 μm, four-spored, clavate, hyaline. Pleurocystidia of two types: macrocystidia, 35-82 × 5-9 μm, nearly clavate to cylindrical, with rounded to tapered or moniliform apices, and with granular hyaline contents; and filamentous pseudocystidia. Cheilocystidia 35-56 × 5-9 μm, in the form of macrocystidia.

Gill trama interwoven to more or less parallel in the center; hyphae 4.0-21.9 μm wide, hyaline. Subhymenium filamentous, interwoven; hyphae 3.7-4.8 μm wide, hyaline. Cap cuticle consisting of a layer of inflated ellipsoidal pedicellate cells 10.5-15.5(-25.0) μm wide, covered by an interwoven turf of tapered hyphal-like pileocystidia 2.3-3.5(-5.1) μm wide; pileocystidia thin-walled to slightly thick-walled, with granular hyaline contents. Cap trama heteromerous, interwoven; hyphae 3.5-9.4 μm wide, hyaline; sphaerocysts 19.1-41.5 × 16.7-29.2 μm, ellipsoidal, hyaline, occurring in rosettes. Stalk cuticle consisting of repent hyphae bearing initially a turf of caulocystidia; caulocystidia (29.2-)35-42(-70) × 5.8-8.2 μm, clavate, with granular yellowish to hyaline contents. Lactiferous hyphae 2.3-7.0 μm wide, yellowish, with granular refractive contents, scattered in the stalk, trama, and cuticle. Clamp connections absent.

Useful references. The following book contains information on *L. piperatus* var. *piperatus* pertinent to North America.

- *Mushrooms of North America* (Miller 1972)—A color photograph and a description and discussion are given.

Lactarius repraesentaneus Britzelm. *sensu* Neuhoff

Figs. 137, 201

Common names. None.

Distinguishing characteristics. Cap pale golden yellow; surface bearded at the edge. Latex white, changing to purple, staining bruised flesh and gills purplish. Stalk spotted with small pits. Spores ornamented but not reticulate, more or less ellipsoidal, 9-11 × 6.5-9 μm in size, yellowish in thick deposit. Fruit bodies often occurring in spruce-fir (*Picea-Abies*) forests in northern or mountainous areas.

Observations. *L. repraesentaneus* was originally described from Europe. It has been reported in several North American publications, but recent studies on *Lactarius* by Hesler and Smith (1979) make its presence in North America doubtful. It is similar to *L. speciosus* (Burlingham) Sacc., which features a zoned cap, and is more southern in distribution than *L. repraesentaneus*. Groves (1979) reported *L. repraesentaneus* as not recommended for eating. Never eat any species of *Lactarius* in which injured areas stain purplish.

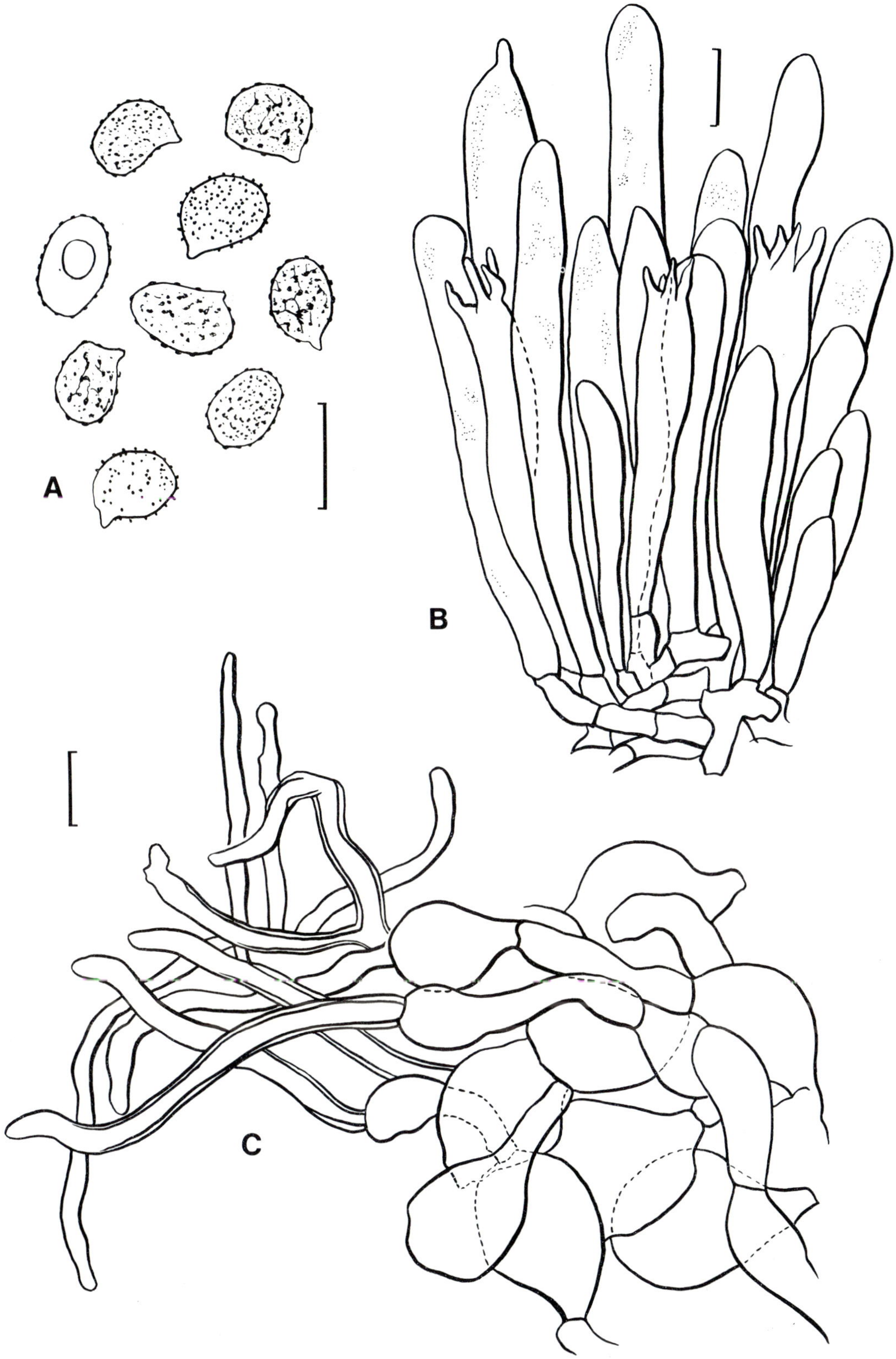

Fig. 136. *Lactarius piperatus* var. *piperatus*: A, spores; B, basidia and macrocystidia; C, cuticular tissue and trichoderm.

Scale line 10 μm

Distribution and seasonal occurrence. This species is reported as common in the Rocky Mountains, occurring in the Pacific northwest, Rocky Mountains, Alaska, central and northeastern United States, and Canada. It fruits from August through September.

Habitat and habit. *L. repraesentaneus* grows solitary to gregarious on soil in moist areas under conifers such as spruce (*Picea sitchensis*), fir (*Abies*), and western hemlock (*Tsuga heterophylla*), and under aspen (*Populus*). It also occurs on the Alaskan tundra, probably associated with dwarf birch (*Betula*) or other dwarf shrubs.

Technical description. Cap 60–150 mm broad, convex-depressed to plano-depressed, broadly funnel-shaped with age; surface viscid, unzoned to more or less zoned with an overlay of fibrils that are more coarse and numerous toward the margin; disc more or less glabrous at times; margin coarsely bearded, enrolled becoming arched; color pale to rich yellow or orange yellow, developing clay to rusty tints over the central area, staining purplish where bruised; fibrils pallid in early stages, becoming yellowish and finally often more or less clay color. Flesh whitish, quickly staining dull lilac to purple, firm, brittle; taste nutmeg-like, bitterish to slightly acrid, or both; odor indistinctive or nutmeg-like; latex copious, viscous in early stages or whey-like to watery, white to cream changing to violet or lilac on the flesh, mild to slightly acrid.

Gills narrow to moderately broad, adnate-subdecurrent, close to crowded, forking near the stalk; color cream to pale ocherous, sometimes developing finally a pale orange cast, soon spotting lavender to purplish where injured.

Stalk 40–120 mm long, 10–30 mm thick, short and thick, equal or clavate; surface viscid in early stages becoming dry, pitted, pruinose at the apex, tomentose at the base; color whitish to more or less similar in color to the cap, staining similar to the gills. Cuticle hard, stuffed, becoming hollow.

Spores yellowish in deposit, 9–11 × 6.5–9 μm, broadly ellipsoidal to ellipsoidal, hyaline in Melzer's reagent or at times with a diffuse coating of amyloid material; ornamentation consisting of warts, ridges, and sparsely branched bands, not reticulate; prominences 0.4–0.8 μm high; plage distinct. Basidia four-spored, 60–70 × 10–14 μm, clavate. Pleurocystidia usually only in the form of macrocystidia; macrocystidia 70–120 × 9–15 μm, fusiform to fusiform-ventricose with a long tapered neck ending in a sharp point, having a more or less small apical protuberance or somewhat apical constrictions, and with more or less granular contents; pseudocystidia absent to rare. Cheilocystidia in the form of macrocystidia, except smaller and scattered.

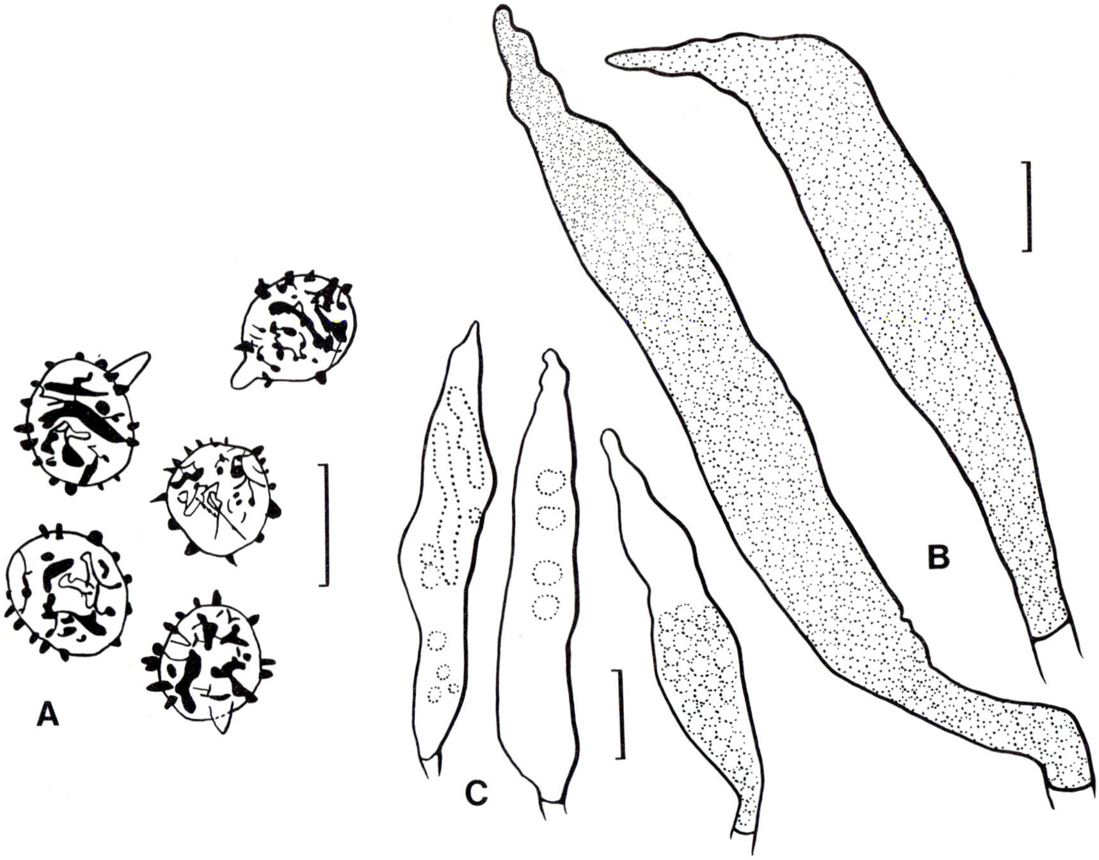

Fig. 137. *Lactarius repraesentaneus*: A, spores; B, macrocystidia; C, cheilocystidia.

Scale line 10 μm

Gill trama with some rosette formation near the junction of the gill and cap trama. Cap cuticle in the form of an ixocutis of narrow hyphae 4-6 μm wide; hyphae loosely aggregated into fascicles, more or less embedded in slime, lacking encrustations and dextrinoid debris. Cap trama heteromerous. Stalk cuticle in the form of a distinct ixotrichoderm originating from a basal zone of longitudinally oriented hyphae; remainder of surface composed of loosely interwoven fibrils; slime not visible. Lactiferous hyphae inconspicuous. Clamp connections absent.

Useful references. The following books contain information on *L. repraesentaneus* pertinent to North America.

- *Edible and Poisonous Mushrooms of Canada* (Groves 1979)—A color photograph and a description and discussion are given.
- *The Mushroom Hunter's Field Guide* (Smith 1974)—A black-and-white photograph, a color photograph, and a description and discussion are given.

Lactarius rufus (Scop. ex Fr.) Fr.

Figs. 138, 202

Common names. Red milky cap, red lactarius.

Distinguishing characteristics. Cap red brown, dry; margin striate to ribbed. Gills pallid, grayish flesh color. Latex white, strongly acrid. Spores pale yellow in thick deposit, subglobose to broadly ellipsoidal, 7-8(-11) × 5.1-7.3(-8) μm. Fruit bodies often growing in well-drained pine woods.

Observations. *L. rufus sensu lato* is a very common species in North America but may be confused with *L. subdulcis* (Pers. ex Fr.) S. F. Gray and relatives because of the similar coloration. *L. subdulcis* is a mild-tasting species rather than acrid, has smaller, more reddish tan fruit bodies, and typically is associated with hardwoods. The many variants of *L. rufus* make it difficult to define. Additional data are needed to determine whether distinct varieties are present. Kauffman (1971) reported this species as poisonous.

Distribution and seasonal occurrence. This species is common and widely distributed, occurring from eastern to western North America. It fruits from late summer to late fall.

Habitat and habit. *L. rufus* grows scattered to gregarious, occurring in sphagnum bogs, moss, pine, duff, and sand under pine (*Pinus*) and larch (*Larix*) in coniferous woods.

Technical description. Cap 40-120 mm broad, convex-umbonate to convex-depressed or more or less plane with a depressed center; margin striate to ribbed, enrolled in early stages, spreading with age; edges even; surface moist to sometimes nearly viscid to dry, not hygrophanous, minutely flocculose-silky in early stages, soon becoming glabrous but sometimes areolate-squamulose to lacerate-squamulose, more or less zoned; color dull reddish, reddish brown, deep reddish brown, or sometimes with slightly deep reddish orange or rufous tints; umbo dark brown; margin grayish red. Flesh thin, 3-8 mm thick, soft; color tan, with a pinkish tinge; latex copious in early stages, white, unchanging; taste very acrid immediately or mild initially with acridity developing; odor fungoid, indistinctive.

Gills 7-32 mm long, 1-5 mm wide, adnate-subdecurrent, close, narrow, sometimes forked; color more or less pallid, grayish flesh color, or slightly yellowish when young, becoming light orange to light pinkish orange or more or less colored similar to the cap.

Stalk 30-110 mm long, 5-17 mm thick, fragile, long, thin, equal or slightly narrowed downward; surface dry, glabrous to pruinose or sometimes hairy at the base; color light pinkish orange to light reddish brown, usually more or less similar in color to the cap surface, paler at the apex and base. Flesh similar to the flesh of the cap, stuffed, becoming hollow.

Spores yellowish white to buff in thin and thick deposit, (7.0-)8.0-11.0 × 4.1-7.3(-8.0) μm, subglobose to broadly elliptic and inequilateral in side view, hyaline; ornamentation amyloid, partially reticulate with isolated ridges, warts, and particles; prominences more or less 0.5 μm high; plage distinct; apiculus prominent. Basidia (36.5-)43.7-46.8 × 7.3-11.7 μm, four-spored, clavate, with granular or globular contents, hyaline. Pleurocystidia only in the form of macrocystidia embedded and projecting from the hymenium, 42.0-70.2 × 5.1-10.5 μm, fusiform-ventricose or clavate with tapering obtuse to sharply acute apices, sometimes with somewhat apical constrictions and encrustations, and with granular to globular hyaline contents. Cheilocystidia embedded and projecting from the hymenium, 23.4-52.0 × 4.6-7.5 μm, fusiform to subcylindrical or nearly ventricose; apices obtuse to more or less acute, sometimes with more or less apical encrustations; contents granular to globular, hyaline.

Gill trama interwoven to cellular; hyphae 2.3-7.3 μm wide, hyaline; other cells also present, 6.6-11.0 μm wide, inflated, more or less isodiametric, hyaline. Subhymenium filamentous; hyphae interwoven, 3.0-6.6 μm wide, thread-like, septate, hyaline. Cap cuticle an appressed layer of longitudinally arranged hyphae with tufts of erect hyphal ends in the form of pileocystidia; hyphae 2.2-5.1(-5.8) μm wide, cylindrical, septate, nongelatinizing, hyaline. Cap trama heteromerous, interwoven; hyphae 3.5-10.2 μm wide, cylindrical, septate, hyaline; sphaerocysts 17.8-46.0 × 8.8-29.2 μm

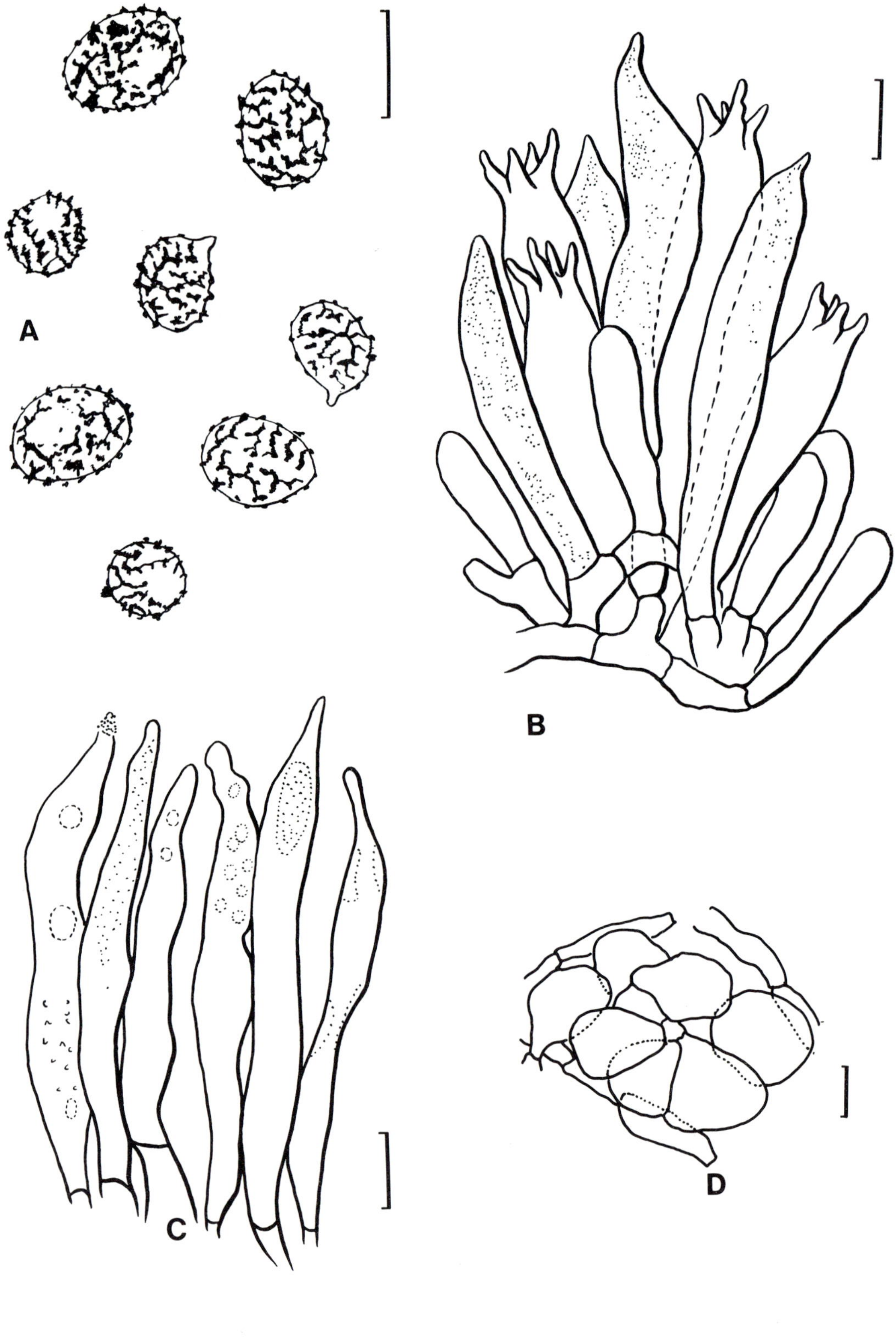

Fig. 138. *Lactarius rufus*: A, spores; B, basidia and macrocystidia; C, cheilocystidia; D, sphaerocysts in tramal tissues.

Scale line 10 μm

wide, cellular, subglobose to ellipsoidal, inflated, hyaline, occurring in rosettes. Stalk cuticle a tangled interwoven layer of hyphae; occasional hyphae grouped into tufts and projecting, or hyphae blunt-ended and singly projecting; caulocystidia 2.2–5.1 μm wide, clavate to cylindrical, septate, nongelatinizing, clustered at the apex of the stalk, hyaline; slime absent. Lactiferous hyphae 3.5–8.2 μm wide, yellowish, refractive, with granular to homogeneous contents, not prominent in the stalk and trama. Clamp connections absent.

Useful references. The following books contain information on *L. rufus* pertinent to North America.

- *Edible and Poisonous Mushrooms of Canada* (Groves 1979)—A color photograph and a description and discussion are given.
- *Mushrooms of North America* (Miller 1972)—A color photograph and a description and discussion are presented.
- *The Mushroom Hunter's Field Guide* (Smith 1974)—A black-and-white and a color photograph, as well as a description and discussion, are presented.
- *The Savory Wild Mushroom* (McKenny 1971)—A black-and-white and a color photograph are presented, along with a description and discussion.

Lactarius scrobiculatus var. *canadensis* (Smith) Hesler & Smith — Fig. 139

Common names. Spotted-stemmed lactarius, pitted milky cap.

Distinguishing characteristics. Cap olive buff or various shades of yellow, or both; surface viscid, becoming dry; margin bearded. Latex scanty, quickly changing to sulfur yellow upon exposure to air. Taste slightly acrid. Gills staining yellowish to more or less clay color. Stalk surface pitted with large yellow glazed spots. Spores white to cream in deposit, more or less ellipsoidal, 7.3–8(–8.8) × 5.1–6.6 μm in size.

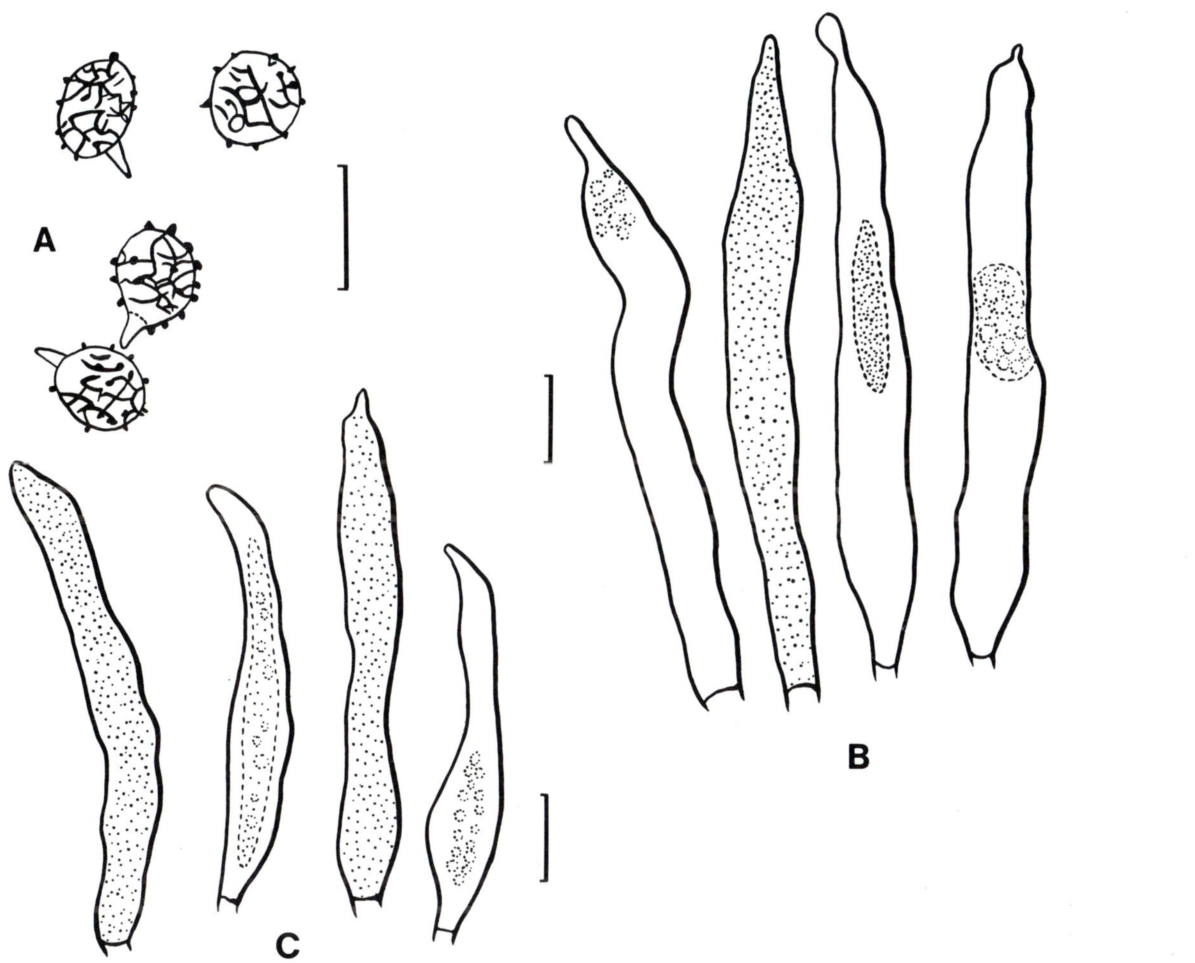

Fig. 139. *Lactarius scrobiculatus* var. *canadensis*: A, spores; B, macrocystidia; C, cheilocystidia.

Scale line 10 μm

Observations. This variety was originally described as *L. payettensis* var. *canadensis* (Smith 1960). According to Hesler and Smith (1979) *L. scrobiculatus* var. *canadensis* is distinguished from var. *scrobiculatus* by its white to cream spore deposit, scanty latex, slightly acrid taste, and less robust stature. The cap color pattern is one of a gradual yellowing with age; usually the cap margin is bearded.

Distribution and seasonal occurrence. This variety's general distribution is uncertain. It occurs mainly in northern and mountainous areas, although it is perhaps even more widely distributed. It fruits in the fall.

Habitat and habit. *L. scrobiculatus* var. *canadensis* occurs solitary to gregarious on soil under conifers such as black spruce (*Picea mariana*).

Technical description. Cap (30-)40-100(-120) mm broad, convex-depressed to plano-depressed; surface viscid, becoming dry, more or less fibrillose to glabrous; margin bearded in early stages, or reduced to marginal pubescence, often naked at maturity, at times with the cuticle near the margin broken into appressed squamules; color whitish becoming olive buff and slowly yellowing with age, or soon turning a dull to brilliant yellow, or remaining yellow from the first; margin white, staining brownish at times. Flesh whitish to pale yellow, firm, brittle becoming lax; latex scanty, white, changing quickly to sulfur yellow upon exposure to air; taste mild to slightly acrid; odor faintly fragrant to indistinctive.

Gills adnate to short-decurrent, crowded, narrow to moderately broad, equal in width, forking at the base; two or three rows of lamellulae present; color whitish, becoming ocherous, staining yellowish, and finally turning more or less clay color where injured.

Stalk 30-110 mm long, 10-30 mm thick at the apex, more or less equal, with short pseudorhizae; surface glabrous, with large glazed spots; color whitish above, with yellow spots circled in white to yellowish shades. Flesh staining yellow to rusty brown where bruised, stuffed, becoming hollow.

Spores white to cream in deposit, 7.3-8.0(-8.8) × 5.1-6.6 μm, elliptic in side view, hyaline; ornamentation amyloid, partially reticulate, consisting of branching lines and ridges with isolated particles and warts; prominences more or less 0.5 μm high, with a prominent hyaline apiculus. Basidia 41.8-52.9 × 8.8-11.7 μm, four-spored, clavate, pale yellow, hyaline; contents granular to globular. Pleurocystidia of two types: macrocystidia, 45-78 × 5-12 μm, scattered, projecting, originating from the subhymenium, fusiform, with acute apices or with constrictions near the apex, hyaline, with granular contents; and pseudocystidia, 1.5-3 μm wide, scattered, filamentous, thin-walled, with obtuse apices, hyaline, refractive. Cheilocystidia 25-45 × 5-8 μm, in the form of macrocystidia and pseudocystidia, but smaller than those of the pleurocystidia.

Gill trama interwoven; hyphae of central portion and subhymenium 2-7 μm wide, tubular, thin-walled, branching, septate, hyaline, with more or less granular contents. Cap cuticle in the form of a well-defined ixocutis; hyphae 2.2-5.1 μm wide, tubular to cylindrical, gelatinous, septate, hyaline, refractive. Cap trama heteromerous, interwoven; hyphae 3-7 μm wide, tubular-cylindrical, branching, septate, hyaline in potassium hydroxide, with more or less granular contents; sphaerocysts 15-30 × 10-19 μm, subglobose to ellipsoidal, thin-walled, inflated, scattered or occurring in rosettes, hyaline. Stalk cuticle interwoven, dry; ascending hyphae in the form of caulocystidia, 2.9-5.8 μm wide, tubular, more or less gelatinous, blunt-ended, septate, hyaline, with granular to refractive contents. Lactiferous hyphae inconspicuous. Clamp connections absent.

Useful references. There are no general books or field guides pertinent to North America that contain information on this variety. Hesler and Smith (1979) provide a technical description and some illustrations.

Lactarius scrobiculatus (Fr.) Fr. var. *scrobiculatus*

Fig. 203

Common names. Spotted-stemmed lactarius, pitted milky cap.

Distinguishing characteristics. Cap ocherous; surface viscid; margin bearded. Latex copious, quickly changing to sulfur yellow upon exposure to air; taste burning acrid. Stalk pitted with spots. Spores bright ocher in thick deposit, subglobose to broadly ellipsoidal, 7-9 × 6-7.5 μm. Cap cuticle consisting of hyphae crowded in bundles.

Observations. The following technical description of *L. scrobiculatus* var. *scrobiculatus* adapted from that of Hesler and Smith (1979) is translated and condensed from the work of Neuhoff (1956). Although this variety has been reported often from North America, it may not be common here. *L. scrobiculatus* var. *scrobiculatus* may be confused with *L. torminosus* and *L. resimus* (Fr.) Fr. *L. torminosus* produces a white, unchanging latex and does not usually have a stalk pitted with spots. *L. resimus* also does not feature a pitted stalk and has a white cap in the early stages.

Kauffman (1971) reported *L. scrobiculatus* var. *scrobiculatus* as poisonous. Never eat species of *Lactarius* that produce a latex which turns bright yellow upon exposure to air.

Distribution and seasonal occurrence. *L. scrobiculatus* var. *scrobiculatus* is generally reported as common in

North America. It fruits from July through September. See *L. scrobiculatus* var. *canadensis*.

Habitat and habit. *L. scrobiculatus* var. *scrobiculatus* grows scattered to gregarious in moist areas in conifer forests of mountainous areas or along mossy margins of swamps.

Technical description. Cap 50-170 mm broad, almost hemispherical in early stages, becoming funnel-shaped; surface viscid, peeling easily along the margin, shiny when dry, more or less reticulate from darker fibrils, becoming more or less scaly from the margin inward; scales arranged in zones; disc remaining more or less glabrous, sometimes with watery zones or rows of squamules; margin strigose-hairy, becoming glabrous with age, scalloped in early stages, long-enrolled; color pale yellow to dull ocher yellow or dark chrome yellow with brownish yellow, orange brown, or dingy brown fibrils and scales, becoming yellow to dingy brown where injured. Flesh pallid, rigid; latex copious, white quickly changing to sulfur yellow upon exposure to air, staining the flesh citron yellow to greenish yellow; taste acrid; odor fruity.

Gills 3-5 mm broad, adnate-decurrent, crowded, rounded at the stalk, toothed; color whitish to yellow, with bruised areas dingy tawny to sordid greenish; lamellulae numerous.

Stalk 30-60 mm long, 10-35 mm thick, equal, at times narrowed to a root-like base; surface overlaid sparsely at the base with white mycelia, pruinose overall except on the glazed spots, pitted above with ocherous to clay spots, bruising pale fulvous. Flesh stuffed, becoming hollow.

Spores bright ocher in thick deposit, 7-9 × 6-7.5 μm, subglobose to broadly ellipsoidal; ornamentation consisting of fine lines emanating from ridges making a sparse reticulum, with isolated warts and broken lines; prominences 0.5-1.0 μm high. Basidia four-spored, 45-50 × 8-10 μm. Pleurocystidia and cheilocystidia 60-100 × 6-11 μm, projecting about 20 μm beyond the basidia, sparse, cylindrical-clavate to fusiform, with aciculate apices, and with granular to refractive contents; pseudocystidia abundant. Cap cuticle in the form of an epicutis; hyphae crowded in bundles, gelatinous, nearly parallel, 2-4 μm wide, forming a layer 20 μm deep, hyaline in potassium hydroxide. Stalk cuticle composed of a covering of hairs 60-80 × 3-4.5 μm. Lactiferous hyphae numerous. Clamp connections absent.

Useful references. The following books contain information on *L. scrobiculatus* var. *scrobiculatus* pertinent to North America.

- *A Field Guide to Western Mushrooms* (Smith 1975)—A color photograph and a description and discussion are given.
- *Edible and Poisonous Mushrooms of Canada* (Groves 1979)—A color photograph and a description and discussion are given.
- *Mushrooms of North America* (Miller 1972)—A description and discussion are given.
- *The Savory Wild Mushroom* (McKenny 1971)—A black-and-white photograph, a color photograph, and a description and discussion are presented.

Lactarius subvellereus var. *subdistans* Hesler & Smith

Figs. 140, 204

Common names. None.

Distinguishing characteristics. Cap and stalk white, velvety, and dry. Gills distant. Latex cream, staining gills and flesh brownish. Spores white in deposit, ovoid to broadly ellipsoidal, (5-)6.5-10 × (4-)5.1-6.6 μm in size.

Observations. *L. subvellereus* var. *subdistans* is often confused with *L. deceptivus*, *L. piperatus*, and *L. vellereus*. *L. deceptivus* at maturity loses its cottony roll on the margin but can still be distinguished by its larger and broader spores that have prominent markings on the walls. *L. piperatus* has narrow gills and the cap is not tomentose. *L. vellereus* is distinguished primarily by its broader spores. Groves (1962) reported *L. vellereus* as suspected of being poisonous.

Distribution and seasonal occurrence. This variety is common, occurring generally east of the Great Plains in North America, specifically from the Great Lakes states eastward in the United States and from Ontario eastward in Canada. It fruits from July to October.

Habitat and habit. *L. subvellereus* var. *subdistans* is gregarious and is often found on pine needles, moss, and grass in open places under pine (*Pinus*), birch (*Betula*), oak (*Quercus*), or beech (*Fagus*) in coniferous, deciduous, or mixed woods.

Technical description. Cap 35-165 mm wide, plano-convex to convex-umbilicate; disc broadly and deeply depressed; margin enrolled, finally becoming plane or elevated; surface dry and velvety to the touch, finely tomentose, without zones; color white with yellowish, grayish, or buff patches. Flesh 3-20 mm thick, firm, compact, white, becoming yellow when cut; latex white and unchanging or becoming creamy yellow, staining the flesh or gills brownish; taste stinging acrid; odor fungoid to pungent.

Gills 10-90 mm long, 1-5 mm deep, adnate to subdecurrent, subdistant to distant, narrow to moderately broad, some forked, pallid cream or pale yellow to creamy yellow, staining brownish where bruised; edges even.

Stalk 15-50 mm long, 12-37 mm thick, equal or narrowed toward the base, at times eccentric, solid, very firm and hard; surface more or less even, unpolished to velvety, dry, white, tinged yellowish to brownish when dried. Flesh firm, white to yellowish, staining pink, stuffed.

Spores chalk white in thick deposit, (5.0-)6.5-10.0 × (4.0-)5.1-6.6 μm, ovate to broadly elliptic in side view, inequilateral in side view, hyaline; ornamentation amyloid, finely warted with some warts faintly connected but not reticulate; prominences about 0.2 μm high; plage indistinct. Basidia 37.0-73.6 × 7.5-11.7 μm, four-spored, narrowly clavate, hyaline. Pleurocystidia only in the form of macrocystidia, 30-94 × 4.4-9.5 μm, fusiform to nearly fusiform or cylindrical, with tapering apices, obtuse to nearly acute or constricted, with granular to globose inclusions, hyaline, embedded, abundant, originating deep within the gill trama. Cheilocystidia 30.0-66.2 × 4.4-7.3 μm, in the form of macrocystidia except shorter than those of pleurocystidia.

Gill trama tightly interwoven; hyphae of the subhymenium and the central portion 2.2-10.5 μm wide, cylindrical, septate, hyaline. Cap cuticle filamentous; subcutis a narrow layer of hyphae just beneath the cutis, giving rise to pileocystidia; these hair-like cells 75-350 × 1.5-5.8 μm, thick-walled, blunt-ended, many-partitioned, hyaline to opaque. Cap trama heteromerous; hyphae interwoven, 3.5-9.4 μm wide, thread-like, septate, hyaline; sphaerocysts occurring in rosettes, 23.4-46.8 × 9.5-41.2 μm, subglobose to many-angled, cellular, inflated, thick-walled, hyaline. Stalk cuticle filamentous, consisting of a layer of interwoven hyphae giving rise to caulocystidia; these filiform cells 70-350 × 1.5-5.8(-7.0) μm, thin- to thick-walled, blunt-ended, septate, hyaline or, in Melzer's reagent, pale yellow to hyaline with encrusted black internal and external debris. Lactiferous hyphae 5.8-10.5 μm wide, granular, refractive, yellowish, scattered in the trama of the cap, gills, and stalk. Clamp connections absent.

Useful reference. The following book contains information on *L. subvellereus* var. *subdistans* pertinent to North America.

- *Edible and Poisonous Mushrooms of Canada* (Groves 1979)—A discussion under *L. vellereus* is presented.

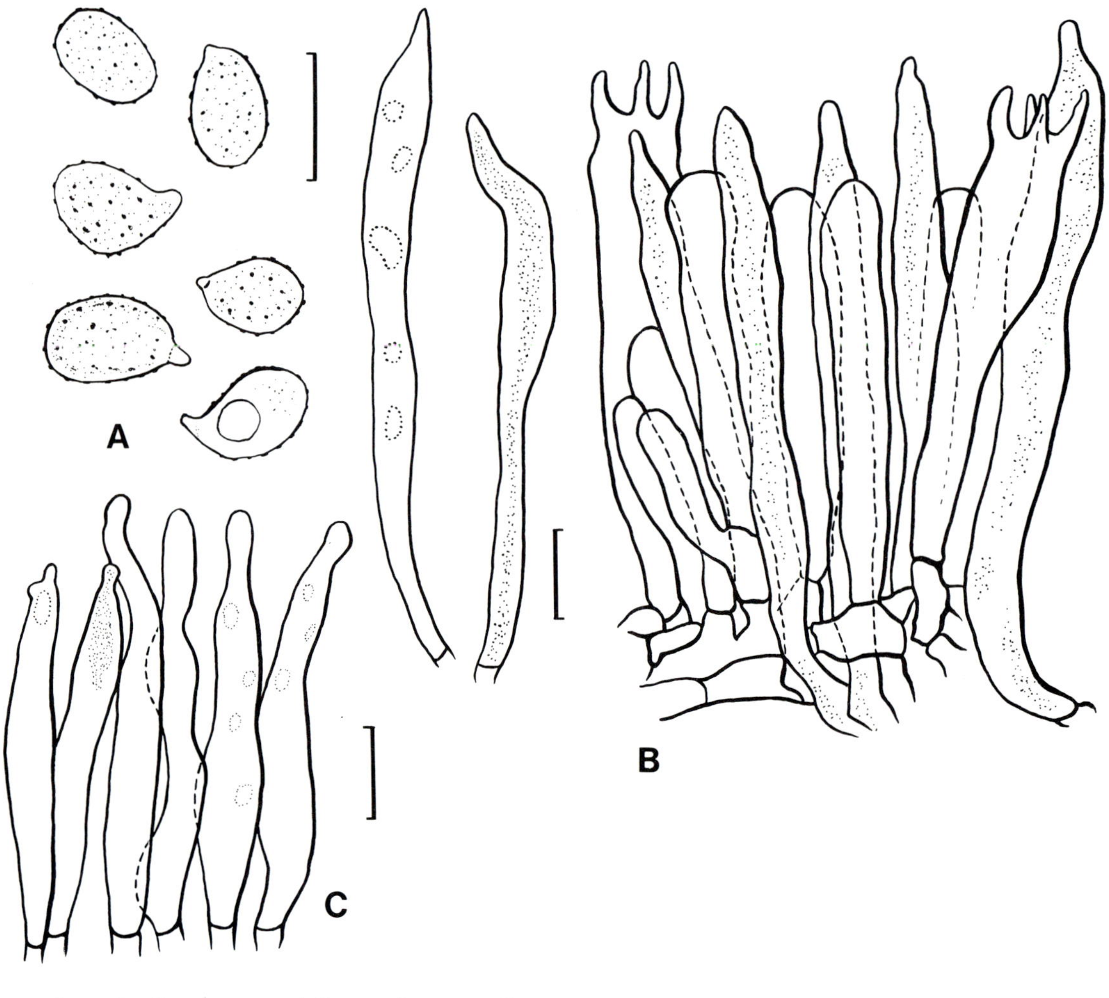

Fig. 140. *Lactarius subvellereus* var. *subdistans*: A, spores; B, basidia and macrocystidia; C, cheilocystidia.

Scale line 10 μm

Lactarius torminosus (Fr.) S. F. Gray

Figs. 141, 205

Common names. Woolly milky cap, woolly lactarius.

Distinguishing characteristics. Cap pinkish orange to pale pinkish, becoming whitish on the margin, more or less zoned; disc viscid, becoming glabrous; margin white-fibrillose-bearded. Latex white, unchanging. Taste acrid. Spores yellowish cream in deposit, subglobose to broadly ellipsoidal, (7-)8-10.2 × 5.8-6.6(-8) µm in size.

Observations. *L. torminosus* is common in North America, nearly always associated with birch (*Betula*). It is often confused with *L. scrobiculatus, L. cilicioides* Fr., and *L. pubescens* Fr. *L. scrobiculatus,* when immature, is similar in appearance to *L. torminosus* but has a white latex that quickly changes to sulfur yellow. *L. cilicioides,* a poorly understood species, is similar in color to *L. torminosus,* but the cap is not zoned and it apparently has smaller spores. *L. pubescens* is almost identical to *L. torminosus* but has smaller spores, 6.0-7.5 × 5.0-6.5 µm. *L. torminosus* var. *nordmanensis* Smith produces latex that changes to yellow and stains the mushroom tissues and white paper yellow. Groves (1979) reported *L. torminosus* as poisonous.

Distribution and seasonal occurrence. This species is common, abundant, and widely distributed, occurring in western, northwestern (Alaska), and eastern North America. It fruits from July to October.

Habitat and habit. *L. torminosus* is terrestrial, occurring scattered to gregarious in grass under birch (*Betula*) and hemlock (*Tsuga*) and in mixed hardwood-coniferous forests. It also grows on urban lawns where birch trees are planted.

Technical description. Cap 20-120 mm broad, convex, convex-depressed, or plano-depressed, expanding to shallowly funnel-shaped; margin incurved, bearded when young, white-fibrillose with age; surface tomentose becoming glabrous, sticky to viscid and glabrous on the central portion in early stages, often somewhat zoned; color pinkish orange to pale dull pink, becoming orange white to whitish on the margin, with the pink gradually fading. Flesh firm, becoming flaccid; color white, at times flesh-colored; latex white to cream, unchanging, not staining the gills; taste acrid; odor slight to pungent.

Gills short-decurrent, close to crowded, narrow, sometimes forked near the stalk; color whitish, becoming pale vinaceous to pale orange or cream tinged vinaceous, turning pale tan with age.

Stalk 15-80 mm long, 6-20 mm thick, fragile, more or less equal, cylindrical or narrowed at the base; surface dry, glabrous to pruinose; color pale light pinkish to yellowish tinged or slightly pinkish orange to orange white, sometimes spotted. Flesh firm, beige white, stuffed, becoming hollow.

Spores cream to pale yellow in thick deposit, (7.0-)8.0-10.2 × 5.8-6.6(-8.0) µm, subglobose to broadly elliptic in side view, hyaline; ornamentation amyloid, partially reticulate, with interrupted ridges and a few isolated warts; prominences more or less 0.5-0.73(-1.5) µm high, with a prominent apiculus. Basidia 30.0-47.7 × 7.3-8.2(-11.0) µm, four-spored, clavate to cylindrical, hyaline. Pleurocystidia only in the form of macrocystidia embedded and originating in the hymenium and the subhymenium, 40.3-80.0 × 5.1-9.5 µm, abundant, fusiform to fusiform-ventricose, gradually tapering, with obtuse apices and zero to three constrictions toward the apex, and with granular hyaline contents. Cheilocystidia 30-52 × 4.5-8.0 µm, in the form of macrocystidia.

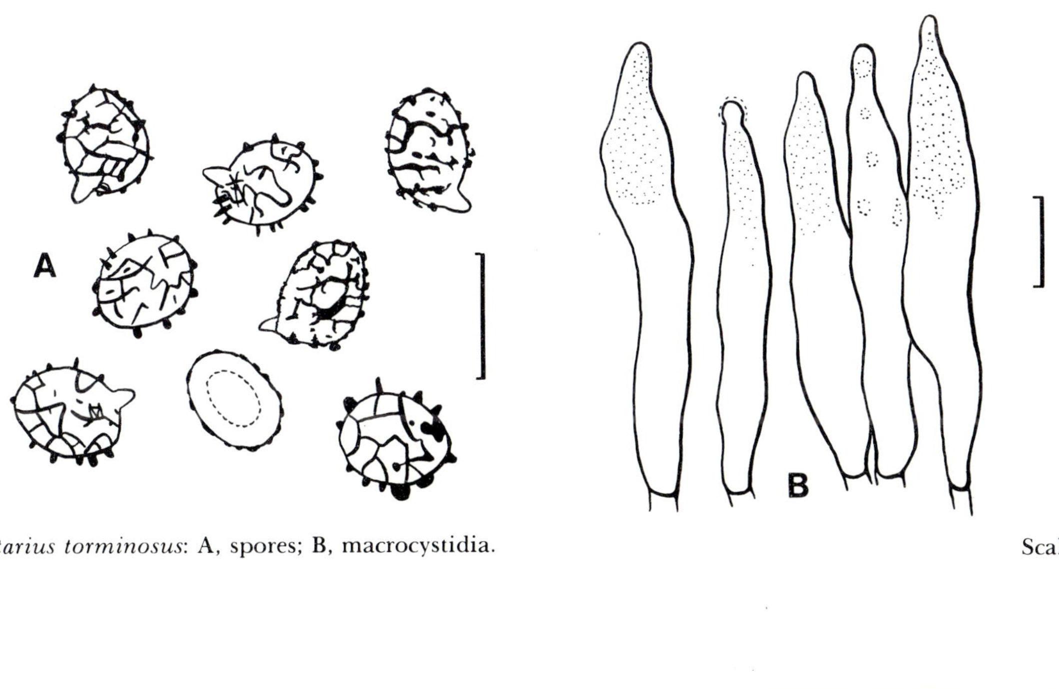

Fig. 141. *Lactarius torminosus*: A, spores; B, macrocystidia.

Scale line 10 µm

Gill trama cellular to filamentous; hyphae more or less parallel to interwoven, 3.7–14.6 μm wide, subglobose to many-angled, hyaline, not occurring in rosettes. Subhymenium hyphae interwoven, 3.7–11.0 μm wide, thread-like, thin-walled, septate, hyaline. Cap cuticle in the form of an ixocutis; hyphae interwoven to more or less radially arranged and parallel in fibrils, 2.5–7.3 μm wide, thread-like, thin-walled, septate, hyaline. Cap trama heteromerous; hyphae interwoven, 3.5–11.7 μm wide, thread-like, thin-walled, septate, hyaline; sphaerocysts 29.2–62.0 × 11.0–48.0 μm, subglobose to ellipsoidal to many-angled, cellular, inflated, thin-walled, hyaline, occurring in rosettes. Stalk cuticle consisting of interwoven hyphae not specialized at the surface; hyphae 3.0–8.0 μm wide, cylindrical, septate, hyaline. Lactiferous hyphae 5.8–10.2 μm wide, yellowish, refractive, with granular to homogeneous contents, inconspicuous to abundant in the stalk, cap and gill tramae, and cap cuticle. Clamp connections absent.

Useful references. The following books contain information on *L. torminosus* pertinent to North America.

- *A Field Guide to Western Mushrooms* (Smith 1975)—This species is discussed under *L. torminosus* var. *nordmanensis*.
- *Edible and Poisonous Mushrooms of Canada* (Groves 1979)—A color photograph and a description and discussion are given.
- *Mushrooms of North America* (Miller 1972)—A color photograph and a description and discussion are given.
- *The Mushroom Hunter's Field Guide* (Smith 1974)—A black-and-white photograph, a color photograph, and a description and discussion are given.
- *The Savory Wild Mushroom* (McKenny 1971)—A black-and-white photograph and a description and discussion are given.

Lactarius uvidus (Fr. ex Fr.) Fr.

Figs. 142, 206

Common names. None.

Distinguishing characteristics. Cap smooth, viscid, soon becoming dry; color pallid to pale lilac drab or grayish lilac, at times whitish. Latex mild to bitter-tasting, staining the flesh lilac on injured areas. Stalk viscid, soon becoming dry; color white to pale yellowish, with light purplish gray toward the base. Gills close; color creamy white, darkening with age, staining lilac to violet. Spore deposit dull white to pale yellow, subglobose to broadly ellipsoidal, 6.6–12.4 × 5.8–8.5 μm in size.

Observations. As described here, this species represents *L. uvidus* as it occurs in the Great Lakes region; however, it may not be the species described originally by Fries (1821). Refer to the work of Hesler and Smith (1979) for comments on *L. uvidus* and its relatives. *L. uvidus* is most likely to be confused with *L. maculatus* Peck, which is larger, with a zoned cap and a spotted stalk. Groves (1979) reported *L. uvidus* as poisonous.

Distribution and seasonal occurrence. This species is reported infrequently but is probably widely distributed, occurring generally from northwestern to northeastern North America. It fruits from late summer through the fall.

Habitat and habit. *L. uvidus* is scattered to gregarious, growing under aspen (*Populus*), in mixed woods of birch (*Betula*) and pine (*Pinus*), under northern conifers, on low ground in swampy or boggy regions, and at times in sphagnum bogs or among liverworts.

Technical description. Cap 21–100 mm broad, convex, convex-umbonate, or nearly plane, finally plano-depressed; margin pruinose and incurved in early stages, at times arched, spreading with age; surface viscid to slimy, becoming glabrous, and without zones, slightly zoned, or, rarely, distinctly zoned; color pallid to very pale lilac drab, darkening to a medium lilac drab with age, or remaining whitish, sometimes with incarnate to avellaneous tinges with age; disc darker. Flesh 4–12 mm thick, firm; color whitish, staining dull lilac when cut, eventually becoming pinkish tan to darker brown; latex bitter, copious, milky white, soon becoming dingy pale cream to lilac or reddish violet, staining broken surfaces dull lilaceous; taste mild, becoming bitter; odor mild to fungoid.

Gills 8–51 mm long, 1–8 mm deep, adnate-subdecurrent to decurrent, close, narrow to broad; color creamy white, staining lilac to violet when bruised, darkening with age.

Stalk 21–70 mm long, 4–16 mm thick, more or less equal, at times slightly enlarged at the base; surface sticky to viscid, naked and shining when dry, glabrous or sometimes tomentose at the base; color cream to pale yellow; apex whitish or slightly yellowish, light purplish gray toward the base. Flesh spongy, white to pale yellow, stuffed, becoming hollow.

Spores dull white to pale yellow in thick deposit, 6.6–12.4 × 5.8–8.5 μm, subglobose to broadly elliptic in side view, hyaline; ornamentation amyloid, partially reticulate, consisting of broken lines and bands, with a few isolated warts; prominences 0.5–1.1 μm high; plage indistinct; apiculus prominent. Basidia 36.5–51.4 × 8.0–14.6 μm, four-spored, clavate, pale yellow with granules or globules, hyaline. Pleurocystidia only in the form of macrocystidia projecting and originating from the hymenium, 40.0–85.0 × 5.8–11.0 μm, nearly ventricose, becoming fusiform-ventricose, one- to three-

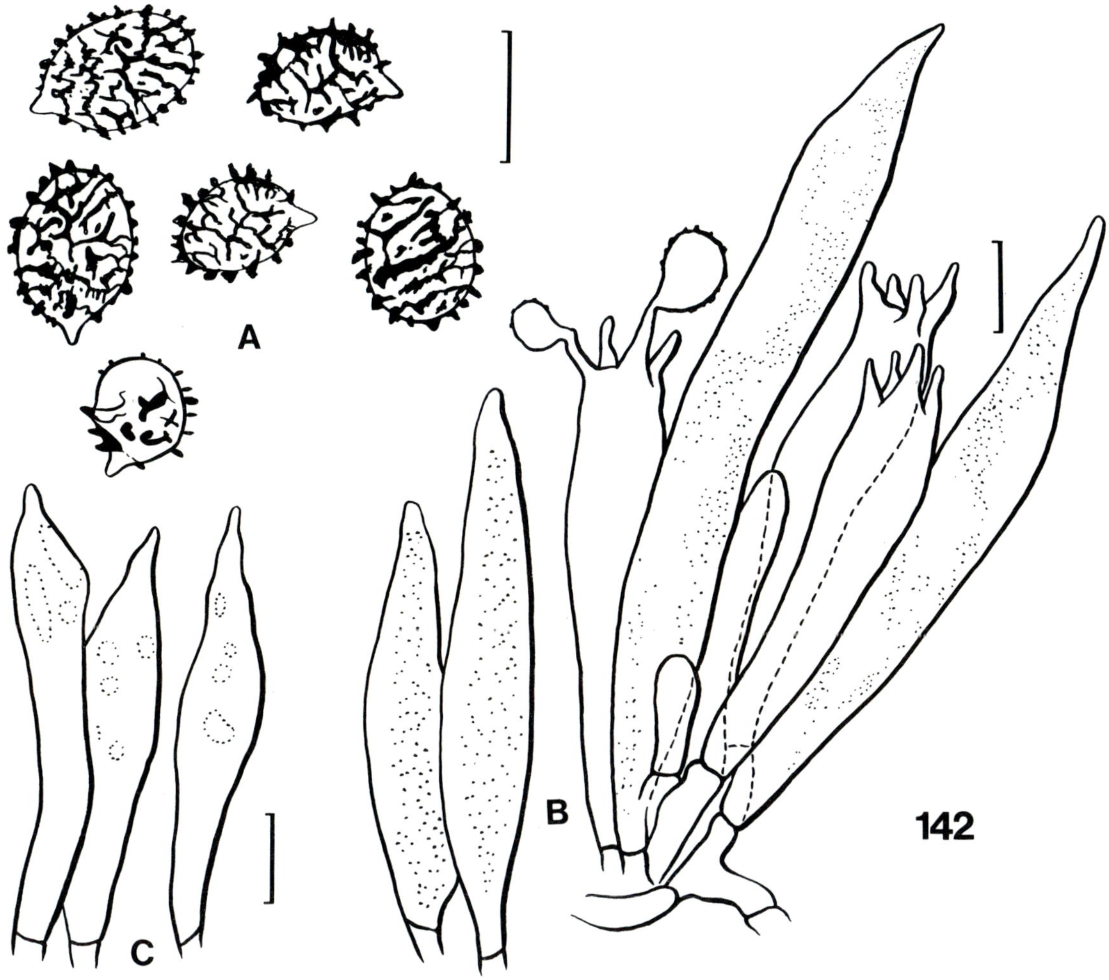

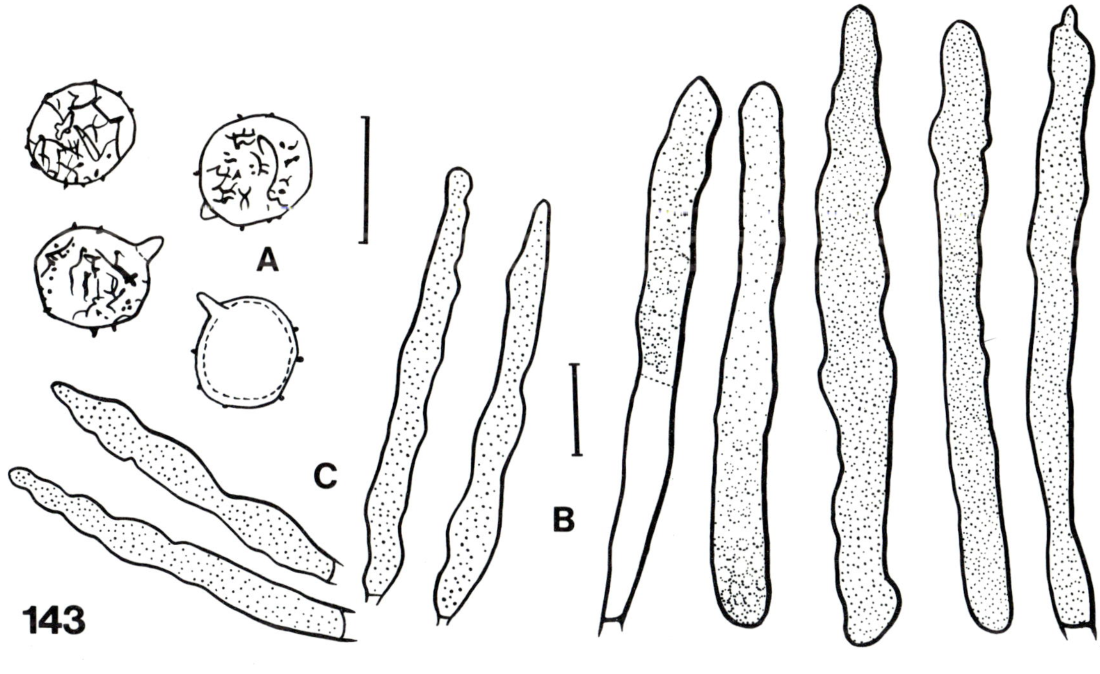

Fig. 142. *Lactarius uvidus*: A, spores; B, basidia and macrocystidia; C, cheilocystidia.

Fig. 143. *Lactarius vellereus* var. *vellereus*: A, spores; B, macrocystidia; C, cheilocystidia.

Scale line 10 μm

celled, with tapering obtuse to acute apices, turning pale yellow with globules or granules in Melzer's reagent, otherwise hyaline. Cheilocystidia projecting from the hymenium, 29.2-47.7 × 5.1-8.8 μm, nearly ventricose to subcylindrical, one- to three-celled, with obtuse, tapering, or at times capitate apices, turning pale yellow with globules or granules in Melzer's reagent, otherwise hyaline.

Gill trama interwoven, gelatinous; subhymenium hyphae tightly interwoven, (2.9-)3.5-5.8(-11.0) μm wide, hyaline. Cap cuticle more or less in the form of an ixocutis; hyphae repent and enmeshed in slime; some hyphae parallel; others cross-latticed, with the slime layer thick and encompassing; surface hyphae (2.0-)4.4-8.0 μm wide, cylindrical, septate, with gelatinized walls, pale yellow with granular and globose amyloid contents in Melzer's reagent. Cap trama heteromerous; hyphae interwoven, (3.5-)8.2-12.9 μm wide, thread-like to tubular, septate, hyaline; sphaerocysts 29.2-46.8 × 11.0-41.1 μm, subglobose to many-angled, more or less thick-walled, hyaline in potassium hydroxide, occurring in rosettes. Stalk cuticle more or less in the form of an ixocutis; hyphae embedded in slime, sometimes parallel, sometimes cross-latticed, with the slime layer thick and encompassing. Caulocystidia lacking; surface hyphae 2.9-8.8 μm wide, thread-like, blunt-ended, thick-walled, septate, hyaline. Lactiferous hyphae 3.5-8.2(-10.5) μm wide, yellowish, refractive; contents granular. Clamp connections absent.

Useful references. The following books contain information on *L. uvidus* pertinent to North America.

- *Edible and Poisonous Mushrooms of Canada* (Groves 1979)—A color photograph and a description and discussion are given.
- *Mushrooms of North America* (Miller 1972)—A color photograph and a description and discussion are given.

Lactarius vellereus (Fr.) Fr. var. *vellereus*

Fig. 143

Common name. Woolly-white lactarius.

Distinguishing characteristics. The European type variant is distinguished by the following features. Gills distant at maturity. Latex changing to and staining tissue yellow, with stains finally becoming more or less clay color to fulvous; taste mild. Cap and stalk cuticles with a turf of long thick-walled hairs. Spores broad, 7-9 μm wide.

The North American variant is distinguished by these characteristics. Gills close to subdistant. Latex white, drying yellowish on the gills; taste mild. Flesh with acrid taste. Cap color basically white; surface velvety-tomentose. Cap and stalk cuticles with a turf of narrow thick-walled hyphae. Spores broad, 7.5-9 μm wide, subglobose to broadly ellipsoidal.

Observations. The distribution of *L. vellereus* on this continent has not been determined. Fries (1821, 1838) stresses that the gills of *L. vellereus* are distant. Hesler and Smith (1979) describe a North American variant (see the technical description below) and also recognize the following taxa: *L. subvellereus* Peck var. *subvellereus* (gills crowded to close); *L. subvellereus* Peck var. *subdistans* Hesler & Smith (gills subdistant); and *L. vellereus* var. *virescens* Hesler & Smith (gills flush green).

L. vellereus var. *vellereus* is often confused with *L. piperatus* and *L. deceptivus*. *L. piperatus* has gills that are narrow, densely crowded, and forked; the cap is not tomentose. *L. deceptivus* has larger spores, 6.0-12.1 × 5-9 μm in size, that are more prominently marked, and a thick and cottony cap margin as compared with *L. vellereus*. Kauffman (1971) reported *L. vellereus* as poisonous. The technical description given below is based in part on the work of Hesler and Smith (1979).

Distribution and seasonal occurrence. *L. vellereus* var. *vellereus* occurs in the Great Lakes states and eastward, including southeastern states, and in Eastern Canada, in Nova Scotia. It fruits from July to October.

Habitat and habit. This species is scattered to gregarious, growing on soil under conifers and in deciduous and mixed woods.

Technical description. Cap 60-150 mm broad, convex-depressed, becoming broadly funnel-shaped, often umbilicate; surface dry, without zones, velvety or tomentose; epicutis matted with age; margin enrolled in early stages, becoming upraised; color white, staining buff to dingy tawny brown. Flesh 4-8 mm thick, white, hard; taste acrid; odor mild. Latex white, drying yellowish on the gills, mild.

Gills 4-8 mm wide, narrow, decurrent, close to subdistant, sometimes forking; color creamy white, developing brown stains, drying a dull vinaceous brown, exuding drops of hyaline liquid.

Stalk 15-65 mm long, 12-30 mm thick, narrowed downward; surface dry, unpolished to velvety; color white. Flesh solid, hard.

Spores white in deposit, 7.5-9 × 7.5-9 μm, subglobose varying to broadly ellipsoidal; ornamentation consisting of fine granules and short lines, forming at most a broken reticulum; prominences 0.3 μm high. Basidia four-spored, 52-60 × 9-11 μm. Pleurocystidia and cheilocystidia 48-75 × 4-8 μm, scattered to numerous, projecting prominently, almost cylindrical but with tapering-acute or hair-like apices, and with more or less yellow contents in potassium hydroxide.

Cap cuticle consisting of three layers: the innermost layer, consisting of inflated cells and somewhat tubular hyphae intermixed; the thin middle layer, consisting of septate matted-down hyphae that are hyaline and narrow (3-6 μm); and the outmost layer, an epicutis, in the form of a turf of narrow (3-5 μm) thick-walled (1-1.5 μm thick) hyphae that are tubular, hyaline, and mostly not septate, and that become matted down with age. Caulocystidia present as a turf of hyaline thick-walled smooth tubular hyphae that are 3-5 μm wide and are not septate. Clamp connections absent.

Useful references. The following books contain information on *L. vellereus* var. *vellereus* pertinent to North America.

- *Edible and Poisonous Mushrooms of Canada* (Groves 1979)—A color photograph and a description and discussion are presented.
- *Mushrooms of North America* (Miller 1972)—A description and discussion are given.
- *The Mushroom Hunter's Field Guide* (Smith 1974)—Black-and-white photographs and a description and discussion are presented.

Lepiota (Pers. ex Fr.) S. F. Gray

Lepiotaceae

There are several interpretations of the genus *Lepiota*. Modern authors have often used a narrow definition of *Lepiota* and in doing so have recognized several segregate genera including *Chlorophyllum*, *Leucocoprinus*, *Leucoagaricus*, and *Macrolepiota* in addition to *Lepiota*. A broad definition of *Lepiota* is used here, however, which includes all the above genera. Species of *Lepiota* have in common distinctly free gills; a spore deposit that is typically white to pale cream, rarely pinkish, and green in one species; a partial veil that usually forms an annulus, although the annulus is sometimes lost in mature specimens; and a more or less fleshy-fibrous to more or less fragile stalk that is often somewhat clavate. Species of *Lepiota* vary from very tiny to large. Often the cap is convex, with a raised knob in the center. There is usually no universal veil and consequently no volva. The structure of the cap cuticle varies but is often composed of filamentous hyphae. The spores are smooth and of various shapes; they sometimes have an apical germ pore and are nonamyloid, dextrinoid, or more rarely amyloid.

In general appearance certain species of *Amanita* resemble the genus *Lepiota*. The presence of a universal veil and consequently a volva in *Amanita* separates it from *Lepiota*. *Amanita* can also be distinguished by a divergent gill trama, as opposed to the parallel to interwoven gill trama of *Lepiota*. Species of *Armillaria* and *Cystoderma* also may resemble *Lepiota* in general appearance. These genera have species with attached gills. Species of *Agaricus* have the same appearance as some *Lepiota* species, but *Agaricus* can be easily distinguished by its blackish brown to dark brown spore deposit.

Species of *Lepiota* that produce amatoxins are presented in Chapter 7. Other species that have been reported as poisonous include *L. clypeolaria* (Bull. ex Fr.) Kummer, *L. cristata* (Bolt. ex Fr.) Kummer, *L. (Leucocoprinus) lutea* (Bolt.) Quél. (=*L. birnbaumi* (Corda) Singer), *L. molybdites*, and *L. naucina*. Lincoff and Mitchel (1977) list *L. americana* Peck, *L. cepaestipes* (Sow. ex Fr.) Kummer, *L. naucina*, *L. procera* (Scop. ex Fr.) S. F. Gray, and *L. rachodes* (Vittad.) Quél. (also spelled *rhacodes* and *racodes*) as good to excellent edible species but as capable of causing mild to severe gastrointestinal upset in some people. Our reports add to the above list *L. brunnea* Farl. & Burt (K. A. Harrison, Personal commun.), *L. naucinoides* Peck (Dearness 1911), which may or may not be distinct from *L. naucina*, and *L. schulzeri* (Kalchbr.) Sacc., which is apparently related to *Lepiota naucina* (Krieger 1967).

Lepiota naucina and *Lepiota molybdites* are the most frequent offenders and are described and discussed below. From the standpoint of poisoning, *L. naucina* is one of the more interesting species. *Lepiota naucina* in many recent publications has been called *Leucoagaricus naucinus* (Fr.) Singer. Because of the broad concept of *Lepiota* used here, however, we place it in the genus *Lepiota*. It is listed as an edible species in most mushroom handbooks and guides. Nevertheless caution is usually recommended for collectors of this species. The greatest danger is the possible confusion of a deadly white *Amanita* species, such as *A. virosa*, with *L. naucina*.

Take care, especially, in the areas where lawns or other grassy places have scattered trees or border on woods, a habitat preferred by both. The white *Amanita* species have a more or less pendent, membranous, skirt-like annulus and a distinct, sac-like or cup-like volva. The annulus in *L. naucina* is somewhat thickened and collar-like, and a volva is lacking. When *A. virosa* or some other deadly *Amanita* species is broken off above the stalk base leaving the volva behind in the soil, the danger of confusing these mushrooms is greatly increased. The spores of *L. naucina* and the deadly white *Amanita* species, however, also differ. *L. naucina* has spores that are thick-walled, with an apical germ pore; they are dextrinoid, turning reddish in Melzer's reagent. *A. virosa* and its relatives have thin-walled spores that lack an apical germ pore and are amyloid, turning grayish to grayish blue in Melzer's reagent.

Smith (1974) reports the confusion of *L. naucina* with young specimens of *L. molybdites*. This confusion apparently arises because the gills of *L. molybdites* remain white until the partial veil breaks or sometimes until the caps are fully expanded. The gills of *L.*

naucina are typically white when young and fresh. Because both species occur in grassy areas, the possibility of confusing the two is fairly high.

Finally, several cases of poisonings by *L. naucina* have been reported. All consisted of gastrointestinal irritation and apparently none were severe. McKenny (1971) suggests that stomach upset is caused by the gray cap form of *L. naucina* (Fig. 209). Several cases of more or less mild poisoning are on record at the North American Poison Mushroom Research Center.

In summary, one cannot list *L. naucina* as a definitely poisonous species. It is apparently eaten by many people without consequence but makes certain individuals sick; therefore, take precautions when trying it for the first time. It is a widely distributed and common species, so perhaps some races or forms of this species are somewhat toxic. The existence of poisonous races would explain why people who know *L. naucina* and have eaten it for years are occasionally poisoned by it.

In conclusion this species is best not recommended for eating, mainly because it can be confused with the deadly white *Amanita* species and white-gilled specimens of *L. molybdites.*

Lepiota molybdites (Meyer ex Fr.) Sacc.

Figs. 144, 207

Common names. Green-spored lepiota, Morgan's lepiota.

Distinguishing characteristics. Cap large, 45–300 mm broad, convex to plano-convex when expanded, ovoid when young, with a broad low umbo; surface even in early stages, then breaking into scales; color variable, tan, dark yellowish brown, light to deep brown, or grayish reddish brown, sometimes more yellowish; cap margin often predominantly white, because as the surface breaks into scales the white flesh is exposed. Flesh on the surface and within usually staining light yellow to yellowish pink, then more orange, finally becoming some shade of reddish brown. Gills free, white in early stages, often shaded with green, yellow green, or grayish yellow green at maturity, sometimes darker with age; edges usually whitish, but in one form brown to almost black. Stalk 6–15 mm thick at the apex, slightly enlarged below to clavate or bulbous; surface white, becoming shaded with grayish reddish brown, brownish pink, or grayish pink. Annulus thick, membranous, persistent, white or similar in color to that of the stalk surface. Spore deposit pale green, pale yellow green, or greenish gray to darker green when air-dried, fading to orange yellow with age. Spores more or less ovoid, smooth, thick-walled with an apical germ pore, (7.7–)9.5–12.4(–13.5) × 5.8–9(–10.9) μm in size.

Observations. *L. molybdites* is placed in the genus *Chlorophyllum* by many authors. Nevertheless it fits within the broad definition of *Lepiota* used here. *L. molybdites* has the following characteristics in common with other *Lepiota* species: distinctly free gills; an annulus; a fleshy-fibrous stalk that is clavate to somewhat bulbous; and the absence of a universal veil. The spore deposit is dull pale green when fresh. The gills are white in the early stages and usually develop various greenish to grayish green hues as the spores mature, usually after expansion of the cap. Smith (1949) described var. *marginata* as having blackish gill edges. The spore deposit of other *Lepiota* species is usually white to whitish, making *L. molybdites* distinct on the basis of spore color alone.

In general appearance *L. molybdites* is similar to *L. rachodes* (Vittad.) Quél., *L. procera* (Fr.) S. F. Gray, and *L. brunnea* Farl. & Burt. The green spore deposit is the best way of separating *L. molybdites* from these species, which all have white spore deposits. Specimens of *L. rachodes* also can be separated from *L. molybdites* by the structure of the cap cuticle. In *L. rachodes* it is composed of a layer of tightly adhering, more or less clavate cells, whereas in *L. molybdites* the cuticle is turf-like, composed of tightly packed more or less upright filamentous hyphae. This feature is particularly helpful in distinguishing young specimens of these species. In the field *L. molybdites* is often confused with other large *Lepiota* species because the gills of young or maturing mushrooms are often white to creamy rather than greenish or grayish green; even large, expanded fruit bodies sometimes lack greenish tones on the gills. Because its gills do not always turn green, *L. molybdites* has been confused with the normally edible large species of *Lepiota* such as *L. procera.* This error has lead to some severe poisonings, even among some experts. Bessey (1939) reports confusion of *Coprinus comatus* with *L. molybdites,* which resulted in two severe poisonings.

Eilers and Barnard (1973) describe a method that utilizes Acid Fuchsine for improving the visibility of *L. molybdites* spores in vomit and stool samples. Congo Red can also be used. They stain the spores of edible, white-spored *L. rachodes* and *L. brunnea* but not the mature spores of *L. molybdites* (Weresub 1971). These procedures may be useful in diagnosing cases of poisoning.

L. molybdites has a long and interesting history. In North America for many years it was called *L. morgani* (Peck) Sacc., and the name *Chlorophyllum molybdites* (Meyer ex Fr.) Massee has also been, and still is, commonly used. More recently the names *L. esculenta* (Massee) Sacc. & Syd. and *Chlorophyllum esculentum* Massee have come into use. The specific name *esculentum,* which means "good to eat", has been applied to this species in tropical and subtropical regions, where it is common and generally considered edible.

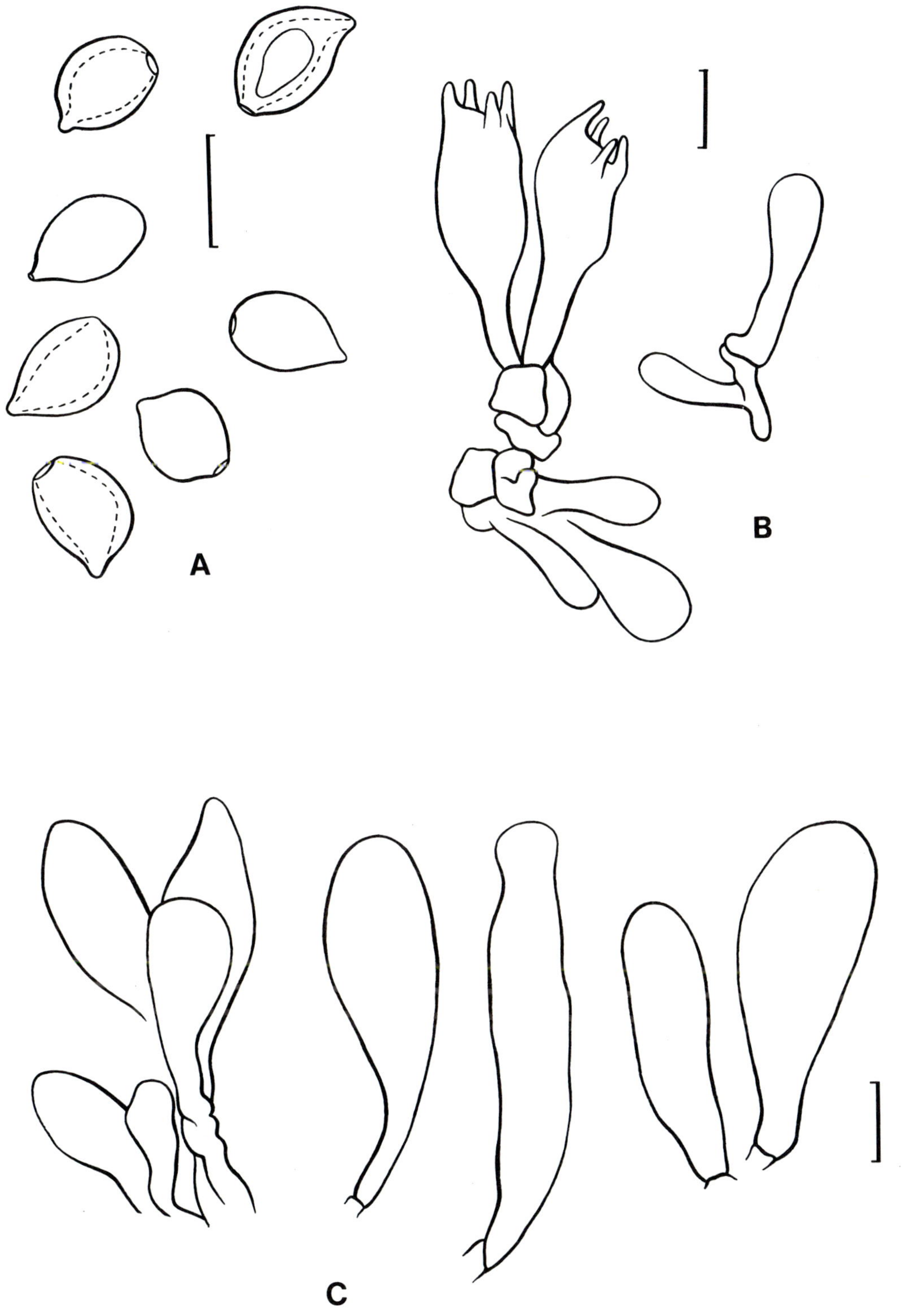

Fig. 144. *Lepiota molybdites*: A, spores; B, basidia; C, cheilocystidia.

Scale line 10 μm

In North America there are numerous reports of severe poisoning by *L. molybdites,* especially in the southern United States, where it is very common (Chesnut 1900, Webster 1915, Dearness 1922, Bessey 1939, Charters 1960). Several other cases are on record at the Poison Fungi Center, Beltsville, MD. In addition, several reports clearly show that some people can eat *L. molybdites* without ill effects (Graff 1922, Horne 1941, Stubbs 1971). Nevertheless, this species should be considered poisonous, and poisoning by *L. molybdites* should not be taken lightly.

The symptoms of poisoning usually begin 1-2 hours after ingestion of the mushroom. Feelings of queasiness and thirst usually develop first, followed by mental haziness, nausea, cold sweats alternating with chills, and intervals of vomiting for 4-5 hours; the victim finally has an attack of copious, watery, or sometimes bloody diarrhea, which persists from several hours to a few days. The degree of abdominal pain varies from mild to intense. Most victims recover within a day or two. Chesnut (1900) reports the only fatality caused by this species in North America: a 2-year-old girl died about 17 hours after she had eaten an undetermined amount of the raw mushroom. Reports indicate that humans can be poisoned by either raw or cooked *L. molybdites.*

Distribution and seasonal occurrence. This species is uncommon to rare in the Great Lakes region and southeastern Canada, and more common to very common southward and across the southern United States. In the Rocky Mountains it has been reported as far north as Colorado. *L. molybdites* fruits in the late spring, summer, and fall.

Habitat and habit. This species is scattered to gregarious or occurs in partial to complete rings. It is terrestrial, found in lawns and gardens and occasionally in open woods.

Technical description. Cap 45-300 mm broad, ovoid to obtuse when young, expanding to convex or plano-convex with a broad low umbo; margin sometimes upturned with age; surface dry, not striate, more or less even when young; disc of expanded cap even to rimose or somewhat scaly; color tan, dark yellowish brown, light brown to deep brown, grayish reddish brown, or somewhat pale orange yellow to moderate orange yellow; margin of unexpanded caps more or less the same color as the disc; surface of expanded caps broken into flattened or upturned scales; scales white to whitish, faintly yellowish or brownish pink to dark reddish brown; flesh between scales white or staining pale yellowish pink, pinkish, or light grayish red to reddish brown. Flesh (2-)5-13 mm thick at the disc, soft, solid, white to pale yellowish pink or light grayish brown to more or less orange yellow, staining light yellow to yellowish pink then darkening to moderate orange, finally becoming reddish brown or grayish red brown; taste mild; odor indistinctive.

Gills white when young, usually becoming pale green to pale yellow to light olive brown with age, free and remote from the apex of the stalk, close, 6-18 mm broad, more or less thin and fragile; edges more or less fimbriate and white in early stages, remaining white or becoming brown to almost black.

Stalk 50-120 mm long, 6-15 mm thick at the apex, equal to slightly enlarged below or clavate to bulbous, central, stuffed, becoming hollow, dry, silky to innately fibrillose, often with white mycelia at the base; color white to off-white or sometimes tinted grayish pink to grayish red brown above the annulus, white to brownish pink below and sometimes tinged grayish pink to grayish red brown, darkening with age to dark grayish red brown. Flesh white to light brown or pinkish, staining pale orange yellow, then becoming moderate orange to brownish orange and finally grayish reddish orange to reddish brown or deep brown. Annulus membranous, fleshy, apical to median on the stalk, sometimes becoming movable, persistent; upper surface white or sometimes spotted pale pink to grayish pink or with age becoming brownish pink to light brown, sometimes greenish from the spores; lower surface moderate brown to dark grayish reddish brown near the edge, white or tinged grayish pink elsewhere. Universal veil absent.

Spores very pale green, pale green, pale yellow green or greenish gray, or slightly darker green in deposit, fading as they age to golden or more or less orange yellow, (7.7-)9.5-12.4(-13.5) × 5.8-9(-10.9) μm, ellipsoidal, with a distinct apical germ pore and a truncate apex in side view, thick-walled, smooth, hyaline to greenish, dextrinoid. Basidia four-spored, short-clavate, 24-33 × 8-11 μm, hyaline. Pleurocystidia lacking. Cheilocystidia clavate, somewhat sac-like or fusiform to ventricose, (18-)22-47.5(-62) × (9.5-)10-25.5 μm, hyaline, thin-walled.

Gill trama compactly interwoven; hyphae more or less cylindrical, 3.5-12.0 μm wide, hyaline. Subhymenium cellular; hyphae 10-16 μm wide, hyaline. Cap cuticle filamentous; hyphae upright and tightly adhering, at times with the apical portions bent over and appearing interwoven and flattened; cells more or less upright, 3.5(-7.3) μm wide, hyaline to yellowish or yellowish brown. Cap trama loosely interwoven; hyphae cylindrical to inflated, 3.5-30 μm wide, hyaline. Caulocystidia lacking; stalk hyphae parallel to interwoven on the surface, (3.5-)5.1-8.8(-11.0) μm wide, hyaline to yellowish. Annulus interwoven; hyphae cylindrical, 3.5-5.8 μm wide, hyaline, with thin to slightly thickened walls. Oleiferous hyphae yellowish to ocherous, refractive, 3.7-8.8(-19.5) μm wide, not septate, contorted, scattered in the cap trama and cuticle, stalk, and annulus. Clamp connections absent to rare in the cap trama.

Useful references. The following books contain information on *L. molybdites* pertinent to North America.

- *Edible and Poisonous Mushrooms of Canada* (Groves 1979)—A black-and-white photograph is given.
- *Mushrooms in their Natural Habitats* (Smith 1949)—A color photograph and a description and discussion are given.
- *Mushrooms of North America* (Miller 1972)—A good color photograph and a description and discussion are presented.
- *The Mushroom Hunter's Field Guide* (Smith 1974)—A black-and-white photograph and a description and discussion are given.

Lepiota naucina (Fr.) Quél.

Figs. 145, 208, 209

Common names. White lepiota, smooth lepiota, chalky lepiota.

Distinguishing characteristics. Cap 30-80(-120) mm broad; shape subglobose to hemispherical in early stages, becoming convex or at times somewhat flattened; surface as smooth as kidskin or rarely somewhat scaly, dry; color usually white to dull white, developing buff to grayish or smoky tones or tinged with gray to smoky tones from the earliest stages. Flesh white and unchanging. Gills free, close, broad, white in early stages, becoming grayish pink although the gills of dried specimens are strongly grayish to grayish brown. Stalk 50-150 mm long, 5-18 mm thick, equal or slightly enlarged toward the base; color white, at times discolored grayish to brownish. Annulus more or less thick, often collar-like and movable; color white or sometimes somewhat buff. Spores white in deposit or sometimes tinted pink, smooth, thick-walled, with a distinct apical germ pore, (5.8-)8-11.7 × (4.4-)5.1-7.3 μm in size.

Observations. *L. naucina* is common and widely distributed, occurring mainly in grassy areas, sometimes very abundant in lawns. In general, this species is fairly easy to recognize when it is studied carefully. However, it can be confused with other *Lepiota* species or the deadly white *Amanita* species. A discussion of the distinctions among these species appears in the introduction to the genus *Lepiota* in this chapter.

The gills often become pinkish to grayish red in *L. naucina,* a feature that is helpful in identifying this species. Smith (1974) reports that some specimens have gills that are strongly olive green at maturity, somewhat similar to the color of the mature gills of *L. molybdites.* The spore deposit for these specimens is white, however, not greenish as in *L. molybdites.* The spore deposit of *L. naucina* is normally white, although we have obtained some pinkish spore deposits from specimens of this species collected in southern Ontario. Both white and pink deposits have been obtained from the same populations, even from the same collection. We have made deposits on both white bond paper and clean glass slides. Pinkish and white deposits have been obtained on paper. Pinkish deposits seem to be more common on paper but have also been obtained on glass slides. Sundberg (1967), in his study of California species of *Lepiota,* listed the deposit for *L. naucina* as white or tinged pinkish in mass. A pinkish spore deposit was obtained from *L. naucina* collected in southeastern Oregon (J. F. Ammirati, Personal observ.). Hesler (1960) reports the spore deposit as white, sometimes tinged cream to pale pinkish. Clearly, the spore deposit from *L. naucina* can be pinkish or white. Why pink deposits are formed is not known, but their occurrence should be kept in mind when identifying this species.

Spore size also appears to be somewhat variable in *L. naucina.* Spore sizes for our specimens are larger than those normally reported, except those by Sundberg (1967), who gives spore measurements of 7.9-10.3(-12.7) × 5.5-6.3 μm. H. V. Smith (1954), Miller (1972), Smith (1975), and Groves (1979) give spore measurements of 7-9 × 5-6 μm.

Distribution and seasonal occurrence. This species is widely distributed across southern Canada and the United States. It fruits during the summer and fall, mostly from August into October, but as late as early November on the Pacific coast where cool, moist weather extends late into the fall.

Habitat and habit. This species is terrestrial, rarely solitary, usually scattered to gregarious. It commonly occurs in lawns, pastures, meadows, waste areas where grass and weeds occur, around compost piles, along roadsides, and occasionally in more or less open wooded areas. It has also been reported from parking strips and plowed fields.

Technical description. Cap 30-80(-120) mm broad, subglobose to ovoid when young, becoming convex and finally broadly convex to plane, sometimes with a slight obtuse umbo; margin incurved in early stages, becoming decurved and finally plane, entire, fibrillose-tomentose to appendiculate with annular fragments, becoming rimose to eroded with age; surface dry, unpolished to appressed-fibrillose, smooth or rarely broken up into scales; color white to dull white, pinkish buff to yellowish buff, or with a smoky gray shade over the disc. Flesh thick, rather firm, white and unchanging; taste mild; odor slight or indistinctive.

Gills white in early stages, slowly changing to sordid grayish dark pink and dingy grayish red to dark grayish brownish pink when dried, free, close, broad, broadest near the margin; edges slightly floccose.

Stalk 50-120(-150) mm long, 5-12(-18) mm thick, equal or slightly enlarged at the base, central, stuffed

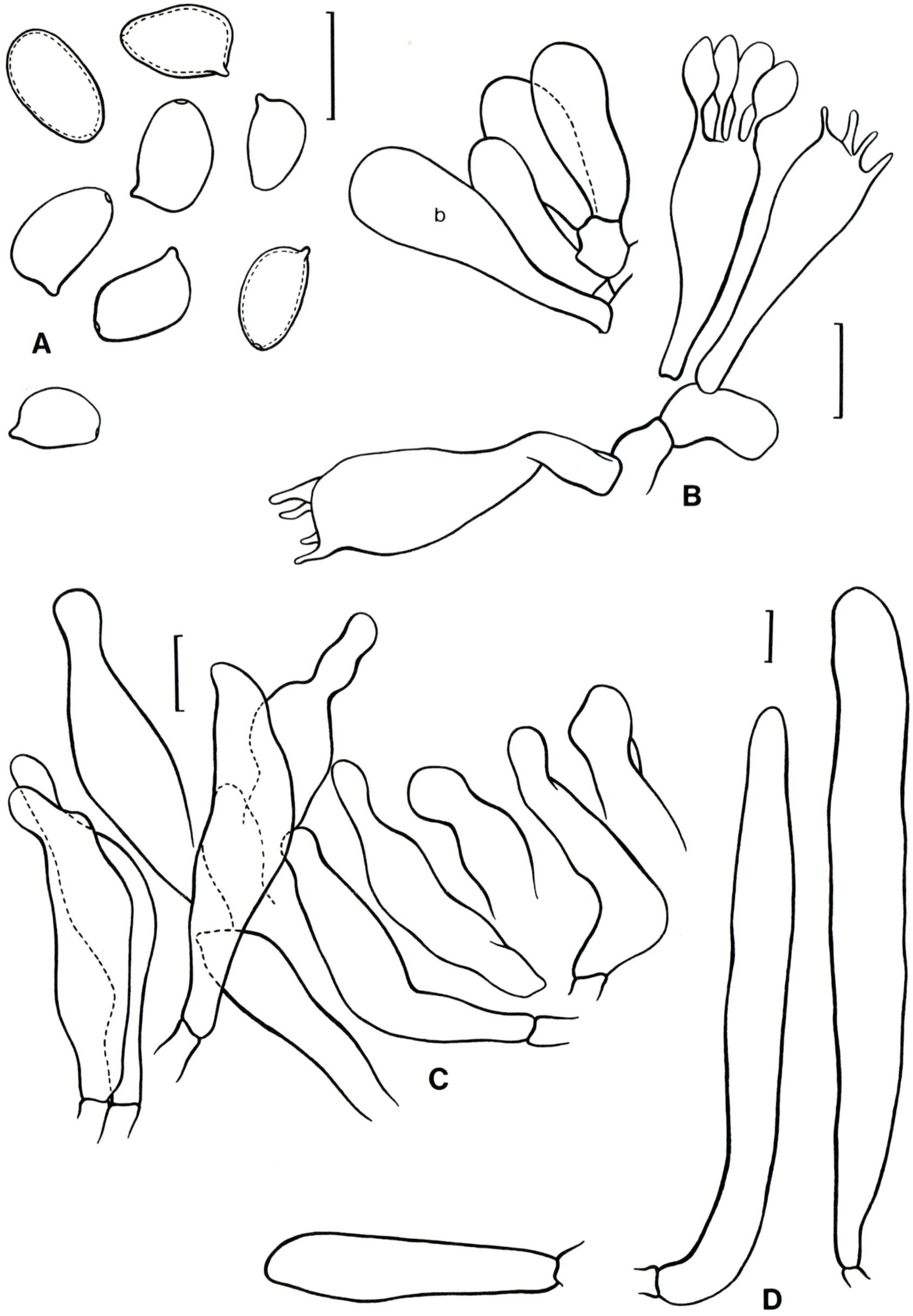

Fig. 145. *Lepiota naucina*: A, spores; B, basidia and basidioles (*b*); C, cheilocystidia; D, pileocystidia.

Scale line 10 μm

but soon becoming hollow, dry, silky to glabrous below the annulus, slightly pruinose above the annulus, white, at times discolored grayish to brownish or occasionally yellowish. Flesh white or creamy, sometimes becoming grayish after prolonged exposure. Annulus membranous with a thick cottony floccose double edge, often collar-like and movable, apical on the stalk, persistent, white and silky on the upper surface, pallid buff at least in a zone near the margin below. Universal veil lacking.

Spores white in deposit, rarely to occasionally tinged pinkish, (5.8-)8.0-11.7 × (4.4-)5.1-7.3 μm, broadly ellipsoidal, inequilateral in side view, ovate in face view, thick-walled, smooth; apex with a small apical germ pore, hyaline, dextrinoid. Basidia four-spored, clavate, 25.6-31.4 × 7.3-8.8 μm, hyaline, thin-walled. Pleurocystidia lacking. Cheilocystidia clavate to narrowly fusiform-ventricose to ventricose; apex nearly capitate to double-forked or irregular, (28-)37-60 × 8.8-14 μm, hyaline, thin-walled.

Gill trama interwoven; hyphae cylindrical, 4.6-7.0 μm wide, hyaline. Subhymenium cellular; hyphae 5.1-8.8 μm wide, thin-walled, hyaline. Cap cuticle in the form of a trichoderm of pileocystidia; these cells elongate and cylindrical, slightly tapered and rounded to almost acute toward the apex; clavate cells also present but not abundant, 56-138 × 6-20 μm, hyaline to yellowish brown. Subcutis present; cells 4-6 μm wide, pale to dark yellowish brown. Cap trama loosely interwoven; hyphae cylindrical to inflated, 4.6-16.7 μm wide, branched, hyaline. Caulocystidia lacking; some recurved hyphal ends present; stalk hyphae parallel in the core to interwoven on the surface, 3.5-11.7 μm wide, hyaline. Annulus interwoven; hyphae cylindrical, 3.5-8.2 μm wide, hyaline to yellowish in potassium hydroxide; some hyphae with tapered or oblong to pear-shaped ends, 28-46(-82) × 12.9-27 μm, hyaline to yellowish, separating at the cross walls. Oleiferous hyphae refractive, greenish yellow, 5.1-11.0 μm wide, scattered in the trama and stalk. Clamp connections lacking.

Useful references. The following books contain information on *L. naucina* pertinent to North America.

- *A Field Guide to Western Mushrooms* (Smith 1975)—A color photograph and a description and discussion are presented.
- *Edible and Poisonous Mushrooms of Canada* (Groves 1979)—A black-and-white photograph and a description and discussion are presented.
- *Mushrooms in their Natural Habitats* (Smith 1949)—A description and discussion, as well as a color photograph on a reel, are presented.
- *Mushrooms of North America* (Miller 1972)—A color photograph and a description and discussion are given.
- *The Mushroom Hunter's Field Guide* (Smith 1974)—A color photograph, a black-and-white photograph, and a description and discussion are presented.
- *The Savory Wild Mushroom* (McKenny 1971)—A black-and-white photograph and a description and discussion are given.

Morchella Dill. ex Fr. — Morchellaceae

The genus *Morchella,* which contains the true morels, is most easily confused with species of *Verpa;* both genera belong to the family Morchellaceae, which is a member of the division Ascomycota. See Chapters 2 and 3, Ascomycota, for a discussion of the main features of this division. Species in both genera have a distinct, hollow to loosely stuffed stalk and a distinct, more or less conic cap that varies from rounded to elongate, depending on the species. Typical morels have a cap surface composed of ridges and distinct pits; the pits are usually more or less rounded but vary to irregular or elongate. The cap usually resembles a honeycomb or a sponge or is pinecone-like in appearance. With the exception of *Morchella semilibera* DC. ex Fr., the half-free morel, the cap of *Morchella* species is attached to the stalk from very near its edge to the stalk apex; thus the cap is not free from the stalk apex. In contrast, the cap of *Verpa* species is skirt-like, attached only to the stalk apex. In *Verpa* the cap surface varies from nearly smooth as in *Verpa conica* to highly ridged by a network of longitudinal grooves or folds as in *Verpa bohemica. Verpa bohemica* is the most easily confused with *Morchella,* especially *M. semilibera. V. bohemica* can be separated from *M. semilibera* by its two-spored asci; *M. semilibera* has eight spores per ascus. See the treatment of *Verpa* later in this chapter (alphabetically listed), and consult the general and technical keys to the genera of the Ascomycota in Chapter 5 for a further comparison of these genera.

The genus *Morchella* is probably the best known and most sought after of all edible fleshy fungi. The species fruit in the spring season but, rarely, occur as late as early summer in the western mountains. In general all are considered edible, although recent surveys and reports indicate that poisonings of the gastrointestinal type are fairly common among morel eaters. Smith (1975) questions the edibility of the black morels, *Morchella conica* Fr. and *Morchella angusticeps,* because of the large number of poisonings reported in the west where these species are more abundant. Several poisonings by morels have also been reported in the Upper Peninsula of Michigan (I. Bartelli, Personal commun.). Lincoff and Mitchel (1977) state that morels are known to cause stomach upsets when eaten raw, even by some people who can eat them cooked without ill effects. They further state that most people can safely eat them, although a few individuals may become allergic and others may experience gastrointestinal upset. Groves (1979) reported stomach upsets caused by combining a meal of morels with alcohol. The compound or compounds causing the gastrointestinal upset are unknown. There is no evidence that *Morchella* species contain gyromitrin or related compounds.

The number of species of morels in North America is still unclear. The names *M. angusticeps, M. conica, M. crassipes* Fr., *M. deliciosa* Fr., *M. elata* Fr., *M. esculenta,* and *M. semilibera* (=*M. hybrida* Sow. ex Grev.) are those most commonly seen in the North American literature. Some of these names may be synonyms for the same species and there are undoubtedly additional species yet unidentified or perhaps undescribed. We have described and illustrated here two species, *M. angusticeps* and *M. esculenta,* as representatives of *Morchella.* Smith and Smith (1973) provide a key to some common species.

Morchella angusticeps Peck

Figs. 146, 210 (*right*)

Common names. Black morel, narrow-capped morel.

Distinguishing characteristics. Cap 50-90 mm long, 25-50 mm broad, usually more or less narrowly conic; pits typically elongated; color pallid to buff or grayish when young; ridges, and sometimes pits, becoming smoky brown to blackish. Stalk 50-150 mm long, 20-30 mm wide, hollow; color whitish or with pinkish to brownish tones. Fungi found in wooded areas, especially in sandy soil. Spores yellowish in deposit, ellipsoidal, (16.1-)18.6-29.2 × (10.2-)12-14 μm in size.

Observations. This species is one of the common black or dark grayish morels that occur in North America. According to Smith (1975) there are at least two species of black morels in North America: a narrow-capped morel that is gray to dull tan in the early stages, with distinctly elongate pits that blacken on the ridges before the fruit body matures, called *M. angusticeps*; and a broadly conic to more or less round-capped morel that is blackish overall when still young, called *M. conica* Fr. (Fig. 211). According to Smith (1975) *M. conica* attains a larger size than *M. angusticeps.* N. Smith Weber (Personal commun.) suggests that spore size might be helpful in separating these species, but this characteristic still requires further study and comparison among the various species. The name *M. elata* Fr. has also been used for the black morel with a broadly conic to more or less rounded cap.

Distribution and seasonal occurrence. This species is widely distributed across North America and fruits in the spring, from April to June.

Habitat and habit. *M. angusticeps* is rarely solitary, usually scattered to gregarious, and terrestrial. It occurs in open woods or at the edges of conifer and deciduous woods. This species is sometimes common in burned areas, especially the first year following a fire.

Technical description. Cap 50-90 mm long, 25-50 mm broad, oblong-conic and somewhat obtuse or narrowly conic and acute, fused with the stalk from the edge of the cap upward; surface pitted; pits elongated, arranged in more or less vertical rows; color pallid to buff or grayish to somewhat brownish when young; ridges and pit borders becoming smoky brown to black at maturity; pits also becoming blackish in some forms.

Stalk 50-150 mm long, 20-30 mm wide, cylindrical but usually expanded slightly above, hollow, fragile, shallowly furrowed, sometimes slit at the base; surface scurfy; color whitish or with pinkish to brownish tones.

Spores yellowish in deposit, (16.1-)18.6-29.2 × (10.2-)12-14 μm, ellipsoidal, smooth, thin-walled, hyaline. Asci eight-spored, 200-300 × 16-22 μm, cylindrical, hyaline. Paraphyses cylindrical, 5.8-7.3(-9.5 μm wide.

Stalk hyphae interwoven in the flesh, 5.8-9.4 μm wide, hyaline; surface hyphae consisting of clavate to sac-like cells, 16.7-46 × 16-35 μm, arranged in a hymeniform-like layer, hyaline.

Useful references. The following books contain information on *M. angusticeps* pertinent to North America.

- *A Field Guide to Western Mushrooms* (Smith 1975)—A color photograph and a description and discussion are presented.
- *Edible and Poisonous Mushrooms of Canada* (Groves 1979)—A black-and-white photograph, a color photograph, and a description and discussion are included.
- *Mushrooms of North America* (Miller 1972)—A color photograph and a description and discussion are given.
- *The Mushroom Hunter's Field Guide* (Smith 1974)—A black-and-white photograph, a color photograph, and a description and discussion are given.
- *The Savory Wild Mushroom* (McKenny 1971)—A black-and-white photograph and a description and discussion are presented.

Morchella esculenta Fr. Figs. 147, 210 (*left*)

Common names. Edible morel, common morel, sponge mushroom.

Distinguishing characteristics. Cap 70-120 mm long and 40-60 mm wide, more or less conic to subglobose or more elongated; surface pitted; pits irregularly shaped, varying from rounded to elongated; color white to gray or yellowish to yellowish brown; ridges lighter than the pits. Stalk 10-100 mm long, 10-40 mm thick, more or less cylindrical although sometimes enlarged above or below, with a more or less scurfy surface, hollow; color white to pallid or yellowish. Fungi fruiting in early spring, occurring in a variety of habitats. Spores lightly yellowish in deposit, ellipsoidal, smooth, (14.6-) 17.5-21.9(-24) × 8.8-11(-14) μm in size.

Observations. *M. esculenta* is probably the most popular of the morels. In contrast to *M. angusticeps* and its relatives, the caps are light-colored throughout development, especially the ridges, which remain paler than the pits. *M. crassipes* is sometimes confused with *M. esculenta*. According to Smith (1975) the two are distinct, but young forms of *M. crassipes* are difficult to separate from *M. esculenta*. The two are similar in color, but *M. crassipes* is larger, often has thin ridges, and sometimes has a stalk base that is enlarged and longitudinally grooved.

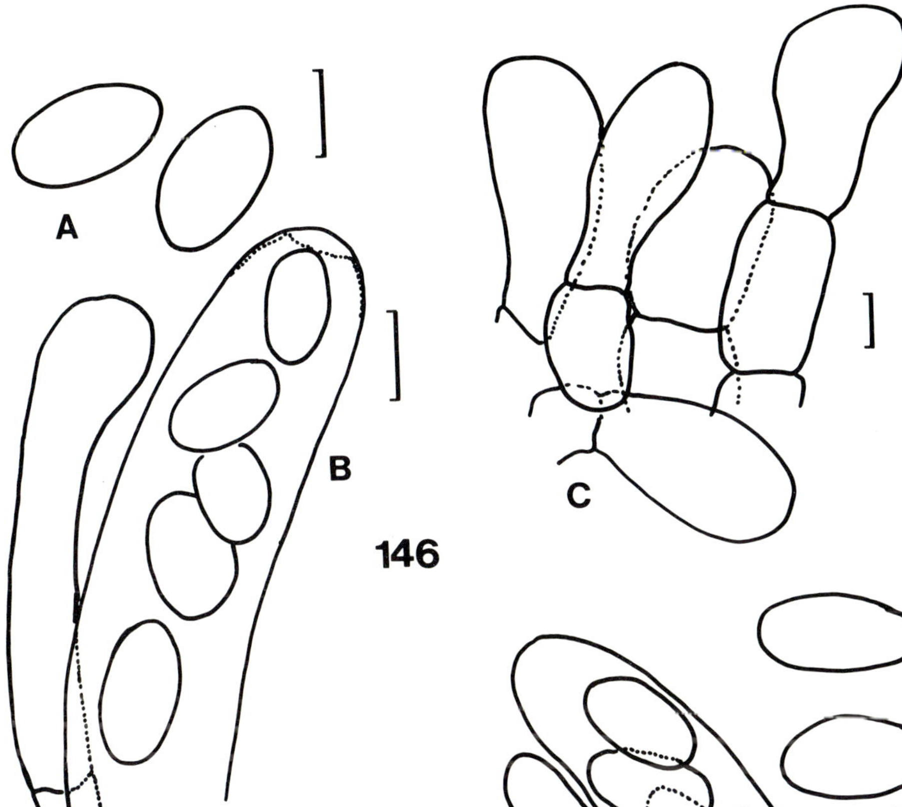

146

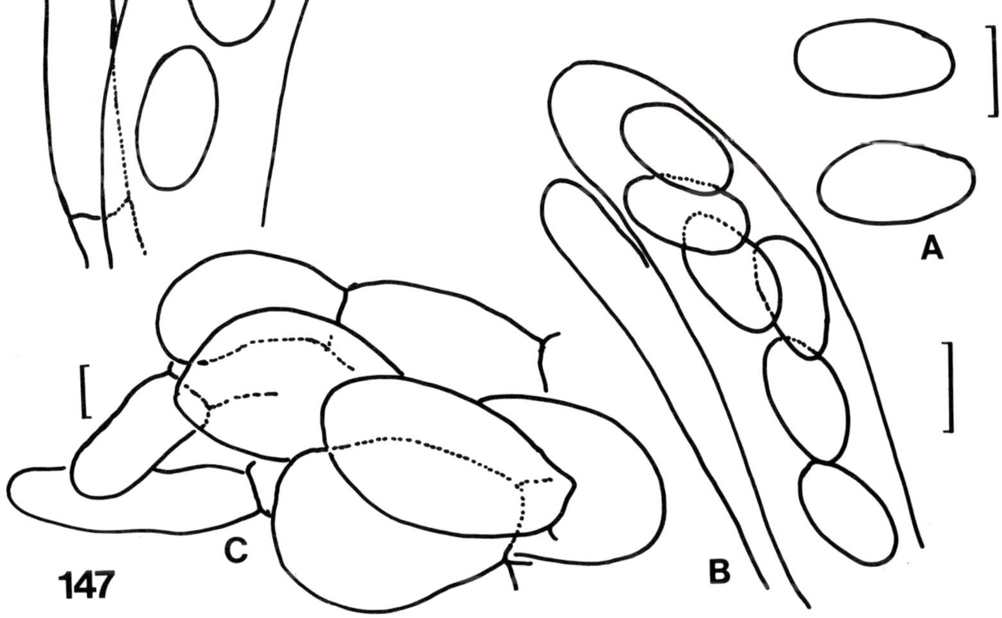

147

Fig. 146. *Morchella angusticeps*: A, spores; B, ascus and paraphyses; C, excipular elements.

Fig. 147. *Morchella esculenta*: A, spores; B, ascus and paraphysis; C, excipular elements.

Scale line 10 μm

Distribution and seasonal occurrence. *M. esculenta* is widely distributed in North America. It typically fruits in the spring, from April to June.

Habitat and habit. This species is rarely solitary, usually scattered to gregarious, and terrestrial. It occurs in various habitats including orchards, deciduous and coniferous forests along the edges of swamps, in open grassy areas, in gardens, and sometimes on burned areas.

Technical description. Cap 70-120 mm long, 40-60 mm wide, more or less conic or nearly globose; apex obtuse when narrowed upward, usually distinctly wider than the stalk; surface pitted; pits quite irregular in shape, rounded to considerably elongated or laterally compressed, 3-10 mm wide, up to 15 mm long; ridges irregularly connecting and not distinctly longitudinally arranged; edges rounded; color white to gray, yellowish, or yellowish brown, with the ridges lighter than the pits.

Stalk 10-100 mm long, 10-40 mm wide, thick, cylindrical or expanded at both the base and apex, sometimes irregularly furrowed; surface scurfy to minutely pubescent, hollow, brittle; color white to pallid or yellowish.

Spores slightly yellowish in deposit, (14.6-)17.5-21.9(-24) × 8.8-11.0(-14) μm, ellipsoidal, smooth, thin-walled, hyaline. Asci eight-spored, 223-300 × 19-20 μm, cylindrical, hyaline. Paraphyses filamentous, cylindrical, 5.8-8.8 μm wide, hyaline.

Stalk hyphae interwoven, 5.8-9.4 μm wide, hyaline; surface hyphae inflated, globose to pear-shaped, 22-44 μm wide, covered by a network of interwoven hyphae 11-16.8 μm wide with recurved cylindrical hyphal ends.

Useful references. The following books contain information on *M. esculenta* pertinent to North America.

- *A Field Guide to Western Mushrooms* (Smith 1975)—A color photograph and a description and discussion are presented.
- *Edible and Poisonous Mushrooms of Canada* (Groves 1979)—A black-and-white photograph, a color photograph, and a description and discussion are included.
- *Mushrooms of North America* (Miller 1972)—A color photograph and a description and discussion are given.
- *The Mushroom Hunter's Field Guide* (Smith 1974)—A black-and-white photograph, a color photograph, and a description and discussion are included.
- *The Savory Wild Mushroom* (McKenny 1971)—A black-and-white photograph and a description and discussion are presented.

Mycena (Fr.) S. F. Gray

Tricholomataceae

The genus *Mycena* contains mostly small to minute species of mushrooms that are not commonly collected for food. The fruit bodies are often quite fragile, with thin, brittle to pliant stalks. The gills are attached to the stalk, often adnate to adnexed. The cap varies from conic to bell-shaped, rarely somewhat convex, with the margin usually straight and often marked with radial striations. There is no partial or universal veil. The spores are white or less commonly pale cream in deposit, lack an apical germ pore, and are typically amyloid, turning grayish to grayish blue in Melzer's reagent, an iodine solution.

Species of *Mycena* often have pleurocystidia and cheilocystidia; these cells are important in identifying species. The cap cuticle often has an epicutis comprising a surface layer of filamentous hyphae, frequently with lateral projections, and a subcutis of inflated cells.

Species of *Collybia* and *Marasmius* are those most likely to be confused with *Mycena. Collybia* species typically have convex caps with the margin enrolled to incurved or decurved, especially when young. The stalks are usually more or less cartilaginous. *Marasmius* is difficult to separate both from *Mycena* and *Collybia*. A useful characteristic of *Marasmius* is the ability of the species to revive after drying; usually, dried specimens of *Marasmius,* soaked in solution, revive to their original size, shape, and color and can again begin to produce spores. Also, some microscopic differences in the structure of the cap are helpful in distinguishing *Mycena, Marasmius,* and *Collybia,* although proper interpretation requires careful study and some training. The filamentous cuticle of *Collybia* differs from that of *Mycena* in that it lacks a subcutis. The hyphae of the epicutis in *Mycena* frequently are filamentous with lateral projections. *Marasmius* species often have a cap cuticle characterized by broom cells.

Mycena pura is the only species of *Mycena* that has been reported as poisonous. It is one of the larger species of the genus and is at times common. It is most likely not intentionally collected for food, but rather, inadvertently collected while gathering other mushrooms. Heim (1963) reported *M. pura* as psychotropic and muscarinic. Other European authors have reported muscarine-type poisoning following its consumption. Minute levels of muscarine have been demonstrated in *M. pura* by Stadelmann et al. (1976). There are no published reports of poisoning by this species in North America. It is sometimes listed in mushroom books as nonpoisonous. Usually it is considered too small to be of value for food.

Mycena pura (Fr.) Quél. Figs. 148, 212

Common name. Pure mycena.

Distinguishing characteristics. Cap 20–40(–65) mm broad, obtuse-expanding to convex or plane; surface naked, moist; margin translucent-striate; color extremely variable, more or less fading with age, rosy red, lilac, purplish to grayish lilac, or sometimes white with tinges of blue or purple, less commonly yellowish. Odor and taste distinctly resembling radish. Gills usually tinted with colors of the cap or rarely white, close to subdistant, broad and ventricose, adnexed to hooked. Stalk 30–100 mm long, 2–6 mm thick, equal or enlarged below, hollow and more or less tough; color white or similar in color to that of the cap but usually a paler shade. Spores white in deposit, amyloid, smooth, subcylindrical to narrowly ellipsoidal, 6.6–9.1(–10.2) × 3.3–4.1 µm in size. Cheilocystidia and pleurocystidia both present.

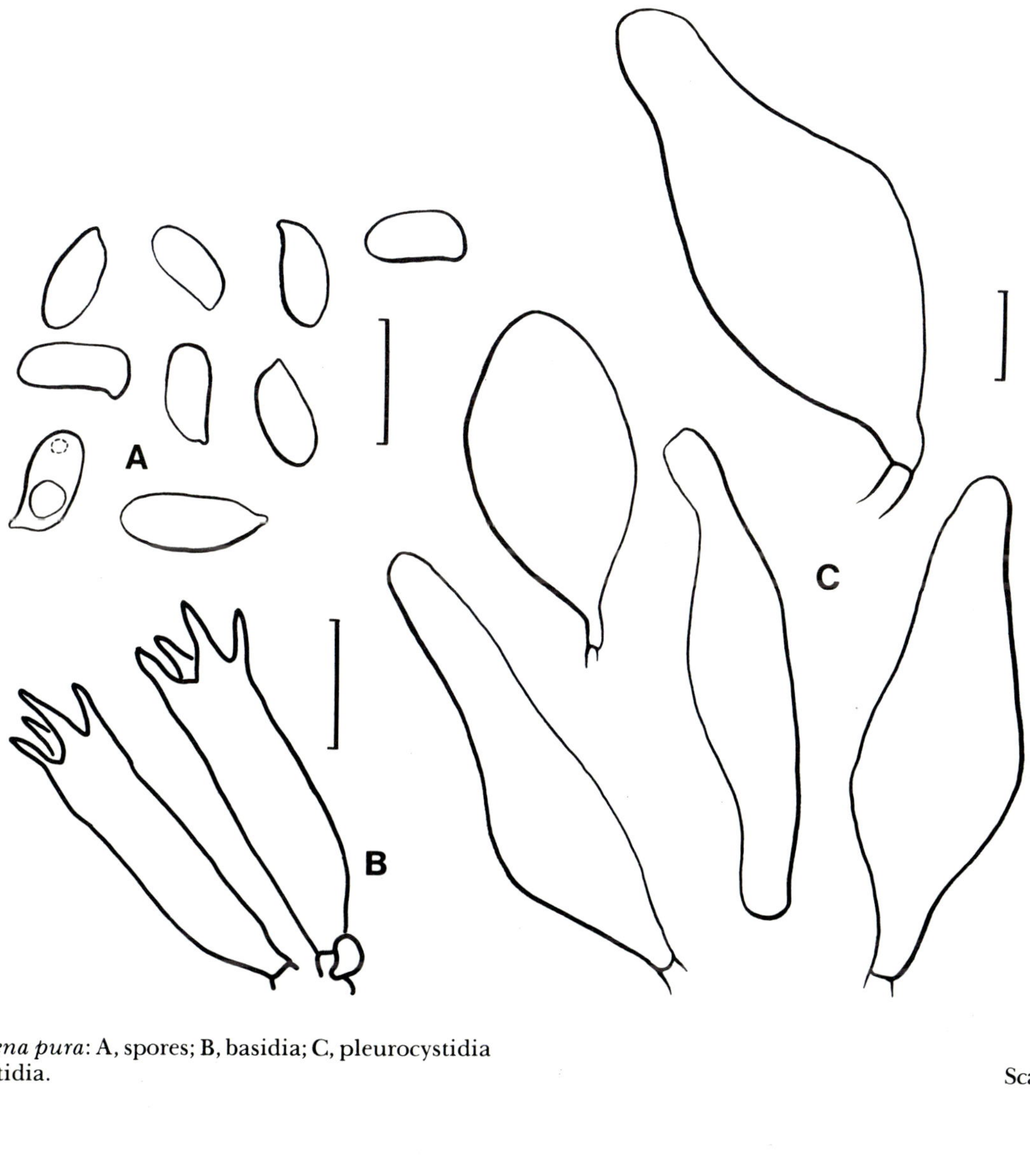

Fig. 148. *Mycena pura*: A, spores; B, basidia; C, pleurocystidia and cheilocystidia.

Scale line 10 µm

Observations. *M. pura* is common and widely distributed, occurring in both hardwood and coniferous forests. Its bright colors make it attractive, and although it is large compared with other species of *Mycena,* it is too small compared with most mushrooms to be of value for food.

Distribution and seasonal occurrence. *M. pura* occurs throughout Canada and the United States and is sometimes abundant. This species fruits from spring to late fall, depending on location and weather, and is common from August to October. It likes cool, moist weather, as do most *Mycena,* and is usually not found in the drier summer months.

Habitat and habit. This species is terrestrial, sometimes solitary, usually scattered to gregarious, very rarely occurring in clusters. It is found on humus in hardwood and coniferous forests, rarely on rotten wood, and is sometimes quite common in conifer plantations.

Technical description. Cap 20–40(–65) mm broad, obtuse when young, becoming umbonate, convex, or plane; disc rarely slightly depressed; margin straight or slightly incurved in early stages, becoming elevated and splitting; surface naked, moist, translucent-striate, hygrophanous, frequently wrinkled with age; color extremely variable, bright rosy red, purplish, lilac gray, yellowish, or white with a faint bluish or purplish tinge on the disc. Flesh moderately thick, tapered to the margin, purplish to pallid or whitish; odor and taste resembling radishes.

Gills tinged purplish lilac or bluish, often shaded more or less gray, occasionally white, adnate, adnexed, or hooked, almost distant, broad, ventricose, interveined; edges whitish.

Stalk 30–100 mm long, 2–6 mm thick, terete or compressed, equal, sometimes enlarged below, hollow, somewhat tough, more or less twisted-striate, glabrous or scabrous-pruinose, sometimes almost scaly; base with a slight growth of mycelia; color white or almost the same as the cap but often paler.

Spores white in deposit, 6.6–9.1(–10.2) × 3.3–4.1 μm, more or less subcylindrical to narrowly ellipsoidal, smooth, hyaline, amyloid. Basidia four-spored, clavate, 22–26 × 6–7.0 μm, hyaline, thin-walled. Pleurocystidia 40–66 × 12.8–18.0(–20) μm, ventricose with obtuse apices or elongated tapered necks, sometimes sac-like to pedicellate, scattered or abundant. Cheilocystidia ventricose to sac-like, 58–66(–100) × 12.8–18.0(–25) μm, hyaline, thin-walled.

Gill trama interwoven to parallel; hyphae cylindrical to inflated, 4.4–18.3 μm wide, hyaline, reddish brown in Melzer's reagent. Subhymenium filamentous; hyphae interwoven, 3.7–5.1 μm wide, hyaline. Cap cuticle composed of an epicutis and a subcutis; hyphae cylindrical, 1.8–5.1 μm wide, repent in the epicutis, thin- to thick-walled with nodulose irregular lateral projections, hyaline to yellowish, somewhat gelatinous; subcutis hyphae inflated, interwoven to subparallel, 30–35 μm wide, hyaline, reddish in Melzer's reagent. Cap trama loosely interwoven; hyphae cylindrical to inflated, 4–23 μm wide, hyaline, reddish in Melzer's reagent. Caulocystidia ventricose to broadly clavate or sac-like, 58–117 × (13.1–)29.2–37.2 μm, thin-walled, hyaline, parallel and overlapping to appressed; stalk hyphae parallel to interwoven near the surface, cylindrical to somewhat inflated, 5.8–23.4 μm wide, hyaline. Oleiferous hyphae refractive, greenish yellow, 4.6–7.0 μm, scattered in the trama and cap cuticle. Clamp connections present, more conspicuous on narrow hyphae.

Useful references. The following books contain information on *M. pura* pertinent to North America.

- *Edible and Poisonous Mushrooms of Canada* (Groves 1979)—A color photograph and a description and discussion are included.
- *Mushrooms in their Natural Habitats* (Smith 1949)—A description and discussion, as well as a color photograph on a reel, are presented.
- *Mushrooms of North America* (Miller 1972)—A good color photograph and a description and discussion are given.

Omphalotus Fayod

Tricholomataceae

Omphalotus is a small genus related to *Clitocybe.* In fact, one of the most common species in North America, *Omphalotus illudens,* was until recently classified as *Clitocybe illudens* (Schw.) Sacc. At present three species of *Omphalotus* have been reported with certainty from North America: *O. illudens, O. olivascens* (Bigelow) Miller & Thiers, and *O. subilludens* (=*Monadelphus subilludens* Murrill). The name *Omphalotus (Clitocybe) olearius* (Fr.) Singer has been applied incorrectly to various North American species of *Omphalotus;* in its modern definition *O. olearius* (=*Pleurotus olearius* (Fr.) Gill.) occurs only in Europe.

The genus *Omphalotus* can be characterized by the large, fleshy, more or less caespitose fruit bodies growing around stumps, at the base of trees, or from buried wood; the bright colors of the cap, often some shade of orange, varying from yellow to saffron yellow, orange yellow, or dull orange to brownish orange, with one species developing strong olive to greenish tones; typically long, decurrent, luminescent gills; a white to pale cream spore deposit; and the smooth spores, lacking an apical germ pore.

All three species of *Omphalotus* that occur in North America are considered poisonous. *O. illudens* appears to have caused most of the poisonings (Farlow 1899, Fischer 1918, Seaver 1939, Gelb 1949, Boston Mycological Club 1963) (several cases on file at the Poison Fungi Center, Beltsville, MD). *O. olivascens* is also poisonous (H. D. Thiers, Personal commun.; Mandell 1978). Symptoms usually include nausea and vomiting and sometimes diarrhea (Farlow 1899, Seaver 1939). Carey (1974) reported several toxic compounds in *O. olearius* (=*O. illudens*), but according to Lincoff and Mitchel (1977), more work is needed to determine the nature and action of the toxin or toxins causing *O. olearius* poisoning. Genest et al. (1968) detected muscarine in *O. olearius*, and Clark and Smith (1913), Maretic (1967), and Maretic et al. (1975) support the presence of muscarine in this species. Lincoff and Mitchel (1977), however, conclude that muscarinic symptoms are not characteristic of poisoning by *O. olearius.*

The main toxins in *O. illudens* may be the sesquiterpenoids illudin-s (lampteral) and illudin-m (McMorris and Anchel 1965). Illudin-s has also been isolated from a related poisonous mushroom *Lampteromyces japonicus* (Kawamura) Singer, a species known only from Japan (Nakanishi et al. 1965).

A description of *Omphalotus illudens* follows, which contains reference to *O. subilludens* and *O. olivascens.*

Omphalotus illudens (Schw.) Sacc.

Figs. 149, 213

Common names. Jack-o-lantern fungus, false chanterelle, deceiving mushroom.

Distinguishing characteristics. Cap 50–200 mm wide, convex to plane when young; margin enrolled; center broadly depressed in later stages; surface dry to moist; color deep orange to moderate orange or orange yellow, the color fading with age. Gills crowded, narrow, long-decurrent, occasionally forked; color yellow orange to more or less the same color as the cap surface, luminescent. Stalk 70–250 mm long, 5–22 mm thick, tapered downward to a narrow more or less pointed base; color about the same as the cap but usually brownish toward the base. Spores sometimes white, usually pale cream to pale yellow in deposit, subglobose to globose, smooth, lacking an apical germ pore, 3.7–7.0(–8.0) × 3.7–5.1 μm in size.

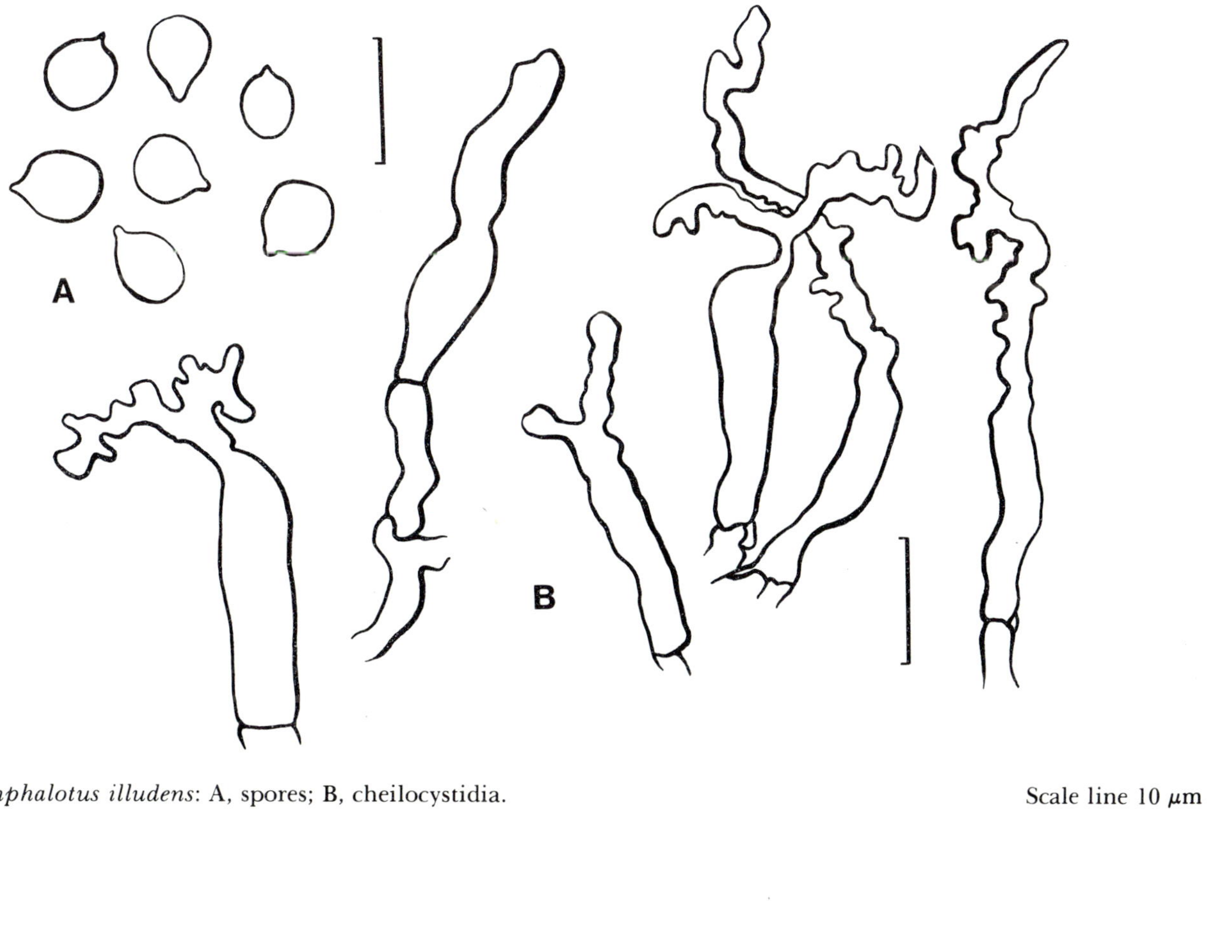

Fig. 149. *Omphalotus illudens*: A, spores; B, cheilocystidia. Scale line 10 μm

Observations. *O. illudens* occurs in Canada from the Great Lakes region and eastward, but it is not common. In the United States it occurs throughout the east and into the south.

The bright colors of *O. illudens* and its habit of fruiting in very large clusters make it especially conspicuous. It is sometimes mistaken for *Cantharellus cibarius* (chanterelle) or yellow forms of *Armillariella mellea. Cantharellus cibarius* is similar in color to *O. illudens* but does not grow in dense, large clusters; on the undersurface of the cap it has distant, more or less gill-like structures that are usually shallow with blunt to obtuse edges. *O. illudens* can be separated from *Armillariella mellea* by its bright orange to orange yellow colors and the lack of an annulus. It may also be confused with *Armillariella* (*Monodelphus*) *tabescens.*

The luminescent gills of *O. illudens* can be plainly seen when fresh fruit bodies are placed in a dark room. After about 5 minutes the luminescence is bright enough that the edges of the gills are visible in the dark. Certain specimens are not luminescent, especially when they are old or partly dried. According to Miller (1972), specimens that do not luminesce may exhibit this character after being wrapped in wax paper for 2-3 hours. The common name jack-o-lantern fungus is derived from the luminescent quality of *O. illudens* and its relatives.

O. olivascens (Fig. 214) and *O. subilludens* have not been reported from Canada. *O. olivascens* has been reported only from California. It has a dull orange to brownish orange or dull bay red cap that develops olive tones as it matures. The gills and stalk also have olive coloration. The spores are subglobose to globose or broadly ellipsoidal and larger, 6.5-8.8(-11) × 5-7.3(-8.8) μm, than those of *O. illudens.*

O. subilludens occurs in the southeastern United States on hardwoods and palms. The cap color is more reddish orange to reddish brown than for *O. illudens* and the spores tend to be more ellipsoidal and larger, 6.6-11 × 4.4-5.8 μm in size.

Distribution and seasonal occurrence. *O. illudens* occurs in southeastern Canada from the Great Lakes region eastward but is not common. In the United States it occurs mainly in the east and into the south. Depending on the weather, it fruits during the summer and early fall, often following a rainfall.

Habitat and habit. *O. illudens* is caespitose, occurring in large, dense clusters that sometimes contain more than 100 fruit bodies. It is usually found at the base of hardwood stumps and living trees or is sometimes terrestrial, fruiting from buried wood. It is most commonly associated with oak (*Quercus*).

Technical description. Cap 59-200 mm broad, convex to nearly plane, with an enrolled margin when young; margin decurved or spreading with age; disc often broadly depressed, sometimes slightly umbonate; margin sometimes irregular or lobed; surface dry to moist, glabrous, occasionally slightly striped; color deep orange or moderately orange to orange yellow, usually fading with age. Flesh thin to moderately thick, firm, white to yellowish or tinted orange; taste strong to disagreeable; odor strong to disagreeable or unpleasantly sweet.

Gills yellow orange, more or less the same color as the cap surface, luminescent, unequally long-decurrent, occasionally forked, crowded, narrow.

Stalk 70-250 mm long, 5-22 mm thick, tapering to a narrow more or less pointed base, often curved to twisted, central or occasionally eccentric, solid, glabrous to minutely downy or occasionally somewhat scaly with age, more or less the same color as the cap surface, brownish toward the base. Flesh firm. Partial and universal veils absent.

Spores white or pale cream to pale yellow in deposit, 3.7-7.0(-8.0) × 3.7-5.1 μm, subglobose to globose, smooth, hyaline, nonamyloid. Basidia four-spored, clavate, 20-29.2 × 5.8-6.6 μm, hyaline. Pleurocystidia not differentiated. Cheilocystidia clavate-cylindrical, with obtuse to filiform or irregular apices, 18-24 × 4-6 μm, hyaline.

Gill trama interwoven to somewhat parallel; hyphae more or less cylindrical, 3.5-12.0 μm wide, hyaline to yellow orange, thin-walled. Subhymenium filamentous, compactly interwoven; hyphae 3.0-5.8 μm wide, hyaline to yellow orange. Cap cuticle filamentous; hyphae interwoven, cylindrical, 2.0-3.0 μm wide, with yellow contents, thin-walled; subcutis hyphae cylindrical, 3.0-9.0 μm wide, yellowish. Cap trama loosely interwoven; hyphae cylindrical to inflated, 3.5-14.0 μm, yellowish. Caulocystidia clavate-cylindrical with obtuse apices to filiform or contorted with tapered apices, 18-23.4 × 4.0-5.8 μm, hyaline; stalk hyphae parallel to interwoven on the surface, more or less cylindrical, 3.5-9.4(-11.0) μm wide, hyaline to yellowish. Oleiferous hyphae refractive, yellowish, thin-walled, not septate, contorted, 3.8-11.0 μm wide, scattered in the trama, cap cuticle, and stalk. Clamp connections present.

Useful references. The following books contain information on *O. illudens* pertinent to North America.

- *A Field Guide to Western Mushrooms* (Smith 1975)—*O. illudens* is discussed under *O. olivascens.*
- *Edible and Poisonous Mushrooms of Canada* (Groves 1979)—A color photograph, a description, and a discussion are included.
- *Mushrooms in their Natural Habitats* (Smith 1949)—A description and discussion, as well as a color photograph on a reel, are presented.
- *Mushrooms of North America* (Miller 1972)—A color photograph and a description and discussion, as *O. olearius,* are presented.
- *The Mushroom Hunter's Field Guide* (Smith 1974)—A black-and-white photograph and a description and discussion, as *Clitocybe illudens,* are presented.

Paxillus Fr.

Paxillaceae

Paxillus is a small genus of mushrooms with both terrestrial and wood-inhabiting species. The species are variable in appearance, making the genus difficult to characterize. In general the species somewhat resemble species of *Clitocybe* or *Pleurotus*. The gills are decurrent and often appear forked, especially near the stalk apex. When care is taken, the gills can be separated as a unit from the flesh of the cap. The spores are yellow brown to vinaceous brown and smooth, and they lack an apical germ pore.

In North America three species of *Paxillus* are commonly encountered: *P. atrotomentosus* (Batsch ex Fr.) Fr., a velvety stalked species that occurs on conifer wood; *P. involutus*, a terrestrial species, described below; and *P. panuoides* (Fr. ex Fr.) Fr., a stalkless species that occurs on coniferous wood. These and other species of *Paxillus* are not recommended for food. In Europe and some areas of North America (Smith 1975) *P. involutus* is eaten and enjoyed by some people, although it can make certain individuals ill.

The toxicity of *Paxillus involutus* is well-documented in Europe and Japan. Grzymala (1965) in a 10-year study in Poland reported 109 cases of poisoning by *Paxillus involutus*, including one death. Three fatalities from eating insufficiently cooked *Paxillus involutus* were reported by Bschor and Mallach (1963) and Bschor et al. (1963); both reports deal with the same poisonings. According to Hatfield and Brady (1975) specific hemolysins were detected in the serum of two previously sensitized patients who developed hemolytic anemia upon ingestion of *Paxillus involutus*. This statement was based on the article by Schmidt et al. (1971). See also comments by Chilton (1972).

Paxillus involutus (Batsch ex Fr.) Fr.

Figs. 150, 215

Common name. Common paxillus.

Distinguishing characteristics. Cap 40–120 mm broad, convex to plane; margin often enrolled, somewhat cottony, often with short rib-like lines; color yellow brown to darker brown; surface viscid. Gills decurrent, close, narrow; color yellow maturing to brown, staining dingy brown where bruised; crossveins usually conspicuous between the gills. Stalk thick, usually short in relation to the width of the cap, dry, about the same color as the cap. Spores yellow brown in deposit, smooth, more or less ellipsoidal, (8–)8.8–10.2(–11) × 4.4–6.2(–7.3) μm in size. Fungi terrestrial, occurring usually near trees.

Observations. *P. involutus* is a fairly common and widely distributed species. The dingy colors, brown stains, enrolled cap margin, close-decurrent gills, and yellow brown spore deposit distinguish this mushroom. *P. vernalis* Watling, a species related to *P. involutus*, has a vinaceous brown spore deposit and other features that distinguish it from *P. involutus*. It's edibility is unknown. Watling (1969) provides a description and discussion.

Distribution and seasonal occurrence. This species is widely distributed, especially in northern forest areas. It fruits in the summer and fall, depending on location and weather.

Habitat and habit. *P. involutus* is solitary or more commonly scattered to gregarious, sometimes caespitose, and terrestrial. It is occasionally found on woody material, often around stumps. It grows in coniferous and mixed woods and is associated with the roots of trees. It is sometimes abundant. In the Pacific northwest it frequently occurs in metropolitan areas where white birch (*Betula papyrifera*) has been planted.

Technical description. Cap 40–120 mm broad, convex when young, expanding to more or less plane and often depressed at the disc; margin persistently enrolled, cottony, becoming smooth or grooved; surface viscid on the disc when wet, downy to cottony, sometimes glabrous; color deep yellowish brown to rusty yellow in early stages with a more or less olivaceous cast, becoming tinged reddish brown or somewhat blotchy red brown when mature or bruised. Flesh thick, soft, yellowish to moderate orange yellow, becoming red brown where cut; taste and odor acidulous.

Gills pale yellowish or ocherous when young, becoming rust yellow, staining brown to red brown where bruised, decurrent, often forked and connecting near the stalk, easily separable from the cap, crowded, broad.

Stalk 40–100 mm long, 14–20 mm thick, equal or enlarged toward the apex, central or slightly off center, solid, dry, glabrous; color paler than or the same color as the cap, soon becoming spotted and streaked darker brown. Flesh soft, brownish, becoming red brown where cut. Annulus or universal veil lacking.

Spores yellowish brown to deep yellow brown in deposit, (8.0–)8.8–10.2(–11.0) × 4.4–6.2(–7.3) μm, elliptic and slightly inequilateral in side view, ovate to broadly elliptic in face view, smooth, thin-walled, dull yellow brown, nonamyloid. Basidia four-spored, clavate, 32.9–41.2 × 8.0–9.5 μm, hyaline to yellowish. Pleurocystidia and cheilocystidia lanceolate to fusiform, 40–65 × 10–15(–18.0) μm, hyaline to yellowish, thin-walled.

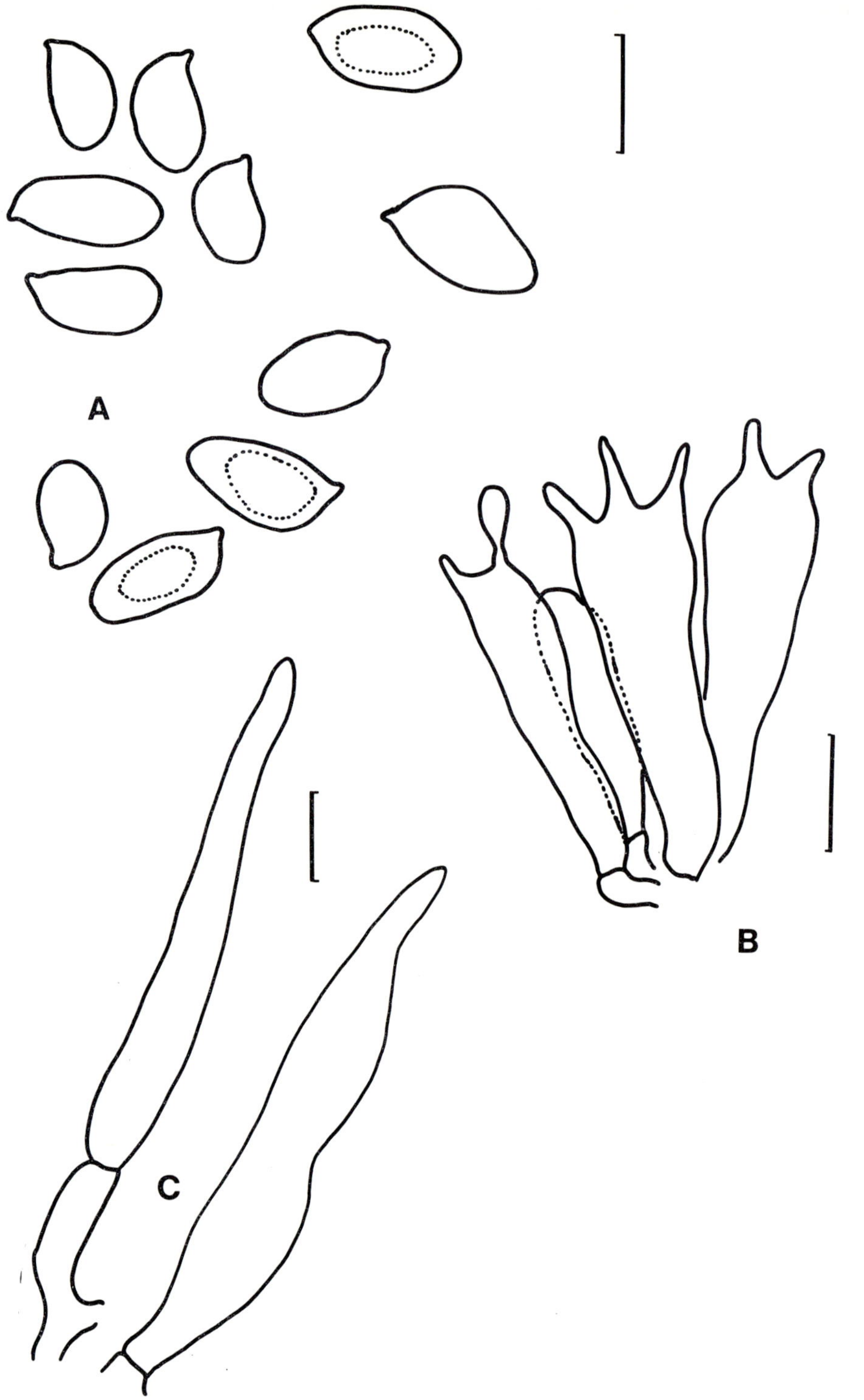

Fig. 150. *Paxillus involutus*: A, spores; B, basidia; C, pleurocystidia.

Scale line 10 μm

Gill trama divergent from a central strand of more or less parallel hyphae; hyphae more or less cylindrical, 4.4-11.3 µm wide, gelatinous, hyaline to yellowish. Subhymenium filamentous, interwoven; hyphae cylindrical, 2.9-3.5 µm wide, hyaline. Cap cuticle filamentous, interwoven; hyphae cylindrical, 3.6-5.4 µm wide; hyphal ends erect to repent, hyaline to yellowish, somewhat gelatinous. Cap trama loosely interwoven; hyphae cylindrical to inflated, 4.6-30.0 µm wide, hyaline. Oleiferous hyphae abundant, refractive, greenish yellow, 5.8-10.2 µm wide, scattered in the trama and cap cuticle. Clamp connections present.

Useful references. The following books provide information on *P. involutus* pertinent to North America.

- *Edible and Poisonous Mushrooms of Canada* (Groves 1979)—A color photograph and a description and discussion are included.
- *Mushrooms of North America* (Miller 1972)—A color photograph and a description and discussion are presented.
- *The Mushroom Hunter's Field Guide* (Smith 1974)—A black-and-white photograph and a description and discussion are given.
- *The Savory Wild Mushroom* (McKenny 1971)—A black-and-white photograph, a color photograph, and a description and discussion are included.

Phaeolepiota Maire — Cortinariaceae

This genus contains only one species, *Phaeolepiota aurea*, in North America. It is sometimes placed in either the genus *Pholiota* or *Togaria*. It is reported as edible in some books but has caused several cases of gastrointestinal upset. Poisonings have been reported by Wells and Kempton (1967) and by D. Barr (Personal commun.). If you are intent on eating it, proceed with caution.

In general appearance *Phaeolepiota aurea* resembles a large species of *Cystoderma*. It may be distinguished, however, by the colored spore deposit, usually light yellow brown to pale ocherous buff or pale orange buff. The fruit bodies are usually large, colored yellowish orange to golden yellow or orange tan. The surfaces of the cap, the veil or annulus, and stalk have a granular to powdery coating that is easily rubbed off. The spores are minutely roughened to nearly smooth, and they lack an apical germ pore. Similar-appearing genera are *Cortinarius* (Ch. 7), *Gymnopilus* (Ch. 12), and *Pholiota* (Ch. 13).

Phaeolepiota aurea (Bull. ex Fr.) Maire ex Konr. & Maubl. — Figs. 151, 216

Common name. Golden pholiota.

Distinguishing characteristics. Cap rounded to convex, up to 200(-300) mm broad; color yellowish orange, light gold, golden yellow to orange tan or leather brown. Granular to powdery coating covering the cap, annulus, and stalk, easily rubbed off and sometimes lost with age. Gills close; color slightly yellowish to yellowish brown, often becoming the same color as the cap surface. Stalk fairly thick, enlarged below, with a granular to powdery coating below the annulus; color about the same as the cap surface. Annulus usually present, distinct, flaring, membranous, powdery to granular on the outer surface, located near the stalk apex. Spores typically minutely roughened, ellipsoidal to subfusiform, 10.2-14.6 × 4.8-5.8 µm in size. Fungi occurring in the west.

Observations. *P. aurea* is a large, beautiful species that in its typical form is not easily confused with other species. When the granular to powdery coating is lost, particularly with age, the species becomes somewhat more difficult to identify.

Distribution and seasonal occurrence. *P. aurea* occurs along the north Pacific coast to Alaska. It is generally uncommon but sometimes fruits in a particular location in great abundance in the late summer and fall.

Habitat and habit. This species has been reported in or on the edge of Douglas-fir (*Pseudotsuga taxifolia*) forests, near alders (*Alnus*). It fruits in clusters or is gregarious.

Technical description. Cap 50-200(-300) mm broad, subovoid to convex when young, becoming obtusely bell-shaped to convex or somewhat plane with age, more or less umbonate; surface dry, not hygrophanous, granulose to floccose-granulose or with age becoming more or less glabrous after most of the granular particles have weathered away; granulose covering generally easily removed by rubbing the surface; margin often ragged with veil fragments; color orange buff or orange ocherous, golden yellow, golden orange brown, or ocherous tawny, usually becoming paler with age, rarely merely ocherous. Flesh moderately thick on the disc, firm, whitish to pallid or yellowish,

Fig. 151. *Phaeolepiota aurea*: A, spores; B, basidia; C, veil elements.

Scale line 10 μm

unchanging; taste indistinctive to somewhat astringent; odor indiscernible to mild, pleasantly aromatic, slightly reminiscent of bitter almonds, or pungent.

Gills pallid to slightly yellowish in early stages, becoming the same color as the cap surface at maturity, adnate to adnexed or occasionally with a slightly decurrent tooth, close, moderately broad; edges entire and the same color as the cap surface.

Stalk 50–150(–250) mm long, 10–50(–60) mm wide, enlarged downward and sometimes almost clavate, central, stuffed, becoming hollow, dry, unpolished, smooth and glabrous above the annulus, granulose to floccose-granulose below the annulus to the base of the stalk; color light orange yellow or more or less the same color as the cap surface, sometimes darker at the apex. Flesh fibrous, whitish or somewhat streaked orange. Annulus flaring, membranous, granular to powdery on the outer surface, large, persistent, apical on the stalk, becoming pendulous and disappearing with extreme age.

Spores light yellow brown to light orange yellow in deposit, 10.2–14.6 × 4.8–5.8 μm, ellipsoidal to subfusiform, inequilateral in side view, elliptic in face view, thin-walled, minutely roughened, yellowish brown, nonamyloid. Basidia two- and four-spored, clavate, 35–48.0 × 7.3–8.0 μm, hyaline. Pleurocystidia lacking or rare, 26–30 × 7.5–8.5 μm, clavate-mucronate, brownish. Cheilocystidia lacking.

Gill trama interwoven to parallel; hyphae more or less cylindrical, 2.9–5.8(–9.0) μm wide; walls thin to slightly thickened, hyaline to yellowish. Subhymenium filamentous; hyphae 2.9–3.7 μm wide, hyaline to yellowish. Cap cuticle cellular; hyphae inflated, often isodiametric, subglobose-pedicellate to broadly fusiform, 16–22 μm across, with finger-like or knob-like projections over the apex, yellowish brown; hyphal walls thin to slightly thickened. Cap trama interwoven; hyphae more or less cylindrical, 4.8–8.0 μm wide, hyaline. Caulocystidia sac-like to globose or ovoid, 12.4–30 × 10–15 μm, occurring in chains, clustered, with minute projections or knobs, brownish; stalk hyphae parallel to interwoven on the surface, more or less cylindrical, 2.3–5.8(–7.3) μm wide, hyaline to brownish, thin-walled to slightly thick-walled. Annulus interwoven; hyphae cylindrical, 2.2–5.8 μm wide, mixed with chains of brownish ellipsoidal to subglobose nodulose cells 14.6–25.6 × 8.8–14.6 μm in size. Oleiferous hyphae refractive, greenish yellow, 5.8–7.0 μm wide, scattered in the trama. Clamp connections present.

Useful references. The following books contain information on *P. aurea* pertinent to North America.

- *Edible and Poisonous Mushrooms of Canada* (Groves 1979)—A black-and-white photograph and a description and discussion are included.
- *Mushrooms in their Natural Habitats* (Smith 1949)—A discussion is found under *Pholiota aurea.*
- *Mushrooms of North America* (Miller 1972)—A description and discussion, as *Pholiota aurea,* are included.
- *The Mushroom Hunter's Field Guide* (Smith 1974)—Black-and-white photographs and a description and discussion are presented.
- *The Savory Wild Mushroom* (McKenny 1971)—A color photograph, a black-and-white photograph, and a description and discussion are presented.

Pholiota (Fr.) Quél.

Cortinariaceae

Most species of *Pholiota* occur on wood, although a few, such as *Pholiota terrestris* Overholts, are terrestrial, possibly fruiting from buried wood. All the species have a brown spore deposit, often dull to dark brown, but sometimes rusty brown to yellow brown. The fruit bodies vary from small to large, with most species being fairly large and fleshy. The fruit bodies typically have a partial veil that leaves a slight to well-developed annulus near the stalk apex. The annulus is usually composed of fibrils, but it may be membranous; it is commonly brownish from spores that have fallen from the gills. The cap is often scaly, and the stalk below the annulus is coated with scales. The fruit bodies are often caespitose but may be scattered or solitary.

The spores of *Pholiota* are typically smooth, and each spore typically has a small to well-developed apical germ pore. Various kinds of cystidia occur in the genus. All species have cheilocystidia. The cap cuticle is usually filamentous.

The wood-inhabiting genus *Gymnopilus* and the wood-inhabiting species of *Galerina* are those most likely to be confused with *Pholiota*. The *Gymnopilus* species typically have a rusty orange brown to ferruginous spore deposit, and the spores are ornamented with small to medium-sized warts and lack an apical germ pore. The cap surface of *Gymnopilus* species is often brightly colored, frequently some shade of orange brown to reddish brown. The wood-inhabiting *Galerina* species are usually fairly small to medium-sized and generally resemble certain smaller species of *Pholiota.* The only sure way to differentiate these species is by studying the spores. The spores are smooth with an apical germ pore in *Pholiota* but are usually roughened and lacking an apical germ pore in *Galerina*. Small, brown-spored mushrooms that occur on wood or are terrestrial should not be eaten. When poisoning results from the ingestion of a small, brown-spored, wood-inhabiting mushroom, immediately check to see if it is *Galerina autumnalis* or related species. These species contain amatoxins and are very dangerous.

To date only a few species of *Pholiota* have caused poisonings. Except for *P. highlandensis*, they are large, scaly species and include *P. hiemalis* Smith & Hesler, *P. abietis*, and *P. squarrosa*. *Phaeolepiota aurea*, which has also caused poisonings, is sometimes named *Pholiota aurea* (Bull. ex. Fr.) Kummer. *Phaeolepiota* is treated earlier in this chapter. *P. abietis, P. highlandensis,* and *P. squarrosa* are described here.

Pholiota abietis Smith & Hesler

Figs. 152, 217

Common names. None.

Distinguishing characteristics. Cap large, convex to plane; surface slimy-viscid, covered with tawny brown to brownish orange spot-like scales that are sometimes arranged in more or less concentric zones; disc usually the same color as the scales, with yellow between the scales and on the margin. Gills close; color pale brownish, becoming light brown. Stalk more or less equal; color yellowish overall, not darkening with age at the base; annulus lacking, although a slight fibrillose zone may occur where the partial veil breaks; surface with scales similar to those on the cap below the annular zone. Spores smooth, with a minute apical germ pore, ellipsoidal, 5.8-7.3(-8.2) × 3.7-4.6 μm in size. Fungi occurring in clusters on coniferous wood.

Observations. *P. abietis* is a western species that grows on coniferous wood. It is one of several large *Pholiota* species that have a slimy cap with conspicuous scales. This species and *P. hiemalis,* the winter pholiota, are apparently poisonous to some individuals, causing gastrointestinal disturbance. In some books on mushrooms *P. abietis* and its relatives are considered edible. In light of the poisoning resulting from this group of species, however, they are not recommended for eating. Farr et al. (1977) consider *P. abietis* to be conspecific with *P. limonella* (Peck) Sacc.

Poisoning by *P. abietis* was reported by Dr. D. L. Largent, of Humboldt State College, California (Personal commun.). Poisoning by *P. hiemalis* was reported by Dr. E. Tylutki of the University of Idaho (Personal commun.).

P. hiemalis, which is not described here, is characterized by a slimy yellow cap covered by broad, dull fulvous scales that finally disintegrate to dark brown discolorations; pallid young gills with yellow edges; scattered rusty brown gelatinous scales on the stalk; and a fibrillose annular zone near the stalk apex. It occurs on coniferous wood and is related to *P. abietis.* The spores are 7-9(-10) × 4-4.5(-5) μm, slightly larger than those of *P. abietis.* Smith (1975) provides a good illustration of *P. hiemalis.*

Distribution and seasonal occurrence. *P. abietis* is widespread in the western United States and Canada, occurring in the fall and winter depending on location and weather.

Habitat and habit. This species occurs on coniferous logs and dead trees, especially fir (*Abies*), and is typically caespitose to gregarious.

Technical description. Cap 40-90(-150) mm wide, broadly convex with an encurved margin, becoming nearly plane, more rarely with an obtuse umbo; margin appendiculate with veil fragments in early stages; surface slimy-viscid, covered by appressed spot-like brownish orange scales sometimes arranged in more or less concentric rows, giving the cap its predominantly brownish orange color; color lemon yellow between the scales and on the margin. Flesh moderately thick, pliant, pale, dull yellow; taste and odor mild.

Gills pallid brownish when young, becoming light brown, turning rusty brown to rusty brownish orange when dried, adnate to slightly adnexed or rounded near the stalk, tapering toward the cap margin, close, moderately broad.

Stalk 80-120 mm long, 9-15 mm wide, equal to slightly enlarged at the base, solid, silky-pruinose to glabrous above the veil line and more densely scaly toward the base; scales fibrillose, dry; ground color of stalk yellowish overall, not distinctly darker near the base. Flesh similar to that of the cap. Partial veil leaving an evanescent zone near the stalk apex but not a distinct annulus. Universal veil absent.

Spores brownish moderate orange in deposit, 5.8-7.3(-8.2) × 3.7-4.6 μm, elliptic and inequilateral in side view, oblong-elliptic to ovate in face view, with a small apical germ pore, thick-walled, dark yellow brown to moderate brown, nonamyloid. Basidia four-spored, narrowly clavate, 17.9-29.2 × 5.8-7.3 μm, hyaline to yellowish. Pleurocystidia clavate to clavate-mucronate, 36-51 × 9.0-11.7 μm, with dark reddish brown contents in potassium hydroxide; chrysocystidia clavate, fusiform, or clavate-mucronate, 26-42 × 6-12 μm, thin-walled, containing yellow refractive bodies in potassium hydroxide. Cheilocystidia subcylindrical-capitate to narrowly fusiform-ventricose, 26-36.5 × 7.3-8.8 μm, hyaline to yellowish, thin-walled to slightly thick-walled.

Gill trama hyphae more or less parallel to somewhat divergent from a central strand, cylindrical to inflated, 5.8-15.5 μm wide, hyaline to yellowish. Subhymenium filamentous, compactly interwoven; hyphae cylindrical, 1.5-3.7 μm wide, hyaline to yellowish, gelatinous. Cap cuticle filamentous, interwoven; hyphae compact in the epicutis, cylindrical, 4.0-9.5 μm wide, gelatinous, yellowish to ocherous; hyphae somewhat wider in the subcutis. Cap trama loosely interwoven; hyphae cylindrical to inflated, 7.0-16.5(-25.5) μm wide, hyaline to yellowish, thin-

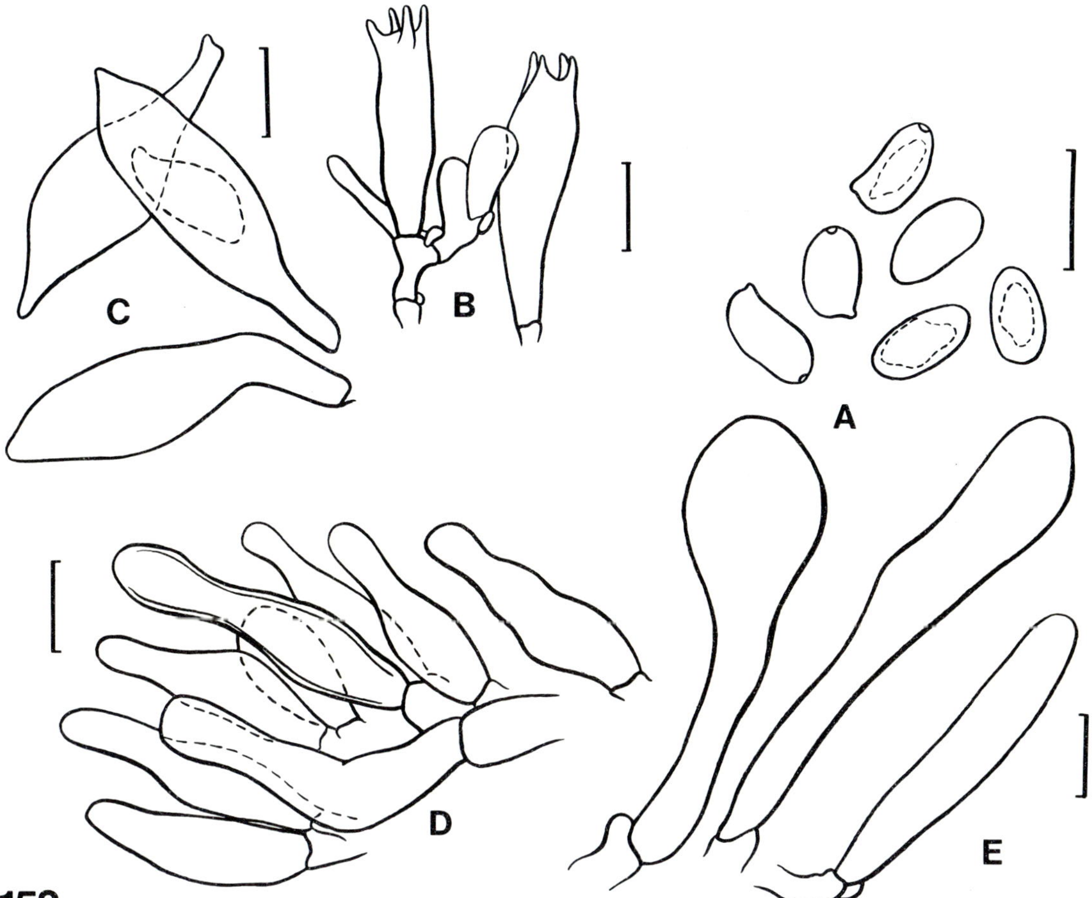

152

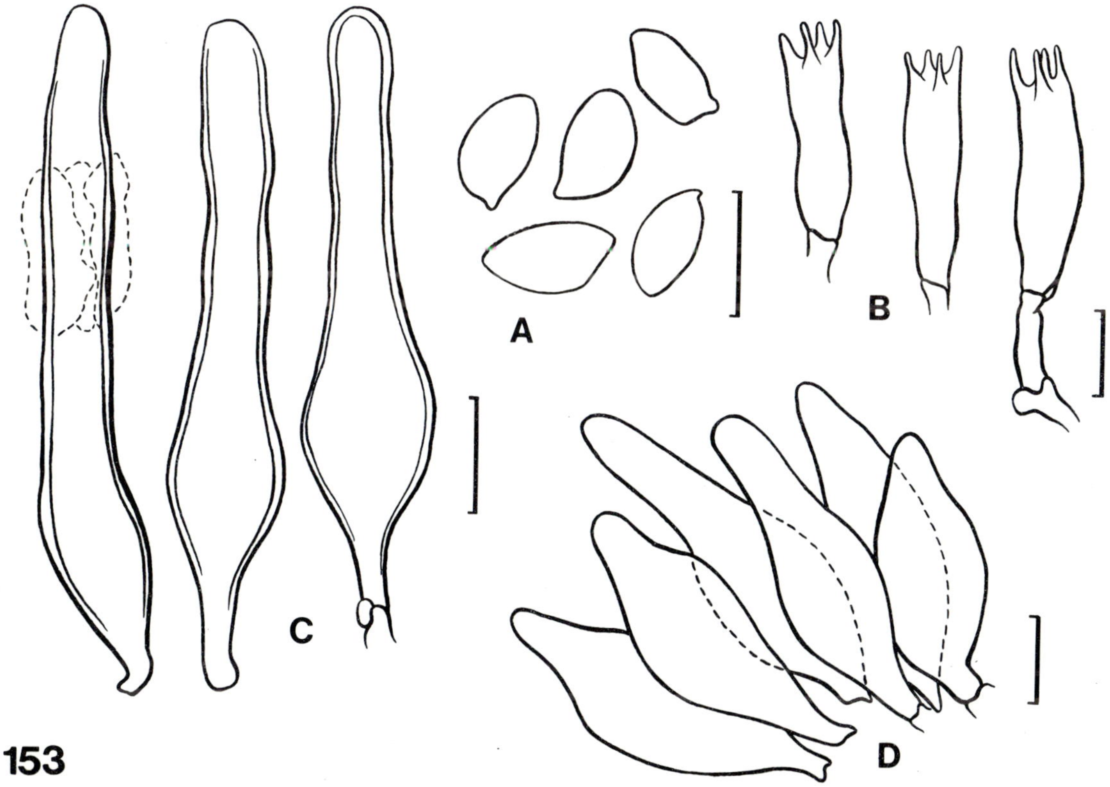

153

Fig. 152. *Pholiota abietis*: A, spores; B, basidia; C, pleurocystidia; D, cheilocystidia; E, caulocystidia.

Fig. 153. *Pholiota highlandensis*: A, spores; B, basidia; C, pleurocystidia; D, cheilocystidia. Scale line 10 μm

walled. Caulocystidia clavate to fusiform, (20-)58-81.9 × 9.4-13.1(-20) μm, thin-walled, occurring clustered to single; contents homogeneous to granular, reddish yellow in potassium hydroxide. Stalk hyphae parallel in the core to more interwoven on the surface, more or less cylindrical, 3.5-11.7 μm wide, hyaline to yellowish, thin-walled to slightly thick-walled and encrusted; scattered thin-walled granular pigmented hyphae also present, 3.5-11.0 μm wide, reddish ocherous in potassium hydroxide. Partial veil interwoven; hyphae cylindrical to more or less inflated, (2.3-)3.5-11.0(-16.7) μm wide, thin-walled to slightly thick-walled, hyaline to yellowish, leaving chains of ellipsoidal cells on the cap. Oleiferous hyphae refractive, greenish yellow, contorted, 4.0-9.5 μm wide, scattered in the cap cuticle, trama, and stalk. Clamp connections present.

Useful reference. The following book contains information on *P. abietis* pertinent to North America.

- *A Field Guide to Western Mushrooms* (Smith 1975)—A color photograph and a description and discussion are included.

Pholiota highlandensis (Peck) Smith & Hesler

Figs. 153, 218

Common names. None.

Distinguishing characteristics. Cap 20-40 mm broad, more or less convex; surface viscid; margin often with pallid to yellowish or buff veil remnants; color brown, usually some shade of reddish brown, fading with age, with the margin usually paler than the center. Gills attached; color pallid to pale yellowish when young, becoming brown when mature. Stalk 3-6 mm thick; color whitish to yellowish above and pallid to dark brown below, with zones and patches of pallid yellow to pale buff veil tissue, sometimes with a slight fibrillose annular zone. Spores smooth, with a small apical germ pore, ellipsoidal, 6.6-8.4 × 3.7-5.8 μm in size. Fungi typically occurring on burned areas, usually on charred wood.

Observations. This small *Pholiota* is usually found where a fire has occurred and usually grows attached to pieces of charred wood. Its small size and general coloration make it similar in appearance to the wood-inhabiting species of *Galerina*, for example, *G. autumnalis*. The best way to distinguish these is on the basis of the spores; in *G. autumnalis* and its relatives the spore surface is roughened and the spores lack an apical germ pore; in *P. highlandensis* the spores are smooth and have an apical germ pore. This species has also been called *Flammula carbonaria* (Fr.) Kummer. It is not the same as *P. carbonaria* Smith.

A case of poisoning by *P. highlandensis*, involving a young girl, was reported in the upper peninsula of Michigan. The child was discovered eating the mushrooms; she had apparently eaten about 20 caps. After eating them, she became sleepy and was difficult to keep awake. The parents took the child to the hospital at about 6 o'clock; by 7 o'clock she complained of a stomachache, perhaps due to the emetic she had been given. At 8 o'clock the doctor pumped her stomach, and the child was retained in the hospital overnight for observation. She developed a severe case of dysentery, and she remained in the hospital for 10 days. Soon after she was released she lost 4.5 kg, so she was returned to the hospital for an additional 2 weeks. She eventually recovered.

The specimens identified as *P. highlandensis* were gathered by the child's father, who collected them from where the child had been found eating the mushrooms. They grew in a lawn that was built in a jack pine plains, where brush piles had been burned. The specimens were sent to Dr. A. H. Smith, University of Michigan, who identified them as *P. highlandensis*. The collection is no longer available, but the notes and illustrations of the collection verify the identification. The child may possibly have eaten mushrooms other than or in addition to *P. highlandensis*, for example, a wood-inhabiting *Galerina* such as *G. autumnalis*. As we have no evidence of this, we have added *P. highlandensis* to the list of poisonous species. Certainly the poisoning was severe enough to deserve consideration in this publication.

Distribution and seasonal occurrence. *P. highlandensis* is fairly common and widely distributed and certainly occurs across Canada and throughout the United States. It has been collected in the late spring, summer, fall, and winter, depending on season and location.

Habitat and habit. This species is scattered to gregarious or caespitose. It is found on pieces of burned or charred wood or on burned soil.

Technical description. Cap 20-40(-60) mm broad, convex becoming somewhat depressed, at times with a low umbo; margin even, bearing veil remnants, otherwise glabrous; surface viscid; color fulvous or reddish cinnamon brown to reddish brown on the disc; margin usually paler, with colors in varying shades of russet, tawny, or pecan brown, or darker brown with age, hygrophanous, fading to ocherous buff. Flesh rather thin, yellow to similar in color to the surface of the cap; odor indistinctive; taste slightly disagreeable to indiscernible. Gills pallid to pale yellowish when young, becoming snuff brown or cinnamon brown, adnate or rounded-adnate, broad, close, with edges even or eroded.

Stalk (10-)30-40 mm long, (2-)3-6 mm thick; apex whitish to yellowish in early stages, becoming dingy yellowish; lower part pallid, becoming dark brown;

lower stalk always darker than above, with pallid yellow to cinnamon buff or pale buff zones or patches of veil tissue that often leave an evanescent fibrillose annular zone, more or less becoming glabrous.

Spores cinnamon brown in deposit, 6.6-8.4 × 3.7-5.8 μm, elliptic and inequilateral in side view, ovate in face view, smooth, thick-walled, with a small apical germ pore, light cinnamon brown, nonamyloid. Basidia four-spored, clavate, 18.3-21.9 × 5.8-6.6 μm, hyaline to yellow, thin-walled. Pleurocystidia fusiform-ventricose, with a long tapered neck, and an obtuse apex, 58-73 × 8.0-14.6 μm, 7-8.0 μm wide at the neck, thin-walled to slightly thick-walled; contents ocherous, sometimes encrusted with an amorphorous crystalline deposit around the neck, arising in the central portion of the gill trama. Cheilocystidia fusiform-ventricose, smaller than pleurocystidia, 30-40(-50) × 7.3-11.7(-15.0) μm, hyaline to yellowish; walls thin to slightly thickened and roughened near the apex.

Gill trama interwoven to parallel; hyphae cylindrical to inflated, 3.7-14.6 μm wide, thin-walled, hyaline to yellowish. Subhymenium filamentous, interwoven; hyphae 3.3-3.7 μm wide, hyaline, gelatinous. Cap cuticle filamentous, interwoven; hyphae more or less radially arranged, narrow in the epicutis, cylindrical, 2.2-4.4 μm wide, hyaline to yellowish, gelatinous, thin- to slightly thick-walled and finely encrusted; subcutis closely interwoven; hyphae cylindrical to inflated, 4.0-13.1 μm wide, thin-walled to slightly thickened, hyaline to yellowish and encrusted. Cap trama loosely to compactly interwoven; hyphae cylindrical to inflated, 4.6-14.3(-19.0) μm wide, hyaline to yellowish, thin-walled. Caulocystidia cylindrical-clavate to fusiform, 23.4-82.9 × 5.8-8.2 μm, with recurved hyphal ends, thin-walled; contents granular, hyaline to yellow; stalk hyphae parallel in the core to interwoven on the surface, more or less cylindrical, 3.5-11.7 μm wide, hyaline to yellowish; walls thin to slightly thickened and encrusted. Oleiferous hyphae greenish yellow, refractive, contorted, not septate, 3.5-7.0(-9.5) μm wide, scattered in the stalk. Clamp connections present.

Useful references. *P. highlandensis* is not discussed or described in any of the common field guides. See the technical work of Smith and Hesler (1968) for further details.

Pholiota squarrosa (Fr.) Kummer

Figs. 154, 219

Common name. Scaly pholiota.

Distinguishing characteristics. Cap conspicuously scaly, dry; scales pale tawny to tawny or light to dark yellow brown, with pale yellow between the scales to pale greenish yellow along the margin. Gills pale yellowish when young, becoming sordid greenish yellow with age, close to crowded. Stalk covered with conspicuous pale tawny to pale yellowish brown scales up to the more or less persistent annular zone. Fruit bodies usually occurring in large clusters, on wood of coniferous and deciduous trees. Spores smooth with an apical germ pore, ellipsoidal to somewhat bean-shaped, 6.6-8.0 × 3.7-4.4 μm in size.

Observations. *P. squarrosa* is commonly reported as edible but is apparently poisonous to some individuals. Shaffer (1965) reported three people poisoned by *P. squarrosa.* All three had consumed alcohol with the mushroom, which suggests a possible mushroom-alcohol reaction. The symptoms included vomiting and diarrhea for all three victims. One victim suffered prolonged vomiting and diarrhea and required hospitalization. The second person went into shock, and the third, who had eaten a smaller quantity, recovered after sleeping all day. The specimens of *P. squarrosa* that caused the poisoning had a garlic-like odor. According to Smith (1975), this variant occurs east of the Mississippi. For other adverse alcohol-mushroom reactions see Chapter 9 and the descriptions of *Coprinus atramentarius* and *Clitocybe clavipes*, as well as Chapter 13, the description of *Morchella angusticeps*.

P. squarrosa, in general appearance, looks like *P. squarrosoides* (Peck) Sacc., which has coarse, tawny to dull rusty brown scales on the cap with a whitish color between the scales, whitish gills with no greenish tints, and a scaly stalk with the scales resembling those of the cap. A thin, sticky layer, present beneath the scales of the cap, is most apparent at maturity or when the cap surface is moistened.

Distribution and seasonal occurrence. *P. squarrosa* is found across Canada and the northern areas of the United States but is most common in the west and generally uncommon in the east. It fruits in the fall.

Habitat and habit. *P. squarrosa* is found on both conifers and hardwoods, on logs and stumps, at the base of both dead and living trees, and at times on living trees. It is often found on aspen (*Populus*) and spruce (*Picea*).

Technical description. Cap 30-120 mm broad, obtuse to convex when young, becoming bell-shaped with a conic umbo and finally convex or nearly plane; margin incurved, conspicuously fringed with veil remnants in early stages; surface dry, broken into numerous innate recurved or squarrose scales, sometimes becoming denuded along the edge leaving a glabrous outer margin; scales cinnamon buff to clay color, becoming dull tawny or light to dark yellowish brown at maturity, with pale yellow between the scales or becoming pale greenish yellow with age along the margin. Flesh moderately thick, pliant, pale yellow; taste mild to

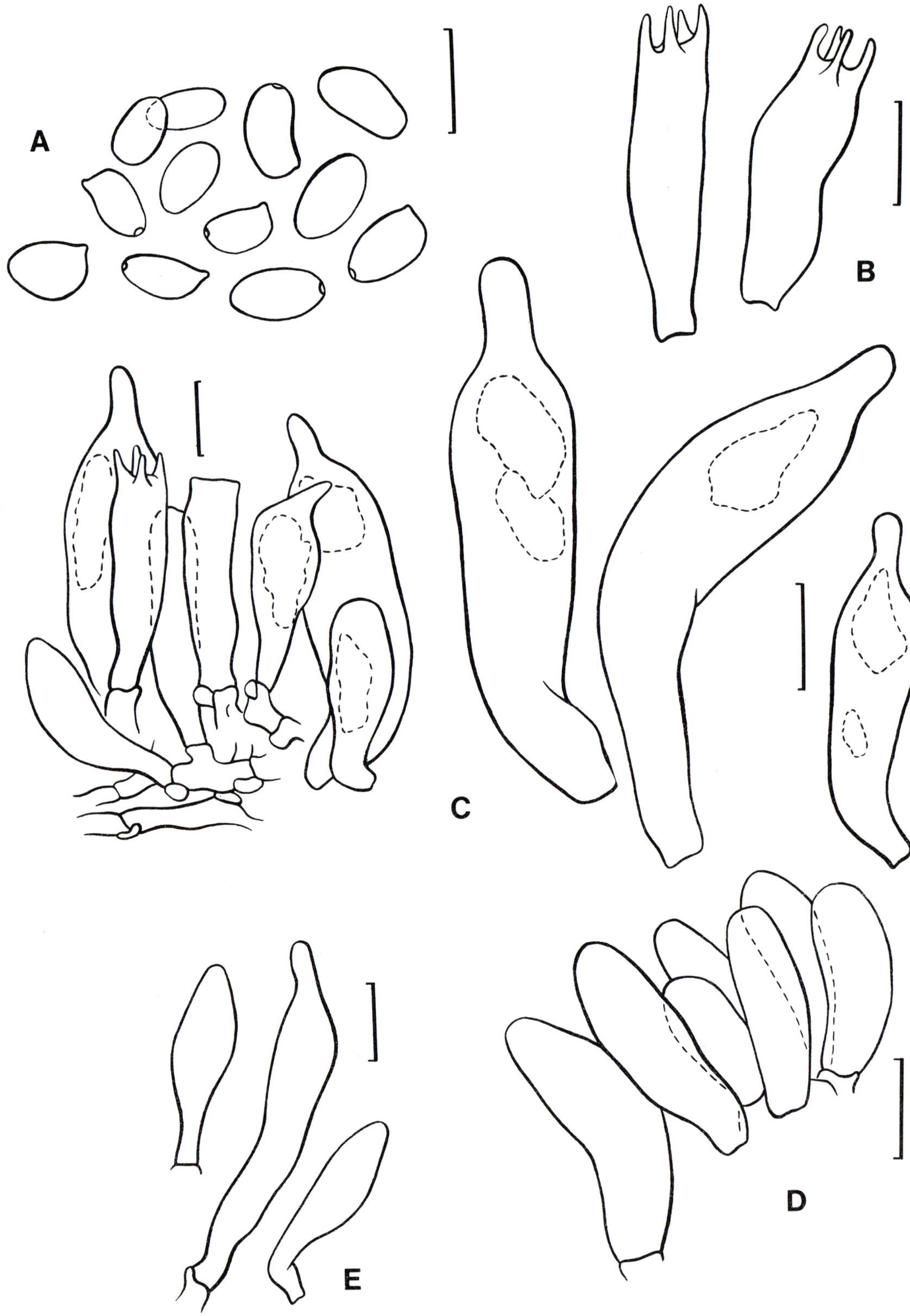

Fig. 154. *Pholiota squarrosa*: A, spores; B, basidia; C, pleurocystidia; D, cheilocystidia; E, caulocystidia.

Scale line 10 μm

slightly rancid; odor indiscernible or like garlic or onion.

Gills pale yellowish when young, soon becoming more or less dark grayish yellow to sordid greenish yellow and finally sordid rusty brown, bluntly adnate to somewhat arcuate with a decurrent line or tooth, close to crowded, narrow; edges even.

Stalk 40-120 mm long, 4-15 mm thick, equal or nearly so, at times tapered at the base, central, solid, dry; surface covered with recurved pale tawny to pale yellowish brown scales up to the annulus or the annular zone. Flesh yellowish. Annulus sometimes membranous and persistent or evanescent. Universal veil lacking.

Spores dull rusty brown in deposit, 6.6-8.0 × 3.7-4.4 μm, elliptic to somewhat reniform or inequilateral in side view, ovate in face view, smooth; walls moderately thick, with an apical germ pore, light cinnamon brown, nonamyloid. Basidia four-spored, clavate, 16-25 × 5.0-7.0 μm, thin-walled, hyaline. Pleurocystidia in the form of chrysocystidia, 26-46.8 × 9.5-11.7(-16.8) μm, clavate to clavate-mucronate or fusiform-ventricose; apex tapered to obtuse or with one or more papillae; contents crystalline, ocherous to yellow, refractive, homogeneous or as a single amorphous body. Cheilocystidia 19-23.4 × 6.6-7.3(-15.0) μm, subfusiform, fusiform-ventricose, clavate, or clavate-mucronate, sometimes with refractive contents, thin-walled, hyaline to yellowish.

Gill trama parallel; hyphae cylindrical to inflated, 3.7-13.1 μm wide, hyaline to yellowish. Subhymenium filamentous; hyphae interwoven, gelatinizing, cylindrical, 3.7-4.4 μm wide, hyaline to yellowish. Cap cuticle filamentous, interwoven; hyphae parallel and radially arranged in fibrils, more or less cylindrical, 5.8-9.4(-14.3) μm wide, thin- to thick-walled, yellowish to brown, encrusted; erect cystidia-like hyphal ends present, 46.8-50.0 × 12.9-15.5 μm, clavate to fusiform-ventricose, with brownish walls. Cap trama loosely interwoven; hyphae cylindrical to inflated, 3.7-19.1 μm wide, thin-walled to slightly thickened, hyaline to yellowish. Caulocystidia clavate or clavate-mucronate to fusiform-ventricose, 38.5-66 × 8.8-13.1 μm, hyaline to yellowish, some with refractive ocherous inclusions, thin-walled; stalk hyphae parallel, cylindrical to somewhat inflated, 3.5-11.7(-16.7) μm wide, thin-walled to slightly thick-walled and encrusted, hyaline to yellowish; end cells recurved, clavate, 5.8-9.4 μm wide. Annulus compactly interwoven; hyphae cylindrical to somewhat inflated, 4.4-7.3(-16.8) μm wide, with thin to slightly thickened and encrusted walls, hyaline to yellowish. Oleiferous hyphae refractive, greenish yellow, contorted, 3.7-6.6(-11.7) μm wide, scattered in the trama, cap cuticle, stalk, and annulus. Clamp connections present.

Useful references. The following books contain information on *P. squarrosa* pertinent to North America.

- *Edible and Poisonous Mushrooms of Canada* (Groves 1979)—A discussion under *P. squarrosoides* is given.
- *Mushrooms of North America* (Miller 1972)—A color photograph and a description and discussion (with an inaccurate gill description) are given.
- *The Mushroom Hunter's Field Guide* (Smith 1974)—A color photograph, some black-and-white photographs, and a brief discussion are given.

Polypores

Polyporaceae

The polypores, which include the conks and bracket fungi, form a large, diverse group of fungi represented in North America by several hundred species and, according to modern systems of classification, a large number of genera. Most polypores and their relatives cause decay of dead wood; however, some are parasitic and attack living trees and other woody plants. Some species are terrestrial, some growing from buried wood and certain others from tree roots. Some polypores and their relatives, such as *Albatrellus* and *Boletopsis*, are bolete-like (see description of *Boletus* earlier in this chapter, alphabetically listed), but they can be distinguished by their firm fruit bodies that do not decay readily and the cream to white spore deposit.

Most of the polypores produce spores in tubes that develop on the undersurface of the cap. A few species have a gill-like spore-producing surface, and more rarely some have tubes that become spine-like with age. Some species have stalked fruit bodies, but often the cap is attached laterally to the substratum, the characteristic from which the common name bracket fungus is derived. The fruit bodies of some species are almost fleshy, but normally they are tough to rubbery, fibrous, leathery, or woody.

The genera and species of polypores are recognized in part by their macroscopic features but they often require a microscopic examination for positive identification. A discussion of the microscopic characteristics is beyond the scope of this treatment. Technical papers by Overholts (1953), Bondartsev (1953), Lowe (1957), and others should be consulted for thorough treatments of both macroscopic and microscopic features. Smith and Smith (1973) provide a key and descriptions to some of the almost fleshy species that are most likely to be collected for food.

Most polypores are inedible because of their tough to woody texture. A few of the almost fleshy species such as *Laetiporus (Polyporus) sulphureus* and *Polypilus*

(Polyporus) frondosus (Fr.) Karst. are sometimes collected for food. Although *L. sulphureus* has often been eaten, especially when young and fresh, it has caused several cases of gastrointestional upset (Orr 1976). Lincoff and Mitchel (1977) list *Bondarzewia (Polyporus) berkeleyi* (Fr.) Singer, *Albatrellus (Polyporus) cristatus* (Pers. ex Fr.) Pouzar, *Grifola (Polyporus) giganteus* (Pers. ex Fr.) Pilát, *Phaeolus (Polyporus) schweinitzii* (Fr.) Pat., and *Laetiporus (Polyporus) sulphureus* as reported or suspected of being poisonous. This list is based mainly on papers by West et al. (1974) and Lee et al. (1975).

Below are described *Laetiporus (Polyporus) sulphureus*, which appears to cause the greatest number of gastrointestinal upsets, and *Albatrellus dispansus*, which, although it has been reported as causing gastrointestinal upset (O. K. Miller, Personal commun.), is not commonly reported in North America (Canfield and Gilbertson 1971).

Albatrellus dispansus (Lloyd) Canfield & Gilbertson

Fig. 155

Common names. None.

Distinguishing characteristics. Fruit bodies firm, brittle, with a short stalk breaking cleanly when fresh, up to 200 mm broad, composed of many petal-like caps densely packed together and sometimes fused at the edges; color yellowish to yellowish buff when fresh. Pores occurring at a density of two or three per millimetre, angular, white. Spores white in deposit, ellipsoidal to subglobose, smooth, 3.7–5.1 × 2.9–3.7 μm in size. Fungi terrestrial, occurring in litter under conifers in western North America.

Observations. Because species of *Albatrellus* are somewhat difficult to identify, consult technical descriptions or someone with a knowledge of this group for verification of identification. The large size and somewhat fleshy nature of *Albatrellus* species and the fact that several are common, particularly in the west, make this genus an attractive edible mushroom. According to Smith (1975), *A. ovinus* (Schaeff. ex Fr.) Kotlaba & Pouzar is eaten in North America, but he notes that there have been recent reports of poisoning by this species in Europe. The edibility of *A. confluens* (Alb. & Schw. ex Fr.) Kotlaba & Pouzar is not known. Smith (1975) provides color photographs and brief descriptions of these two species.

Distribution and seasonal occurrence. *A. dispansus* has been reported from western North America (Arizona to Idaho) and was originally described from Japan. It fruits in late summer and fall.

Habitat and habit. *A. dispansus* is terrestrial. It is found in litter under conifers (*Pinus, Pseudotsuga*) in North America.

Technical description. Fruit bodies annual, up to 200 mm in diameter, with a very short attachment, developing from a broad shallowly rooted base up to 70 mm thick, ultimately branching into a large number of petal-like caps. Caps densely overlapping, confluent, 10–40 mm broad; upper surface finely tomentose or scurfy, becoming coarsely wrinkled with age and from drying, without zones; color yellow to yellow buff when fresh.

Pore surface white when fresh, drying faintly vinaceous; pores angular, with thin entire walls, and occurring at a density of two or three per millimetre.

Flesh white when fresh, firm, brittle, easily sectioned, breaking cleanly with no fibrous ends when fresh but very hard and difficult to section after drying, tapering from less than 1 mm thick at the edge of the cap margin to 10 mm thick near the base of the branches.

Spores white in deposit, 3.7–5.1 × 2.9–3.7 μm, broadly ellipsoidal to subglobose, slightly inequilateral in side view, guttulate, thin-walled, hyaline, nonamyloid. Basidia four-spored, clavate, 19.0–22.0 × 5.8–6.6 μm wide, hyaline. Cystidia lacking.

Entire fruit body basically with a single type of hyphae. Tube trama interwoven; hyphae parallel, 2.2–2.9 μm wide at the edge, 3.5–4.7 μm wide behind the margin, regularly cylindrical, thin-walled. Cap trama interwoven; hyphae cylindrical or zigzagged to irregularly inflated, 4.7–23.4(–82) μm wide; walls thin to slightly thickened, with crystalline contents, turning cherry to blood red in potassium hydroxide. Hyphae at cap margin parallel, radially arranged, 4.6–5.9 μm wide, thin-walled. Cap cuticle interwoven; hyphae 2.9–3.8 μm wide; hyphal ends erect and projecting to form a felt-like trichoderm, refractive, thin-walled, greenish yellow in potassium hydroxide. Clamp connections absent.

Useful references. There are no books or field guides pertinent to North America that contain information on this species.

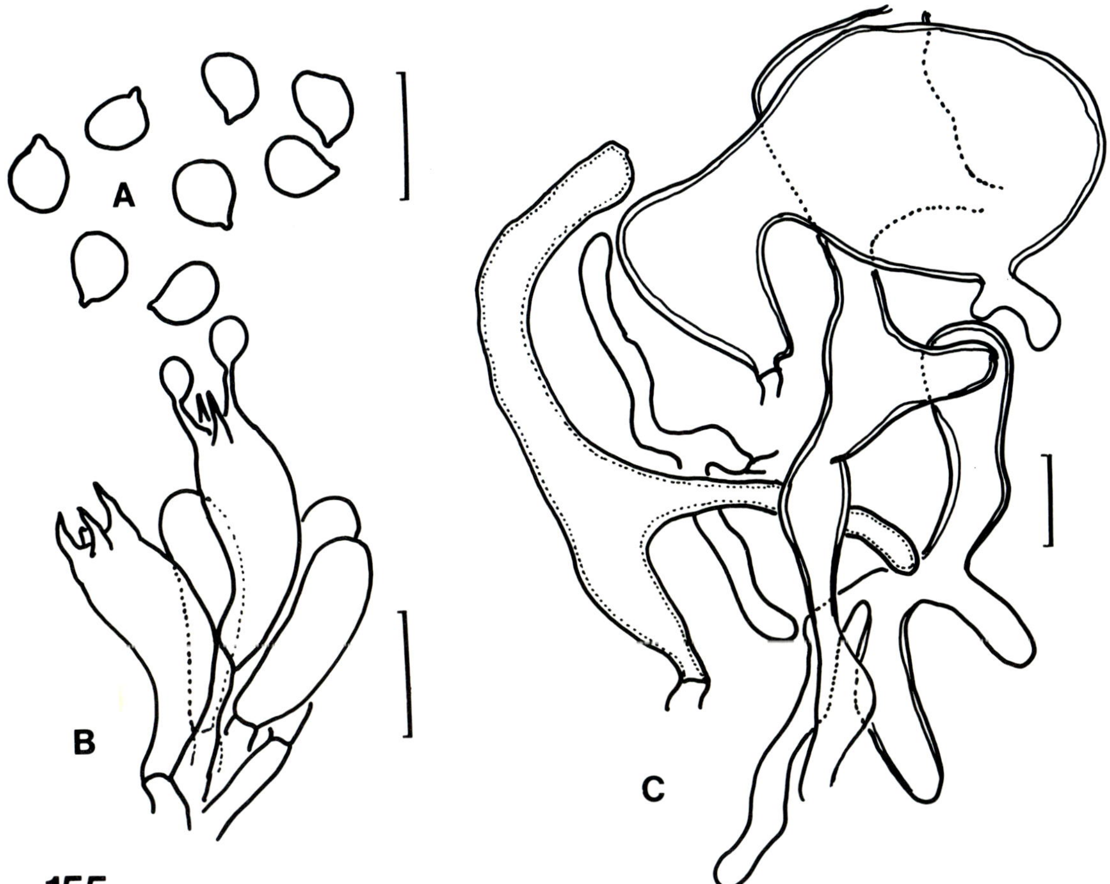

155

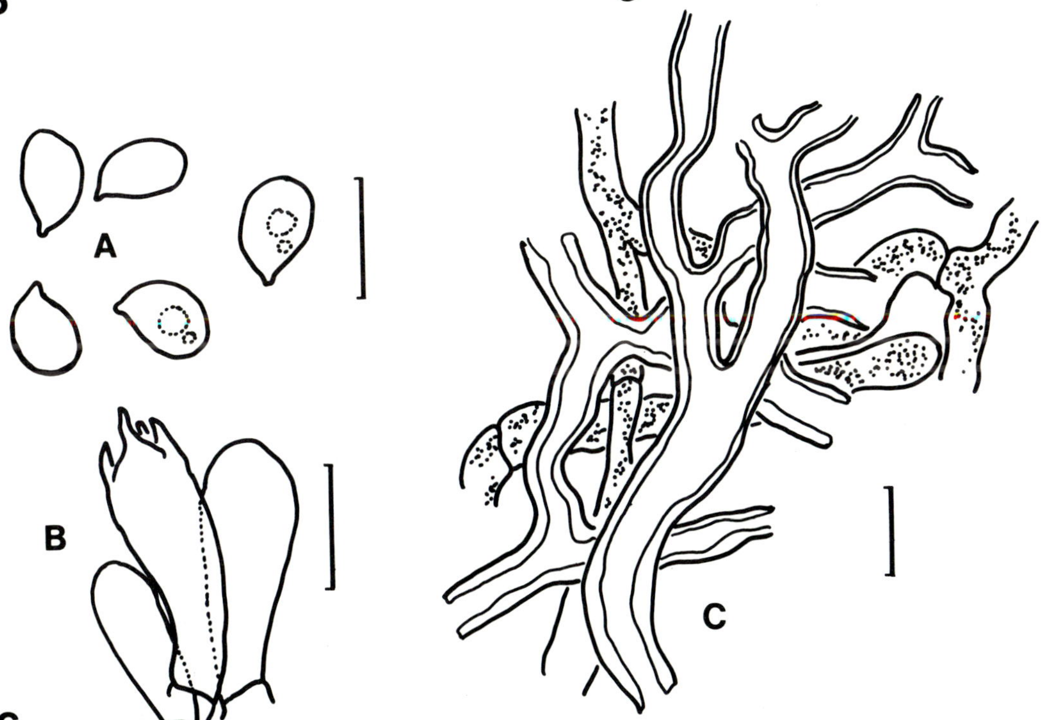

156

Fig. 155. *Albatrellus dispansus*: A, spores; B, basidia; C, tramal elements.

Fig. 156. *Laetiporus sulphureus*: A, spores; B, basidia; C, tramal elements in the tube walls.

Scale line 10 μm

Laetiporus (Polyporus) sulphureus (Bull. ex Fr.) Murrill

Figs. 156, 220

Common names. Sulfur-shelf, chicken-of-the-woods.

Distinguishing characteristics. Fruit bodies annual, typically occurring in large rosettes or clusters; individual caps 100–400(–600) mm broad, usually flat to shelf-like; color usually yellow orange to pale yellow, at times with reddish colors on the margin, or generally more pinkish orange over all. Flesh soft to cheesy and fleshy to watery when moist and young, more fibrous to rubbery with age. Pore surface sulfur yellow to pale yellow, white in one form; pores small and circular. Stalk usually lacking to slight. Spores white to pale yellow in deposit, ellipsoidal to ovoid, smooth, 4.6–7.3 × 3.5–4.4 μm in size. Fungi occurring on stumps, logs, and dead trees of both conifers and hardwoods.

Observations. *L. sulphureus* is a beautiful and easily recognized species. A form with white pores is sometimes called var. *semialbinus* Peck; typically the pore surface is yellow. The odor is often strongly musky to strongly fungoid and the taste is at times acidulous.

Distribution and seasonal occurrence. This species is widely distributed in North America. It fruits most commonly in the late summer and fall but has also been reported as early as June.

Habitat and habit. This species occurs on wood of conifers and hardwoods, on stumps, logs, and dead trees, and in wounds of living trees. The mycelia are perennial, producing fruit bodies each year. In some cases, hundreds of fruit bodies may grow on one log or tree.

Technical description. Cap annual, 100–400(–600) mm broad, horizontally expanded, flat, sometimes irregular in shape, sessile or gradually narrowed at the base, at times more or less with a very short attachment, usually occurring in large rosette-like or overlapping clusters; margin thick to thin, wavy, zigzagged, slightly enrolled, or upturned; surface uneven, coarsely wrinkled with the lines spreading from a common center, nodulose, matted-tomentose; color yellow orange to pale yellow or pale muddy ocher, with the margin paler yellow or reddish when young, sometimes more pinkish orange. Flesh watery, thick, fibrous when dry, soft to cheesy, fleshy to succulent, watery when moist, creamy yellow to apricot colored, straw yellow, or almost white when dry. Odor intensive, strongly musky to fungoid; taste acidulous, especially with age.

Tubes sulfur yellow, rarely white, avellaneous to pale straw color after drying, short, 1.5–3.0 mm long; pores small, circular, occurring at a density of one to four per millimetre, and with regular edges becoming somewhat toothed with age.

Spores white to pale yellow in deposit, 4.6–7.3 × 3.5–4.4 μm, broadly ellipsoidal or ovoid, hyaline, guttulate, nonamyloid; walls thin to slightly thickened. Basidia four-spored, clavate, 15.5–19.1 × 6.6–8.2 μm. Cystidia absent.

Entire fruit body having two types of hyphae, generative and binding hyphae. Tube trama interwoven; hyphae more or less parallel; generative hyphae 3.5–4.6(–6) μm wide, thin-walled; binding hyphae 3.5–4.6(–6) μm wide, hyaline, refractive, thick-walled, with walls 0.7–1.5 μm wide. Subhymenium filamentous, interwoven; hyphae 3.5–4.6 μm wide, hyaline, thin-walled. Cap cuticle filamentous, interwoven; hyphae matted, appressed, cylindrical, 3.5–7.0 μm wide, branched, thin-walled. Cap trama interwoven; generative hyphae 4.4–7.3 μm wide, hyaline, thin-walled; binding hyphae 3.7–11.7(–20) μm wide, thick-walled, refractive; cap margin composed of cylindrical thin-walled hyphae 3.5–6.6 μm wide with parallel hyphal tips. Clamp connections absent.

Useful references. The following books contain information on *L. sulphureus* pertinent to North America.

- *A Field Guide to Western Mushrooms* (Smith 1975)—A color photograph and a description and discussion of var. *semialbinus* are presented.
- *Edible and Poisonous Mushrooms of Canada* (Groves 1979)—A color photograph and a description and discussion are given.
- *Mushrooms of North America* (Miller 1972)—A color photograph and a description and discussion are given.
- *The Mushroom Hunter's Field Guide* (Smith 1974)—A black-and-white photograph and a description and discussion are given.
- *The Savory Wild Mushroom* (McKenny 1971)—A black-and-white photograph, a color photograph, and a description and discussion are given.

Ramaria Holmskjold ex S. F. Gray

Clavariaceae

Ramaria is a member of the family Clavariaceae, commonly called the coral fungi. The coral fungi in modern taxonomic treatments are divided into several genera, whereas in older literature often the single genus *Clavaria* was used for most of the species. The commonly encountered modern genera include *Clavaria, Clavariadelphus, Clavulina, Clavicorona, Ramariopsis,* and *Ramaria. Sparassis,* commonly called the cauliflower fungus, is also in this group. Although many species have not been tested for edibility, several are known to be edible and others poisonous. Those species reported as poisonous are in the genus *Ramaria*, which also contains some of the best edible coral fungi.

The genus *Ramaria* is characterized by often large, always profusely branched fruit bodies; a cream color to yellow, golden yellow, grayish yellow, yellow brown, or brown spore deposit; nonamyloid spores in Melzer's reagent, usually ornamented with warts or ridges best seen in preparations stained with Cotton Blue; and the dull olive to blue or dull to bright green reaction of the upper branches of the fruit bodies when a drop of iron sulfate is applied to their surface.

The gastrointestinal upset caused by eating certain species of *Ramaria* typically results in diarrhea. The toxin or toxins responsible for the poisonings have not been identified. Exactly which species are poisonous has not yet been clearly determined. Almost all handbooks on coral fungi, except for that by McKenny (1971), list *R. formosa* as poisonous.

Dr. Currie Marr, however, a specialist on *Ramaria*, questions the toxicity of *R. formosa*. He states that *R. formosa* is rarely accurately identified in the popular mushroom guides. He finds that the descriptions are usually too general for accurate identification. The main discrepancy concerns the taste of *R. formosa*, often described as strongly acrid or bitter. Marr (Personal commun.) believes that *R. formosa* has an indistinctive taste, and he states that specimens exhibiting a bitter taste are in fact quite possibly members of the *R. pallida* complex. This complex occurs more frequently than *R. formosa*, at least in the Pacific northwest. *R. formosa* and *R. pallida* are described below. The description of *R. formosa* given below, however, does not agree entirely with those given in most of the popular literature on mushrooms. Rather, it agrees with the modern, more correct concept of Marr and Stuntz (1973) and Petersen (1976). Quite possibly this species as now described may prove to be not at all poisonous. At present, though, it should not be eaten.

McKenny (1971) also does not list *R. formosa* as poisonous. Although she does describe its taste as being bitter or astringent, she suggests that species with a bitter or astringent taste and a tough and stringy consistency are generally edible, whereas those with a semitransparent flesh of gelatinous to rubbery consistency, such as the *R. gelatinosa* complex, are generally poisonous. The *R. gelatinosa* complex is characterized by medium- to large-sized fruit bodies branched from a large base; a white to light yellow or light orange base, with pinkish orange, orange buff, or yellowish orange branches that become darker pinkish cinnamon, grayish orange, or more brownish with age; and a gelatinous consistency caused by pockets of gelatinous material, marbled throughout the flesh, visible upon cutting through the center of the fruit body. Because many corals have a similar appearance externally, she advises that edible and poisonous species can easily be distinguished by cutting each specimen through the base and examining the interior carefully before they are cooked. Two varieties of *R. gelatinosa*, var. *gelatinosa* and var. *oregonensis*, are described below. Descriptions of other species with a gelatinous consistency can be found in the technical work of Marr and Stuntz (1973).

A description of *R. flavobrunnescens* is also included here, along with the other four, because it is widely distributed in North America and is reported to have caused the death of livestock.

Ramaria flavobrunnescens (Atk.) Corner

Fig. 157

Common names. None.

Distinguishing characteristics. Fruit bodies 50–70 mm high, 40–60 mm wide; stalk single or with up to three converging at one point, branching five or six times above the base; branch tips fine and rounded or with small knobs; color uniformly yellow except for white at the base, staining brown where bruised. Flesh hard and brittle when dry, probably somewhat cartilaginous when fresh. Spores yellow in deposit, cylindrical to ellipsoidal, 8–10 × 3–4.5 μm in size, with the surface roughened with warts.

Observations. There are several varieties of *R. flavobrunnescens* (Marr and Stuntz 1973). It is a widely distributed species, occurring in hardwood and coniferous forests.

R. flavobrunnescens has not yet been implicated in human poisonings. Fidalgo and Fidalgo (1970) reported it as the cause of several deaths in livestock (it had been eaten raw).

Distribution and seasonal occurrence. *R. flavobrunnescens* seems widely distributed in North America; however, there are several varieties, and the distribution of each has not been determined. *R. flavobrunnescens* var. *aromatica* Marr & Stuntz was described from the Pacific northwest. The type variety was described from the southeastern United States. The species seems to fruit in the late summer and fall, depending on location, weather, and variety involved.

Habitat and habit. This species is terrestrial, occurring under hardwoods and conifers.

Technical description. Fruit bodies 50–70 mm high, 40–60 mm wide, hard and brittle when dry, probably somewhat cartilaginous when fresh; stalk single and 5–10 mm thick, or consisting of three stalks bundled together at one point, branching five or six times above; lower branches divided at the apex into several more branches that are channeled, 7–9 mm in diameter, and up to 15 mm long; upper branches mostly dividing at the apex into only two branches that are 1-3 mm in diameter and have fine rounded to polynodulose apices; color uniformly yellow except for some white occurring at the extreme base, turning brown and becoming water-soaked when bruised. Flesh of dried specimens slightly acid in taste; taste of fresh specimens not recorded by Atkinson (1909).

Spores yellow in deposit, pale yellow under the microscope, 8-10 × 3-4.5 μm, cylindrical to ellipsoidal; ornamentation consisting of small to moderately large discrete lobed warts sometimes arranged linearly. Basidia clavate, 46-58 × 8-8.5 μm, mostly four-spored.

Surface hyphae of the stalk 2-3 μm wide, thin-walled, cylindrical. Flesh hyphae interwoven in the stalk and parallel in the branches; hyphae mostly thin-walled, 2.5-10 μm wide, basically cylindrical. Clamp connections present. Gloeoplerous hyphae not observed.

Useful references. There are no books or field guides pertinent to North America that contain information on this species. See the technical papers by Marr and Stuntz (1973) and Petersen (1976).

Ramaria formosa (Pers. per Fr.) Quél.

Figs. 158, 221

Common names. Beautiful clavaria, pink coral fungus, yellow-tipped coral.

Distinguishing characteristics. Fruit bodies medium to large, 70-200 mm high, 70-140 mm wide, branching up to six times from the base; terminal branches finely divided, with acute to rounded apices; color of underground portion white to brownish white; branches peach or salmon pink to light coral red with light yellow tips when young; tips similar in color to branches with age; flesh staining brown when handled or bruised. Flesh more or less spongy at the base, fleshy-fibrous above; dried specimens brittle and easily crumbled to powder when rolled between the thumb and forefinger. Odor and taste indistinctive. Spores golden yellow in deposit, ornamented with warts on the surface, ellipsoidal, 9-12 × 4.5-6 μm in size.

Observations. *R. formosa* apparently has been frequently interpreted incorrectly in North America. Most authors report it as having a bitter, acid taste, but according to Marr and Stuntz (1973) the taste is indistinctive. According to these authors, the mistaken notion that any large, fruiting *Ramaria* with pinkish branches and yellowish tips is *R. formosa* would in western Washington result in incorrect identification in most cases. Species that fit this general description and have a bitter, acid taste are probably members of the *R. pallida* complex. It is likely that errors in identification of species of *Ramaria* have occurred throughout North America, making it impossible to determine which ones are poisonous. It may be that *R. formosa* as described in this publication is not a poisonous species. *R. acrisiccescens* Marr & Stuntz is similar to *R. formosa*, but it has a bitter, acid taste and lacks clamp connections.

Some of the characteristics that help to identify *R. formosa* are generally massive, salmon pink to coral pink fruit bodies with yellow tips when young, bruising and becoming dark when handled; coarsely warted spores, about 10.4 × 5.4 μm in size; and the presence of clamp connections. There are other characteristics that are helpful in identification, but they require the use of special stains and chemicals that are not readily available. For further information, see technical publications by Marr and Stuntz (1973) and Petersen (1976).

Distribution and seasonal occurrence. Because of the confusion surrounding the concept of *R. formosa*, its distribution in Canada or the United States is difficult to determine. Nevertheless, it is probably widely distributed. It has been collected in eastern and western North America and it occurs also in Europe. Apparently it fruits in the fall in most areas, but it may occur into the early winter on the west coast, where location and weather are suitable.

Habitat and habit. *R. formosa* is terrestrial, occurring in hardwood or mixed forests; it is known to grow under conifers in the west.

Technical description. Fruit bodies 70-200 mm high, 70-140 mm wide, medium or large, generally broad; base single and massive, or consisting of two to six axes connected basally, covered by a thin white tomentum; small whitish abortive or primordial branches frequently present, branching up to six times from the base, dividing at the apices into two to many other branches that are sometimes brush-like; axils subacute to U-shaped; branches slightly divergent, with the diameter and internodal length decreasing upward; terminal branches less than 5 mm wide, finely divided or with many lobes near the apices; apices acute to rounded; color of underground portion white or brownish white; branches peach or light red; apices distinctly light yellow when young, similar in color to branches with age, becoming brown when handled, turning grayish orange to cinnamon brown, drying grayish yellow or orange. Flesh of branches similar in color or more intensely colored than the surface; flesh at the base yellowish white and often watery; consistency fleshy, fibrous, more or less spongy at the base, drying brittle to chalky and easily powdered; taste and odor indistinctive.

Spores golden yellow in deposit, 9-12 × 4.5-6 μm, ellipsoidal; surface ornamented with large irregularly lobed warts. Basidia 40-82 × 8-13 μm, with one to four sterigmata; contents granular, cyanophilous. Flesh hyphae interwoven in the stalk, parallel in the

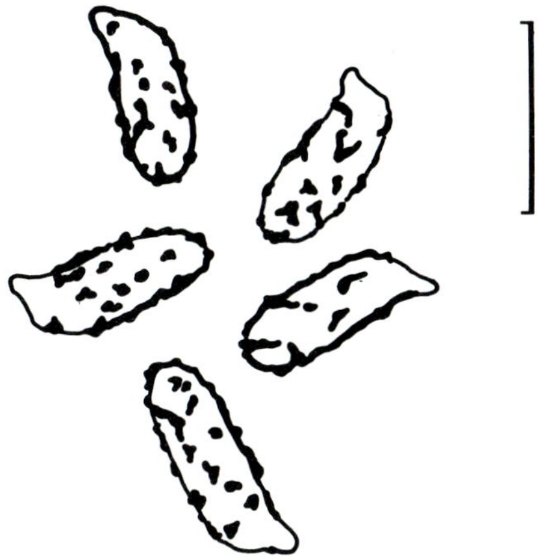

157

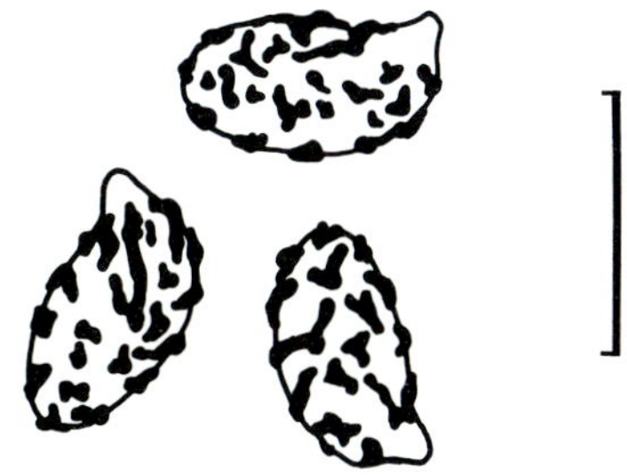

158

159

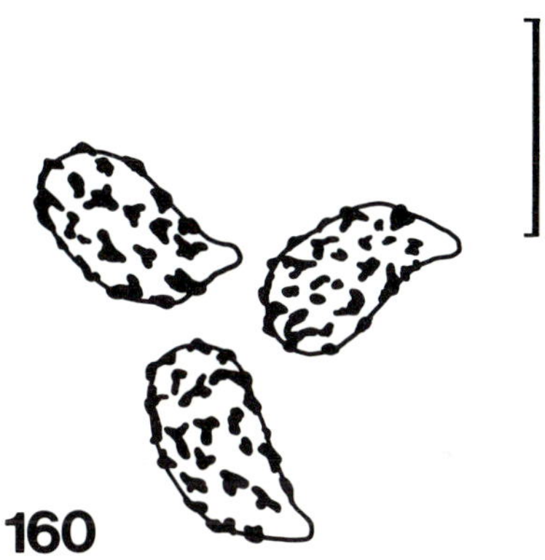

160

161

Fig. 157. *Ramaria flavobrunnescens*: spores.

Fig. 158. *Ramaria formosa*: spores.

Fig. 159. *Ramaria gelatinosa* var. *gelatinosa*: spores.

Fig. 160. *Ramaria gelatinosa* var. *oregonensis*: spores.

Fig. 161. *Ramaria pallida*: spores.

Scale line 10 μm

branches, noninflated or slightly so, 4–13 μm wide, thin-walled. Gloeoplerous hyphae present but not abundant.

Useful references. The following books contain information on *R. formosa* pertinent to North America. In some publications the description can be found under *Clavaria formosa* Pers. per Fr. Because of confusion over the concept of *R. formosa*, the references listed below probably illustrate and describe other species. It is usually impossible to make an accurate identification of species from the descriptions given.

- *Mushrooms of North America* (Miller 1972)—A color photograph and a description and discussion are given, although the species is described as having a bitter taste.
- *The Mushroom Hunter's Field Guide* (Smith 1974)—A black-and-white photograph and a description and discussion are given under *Clavaria formosa*, although the raw flesh is described as mild but not staining.
- *The Savory Wild Mushroom* (McKenny 1971)—A black-and-white photograph and a description and discussion are given.

Ramaria gelatinosa (Coker) Corner var. *gelatinosa*

Fig. 159

Common name. Gelatinous coral.

Distinguishing characteristics. Fruit body medium-sized; stalk broad and compound, composed of fused branched axes, branched four to six times; color of branches pale creamy white to buff pink becoming brownish with age, creamy white at the base. Fresh flesh gelatinous, with gelatinous lines mottling the flesh in cross section; dried flesh with mottling resembling dried agar. Spores grayish yellow in deposit, coarsely warted, 8.5–10 × 4.5–6 μm in size.

Observations. The distribution of *R. gelatinosa* var. *gelatinosa* is unclear. It occurs in eastern North America. The var. *oregonensis* occurs on the west coast. Several species have a gelatinous consistency. They all must be identified and the distribution of each determined before the general distribution of the *gelatinosa* complex is known. Avoid all species of *Ramaria* with a gelatinous consistency because they are known to be poisonous, at least to some individuals.

Distribution and seasonal occurrence. The var. *gelatinosa* occurs in eastern North America, but its general distribution is unclear. It may be widely distributed. Apparently it fruits in the late summer and fall.

Habitat and habit. This species is terrestrial and is found in hardwood forests and perhaps in mixed conifer forests. See also var. *oregonensis*.

Technical description. Fruit body medium-sized, 50–60 mm high and 20–70 mm wide when dried; base 10–20 mm high and 10–35 mm wide when dried, composed of fused axes creamy white in color; branches arising at about the same level from the base, dividing at the apex into two or more branches, producing four to six nodes above the base; lower branches often flattened and connate; middle and upper branches ascending parallel; upper angles almost acute to U-shaped and divergent in the lower nodes, minutely divided near the apices, with apices nearly acute; color pale creamy white when young, maturing buff pink, becoming fleshy brown with age; apices similar in color or lighter. Flesh drying hard; branches brittle; flesh of dried specimens appearing like dried agar; fresh flesh with a gelatinous consistency.

Spores grayish yellow in deposit, 8.5–10 × 4.5–6 μm, broadly ovoid to broadly cylindrical with a prominent eccentric apiculus, ornamented with conspicuous lobed cyanophilous warts. Basidia clavate, 48 × 10 μm, four-spored; contents strongly cyanophilous and granular.

Trama hyphae underlying the subhymenium parallel, interwoven in the center of the branches, noninflated, 3–7 μm wide, with gelatinizing walls; hyphae in the base more completely gelatinous, forming an amorphous agar-like matrix; gloeoplerous hyphae 3–4(–13) μm wide; contents strongly cyanophilous. Clamp connections present at the base of the basidia and at the hyphae septa.

Useful references. The following books contain information on *R. gelatinosa* var. *gelatinosa* pertinent to North America.

- *The Mushroom Hunter's Field Guide* (Smith 1974)—An excellent black-and-white photograph and a description and discussion on the western form are presented.
- *The Savory Wild Mushroom* (McKenny 1971)—A color photograph, a black-and-white photograph, and a description and discussion on the western form are presented.

Ramaria gelatinosa var. *oregonensis* Marr & Stuntz

Figs. 160, 222, 223

Common name. Oregon gelatinous coral.

Distinguishing characteristics. Fruit body medium-sized, broad, 80-150 mm high, 50-120 mm wide. Base compound, 40-70 mm long, 30-80 mm wide, composed of fused axes; color white, light yellow, or light orange, with any one of these colors predominating. Branches arising from about the same level in the compound base, branching up to seven times from the base; apices narrow-rounded; color light orange in early stages, darker with age becoming more grayish to brownish, at times with a violet gray cast, similar in color at the apices or paler. Consistency gelatinous; base resembling stiff agar when fresh, drying hard and brittle. Taste indistinctive. Spores golden yellow in deposit, broadly ovoid to broadly cylindrical, coarsely warted, 7-10 × 4.5-6 µm in size.

Observations. *R. gelatinosa* var. *oregonensis* is extremely gelatinous in consistency. The main difference between var. *gelatinosa* and var. *oregonensis* is the color. The var. *oregonensis* has orange to orange brown fruit bodies that become grayish violet with age. The var. *gelatinosa* is creamy white when young, becoming buff pink and finally fleshy brown with age.

Distribution and seasonal occurrence. The var. *oregonensis* is known from the Pacific northwest. It fruits in the late summer and fall.

Habitat and habit. This species is terrestrial and is found under conifers such as western hemlock (*Tsuga heterophylla*).

Technical description. Fruit body usually medium-sized, broad, 80-150 mm high, 50-120 mm wide. Base compound, broad, 40-70 mm long, 30-80 mm wide, consisting of a wrinkled gelatinous mass of fused axes, with the component parts delimited by the convoluted surface in longitudinal section; surface covered with a thin white tomentum; color white, light yellow, and light orange, with any one of these colors predominating. Branches characteristically numerous, arising at about the same level from the compound basal mass, branching up to seven times from the base; nodes commonly but not always dividing into two at the apex; upper angles nearly acute to U-shaped; secondary branches nearly parallel, sometimes slightly flattened, split into two in the middle or finely divided near the apices, with narrowly rounded apices; color light orange, developing darker shades with age such as grayish orange, Pompeian yellow, or agate brown, sometimes with a definite violet gray cast, nearly the same color or distinctly paler at the apices. Flesh translucent, a shade lighter than grayish orange; consistency gelatinous, resembling stiff agar at the base when fresh, drying hard and brittle; odor musty sweet; taste indistinctive.

Spores golden yellow in deposit, 7-10 × 4.5-6 µm, broadly ovoid to broadly cylindrical with a prominent lateral apiculus, coarsely ornamented with lobed cyanophilous warts. Basidia 43-65 × 5.5-11 µm, clavate, two- or four-spored, often soon collapsing after spore release, containing cyanophilous granules.

Hyphae of the stalk trama almost completely gelatinized making structure practically indistinguishable, although crystalloid clusters are prominent in the gelatinous mass. Hyphae of the branches less gelatinized, parallel to subparallel; cells noninflated or slightly so, 4-11 µm wide, thin-walled. Gloeoplerous hyphae rare, 3-3.5 µm wide. Clamp connections present in the trama and hymenium.

Useful references. See the technical paper by Marr and Stuntz (1973); also see references under *R. gelatinosa* var. *gelatinosa*, which treat the western forms.

Ramaria pallida (Schaeff. per Schulz.) Ricken

Fig. 161

Common names. None.

Distinguishing characteristics. Fruit body up to 140 mm high, 80 mm broad, profusely branched. Stalk color off-white to pale ocherous tan, staining brownish to watery brown on handling, although at times hardly staining. Branches numerous; color pallid tan with a tint of flesh color; tips often light dull violaceous or faded to the same color of the branches. Flesh off-white, slowly turning brownish on cutting or exposure. Taste faintly to mildly sour or bitter, but quickly disappearing. Spores more or less ocherous in deposit, broadly ovoid, 10.7-13.7 × (5.2-)5.6-6.3(-6.7) µm in size; surface ornamented with delicate longitudinal ridges and low warts.

Observations. The *R. pallida* complex is widely distributed in North America and Europe, although the distributions of individual species are still imperfectly known. This complex can be distinguished from the *R. formosa* complex by the cream to brown fruit bodies (some forms are tinted pink or lavender, but the tips of the branches are not yellow when young); the decidedly bitter taste; the absence of clamp connections; and the presence of basidia that are not cyanophilous.

Some of the other species in this group are *R. spinulosa* (Fr.) Quél. and *R. subspinulosa* (Coker) Corner. *R. acrisiccescens* Marr & Stuntz is probably the same as *R. pallida* (C. Marr, Personal commun.), as is *R. mairei* Donk. For further information on this group, see the technical works of Petersen (1974) and Marr and Stuntz (1973). Marr and Stuntz (1973) referred to this group as the *R. subspinulosa* complex.

Distribution and seasonal occurrence. *R. pallida* occurs in Europe and North America. It fruits in late summer and fall.

Habitat and habit. *R. pallida* is terrestrial; it grows under conifers, in mixed woods and in hardwood forests.

Technical description. Fruit body up to 140 mm high, up to 80 mm broad, profusely branched, obovoid when mature. Stalk large when young compared with the branches but not expanding at the same rate and thus often comparatively small at maturity, solid, off-white to pale ocherous tan, smooth usually but occasionally almost tomentose below the substrate, quickly turning brown to watery brown on handling, but occasionally hardly changing. Major branches numerous, similar in color to the apex of the stalk, pallid tan with a tint of flesh color, terete; length between nodes long at maturity, giving an open lax appearance; upper angles rounded to narrowly rounded, often with a rusty ocherous coloration from the spores; apices minute when young, elongate and delicate when mature, light dull violaceous when young, often retaining this color through maturity or fading to the same color as the branches. Flesh off-white, slowly browning on cutting and exposure, punky or somewhat chalky when dry; taste faintly to mildly sour or bitter, but quickly disappearing; odor faintly aromatic, perhaps resembling fenugreek with age.

Spores more or less ocherous in deposit, 10.7–13.7 × (5.2–)5.6–6.3(–6.7) μm, broadly ovoid, weakly cyanophilous, with an eccentric apiculus; ornamentation consisting of delicate longitudinal ridges and scattered low warts that are randomly placed to convergent and strongly cyanophilous. Basidia 75 × 12.5 μm, clavate, hyaline, four-spored.

Hyphae of stalk trama 3.7–18.5 μm wide, densely interwoven, hardly inflated, hyaline. Hyphae of stalk surface densely packed, similar to those of stalk trama, but at the surface 1.5–3 μm wide and more or less perpendicular to the surface. Hyphae of upper branches parallel, often somewhat inflated to barrel-shaped, 2–10 μm wide, hyaline. Gloeoplerous hyphae not observed. Clamp connections absent.

Useful references. There are no books or field guides pertinent to North America that contain information on this species. See the description by Petersen (1974). More detailed descriptions for North American material are needed.

Rhodophylloid mushrooms

Rhodophyllaceae (= Entolomataceae)

The rhodophylloid mushrooms vary in size, shape, and color but have in common attached gills and a pinkish spore deposit, specifically fleshy brown or pinkish brown in color. Also, all these mushrooms lack a partial and a universal veil. Under the microscope the spores of most rhodophylloid mushrooms are usually more or less angular in outline. The cap cuticle is usually composed of filamentous hyphae in the form of a cutis or trichoderm. This group of mushrooms has been divided by some workers into several genera, including *Claudopus, Clitopilus, Pouzarella* (=*Pouzaromyces*), *Alboleptonia, Entoloma, Nolanea, Leptonia,* and *Eccilia.* Others have used or proposed the use of a single genus for the entire group. Three genera that have been used in this way are *Acurtis, Entoloma,* and *Rhodophyllus;* only *Entoloma* and *Rhodophyllus* have been widely used. The more fleshy of the rhodophylloid mushrooms are likely to be confused with species of *Tricholoma. Tricholoma* can be distinguished from the rhodophylloids, however, by its pure white spore deposit. The genus *Pluteus* may also be confused with *Entoloma* or related genera. *Pluteus* has free gills, a pink spore deposit, and smooth spores that are not angular in outline and it usually fruits on wood.

Some rhodophylloid mushrooms are edible, for example *Clitopilus prunulus* (Berk. & Curt.) Sacc. and *Entoloma abortivum* (Berk. & Curt.) Donk (=*Rhodophyllus abortivus* (Berk. & Curt.) Singer and *Clitopilus abortivus* (Berk. & Curt.) Sacc.), whereas others are reported as poisonous. The edibility of many other species is unknown. Some of these are probably edible but several will no doubt prove to be poisonous. The best policy is to avoid eating all pink-spored mushrooms that have attached gills. Never eat one of these unless you are absolutely sure of the identification of your specimens and know the species is edible. The probability of poisoning is great and the consequences can be severe.

Poisoning by rhodophylloid mushrooms produces a variety of symptoms. Headache, vomiting, and diarrhea usually begin 30 minutes to 2 hours after ingestion of the mushrooms and may continue for up to 48 hours. Mild liver damage has occasionally been reported (Tyler 1963). The victim usually does not fully recover from the poisoning for several days. Alder (1953, 1964) provides more information on symptoms.

Entoloma lividum (=*Entoloma sinuatum* (Bull. ex Fr.) Kummer) is the species of rhodophylloid mushrooms most commonly reported in Europe as the cause of human poisoning (Pilát 1951, Alder 1960). According to Benedict (1972) it has caused fatalities in both adults and children. Poisonings by *Entoloma nidoro-*

sum (Alder 1953), *Entoloma rhodopolium* (Alder 1961, 1963, 1966), *Entoloma aprile* (Britzelmayr) Sacc. (Alder 1966), and *Entoloma verna* (=*Nolanea verna*) (Ayer 1974) have been reported. *Entoloma albidum* was reported to have poisoned five people (Murrill 1921). Besides these species Lincoff and Mitchel (1977) list *Entoloma mammosum* (Fr.) Hesler, *Nolanea pascua* (L. ex Fr.) Kummer, *Entoloma salmoneum* (Peck) Sacc., and *Entoloma strictius* (Peck) Sacc. as poisonous.

Only four of the above species of rhodophylloid mushrooms, those that are commonly encountered or proven dangerous, are described below. It is best, however, to avoid the entire group. When you are collecting edible species of *Tricholoma* for food, note especially the spore deposit color, because the dangerous rhodophylloid mushrooms can resemble the edible *Tricholoma* species in size, stature, and coloration.

Entoloma lividum (Bull. ex Mérat) Quél.

Figs. 162, 224

Common names. Leaden entoloma, livid entoloma.

Distinguishing characteristics. Cap 80–150 mm broad, broadly bell-shaped to convex or plano-convex, with the center becoming depressed or sometimes raised; surface usually smooth, tacky giving a soapy feeling when moist, silky when dry; color at times whitish yellow to very pale ashy brown or more commonly dingy brown or brownish gray. Taste and odor sweet to resembling fresh meal. Gills white to whitish, becoming pinkish or somewhat grayish at maturity, broad, subdistant to crowded, attached, sinuate-rounded to adnexed, decurrent by a tooth. Stalk 40–120 mm long, 10–40 mm thick, robust, tending to be enlarged at the base; surface silky to fibrillose-scaly; color white to somewhat grayish. Spores pinkish in deposit, angular in outline with five to eight sides, 8.8–11.0 × 5.5–8.0 μm in size.

Observations. *E. lividum* has also been called *E. sinuatum*, and the names *lividum* and *sinuatum* have both been used in the genus *Rhodophyllus* as *R. lividus* (Bull. ex Mérat) Quél. and *R. sinuatus* (Bull. ex Fr.) Singer. The large size and more or less robust stature of *E. lividum* give it the general appearance of a *Tricholoma*. However, *Entoloma* species have a pinkish to pinkish brown spore deposit, whereas *Tricholoma* species have a white spore deposit. Also, under the microscope the spores of *Entoloma* are angular in outline, whereas those of *Tricholoma* have an even outline. In the Pacific northwest, *Clitocybe* (*Lyophyllum*) *multiceps* Peck, a species closely related to *Lyophyllum decastes* (Fr. ex Fr.) Singer, is apparently often mistaken for *Entoloma lividum* (McKenny 1971). *C. multiceps* typically grows in clusters of several mushrooms and has a white spore deposit; also, the outline of the spores under the microscope is even, rather than angular as in *Entoloma*. Small groups or solitary fruit bodies of *C. multiceps* could easily be confused with *Entoloma*.

E. albidum Murrill, a species similar to *E. lividum*, is described here for comparison with *E. lividum*. Little else is known about *E. albidum*. Its cap is 30–60 mm broad, convex to plane or slightly depressed, not umbonate, hygrophanous, glabrous, shining, with the margin entire or slightly lobed, opaque when dry, minutely striate when moist, white becoming grayish when moist, and tinged light grayish pinkish brown with age or on drying. The flesh is white, with a taste slightly like fresh meal to bitter and with a mild odor. The gills are sinuate, close to nearly crowded, rather narrow, white becoming pinkish orange or salmon color, with the edges nearly even. The stalk is 5–9 cm long, 6–10 mm thick, equal or slightly tapered downward, central, more or less curved at the base, stuffed to hollow, dry, white, and fibrillose except at the apex where it is squamulose. The flesh of the stalk is white. The partial and universal veils are absent. The spores are more or less light cinnamon pink in deposit, 8.5–11 × 7–8 μm, globose to subglobose, ovoid or ellipsoidal, angular and five- or six-sided or sometimes obscurely five- to seven-sided, lacking an apical germ pore, and hyaline. Pleurocystidia are lacking. The cheilocystidia are 30–48 × 3–7 μm, cylindrical, clavate or sometimes nearly capitate, thin-walled and often collapsed, and hyaline. The cap cuticle is composed of repent filamentous hyphae, with a few hyphal ends more or less upturned; the terminal cells are clavate and hyaline.

Distribution and seasonal occurrence. *E. lividum* is widely distributed in North America. It appears to be most common in the fall but is reported also in the summer.

Habitat and habit. This species is scattered to somewhat gregarious and at times is fairly common. Apparently it occurs in various habitats and has been reported in both hardwood and conifer forests.

Technical description. Cap 80–150 mm broad, broadly bell-shaped to convex or plano-convex, becoming depressed or sometimes umbonate, striate, splitting with age; margin enrolled in early stages, becoming upturned; surface smooth, glabrous to silky or fibrillose on the umbo, tacky or almost viscid to dry; color whitish yellow or very pale ashy brown to dingy brown or brownish gray. Flesh spongy to fibrous, more or less fragile; taste and odor sweet to resembling that of fresh meal.

Gills white, becoming pinkish or somewhat grayish at maturity, strongly sinuate-rounded to somewhat adnexed, decurrent by a tooth, broad, subdistant to crowded.

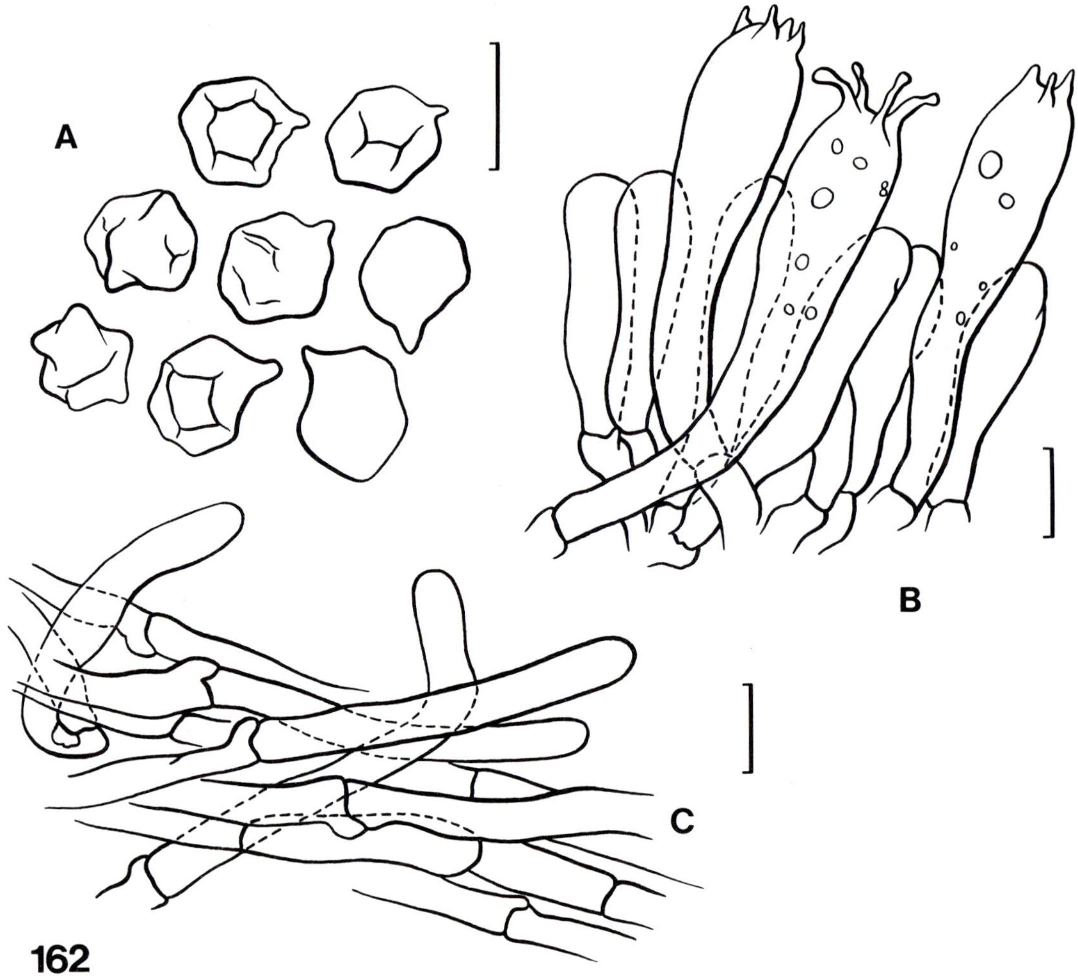

162

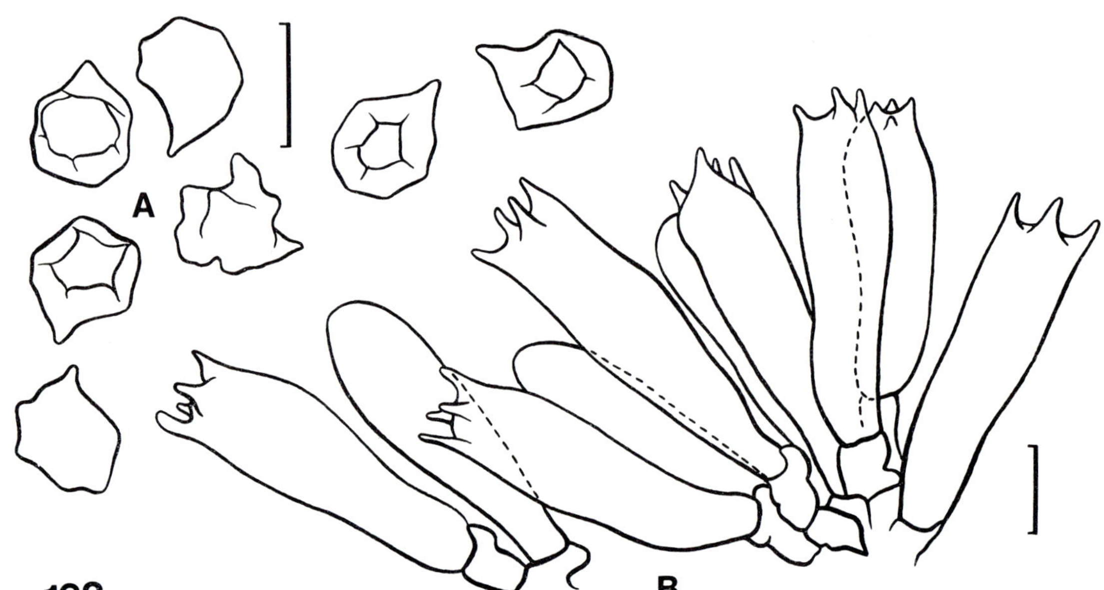

163

Fig. 162. *Entoloma lividum*: A, spores; B, basidia; C, cuticle.

Fig. 163. *Entoloma nidorosum*: A, spores; B, basidia.

Scale line 10 μm

Stalk 40-120 mm long, 10-40 mm thick, robust, sometimes curved, equal to more or less enlarged at the base, solid, fibrous; surface dry, silky to fibrillose-scaly, white to tinted grayish. Flesh white. Partial veil and universal veil lacking.

Spores pinkish in deposit, 8.8-11.0 × 6.6-8.0 μm, subglobose, angular, five- to eight-sided, wrinkled when dry, hyaline, nonamyloid. Basidia four-spored, narrowly long and clavate, 46-58.5 × 9.4-11.7 μm, hyaline; contents refractive. Pleurocystidia absent. Cheilocystidia not well-differentiated, but the gill edges are lined with more or less clavate cells.

Gill trama interwoven to parallel; hyphae cylindrical, 3.5-5.8 μm wide, hyaline. Subhymenium filamentous, compactly interwoven; hyphae cylindrical, 2.9-4.6 μm wide, hyaline. Cap cuticle filamentous; hyphae interwoven, radially arranged, parallel in fibrils, somewhat gelatinous, cylindrical, 3.5-8.2 μm wide, hyaline. Cap trama compactly interwoven; hyphae more or less cylindrical to inflated, 3.5-11.7 μm wide, hyaline to yellowish; hyphae of subcutis compactly interwoven, 14.3-23.4 μm wide, cylindrical to inflated; contents yellowish brown and granular. Caulocystidia cylindrical to clavate, occurring as recurved hyphal ends, 16.7-46.8 × 3.5-7.0 μm; stalk hyphae parallel to interwoven, more or less cylindrical, 3.5-9.4(-19.1) μm wide, hyaline to yellowish; walls thin to slightly thickened; hyphae of basal tomentum interwoven, cylindrical, 2.9-4.5 μm wide, hyaline to yellowish, thin-walled to slightly thick-walled. Oleiferous hyphae refractive, greenish yellow, 3.5-7.0(-9.4) μm wide, scattered in the trama and stalk. Clamp connections present.

Useful references. The following books contain information on *E. lividum* pertinent to North America.

- *Mushrooms in their Natural Habitats* (Smith 1949)—A color photograph on a reel and a description and discussion as *Rhodophyllus lividus* are presented.
- *Mushrooms of North America* (Miller 1972)—A color photograph and a description and discussion are given.
- *The Savory Wild Mushroom* (McKenny 1971)—A black-and-white photograph and a description and discussion are given.

Entoloma nidorosum (Fr.) Quél.

Fig. 163

Common name. Foul-smelling entoloma.

Distinguishing characteristics. Cap 20-50 mm broad, convex; surface finely tomentose-silky; color grayish brown when fresh and moist, fading with age or on drying. Gills subdistant, broad, attached (adnexed); color pale flesh-toned or pinkish. Stalk 40-70 mm long, 4-8 mm thick, more or less equal; surface slightly fibrillose; color whitish. Spores pinkish in deposit, angular in outline, six- to eight-sided, 8.8-10.2 × 6.6-7.3 μm in size.

Observations. *E. nidorosum* is difficult to identify because it is similar in size and coloration to several other species of *Entoloma*. Consult an expert to be sure of identification of species. The strongly acid to alkaline odor may be helpful in identifying this species. The species name, *nidorosum*, means foul-smelling.

Distribution and seasonal occurrence. This species has been collected in the Great Lakes region and in southeastern Canada. Its general distribution seems to be poorly known. According to our records, it fruits in late summer to early fall.

Habitat and habit. This species is terrestrial and is found in mixed woods.

Technical description. Cap 20-50 mm broad, convex, obtuse; margin incurved; surface hygrophanous, minutely tomentose-silky; color grayish brown when moist. Flesh thin, fragile, white; taste not recorded; odor strongly nitrous.

Gills pale flesh color, adnexed, broad, subdistant, more or less wavy.

Stalk 40-70 mm long, 4-8 mm thick, equal or nearly equal, stuffed, soon becoming hollow, pruinose at the apex, slightly fibrillose, whitish.

Spores pinkish to flesh-colored in deposit, 8.8-10.2 × 6.6-7.3 μm, ellipsoidal to ovoid in outline, angular, six- to eight-sided, thin-walled, hyaline, nonamyloid. Basidia four-spored, clavate, 31.6-37.3 × 9.4-10.5 μm, hyaline. Pleurocystidia and cheilocystidia absent.

Gill trama interwoven, more or less parallel; hyphae cylindrical to inflated, 5.8-21.9 μm wide, hyaline. Subhymenium filamentous, interwoven; hyphae more or less cylindrical, 3.5-5.1 μm wide, hyaline. Cap cuticle filamentous, interwoven, more or less radially arranged; hyphae appressed, cylindrical to inflated, 3.5-15.0(-23.4) μm wide, hyaline to brownish, thin-walled to slightly encrusted; contents pale yellow brown, granular. Cap trama loosely interwoven; hyphae cylindrical to inflated, 3.5-29.2 μm wide, hyaline. Caulocystidia clavate, 23.4-46.0 × 7.3-11.7 μm, hyaline; stalk hyphae parallel to interwoven at the surface, cylindrical to inflated, 3.5-18.0 (-23.4) μm wide, hyaline; hyphae of basal tomentum interwoven, cylindrical, 3.5-9.4 μm wide, hyaline. Oleiferous hyphae not observed. Clamp connections present.

Useful references. *E. nidorosum* is described by Kauffman (1971). It is not referred to in any of the recent general floras or mushroom books published in North America.

Entoloma rhodopolium (Fr.) Kummer

Fig. 164

Common name. Rosy entoloma.

Distinguishing characteristics. Cap 40-80 mm broad, bell-shaped expanding to plane; surface silky when dry, lubricous when fresh and moist; color umber brown to smoky brown, fading upon drying to pale livid gray. Odor and taste mild. Gills attached, adnate becoming emarginate, somewhat subdistant; color whitish in early stages, becoming pinkish or deep rose. Stalk 40-100 mm long, 6-12 mm thick, more or less equal; color white. Spores deep rose or pinkish in deposit, angular in outline, six- or seven-sided, 8-11 × 6.6-8(-8.8) μm in size.

Observations. *E. rhodopolium* apparently occurs across southern Canada, wherever a suitable habit is found. Because its general stature and coloration are similar to several other *Entoloma* species, correct identification requires study by an expert on this group. Any specimen of *Entoloma* resembling this species should be considered dangerous.

Distribution and seasonal occurrence. This species has been collected both in southeastern and southwestern Canada; however, its actual distribution is poorly known. Our records indicate that it fruits in late summer and fall; Groves (1979) reports it from July to October.

Habitat and habit. This species can be found solitary, scattered, or in groups, and at times in clusters of two or three (Groves 1979). It is terrestrial, growing in mixed woods and in hardwood forests.

Technical description. Cap 40-80 mm broad, bell-shaped expanding to plane; margin even, wavy; surface hygrophanous, dry, and silky-shining glabrous, lubricous but not viscid; color umber brown or smoky brown when moist, fading to pale livid gray. Flesh watery white; odor and taste mild.

Gills whitish to deep rose when mature, adnate, becoming emarginate, somewhat subdistant, sometimes veined, moderately broad; edges minutely eroded.

Stalk 40-100 mm long, 6-12 mm wide, nearly equal, tapering up or down, sometimes curved, stuffed becoming hollow, glabrous, fibrous to almost cartilaginous, readily splitting longitudinally; apex floccose or scurfy; color white; base tomentose.

Spores deep rose in deposit, 8.0-11.0 × 6.6-8.0(-8.8) μm, subglobose in general outline, angular, six- or seven-sided, sometimes wrinkled on the outer layer, thin-walled, hyaline, nonamyloid. Basidia four-spored, short-clavate, 25-35.1 × 9.5-11.0 μm, hyaline. Pleurocystidia and cheilocystidia not differentiated.

Gill trama interwoven to subparallel; hyphae cylindrical to inflated, 3.7-10.5(-30.0) μm, hyaline. Subhymenium filamentous to almost cellular, interwoven; hyphae more or less cylindrical, 5.8-7.3 μm wide, hyaline. Cap cuticle filamentous, interwoven to radially arranged; hyphae cylindrical to inflated, 3.5-19.5 μm wide, thin-walled, hyaline or with pale brownish contents. Cap trama interwoven, compact; hyphae cylindrical to inflated, 3.7-25.6 μm wide, hyaline. Caulocystidia clavate with recurved hyphal ends, 25-46.8 × 8.0-12.9 μm, hyaline, single to clustered at the apex; stalk hyphae parallel to compactly interwoven, cylindrical to somewhat inflated, 5.1-11.7(-16.7) μm wide, hyaline to yellowish, thin-walled to slightly thick-walled; hyphae of basal tomentum interwoven, cylindrical, 3.5-8.2 μm wide, hyaline to yellowish. Oleiferous hyphae refractive, yellowish green, 3.5-10.6 μm wide, scattered in the trama and cap cuticle. Clamp connections present.

Useful references. The following book contains information on *E. rhodopolium* pertinent to North America.

- *Edible and Poisonous Mushrooms of Canada* (Groves 1979)—A color photograph and a description and discussion are presented.

This species is also described and discussed by Kauffman (1971).

Nolanea verna (Lundell) Mazzer

Figs. 165, 225

Common names. Spring nolanea, spring entoloma.

Distinguishing characteristics. Cap 20-60 mm broad, conic to bell-shaped with a distinct small knob in the center, usually expanding somewhat with age; surface moist and glabrous, at times streaked with silvery gray; color variable, from dark gray brown to near chocolate brown, fading to an opaque shiny grayish tan on the margin, usually remaining darker in the center. Odor varying from indiscernible to faintly nitrous. Gills subdistant, attached, sinuate to deeply adnexed or nearly free; color grayish buff to pinkish brown. Stalk 30-110 mm long, 3-10 mm thick, more or less equal, usually longitudinally streaked from silvery fibrils; color brown to gray brown, with white mycelia on the base. Spores pinkish light brown in deposit, angular in outline, five- to seven-sided, 8-11.0 × 7.3-8.8 μm in size.

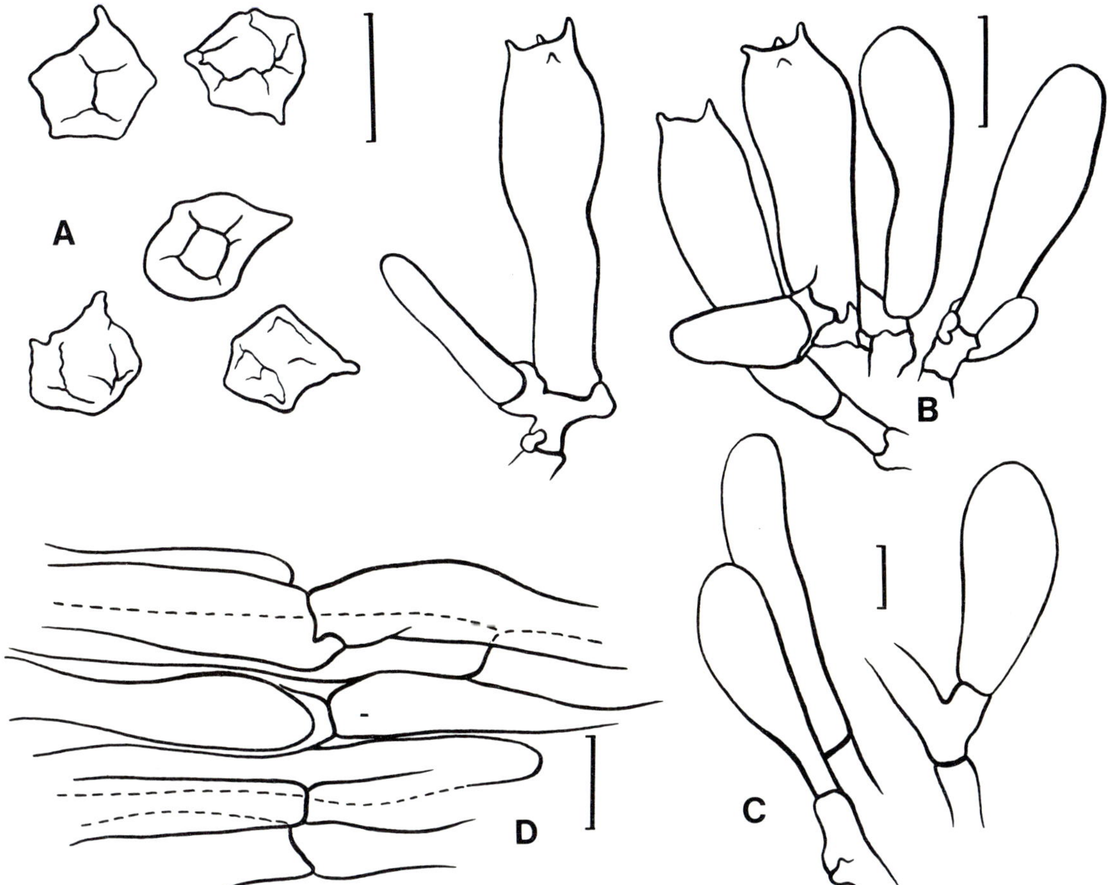

164

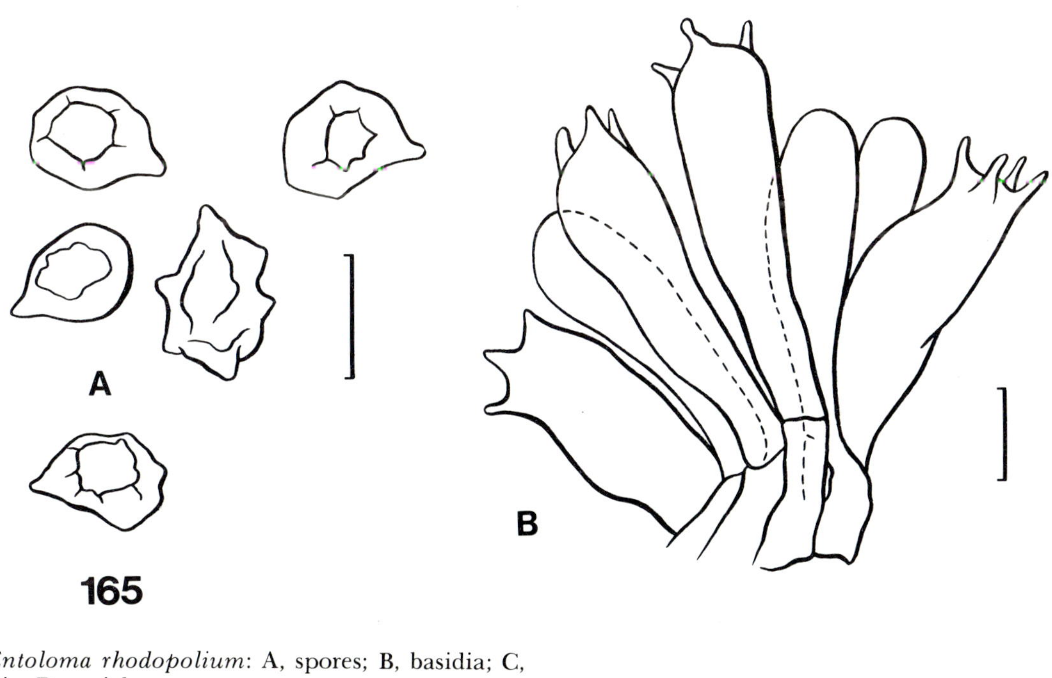

165

Fig. 164. *Entoloma rhodopolium*: A, spores; B, basidia; C, caulocystidia; D, cuticle.

Fig. 165. *Nolanea verna*: A, spores; B, basidia.

Scale line 10 μm

Observations. *N. verna* (=*Entoloma vernum* Lundell) occurs in Europe and in the Great Lakes region of North America (Mazzer 1977). It fruits during the spring months and is one of a few dark brown to grayish brown species of *Nolanea* that occurs in the Great Lakes region in the spring. *N. verna* has not been reported from elsewhere in North America; however, it and similar species probably fruit in spring or early summer in other areas of North America. For example, *N. sericea* (Bull. ex Mérat) Orton sometimes fruits during the spring in the Pacific northwest. All species of *Nolanea* should be considered dangerous. Because the gills of *N. verna* as well as other *Nolanea* species are at times nearly free, there is the possibility of mistaking *Nolanea* for the edible wood-inhabiting genus *Pluteus*. This possibility is further increased when *Nolanea* occurs on rotten wood. Because the spore deposits of both genera are pinkish, the best way to separate these genera is by the shape of the spores. In *Pluteus* the spores have an even, smooth outline, whereas the spores of *Nolanea* and most other rhodophylloid mushrooms are angular.

Distribution and seasonal occurrence. This species is known from the Great Lakes region and no doubt occurs elsewhere in eastern Canada. Similar or related species may be found in other regions. It fruits in spring, from mid-April through early June. It fruits during the morel season and is likely to be seen growing near them. It is one of a few mushrooms that fruits early; therefore, it is likely to be noticed and collected as a possible edible mushroom, possibly as a species of the edible genus *Pluteus*, which may also fruit early in the year.

Habitat and habit. This species is terrestrial, growing scattered to gregarious, at times occurring in large numbers. It is often found in pine (*Pinus*) plantations, oak-hickory (*Quercus-Carya*) forests, and mixed woods.

Technical description. Cap 20–60 mm broad, conic to bell-shaped in early stages, often with a distinct umbo in the center, expanding to umbonate or broadly convex but nearly always retaining the central umbo; margin incurved in early stages, often exceeding the gills, often lobed or uneven and broadly undulate; translucent-striate; surface hygrophanous, moist, and glabrous, except for the disc, which is faintly finely wrinkled to striate, at times streaked with innate silvery gray fibrils when young, fading with age; color varying from dark gray brown to near chocolate brown, fading to an opaque shiny grayish tan with a slightly darker disc. Flesh watery, similar in color to the surface of the cap when moist, fading to whitish tan when dry; odor indiscernible to faintly nitrous; taste indistinguishable to faintly like fresh meal.

Gills grayish buff to pinkish brown, staining brown, sinuate or deeply adnexed to nearly free, 3–14 mm broad, broadest near the stalk, subdistant.

Stalk 30–110 mm long, 3–10 mm thick, equal or slightly enlarged downward, in early stages longitudinally striate from innate silvery fibrils, scurfy near the apex, stuffed to hollow, central or nearly so, snuff brown or gray brown, with white mycelia at the base.

Spores vinaceous cinnamon in deposit, 8–11.0 × 7.3–8.8 μm, subglobose in outline, angular, five- to seven-sided, sometimes wrinkled when revived from old specimens, hyaline, nonamyloid. Basidia four-spored, clavate, 29.2–41.2 × 10.5–15 μm, hyaline. Pleurocystidia and cheilocystidia lacking.

Gill trama interwoven to more or less parallel; hyphae cylindrical to inflated, 3.5–20 μm wide, hyaline. Subhymenium filamentous to almost cellular, interwoven; hyphae more or less cylindrical, 4.7–7.0 μm wide, hyaline. Cap cuticle filamentous, interwoven; hyphae more or less radially arranged, erect in fibrils, more or less cylindrical, 3.5–10.5(–16.7) μm wide, yellowish brown, conspicuously encrusted. Cap trama loosely interwoven; hyphae cylindrical to inflated, 5.8–30 μm wide, hyaline to brownish. Caulocystidia clavate with recurved hyphal ends, 4.6–5.8 μm wide; stalk hyphae parallel to more or less interwoven near the surface, cylindrical to inflated, 3.5–16.7(–29.2) μm wide, hyaline to yellowish; hyphae of basal tomentum interwoven, cylindrical, 3.5–5.8 μm wide, hyaline, encrusted. Oleiferous hyphae greenish-refractive, not septate, contorted, 3.5–7.1 μm wide, scattered in the trama, stalk, and cap cuticle. Clamp connections rare at the base of the basidia, present on the hyphae at the base of the stalk.

Useful references. See the technical paper by Mazzer (1977). This species is not illustrated or described in the common mushroom floras or handbooks referred to in this publication.

Russula Pers. ex S. F. Gray — Russulaceae

Russula is a large and widely distributed genus of mushrooms. The species occur in forested areas throughout North America. Most species fruit during the summer. The fruit bodies are usually large to medium-sized, and the cap is often a brightly colored blue, purple, green, yellow, or red, a mixture of these colors, or less commonly some shade of brown, gray, or dull white. Therefore, they are a conspicuous part of the mushroom flora and attract people collecting mushrooms for food.

Many publications deal with the classification of *Russula* species. For North America the publications by Shaffer (1962, 1964, 1970, 1975) include several species and are excellent treatments. Bibliographies in his publications name other useful references.

The genus *Russula* is fairly easy to recognize. Besides the stature and color mentioned above, the genus may be distinguished by the short, thick stalk, in comparison with the size of the cap; the brittle texture of the fruit bodies, such that they are easily broken; the pale spore deposit, pure white to cream, pale yellow, or pale to bright ocher (ocherous orange) to orange; the typical absence of the partial and universal veils in North American species; the absence of latex; and the often white to whitish stalk and gills, contrasting with the colored cap surface. The spores of *Russula* have a roughened surface and are broadly ellipsoidal to globose. When the spores are mounted in Melzer's reagent, the ornamentation turns blackish to grayish and can be clearly seen. The brittle to fragile nature of the flesh is imparted by groups of inflated to spherical cells, called sphaerocysts, distributed among the more cylindrical hyphae. Special cystidia called pseudocystidia are present in the hymenium among the basidia. The taste of the fruit bodies is also helpful in identification and must be noted when the specimens are fresh. Some species have a mild taste; others are acrid or hot, bitter, or nauseous to fetid.

The genus *Lactarius* is the one most likely to be confused with *Russula*. The main feature separating them is the presence of latex in *Lactarius*, and its absence in *Russula*. Some *Lactarius* species, however, produce so little latex that even this character can be misleading. However, the presence or absence of sphaerocysts in the gill trama can also be used to distinguish these genera. Sphaerocysts are usually numerous in the gill trama of *Russula* whereas they are usually lacking in that of *Lactarius*.

Controversy exists over the edibility of *Russula*. Some specialists consider the entire genus safe to eat whereas others consider certain species to be poisonous. Generally, those species with an acrid to peppery taste are considered dangerous, especially when eaten raw, and should be avoided. Groves (1979) reports that the acrid *R. vesicatoria* Burlingham causes blistering of the lips and tongue when tasted. The method of preparing the mushrooms for eating may influence the edibility of acrid species of *Russula*. Some collectors who eat *Russula* may also simply discard the acrid specimens, although McKenny (1971) provides evidence that they do not always do so (under *R. cascadensis* Shaffer). Other species such as *Russula foetens* Fr. are not usually eaten because of their nauseous odor and taste. *Russula subnigricans* Hongo, which occurs in Japan, has been reported to have caused deaths of several individuals. We have no reports of deaths in North America by species related to *R. subnigricans*.

The genus *Russula* cannot be fairly labeled as poisonous because the species are so commonly eaten. We recommend that acrid or offensive-tasting species be avoided and that caution be used in general when eating *Russula* species. Never eat them raw. Avoid *Russula subnigricans* and related species such as *R. densifolia* (Secr.) Gill., which Miller (1972) lists as poisonous but without reference to the source of this information. Groves (1979) and McKenny (1971) report it and related species as edible, although they indicate that the species are of poor quality. The species in this group, commonly called the *Compactae*, are often white to whitish in color, at least when young. Eventually they become, at least in part, brown, gray, black, or red when bruised or cut, or with age. The fruit bodies have short stalks and broad caps, and the flesh is hard and compact.

We describe only *Russula emetica* here. It and related species are often very acrid and many say that *R. emetica* is poisonous. Some specialists report that cooking *R. emetica* and *R. sardonia* Fr. does not destroy their irritant properties.

Russula emetica (Schaeff. ex Fr.) Pers. ex S. F. Gray

Figs. 166, 226

Common name. Emetic russula.

Distinguishing characteristics. Cap 25–85 mm broad, convex with the margin incurved in early stages, expanding to plane or more or less depressed in the center; surface glabrous, viscid when moist; color some shade of red to deep pink or reddish orange, sometimes more orangish. Taste of flesh acrid. Gills attached, close; color white to yellowish white. Stalk 20–105 mm long, 7–24 mm thick, more or less equal; color white to yellowish white, at times discolored grayish yellow. Spores yellowish white to white in deposit, ellipsoidal to ovoid or nearly subglobose, 8.0–11.3 × 6.7–9.0 μm in size, ornamented with amyloid warts and spines.

Observations. *R. emetica* is a widely used name and has been applied to almost any red *Russula* with a whitish stalk and gills and an acrid taste. Shaffer (1975) clearly defines *R. emetica* and separates it from similar-appearing, related species. It is distinguished by the medium to large fruit bodies with strong to moderate red or reddish orange to deep pink caps, the strongly acrid taste, the larger spores, and its habit of growing in sphagnum. Cystidial characteristics are also helpful in distinguishing it. According to the number of collections reported in Shaffer's paper, it is not as common as generally believed. Apparently *R. fragilis* (Pers. ex Fr.) Fr. and *R. silvicola* Shaffer are more common than *R. emetica* in eastern North America. In the west *R.*

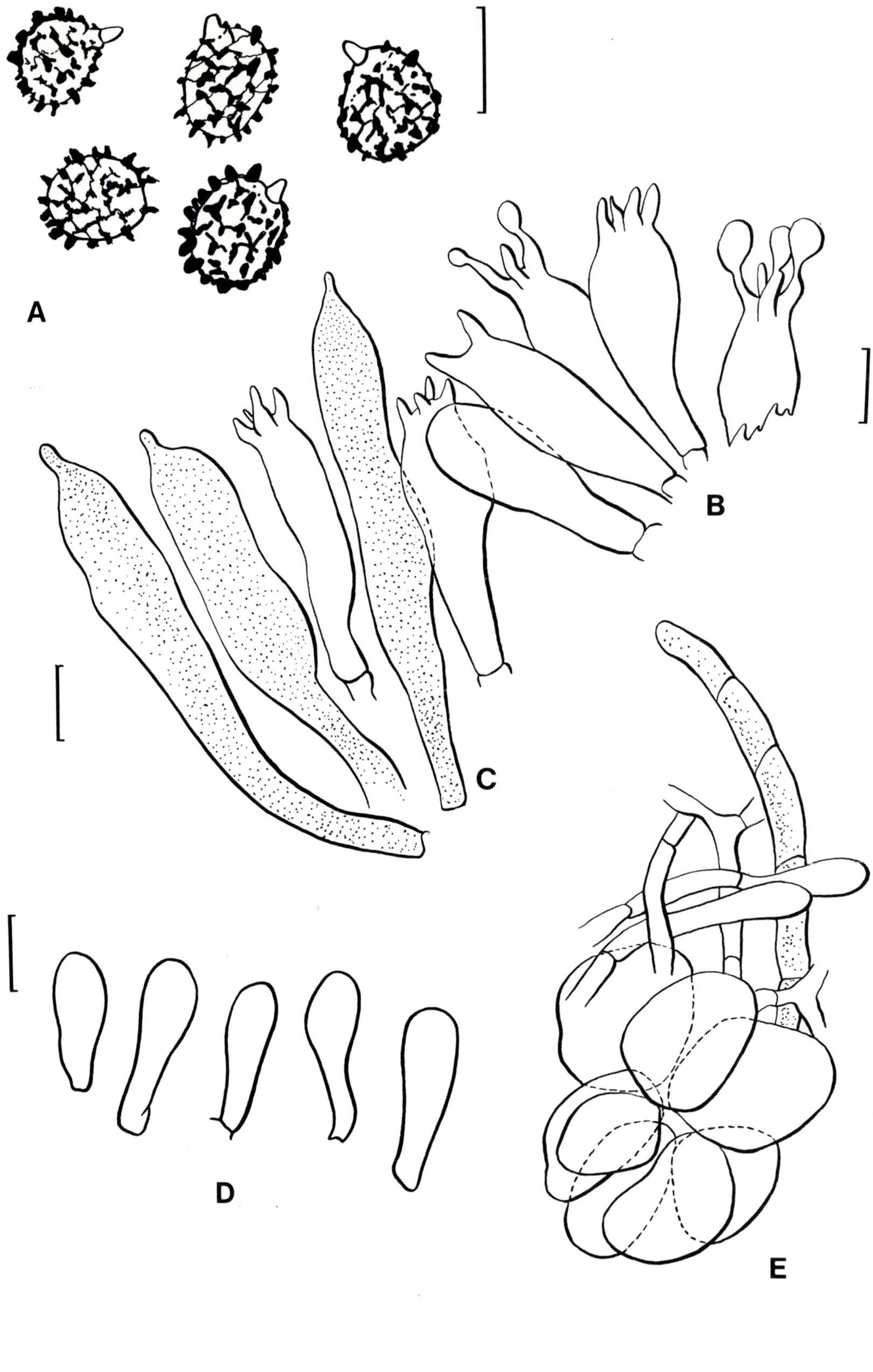

Fig. 166. *Russula emetica*: A, spores; B, basidia; C, macrocystidia; D, cheilocystidia; E, sphaerocysts and filamentous elements of the cap cuticle.

Scale line 10 μm

montana Shaffer is another fairly common species in this group. There are thus several acrid, reddish-colored species of *Russula* in the *emetica* group. Correct identification requires a good knowledge of *Russula*. Specimens reported in the general literature as *R. emetica* could be one of the above species, one of several other red species of *Russula*, or even *R. emetica* itself. These species should all be avoided for food. *R. mairei* Singer (=*R. emetica* ssp. *mairei* (Singer) Romagnesi) has also been reported as poisonous. According to Shaffer (1975) it has not been documented from North America.

Distribution and seasonal occurrence. This species is generally reported as widely distributed, but in light of the new species described in this group by Shaffer (1975), more study is needed before the distribution of *R. emetica* can be determined. Certainly the group to which it belongs is widely distributed. It fruits during the summer and fall.

Habitat and habit. *R. emetica* grows singly or scattered, or it is gregarious or rarely caespitose. It is usually found on sphagnum or on humus or very rotten wood, often in boggy places, and in coniferous and deciduous-coniferous woods.

Technical description. Cap 25-85 mm broad, cushion-shaped to convex when young with the margin incurved, expanding to plano-convex, sometimes more or less depressed, more or less umbonate; margin not striate when young, then becoming tuberculate-striate for 2-7 mm; surface viscid to more or less shiny, glabrous, finely wrinkled outward from the disc, more or less areolate with age; surface layer easily separable from the flesh; color strongly vivid or moderate red, strong to deep reddish orange or deep pink, occasionally spotted deep red near the center, sometimes partly moderate to pale orange yellow or moderate orange. Flesh moderately thick, 4-9 mm, firm-brittle to soft-brittle when old, white or tinged red under the cuticle; taste acrid; odor faintly fruity, spicy, or more or less fungoid.

Gills white to yellowish white, unchanging, 2-10 mm broad, fragile, unequal, rounded, adnexed to adnate, seceding, close, occasionally forked near the stalk, interveined, entire; lamellulae rare.

Stalk (20-)45-105 mm long, 7-24 mm wide, almost equal, flared at the apex, enlarged at the base, rarely tapered at the base, dry, dull, glabrous to longitudinally finely wrinkled, stuffed to partly hollow, white to yellowish white, sometimes discolored grayish yellow at the base. Annulus and universal veil lacking.

Spores yellowish white to white in deposit, (6.0-)8.8-11.0 × 6.6-8.0(-9.0) μm, broadly ellipsoidal to ovoid, rarely subglobose; surface ornamented, strongly warted, partially reticulate; warts conical to cylindrical, amyloid. Basidia four-spored, clavate, 32.9-50 × 9.0-11.7(-13.6) μm, hyaline. Pleurocystidia in the form of pseudocystidia, 35-88 × 7.3-12.4(-15.0) μm, arising in the subhymenium and trama, subcylindrical to clavate or subfusiform; apex often mucronate, capitate, rounded, or acute; contents granular, yellowish. Cheilocystidia like pleurocystidia or subcylindrical to clavate, 14-24 × 4.4-7.3 μm, thin-walled, hyaline.

Gill trama heteromerous; hyphae interwoven, more or less cylindrical, 2.2-8.8 μm wide, hyaline; sphaerocysts present, globose to subglobose or broadly ellipsoidal, 29-58 × 18.7-50.0 μm, hyaline. Subhymenium cellular, compact; cells 7.3-10.2(-25) μm wide, hyaline to yellowish. Cap cuticle gelatinous, filamentous; hyphae interwoven, cylindrical, 2.3-3.7 μm wide; contents granular, reddish in potassium hydroxide; hyphal ends projecting in the form of pseudocystidia, erect, 29.2-220 × 3.7-7.3(-9.0) μm, subcylindrical to clavate, 1-7(-10) μm, septate. Cap trama heteromerous, filamentous; hyphae interwoven, more or less cylindrical, 2.2-10.0 μm wide, sometimes giving rise to pseudocystidia in the form of pileocystidia; contents granular, hyaline; sphaerocysts present, 29-58 × 18.7-50 μm, globose to subglobose and broadly ellipsoidal, hyaline. Caulocystidia cylindrical to clavate, consisting of the hyphal ends of the vascular hyphae, 35-70.2(-117) × 7.0-8.2(-9.6) μm, septate; contents granular, yellowish. Stalk tissue heteromerous; hyphae interwoven, cylindrical, 2.3-4.6 μm wide, hyaline; sphaerocysts 15-35.1 × 10.5-29.2 μm, subglobose to globose or ellipsoidal, hyaline, occurring in rosettes. Oleiferous hyphae not observed. Clamp connections absent.

Useful references. The following books contain information on *R. emetica* and related species pertinent to North America.

- *Edible and Poisonous Mushrooms of Canada* (Groves 1979)—A color photograph and a description and discussion are given.
- *Mushrooms in their Natural Habitats* (Smith 1949)—A description and discussion, as well as a color photograph on a reel, are presented, representing *R. emetica* var. *gregaria* Kauffman.
- *Mushrooms of North America* (Miller 1972)—A color photograph and a description and discussion are given.
- *The Mushroom Hunter's Field Guide* (Smith 1974)—A color photograph and a description and discussion are given.
- *The Savory Wild Mushroom* (McKenny 1971)—A black-and-white photograph and a description and discussion are given.

The best technical description of *R. emetica* and related species is given by Shaffer (1975).

Sarcosphaera Auersw.

Pezizaceae

The genus *Sarcosphaera* is one of several fleshy fungi of the Ascomycota, in the class Discomycetes. See Chapters 2 and 3, "Ascomycota", for a discussion of the major features of this division. The genus *Sarcosphaera* belongs to a group commonly called the cup fungi, because of the medium to fairly large, cup-like fruit bodies they form. Some additional common genera include *Aleuria, Discina, Disciotis, Peziza,* and *Sarcoscypha.* The edibility of most of these genera is not well known. Some general reports state that *Discina* and *Peziza* species are eaten. *Aleuria aurantia* (Fr.) Fuckel is reported as edible by McKenny (1971). Lincoff and Mitchel (1977) report that certain species of *Peziza, Disciotis,* and *Sarcoscypha* are poisonous when eaten raw and remain poisonous to some individuals even when the fungi are cooked. The only consistent reports of poisoning by any of these cup fungi in North America have involved *Sarcosphaera crassa.* Several poisonings by this species also have been reported from Europe (Pilát 1951). Smith and Smith (1973) and Tylutki (1979) provide an additional key to the above genera, including brief descriptions of several common species. Tylutki (1979) says *S. crassa* is not edible and is poisonous to some people.

Sarcosphaera forms large, deep, cup-like fruit bodies that often are wholly immersed in the soil, at least when young. The deep cups expand somewhat as they mature, and the edge of the margin often splits and appears crown-like. The interior of the cup is tinted violaceous or is somewhat pinkish tinted.

Sarcosphaera crassa (Santi ex Steudl) Pouzar

Figs. 167, 227

Common names. Pink crown, violet star cup.

Distinguishing characteristics. Fruit body cup-shaped, 40–120 mm broad, deep and rounded, buried in the soil at least when young; margin incurved in early stages, splitting more or less radially on expanding with maturity to become crown-like; color pinkish to dull lilac or lilac brown on the interior, whitish to grayish on the exterior. Flesh white and brittle. Stalk absent, although a short stalk-like base may be present in some specimens. Spores smooth, ellipsoidal to cylindrical, 15.5–17.8 × 8.2–9.0 μm in size.

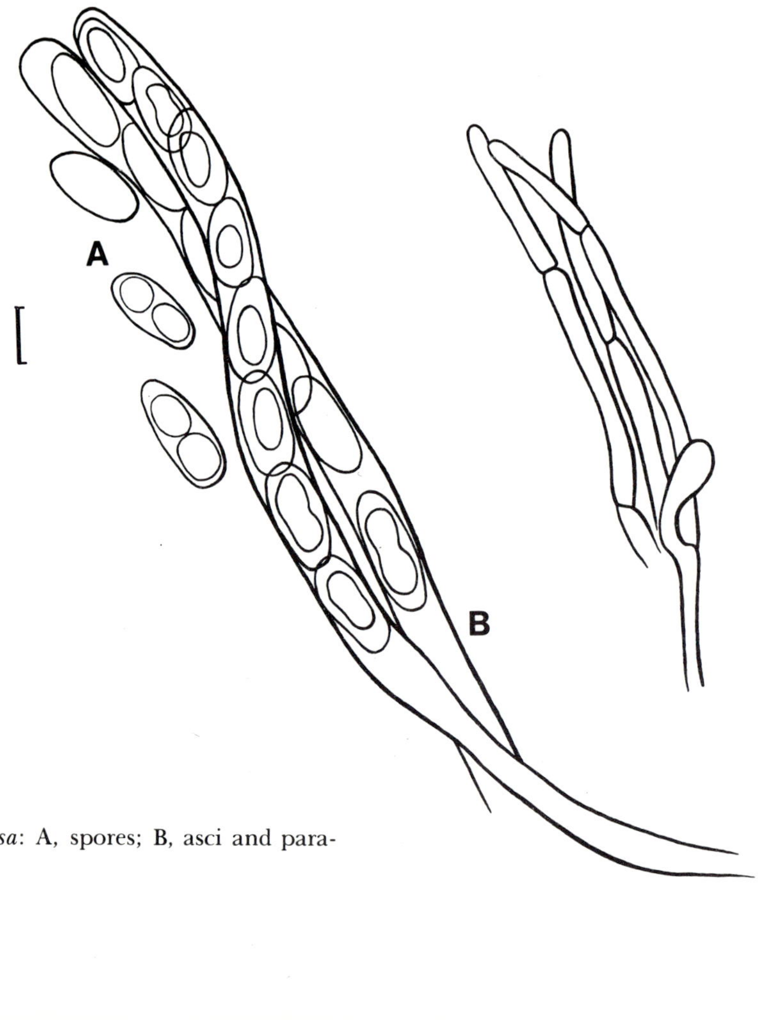

Fig. 167. *Sarcosphaera crassa*: A, spores; B, asci and paraphyses.

Scale line 10 μm

Observations. This species has been described as *S. eximia* (Dur. & Lév.) Maire and *S. coronaria* (Jacquin ex Cooke) Boud. *S. crassa* has been reported from across North America. It is rare in the east but more frequent in the west. At times it can be fairly common in the Rocky Mountains and Cascade Mountains.

Distribution and seasonal occurrence. This species occurs across North America. It fruits in the spring, summer, and fall, but in the northwest it is most common in the spring and early summer.

Habitat and habit. *S. crassa* grows singly, scattered, or in clusters of several to numerous specimens, buried in soil and litter. It is often found in coniferous forests but is also reported under hardwoods and in shrubby and grass-covered areas.

Technical description. Fruit body 40-120 mm broad, cup-shaped, deep, rounded, buried in soil in early stages, sessile to substipitate; margin incurved, soon splitting and radiating crown-like, with the interior surface of the cup smooth; color of interior sometimes whitish in early stages, usually pinkish to dull lilac or lilac brown; exterior surface of the cup white to grayish, smooth. Flesh white, brittle.

Spores pale yellowish, 15.5-17.8 × 8.2-9.0 μm, ellipsoidal to cylindrical, with slightly truncate ends, smooth, thin-walled, hyaline, biguttulate. Asci eight-spored, narrowly cylindrical, 7.0-11.7 μm wide; walls amyloid, especially at the apex. Paraphyses narrow, filamentous, 5-8 μm wide, hyaline, septate, constricted at the septa. Subhymenium filamentous, compactly interwoven; hyphae 3.5-4.7 μm wide, hyaline. Flesh interwoven; hyphae inflated, ellipsoidal to fusiform, 16.7-46.8(-70.0) μm wide, hyaline. Outer excipulum compactly interwoven; hyphae 3.5-8.2 μm wide, hyaline.

Useful references. The following books contain information on *S. crassa* pertinent to North America.

- *A Field Guide to Western Mushrooms* (Smith 1975)—A color photograph and a description and discussion are given.
- *Mushrooms of North America* (Miller 1972)—A color photograph and a description and discussion are given.
- *The Mushroom Hunter's Field Guide* (Smith 1974)—Black-and-white photographs and a description and discussion are given.
- *The Savory Wild Mushroom* (McKenny 1971)—A black-and-white photograph and a description and discussion are given.

Scleroderma Pers.

Sclerodermataceae

The genus *Scleroderma,* often called the thick-skinned puffballs, is common and widely distributed in North America. *Scleroderma* is easy to recognize and separate from the other puffball genera. The fruit bodies have a thick, hard, persistent rind or wall that encloses the spore-producing tissue. The spore-producing tissue, or gleba, is firm and sometimes white in very young specimens but soon turns grayish violet to deep purple or blackish violet, with a fine mottling of small, white lines. With age the spore tissue typically forms a violet black to blackish brown powder. The fruit bodies often have a disagreeable to bitter taste.

The true puffballs, for example *Calvatia* and *Lycoperdon,* often have a thin rind or outer wall and a soft to spongy, white spore-producing tissue when young. This tissue gradually becomes yellowish and often forms an olive brown, powdery spore mass at maturity. The spore-producing tissue in these puffballs is never hard, firm, and purplish with white mottling as in *Scleroderma.* Those true puffballs that have a purplish spore mass when mature, such as *Calvatia cyathiformis* (Bosc) Morgan, are thin-skinned in comparison to *Scleroderma* and the spore mass changes from white to yellowish before changing to purple brown.

Puffballs are generally considered edible, but not all are safe to eat. At least two species of *Scleroderma, S. citrinum* and *S. cepa,* have been reported in North America as poisonous (Stevenson and Benjamin 1961). These are described below. Two cases of poisoning by *Scleroderma* were reported by A. Funk, Forest Pathologist, Victoria, B.C. (Personal commun.). Both victims said they had read that all puffballs were edible. Little is known about the edibility of other *Scleroderma* species, but they should all be avoided because the genus has a bad reputation. Genera such as *Lycoperdon* and *Calvatia* are generally considered edible when young and fresh. Krieger (1967) reported *Lycoperdon subincarnatum* Peck as unwholesome, and not to be eaten, because prolonged diarrhea would likely result. He reported the only case of poisoning by this genus in Canada and the United States. Schultes and Hofmann (1973) reported that *Lycoperdon marginatum* Vittad. and *Lycoperdon mixtecorum* Heim contain hallucinogens, but their finding has not been fully documented. According to Vergeer (1977), a study done in Mexico (Ott et al. 1975) disproved the presence of psychoactive puffballs in that country.

Scleroderma cepa Pers.

Fig. 168

Common name. Thick-skinned puffball.

Distinguishing characteristics. Fruit body 15-120 mm wide, nearly spherical to somewhat lobed, typically sessile; surface smooth to finely cracked, at times with small scales; surface color nearly pure white, becoming straw color and gradually brownish to dull leather brown, quickly staining deep vinaceous on the surface when rubbed. Wall about 1.5 mm thick when fresh; color white, staining vinaceous when cut. Spore-producing tissue white, becoming blackish with purple tints, forming a blackish brown to somewhat yellow brown, powdery spore mass when mature. Taste bitter. Spores covered with spines, spherical, 7-11(-13.1) μm in diameter.

Observations. *S. cepa* is not reported in common mushroom guides or floras. The surface of the fruit body is not conspicuously warted as it is in *S. citrinum*, and the spores are distinctly spiny and lack the net-like surface pattern exhibited by *S. citrinum*. However, when the surface of the fruit body is quite smooth, *S. cepa* may be mistaken for *S. bovista* Fr., especially when the color of mature fruit bodies are similar. *S. bovista*, however, can be distinguished by the net-like surface pattern on the spores. All three species should be avoided. The technical description of *S. cepa* given below is taken from that by Coker and Couch (1928).

Distribution and seasonal occurrence. The distribution of *S. cepa* is poorly known. There are collections on file from both eastern and western North America. It seems to be most common in the east. It fruits in late summer and fall.

Habitat and habit. *S. cepa* is reported from forested areas and grassy areas, and under shrubs and bushes.

Technical description. Fruit body 15-60(-120) mm wide, subglobose or lobed, often flattened, sessile or nearly so, and attached by a thick mass of fibrous mycelia; surface finely areolated, or covered over most of the area with very small inherent scales, or in large part smooth; color nearly pure white when young, soon becoming straw-colored or yellowish ocherous to dull leather brown, turning quickly to deep vinaceous when rubbed. Peridium up to 1.5 mm thick when fresh, less than 0.5 mm thick when dry, firm, not very brittle, white in section when young, turning vinaceous on exposure. Gleba watery white, at least when young, becoming nearly black with purple tints, finally less dark, blackish brown to deep yellowish brown; taste bitter; odor indiscernible or resembling fresh meal in the western specimens. Spores blackish brown in mass, globose, 7-11(-13.1) μm, covered overall with sharp spines up to 2 μm long, not reticulate, blackish brown.

Useful references. There are no books or field guides pertinent to North America that contain information on this species.

168

169

Fig. 168. *Scleroderma cepa*: spores.

Fig. 169. *Scleroderma citrinum*: spores.

Scale line 10 μm

Scleroderma citrinum Pers.

Figs. 169, 228

Common name. Thick-skinned puffball.

Distinguishing characteristics. Fruit body 20-120 mm wide, generally spherical, but at times depressed to slightly irregularly shaped, sessile or with a short stalk; surface cracked, forming a pattern of raised flattened scales, often with a small central dark wart on the individual scales; color usually some shade of yellow brown. Walls of fresh specimens whitish, turning pinkish when cut, at maturity breaking irregularly, not opening in a definite star-like pattern. Spore-producing tissue grayish to purplish gray and mottled with whitish lines, forming a blackish powdery mass when mature. Taste bitter, although it may be mild in very early stages. Spores globose; surface roughened with warts that are interconnected to form a net-like pattern, 8-11(-13) μm in diameter.

Observations. *S. citrinum* has commonly been called *S. aurantium* Pers. in much of the older literature. *S. vulgare* Fr. is another name for this species. It is one of the most common species of *Scleroderma* in North America. It is at times parasitized by *Boletus parasitius* Fr.

The recommendations on the edibility of *S. citrinum* vary. Some report it as nonpoisonous and edible, whereas others do not recommend it for eating. We do not recommend it or any other species of *Scleroderma* for food. They are definitely poisonous and have made some people very sick. The technical descriptions provided below are taken from those of Coker and Couch (1928) and Smith and Smith (1973).

Distribution and seasonal occurrence. *S. citrinum* seems to be widely distributed, but the genus *Scleroderma* has not been thoroughly studied in North America. This lack of study, together with the confusion concerning the names of *Scleroderma* species, makes the actual distribution of this species uncertain. It fruits in late summer and fall.

Habitat and habit. *S. citrinum* has been reported from both coniferous and hardwood forests. It grows solitary to scattered, in groups, or in caespitose clusters on rotten wood, humus, or soil.

Technical description. Fruit body 20-120 mm wide, globose to depressed-globose, plicate below, rarely lobed, sessile or substipitate, attached by a thick mass of fibers; surface cracked into distinct flattened scales, each usually having a central wart, with scales and warts typically arranged in a distinct pattern; color yellow brown, yellowish ocherous, or brownish. Peridium about 2 mm thick when fresh, less than 1 mm thick when dry, white when fresh, turning pink when cut and becoming more brownish with age, slowly cracking into irregular lobes but not opening in a star-like pattern. Gleba gray, becoming purplish gray to nearly black, mottled with white to pallid lines, blackish and powdery when mature; taste mild to bitter with age; odor not observed.

Spores purplish to blackish brown in mass, 8-11(-13) μm in diameter, globose, prominently warted-reticulate, blackish to dark brown.

Useful references. The following books contain information on *S. citrinum* pertinent to North America.

- *Mushrooms in their Natural Habitats* (Smith 1949)—A description and discussion as *S. aurantium*, as well as a color photograph on a reel, are presented.
- *Mushrooms of North America* (Miller 1972)—A color photograph and a description and discussion as *S. aurantium* are given.
- *The Mushroom Hunter's Field Guide* (Smith 1974)—A black-and-white photograph and a description and discussion are given.

Stropharia (Fr.) Quél.

Strophariaceae

Stropharia is a fairly large genus, but there is no treatment of it for Canada or the United States. The species are in general widely distributed and there are a variety of kinds. To date only a few have been reported as poisonous. Certain recent treatments place *Stropharia* species in the genus *Psilocybe*.

Stropharia in general appearance is similar to certain species of *Agaricus* (Ch. 13), *Naematoloma* (Ch. 7), *Pholiota* (Ch. 13), and *Psilocybe* (Ch. 12). Compare all four carefully before making a final generic determination.

The characteristics that help to define *Stropharia* include a lilac to sooty lilac or purplish smoky gray to purplish brown spore deposit; the presence of a partial veil, often leaving a distinct annulus or an annular zone and sometimes leaving veil material on the cap margin; a moist to viscid cap, often becoming more or less convex; attached, adnexed to adnate, usually not decurrent gills, not mottled as in *Panaeolus*; a variable habitat, usually terrestrial and humicolous, with fruit bodies occurring at times on wood or woody debris or sawdust and less commonly on dung; smooth spores, with an apical germ pore; the presence of chrysocystidia in the hymenium; and a filamentous cap surface composed of cylindrical hyphae that are more or less interwoven. The filamentous cap surface separates *Stropharia* from *Panaeolus* and most *Psathyrella* species; these genera typically have a cellular cap surface, as viewed under the microscope.

For a comparison of *Stropharia* with *Naematoloma* and *Psilocybe*, see the discussions under those genera in Chapters 7 and 12, respectively. Certain *Pholiota* species are similar in appearance to some species of *Stropharia*. Usually the two genera can be separated by the spore deposit color; the spores of *Pholiota* are typically dull brown to dark brown in deposit, or at times rusty brown to yellow brown. *Stropharia coronilla* is discussed and described below.

Stropharia coronilla (Bull. ex Fr.) Quél.

Figs. 170, 229

Common names. None.

Distinguishing characteristics. Cap 10–40 mm broad, usually more or less convex to plane, when moist usually somewhat viscid; color pale tan to yellowish brown or more yellowish. Gills close to crowded; color grayish to purplish gray (drab) to more brownish and purple black with age. Stalk 3–7 mm thick, more or less equal, white. Partial veil leaving a persistent ring that is white to pallid and often with radial lines on the upper surface. Fruit bodies typically occurring in lawns or grassy places. Spores smooth, with an apical germ pore, 7.3–11 × 4.8–5.8 μm in size.

Observations. *S. coronilla* and similar species such as *S. hardii* Atk. are most likely to be confused with species of *Agaricus* that grow in lawns and other grassy areas. *Agaricus* species have free to nearly free gills, although they may be slightly attached at times, whereas *Stropharia* species have attached gills; these in *S. coronilla* are usually adnexed.

Species of *Stropharia* are often listed as poisonous or suspected of being poisonous, but little is actually known about the edibility of most of them. *S. semiglobata* is discussed in Chapter 12. *S. aeruginosa* (Curt. ex Fr.) Quél. is reported as poisonous in several publications, but there are no reports of poisoning by this species in North America. Miller (1972) provides a description and a color photograph.

S. coronilla is listed as poisonous by Miller (1972) and as suspected by Groves (1979). A case of poisoning was reported by Thomas et al. (1977). This poisoning is also discussed by Lincoff and Mitchel (1977) in their chapter on hallucinogenic poisoning. The symptoms in this case began within an hour after eating the

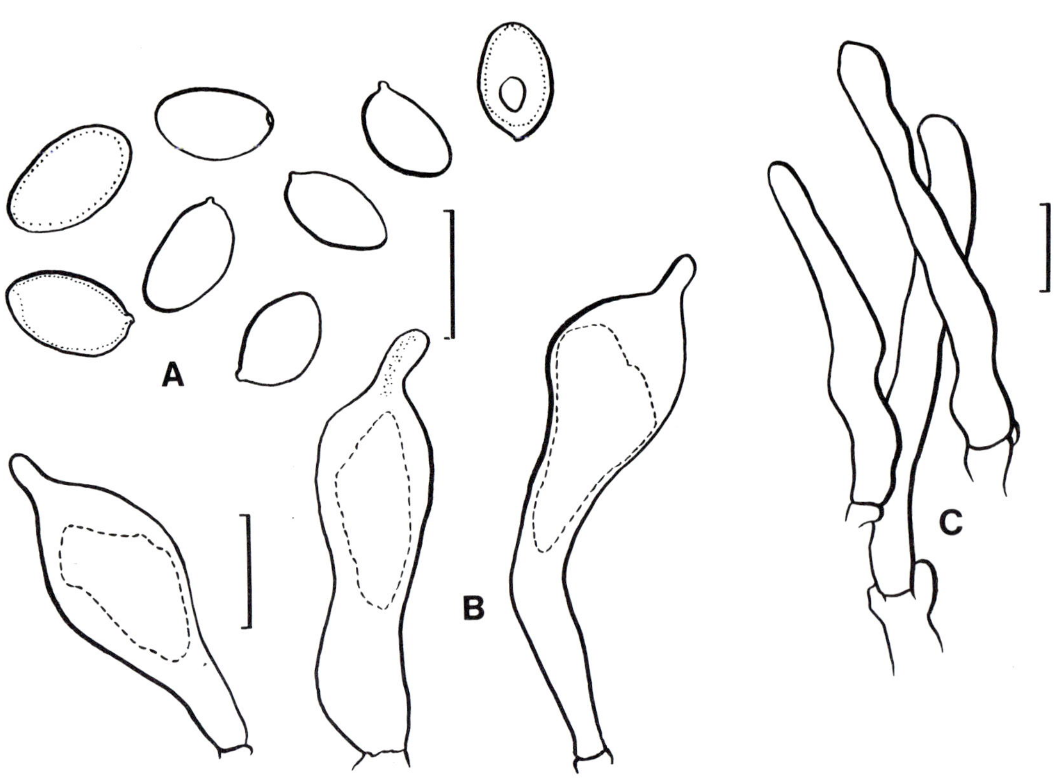

Fig. 170. *Stropharia coronilla*: A, spores; B, pleurocystidia; C, caulocystidia.

mushroom meal. Headaches and boneaches were followed by dizziness, and the two victims had difficulty walking. Vomiting, cramps, and in one case brief diarrhea occurred. Both had visions and one became delirious. The victims recovered within approximately 24 hours, except for some residual pain.

Distribution and seasonal occurrence. *S. coronilla* is widely distributed in North America. It is infrequent to fairly common and at times can be abundant. It fruits in the summer and fall, depending on location and weather.

Habitat and habit. *S. coronilla* is terrestrial and grows scattered to gregarious in lawns, meadows, pastures, and other grassy areas. It also grows under open stands of trees.

Technical description. Cap 10-40 mm broad, broadly bell-shaped to convex and umbonate; margin often appendiculate; surface smooth, glabrous, viscid when moist; color pale tan to yellowish brown, straw yellow, dull yellow, or ocherous, at times whitish. Flesh thick, firm, white; taste not noted; odor disagreeable.

Gills grayish to purplish drab or brownish to brownish violet or purplish black with age, adnate to adnexed, broad, close to crowded; edges more or less pale.

Stalk 30-45 mm long, 3-7 mm wide, equal to slightly enlarged at the apex, solid, stuffed, scurfy above the annulus, shiny, fibrillose-scaly below, entirely white. Flesh white. Annulus median, membranous, floccose, easily moved, leaving a persistent ring, grooved-striate above, white to pallid.

Spores purplish brown in deposit, 7.3-11.0 × 4.8-5.8 μm, ellipsoidal, unequal in side view, ovate in face view, thick-walled, smooth, dull rusty brown to ocherous; apex with a narrow apical germ pore. Basidia four-spored, clavate. Pleurocystidia and cheilocystidia similar, fusiform-ventricose to broadly clavate; apex obtuse to acute or mucronate, 32-48 × 8.0-12.4 μm, thin-walled; these cells often modified as chrysocystidia, namely containing an ocherous crystalline inclusion.

Gill trama interwoven, more or less parallel; hyphae cylindrical, 2.2-8.8 μm wide, hyaline. Subhymenium subcellular to filamentous, interwoven; hyphae more or less cylindrical, 3.5-7.5 μm wide, hyaline. Cap cuticle filamentous, gelatinous; hyphae interwoven, cylindrical, 3.0-5.1 (-9.9) μm wide, hyaline in potassium hydroxide. Cap trama interwoven, compact just below the surface; hyphae cylindrical to inflated, 5.1-21.9 μm wide, yellowish to hyaline; walls slightly thickened, sometimes spirally encrusted. Caulocystidia filiform-cylindrical, with clustered to single recurved hyphal ends at the apex, 3.5-9.0 μm wide, hyaline; stalk hyphae parallel to interwoven, more or less cylindrical, 3.5-9.0(-11.0) μm wide, hyaline. Annulus interwoven; hyphae cylindrical, 3.5-9.4 μm wide, hyaline. Oleiferous hyphae not observed. Clamp connections present.

Useful references. The following books contain information on *S. coronilla* pertinent to North America.

- *Edible and Poisonous Mushrooms of Canada* (Groves 1979)—A color photograph and a description and discussion are given.
- *Mushrooms in their Natural Habitats* (Smith 1949)—A description and discussion, as well as a color photograph on a reel, are presented.
- *Mushrooms of North America* (Miller 1972)—A description and discussion are given.

Tricholoma (Fr.) Quél.

Tricholomataceae

Tricholoma is a fairly distinct genus and can sometimes be recognized by its major macroscopic characteristics, which include a fleshy stalk and usually fairly large to robust fruit bodies, although some species are small; sinuate to notched gills; and a white spore deposit. Some species have a partial veil and some an annular zone.

Tricholoma spores are smooth and nonamyloid. Cystidia are present in the genus but are not well differentiated in most species. The cap cuticle is filamentous and several of the species lack clamp connections. Certain toxic species, such as *T. pardinum*, have clamp connections.

Most species of *Tricholoma* fruit in the fall, although several occur in the western mountains during the summer and they can occasionally be found in the spring. They are terrestrial and typically occur in forested areas.

Tricholoma may be confused with several similar-appearing genera, and at times a microscope is required to be sure of genus identification. *Melanoleuca* species usually have a more or less slender stalk that breaks easily and cleanly, a usually plane cap at maturity, and spores ornamented with amyloid warts. *Leucopaxillus* is very similar in stature to *Tricholoma*. Often a mass of white mycelia is attached to the base of the stalk and the spores are amyloid and roughened. *Tricholomopsis* can usually be separated from *Tricholoma* by its lignicolous fruiting habit, the species usually occurring on logs or around stumps, and by the well-differentiated cheilocystidia. Goos and Shoop (1980) report *Tricholomopsis platyphylla* (Fr.) Singer as causing poisoning. The genus *Tricholomopsis* and the species *T. platyphylla* are described later in this chapter, placed alphabetically.

Clitocybe species can usually be separated from those of *Tricholoma* by their decurrent gills, although the larger, more robust species, sometimes placed in the genus *Lepista*, are similar in size and stature to *Tricholoma*. This group of *Clitocybe* species usually has slightly pinkish to buff spore deposits and usually slightly roughened spores. The genera *Lyophyllum* and *Calocybe* are difficult to separate from *Tricholoma*. They often are very similar in appearance and spore deposit color. They can be distinguished, however, by the siderophilous granules in the basidia, detectable using a special technique described in Chapter 4. Finally, *Russula* and *Lactarius* may be confused with *Tricholoma*. *Tricholoma* can be separated from these genera by the texture of the flesh, which is chalky to brittle in *Lactarius* and *Russula*, snapping or breaking much like a piece of chalk.

Lincoff and Mitchel (1977) list the following *Tricholoma* species as being poisonous: *T. album* (Schaeff. ex. Fr.) Kummer, *T. muscarium* Kawamura ex Hongo (Japan), *T. nudum* (Bull. ex Fr.) Kummer (=*Clitocybe nuda* (Bull. ex Fr.) Bigelow & Smith) eaten raw, *T. pardinum*, *T. pessundatum*, *T. saponaceum* (Fr.) Staude, *T. sejunctum* (Sow. ex Fr.) Quél., *T. sulphureum*, and *T. venenatum*. Of these species, *T. pardinum* causes the most poisonings, apparently because it is fleshy, fairly large, and attractive to collectors of edible mushrooms (Pilât 1951, 1961; Alder 1950, 1960). Poisonings by this species may be very severe. The symptoms usually begin 15 minutes to 2 hours after the meal and may persist for several hours. Main symptoms include malaise, dizziness, vomiting, and diarrhea. Some mild liver damage has been reported. Victims usually recover in 4-6 days. A case of poisoning occurred near Arnprior, Ont., in September 1975 (S. C. Thompson, Personal commun.).

Poisonings by *T. inamoenum* (Alder 1967) and *T. venenatum* (Fischer 1918) have been reported. Stadelmann et al. (1976) have reported trace amounts of muscarine and its stereoisomers in several mushrooms, including *T. sulphureum* (Fr.) Staude. McKenny (1971) report *T. pessundatum* as causing gastric disturbances. Pilât (1961) also reports it as poisonous. McKenny (1971) reports that *T. sulphureum* and *T. inamoenum* are inedible because of the repulsive odor, like coal-tar gas. Pilât (1961) reports cases of poisoning by *T. sejunctum*, but this species is considered edible in North America.

The toxic properties of *Tricholoma* are in need of further study. To date, the toxin or toxins responsible for poisonings in this genus have not been elucidated. When collecting *Tricholoma* species for food, totally avoid the poisonous or suspected species and be absolutely sure of the identification of those you eat. When collecting species such as *T. flavovirens* (Pers. ex Fr.) Lundell (=*T. equestre* (L. ex Fr.) Kummer), be sure to check each specimen carefully. Because of similarity in color, it and similar species of *Tricholoma* can be confused with the deadly *Amanita phalloides* (Ch. 7).

The four *Tricholoma* species that present the most serious danger in North America are described below.

Tricholoma inamoenum (Fr.) Gill.

Figs. 171, 230

Common names. None.

Distinguishing characteristics. Fruit body medium to small. Cap more or less convex; color white to yellowish, with white, gray buff, or clay color in the center; margin usually paler than the center. Odor strong, repulsive, resembling coal tar. Gills white with yellowish tints, distant. Stalk equal; color white, sometimes brownish at the base. Spores smooth, large, 10.5-12.4 × 6.7-7.6 μm.

Observations. *T. inamoenum* is apparently uncommon in North America and has only been reported from the Pacific northwest (D. E. Stuntz, Personal commun.). It grows in Europe, where poisonings have been reported. *T. inamoenum* belongs to a complex of species that centers around *T. sulphureum*. These species all possess the characteristic repulsive, coal-tar odor and have large spores. *T. sulphureum* is a yellow species but has several variations, especially in Europe. *T. ódorum* Peck is a North American species that is yellow when young, fading to nearly buff at maturity. *T. platyphyllum* (Murrill) Murrill is entirely buff and is smaller than *T. inamoenum*; these two species seem to be closely related, and are perhaps the same.

Distribution and seasonal occurrence. This species occurs in the fall in coniferous woods in the Pacific northwest and in Europe. Its distribution is not well known in North America.

Habitat and habit. *T. inamoenum* grows under conifers.

Technical description. Cap 40-60 mm broad, convex to plano-convex, obtusely umbonate; margin thin, irregular; surface dry, smooth to weakly matted-tomentose or filamentous-silky; color white to yellowish, white, gray buff, or clay color on the disc, paler on the margin. Flesh white; taste mild, salty becoming acrid; odor repulsive, like coal tar.

Gills white with a yellowish tint, bluntly adnate, arcuate, distant, broad.

Stalk 90-110 mm long, 8-10 mm wide, equal, slightly enlarged at the base, solid, firm; surface dry, smooth to finely fibrillose, white, sometimes browning at the base. Flesh white.

Spores white in deposit, 10.5-12.4 × 6.7-7.6 μm, elliptic to slightly ovate in side view, elliptic in face view, smooth, hyaline, nonamyloid. Basidia four-spored, clavate, 41-52 × 10.5-11.4 μm, hyaline. Pleurocystidia and cheilocystidia lacking.

Gill trama parallel; hyphae 3.8-11.4 μm wide, cylindrical to slightly inflated, hyaline. Subhymenium filamentous; hyphae interwoven, 1.9-2.9 μm wide, hyaline. Cap cuticle filamentous; hyphae radially appressed to slightly interwoven, cylindrical, 4.8-9.5 μm wide, thin-walled, hyaline to light golden yellow. Cap trama interwoven to radially arranged near the surface; hyphae cylindrical to inflated, 3.8-14 μm wide, hyaline. Caulocystidia cylindrical-clavate, zigzagged, occurring as recurved hyphal end cells at the stalk apex, 33-71 × 3.8-4.8 μm, clustered or in a turf, hyaline; stalk hyphae longitudinally appressed on the surface, somewhat interwoven to parallel in the trama, cylindrical, 3.8-9.5 μm wide, thin-walled, smooth to slightly roughened, hyaline to pale yellowish green. Oleiferous hyphae not seen. Clamp connections present.

Useful references. There are no books or field guides pertinent to North America that contain information on this species.

Tricholoma pardinum Quél.

Figs. 172, 231

Common name. Tiger tricholoma.

Distinguishing characteristics. Fruit body medium to large, robust. Cap 20-150 mm broad, hemispherical to convex or plane; surface scaly to squamulose; scales light to dark gray or grayish brown, paler between the scales. Taste like fresh meal. Gills white, close to crowded. Stalk 10-35 mm thick, equal or slightly tapered upward. Spores smooth, broadly ellipsoidal, 7.5-9.8(-10.5) × 5-6.5 μm in size. Clamp connections present. Cheilocystidia present.

Observations. *T. pardinum* in general appearance is similar to *T. venenatum*, to which it is closely related. Both species are reported to have caused severe gastrointestinal poisonings in North America. *T. pardinum* also occurs in Europe, where it is considered poisonous. Although most species of *Tricholoma* lack clamp connections, *T. pardinum*, *T. venenata*, and certain other species of this genus have them.

Distribution and seasonal occurrence. This species occurs throughout temperate North America. It usually fruits in the fall but can be found in the western mountains during the late summer.

Habitat and habit. *T. pardinum* is terrestrial and grows solitary to gregarious or caespitose under conifers or in mixed woods.

Technical description. Cap 20-150 mm broad, hemispherical or convex, becoming broadly convex to nearly plane, sometimes with a low umbo; margin incurved or enrolled in early stages, becoming straight; surface dry,

171

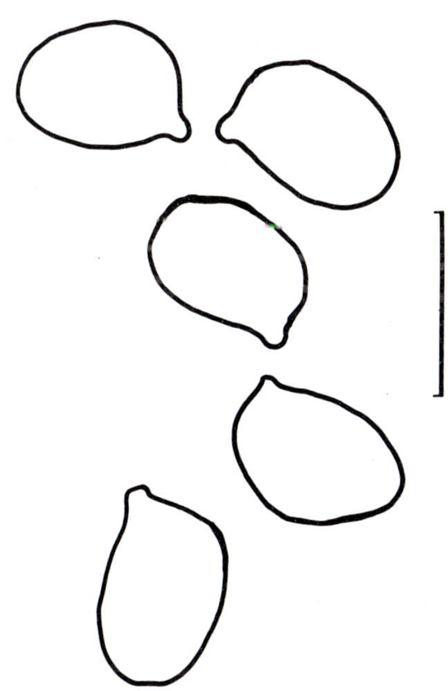

172

Fig. 171. *Tricholoma inamoenum*: spores.

Fig. 172. *Tricholoma pardinum*: spores.

Scale line 10 μm

covered with short fine fibrillose scales; color drab gray to grayish brown because of the drab to brownish black fibrils or squamules against the lighter grayish buff background. Flesh firm, white; taste like fresh meal; odor mild.

Gills white in early stages, unchanging or discoloring dingy brown, adnexed to emarginate-sinuate or with a slight decurrent tooth, close to crowded; edges eroded with age; lamellulae present.

Stalk 25-120 mm long, 10-35 mm thick, equal or slightly tapered to the apex, dry, dull, slightly fibrillose to glabrous, white, unchanging or becoming dingy brown with age toward the base.

Spores white in deposit, 7.5-9.8(-10.5) × 5.0-6.5 μm, broadly ellipsoidal, smooth, hyaline, nonamyloid. Basidia four-spored, clavate, 41-52(-67.5) × 7.5-9.8(-11.3) μm, hyaline. Pleurocystidia lacking. Cheilocystidia with a short stalk and an enlarged spherical apical portion, 29.0-41.0 × 12.0-21.0 μm, thin-walled, hyaline.

Gill trama parallel to slightly interwoven; hyphae more or less cylindrical, 3.8-19.0 μm wide, hyaline. Subhymenium filamentous. Cap cuticle filamentous; hyphae parallel, radially arranged, agglutinated, cylindrical, 2.9-6.7 μm wide, hyaline to golden yellow. Cap trama interwoven; hyphae cylindrical to inflated, 4.5-15.0 μm wide, hyaline. Caulocystidia cylindrical, approximately 48 × 3.3-4.8 μm, occurring as recurved hyphal end cells, hyaline; stalk hyphae parallel in the core to interwoven on the surface or longitudinally appressed, more or less cylindrical, 3.8-15.0 μm wide, hyaline. Oleiferous hyphae not observed. Clamp connections present.

Useful references. The following books contain information on *T. pardinum* pertinent to North America.

- *A Field Guide to Western Mushrooms* (Smith 1975)—A color photograph and a description and discussion are given.
- *Mushrooms of North America* (Miller 1972)—A color photograph and a description and discussion are given.
- *The Mushroom Hunter's Field Guide* (Smith 1974)—A black-and-white photograph and a description and discussion are given.
- *The Savory Wild Mushroom* (McKenny 1971)—A black-and-white photograph and a description and discussion are given.

Tricholoma pessundatum (Fr.) Quél.

Figs. 173, 232

Common name. Red brown tricholoma.

Distinguishing characteristics. Cap 25-100 mm broad, convex to broadly convex or plane, often raised in the center; surface viscid when fresh, becoming dry with age; color dark reddish brown in the center, dark buff to light brown on the margin. Odor and taste like fresh meal. Gills close; color pale whitish buff in early stages, staining or discoloring yellowish brown or brown with age. Stalk 20-70 mm long, 6-20 mm thick, equal, rounded or tapered at the base; surface silky-fibrillose to rarely somewhat scaly; color whitish buff at the apex, elsewhere light brown to brown, darkest below. Spores white in deposit, smooth, ellipsoidal, 5.7-6.7(-7.6) × 4.3-4.8 μm in size.

Observations. *T. pessundatum* is one of several species of *Tricholoma* that have reddish brown and viscid caps. It is best characterized by these cap features and the light whitish buff gills that discolor brown. The stalk is also whitish buff and discolors brown. In addition, the odor and taste are like fresh meal. It occurs under both hardwood and coniferous trees. The concept for *T. pessundatum* as reported here corresponds closely to the European concept.

The brown, viscid species of *Tricholoma* are in need of critical study in North America. They vary with respect to the shade of brown of the cap, cap vestiture, gill color, and spore size. A change in the interpretation of *T. pessundatum* may occur as more data are gathered.

Distribution and seasonal occurrence. This and related species occur across North America, fruiting in late summer and fall.

Habitat and habit. *T. pessundatum* is terrestrial and grows scattered to gregarious or caespitose in coniferous and hardwood forests.

Technical description. Cap 25-100 mm broad, convex when young, expanding to broadly convex, cushion-shaped or plane when mature, often umbonate; margin incurved when young, straight and often wavy or lobed when mature; surface viscid, becoming dry with age, matted-fibrillose on the disc, with few to abundant minute radially appressed fibrils elsewhere, generally naked on the extreme margin, occasionally with scattered minute squamules, dark reddish brown on the disc, dark buff or light brown on the margin; flesh 7-18 mm thick, pale white, often light brown near the surface or watery gray above the gills; odor and taste like fresh meal.

Gills 2-8 mm broad, sinuate, rarely adnexed or adnate, close; color pale whitish buff, with yellowish brown or brown discolorations occurring on the edge, in patches, or when overmature, occasionally nearly overall, generally not apparent on extremely young specimens; edges entire or often eroding or splitting transversely; lamellulae numerous but not arranged in tiers.

Stalk 20-70 mm long, 6-20 mm thick, equal, rounded or abruptly tapered at the base; surface roughened silky-fibrillose, with surface fibrils breaking loose, rarely with scattered squamules, whitish buff at the apex (the same color as the gills), elsewhere light brown to brown with the base the darkest, with the whitish ground color generally visible beneath the colored surface fibrils; flesh pale to white, discoloring brown around wormholes, solid or hollow.

Spores white in deposit, 5.7-6.7(-7.6) × 4.3-4.8 μm, elliptic in side view and face view, smooth, hyaline, nonamyloid. Basidia 25-33 × 5.7-6.7 μm, four-spored, clavate, hyaline to light brown. Cheilocystidia and pleurocystidia lacking.

Gill trama parallel; hyphae cylindrical to slightly inflated, 3.8-14 μm wide. Subhymenium filamentous; hyphae 1.9-3.8 μm wide, cylindrical, hyaline. Cap cuticle radially interwoven; hyphae embedded in a gelatinous matrix, cylindrical, 3.8-6.9 μm wide, with hyaline to light brown encrustations; contents hyaline to reddish brown. Cap trama radially arranged, interwoven; hyphae cylindrical to inflated, 5.5-19 μm wide, with slightly refractive walls, hyaline. Stalk surface longitudinally appressed to slightly interwoven; hyphae cylindrical, 2.9-5.7 μm wide, smooth and thin-walled or slightly roughened, hyaline to light brown. Caulocystidia in the form of recurved hyphal end cells at the stalk apex and occasionally elsewhere, 16-33(-44) × 3.8-6.7 μm, cylindrical to zigzagged, clavate, or irregularly contracted and expanded, often multicellular, smooth, thin-walled, hyaline, single, occurring in fascicles, pyramidal clusters, or as a turf. Oleiferous hyphae present, light yellow or light cinnamon. Clamp connections absent.

Useful references. The following books contain information on *T. pessundatum* pertinent to North America.

- *Edible and Poisonous Mushrooms of Canada* (Groves 1979)—A color photograph and a description and discussion are given.
- *The Savory Wild Mushroom* (McKenny 1971)—A black-and-white photograph, a color photograph, and a description and discussion are given.

Tricholoma venenatum Atk. Fig. 174

Common names. None.

Distinguishing characteristics. Fruit body medium-sized. Cap 25-70 mm broad, conic, expanding to convex; margin squamulose; color light buff to light tan on the disc, tan to nearly white at the edge. Odor and taste like fresh meal. Gills white, becoming pale ivory buff with age, close. Stalk 7-17 mm thick, equal to slightly clavate or slightly bulbous; color white, becoming dingy below or overall with age. Spores smooth, ellipsoidal, 6.9-8.2 × 5.5 μm in size. Clamp connections present.

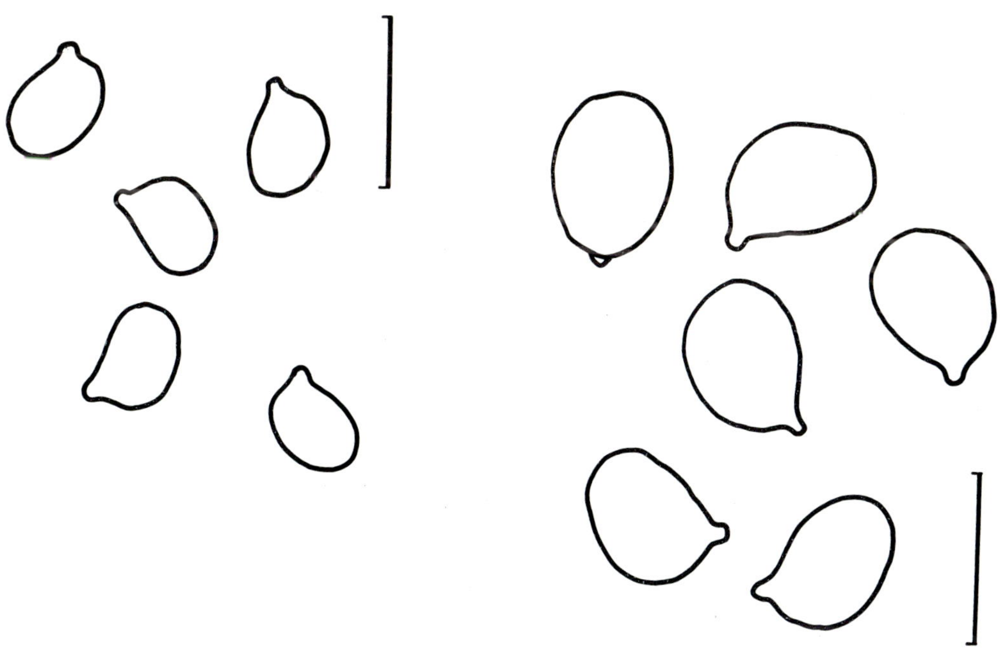

173 174

Fig. 173. *Tricholoma pessundatum*: spores.
Fig. 174. *Tricholoma venenatum*: spores.

Scale line 10 μm

Observations. *T. venenatum* is rare, known only from southern Michigan. Related to *T. pardinum,* it differs by its tan color, smaller spores, and absence of cheilocystidia. It grows under hardwoods. *T. vaccinum* (Pers. ex Fr.) Quél. is another brown, scaly species; however, it is rusty brown and occurs under conifers.

Distribution and seasonal occurrence. This species is known only from southern Michigan but is probably more widely distributed. It fruits in the fall.

Habitat and habit. *T. venenatum* is terrestrial and grows gregarious to caespitose under hardwoods.

Technical description. Cap 25-70 mm broad, obtusely conic when young, expanding to convex-umbonate when old; margin incurved when young, straight and often wavy when mature; surface dry, densely matted-fibrillose on the disc, squamulose elsewhere; squamules becoming less distinct and lighter toward the margin; color light buff to light tan on the disc, becoming lighter toward the margin, nearly white on the edge. Flesh 3-10 mm thick, white or watery gray; odor and taste like fresh meal.

Gills white when young, fading to pale ivory buff with age, not discoloring, sinuate, 4-6 mm wide, entire, close; lamellulae numerous but not in tiers.

Stalk 30-60 mm long, 7-17 mm wide, equal or slightly clavate, occasionally slightly bulbous; base rounded; surface silky-fibrillose, often with loosened surface fibrils, white becoming dingy over the lower half and overall with age. Flesh white, solid.

Spores white in deposit, 6.9-8.2 × 5.5 μm, ellipsoidal, smooth, hyaline, nonamyloid. Basidia four-spored, clavate, 30-36 × 8.2 μm, hyaline. Pleurocystidia and cheilocystidia lacking.

Gill trama parallel; hyphae cylindrical to slightly inflated, 4.1-21 μm wide, hyaline. Subhymenium filamentous; hyphae more or less 2.9 μm wide, hyaline. Cap cuticle filamentous; hyphae radially arranged to slightly interwoven, cylindrical, 3.8-4.5 μm wide, often agglutinated and recurved at the tips, hyaline to yellowish, thin-walled. Cap trama interwoven to radially arranged near the surface; hyphae cylindrical to inflated, 4.8-14 μm wide, hyaline, with often slightly refractive walls. Caulocystidia cylindrical to clavate, 28.0-40.0 × 3.8-8.2 μm, occurring as recurved hyphal end cells at the stalk apex, hyaline; stalk hyphae appressed, parallel in the core, thin-walled, cylindrical, 3.8-9.5 μm wide, hyaline. Oleiferous hyphae not observed. Clamp connections present.

Useful reference. The following book contains information on *T. venenatum* pertinent to North America.

- *The Mushroom Hunter's Field Guide* (Smith 1974)—A black-and-white photograph and a description and discussion are given.

Tricholomopsis Singer — Tricholomataceae

The genus *Tricholomopsis* is fairly common in North America, usually fruiting on rotting wood, sometimes from thick humus deposits or from the base of stumps or trees. In general appearance it is similar to certain species of *Tricholoma*; however, *Tricholoma* species are usually terrestrial and mycorrhizal. The best combination of diagnostic characteristics for identifying *Tricholomopsis* species is the habit of fruiting on rotting wood, nonamyloid spores, well-differentiated cheilocystidia, the presence of clamp connections, and the fibrillose to scaly cap surface that is evident in most species. The cap surface ranges from grayish to blackish or reddish to purplish or yellowish, with yellow being the predominant color in the genus. *Tricholomopsis platyphylla,* the only species in this genus described here, has a blackish to grayish cap. Smith (1960) provides a treatment of *Tricholomopsis*; most of the information on *T. platyphylla* given below comes from this publication.

Tricholomopsis platyphylla (Pers. ex Fr.) Singer

Common name. Broad-gilled collybia.

Distinguishing characteristics. Cap 50-120(-200) mm broad, convex to plane; margin enrolled, becoming decurved and frequently frayed with age; surface moist to dry, streaked with fibrils; color usually blackish to grayish or whitish gray. Gills attached, broad, subdistant, white to somewhat grayish. Stalk 60-120 mm long, 10-30 mm thick; color usually whitish to grayish. Partial veil and universal veil absent. Spores white in deposit, 7-9(-10) × 4.5-6 μm, smooth, nonamyloid. Fruit bodies fairly common on rotten wood or thick humus layer.

Observations. In some of the older literature *Tricholomopsis platyphylla* is called *Collybia platyphylla* (Pers. ex Fr.) Kummer. In general appearance it looks much like a species of *Tricholoma,* especially since the stalk is more fleshy than cartilaginous. The presence of well-differentiated cheilocystidia helps to separate it from *Collybia,* and the cheilocystidia along with its habit of fruiting on or near decaying wood or rich humus help to separate it from *Tricholoma.*

According to Smith (1949) it is edible but not highly rated. Groves (1979) lists it as edible and Miller (1972) lists it as nonpoisonous. Goos and Shoop (1980) report it as being poisonous. In the case they reported, two people were involved, a woman and her brother. The symptoms appeared about 2 hours after eating the cooked mushrooms. The woman experienced symptoms of light-headedness and extreme fatigue, followed shortly by stomach cramps and severe diarrhea that persisted most of the night. By morning the diarrhea had subsided, but she felt generally weak and had a queasy stomach most of the following day. The brother experienced similar symptoms but, instead of diarrhea, had recurring attacks of nausea and vomiting throughout the night.

Distribution and seasonal occurrence. *T. platyphylla* fruits from June to October in the central United States and in spring and fall along the Pacific coast. It is known from Texas north into Canada and across the United States and much of Canada. It is frequently common or even more abundant in some seasons.

Habitat and habit. *T. platyphylla* grows solitary to scattered or in small groups. It occurs on wood of both conifers and hardwoods but is most abundant on hardwoods. It usually fruits along rotting logs or on very rich humus.

Technical description. Cap 50-120(-200) mm broad, convex with an enrolled margin but soon becoming broadly convex to plane or with the margin recurved, sometimes retaining a low broad umbo; disc sometimes slightly depressed; margin opaque or when wet finely striate, very thin, and frequently becoming frayed with age; surface moist to dry and streaked with fibrils, at times almost glabrous but usually scurfy to subscaly; color blackish to avellaneous on the disc, paler and grayish toward the margin, sometimes very pale whitish gray overall except on the slightly darker disc. Context thin and moderately pliant, watery gray; odor and taste not distinctive.

Gills broad (10-20 mm), subdistant with two or three tiers of lamellulae present, adnate but becoming adnexed to nearly sinuate, often interveined, white to grayish but usually darker in forms that have deeply colored caps; edges even to eroded.

Stalk 60-120 mm long, 10-20(-30) mm thick, equal or sometimes slightly enlarged downward, stuffed, becoming hollow, with a somewhat cartilaginous cortex; surface glabrous to fibrillose, striate, usually appearing merely unpolished, whitish or tinged grayish and sometimes more or less similar in color to that of the cap surface. Partial and universal veils absent.

Spores white in deposit, 7-9(-10) × 4.5-6 μm, smooth, ellipsoidal, nonamyloid, hyaline, without an apical germ pore. Basidia four-spored, 34-36 × 5-7 μm, narrowly clavate. Pleurocystidia not differentiated or rare, and basidiole-like to nearly fusiform. Cheilocystidia abundant, 24-48 × 2.5-38 μm, variously shaped: sometimes basidiole-like but usually broader than basidioles, sometimes somewhat irregularly shaped or with one or two small projections, or sometimes fusiform-ventricose or hair-like. Gill trama parallel to somewhat interwoven; hyphae hyaline. Cap cuticle an irregularly arranged layer of pileocystidia with smoky brown contents; pileocystidia 30-60 × 8-12 μm, narrowly clavate with rounded apices. Cap trama homogeneous. Clamp connections present.

Useful references. The following books contain information on *T. platyphylla* pertinent to North America.

- *Edible and Poisonous Mushrooms of Canada* (Groves 1979)—A color photograph and a description and discussion are presented under the name *Collybia platyphylla.*
- *Mushrooms in their Natural Habitats* (Smith 1949)—A color photograph on a reel and a description and discussion are given under *Collybia platyphylla.*
- *Mushrooms of North America* (Miller 1972)—A color photograph and description and discussion are given.

Tylopilus Karst. — Boletaceae

The genus *Tylopilus* is one of several bolete genera in the family Boletaceae, a general discussion of which can be found under *Boletus* in this chapter. *Tylopilus* species are medium to large boletes with a spore deposit color ranging from pinkish to vinaceous, vinaceous brown, dark reddish brown to purplish brown, or more or less rust color. The tubes are often white to buff or pallid in the early stages or may be pinkish to vinaceous or even darker, for example, dark reddish brown in *Tylopilus pseudoscaber* (Secr.) Smith & Thiers. Certain species of *Leccinum* have a spore deposit color that is similar to *Tylopilus*; however, the stalk surface in *Leccinum* has dark ornamentation in the form of small scales or points. Compare this species with *Boletus* treated earlier in this chapter and refer to the keys to bolete genera (Ch. 5) to confirm generic determination.

Tylopilus felleus is described below. It is treated here as an inedible species because of its bitter taste but cannot be considered as truly poisonous. It is most likely to be confused with *Boletus edulis* Bull. ex Fr., an excellent edible species.

Tylopilus felleus (Fr.) Karst.

Figs. 175, 233

Common name. Bitter tylopilus.

Distinguishing characteristics. Cap 50–150(–300) mm broad, convex, often becoming plane by maturity; surface dry or sometimes sticky with age when moist; color pale to light leather color to vinaceous brown, unchanging or somewhat cinnamon with age. Flesh thick and white, unchanging or changing to pinkish brown where exposed; taste usually extremely bitter. Tubes usually white to pinkish when young, developing vinaceous to cinnamon pinkish tones with age, staining brown when bruised; pores whitish to pallid or about the same color as the tubes when viewed at an angle or with age, also staining brown when bruised. Stalk 49–100 mm long, 10–30 mm thick at the apex, clavate to bulbous, up to 60 mm thick at the base; surface reticulate over the upper portion; color pallid above and pale brown or leather color to darker below, often discolored or stained with olive tones. Spores pinkish to deep vinaceous or vinaceous brown in deposit, ellipsoidal to subfusiform, smooth, 11.0–14.6(–20) × 3.7–4.4(–5.0) μm in size.

Observations. *T. felleus* is a common species east of central North America. The exceedingly bitter taste, the pinkish to vinaceous spore deposit, the reticulate stalk, and the habit of growing around rotten stumps and woody debris, especially of conifers, are helpful features in identifying this species. A closely related species, *T. rubrobrunneus* Mazzer & Smith, occurs in eastern hardwood forests. It is also bitter and, like *T. felleus*, is inedible because of the bitter taste.

Because of the large, fleshy fruit bodies produced by *T. felleus* and *T. rubrobrunneus*, they are attractive to collectors looking for edible mushrooms. The reticulate stalk and general stature, together with the brownish caps of these species, cause collectors to confuse them with edible boletes, especially *Boletus edulis*. There are reports of collectors filling a bushel basket with *T. felleus* and finding after cooking them that they are bitter and inedible. Certain people are unable to detect the bitter taste of *T. felleus* and would probably be able to eat this species. Because we have no actual cases of poisoning by *T. felleus* it cannot really be labeled poisonous.

Distribution and seasonal occurrence. *T. felleus* grows in the region east of the Great Plains. It is especially common in the northern part of the Great Lakes region. It fruits in the summer and fall, and occasionally in late spring.

Habitat and habit. *T. felleus* grows singly to gregarious, and at times in clusters. It is typically as-

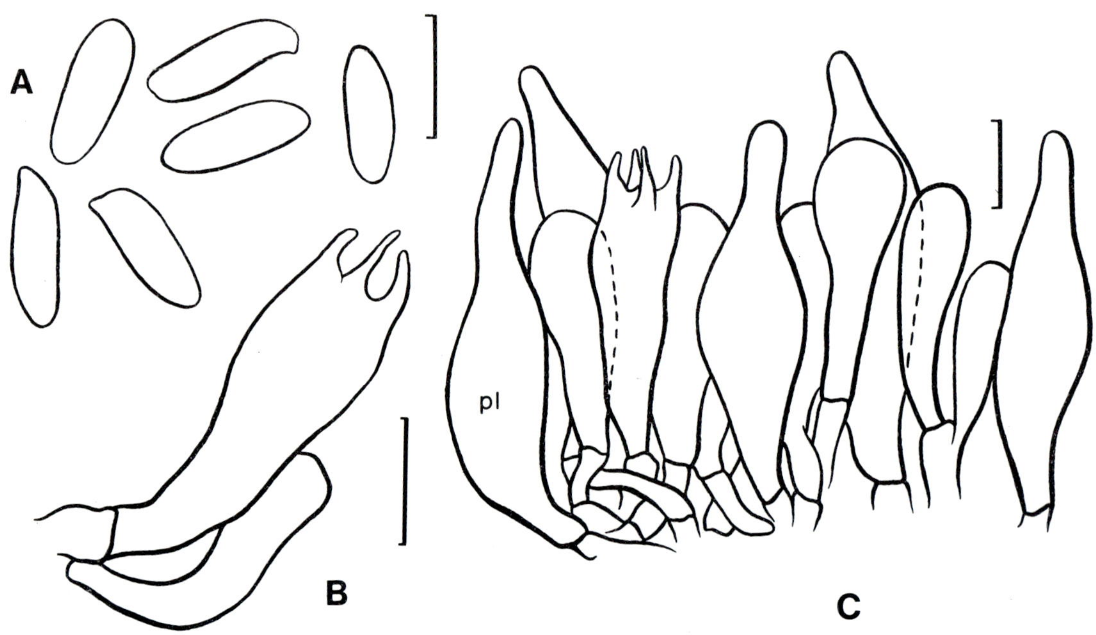

Fig. 175. *Tylopilus felleus*: A, spores; B, basidium; C, basidia and pleurocystidia (*pl*).

Scale line 10 μm

sociated with conifer stumps or debris, especially hemlock (*Tsuga*). It also grows in humus of hardwood forests containing oak (*Quercus*). In hardwood areas this species can be confused with *T. rubrobrunneus.*

Technical description. Cap 50-150(-300) mm broad, hemispherical to convex, becoming broadly convex to plano-convex to plane, occasionally depressed when old; margin entire, incurved, becoming decurved to straight or flared; surface glabrous or pruinose to distinctly tomentose, glabrous with age, dry, sometimes sticky with age when moist, smooth, occasionally split or rimose, especially toward the margin, sometimes pitted near the margin; color pale to dark leather color to vinaceous brown, unchanging or becoming cinnamon with age, often paler on the margin. Flesh very thick, up to 30 mm, white, unchanging or becoming pinkish brown when exposed; taste very bitter; odor indiscernible or occasionally like fresh meal.

Tubes 10-20 mm long, white to pale pinkish when young, soon more vinaceous to deep flesh-colored at times, becoming pale cinnamon pinkish to vinaceous red at maturity, staining brownish when bruised, adnate to subdecurrent, becoming deeply depressed around the stalk; pores about 1 mm across, whitish to dingy pallid, staining dingy brown when bruised.

Stalk 40-100 mm long, 10-30 mm thick at the apex, rarely up to 60 mm below, clavate to bulbous when young, rarely equal, becoming ventricose or remaining bulbous to clavate, usually reticulate over the upper portion, often smooth lower down, at times entirely reticulate, solid, firm, dry, fleshy, pallid upward, and becoming pale brown or leather-colored to darker below, often stained olivaceous where handled.

Spores pinkish to deep vinaceous or vinaceous brown in deposit, 11.0-14.6(-20) × 3.7-4.4(-5.0) μm, ellipsoidal to subfusiform, inequilateral in side view, with a plage, cylindrical-elliptic in face view, smooth, thin-walled, hyaline, nonamyloid to dextrinoid. Basidia four-spored, short-clavate, 18-25.6 × 7.0-10.2 μm, hyaline. Pleurocystidia fusiform-ventricose with mucronate to tapered apices, 36-44(-50) × 8.0-11.0(-14.0) μm, thin-walled; contents granular, yellowish, dextrinoid. Cheilocystidia similar to pleurocystidia, 24.8-44.0 × 7.3-11.0 μm, hyaline to yellowish.

Tube trama divergent from a central strand of subparallel hyphae; hyphae cylindrical, 3.7-8.0 μm wide, hyaline to yellowish, gelatinous, nonamyloid. Subhymenium filamentous, interwoven; hyphae cylindrical, 3.7-4.4 μm, hyaline. Cap cuticle filamentous, interwoven; hyphae cylindrical, 3.7-7.0(-9.4) μm wide, hyaline to yellowish brown, thin-walled to slightly encrusted, with cylindrical to clavate hyphal end cells, projecting from a matted trichoderm. Cap trama loosely interwoven; hyphae more or less cylindrical, 4.6-14.3 μm wide, hyaline, thin-walled; lower layer of cap trama more compactly interwoven, almost gelatinous; hyphae cylindrical, 3.7-7.0 μm wide. Caulocystidia clavate to clavate-mucronate, or fusiform-ventricose with tapered apices, 20-32.9 × 7.3-8.8(-12.0) μm, hyaline to ocherous, thin-walled, clustered on reticulations; stalk hyphae parallel to compactly interwoven, cylindrical, hyaline, 3.7-15.5 μm, narrower on the surface (3.7-5.1 μm). Oleiferous hyphae refractive, yellowish, 4.0-8.8 μm wide, scattered in the tube trama and subcutis of the cap. Clamp connections not seen.

Useful references. The following books contain information on *T. felleus* pertinent to North America.

- *Edible and Poisonous Mushrooms of Canada* (Groves 1979)—A color photograph and a description and discussion are given.
- *Mushrooms of North America* (Miller 1972)—A color photograph and a description and discussion are given.
- *The Mushroom Hunter's Field Guide* (Smith 1974)—A black-and-white photograph, a description and discussion, and a color photograph (=*T. felleus* var. *rubrobrunneus*) are presented.

Verpa Swartz ex Fr. — Morchellaceae

The genus *Verpa* belongs to the Ascomycota, in the class Discomycetes. It closely resembles *Morchella,* known as the true morels, and has the general appearance of *Gyromitra* and *Helvella,* two additional genera of false morels. *Gyromitra* is discussed in Chapter 8, and *Helvella* and *Morchella* are treated earlier in this chapter, alphabetically listed. See Chapters 2 and 3, "Ascomycota," for a discussion of the main features of this division.

The stalked, more or less conic cap of *Verpa* species causes some confusion with the genus *Morchella.* In addition, the caps of *Verpa bohemica* have longitudinal folds and ridges forming somewhat irregular, longitudinal pits or grooves. In *Verpa conica* Swartz ex Fr. the cap surface is almost smooth to slightly wrinkled. The best field characteristic for separating *Verpa* and *Morchella* is the attachment of the cap to the stalk. The cap in *Verpa* is skirt-like, attached at the stalk apex only, leaving the margin of the cap free. In *Morchella* the stalk and cap are attached from very near the edge of the cap margin to the apex of the stalk so that the cap margin is almost continuous with the stalk, not skirt-like. In the half-free morel, *Morchella semilibera* DC. ex Fr. (=*Morchella hybrida* Sow. ex Grev.), more or less half the margin of the cap is free from the stalk, and this species is the one most likely to be confused with *Verpa*

Fig. 176. *Verpa bohemica*: A, spores; B, asci and paraphysis. Scale line 10 μm

bohemica. Because they fruit at the same time of year, collectors should learn both these species. The best way to separate them is by the number of spores in the ascus. *Verpa bohemica* typically has two, but sometimes three, spores per ascus, whereas *Morchella semilibera* has eight spores per ascus. *Verpa bohemica* is sometimes placed in the genera *Mitrophora* or *Ptychoverpa*, because of the two-spored asci.

Verpa conica has not been reported as poisonous. However, some people cannot tolerate *Verpa bohemica*, which was reported by Wells and Kempton (1967) to have caused gastrointestinal upset in some individuals.

Simons (1971) states that a few hours after eating *V. bohemica*, he became so uncoordinated that even simple tasks like writing proved difficult. Smith (1974) says most people find *V. bohemica* edible but he suggests it should not be eaten in large quantities or for several days in sequence. He reports lack of muscular coordination 4-5 hours after eating *V. bohemica.*

Verpa bohemica (Krombh.) Schroet.

Figs. 176, 234

Common names. Early morel, two-spored morel.

Distinguishing characteristics. Cap 20-35 mm long, 9-30 mm wide, conic to bell-shaped, attached at the stalk apex; margin free, hanging down like a skirt; surface with longitudinal wrinkles, ridges, or folds that have corresponding longitudinal depressions or pits usually formed from crossveins that connect ridges; color yellowish brown to somewhat reddish brown. Flesh fragile, brittle. Stalk 60-80 mm long, 10-25 mm thick, usually more or less cylindrical although sometimes enlarged somewhat below or tapered upward, sometimes compressed and slightly furrowed; color white to cream or at times becoming pale tan with age, hollow and with a single channel when mature. Two or rarely three spores per ascus, ellipsoidal to subcylindrical, 45-60.8(-84) × 15-22 μm in size.

Observations. *V. bohemica* is one of the first fleshy fungi to appear in the spring. *V. conica*, another spring species, has a tapered, more or less slender stalk and a smooth to slightly wrinkled, dark brown to olive-tinged cap surface. The spores are smaller, 22-26 × 12-16 μm (Smith and Smith 1973), and there are eight spores per ascus.

Distribution and seasonal occurrence. This species is widely distributed in North America and at times is locally abundant. It fruits in the spring, often before the trees have leafed out, and is among the first fleshy fungi to appear, as early as March in some areas. Smith (1975) reports it from late winter through early spring in the west.

Habitat and habit. *Verpa bohemica* grows scattered or in groups, rarely solitary, and is terrestrial. It seems to favor moist to wet, rich soil and often occurs along the edges of swamps or streams. It seems to be most common in deciduous forests and is often hidden among dead leaves.

Technical description. Cap 20-35 mm long, 9-20(-30) mm wide, conic to nearly conic or bell-shaped, attached to the apex of the stalk and hanging down like a skirt; margin free and often upturned; surface of hymenium wrinkled with longitudinal folds or shallowly reticulate with irregular pits, sometimes smooth, somewhat granulose; color yellowish tan, yellowish brown, or moderate reddish brown, whitish on the undersurface. Flesh brittle, fragile.

Stalk 60-80 mm long, 10-25 mm wide, cylindrical to narrowly clavate, tapered upward, glabrous to granulose or somewhat floccose-scaly, especially toward the base, stuffed, becoming hollow, somewhat compressed and slightly furrowed; color whitish to yellowish or cream, at times becoming pale tan. Flesh brittle, whitish.

Spores yellowish in deposit, 45.0-60.8(-84) × 15-22 μm, ellipsoidal to subcylindrical, smooth, hyaline. Asci two-spored, rarely three-spored, cylindrical, 144-250 × 14.0-23.4 μm wide, hyaline. Stalk hyphae interwoven in the interior, more loosely interwoven on the stalk surface, 5.4-9.4(-14.7) μm wide, hyaline.

Useful references. The following books contain information on *V. bohemica* pertinent to North America.

- *A Field Guide to Western Mushrooms* (Smith 1975)—A color photograph and a description and discussion are given.
- *Edible and Poisonous Mushrooms of Canada* (Groves 1979)—A color photograph and a description and discussion are given.
- *Mushrooms of North America* (Miller 1972)—A color photograph and a description and discussion are given.
- *The Mushroom Hunter's Field Guide* (Smith 1974)—Black-and-white and color photographs and a description and discussion are given.
- *The Savory Wild Mushroom* (McKenny 1971)—A black-and-white photograph and a description and discussion are given.

Literature

Abdel-Malak, S. Chemotaxonomic significance of alkaloids and cyclopeptides in *Amanita* species. Ph.D. Thesis. Ann Arbor, MI: University of Michigan microfilms; 1974.

Alder, A. E. Die Pilzvergiftungen in der Schweiz im Jahre 1948. Schweiz. Z. Pilzkd. 28:122-132; 1950.

Alder, A. E. Die Pilzvergiftungen in der Schweiz in den Jahren 1950 und 1951. Schweiz. Z. Pilzkd. 31:111-119; 1953.

Alder, A. E. Die Pilzvergiftungen in der Schweiz in den Jahren 1952 und 1953. Schweiz. Z. Pilzkd. 34:4-11; 1956.

Alder, A. E. Die Pilzvergiftungen in der Schweiz in den Jahren 1956 und 1957. Schweiz. Z. Pilzkd. 37:75-82; 1959.

Alder, A. E. Die Pilzvergiftungen in der Schweiz während 40 Jahren. Schweiz. Z. Pilzkd. 38:65-73; 1960.

Alder, A. E. Die Pilzvergiftungen in der Schweiz in den Jahren 1958 und 1959. Schweiz. Z. Pilzkd. 39:185-194; 1961.

Alder, A. E. Die Pilzvergiftungen in der Schweiz in den Jahren 1960 und 1961. Schweiz. Z. Pilzkd. 41:149-158; 1963.

Alder, A. E. Die Pilzvergiftungen in der Schweiz im Jahre 1962. Schweiz. Z. Pilzkd. 42:97-101; 1964.

Alder, A. E. Die Pilzvergiftungen in der Schweiz im Jahre 1963. Schweiz. Z. Pilzkd. 44:33-44; 1966.

Alder, A. E. Die Pilzvergiftungen in der Schweiz in den Jahren 1964-65. Schweiz. Z. Pilzkd. 45:33-37; 1967.

Anderson, J. B.; Ullrich, R. C. Biological species of *Armillariella mellea* in North America. Mycologia 71:402-414; 1979.

Atkinson, G. F. Preliminary notes on some new species of Agaricaceae and *Clavaria*. Ann. Mycol. 7:367-368; 1909.

Ayer, F. *Rhodophyllus vernus* (Lund.) Romagn. Note critique, toxicité. Schweiz. Z. Pilzkd. 52:17-19; 1974.

Bas, C. Morphology and subdivision of *Amanita* and a monograph on its section *Lepidella*. Persoonia 5(4): 285-579; 1969.

Benedict, R. G.; Tyler, V. E., Jr.; Brady, L. R. Chemotaxonomic significance of isoxazole derivatives in *Amanita* species. Lloydia 29:333-342; 1966.

Benedict, R. G. Mushroom toxins other than *Amanita*. Kadis, S.; Ciegler, A.; Ajl, S. J., eds. Microbiological toxins. Vol. 8. New York, NY: Academic Press; 1972.

Bessey, E. A. A case of poisoning by *Lepiota morgani*. Mycologia 31(1):109-110; 1939.

Bondartsev, A. S. The Polyporaceae of the European USSR and Caucasia. Leningrad, USSR: Acad. Sci.; 1953 [English translation published by the U.S. Department of Agriculture 1971].

Boston Mycological Club. The *Clitocybe illudens* affair. Boston Mycol. Club Bull. 4(1):3-4; 1963.

Bschor, F.; Kohlmeyer, J.; Mallach, H. J. Neue Vergiftungsfälle durch *Paxillus involutus*. Schweiz. Z. Pilzkd. 29:1-3; 1963.

Bschor, F.; Mallach, H. J. Vergiftungen durch den Kahlen Kremplinge (*Paxillus involutus*), eine geniessbare Pilzart. Arch. Toxicol. 20:82-95; 1963.

Canfield, E. R.; Gilbertson, R. L. Notes on the genus *Albatrellus* in Arizona. Mycologia 63:964-971; 1971.

Carey, S. *Clitocybe illudens*: its cultivation, chemistry and classification. Mycologia 66:951-968; 1974.

Carrano, R. A.; Malone, M. H. Pharmacologic study of norcaperatic and agaric acids. J. Pharm. Sci. 56:1611-1614; 1967.

Charles, V. K. Mushroom poisoning caused by *Lactarius glaucescens*. Mycologia 34:112-113; 1942.

Charters, A. D. Mushroom poisoning. Cent. Afr. J. Med. 6:213-214; 1960.

Chesnut, V. K. Poisonous properties of the green-spored *Lepiota*—Plate V. Asa Gray Bull. 8(5):87-93; 1900.

Chilton, W. S. Poisonous mushrooms. Pac. Search, March; 1972.

Chilton, W. S.; Ott, J. Toxic metabolites of *Amanita pantherina*, *A. cothurnata*, *A. muscaria* and other *Amanita* species. Lloydia 39:150-157; 1976.

Clark, E.; Smith, C. Toxicological studies on the mushrooms *Clitocybe illudens* and *Inocybe infida*. Mycologia 5:224-233; 1913.

Coker, W. C.; Couch, J. N. The gasteromycetes of the Eastern United States and Canada. Chapel Hill, NC: University of North Carolina Press; 1928.

Collins, F. S. A case of *Boletus* poisoning. Rhodora 1(2):21-23; 1899.

Dearness, J. The personal factor in mushroom poisoning. Mycologia 3:75-78; 1911.

Dearness, J. Fleshy fungi. Mycologia 14(4):228-229; 1922.

Eilers, F. I.; Bernard, B. L. A rapid method for the diagnosis of poisoning by the mushroom *Lepiota morgani*. Am. J. Clin. Pathol. 60:823-825; 1973.

Eugster, C. H. Wirkstoffe aus dem Fliegenpilz. Naturwissenschaften 55:305-313; 1968.

Farlow, W. G. Poisoning by *Agaricus illudens*. Rhodora 1:43-45; 1899.

Farr, E. R.; Miller, O. K.; Farr, D. F. Biosystematics studies in the genus *Pholiota*, stirps *Adiposa*. Can. J. Bot. 55:1167-1180; 1977.

Fidalgo, O.; Fidalgo, M. E. P. A poisonous *Ramaria* from southern Brazil. Rickia Ser. Criptogam Arq. Bot. Estado Sao Paulo 5:71-91; 1970.

Fischer, O. E. Mushroom poisoning. Kauffman, C. H., ed. The Agaricaceae of Michigan. Michigan Geol. Biol. Surv., Biol. Ser. 1:825-875; 1918.

Fries, E. M. Systema Mycologicum. Vol. I. Greifswald, Sweden: Moritz; 1821.

Fries, E. M. Epicrisis Systematis Mycologici. Uppsala, Sweden: Typographia Academica; 1838.

Gelb, A. Wild, deadly fungi lure city hunters. The New York Times, Sat. Oct. 22; 1949.

Genest, K.; Hughes, D. W.; Rice, W. B. Muscarine in *Clitocybe* species. J. Pharm. Sci. 57:331-333; 1968.

Ginns, J.; Sunhede, S. Three species of *Christiansenia* (Corticiaceae) and the teratological gall on *Collybia dryophila*. Bot. Notiser 131:167-173; 1978.

Goos, R. D.; Shoop, C. R. A case of mushroom poisoning caused by *Tricholomopsis platyphylla*. Mycologia 72:433-435; 1980.

Graff, P. W. The green-spored *Lepiota*. Mycologia 19(6):322-326; 1922.

Groves, J. W. Poisoning by morels when taken with alcohol. Mycologia 56:779-780; 1964.

Groves, J. W. Edible and poisonous mushrooms of Canada. Ottawa, Ont.: Agriculture Canada. Agric. Can. Publ. 1112; 1979 [first printed in 1962].

Grund, D. W.; Harrison, K. A. Nova Scotian boletes. Bibl. Mycol. Bd. 47. Lehre, Germany: J. Cramer; 1976.

Grzymala, S. Les recherches sur la fréquence des intoxications par les champignons. Bull Med. Leg. 8(2):200-210; 1965.

Haines, J. H. On the edibility of *Amanita flavorubens*. Mycophile 18(1):3; 1977.

Hatfield, G. M.; Brady, L. R. Toxins of higher fungi. Lloydia 38(1):36-55; 1975.

Heim, R. Les champignons toxiques et hallucinogènes. Paris, France: N. Boubee and Cie.; 1963.

Henry, E. D.; Sullivan, G. Phytochemical evolution of some cantharelloid fungi. J. Pharm. Sci. 58:1497-1500; 1969.

Hesler, L. R. Mushrooms of the Great Smokies. Knoxville, TN: University of Tennessee Press; 1960.

Hesler, L. R.; Smith, A. H. The North American species of *Lactarius*. Ann Arbor, MI: University of Michigan Press; 1979.

Horne, W. T.; Condit, I. J. 1941. *Lepiota morgani*, an unwholesome fungus. Mycologia 33(6):666-667; 1941.

Kauffman, C. H. The Agaricaceae of Michigan. Vol. 1-2. New York, NY: Dover Publ. Inc.; 1971 [first printed in 1918].

Kerrigan, R. Poisonous species in the genus *Agaricus*. Mycena News 28:46; 1978.

Kienholz, J. R. A poisonous *Boletus* from Oregon. Mycologia 26:275-276; 1934.

Krieger, L. C. C. The mushroom handbook. New York, NY: Dover Publ. Inc.; 1967 [first printed in 1936].

Lee, T. M.; West, L. G.; McLaughlin, J. L. Screening for *N*-methylated tyramines in some higher fungi. Lloydia 38(5):450-452; 1975.

Lincoff, G.; Mitchel, D. H. Toxic and hallucinogenic mushroom poisoning—A handbook for physicians and mushroom hunters. New York, NY: Van Nostrand Reinhold Co.; 1977.

List, P. H.; Hackenberg, H. Die Scharf Schmeckenden Stoffe von *Lactarius vellereus* Fries. Schweiz. Z. Pilzkd. 39:97-102; 1973.

Lowe, J. L. Polyporaceae of North America. The genus *Fomes*. New York, NY: Technic. Publ. N.Y. State Coll. For. 80; 1957.

Magnusson, G.; Thoren, S.; Wickberg, B. Fungal extractives I. Structure of sesquiterpene dialdehyde from *Lactarius* by computer simulation of the NMR spectrum. Tetrahedron Lett. 12:1105-1108; 1972.

Mandell, J. Jack-o-lantern. Mycena News 28(11):41; 1978.

Maretic, Z. Poisoning by the mushroom *Clitocybe olearia* Maire. Toxicon 4:263; 1967.

Maretic, Z.; Russell, F. E.; Golobič, V. Twenty-five cases of poisoning by the mushroom *Pleurotus olearius*. Toxicon 13:379-381; 1975.

Marr, C. D.; Stuntz, D. E. *Ramaria* of western Washington. Bibl. Mycol. 38:1-232; 1973.

Matzinger, P.; Catalfomo, P.; Eugster, C. H. Isolierung von (2*S*,4*S*)-(+)-γ-Hydroxynorvalin und (2*S*,4*R*)-(−)-γ-Hydroxynorvalin aus *Boletus satanas* Lenz. Helv. Chim. Acta 55:1478-1490; 1972.

Mazzer, S. J. *Nolanea verna* in North America. Mich. Bot. 16(4):195-200; 1977.

McKenny, M. The savory wild mushroom (revised and enlarged by Stuntz, D. E.). Seattle, WA: University of Washington Press; 1971.

McMorris, T. C.; Anchel, M. Fungal metabolites. The structures of the novel sesquiterpenoids illudin-S and -M. J. Am. Chem. Soc. 87:1594-1600; 1965.

Miller, O. K., Jr. Mushrooms of North America. New York, NY: E. P. Dutton & Co., Inc.; 1972.

Miyata, J. T.; Tyler, V. E., Jr.; Brady, L. R.; Malone, M. H. The occurrence of norcaperatic acid in *Cantharellus floccosus*. Lloydia 29:43-49; 1966.

Möller, F. H. Danish *Psalliota* species. Friesia 4:1-60; 135-220; 1950.

Murrill, W. A. Notes and brief articles. Mycologia 13:338; 1921.

Nakanishi, K.; Ohashi, M.; Tada, M.; Yamada, Y. Illudin S (lampterol). Tetrahedron Lett. 21:1231-1246; 1965.

Neuhoff, W. Die Milchling (*Lactarii*). Die Pilze Mitkleuropas Bd. IIb. Bad Heilbruon Obb.: J. Klinkhardt; 1956.

Orr, D. B. Poisoning by *Laetiporus sulphureus*. Mycena News 17(5):4; 1976.

Ott, J.; Guzmán, G.; Romano, J.; Luiz Diaz, J. Nuevos datos sobre los supuestos Licoperdaceos psicotropicos y dos caso de intoxicacicn provocodos por hongos del gaero *Scleroderma* en Mexico. Bol. Soc. Mex. Micol. 9:67-76; 1975.

Overholts, L. O. The Polyporaceae of the United States, Alaska and Canada. Ann Arbor, MI: University of Michigan Press; 1953.

Petersen, R. H. The genera *Gomphus* and *Gloeocantharellus* in North America. Nova Hedwigia 21:1-112; 1971.

Petersen, R. H. Contribution toward a monograph of *Ramaria* I. Some classic species redescribed. Am. J. Bot. 61(7):739-748; 1974.

Petersen, R. H. Contribution toward a monograph of *Ramaria* III. *R. sanguinea*, *R. formosa*, and two new species from Europe. Am. J. Bot. 63(3):309-316; 1976.

Pilát, A. Mushrooms. London, England: Spring Books; 1951.

Pilát, A. Mushrooms and other fungi. London, England: P. Nevill; 1961.

Price, H. W. Mushroom poisoning due to *Hebeloma crustuliniforme*. Am. J. Dis. Child. 34:441-442; 1927.

Schultes, R. E.; Hofmann, A. The botany and chemistry of hallucinogens. Springfield, IL: Charles E. Thomas; 1973.

Schmidt, J.; Hartmann, W.; Würstlin, A.; Deicher, H. Acute renal failure due to immune-haemolytic anaemia after ingestion of *Paxillus involutus*. Dtsch. Med. Wochenschr. 96(28):1188-1191; 1971 [in German].

Seaver, F. J. Recent mushroom poisonings. J. N.Y. Bot. Gard. 40:236-237; 1939.

Shaffer, R. L. The subsection *Compactae* of *Russula*. Brittonia 14:254-284, 18 figs., plates 7-14; 1962.

Shaffer, R. L. Poisoning by *Pholiota squarrosa*. Mycologia 57:318-319; 1965.

Shaffer, R. L. The subsection *Lactarioideae* of *Russula*. Mycologia 56:202-231; 1964.

Shaffer, R. L. Notes on the subsection *Crassotunicatinae* and other species of *Russula*. Lloydia 33:49-96; 1970.

Shaffer, R. L. Some common North American species of *Russula* subsect. Emeticinae. Beih. Nova Hedwigia 51:207-237, plates 49-54; 1975.

Simons, D. M. The mushroom toxins. Del. Med. J. 43(7):177-187; 1971.

Singer, R. The Agaricales in modern taxonomy. Vaduz, Germany: J. Cramer; 1975.

Smith, A. H. Studies in the genus *Agaricus*. Pap. Mich. Acad. Sci. Arts Lett. 25(1939):107-138; 1940.

Smith, A. H. Mushrooms in their natural habitats. 2 Vols. Portland, OR: Sawyer's, Inc.; 1949.

Smith, A. H. *Tricholomopsis* (Agaricales) in the western hemisphere. Brittonia 12(4):41-70; 1960.

Smith, A. H. *Tricholoma irinum* in Michigan. Bot. 1:51-53; 1962.

Smith, A. H. The mushroom hunter's field guide. Ann Arbor, MI: University of Michigan Press; 1974 [first printed in 1963].

Smith, A. H. A field guide to western mushrooms. Ann Arbor, MI: University of Michigan Press; 1975.

Smith, A. H.; Hesler, L. R. The North American species of *Pholiota*. New York, NY: Hafner Publ. Co.; 1968.

Smith, A. H.; Morse, E. E. The genus *Cantharellus* in the western Untied States. Mycologia 39:497-534; 1947.

Smith, A. H.; Thiers, H. D. The boletes of Michigan. Ann Arbor, MI: University of Michigan Press; 1971.

Smith, H. V. A revision of the Michigan species of *Lepiota*. Lloydia 17:307-328; 1954.

Smith, H. V.; Smith, A. H. How to know the non-gilled fleshy fungi. Dubuque, IA: William C. Brown Co.; 1973.

Snell, W. H.; Dick, E. A. The boleti of northeastern North America. Lehre, Germany: J. Cramer; 1970.

Stadelmann, R. J.; Müller, E.; Eugster, C. H. Über die Verbreitung der stereomeren Muscarine innerhalb der Ordnung der Agaricales. Helv. Chim. Acta 59:2432-2436; 1976.

Stevenson, J. A.; Benjamin, C. R. *Scleroderma* poisoning. Mycologia 53:438-439; 1961.

Stubbs, A. H. 1971. Wild mushrooms of the central midwest. Lawrence, KS: University of Kansas Press; 1971.

Sundberg, W. The Family Lepiotaceae in California. M.A. Thesis. San Francisco, CA: San Francisco State University; 1967.

Thiers, H. D. California boletes. I. Mycologia 57:524-534; 1965.

Thiers, H. D. California mushrooms, a field guide to the boletes. New York, NY: Hafner Publ. Co.; 1975.

Thiers, H. D.; Halling, R. E. California boletes. V. Two new species of *Boletus*. Mycologia 68(5):976-983; 1976.

Thomas, H. N.; Mitchel, D. H.; Rumack, B. W. Poisoning from the mushroom *Stropharia coronilla* (Bull. ex Fr.) Quél. J. Arkansas Med. Soc. 73:311-312; 1977.

Tyler, V. E., Jr. Poisonous mushrooms. Stolman, A., ed. Progress in chemical toxicology. Vol. 1. New York, NY: Academic Press, pp. 339-384; 1963.

Tylutki, E. E. Mushrooms of Idaho and the Pacific Northwest. Discomycetes. Moscow, ID: University Press of Idaho; 1979.

Vergeer, P. Book review. Mycena News 28(1):2; 1977.

Watling, R. New fungi from Michigan. Notes Roy. Bot. Gard. Edinb. 29(1):59-66; 1969.

Webber, N. S. The genus *Helvella* in Michigan. Mich. Bot. 11:147-201; 1972.

Webster, H. A rash mycophagist. Rhodora 17:30-32; 1915.

Wells, V. L.; Kempton, P. E. Studies on the fleshy fungi of Alaska I. Lloydia 30:258-268; 1967.

Weresub, L. K. Congo red for instant distinction between poisonous *Lepiota molybdites* and edible *L. brunnea*. Can. J. Bot. 49:2059-2060, plate; 1971.

West, L. G.; Johnson, I. T.; McLaughlin, J. L. Hordenine from *Polyporus berkeleyi*. Lloydia 37(4):633-635; 1974.

Zeller, S. M. New or noteworthy agarics from the Pacific coast states. Mycologia 30(4):468-474; 1938.

Fig. 177. *Agaricus xanthodermus.*

Fig. 178. *Agaricus arvensis.*

Fig. 179. *Agaricus hondensis.*

Fig. 180. *Agaricus placomyces*

Fig. 181. *Agaricus meleagris.*

Fig. 182. *Agaricus silvaticus.*

Fig. 183. *Agaricus sylvicola.*

Fig. 184. *Armillariella mellea.*

Fig. 185. *Boletus erythropus.*

Fig. 186. *Boletus pulcherrimus.*

Fig. 187. *Boletus satanas.*

Fig. 188. *Boletus sensibilis.*

Fig. 189. *Clitocybe irina.*

Fig. 190. *Clitocybe nebularis.*

Fig. 191. *Collybia dryophila.*

Fig. 192. *Gomphus floccosus.*

Fig. 193. *Hebeloma crustuliniforme.*

Fig. 194. *Hebeloma sinapizans.*

Fig. 195. *Helvella crispa.*

Fig. 196. *Helvella elastica.*

Fig. 197. *Helvella lacunosa.*

Fig. 198. *Lactarius chrysorheus.*

Fig. 199. *Lactarius deceptivus.*

Fig. 200. *Lactarius piperatus* var. *piperatus.*

Fig. 201. *Lactarius repraesentaneus.*

Fig. 202. *Lactarius rufus.*

Fig. 203. *Lactarius scrobiculatus* var. *scrobiculatus.*

Fig. 204. *Lactarius subvellereus* var. *subdistans.*

Fig. 205. *Lactarius torminosus.*

Fig. 206. *Lactarius uvidus.*

Fig. 207. *Lepiota molybdites.*

Fig. 208. *Lepiota naucina,* white form.

Fig. 209. *Lepiota naucina,* gray form.

Fig. 210. *Morchella esculenta* (*left*) and *Morchella angusticeps* (*right*).

Fig. 211. *Morchella conica.*

Fig. 212. *Mycena pura.*

Fig. 213. *Omphalotus illudens.*

Fig. 214. *Omphalotus olivascens.*

Fig. 215. *Paxillus involutus.*

Fig. 216. *Phaeolepiota aurea.*

Fig. 217. *Pholiota abietis.*

Fig. 218. *Pholiota highlandensis.*

Fig. 219. *Pholiota squarrosa.*

Fig. 220. *Laetiporus sulphureus.*

Fig. 221. *Ramaria formosa.*

Fig. 222. *Ramaria gelatinosa* var. *oregonensis.*

Fig. 223. *Ramaria gelatinosa* var. *oregonensis.*

Fig. 224. *Entoloma lividum.*

Fig. 225. *Nolanea verna.*

Fig. 226. *Russula emetica.*

Fig. 227. *Sarcosphaera crassa.*

Fig. 228. *Scleroderma citrinum.*

Fig. 229. *Stropharia coronilla.*

Fig. 230. *Tricholoma inamoenum.*

Fig. 231. *Tricholoma pardinum.*

Fig. 232. *Tricholoma pessundatum.*

Fig. 233. *Tylopilus felleus.*

Fig. 234. *Verpa bohemica.*

Appendix 1 The Collections

The collections used for describing and illustrating the fungi included in this text are listed below, arranged alphabetically by genus and species. The institutes where the collections are housed are abbreviated in the second column, with their full names and addresses given following this list. The name of the collector, the collection number when available, and the date and location of the collection are included for each entry. A list of additional collections used for determining distributions and habitats is on file at the National Mycological Herbarium, Biosystematics Research Institute, Research Branch, Agriculture Canada, Ottawa, Ont.

Agaricus arvensis	UBC	L. Broome and M. A. Waugh; 27 Oct. 1966; University of British Columbia Endowment Lands, Vancouver, B.C.
	WTU	J. F. Ammirati and M. McAskie; 7996; 26 Oct. 1977; Erindale College Campus, Mississauga, Ont.
Agaricus hondensis	UBC	M. A. Waugh; 172; 19 Nov. 1966; San Juan Islands, WA
		R. J. Bandoni; 2279; 17 Nov. 1961; Colwood, Vancouver Island, B.C.
Agaricus placomyces	DAOM	S. C. Hoare; 34692; 5 Oct. 1952; Ottawa, Ont.
	MICH	A. H. Smith; 14921; 15 Sept. 1939; Saginaw Forest, Ann Arbor, MI
	WTU	J. F. Ammirati; 7277; 9 Sept. 1977; Erindale College Campus, Mississauga, Ont.
Agaricus placomyces (western form, affin. *Agaricus meleagris)*	UBC	M. McKenny; 2769; 24 Oct. 1962; Olympic Park, WA
		A. W. L. Stewart; 5 Oct. 1966; University of British Columbia Campus, Vancouver, B.C.
Agaricus silvaticus	MICH	A. H. Smith; 14813; 6 July 1939; Joyce, WA
Agaricus sylvicola	UBC	R. Pillsbury; 13 Nov. 1960; Saturna Island, B.C.
	WTU	J. F. Ammirati; 7065; 29 July 1975; Booth's Rock Trail, Rock Lake, Algonquin Park, Ont.
		J. F. Ammirati; 7221; 4 Aug. 1975; Hemlock Bluff Trail, Algonquin Park, Ont.
Albatrellus dispansus	MICH	A. H. Smith; 82806; 9 Sept. 1972; Priest Lake, ID
Amanita bisporigera	DAOM	R. Shoemaker; 48137; 28 Aug. 1955; Port Elgin, Bruce County, Ont.
		R. T. Pennoyer; 51934; 13 Sept. 1956; Burnet, Que.
		M. Elliot and A. Sarkar; 56596; 1 Aug. 1957; Gatineau Park, Que.
		M. Pantidou and M. Elliot; 62849; 3 Sept. 1958; Bell's Corners, Ottawa, Ont.
		K. Harrison; 111769; 24 Sept. 1952; Kentville, N.S.
	WTU	D. M. Simons; 1496, 1497, 1498; 29 Aug. 1971; Fahrney's Woods, Chester County, PA
		D. M. Simons; 1740; 22 Sept. 1974; Cumberland County, NJ
		D. M. Simons; 1784; 20 July 1975; Woodlawn, New Castle County, DE
	WTU	J. A. Traquair; 856; 1 Aug. 1977; Port Frank, Ont.
Amanita cothurnata	BPI	K. H. McKnight; 14180; 21 Aug. 1974; Prince George County, MD
	DAOM	S. C. Hoare (Thompson); 28896; 19 July 1952; Cantley, Que.
	MICH	R. L. Shaffer; 5630; 2 Aug. 1967; Sillery, Quebec County, Que.
Amanita crenulata	DAOM	E. Blackford; 108235 (isotype), 1899; near Boston, MA
Amanita frostiana	DAOM	K. A. Harrison; 44445; 26 Aug. 1953; Kentville, N.S.
		R. T. Pennoyer; 54099; 13 Sept. 1956; Burnet, Que.

	WTU	J. F. Ammirati; 7122; 30 July 1975; Hemlock Bluff Trail, Algonquin Park, Ont.
		J. F. Ammirati; 7917; 18 Sept. 1977; Temagami, Ont.
Amanita gemmata	DAOM	H. Güssow et al.; F5581; 14 July 1935; Mount Burnet, Que.
		J. W. Groves; 16754; 25 June 1946; Ile Perrot, Que.
		K. A. Harrison; 44455; 25 Aug. 1953; Kentville, N.S.
		J. Ginns; 142201; 20 July 1971; Cantley, Que.
		J. Ginns; 143180; 20 July 1972; Cantley, Que.
		J. Ginns; 144194; 13 July 1972; Cantley, Que.
	UBC	R. J. Bandoni; 9 Nov. 1959; University of British Columbia Campus, Vancouver, B.C.
		F. Waugh; 29 Oct. 1958; University of British Columbia Endowment Lands, Vancouver, B.C.
Amanita muscaria	DAOM	E. Kuyt; 56681; 17 Aug. 1957; Mosquitoh, N.W.T.
		K. Harrison; 112673; 9 Sept. 1958; Beaver Bank, Halifax County, N.S.
		G. H. Davies; 115661; Aug. 1966; Carlea, Sask.
		T. V. Clark; 130893; 4 Sept. 1970; Duncan Creek Road between Mayo and Keno, Y.T.
		J. Ginns; 141998; 28 June 1971; Cantley, Que.
		158685 and 158686; 25 Aug. 1976; Staveley (near Fort McLeod), Alta.
	UBC	Chalette; 6 Oct. 1974; Blackwater Lake (near Darey), B.C.
		J. Lanko; 28 Oct. 1960; University of British Columbia Campus, Vancouver, B.C.
		F. Waugh; 18 Oct. 1958; Lynn Valley Park, North Vancouver, B.C.
	WTU	J. F. Ammirati; 6715; 8 July 1975; Algonquin Park, Ont.
		J. F. Ammirati; 7126; 30 July 1975; Algonquin Park, Ont.
		J. F. Ammirati; 7161; 1 Aug. 1975; South Lookout Trail, Algonquin Park, Ont.
		J. F. Ammirati; 7450; 21 Aug. 1976; Harquail Lake, Chibougamau Park, Que.
		M. McAskie; 7879; 17 Aug. 1977; Temagami, Ont.
		J. F. Ammirati; 7971 and 7972; Oct. 1977; Macey Lake Bog, Simcoe County, Ont.
Amanita ocreata	BPI	C. F. Baker; 5136 (isotype), 1909; Claremont, CA
	SFSU	G. Breckon; 227; 1 March 1965; San Francisco Watershed, San Mateo County, CA
		G. Breckon; 359; 16 April 1966; San Francisco Watershed, San Mateo County, CA
		G. Breckon; 696; 22 April 1967; Oakhurst, Madera County, CA
		H. D. Thiers; 7574; 25 March 1960; Lake Merced, San Francisco, CA
Amanita pantherina	BPI	K. H. McKnight; 13840; 26 Aug. 1973; Summit County, UT
	DAOM	A. F. Porsild and A. J. Breitung; 48918; 8 Aug. 1944; Sheldon Lake, Y.T.
		J. Y. Tsukamoto; 89850; 8 Sept. 1962; Alaskan Highway, Mile 1016, Y.T.
		J. W. Groves et al.; 109783; 12 June 1965; Francis Park, Victoria, B.C.
		H. M. E. Schalkwyle; 149370; 14 Aug. 1974; Manning Provincial Park, B.C.
	UBC	N. Corfman; 12 July 1964; University of British Columbia Campus, Vancouver, B.C.
		J. MacDonald; 24 April 1973; Downer Point, Hornsby Island, B.C.
Amanita phalloides	DAOM	M. Concannon; 144075; 5 Nov. 1972; Marin County, CA
	SFSU	H. D. Thiers; 28598; Marin County, CA

	WTU	D. M. Simons; 1816a; 11 Oct. 1975; Dennisville, NJ
		D. M. Simons; 1817a; 12 Oct. 1975; Berk County, PA
		J. Wolf; 28 Sept. 1975, 6 and 10 Oct. 1977; Durand Eastman Park, Rochester, NY
Amanita verna	WTU	D. M. Simons; 1834; 6 Aug. 1976; Lubob Farm, Upshur County, WV
		D. M. Simons; 1895; 21 Aug. 1977; Woodlawn Tract, New Castle County, DE
Amanita virosa	DAOM	J. W. Groves; 71857; 5 Sept. 1960; Isle Madame, Cape Breton, N.S.
		K. Harrison; 112855; 9 Oct. 1959; Kentville, N.S.
	WTU	J. F. Ammirati; 7808, 7809, 7810; 7 Aug. 1977; High Park, Toronto, Ont.
		J. F. Ammirati; 7914; 18 Sept. 1977; Temagami, Ont.
		D. M. Simons; 1490; 26 Aug. 1971; Woodlawn, New Castle County, DE
		D. M. Simons; 1493; 28 Aug. 1971; Woodlawn, New Castle County, DE
		D. M. Simons; 1504; 7 Sept. 1971; Woodlawn, New Castle County, DE
		D. M. Simons; 1519; 16 Sept. 1971; Alapocas Park, New Castle County, DE
		D. M. Simons; 1708; 13 Oct. 1973; Hoopes Reservoir, New Castle County, DE
Armillariella mellea	ALTA	I. Tsuneda; 7328; Sept. 1975; Edmonton, Alta.
	DAOM	R. F. Cain; 80307; 17 Oct. 1931; Aurora, Ont.
		K. A. Harrison; 110626; 1 Sept. 1930; Kentville, N.S.
	UBC	R. J. Bandoni; 5378; 19 Oct. 1970; University of British Columbia Campus, Vancouver, B.C.
	WTU	J. F. Ammirati; 7315; 23 Sept. 1975; Halton Forest, Ont.
Boletus erythropus	MICH	M. Wills; 77780 (A. H. Smith No.); 10 Aug. 1969; Pellston Hills, Emmet County, MI
		A. H. Smith; 77834; 12 Aug. 1969; Tahquamenon Falls State Park, Upper Luce County, MI
Boletus luridus	MICH	M. Wells; 75716 (A. H. Smith No.); 25 July 1968; Pellston Hills, Emmet County, MI
Boletus pulcherrimus	MICH	A. H. Smith; 30521; 23 Aug. 1948; Mount Rainier National Park, Pierce County, WA
		A. H. Smith; 31397; 18 Sept. 1948; Mount Rainier National Park, Pierce County, WA
Boletus satanas	MICH	A. H. Smith; 79143; 18 Oct. 1970; Aloha, Washington County, OR
Boletus sensibilis	MICH	A. H. Smith; 74479; 10 July 1967; Berry Creek, Cheboygan County, MI
		A. H. Smith and H. D. Thiers; 75624; 22 July 1968; Berry Creek, Cheboygan County, MI
Clitocybe clavipes	WTU	J. F. Ammirati; 7556; 26 Aug. 1976; south end of Lake Mistassini, Que.
Clitocybe dealbata	DAOM	D. Malloch; 161196; Forest Technical School, Dorset, Ont.
	MICH	A. H. Smith; 58027; 8 Sept. 1957; Betsy Lake area, Luce County, MI
		A. H. Smith; 75181; 12 Oct. 1967; Proud Lake, Oakland County, MI
		L. J. Tanghe; 14 Oct. 1977; Rochester, NY
Clitocybe dilatata	DAOM	J. A. Calder and L. G. Billard; 55727; 18 Aug. 1949; Dawson, Y.T.
	UBC	M. McKenny; 24 Oct. 1962; Olympic National Park, WA
		M. A. Waugh; 2 Oct. 1965; Vancouver, B.C.
Clitocybe irina	DAOM	R. F. Cain; 115058; 30 Sept. 1962; Nashville, York County, Ont.
		Maxwell; 161194; 7 Oct. 1971; east of Ottawa, Ont.
	WTU	J. F. Ammirati; 7951, 7956; 28 Sept. 1977; Eugenia Falls, Ont.

Clitocybe morbifera	MICH	H. Bishop; Nov. 1941; Louisville, KY
Clitocybe nebularis	MICH	A. H. Smith; 73549; 20 Sept. 1966; Upper Priest Lake, Bonner County, ID
		H. E. Bigelow; 2275; 16 Oct. 1954; Lower Tahoma Creek, Mount Rainier National Park, WA
Clitocybe rivulosa	MASS	H. E. Bigelow; 14592; 20 Sept. 1965; Amherst, Hampshire County, MA
	MICH	A. H. Smith; 17026; Hurricane Ridge Road, Olympics, WA
		A. H. Smith; 63394; 15 Oct. 1960; Wilderness State Park, Emmet County, MI
Collybia dryophila	DAOM	T. McCabe; 114825; 29 June 1940; Fort St. James, B.C.
	UBC	S. Redhead; 48; 28 July 1970; University of British Columbia Campus, Vancouver, B.C.
	WTU	J. F. Ammirati; 6797; July 1975; Wildlife Research Station, Algonquin Park, Ont.
		O. Zurba; 30
Conocybe cyanopus	MICH	V. E. Tyler; 4 Sept. 1961; Seattle, King County, WA
Conocybe filaris	DAOM	M. C. Melburn; 87392; 16 Oct. 1961; John Dean Park, Victoria, B.C.
	MICH	A. H. Smith; 16537; 1 Sept. 1941; Baker Lake, WA
Conocybe smithii	DAOM	D. Malloch and E. Ohenoja; 148924; 24 July 1974; Central Experimental Farm, Carlton County, Ont.
	MICH	A. H. Smith; 43206; 24 Sept. 1953; Wilderness Park, Emmet County, MI
Coprinus atramentarius	UBC	G. Rouse; 11 Nov. 1958; University of British Columbia Endowment Land, Vancouver, B.C.
	WTU	J. F. Ammirati; 7276; 9 Sept. 1975; Sheridan Park, Mississauga, Ont.
Cortinarius gentilis	WTU	M. Moser; 70181; 26 Aug. 1970; near Ekornahult, Femsjö, Småland, Sweden
Cortinarius orellanus	WTU	M. Moser; Sept. 1974; Hartwald near Hombourg, Elsass, Frankreich, Germany
Cortinarius speciosissimus	WTU	M. Moser; 72/217; 11 Aug. 1972; near St. Kulkagölen, Femsjö, Småland, Sweden
Entoloma lividum	MICH	A. H. Smith; 872; 4 Sept. 1931; Upper Brookside, Colchester County, N.S.
		F. Hoseney; 84434 (A. H. Smith No.); Eberwhite Woods, Ann Arbor, MI
Entoloma nidorosum	MICH	K. A. Harrison; 7874; 7 Sept. 1968; Kentville, N.S.
		A. H. Smith; 43526; 28 Sept. 1953; Wilderness State Park, Emmet County, MI
Entoloma rhodopolium	MICH	S. J. Mazzer; 2561; 1 Sept. 1964; Eberwhite Woods, Washtenau County, MI
		S. J. Mazzer; 4469; 30 Aug. 1966; Bower Station Woods, Wayne County, MI
Galerina autumnalis	UBC	D. Stuntz; 20 Nov. 1966; Friday Harbor, San Juan Island, WA
	WTU	J. A. Traquair; 1242; 20 May 1978; Gilford area, Simcoe County, Ont.
Galerina marginata	DAOM	J. W. Groves; 27250
	MICH	A. H. Smith; 87160
Galerina venenata	MICH	W. B. Gruber; 1952, 1953; Portland, OR
		A. H. Smith; 83255; 29 Oct. 1972; Cispus Environmental Center, Randle, Lewis County, WA
Gomphus floccosus	MICH	A. H. Smith; 40986
	WTU	J. F. Ammirati; 6994; 28 July 1975; Found Lake, Algonquin Park, Ont.
		J. F. Ammirati; 7242; 5 Aug. 1975; Wildlife Research Station, Algonquin Park, Ont.
Gymnopilus aeruginosus	MICH	A. H. Smith; 62277; 1977; Michigan
Gymnopilus luteus	MICH	J. F. Ammirati; 2582; 1968; Michigan

Gymnopilus spectabilis	MICH	A. H. Smith; 54696; 14 Oct. 1956; Kaniksu National Forest, Bonner County, ID
	WTU	J. F. Ammirati; 7281; 14 Sept. 1975; Sheridan Park, Mississauga, Ont.
Gymnopilus validipes	MICH	A. H. Smith; 87285; 1976; Michigan
Gymnopilus viridans	MICH	J. F. Ammirati; 7751, 7752; 25 Sept. and 28 Sept., respectively, 1976; Mississauga, Ont.
Gyromitra ambigua	WTU	J. F. Ammirati; 7652; 30 Aug. 1976; Lac Nicabau, Chibougamau Park, Que.
		J. F. Ammirati; 7742; 5 Sept. 1976; Lac Dufresne, Abitibi Est County, Que.
Gyromitra esculenta	DAOM	W. T. Blair; 112933; 25 May 1954; Aylesford Lake, Kings County, N.S.
		J. Ginns; 155442; 13 May 1967; Victoria, B.C.
	WTU	J. F. Ammirati; 7770; 19 May 1977; Pog Lake, Algonquin Park, Ont.
		J. F. Ammirati; 7777; 18 May 1977; Lake of Two Rivers, Algonquin Park, Ont.
		J. F. Ammirati; 7783; 19 May 1977; Wildlife Research Station, Algonquin Park, Ont.
		J. F. Ammirati; 8369; 15–16 May 1979; Algonquin Park, Ont.
Gyromitra fastigiata	WTU	J. F. Ammirati; 8363; 9 May 1979; Halton Forest, Campbellville Road, Ont.
Gyromitra infula	WTU	J. A. Traquair; 6 Sept. 1977; Slave Lake, Alta.
	ALTA	R. Currah; 7538; 26 Aug. 1977; Carson Lake, Alta.
Hebeloma crustuliniforme	UBC	M. A. Waugh; 168; 19 Nov. 1968; San Juan Islands, WA
Hebeloma sinapizans	WTU	J. F. Ammirati; 7977; 14 Oct. 1977; Macey Lake Bog, Ont.
Inocybe fastigiata	MICH	R. L. Shaffer; 5843; 14 Aug. 1967; Villieu, Leirs County, Que.
Inocybe geophylla	MICH	A. H. Smith; 4916; 20 Sept. 1936; George Reserve, Pickney, MI
	UBC	R. J. Bandoni; 746; 10 Oct. 1959; Saturna Island, B.C.
Inocybe lacera	MICH	F. Hoseney; 770; 31 May 1968; George Reserve, Livingston County, MI
Inocybe mixtilis	MICH	R. L. Shaffer; 1344; 8 July 1957; Tahquamenon Falls State Park, Luce County, MI
Inocybe napipes	UBC	D. Stuntz; 30 Oct. 1965; Haney Forest, B.C.
Inocybe patouillardi	MICH	R. A. Maas Geesteranus; 11573; 24 June 1956; Wassenaar, north of The Hague, Netherlands
Inocybe pudica	UBC	M. A. Waugh; 166; 19 Nov. 1966; San Juan Islands, WA
Inocybe sororia	MICH	W. R. Gruber; 13 Nov. 1942; Apale Station, OR
Inocybe subdestricta	UBC	D. Stuntz; 30 Oct. 1965; University of British Columbia Forest, Haney, B.C.
Lactarius chrysorheus	MICH	A. H. Smith; 86232; 6 Sept. 1975; Half Moon Lake, Livingston County, MI
Lactarius deceptivus	MICH	F. Hoseney; 2355; 22 Aug. 1972; George Reserve, Livingston County, MI
Lactarius piperatus var. *glaucescens*	MICH	A. H. Smith; 64370; 18 Sept. 1961; Ann Arbor, Washtenaw County, MI
Lactarius piperatus var. *piperatus*	MICH	F. Hoseney; 2643; 15 Aug. 1973; Ann Arbor, Washtenaw County, MI
Lactarius repraesentaneus	MICH	J. F. Ammirati; 2244; 12 Aug. 1968; Pine Lake area, Huron Mountain Club, Marquette County, MI
Lactarius rufus	MICH	A. H. Smith; 68528; 20 July 1964; Priest River, Bonner County, ID
Lactarius scrobiculatus var. *canadensis*	MICH	J. F. Ammirati; 4817; 7 Aug. 1970; Canyon Lake Bog, Huron Mountain Club, Marquette County, MI

Lactarius subvellereus var. *subdistans*	MICH	N. S. Weber; 4081; 6 Aug. 1974; Gorge, University of Michigan Biological Station, Cheboygan County, MI
Lactarius torminosus	MICH	A. H. Smith; 36962; 16 July 1951; Tahquamenon Falls, MI
Lactarius uvidus	MICH	A. H. Smith; 84526; 23 Aug. 1973; Chippewa Lake, MI
Lactarius vellereus var. *vellereus*	MICH	K. A. Harrison; 11690; 3 Aug. 1973; Kentville, Kings County, N.S.
Laetiporus sulphureus	WTU	S. Martin; 54; 15 Sept. 1975; Mississauga, Peel County, Ont.
Lepiota clypeolarioides	MICH	A. H. Smith; 8885; 19 Nov. 1937; Smith River, CA
Lepiota helveola	MICH	A. H. Smith; 5714; 26 Nov. 1930; Alsae Mountain, OR
Lepiota molybdites	BPI	J. F. Ammirati; 6470; 25 Aug. 1973; Washington, D.C.
	DAOM	D. C. McArthur; 59769; 11 Aug. 1958; Eardley Road, Que.
		D. C. McArthur; 63893; 2 Sept. 1959; near Aylmer, Que.
Lepiota naucina	ACAD	K. A. Harrison; 372; 27 Sept. 1930; Research Station, Kentville, N.S.
	DAOM	157683
	UBC	M. A. F. Waugh; 13 Oct. 1963; Crescent Beach, B.C.
	WTU	J. F. Ammirati; 7930; 21 Sept. 1977; Erindale College Campus, Mississauga, Ont.
Lepiota rufescens	MICH	A. H. Smith; 24399; 10 Oct. 1946; Salmon River, Welches, OR
Morchella angusticeps	ACAD	K. A. Harrison; 9078; 21 May 1969; Harrington Road, King County, N.S.
	WTU	J. F. Ammirati; 8365; 9 May 1979; Halton Forest, Campbellville Road, Ont.
Morchella esculenta	WTU	J. F. Ammirati; 7784; 11 May 1977; Halton Forest, Campbellville Road, Ont.
Mycena pura	WTU	J. F. Ammirati; 7162; 1 Aug. 1975; South Lookout Trail, Algonquin Park, Ont.
		J. F. Ammirati; 7420; 20 Aug. 1976; Lake Nicabau, Chibougamau Park, Que.
Naematoloma fasciculare	UBC	M. A. F. Waugh; 30 Oct. 1952; Burnaby Lake, B.C.
Nolanea verna	MICH	S. J. Mazzer; 3998; Washtenaw County, MI
Omphalotus illudens	MASS	H. E. Bigelow; 16868; 5 Oct. 1971; Haydenville, MA
	WTU	J. F. Ammirati; 7880, 7881; 22 Aug. 1977; Mimico, Ont.
Panaeolus foenisecii	DAOM	158303; 31 Aug. 1927; Winnipeg, Man.
	UBC	S. Redhead; 332; 25 May 1971; Vancouver, B.C.
	WTU	J. A. Traquair; 819; 10 July 1977; Mississauga, Ont.
		J. A. Traquair; 820; 10 July 1977; Georgetown, Ont.
Panaeolus sphinctrinus	DAOM	113269; 15 Oct. 1965; Lakefield, Que.
		117570; 24 Aug. 1967; South March, Ont.
Panaeolus subbalteatus	MICH	A. H. Smith; 84164
	UBC	L. Broome and M. A. Waugh; 27 Aug. 1965; University of British Columbia Endowment Lands, Vancouver, B.C.
Paxillus involutus	ACAD	K. A. Harrison; 422; 2 Sept. 1930; Research Station, Kentville, N.S.
	DAOM	Kraschewski; 133841; 24 Sept. 1970; Ottawa, Ont.
	UBC	E. King; 25 Sept. 1958; University of British Columbia, Vancouver, B.C.
Phaeolepiota aurea	DAOM	Hume; 113986; Nov. 1965; Revelstoke, B.C.
	MICH	M. Wells and P. Kempton; 4967; 20 Sept. 1970; mile 40, Haines County cutoff road, AK
	UBC	S. Brough; 21 Sept. 1968; 10 miles north of Prince George, B.C.
		R. Stace-Smith; 9 Oct. 1961; Mount Baker, WA

Pholiota abietis	MICH	A. H. Smith; 16625; 5 Sept. 1941; Baker Lake, WA
		A. H. Smith; 73371; 14 Sept. 1966; Usk, Pend Oreille County, WA
Pholiota highlandensis	MICH	A. H. Smith; 73152; 24 Aug. 1966; near Manchester, Washtenaw County, MI
		A. H. Smith; 73532; 20 Sept. 1966; Upper Priest River, Boundary County, ID
Pholiota squarrosa	MICH	K. A. Harrison; 29 Aug. 1966; Beaver Road, Glacier National Park, B.C.
		A. H. Smith; 72388; 23 Aug. 1965; Tahquamenon, Chippewa County, MI
Psilocybe baeocystis	MICH	V. P. Shoemaker; 30 Oct. 1960; Milwaukee, Clackamas County, OR
		F. P. Sipe; 1281; 12 Nov. 1958; Eugene, OR
Psilocybe caerulipes	MICH	H. E. Bigelow; 3531; 26 July 1956; near Geurette, ME
		A. H. Smith; 49967; 5 Aug. 1955; Tahquamenon Falls State Park, Chippewa County, MI
Psilocybe cyanescens	WTU	M. Beug; 20 Nov. 1979; Olympia, WA
Psilocybe pelliculosa	MICH	A. H. Smith; 54268; 8 Oct. 1956; Nordman, ID
Psilocybe semilanceata	MICH	A. H. Smith; 18020; 19 Oct. 1941; Cape Flattery, WA
Psilocybe strictipes	MICH	A. H. Smith; 24554; 13 Oct. 1946; above Welches, Salmon River, OR
Psilocybe stuntzii	WTU	D. T. Leslie; 2709; 26 Oct. 1975; Randle, Lewis County, WA
		D. T. Leslie; 3996; 22 Nov. 1976; Tigard, Washington County, OR
Ramaria flavobrunnescens	CUP-A	W. C. Coker; 22639 (holotype); Battle's Park, NC
Ramaria formosa	WTU	C. Marr; M-513B; 1 Nov. 1966; Mason Lake, near Shelton, WA
Ramaria gelatinosa var. *gelatinosa*	NCU	W. C. Coker; C-2413 (holotype); Chapel Hill, NC
Ramaria gelatinosa var. *oregonensis*	WTU	C. Marr; M-250 (holotype); 10 Oct. 1965; Silver Springs, Pierce County, WA
Ramaria pallida (probably the same as *Ramaria acrisiccescens*)	WTU	C. Marr; M-535 (holotype, *Ramaria acrisiccescens*); Pleasant Valley, about 5 miles south of Elba, Lewis County, WA
Russula emetica	MICH	R. L. Shaffer; 1955; 16 Sept. 1957; Trout Lake, Upper Peninsula, Chippewa County, MI
		R. L. Shaffer; 2032; 20 Sept. 1957; Mud Lake, Cheboygan County, MI
Sarcosphaera crassa (as *S. eximia*)	MICH	K. A. Harrison; 7011; 9 Sept. 1967; Sandia Mountain near the crest, Sandoval County, NM
Scleroderma cepa	BPI	K. H. McKnight; 14041; 23 Nov. 1973; Prince George County, VA
Scleroderma citrinum	DAOM	Jackson; 87097; 24 Aug. 1938; Duchesnay, Que.
		S. Redhead; 151874; Grenadur Island, St. Lawrence Island National Park, Ont.
	WTU	J. F. Ammirati; 7215; 4 Aug. 1975; Algonquin Park, Ont.
Stropharia coronilla	DAOM	K. A. Harrison; 112610; 27 Sept. 1957; Kentville, N.S.
		155546; 1975; Ottawa, Ont.
	MICH	K. A. Harrison; 15 Oct. 1968; Kentville, N.S.
	UBC	R. J. Bandoni; 2770; 14 Oct. 1962; Spanish Banks, Vancouver, B.C.
Stropharia semiglobata	MICH	R. L. Shaffer; 5996; 29 Aug. 1967; St. Jean-Post-Joli, L'Islet County, Que.
		A. H. Smith; 2506; 20 Sept. 1935; Lake Crescent, WA
Tricholoma inamoenum	MICH	A. H. Smith; 80411

	HSC	C. Ardrey; 61 (Largent 5708); Coos County, OR
Tricholoma pardinum	WTU	C. Ovrebo; 496; Clearwater County, MN
Tricholoma pessundatum	WTU	C. Ovrebo; 362; Washtenaw County, MI
Tricholoma venenatum	WTU	C. Ovrebo; 296; Oakland County, MI
Tylopilus felleus	WTU	J. F. Ammirati; 7075a; 30 July 1975; Algonquin Park, Ont.
Verpa bohemica	UBC	R. J. Bandoni; 910; 19 April 1960; University of British Columbia, Vancouver, B.C.
		V. J. Krajina; 2 April 1958; Vancouver, B.C.
		M. Murry; 1618; Crawford Bay, B.C.
	WTU	J. F. Ammirati; 7764; 24 April 1977; Cap Dundas near Barrow Bay, Bruce Peninsula, Ont.

INSTITUTES HOUSING THE COLLECTIONS

ACAD	E. C. Smith Herbarium of Acadia University Wolfville, N.S. B0P 1X0 Curator: Dr. S. P. Vander Kloet
ALTA	University of Alberta Edmonton, Alta. T6G 2E9 Curator: Dr. L. L. Kennedy
BPI	National Fungus Collections Agricultural Research Center—West Beltsville, MD 20705 Curator: Dr. P. L. Lentz
CUP-A	Cornell University Department of Plant Pathology Ithaca, NY 14853
DAOM	National Mycological Herbarium Biosystematics Research Institute Research Branch, Agriculture Canada Central Experimental Farm Ottawa, Ont. K1A 0C6 Curator: Dr. J. A. Parmelee
HSC	Herbarium Department of Biology Humboldt State University Arcata, CA 95521 Curator (fungi): Dr. D. Largent
MASS	Herbarium of the University of Massachusetts Department of Botany University of Massachusetts Amherst, MA 01003 Curator: Dr. H. E. Bigelow
MICH	University of Michigan Herbarium University of Michigan Ann Arbor, MI 48109 Curator: Dr. R. Fogel
NCU	Herbarium of the University of North Carolina Chapel Hill, NC 27514
SFSU	San Francisco State University Herbarium Department of Biology San Francisco State University 1600 Holloway Ave. San Francisco, CA 94132 Curator: Dr. H. D. Thiers
UBC	University of British Columbia Vancouver, B.C. V6T 1W5 Curator: Dr. R. J. Bandoni
WTU	Herbarium Department of Botany, AK-10 University of Washington Seattle, WA 98195 Curator: Dr. M. F. Denton

Appendix 2 Abbreviations of Authorities

The abbreviations used for the names of authors of genera and species in this book are in accordance with the list published in *Plant Dis. Surv. Special Publ. 1* (Stevenson 1953; pp. 1233–1263). Names of authors that are not listed below are written out completely in this book.

Adans. Adanson
Alb. Albertini, von
Atk. Atkinson, G. F.
Auersw. Auerswald
Berk. Berkeley
Bolt. Bolton
Bon. Bonorden
Boud. Boudier
Bres. Bresadola
Britzelm. Britzelmayr
Br. Broome
Bull. Bulliard
Curt. Curtis
DC. de Candolle
Dill. Dillenius
Dur. Durieu
Farl. Farlow
Fr. Fries, E. M.
Gill. Gillet
Grev. Greville
Hoffm. Hoffmann
Hook. Hooker
Huds. Hudson
Kalchbr. Kalchbrenner
Karst. Karsten
Konr. Konrad
Krombh. Krombholz
Lév. Léveillé
L. Linnaeus
Lovej. Lovejoy
Maubl. Maublanc
Mich. Micheli
Mont. Montagne
Pat. Patouillard
Pers. Persoon
Quél. Quélet
Sacc. Saccardo, P. A.
Schaeff. Schaeffer
Schroet. Schroeter
Schulz. Schulzer
Schw. Schweinitz, von
Scop. Scopoli
Secr. Secretan
Sow. Sowerby
Speg. Spegazzini
Syd. Sydow, P.
Trott. Trotter
Tul. Tulasne, L. R.
Underw. Underwood
Velen. Velenovsky
Vittad. Vittadini
Weinm. Weinmann

Appendix 3 Ridgway Colors

The following color terms, used mainly in the technical descriptions of species in Part Two, are defined by R. Ridgway in *Color Standards and Color Nomenclature* (Washington, D.C.; 1912). Note, however, that in this book we use the spelling ocherous instead of Ridgway's ochraceous.

The other terms used throughout the text are general descriptions of colors and as such are more approximate and subjective than are the Ridgway colors given below. Compare color carefully with that in the photographs when it is a vital character in identification. Keep in mind that the particular shade of color of any species varies with the age and growing conditions of the specimen and, as well, is dependent on the perception of the observer.

amber yellow
antimony yellow
antique brown
apricot orange
argus brown
avellaneous
bittersweet pink
brick red
buffy brown
cadmium yellow
cinnamon
cinnamon brown
cinnamon buff
cinnamon rufous
clay color
fawn color
flame scarlet
flesh color
fuscous
heliotrope gray
honey yellow
madder brown
Mars yellow
ocher
ocher yellow
ocherous (=ochraceous)
ocherous buff (=ochraceous buff)
ocherous orange (=ochraceous orange)
ocherous tawny (=ochraceous tawny)
pale cinnamon buff
pale fawn
pale pinkish buff
pecan brown
pinkish buff
Pompeian yellow
raw sienna
raw umber
rufous
russet
salmon
salmon orange
sea green
snuff brown
Sudan brown
tawny
tawny olive
tea green
umber brown
vinaceous
vinaceous brown
vinaceous buff
vinaceous cinnamon
warm buff
wax yellow
xanthine orange

Glossary

acuminate (of cystidia) Gradually narrowed to a point (Fig. 138C).

acutely umbonate (of the cap) Having a sharply pointed umbo (Fig. 5C).

acyanophilous (e.g. of spore walls, ornamentation, or hyphae) Not staining dark blue with Cotton Blue. See Ch. 4, "Chemical Reagents".

adnate (of gills or tubes) Broadly or squarely attached to the stalk along all or most of the gill width or tube layer (Fig. 7D).

adnexed (of gills or tubes) Narrowly attached to the stalk, appearing as if a triangular piece has been removed where the gill or tube mass meets the stalk (Fig. 7 E and H).

agate (e.g. of the cap, stalk, or gills) A blend of colors resembling that in the precious stone, defined in *A Dictionary of Color* by A. Maerz and M. R. Paul (McGraw-Hill, NY; 1930).

alutaceous (e.g. of the cap) Light leather-colored, pale tan, pale brown, or pinkish cinnamon.

amygdaliform (e.g. of spores) Almond-shaped (Fig. 28A).

amyloid (e.g. of spores, walls, spore ornamentation, hyphal walls, or ascus tips) Staining grayish to blackish violet or bluish in Melzer's reagent. See Ch. 4, "Chemical Reagents".

angular (of spores) Having an outline exhibiting distinct angles (Fig. 162A).

annular (of the remains of the veil on the stalk) Resembling a ring. See annulus.

annulate (of the stalk) Bearing a ring.

annulus, pl. **annuli** A ring, formed from the remains of the partial veil, encircling the stalk (Fig. 4J).

apex The distal end of the stalk or spore; the end furthest from the base or point of attachment to the substrate in the case of the stalk or from the point of attachment to the hymenium in the case of the spore.

apical annulus A ring, formed from the remains of the partial veil, found at or near the apex of the stalk.

apical germ pore A pore on or near the apex of a basidiospore (Fig. 104A).

apiculate (of spores) Having a short projection at one or both ends.

apiculus, pl. **apiculi** A short projection on basidiospores near the point where the spore is attached to the sterigma (Fig. 15a); a short projection on ascospores at each end.

apothecium, pl. **apothecia** The fruit body of the Discomycetes (Fig. 2A).

appendiculate (of the margin of the cap) Hung with pieces of the partial veil.

appressed-fibrillose (of the surface of the cap or stalk) Having filaments flattened close to the surface.

appressed-squamulose (of the cap or stalk) Having scales flattened close to the surface.

arcuate (of gills) Curved upward like a bow.

arcuate-decurrent (of gills) Curved upward then running down the stalk for a short distance.

areolate (of the cap surface) Cracked into block-like areas similar to those formed when a mud flat dries.

ascocarp The fruit body of the Ascomycota (Figs. 2A, 10H, 12 A–D).

ascospore A sexual spore that is produced in an ascus (Fig. 2C).

ascus, pl. **asci** A reproductive cell, often sac-like or club-shaped, in which ascospores are produced (Fig. 2C).

basal annulus A ring, formed toward the base of the stalk from the remains of a veil or the proliferation of the cap margin.

basal mycelium Mass of hyphae associated with the base of the stalk.

basal tomentum The densely matted or wooly covering sometimes associated with the base of the stalk.

basidiocarp The fruit body of the Basidiomycota (Fig. 3 A–G).

basidiole An immature basidium, always present in the Basidiomycota.

basidiospore A spore that is produced on a basidium (Fig. 1B).

basidium, pl. **basidia** A reproductive cell, often club-shaped, on which basidiospores are produced (Fig. 1B).

biguttulate (of spores or other cells) Having two internal droplets (Fig. 43A).

bilateral See **divergent.**

boletinoid (of the hymenophore) Having rather long and radially arranged pores with the radial walls of the tubes at times gill-like and often forming prominent lines near the apex of the stalk.

brachybasidiole An inflated basidiole acting as a spacing cell, as found in the genus *Coprinus* (Fig. 51B).

brachycystidium See **brachybasidiole.**

bracket fungus A common name for polypores having bracket-like fruit bodies.

broadly umbonate (of the cap) Having a broad or rounded umbo.

broom cell A special cell found on the cap surface or at the edge of a gill, usually swollen and sometimes darkly colored, with numerous protuberances located over the apex or over the entire cell surface.

bulbous (of the stalk) Enlarged at the base, something like the base of a green onion (Fig. 10 E–G).

button A fruit body of the Basidiomycota during the early stage of its development.
caespitose (of fruit bodies) Growing close together with their bases joined (Fig. 9E).
calyptrate (of basidiospores) With the outer layer of the spore wall loosening to form wing-like folds, usually at the proximal end.
campanulate (of the cap) Bell-shaped, with a flared margin (Fig. 4C).
cap The umbrella-shaped upper structure that bears the hymenium in many Basidiomycota.
capillitium, pl. **capillitia** A mass of differentiated hyphae in puffballs, found among the spores.
capitate (of cystidia) With a distinctive swelling at the apex (Fig. 90C).
carminophilous See **siderophilous.**
cartilaginous (of the stalk) Usually fairly thin, often less than 5 mm, firm and tough but readily bent, often cleanly splitting when bent in half.
caulocystidium A special cystidium occurring on the surface of the stalk.
cellular (of the cap or stalk surface) Composed of globose to sac-like cells, arranged in a single layer (Fig. 13B).
central (of the stalk) Attached in the center of the cap (Fig. 3A).
cheilocystidium A special cystidium occurring on the edge of a gill or tube.
chrysocystidium A type of gloeocystidium with an internal body or contents that turns yellow in an aqueous alkali solution.
clamp connection (of Basidiomycota) A small, semicircular hyphal branch laterally attached to the walls of two adjacent cells and arching over the septum between them (Fig. 50D).
clavate (of the stalk) Thickened like a club toward the base (Fig. 10 C and D); (of basidia or cystidia) Thickened and club-shaped at the apex (Fig. 56B).
close (of gills) Having the individual gills spaced intermediate between subdistant and crowded (Fig. 7L).
compressed (of the stalk) Flattened lengthwise in cross section.
conic (of the cap) More or less cone-shaped (Fig. 4 A and B).
conk The woody fruit body of some polypores.
connate (of fruit bodies) With the stalks grown together for a considerable distance from the base upward.
context See **flesh.**
convergent (of the trama of gills or tubes as seen in section) Having hyphae that converge toward the center of the trama, giving the impression of a series of V's, sometimes with a thin central strand visible (Fig. 14D).
convex (of the cap) Rounded, with a shape much like an inverted bowl (Fig. 4G).
coprophilous (of fruit bodies) Growing on dung or manure.
corrugate (of a surface) Coarsely wrinkled or ridged radially.
cortina (of the stalk) A partial veil composed of a cobwebby mass of filaments.
cortinate (of the stalk) Having a partial veil composed of a cobwebby mass of filaments.
crenate (of the edges of gills or tube mouths, or the margin of the cap) Scalloped, wavy, or round-toothed (Fig. 8B).
crenulate (of the edges of gills or tube mouths, or the margin of the cap) Very finely scalloped or wavy.
crisped (of gill edges) Finely curled or crinkled (Fig. 8E).
cross section A transverse section of the stalk made perpendicular to the long axis of the stalk; a transverse section across the long axis of a gill or tube.
crowded (of gills) So close together that spaces between the gills cannot be seen (Fig. 7M).
cuticle A differentiated covering tissue of the cap or stalk consisting of one to several layers of hyphae, visible at the microscopic level.
cutis A specialized cuticle in which the hyphae are arranged parallel to the surface.
cyanophilous (e.g. of spore walls, ornamentation, or hyphae) Staining dark blue with Cotton Blue. See Ch. 4, "Chemical Reagents".
cystidium, pl. **cystidia** A specialized, sterile cell occurring in the hymenium or as an end cell in the hyphae of the surface of the cap or stalk; it can be described on the basis of position, morphology, contents, function, or origin.
dacryoid (e.g. of spores) More or less tear-shaped; nearly round but tapering at one end.
decurrent (of gills) Extending down the stalk beyond their width (Fig. 7 I).
decurved (of the cap) Having the margin turned downward so that it points toward the stalk (Fig. 5 I).
deliquescent (of gills) Capable of autodigestion, namely of dissolving or liquefying at maturity, as in the genus *Coprinus.*
depressed (of the cap) Having the central portion lower than the margin (Fig. 5 D and E).
derm A specialized cuticle in which the hyphae are arranged perpendicular to the surface.
dextrinoid (e.g. of spore walls or hyphal walls) Staining reddish to purplish brown in Melzer's reagent. See Ch. 4, "Chemical Reagents".
dichotomous (e.g. of gills) Forking in pairs.
disc Center of the cap (Fig. 4H).
distant (of gills) Spaced far apart (Fig. 7J).
divergent (of the trama of gills or tubes as seen in section) Consisting of a central strand of parallel to interwoven hyphae from which other hyphae diverge in an oblique fashion; the diverging hyphae typically branch toward the subhymenium from both sides of the central strand (Fig. 14C).

eccentric (of the stalk) Off-center, not centrally attached (Fig. 9C).
echinate (e.g. of spores) Covered with sharply pointed spines on the surface.
echinulate (e.g. of spores) Covered with small, finely pointed spines on the surface.
ectal excipulum Tissue making up the outer surface layer of the apothecium.
eguttulate (of spores or cells) Lacking small internal droplets.
ellipsoidal (of spores or cells) Having the three-dimensional form of an ellipse, namely with the sides curved and the ends more or less equally rounded (Fig. 146A).
elliptic (of surfaces) Having an outline in the form of an ellipse.
emarginate (of gills) Notched near the stalk.
encrusted (of the surface of hyphae) Covered with a crust in the form of spirals, rings, pegs, or irregular clumps (Fig. 28D).
enrolled (of the cap margin) Rolled inward so that it points toward itself (Fig. 5G).
entire (e.g. of gills or the margin of the cap) Having the edges even (Fig. 8A).
epicutis The outer, surface layer of the cuticle when the cuticle is differentiated into two or more layers. Compare **subcutis**.
epithelium, pl. **epithelia** A specialized surface of the cap or stalk, composed of isodiametric cells, usually in many layers (Fig. 13C).
equal (of the stalk) Of uniform diameter (Fig. 10A).
eroded (of gills) Having the edges uneven, as if gnawed (Fig. 8D).
evanescent (of the annulus or veil) Only slightly developed and soon disappearing.
even (of the surface of the cap or stalk) Smooth, lacking protuberances or irregularities in the topography.
excipulum A specialized name for the fleshy tissue of the apothecium (Fig. 2B).
face The side of a gill, as opposed to the edge.
face view The view of a basidiospore from the dorsal side, namely the side facing the imaginary axis of the basidium, or the ventral side, the side facing outward from the axis of the basidium; from this view the spore can be divided into two halves that are mirror images of one another (Fig. 15, spore 4).
farinaceous (of taste and odor) Like fresh meal.
fawn (e.g. of the cap) Pinkish cinnamon.
ferruginous (e.g. of the cap) Rust-colored.
fibrillose (of the surface of the cap or stalk) With visible filaments arranged in a parallel or interwoven manner.
fibrous (of the flesh of the stalk or cap) Composed of toughish, string-like tissue.
filamentous (of hyphae) Thread-like.
filiform (of a cystidium, spore, or stalk) Thin to thread-like.
fimbriate (of gills) Minutely fringed along the edges.
flesh The fleshy parts of any fruit body, as distinguished from the hymenium.
fleshy (e.g. of the flesh of mushrooms or boletes) Having a rather soft consistency and typically putrescent.
fleshy-fibrous (of the flesh of the stalk or cap) Having a condition between fleshy and fibrous.
floccose (of the cap or stalk) Having a surface like cotton flannel or downy wool (Fig. 6D).
free (e.g. of gills or tubes) Not attached to the stalk (Fig. 7A and B).
fruit body The organized reproductive structure of fungi that produces spores.
fulvous (e.g. of the cap) Reddish cinnamon brown.
furfuraceous See **scurfy.**
fusiform (e.g. of cystidia, spores, or stalk) Spindle-shaped, broad in the middle and tapered toward each end.
fusiform-ventricose (e.g. of cystidia, spores, or the stalk) Midway between being spindle-shaped and broadly swollen in the middle.
gill The reproductive, blade-like structure of gill mushrooms, typically arranged radially on the undersurface of the cap.
gill trama The fleshy tissue of the gills between the two layers of the hymenium (Fig. 14 A–D).
glabrous (of the surface of the cap or stalk) Smooth, lacking any scales, hairs, or fibrils. Compare **even.**
glandular dot (of the stalk) A color spot, drop, or glandule on the surface of the stalk, as in the bolete *Suillus.*
gleba, pl. **glebae** A specialized name for the spore-producing tissue within the peridium in puffballs and tubers (Fig. 11 D and F); the spore mass.
globose (e.g. of spores) Spherical or nearly so (Fig. 18A).
gloeocystidium A special type of cystidium, variable in form, usually staining readily with chemical reagents, and often containing conspicuous amorphous or granular contents.
gloeoplerous (of hyphae) Containing an oily, resinous, or granular material.
gluten A jelly-like material imparting a tacky to sticky texture to the surface of the cap or stalk when moist.
glutinous (of the surface of the cap or stalk) Distinctly coated in a jelly-like material.
granulose (e.g. of the surface of the cap or stalk) Covered with granules like grains of salt.
gregarious (of fruit bodies) Occurring close together or in groups. Compare **caespitose.**
guttulate (of spores or cells) Having one or more small internal droplets.
guttula, pl.**guttulae** A small internal droplet; the oil globule in spores.
heteromerous (of trama) Containing clusters of inflated hyphae called sphaerocysts surrounded by ordinary thread-like hyphae; found in *Russula* and *Lactarius.*

hirsute (of the cap or stalk) Covered with rather long, stiff hairs.

hoary (of the cap or stalk) Covered with dense, silky down.

horny (e.g. of the stalk) Having the consistency of animal horn.

humicolous (of fruit bodies) Growing in or on soil or humus.

hyaline (e.g. of spores, as seen under the microscope) Transparent, clear, colorless.

hygrophanous (of the cap surface or flesh) Watery in appearance at first, but with the moisture disappearing rapidly and the color fading or otherwise changing markedly.

hymeniform (e.g. of the cap surface) Composed of club-shaped to pear-shaped cells arranged in vertical rows (Fig. 13D).

hymenium, pl. **hymenia** The spore-producing surface of the Ascomycota or Basidiomycota, visible microscopically, usually composed of sterile as well as spore-producing cells (Figs. 1B, 2C).

hymenophore That part of the fruit body which visibly produces or supports the hymenium and the subhymenium; it takes the form of gills, tubes, or spines, or it may be convoluted, wrinkled, unpatterned, or smooth.

hypha, pl. **hyphae** A filament or thread-like structure of the vegetative mycelium and fruit body of fungi. Compare **mycelium**.

hypoderm See **subcutis**.

imbricate-scaly (of the cap surface) With scales overlapping like the shingles of a roof.

inamyloid See **nonamyloid**.

incurved (of the cap margin) Pointed inward toward the gills (Fig. 5H).

inferior annulus See **basal annulus**.

infundibuliform (of the cap) Funnel-shaped.

innate (e.g. of scales or fibrils) Forming a part of the surface layer of the cap or stalk; not superficial.

inner veil See **partial veil.**

inserted (of the stalk base) Attached to the substratum directly, without means of rhizomorphs, rhizoids, or hyphae; hence stalk base typically naked (Fig. 9F).

interveined (of gills) Having surface veins that extend various distances into the interspace, or sometimes crossing the interspace to the next gill.

interwoven (of hyphae) Intermingled or irregularly arranged; (of the trama) Composed of intermingled hyphae (Fig. 14B).

inverse (of gill trama) See **convergent**.

irregular (of gill trama) See **interwoven**.

isodiametric (of cells) Having dimensions that are equal in all directions.

ixocutis A specialized cutis in which the hyphae are embedded in or produce a jelly-like material.

ixotrichoderm A specialized trichoderm in which the hyphae are embedded in or produce a jelly-like material.

lacerated (e.g. of the annulus or the cap surface) Appearing as if torn.

laciniated (e.g. of the cap margin or annulus) Torn into lobes.

lacrimoid (of spores or cells) Tear-shaped.

lactiferous (of hyphae) Containing a latex-like material.

lacunary (of the cap or stalk) Pitted or indented, with the cavities surrounded by ridges (Fig. 10H).

lamella See **gill.**

lamellula, pl. **lamellulae** A short gill not reaching the stalk, usually alternating with longer gills of various lengths; each length category is called a series (Fig. 9B).

lamprocystidium See **metuloid**.

lanceolate (e.g. of cystidia) Shaped like a lance, slightly swollen in the middle and tapered at both ends (Fig. 142B).

lateral (of the stalk) Attached to one side of the cap (Fig. 9B).

latex A milky juice that exudes from the fruit body when it is cut or broken, as in the genus *Lactarius*.

laticiferous See **lactiferous**.

leptocystidium A smooth, thin-walled cystidium, often longer than the basidia, having no distinguishing contents and originating in the hymenium (Fig. 27C).

lignicolous (of fruit bodies) Growing on wood.

lubricous (of the surface of the stalk or cap) Covered with a thin, oily layer, feeling slippery to the touch.

macrocystidium A type of gloeocystidium that originates from the gill trama and projects into or beyond the level of the hymenium, relatively large, and usually not staining with acid aldehyde reagents. Compare **pseudocystidium**.

margin The periphery of the cap; in general, that area occurring between the center and the edge of the cap (Fig. 4H).

marginate (of the gills) Having the edge of the gill colored differently from the side of the gill; (of the bulb of the stalk) Having a ridge on the upper part.

median annulus A ring, formed from the remains of a veil, found near the midpoint of the stalk.

median longitudinal section A cut along the length of the stalk and passing through the center.

medullary excipulum The tissue of the apothecium making up the inner layer of the flesh.

Melzer's reagent A mixture of chemicals used for testing for the amyloid or dextrinoid characteristic in structures such as spore walls. See. Ch. 4, "Chemical Reagents".

membranous (e.g. of the cap) Thin and pliant, like a membrane.

metachromatic (of spores and hyphae) Staining red to violet in Cresyl Blue. See Ch. 4, "Chemical Reagents".

metuloid A special cystidium that is thick-walled, variable in shape, and sometimes encrusted, pigmented, or giving a particular color reaction with certain mounting media (Fig. 61 C and D).

micaceous (of the surface of the cap or stalk) Covered with glistening, mica-like particles.

micrometer A tool used to measure small distances, for example, an ocular micrometer.

micrometre (symbol μm) A unit of distance measuring one millionth of a metre; the term micron (μ) is often also used to indicate one millionth of a metre, but the micron is not an accepted SI unit of measurement.

micron (symbol μ) See **micrometre**.

mucronate (e.g. of cystidia) Tipped with an abrupt, short, sharp point.

mycelium, pl. **mycelia** The collective term for a mass of hyphae.

mycorrhiza, pl. **mycorrhizae** The symbiotic association between a fungus and the roots of trees or other plants.

mycorrhizal (of fruit bodies) Surviving by means of a symbiotic association between the fungus and the roots of trees or other plants.

nodulose (of spores) With large outgrowths that make the spore appear as if decorated with knobs (Fig. 62A).

nonamyloid (e.g. of spore walls) Remaining hyaline or merely yellowish in Melzer's reagent. See Ch. 4, "Chemical Reagents".

notched (of gills) See **emarginate**.

oblong (e.g. of spores) Long in relation to width, with the sides almost parallel and the ends more or less squarish.

obovoid (e.g. of spores) With the broader end uppermost, opposite in three-dimensional shape to ovoid.

oleiferous (of hyphae) Containing resinous or oily material.

operculate (of asci) Having a hinged lid (Fig. 2C).

operculum, pl. **opercula** A hinged lid.

outer veil See **universal veil**.

ovate (of surfaces) Having an outline like the longitudinal shape of an egg.

ovoid (of spores) Three dimensionally egg-shaped, namely round at each end but with the ends unequal in width.

parabolic (of the cap) Having the height greater than the width but with the cap still regularly rounded (Fig. 4D).

parallel (of the gill trama) With the hyphae arranged parallel to one another (Fig. 14A).

paraphysis, pl. **paraphyses** A sterile filament found among asci in the hymenium of certain Ascomycota, branched or unbranched, and varying in shape from clavate to filiform (Fig. 2C).

parasitic (of fruit bodies) Living on or in another living organism and deriving nourishment from it.

partial veil A veil that extends from the cap edge to the stalk at a point usually near the apex. In young stages the partial veil often covers the developing gills or tubes (Fig. 1H).

pedicel A modified sterigma occurring on basidiospores of certain puffballs and relatives.

pedicellate (of cells) Having a pedicel or attached by a pedicel.

pellis See **cuticle**.

peridium, pl. **peridia** The outer, enveloping coat of the fruit body of puffballs and relatives (Fig. 11F).

pileocystidium A cystidium occurring on the cap surface (Fig. 91D).

pileus See **cap**.

plage A smooth area near the proximal end of certain basidiospores, located just above the point of attachment of the spore to the sterigma (Fig. 27A).

plane (of the cap) Having a flat surface (Fig. 4H).

pleurocystidium A special cystidium occurring on the side of a gill or tube.

pliant (of the cap or stalk) Easily bent.

plicate (of the cap margin) Folded like a fan.

pore A small opening indicating the mouth of a tube in boletes and polypores.

profile view See **side view**.

proximal end The base of a basidiospore, namely the end that is attached to the sterigma (Fig. 15A).

pruinose (of a surface) Finely powdered, as if sprinkled with flour (Fig. 6A).

pseudoamyloid See **dextrinoid**.

pseudocystidium A type of gloeocystidium that originates from the gill trama and projects into or beyond the level of the hymenium, relatively small, and staining with various acid aldehyde reagents. Compare **macrocystidium**.

pseudorhiza, pl. **pseudorhizae** A root-like extension of the stalk base (Fig. 9D).

pubescent (of a surface) With a covering of short, often soft, downy hairs (Fig. 6C).

pulverulent (e.g. of the surface of the cap or stalk) Powdery.

punctate (e.g. of spores) Marked with very small dots.

pyriform (e.g. of spores, cells, or the stalk) Pear-shaped.

radial section A cut following the radius of the cap, namely projecting from the center toward the edge.

radicating See **rooting**.

receptacle The stalk or cap, in stinkhorns, which supports the spore-producing tissue.

recurved (e.g. of the margin of the cap) Curved backward.

regular See **parallel**.

repent (of hyphae) Appressed to the cap surface.

reticulate (e.g. of the surface of the cap, stalk, or spores) Marked with lines, veins, or ridges that cross one another as in a net (Fig. 10K).

rhizoid A distinct strand of hyphae attached to the stalk base.

rhizomorph A visible strand of compacted mycelium, light or often dark colored, penetrating a soft substratum such as rotten wood (Fig. 9E).

rimose (of the cap surface) Split usually in a radial manner, with the splits normally extending through the cap surface.

rivulose (of the cap surface) Having lines resembling a river and its tributaries.

rooting (of the stalk) Having a root-like extension into the soil (Fig. 9D).
rufous (e.g. of the cap, stalk, or gills) Reddish or dull red.
rugose (of a surface) Coarsely wrinkled.
rugulose (of a surface) Finely wrinkled.
saccate (of cells) Sac-like.
saprobic (of fruit bodies) Living on dead organic material.
scabrous (of the surface) Roughened with short, rigid projections such as small scales or points, and rough to the touch.
scalp section A cut along the slope or surface of the cap or stalk.
scattered (of fruit bodies) Widely separated, 30–60 cm apart.
scurfy (of the surface of the cap or stalk) Covered with bran-like particles.
seceding (of gills) In early stages attached to the stem, e.g. adnate or adnexed, but separating from it at maturity (Fig. 7C).
septate (e.g. of hyphae or paraphyses) Divided by cross walls.
septum A wall separating two cells.
serrate (of gills) Notched on the edge like a saw (Fig. 8G).
serrulate Minutely notched on the edge (Fig. 8F).
sessile (of the cap) Without a stalk; (of basidiospores) Without a sterigma.
siderophilous (of basidia) Having internal particles that stain blackish purple or violet black in aceto-carmine. See Ch. 4, "Chemical Reagents".
side view The profile view of spores, as seen with the concave side of the spore facing the longitudinal axis of the basidium (Fig. 15, spore 2); the view in which accurate spore measurements are taken.
sinuate (of gills) Having a concave indentation of the gill edge where it meets the stalk (Fig. 7F).
solitary (of fruit bodies) Occurring singly.
sphaerocyst An inflated cell, more or less spherical in shape, found in clusters in the trama of *Russula* and *Lactarius* (Fig. 166E), but also found in veils and other structures.
squamous (of the cap and stalk) Covered with scales.
squamulose (of the cap and stalk) Covered with minute scales.
squarrose (of the cap and stalk) Covered with upright to recurved scales (Fig. 6 H and I).
stalk The stem of mushrooms, boletes, polypores, and other fungi.
sterigma, pl. **sterigmata** The tiny, spicule-like pedicel upon which a basidiospore is borne and from which it may be forcibly discharged (Fig. 15).
sterile base A stalk-like structure occurring in certain puffballs.
strangulate (of cystidia) Having wavy walls, alternately contracted and expanded along the length.
stipe See **stalk**.
stipitate (of fruit bodies) Having a stalk.
striate (of the cap margin) Having minute, radiating furrows or lines; (of spores or the stalk) Having longitudinal lines or minute furrows.
strigose (of the cap or stalk) Having long, coarse hairs resembling bristles.
stuffed (of the stalk) Filled in the center with a differentiated, often soft pith that sometimes disappears with age leaving a hollow space.
subcaespitose (of fruit bodies) Growing so close together their bases appear nearly joined.
subcutis A differentiated layer of the cuticle sometimes occurring between the epicutis and the cap trama. Compare **epicutis**.
subcylindrical (of spores) Almost cylindrical.
subdecurrent (of gills) Extending down the stalk for a short distance.
subdistant (of gills) Having the individual gills spaced intermediate between close and distant (Fig. 7K).
subfusiform (e.g. of spores) Nearly fusiform, namely somewhat elongate and usually tapered at one end and rounded at the other (Fig. 127A).
subglobose (e.g. of spores) Nearly spherical, but not perfectly round in outline (Fig. 22A).
subhymenium The tissue, visible only microscopically, that produces the hymenium.
subparallel (of the gill trama) With the hyphae arranged nearly parallel to one another or slightly interwoven.
substipitate (of the cap) Hardly having a stalk, but with a very short attachment.
sulcate (of the cap or stalk) Grooved, more deeply than striate but less so than plicate.
superior annulus See **apical annulus**.
superficial (e.g. of fibrils, scales, or warts) On the surface and easily removable, as opposed to innate.
surface The outermost layer of the cap and stalk, visible macroscopically.
tangential section A cut from margin to margin without passing through the center of the cap.
tapered (of the stalk) Narrowing either toward the apex or the base (Fig. 10B).
terete (of the stalk) Perfectly round as seen in end view or in cross section.
terrestrial (of fruit bodies) Growing on the ground, often in a mixture of soil and dead organic material.
thick-walled (e.g. of spores or hyphae) With a wall thickness greater than 0.5 μm.
thin-walled (e.g. of spores or hyphae) With a wall thickness of 0.5 μm or less.
tomentose (of the surface of the cap or stalk) Densely matted and wooly like a woolen blanket (Fig. 6 D and E).
trama, pl. **tramae** The fleshy tissue of the gills or tubes supporting the hymenium and found between adjacent hymenia; the fleshy part of the cap or stalk of the Basidiomycota and Ascomycota.
translucent-striate (of the cap margin) Having the flesh of the cap thin enough that lines representing

an image of the gills can be seen through the top when the cap is moist or wet.

transverse section A cut in a crosswise manner; cross-sectional.

trichoderm A specialized cuticle in which the hyphae are arranged perpendicular to the surface though they may be somewhat intertwined; when sufficiently dense the trichoderm imparts a matted, wooly texture to the surface.

tube The cylindrical, perforation-like hollow that bears the hymenium and produces the spores in boletes and polypores.

tuberculate (e.g. of spores) Covered with prominent, knob-like outgrowths often more than 1 μm high.

umbilicate (of the cap) Having a central, navel-like depression (Fig. 5F).

umbo A raised, conical to rounded knob at the center of the cap (Fig. 5 A and B).

umbonate (of the cap) Having a raised, conical to rounded knob at the center of the cap (Fig. 5 A and B).

uncinate (of the gills) Hooked, having a narrow, decurrent extension at the stalk (Fig. 7G).

undulate (of the cap margin) Wavy.

universal veil A veil that completely covers certain mushrooms during their young stages, as in *Amanita*, eventually breaking away leaving remnants, the volva, on the stalk and often leaving patches or scales on the cap surface (Fig. 1 F–H).

uplifted See **upturned**.

upturned (of the cap margin) Pointed upward.

vegetative mycelium The food-absorbing part of the fungus, as opposed to the spore-producing fruit body of the fungus.

ventricose (e.g. of the stalk, spores, or cystidia) Swollen in the middle.

verrucose (of the spore surface) Covered with rounded, wart-like processes (Fig. 26A).

verruculose (of the spore surface) Covered with minute, wart-like processes.

vesiculate (e.g. of cystidia or cells) Vesicle-shaped, inflated like a large sac or bladder (Fig. 51C).

villous (of the surface of the cap or stalk) Coated with rather long, weak hairs (Fig. 6G).

virgate (of the surface of the cap or stalk) Striped or streaked, with filaments.

viscid (of the surface of the cap or stalk) Sticky, slippery, or tacky when moist.

volva The remains of the universal veil left in various forms at or on the base of the stalk of mushrooms (Fig. 1 A and H).

volvate (of the stalk) Having the remains of the universal veil at or on the base.

Index

This index includes primarily all the taxonomic and common names of fungi that appear in the book. The morphological and technical terms that are used in this book for fungi are also given, with the page references restricted to those places in the text where these terms are defined in Part One; you may refer also to the Glossary for an alphabetic listing of many of these terms. Toxins, symptoms of poisoning, common and scientific names of plants, and a few other key-subject words are indexed here as well.

Accepted taxonomic names for all categories of fungi (kingdoms, divisions, classes, orders, families, genera, subgenera, species, and subspecies) are given in bold type. Synonyms appear in lightface. Morphological and technical terms, common names of fungi, toxins and symptoms related to fungal poisoning, scientific and common names of plants, and other key-subject words are in small capitals. Page numbers in boldface indicate a significant citation. Italicized page numbers indicate an illustrated citation.